D0142651

Atmospheric Science

Second Edition

This is Volume 92 in the
INTERNATIONAL GEOPHYSICS SERIES
A series of monographs and textbooks
Edited by RENATA DMOWSKA, DENNIS HARTMANN, and H. THOMAS ROSSBY
A complete list of books in this series appears at the end of this volume.

Atmospheric Science

An Introductory Survey

Second Edition

John M. Wallace • Peter V. Hobbs
University of Washington

AMSTERDAM • BOSTON • HEIDELBERG • LONDON
NEW YORK • OXFORD • PARIS • SAN DIEGO
SAN FRANCISCO • SINGAPORE • SYDNEY • TOKYO
Academic Press is an imprint of Elsevier

Acquisitions Editor: Jennifer Helé
Project Manager: Jeff Freeland
Marketing Manager: Linda Beattie
Marketing Coordinator: Francine Ribeau
Cover Art Direction: Cate Rickard Barr
Text Design: Julio Esperas
Composition: Integra Software Services Private Limited
Cover Printer: Transcontinental Printing
Interior Printer: Transcontinental Printing

Academic Press is an imprint of Elsevier
30 Corporate Drive, Suite 400, Burlington, MA 01803, USA
525 B Street, Suite 1900, San Diego, California 92101-4495, USA
84 Theobald's Road, London WC1X 8RR, UK

This book is printed on acid-free paper. ♾

Cover Photo Credits:
Noboru Nakamura's drawing, bearing the Latin inscription "*Here watch the powerful machine of nature, formed by the air and other elements*", was inspired by mankind's long-standing fascination with weather and climate. Some of the products of this machine are represented by the other images (left to right): annual mean sea surface temperature (courtesy of Todd Mitchell), a single cell thunderstorm cell over a tropical Pacific atoll (courtesy of Art Rangno), a supercell thunderstorm over Kansas (courtesy of Chris Kridler), and Antarctic sea ice (courtesy of Miles McPhee).

Library of Congress Cataloging-in-Publication Data
Wallace, John M. (John Michael), 1940–
Atmospheric science : an introductory survey / John M. Wallace, Peter V. Hobbs.—2nd ed.
p. cm.
ISBN 0-12-732951-X
1. Atmosphere—Textbooks. 2. Atmospheric physics—Textbooks.
3. Atmospheric chemistry—Textbooks. 4. Meteorology—Textbooks.
I. Hobbs, Peter Victor, 1936–2005 II. Title.
QC861.3.W35 2006
551.5—dc22

2005034642

British Library Cataloguing-in-Publication Data
A catalogue record for this book is available from the British Library.

ISBN 13: 978-0-12-732951-2
ISBN 10: 0-12-732951-X

For information on all Academic Press Publications visit our Web site at www.books.elsevier.com

Printed in Canada
06 07 08 09 10 8 7 6 5 4 3 2 1

In Memory of Peter V. Hobbs (1936–2005)

I am the daughter of Earth and Water,
And the nursling of the Sky;
I pass through the pores of the ocean and shores;
I change, but I cannot die.
For after the rain when with never a stain
The pavilion of Heaven is bare,
And the winds and sunbeams with their convex gleams
Build up the blue dome of air,
I silently laugh at my own cenotaph,
And out of the caverns of rain,
Like a child from the womb, like a ghost from the tomb,
I arise and unbuild it again.

PERCY BYSSHE SHELLEY
The Cloud

Contents

Preface to the Second Edition

In the 30 years that have passed since we embarked on the first edition of this book, atmospheric science has developed into a major field of study with far-reaching scientific and societal implications. Topics such as climate and atmospheric chemistry, which were not deemed sufficiently important to warrant chapters of their own 30 years ago, are now major branches of the discipline. More traditional topics such as weather forecasting, understanding the processes that lead to severe storms, and the radiation balance of the Earth have been placed on firmer foundations. Satellite-borne sensors that were in the early stages of development 30 years ago are now providing comprehensive observations of Earth's atmosphere. Those who have witnessed these accomplishments and contributed to them, if only in minor ways, have been fortunate indeed.

As we drafted new section after new section describing these exciting new developments, we began to wonder whether we would still be capable of cramming a summary of the entire field of atmospheric science into a book light enough to be carried in a student's backpack. This second edition does, in fact, contain much more material than its predecessor, but thanks to the double column formatting and the supplementary Web site, it is not correspondingly heavier. In deciding which of the recent developments to include and which ones to leave out, we have elected to emphasize fundamental principles that will stand students in good stead throughout their careers, eschewing unnecessary details, interesting though they may be, and important to the specialist.

The second edition contains new chapters on atmospheric chemistry, the atmospheric boundary layer, the Earth system, and climate dynamics. The chapters in the first edition entitled *Clouds and Storms, the Global Energy Balance, and the General Circulation* have been dropped, but much of the material that was contained in them has been moved to other chapters. The coverage of atmospheric dynamics, radiative transfer, atmospheric electricity, convective storms, and tropical cyclones has been expanded. The treatment of atmospheric thermodynamics has been modernized by using the skew $T - \ln p$ chart as the primary format for plotting soundings. The second edition contains many more illustrations, most of which are in color.

A popular feature of the first edition that is retained in the second edition is the inclusion of quantitative exercises with complete solutions embedded in the text of each chapter, as well as additional exercises for the student at the end of each chapter. The second edition retains these features. In addition, we have included many new exercises at the end of the chapters, and (available to instructors only) a nearly complete set of solutions for the exercises. Boxes are used in the second edition as a vehicle for presenting topics or lines of reasoning that are outside the mainstream of the text. For example, in Chapter 3 a qualitative statistical mechanics interpretation of the gas laws and the first law of thermodynamics is presented in a series of boxes.

Academic Press is providing two web sites in support of the book. The first web site includes information and resources for all readers, including a printable, blank skew T-$\ln p$ chart, answers to most of the exercises, additional solved exercises that we did not have space to include in the printed text, errata, an appendix on global weather observations and data assimilation, and climate data for use in the exercises. The second web site, which will be accessible only to instructors, contains a short instructor's guide, solutions to most of the exercises, electronic versions of most of the figures that appear in the book, and electronic versions of a set of supplementary figures that may be useful in customizing classroom presentations.

To use the book as a text for a broad survey course, the instructor would need to be selective, omitting much of the more advanced material from the quantitative Chapters 3, 4, 5, and 7, as well as sections of other, more descriptive chapters. Selected chapters of the book can be used as a text for several different kinds of courses. For example, Chapters 3–6 could be used in support of an atmospheric physics and chemistry course; Chapters 1, 3, 7, and 8 for a course emphasizing weather; and Chapters 1 and 2 and parts of 3, 4, and 9 and Chapter 10 in support of a course on climate in a geosciences curriculum.

Corrigenda and suggestions for the instructor's guide will be gratefully received.

John M. Wallace
Peter V. Hobbs
Seattle, January 2005

Acknowledgments

In 1972, I accepted Peter Hobbs' invitation to collaborate with him in writing an introductory atmospheric science textbook. We agreed that he would take the lead in drafting the thermodynamics and cloud physics chapters and I would be primarily responsible for the chapters dealing with radiative transfer, synoptic meteorology, and dynamic meteorology. Over the course of the following few years we struggled to reconcile his penchant for rigor and logic with my more intuitive, visually based writing style. These spirited negotiations tested and ultimately cemented our friendship and led to a text that was better than either of us could have produced working in isolation.

Three years ago, on a walk together in the rain, Peter warned me that if I wanted to produce that long overdue second edition, we needed to get started soon because he was contemplating retirement in a few years. When I agreed, he immediately set to work on his chapters, including an entirely new chapter on atmospheric chemistry, and completed drafts of them by the end of 2003. Soon afterward, he was diagnosed with pancreatic cancer.

Despite his illness, Peter continued to revise his chapters and offer helpful feedback on mine. Even after he was no longer able to engage in spirited debates about the content of the book, he continued to wield his infamous red pen, pointing out grammatical mistakes and editorial inconsistencies in my chapters. A few months before his death, July 25, 2005, we enjoyed a party celebrating (albeit a bit prematurely) the completion of the project that he had initiated.

For the dedication of the first edition, it was Peter's choice to use Shelley's poem, "Clouds," a visual metaphor for life, death, and renewal. For the second edition, I have chosen the same poem, this time in memory of Peter.

Several members of Peter's "Cloud and Aerosol Research Group" were instrumental in preparing the book for publication. Debra Wolf managed the manuscript and produced many of the illustrations, Judith Opacki obtained most of the permissions, Arthur Rangno provided several cloud photos, and he and Mark Stoelinga provided valuable scientific advice.

Peter and I are indebted to numerous individuals who have generously contributed to the design, content, and production of this edition. Roland Stull at the University of British Columbia is the primary author of Chapter 9 (the Atmospheric Boundary Layer). Three of our colleagues in the Department of Atmospheric Sciences at the University of Washington served as advisors for portions of other chapters. Qiang Fu advised us on the design of Chapter 4 (Radiative Transfer) and provided some of the material for it. Lynn McMurdie selected the case study presented in Section 8.1 (Extratropical Cyclones) and advised us on the content for that section. Robert A. Houze advised us on the design and content of Section 8.3 (Convective Storms) and Section 8.4 (Tropical Cyclones). Other colleagues, Stephen Warren, Clifford Mass, Lyatt Jaegle, Andrew Rice, Marcia Baker, David Catling, Joel Thornton, and Greg Hakim, read and provided valuable feedback on early drafts of chapters. Others who provided valuable feedback and technical advice on specific parts of the manuscript include Edward Sarachik, Igor Kamenkovich, Richard Gammon, Joellen Russell, Conway Leovy, Norbert Untersteiner, Kenneth Beard, William Cotton, Hermann Gerber, Shuyi Chen, Howard Bluestein, Robert Wood, Adrian Simmons, Michael King, David Thompson, Judith Lean, Alan Robock, Peter Lynch, Paquita Zuidema, Cody Kirkpatrick, and J. R. Bates. I also thank the graduate students who volunteered their help in identifying errors, inconsistencies, and confusing passages in the numerous drafts of the manuscript.

Jennifer Adams, a research scientist at the Center for Ocean–Land–Atmosphere Studies (COLA), produced most of the illustrations that appear in Section 8.1, under funding provided by COLA and using graphics software (GrADS) developed at COLA. Some of the design elements in the illustrations were provided by David W. Ehlert. Debra Wolf, Candace Gudmundson, Kay Dewar, and Michael Macaulay and Beth Tully prepared many of the illustrations. Steven Cavallo and Robert Nicholas provided the table of units and numerical values. Most of the photographs of clouds and other atmospheric phenomena that appear in the book were generously provided free of charge.

I am deeply indebted to Qiang Fu and Peter Lynch who generously volunteered their time to correct errors in the equations, as well as James Booth, Joe Casola, Ioana Dima, Chaim Garfinkel, David Reidmiller Kevin Rennert, Rei Ueyama, Justin Wettstein and Reddy Yatavelli who identified many errors in the cross-referencing.

Finally, I thank Peter's wife, Sylvia, and my wife, Susan, for their forbearance during the many evenings and weekends in which we were preoccupied with this project.

Preface to the First Edition

This book has been written in response to a need for a text to support several of the introductory courses in atmospheric sciences commonly taught in universities, namely introductory survey courses at the junior or senior undergraduate level and beginning graduate level, the undergraduate physical meteorology course, and the undergraduate synoptic laboratory. These courses serve to introduce the student to the fundamental physical principles upon which the atmospheric sciences are based and to provide an elementary description and interpretation of the wide range of atmospheric phenomena dealt with in detail in more advanced courses. In planning the book we have assumed that students enrolled in such courses have already had some exposure to calculus and physics at the first-year college level and to chemistry at the high school level.

The subject material is almost evenly divided between physical and dynamical meteorology. In the general area of physical meteorology we have introduced the basic principles of atmospheric hydrostatics and thermodynamics, cloud physics, and radiative transfer (Chapters 2, 4, and 6, respectively). In addition, we have covered selected topics in atmospheric chemistry, aerosol physics, atmospheric electricity, aeronomy, and physical climatology. Coverage of dynamical meteorology consists of a description of large-scale atmospheric motions and an elementary interpretation of the general circulation (Chapters 3, 8, and 9, respectively). In the discussion of clouds and storms (Chapter 5) we have attempted to integrate material from physical and dynamical meteorology. In arranging the chapters we have purposely placed the material on synoptic meteorology near the beginning of the book (Chapter 3) in order to have it available as an introduction to the daily weather map discussions, which are an integral part of many introductory survey courses.

The book is divided into nine chapters. Most of the basic theoretical material is covered in the even-numbered chapters (2, 4, 6, and 8). Chapters 1 and 3 are almost entirely descriptive, while Chapters 5, 7, and 9 are mainly interpretive in character. Much of the material in the odd-numbered chapters is straightforward enough to be covered by means of reading assignments, especially in graduate courses. However, even with extensive use of reading assignments we recognize that it may not be possible to completely cover a book of this length in a one-semester undergraduate course. In order to facilitate the use of the book for such courses, we have purposely arranged the theoretical chapters in such a way that certain of the more difficult sections can be omitted without serious loss of continuity. These sections are indicated by means of footnotes.

Descriptive and interpretive material in the other chapters can be omitted at the option of the instructor.

The book contains 150 numerical problems and 208 qualitative problems that illustrate the application of basic physical principles to problems in the atmospheric sciences. In addition, the solutions of 48 of the numerical problems are incorporated into the text. We have purposely designed problems that require a minimum amount of mathematical manipulation in order to place primary emphasis on the proper application of physical principles. Universal constants and other data needed for the solution of quantitative problems are given on pages xvi–xvii.

It should be noted that many of the qualitative problems at the ends of the chapters require some original thinking on the part of the student. We have found such questions useful as a means of stimulating classroom discussion and helping the students to prepare for examinations.

Throughout the book we have consistently used SI units, which are rapidly gaining acceptance within the atmospheric sciences community.

A list of units and symbols is given on pages xv–xvi.

The book contains biographical footnotes that summarize the lives and work of scientists who have made major contributions to the atmospheric sciences. Brief as these are, it is hoped that they will give the student a sense of the long history of meteorology and its firm foundations in the physical sciences. As a matter of policy we have included footnotes only for individuals who are deceased or retired.

We express our gratitude to the University of Washington and the National Science Foundation for their support of our teaching, research, and other scholarly activities that contributed to this book. While working on the book, one of us (J.M.W.) was privileged to spend 6 months on an exchange visit to the Computer Center of the Siberian Branch of the Soviet Academy of Sciences, Novosibirsk, USSR, and a year at the US National Center for Atmospheric Sciences under the auspices of the Advanced Study Program. The staff members and visitors at both of these institutions made many important contributions to the scientific content of the book. Thanks go also to many other individuals in the scientific community who provided help and guidance.

We wish especially to express our gratitude to colleagues in our own department who provided a continuous source of moral support, constructive criticism, and stimulating ideas. Finally, we acknowledge the help received from many individuals who aided in the preparation of the final manuscript, as well as the many interim manuscripts that preceded it.

Introduction and Overview

1

1.1 Scope of the Subject and Recent Highlights

Atmospheric science is a relatively new, applied discipline that is concerned with the structure and evolution of the planetary atmospheres and with the wide range of phenomena that occur within them. To the extent that it focuses mainly on the Earth's atmosphere, atmospheric science can be regarded as one of the *Earth* or *geo*sciences, each of which represents a particular fusion of elements of physics, chemistry, and fluid dynamics.

The historical development of atmospheric sciences, particularly during the 20th century, has been driven by the need for more accurate weather forecasts. In popular usage the term "meteorologist," a synonym for atmospheric scientist, means "weather forecaster." During the past century, weather forecasting has evolved from an art that relied solely on experience and intuition into a science that relies on numerical models based on the conservation of mass, momentum, and energy. The increasing sophistication of the models has led to dramatic improvements in forecast skill, as documented in Fig. 1.1. Today's weather forecasts address not only the deterministic, day-to-day evolution of weather patterns over the course of the next week or two, but also the likelihood of hazardous weather events (e.g., severe thunderstorms, freezing rain) on an hour-by-hour basis (so called "nowcasting"), and departures of the climate (i.e., the statistics of weather) from seasonally adjusted normal values out to a year in advance.

Weather forecasting has provided not only the intellectual motivation for the development of atmospheric science, but also much of the infrastructure. What began in the late 19th century as an assemblage of regional collection centers for real time teletype transmissions of observations of surface weather variables has evolved into a sophisticated *observing system* in which satellite and in situ measurements of many surface and upper air variables are merged (or *assimilated*) in a dynamically consistent way to produce optimal estimates of their respective three-dimensional fields over the entire globe. This global, real time atmospheric dataset is the envy of oceanographers and other geo- and planetary scientists: it represents both an extraordinary technological achievement and an exemplar of the benefits that can derive from international cooperation. Today's global weather observing system is a vital component of a broader Earth observing system, which supports a wide variety of scientific endeavors, including climate monitoring and studies of ecosystems on a global scale.

A newer, but increasingly important organizing theme in atmospheric science is *atmospheric chemistry*. A generation ago, the principal focus of this field was urban air quality. The field experienced

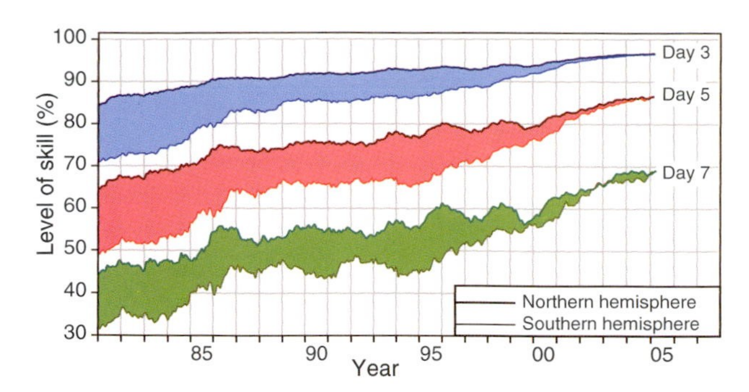

Fig. 1.1 Improvement of forecast skill with time from 1981 to 2003. The ordinate is a measure of forecast skill, where 100% represents a perfect forecast of the hemispheric flow pattern at the 5-km level. The upper pair of curves is for 3-day forecasts, the middle pair for 5-day forecasts, and the lower pair for 7-day forecasts. In each pair, the upper curve that marks the top of the band of shading represents the skill averaged over the northern hemisphere and the lower curve represents the skill averaged over the southern hemisphere. Note the continually improving skill levels (e.g., today's 5-day forecasts of the northern hemisphere flow pattern are nearly as skillful as the 3-day forecasts of 20 years ago). The more rapid increase in skill in the southern hemisphere reflects the progress that has been made in assimilating satellite data into the forecast models. [Updated from *Quart. J. Royal Met. Soc.*, **128**, p. 652 (2002). Courtesy of the European Centre for Medium-Range Weather Forecasting.]

Fig. 1.2 The Antarctic ozone hole induced by the buildup of synthetic chlorofluorocarbons, as reflected in the distribution of vertically integrated ozone over high latitudes of the southern hemisphere in September, 2000. Blue shading represents substantially reduced values of total ozone relative to the surrounding region rendered in green and yellow. [Based on data from NASA TOMS Science Team; figure produced by NASA's Scientific Visualization Studio.]

a renaissance during the 1970s when it was discovered that forests and organisms living in lakes over parts of northern Europe, the northeastern United States, and eastern Canada were being harmed by *acid rain* caused by sulfur dioxide emissions from coal-fired electric power plants located hundreds and, in some cases, thousands of kilometers upwind. The sources of the acidity are gaseous oxides of sulfur and nitrogen (SO_2, NO, NO_2, and N_2O_5) that dissolve in microscopic cloud droplets to form weak solutions of sulfuric and nitric acids that may reach the ground as raindrops.

There is also mounting evidence of the influence of human activity on the composition of the global atmosphere. A major discovery of the 1980s was the *Antarctic "ozone hole"*: the disappearance of much of the stratospheric ozone layer over the southern polar cap each spring (Fig. 1.2). The ozone destruction was found to be caused by the breakdown of chlorofluorocarbons (CFCs), a family of synthetic gases that was becoming increasingly widely used for refrigeration and various industrial purposes. As in the acid rain problem, heterogeneous chemical reactions involving cloud droplets were implicated, but in the case of the "ozone hole" they were taking place in wispy polar stratospheric clouds. Knowledge gained from atmospheric chemistry research has been instrumental in the design of policies to control and ultimately reverse the spread of acid rain and the ozone hole. The unresolved scientific issues surrounding *greenhouse warming* caused by the buildup of carbon dioxide (Fig. 1.3) and other trace gases released into the atmosphere by human activities pose a new challenge for atmospheric chemistry and for the broader field of geochemistry.

Atmospheric science also encompasses the emerging field of *climate dynamics*. As recently as a generation ago, climatic change was viewed by most atmospheric scientists as occurring on such long timescales that, for most purposes, today's climate could be described in terms of a standard set of

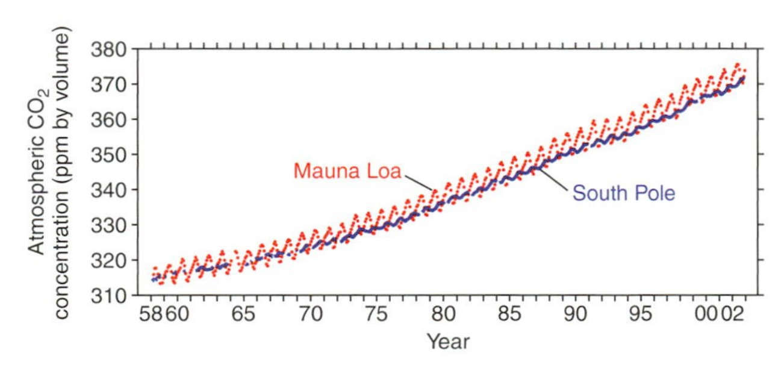

Fig. 1.3 Time series showing the upward trend in monthly mean atmospheric CO_2 concentrations (in parts per million by volume) at Mauna Loa and the South Pole due to the burning of fossil fuels. A pronounced annual cycle is also evident at Mauna Loa, with minimum values in the summer. [Based on data of C. D. Keeling. Courtesy of Todd P. Mitchell.]

statistics, such as January climatological-mean (or "normal") temperature. In effect, climatology and climate change were considered to be separate subfields, the former a branch of atmospheric sciences and the latter largely the province of disciplines such as geology, paleobotany, and geochemistry. Among the factors that have contributed to the emergence of a more holistic, dynamic view of climate are:

- documentation of a coherent pattern of year-to-year climate variations over large areas of the globe that occurs in association with El Niño (Section 10.2).
- proxy evidence, based on a variety of sources (ocean sediment cores and ice cores, in particular), indicating that large, spatially coherent climatic changes have occurred on time scales of a century or even less (Section 2.5.4).
- the rise of the global-mean surface air temperature during the 20th century and projections of a larger rise during the 21st century due to human activities (Section 10.4).

Like some aspects of atmospheric chemistry, climate dynamics is inherently multidisciplinary: to understand the nature and causes of climate variability, the atmosphere must be treated as a component of the *Earth system*.

1.2 Some Definitions and Terms of Reference

Even though the Earth is not perfectly spherical, atmospheric phenomena are adequately represented in terms of a spherical coordinate system, rotating with the Earth, as illustrated in Fig. 1.4. The coordinates are latitude ϕ, longitude λ, and height z above sea level, z.[1] The angles are often replaced by the distances

$$dx \equiv r\, d\lambda \cos\phi \tag{1.1}$$

and

$$dy \equiv r\, d\phi$$

where x is distance east of the Greenwich meridian along a latitude circle, y is distance north of the

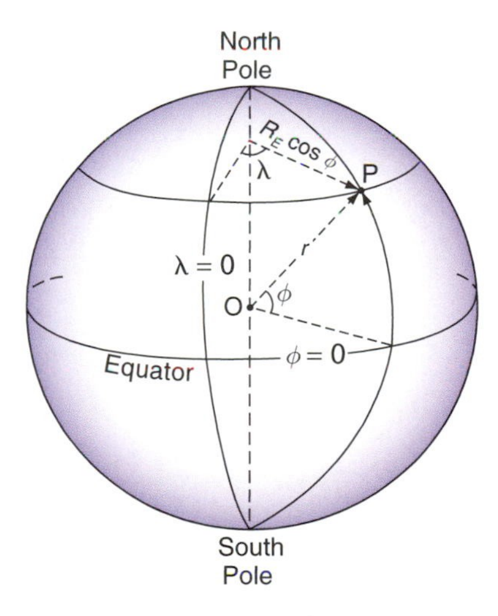

Fig. 1.4 Coordinate system used in atmospheric science. Angle ϕ is latitude, defined as positive in the northern hemisphere and negative in the southern hemisphere, and λ is longitude relative to the Greenwich meridian, positive eastward. The radial coordinate (not shown) is height above sea level.

[1] Oceanographers and applied mathematicians often use the colatitude $\theta = \pi/2 - \phi$ instead of ϕ.

Fig. 1.5 The limb of the Earth, as viewed from space in visible satellite imagery. The white layer is mainly light scattered from atmospheric aerosols and the overlying blue layer is mainly light scattered by air molecules. [NASA Gemini-4 photo. Photograph courtesy of NASA.]

equator, and r is the distance from the center of the Earth. At the Earth's surface a degree of latitude is equivalent to 111 km (or 60 nautical miles). Because 99.9% of the mass of the atmosphere is concentrated within the lowest 50 km, a layer with a thickness less than 1% of the radius of the Earth, r, is nearly always replaced by the mean radius of the Earth (6.37×10^6 m), which we denote by the symbol R_E. Images of the limb of the Earth (Fig. 1.5) emphasize how thin the atmosphere really is.

The three velocity components used in describing atmospheric motions are defined as

$$u \equiv \frac{dx}{dt} = R_E \cos\phi \frac{d\lambda}{dt} \quad \text{(the } \textit{zonal velocity component}\text{)}, \tag{1.2}$$

$$v \equiv \frac{dy}{dt} = R_E \frac{d\phi}{dt} \quad \text{(the } \textit{meridional velocity component}\text{)},$$

and

$$w \equiv \frac{dz}{dt} = \frac{dr}{dt} \quad \text{(the } \textit{vertical velocity component}\text{)}.$$

where z is height above mean sea level. The adjectives *zonal* and *meridional* are also commonly used in reference to averages, gradients, and cross sections. For example, a *zonal average* denotes an average around latitude circles; a *meridional cross section* denotes a north–south slice through the atmosphere. The *horizontal velocity vector* **V** is given by $\mathbf{V} \equiv u\mathbf{i} + v\mathbf{j}$, where **i** and **j** are the unit vectors in the zonal and meridional directions, respectively. Positive and negative zonal velocities are referred to as *westerly* (from the west) and *easterly* (from the east) winds, respectively; positive and negative meridional velocities are referred to as *southerly* and *northerly* winds (in both northern and southern hemispheres, respectively.[2] For scales of motion in the Earth's atmosphere in excess of 100 km, the length scale greatly exceeds the depth scale, and typical magnitudes of the horizontal velocity component **V** exceed those of the vertical velocity component w by several orders of magnitude. For these scales the term *wind* is synonymous with *horizontal velocity component*. The SI unit for velocity (or speed) is m s^{-1}. One meter per second is equivalent to 1.95 knots (1 knot = 1 nautical mile per hour). Vertical velocities in large-scale atmospheric motions are often expressed in units of cm s^{-1}: 1 cm s^{-1} is roughly equivalent to a vertical displacement of 1 kilometer per day.

Throughout this book, the local derivative $\partial/\partial t$ refers to the rate of change at a fixed point in rotating (x, y, z) space and the total time derivative d/dt refers to the rate of change following an air parcel as it moves along its three-dimensional trajectory through the atmosphere. These so-called *Eulerian*[3]

[2] Dictionaries offer contradictory definitions of these terms, derived from different traditions.

[3] **Leonhard Euler** (1707–1783) Swiss mathematician. Held appointments at the St. Petersburg Academy of Sciences and the Berlin Academy. Introduced the mathematical symbols e, i, and $f(x)$. Made fundamental contributions in optics, mechanics, electricity, and magnetism, differential equations, and number theory. First to describe motions in a rotating coordinate system. Continued to work productively after losing his sight by virtue of his extraordinary memory.

and *Lagrangian*[4] rates of change are related by the chain rule

$$\frac{d}{dt} = \frac{\partial}{\partial t} + u\frac{\partial}{\partial x} + v\frac{\partial}{\partial y} + w\frac{\partial}{\partial z}$$

which can be rewritten in the form

$$\frac{\partial}{\partial t} = \frac{d}{dt} - u\frac{\partial}{\partial x} - v\frac{\partial}{\partial y} - w\frac{\partial}{\partial z} \tag{1.3}$$

The terms involving velocities in Eq. (1.3), including the minus signs in front of them, are referred to as *advection terms*. At a fixed point in space the Eulerian and Lagrangian rates of change of a variable ψ differ by virtue of the advection of air from upstream, which carries with it higher or lower values of ψ. For a hypothetical *conservative tracer*, the Lagrangian rate of change is identically equal to zero, and the Eulerian rate of change is

$$\frac{\partial}{\partial t} = -u\frac{\partial}{\partial x} - v\frac{\partial}{\partial y} - w\frac{\partial}{\partial z}$$

The fundamental thermodynamic variables are pressure p, density ρ, and temperature T. The SI unit of pressure is $1\ \text{N m}^{-2} = 1\ \text{kg m}^{-1}\ \text{s}^{-2} = 1$ pascal (Pa). Prior to the adoption of SI units, atmospheric pressure was expressed in millibars (mb), where $1\ \text{bar} = 10^6\ \text{g cm}^{-1}\ \text{s}^{-2} = 10^6$ dynes. In the interests of retaining the numerical values of pressure that atmospheric scientists and the public have become accustomed to, atmospheric pressure is usually expressed in units of hundreds of (i.e., hecto) pascals (hPa).[5] Density is expressed in units of kg m^{-3} and temperature in units of °C or K, depending on the context, with °C for temperature differences and K for the values of temperature itself. Energy is expressed in units of joules ($\text{J} = \text{kg m}^2\ \text{s}^{-2}$).

Atmospheric phenomena with timescales shorter than a few weeks, which corresponds to the theoretical limit of the range of deterministic (day-by-day) weather forecasting, are usually regarded as relating to *weather*, and phenomena on longer timescales as relating to *climate*. Hence, the adage (intended to apply to events a month or more in the future): "Climate is what you expect; weather is what you get." Atmospheric variability on timescales of months or longer is referred to as *climate variability*, and statistics relating to conditions in a typical (as opposed to a particular) season or year are referred to as *climatological-mean* statistics.

1.1 Atmospheric Predictability and Chaos

Atmospheric motions are inherently unpredictable as an initial value problem (i.e., as a system of equations integrated forward in time from specified initial conditions) beyond a few weeks. Beyond that time frame, uncertainties in the forecasts, no matter how small they might be in the initial conditions, become as large as the observed variations in atmospheric flow patterns. Such exquisite *sensitivity to initial conditions* is characteristic of a broad class of mathematical models of real phenomena, referred to as *chaotic nonlinear systems*. In fact, it was the growth of errors in a highly simplified weather forecast model that provided one of the most lucid early demonstrations of this type of behavior.

In 1960, Professor Edward N. Lorenz in the Department of Meteorology at MIT decided to rerun an experiment with a simplified atmospheric model in order to extend his "weather forecast" farther out into the future. To his surprise, he found that he was unable to duplicate his previous forecast. Even though the code and the prescribed initial conditions in the two experiments were identical, the states of the model in the two forecasts

Continued on next page

[4] **Joseph Lagrange** (1736–1813) French mathematician and mathematical physicist. Served as director of the Berlin Academy, succeeding Euler in that role. Developed the calculus of variations and also made important contributions to differential equations and number theory. Reputed to have told his students "Read Euler, read Euler, he is our master in everything."

[5] Although the pressure will usually be expressed in hectopascals (hPa) in the text, it should be converted to pascals (Pa) when working quantitative exercises that involve a mix of units.

1.1 Continued

diverged, over the course of the first few hundred time steps, to the point that they were no more like one another than randomly chosen states in experiments started from entirely different initial conditions. Lorenz eventually discovered that the computer he was using was introducing round-off errors in the last significant digit that were different each time he ran the experiment. Differences between the "weather patterns" in the different runs were virtually indistinguishable at first, but they grew with each time step until they eventually became as large as the range of variations in the individual model runs.

Lorenz's model exhibited another distinctive and quite unexpected form of behavior. For long periods of (simulated) time it would oscillate around some "climatological-mean" state. Then, for no apparent reason, the state of the model would undergo an abrupt "regime shift" and begin to oscillate around another quite different state, as illustrated in Fig. 1.6. Lorenz's model exhibited two such preferred "climate regimes." When the state of the model resided within one of these regimes, the "weather" exhibited quasi-periodic oscillations and consequently was predictable quite far into the future. However, the shifts between regimes were abrupt, irregular, and inherently unpredictable beyond a few simulated days. Lorenz referred to the two climates in the model as *attractors*.

The behavior of the real atmosphere is much more complicated than that of the highly simplified model used by Lorenz in his experiments. Whether the Earth's climate exhibits such regime-like behavior, with multiple "attractors," or whether it should be viewed as varying about a single state that varies in time in response to solar, orbital, volcanic, and anthropogenic forcing is a matter of ongoing debate.

Fig. 1.6 The history of the state of the model used by Lorenz can be represented as a trajectory in a three-dimensional space defined by the amplitudes of the model's three dependent variables. Regime-like behavior is clearly apparent in this rendition. Oscillations around the two different "climate attractors" correspond to the two, distinctly different sets of spirals, which lie in two different planes in the three-dimensional phase space. Transitions between the two regimes occur relatively infrequently. [Permission to use figure from *Nature*, **406**, p. 949 (2000). © Copyright 2000 Nature Publishing Group. Courtesy of Paul Bourke.]

1.3 A Brief Survey of the Atmosphere

The remainder of this chapter provides an overview of the optical properties, composition, and vertical structure of the Earth's atmosphere, the major wind systems, and the climatological-mean distribution of precipitation. It introduces some of the terminology that will be used in subsequent chapters and some of the conventions that will be used in performing calculations involving amounts of mass and rates of movement.

1.3.1 Optical Properties

The Earth's atmosphere is relatively transparent to incoming solar radiation and opaque to outgoing radiation emitted by the Earth's surface. The blocking of outgoing radiation by the atmosphere, popularly referred to as the *greenhouse effect*, keeps the surface of the Earth warmer than it would be in the absence of an atmosphere. Much of the absorption and reemission of outgoing radiation are due to air molecules, but cloud droplets also play a significant role. The radiation emitted to space by air molecules and cloud droplets provides a basis for remote sensing of the three-dimensional distribution of temperature and various atmospheric constituents using satellite-borne sensors.

The atmosphere also scatters the radiation that passes through it, giving rise to a wide range of optical effects. The blueness of the outer atmosphere in Fig. 1.5 is due to the preferential scattering of incoming short wavelength (solar) radiation by air molecules, and the whiteness of lower layers is due to scattering from cloud droplets and atmospheric aerosols (i.e., particles). The backscattering of solar radiation off the top of the deck of low clouds off the California coast in Fig. 1.7 greatly enhances the whiteness (or reflectivity) of that region as viewed from space. Due to the presence of clouds and aerosols in the Earth's atmosphere, ~22% of the incoming solar radiation is backscattered to space without being absorbed. The backscattering of radiation by clouds and aerosols has a cooling effect on climate at the Earth's surface, which opposes the greenhouse effect.

Fig. 1.7 A deck of low clouds off the coast of California, as viewed in reflected visible radiation. [NASA MODIS imagery. Photograph courtesy of NASA.]

1.3.2 Mass

At any point on the Earth's surface, the atmosphere exerts a downward force on the underlying surface due to the Earth's gravitational attraction. The downward force (i.e., the *weight*) of a unit volume of air with density ρ is given by

$$F = \rho g \tag{1.4}$$

where g is the acceleration due to gravity. Integrating Eq. (1.4) from the Earth's surface to the "top" of the atmosphere, we obtain the atmospheric pressure on the Earth's surface p_s due to the weight (per unit area) of the air in the overlying column

$$p_s = \int_0^\infty \rho g dz \tag{1.5}$$

Neglecting the small variation of g with latitude, longitude and height, setting it equal to its mean value of $g_0 = 9.807$ m s^{-2}, we can take it outside the integral, in which case, Eq. (1.5) can be written as

$$p_s = m g_0 \tag{1.6}$$

where $m = \int_0^\infty \rho dz$ is the vertically integrated mass per unit area of the overlying air.

Exercise 1.1 The globally averaged surface pressure is 985 hPa. Estimate the mass of the atmosphere.

Solution: From Eq. (1.6), it follows that

$$\overline{m} = \frac{\overline{p_s}}{g_0}$$

where the overbars denote averages over the surface of the Earth. In applying this relationship the pressure must be expressed in pascals (Pa). Substituting numerical values we obtain

$$\overline{m} = \frac{985 \times 10^2 \text{ Pa/hPa}}{9.807} = 1.004 \times 10^4 \text{ kg m}^{-2}$$

The mass of the atmosphere[6] is

$$\begin{aligned} M_{atm} &= 4\pi R_E^2 \times \overline{m} \\ &= 4\pi \times (6.37 \times 10^6)^2 \text{ m}^2 \times 1.004 \times 10^4 \text{ kg m}^{-2} \\ &= 5.10 \times 10^{14} \text{ m}^2 \times 1.004 \times 10^4 \text{ kg m}^{-2} \\ &= 5.10 \times 10^{18} \text{ kg} \end{aligned}$$

■

1.3.3 Chemical Composition

The atmosphere is composed of a mixture of gases in the proportions shown in Table 1.1, where fractional concentration *by volume* is the same as that based on numbers of molecules, or partial pressures exerted by the gases, as will be explained more fully in Section 3.1. The fractional concentration *by mass* of a constituent is computed by weighting its fractional concentration *by volume* by its molecular weight, i.e.,

$$\frac{m_i}{\Sigma m_i} = \frac{n_i M_i}{\Sigma n_i M_i} \tag{1.7}$$

where m_i is the mass, n_i the number of molecules, and M_i the molecular weight of the ith constituent, and the summations are over all constituents.

Table 1.1 Fractional concentrations by volume of the major gaseous constituents of the Earth's atmosphere up to an altitude of 105 km, with respect to dry air

Constituent[a]	Molecular weight	Fractional concentration by volume
Nitrogen (N_2)	28.013	78.08%
Oxygen (O_2)	32.000	20.95%
Argon (Ar)	39.95	0.93%
Water vapor (H_2O)	18.02	0–5%
Carbon dioxide (CO_2)	44.01	380 ppm
Neon (Ne)	20.18	18 ppm
Helium (He)	4.00	5 ppm
Methane (CH_4)	16.04	1.75 ppm
Krypton (Kr)	83.80	1 ppm
Hydrogen (H_2)	2.02	0.5 ppm
Nitrous oxide (N_2O)	56.03	0.3 ppm
Ozone (O_3)	48.00	0–0.1 ppm

[a] So called *greenhouse gases* are indicated by bold-faced type. For more detailed information on minor constituents, see Table 5.1.

Diatomic nitrogen (N_2) and oxygen (O_2) are the dominant constituents of the Earth's atmosphere, and argon (Ar) is present in much higher concentrations than the other noble gases (neon, helium, krypton, and xenon). Water vapor, which accounts for roughly 0.25% of the mass of the atmosphere, is a highly variable constituent, with concentrations ranging from around 10 parts per million by volume (ppmv) in the coldest regions of the Earth's atmosphere up to as much as 5% by volume in hot, humid air masses; a range of more than three orders of magnitude. Because of the large variability of water vapor concentrations in air, it is customary to list the percentages of the various constituents in relation to dry air. Ozone concentrations are also highly variable. Exposure to ozone concentrations >0.1 ppmv is considered hazardous to human health.

For reasons that will be explained in Section 4.4, gas molecules with certain structures are highly effective at trapping outgoing radiation. The most important of these so-called *greenhouse gases* are water vapor, carbon dioxide, and ozone. Trace constituents CH_4, N_2O, CO, and chlorofluorocarbons (CFCs) are also significant contributors to the greenhouse effect.

Among the atmosphere's trace gaseous constituents are molecules containing carbon, nitrogen, and sulfur atoms that were formerly incorporated into the cells of living organisms. These gases enter the atmosphere through the burning of plant matter and fossil fuels, emissions from plants, and the decay of plants and animals. The chemical transformations that remove these chemicals from the atmosphere involve oxidation, with the hydroxyl (OH) radical playing an important role. Some of the nitrogen and sulfur compounds are converted into particles that are eventually "scavenged" by raindrops, which contribute to acid deposition at the Earth's surface.

[6] When the vertical and meridional variations in g and the meridional variations in the radius of the Earth are accounted for, the mass per unit area and the total mass of the atmosphere are ~0.4% larger than the estimates derived here.

Although aerosols and cloud droplets account for only a minute fraction of the mass of the atmosphere, they mediate the condensation of water vapor in the atmospheric branch of the hydrologic cycle, they participate in and serve as sites for important chemical reactions, and they give rise to electrical charge separation and a variety of atmospheric optical effects.

1.3.4 Vertical structure

To within a few percent, the density of air at sea level is 1.25 kg m^{-3}. Pressure p and density ρ decrease nearly exponentially with height, i.e.,

$$p \simeq p_0 e^{-z/H} \tag{1.8}$$

where H, the e-folding depth, is referred to as the *scale height* and p_0 is the pressure at some reference level, which is usually taken as sea level ($z = 0$). In the lowest 100 km of the atmosphere, the scale height ranges roughly from 7 to 8 km. Dividing Eq. (1.8) by p_0 and taking the natural logarithms yields

$$\ln \frac{p}{p_0} \simeq -\frac{z}{H} \tag{1.9}$$

This relationship is useful for estimating the height of various pressure levels in the Earth's atmosphere.

Exercise 1.2 At approximately what height above sea level $\overline{z}_m$ does half the mass of the atmosphere lie above and the other half lie below? [Hint: Assume an exponential pressure dependence with $H = 8$ km and neglect the small vertical variation of g with height.]

Solution: Let $\overline{p}_m$ be the pressure level that half the mass of the atmosphere lies above and half lies below. The pressure at the Earth's surface is equal to the weight (per unit area) of the overlying column of air. The same is true of the pressure at any level in the atmosphere. Hence, $\overline{p}_m = \overline{p}_0/2$ where $\overline{p}_0$ is the global-mean sea-level pressure. From Eq. (1.9)

$$\overline{z}_m = -H \ln 0.5 = H \ln 2$$

Substituting $H = 8$ km, we obtain

$$\overline{z}_m = 8 \text{ km} \times 0.693 \sim 5.5 \text{ km}$$

Because the pressure at a given height in the atmosphere is a measure of the mass that lies above that level, it is sometimes used as a vertical coordinate in lieu of height. In terms of mass, the 500-hPa level, situated at a height of around 5.5 km above sea level, is roughly halfway up to the top the atmosphere. ■

Density decreases with height in the same manner as pressure. These vertical variations in pressure and density are much larger than the corresponding horizontal and time variations. Hence it is useful to define a *standard atmosphere*, which represents the horizontally and temporally averaged structure of the atmosphere as a function of height only, as shown in Fig. 1.8. The nearly exponential height dependence of pressure and density can be inferred from the fact that the observed vertical profiles of pressure and density on these semilog plots closely resemble straight lines. The reader is invited to verify in Exercise 1.14 at the end of this chapter that the corresponding 10-folding depth for pressure and density is ~17 km.

Exercise 1.3 Assuming an exponential pressure and density dependence with $H = 7.5$ km, estimate the heights in the atmosphere at which (a) the air density is equal to 1 kg m^{-3} and (b) the height at which the pressure is equal to 1 hPa.

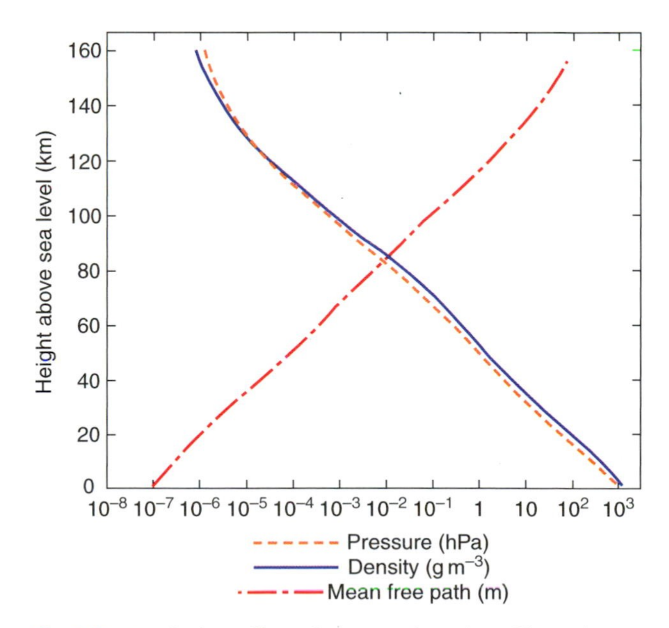

Fig. 1.8 Vertical profiles of pressure in units of hPa, density in units of kg m^{-3}, and mean free path (in meters) for the U.S. Standard Atmosphere.

Solution: Solving Eq. (1.9), we obtain $z = H \ln(p_0/p)$, and similarly for density. Hence, the heights are (a)

$$7.5 \text{ km} \times \ln\left(\frac{1.25}{1.00}\right) = 1.7 \text{ km}$$

for the 1-kg m^{-3} density level and (b)

$$7.5 \text{ km} \times \ln\left(\frac{1000}{1.00}\right) = 52 \text{ km}$$

for the 1-hPa pressure level. Because H varies with height, geographical location, and time, and the reference values ρ_0 and p_0 also vary, these estimates are accurate only to within ~10%. ■

Exercise 1.4 Assuming an exponential pressure and density dependence, calculate the fraction of the total mass of the atmosphere that resides between 0 and 1 scale height, 1 and 2 scale heights, 2 and 3 scale heights, and so on above the surface.

Solution: Proceeding as in Exercise 1.2, the fraction of the mass of the atmosphere that lies between 0 and 1, 1 and 2, 2 and 3, and so on scale heights above the Earth's surface is $e^{-1}, e^{-2}, \ldots e^{-N}$ from which it follows that the fractions of the mass that reside in the 1st, 2nd . . ., N^{th} scale height above the surface are $1 - e^{-1}, e^{-1}(1 - e^{-1}), e^{-2}(1 - e^{-1}) \ldots, e^{-N}(1 - e^{-1})$, where N is the height of the base of the layer expressed in scale heights above the surface. The corresponding numerical values are 0.632, 0.233, 0.086 . . . ■

Throughout most of the atmosphere the concentrations of N_2, O_2, Ar, CO_2, and other long-lived constituents tend to be quite uniform and largely independent of height due to mixing by turbulent fluid motions.[7] Above ~105 km, where the mean free path between molecular collisions exceeds 1 m (Fig. 1.8), individual molecules are sufficiently mobile that each molecular species behaves as if it alone were present. Under these conditions, concentrations of heavier constituents decrease more rapidly with height than those of lighter constituents: the density of each constituent drops off exponentially with height, with a scale height inversely proportional to molecular weight, as explained in Section 3.2.2. The upper layer of the atmosphere in which the lighter molecular species become increasingly abundant (in a relative sense) with increasing height is referred to as the *heterosphere*. The upper limit of the lower, well-mixed regime is referred to as the *turbopause*, where *turbo* refers to turbulent fluid motions and *pause* connotes limit of.

The composition of the outermost reaches of the atmosphere is dominated by the lightest molecular species (H, H_2, and He). During periods when the sun is active, a very small fraction of the hydrogen atoms above 500 km acquire velocities high enough to enable them to escape from the Earth's gravitational field during the long intervals between molecular collisions. Over the lifetime of the Earth the leakage of hydrogen atoms has profoundly influenced the chemical makeup of the Earth system, as discussed in Section 2.4.1.

The vertical distribution of temperature for typical conditions in the Earth's atmosphere, shown in Fig. 1.9, provides a basis for dividing the atmosphere into four layers (*troposphere*, *stratosphere*,

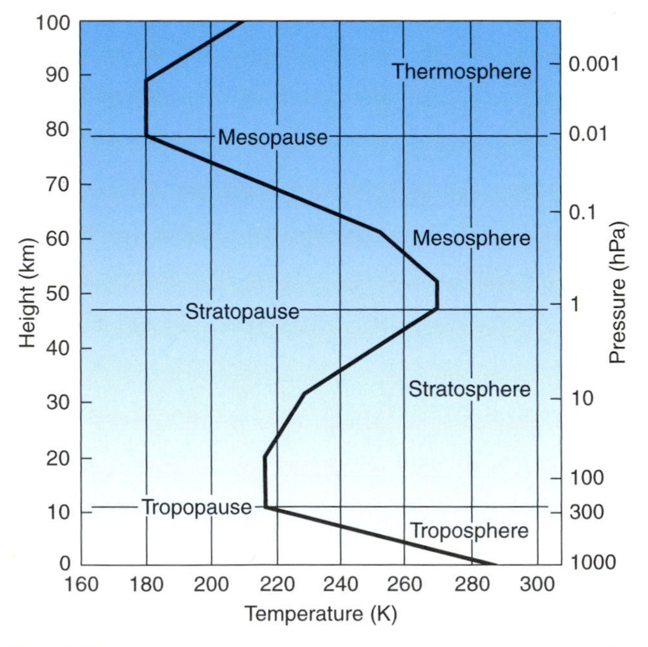

Fig. 1.9 A typical midlatitude vertical temperature profile, as represented by the U.S. Standard Atmosphere.

[7] In contrast, water vapor tends to be concentrated within the lowest few kilometers of the atmosphere because it condenses and precipitates out when air is lifted. Ozone are other highly reactive trace species exhibit heterogeneous distributions because they do not remain in the atmosphere long enough to become well mixed.

mesosphere, and *thermosphere*), the upper limits of which are denoted by the suffix *pause*.

The *tropo*(turning or changing)*sphere* is marked by generally decreasing temperatures with height, at an average *lapse rate*, of ~6.5 °C km^{-1}. That is to say,

$$\Gamma \equiv \frac{\partial T}{\partial z} \sim 6.5\ ^{\circ}\mathrm{C\ km}^{-1} = 0.0065\ ^{\circ}\mathrm{C\ m}^{-1}$$

where T is temperature and Γ is the lapse rate. Tropospheric air, which accounts for ~80% of the mass of the atmosphere, is relatively well mixed and it is continually being cleansed or scavenged of aerosols by cloud droplets and ice particles, some of which subsequently fall to the ground as rain or snow. Embedded within the troposphere are thin layers in which temperature increases with height (i.e., the lapse rate Γ is negative). Within these so-called *temperature inversions* it is observed that vertical mixing is strongly inhibited.

Within the *strato*-(layered)-*sphere*, vertical mixing is strongly inhibited by the increase of temperature with height, just as it is within the much thinner temperature inversions that sometimes form within the troposphere. The characteristic anvil shape created by the spreading of cloud tops generated by intense thunderstorms and volcanic eruptions when they reach the tropopause level, as illustrated in Fig. 1.10, is due to this strong stratification.

Cloud processes in the stratosphere play a much more limited role in removing particles injected by volcanic eruptions and human activities than they do in the troposphere, so residence times of particles tend to be correspondingly longer in the stratosphere. For example, the hydrogen bomb tests of the 1950s and early 1960s were followed by hazardous radioactive fallout events involving long-lived stratospheric debris that occurred as long as 2 years after the tests.

Fig. 1.10 A distinctive "anvil cloud" formed by the spreading of cloud particles carried aloft in an intense updraft when they encounter the tropopause. [Photograph courtesy of Rose Toomer and Bureau of Meteorology, Australia.]

Stratospheric air is extremely dry and ozone rich. The absorption of solar radiation in the ultraviolet region of the spectrum by this *stratospheric ozone layer* is critical to the habitability of the Earth. Heating due to the absorption of ultraviolet radiation by ozone molecules is responsible for the temperature maximum ~50 km that defines the stratopause.

Above the ozone layer lies the *mesosphere* (meso connoting "in between"), in which temperature decreases with height to a minimum that defines the *mesopause*. The increase of temperature with height within the *thermosphere* is due to the absorption of solar radiation in association with the dissociation of diatomic nitrogen and oxygen molecules and the stripping of electrons from atoms. These processes, referred to as *photodissociation* and *photoionization*, are discussed in more detail in Section 4.4.3. Temperatures in the Earth's outer thermosphere vary widely in response to variations in the emission of ultraviolet and x-ray radiation from the sun's outer atmosphere.

At any given level in the atmosphere temperature varies with latitude. Within the troposphere, the *climatological-mean* (i.e., the average over a large number of seasons or years), *zonally averaged* temperature generally decreases with latitude, as shown in Fig. 1.11. The meridional temperature gradient is substantially stronger in the winter hemisphere where the polar cap region is in darkness. The tropopause is clearly evident in Fig. 1.11 as a discontinuity in the lapse rate. There is a break between the tropical tropopause, with a mean altitude ~17 km, and the extratropical tropopause, with a mean altitude ~10 km. The tropical tropopause is remarkably cold, with temperatures as low as −80 °C. The remarkable dryness of the air within the stratosphere is strong evidence that most of it has entered by way of this "cold trap."

Exercise 1.5 Based on data shown in Fig. 1.11, estimate the mean lapse rate within the tropical troposphere.

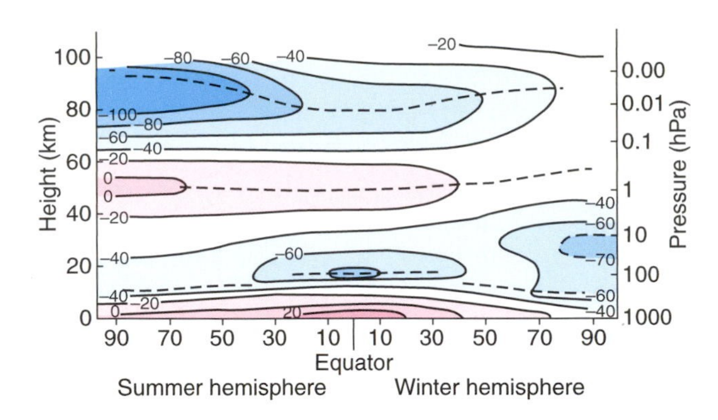

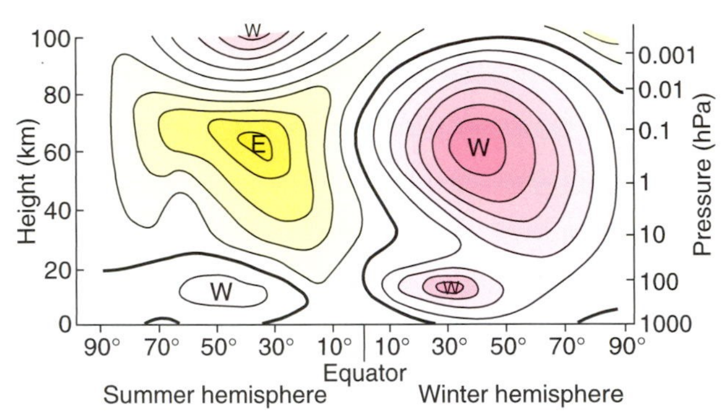

Fig. 1.11 Idealized meridional cross sections of zonally averaged temperature (in °C) (Top) and zonal wind (in m s^{-1}) (Bottom) around the time of the solstices, when the meridional temperature contrasts and winds are strongest. The contour interval is 20 °C; pink shading denotes relatively warm regions, and cyan shading relatively cold regions. The contour interval is 10 m s^{-1}; the zero contour is bold; pink shading and "W" labels denote westerlies, and yellow shading and "E" labels denote easterlies. Dashed lines indicate the positions of the tropopause, stratopause, and mesopause. This representation ignores the more subtle distinctions between northern and southern hemisphere climatologies. [Courtesy of Richard J. Reed.]

Solution: At sea level the mean temperature of the tropics is ~27 °C, the tropopause temperature is near −80 °C, and the altitude of the tropopause altitude is ~17 km. Hence the lapse-rate is roughly

$$\frac{[27 - (-80)]\ ^\circ\text{C}}{17\ \text{km}} = 6.3\ ^\circ\text{C km}^{-1}$$

Note that a decrease in temperature with height is implicit in the term (and definition of) *lapse rate*, so the algebraic sign of the answer is positive. ■

1.3.5 Winds

Differential heating between low and high latitudes gives rise to atmospheric motions on a wide range of scales. Prominent features of the so-called *atmospheric general circulation* include planetary-scale west-to-east (westerly) midlatitude *tropospheric jet streams*, centered at the tropopause break around 30° latitude, and lower *mesospheric jet streams*, both of which are evident in Fig. 1.11. The winds in the tropospospheric jet stream blow from the west throughout the year; they are strongest during winter and weakest during summer. In contrast, the mesospheric jet streams undergo a seasonal reversal: during winter they blow from the west and during summer they blow from the east.

Superimposed on the tropospheric jet streams are eastward propagating, *baroclinic waves* that feed upon and tend to limit the north–south temperature contrast across middle latitudes. Baroclinic waves are one of a number of types of *weather systems* that develop spontaneously in response to *instabilities* in the large-scale flow pattern in which they are embedded. The low level flow in baroclinic waves is dominated by *extratropical cyclones*, an example of which is shown in Fig. 1.12. The term *cyclone* denotes a closed circulation in which the air spins in the same sense as the Earth's rotation as viewed from above (i.e., counterclockwise in the northern hemisphere). At low levels the air spirals inward toward the center.[8] Much of the significant weather associated with extratropical cyclones is concentrated within narrow *frontal zones*, i.e., bands, a few tens of kilometers in width, characterized by strong horizontal temperature contrasts. Extratropical weather systems are discussed in Section 8.1.

Tropical cyclones (Fig. 1.13) observed at lower latitudes derive their energy not from the north–south temperature contrast, but from the release of latent heat of condensation of water vapor in deep convective clouds, as dicussed in Section 8.3. Tropical cyclones tend to be tighter and more axisymmetric than extratropical cyclones, and some of them are much more intense. A distinguishing feature of a well-developed tropical cyclone is the relatively calm, cloud-free *eye* at the center.

[8] The term *cyclone* derives from the Greek word for "coils of a snake."

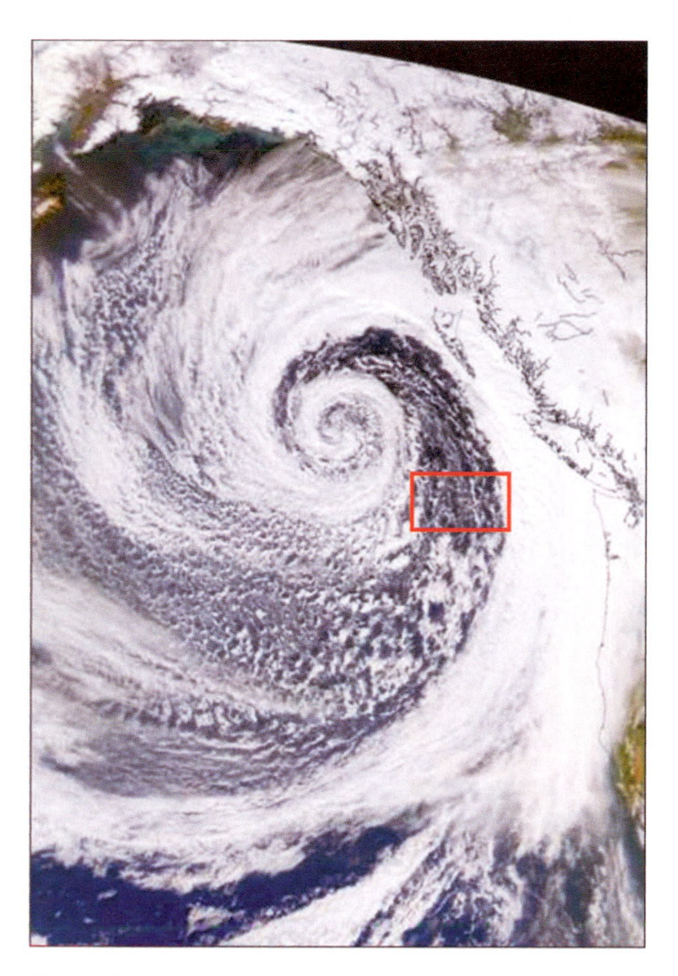

Fig. 1.12 An intense extratropical cyclone over the North Pacific. The spiral cloud pattern, with a radius of nearly 2000 km, is shaped by a vast counterclockwise circulation around a deep low pressure center. Some of the elongated cloud bands are associated with frontal zones. The region enclosed by the red rectangle is shown in greater detail in Fig. 1.21. [NASA MODIS imagery. Photograph courtesy of NASA.]

Fig. 1.13 The cloud pattern associated with Hurricane Floyd September 14, 1999. The eye of the hurricane is clearly visible. The radius of the associated cloud system is ~600 km. Data from NOAA GOES satellite imagery. [Photograph courtesy of Harold F. Pierce, Laboratory of Atmosphere, NASA Goddard Space Flight Center.]

a. Wind and pressure

The pressure field is represented on weather charts in terms of a set of *isobars* (i.e., lines or contours along which the pressure is equal to a constant value) on a horizontal surface, such as sea level. Isobars are usually plotted at uniform increments: for example, every 4 hPa on a sea-level pressure chart (e.g., . . . 996, 1000, 1004 . . . hPa). Local maxima in the pressure field are referred to as *high pressure centers* or simply *highs*, denoted by the symbol **H**, and minima as *lows* (**L**). At any point on a pressure chart the local *horizontal pressure gradient* is oriented perpendicular to the isobars and is directed from lower toward higher pressure. The strength of the horizontal pressure gradient is inversely proportional to the horizontal spacing between the isobars in the vicinity of that point.

With the notable exception of the equatorial belt (10 °S–10 °N), the winds observed in the Earth's atmosphere closely parallel to the isobars. In the northern hemisphere, lower pressure lies to the left of the wind (looking downstream) and higher pressure to the right.[9,10] It follows that air circulates counterclockwise around lows and clockwise around highs, as shown in the right-hand side of Fig. 1.14. In the southern hemisphere the relationships are in the opposite sense, as indicated in the left-hand side of Fig. 1.14.

This seemingly confusing set of rules can be simplified by replacing the words "clockwise" and

[9] This relationship was first noted by Buys-Ballot in 1857, who stated: If, in the northern hemisphere, you stand with your back to the wind, pressure is lower on your left hand than on your right.

[10] **Christopher H. D. Buys-Ballot** (1817–1890) Dutch meteorologist, professor of mathematics at the University of Utrecht. Director of Dutch Meteorological Institute (1854–1887). Labored unceasingly for the widest possible network of surface weather observations.

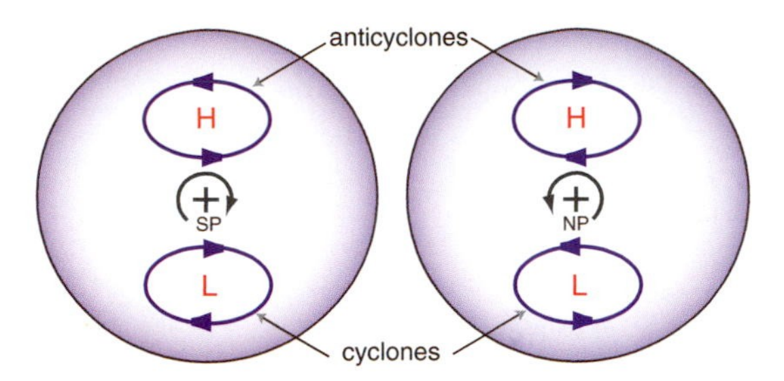

Fig. 1.14 Blue arrows indicate the sense of the circulation around highs (H) and lows (L) in the pressure field, looking down on the South Pole (left) and the North Pole (right). Small arrows encircling the poles indicate the sense of the Earth's rotation.

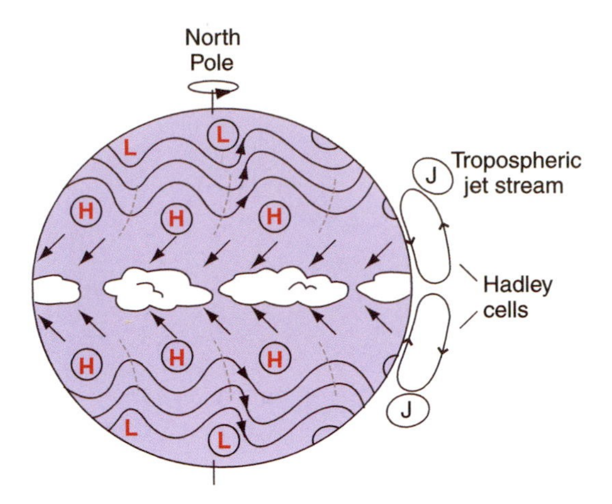

Fig. 1.15 Schematic depiction of sea-level pressure isobars and surface winds on an idealized *aqua planet*, with the sun directly overhead on the equator. The rows of H's denote the subtropical high-pressure belts, and the rows of L's denote the subpolar low-pressure belt. Hadley cells and tropospheric jet streams (J) are also indicated.

"counterclockwise" with the terms *cyclonic* and *anticyclonic* (i.e., in the same or in the opposite sense as the Earth's rotation, looking down on the pole). A *cyclonic circulation* denotes a counterclockwise circulation in the northern hemisphere and a clockwise circulation in the southern hemisphere. In either hemisphere the circulation around low pressure centers is cyclonic, and the circulation around high pressure centers is anticyclonic: that is to say, in reference to the pressure and wind fields, the term *low* is synonymous with *cyclone* and *high* with *anticyclone*.

In the equatorial belt the wind tends to blow straight down the pressure gradient (i.e., directly across the isobars from higher toward lower pressure). In the surface wind field there is some tendency for *cross-isobar flow* toward lower pressure at higher latitudes as well, particularly over land. The basis for these relationships is discussed in Chapter 7.

b. The observed surface wind field

This subsection summarizes the major features of the geographically and seasonally varying *climatological-mean* surface wind field (i.e., the background wind field upon which transient weather systems are superimposed). It is instructive to start by considering the circulation on an idealized ocean-covered Earth with the sun directly overhead at the equator, as inferred from simulations with numerical models.

The main features of this idealized "aqua-planet, perpetual equinox" circulation are depicted in Fig. 1.15. The extratropical circulation is dominated by *westerly wind belts*, centered around 45 °N and 45 °S. The westerlies are disturbed by an endless succession of eastward migrating disturbances called *baroclinic waves*, which cause the weather at these latitudes to vary from day to day. The average wavelength of these waves is ~4000 km and they propagate eastward at a rate of ~10 m s^{-1}.

The tropical circulation in the aqua-planet simulations is dominated by much steadier *trade winds*,[11] marked by an easterly zonal wind component and a component directed toward the equator. The *northeasterly trade winds* in the northern hemisphere and the *southeasterly trade winds* in the southern hemisphere are the surface manifestation of overturning circulations that extend through the depth of the troposphere. These so-called *Hadley*[12] *cells* are characterized by (1) equatorward flow in the boundary layer, (2) rising motion within a few degrees of the equator, (3) poleward return flow in the tropical upper troposphere, and (4) sinking motion in the

[11] The term *trade winds* or simply *trades* derives from the steady, dependable northeasterly winds that propelled sailing ships along the popular trade route across the tropical North Atlantic from Europe to the Americas.

[12] **George Hadley** (1685–1768) English meteorologist. Originally a barrister. Formulated a theory for the trade winds in 1735 which went unnoticed until 1793 when it was discovered by John Dalton. Hadley clearly recognized the importance of what was later to be called the Coriolis force.

subtropics, as indicated in Fig. 1.15. Hadley cells and trade winds occupy the same latitude belts.

In accord with the relationships between wind and pressure described in the previous subsection, trade winds and the extratropical westerly wind belt in each hemisphere in Fig. 1.15 are separated by a *subtropical high-pressure belt* centered ~30° latitude in which the surface winds tend to be weak and erratic. The jet streams at the tropopause (12 km; 250 hPa) level are situated directly above the subtropical high pressure belts at the Earth's surface. A weak minimum in sea-level pressure prevails along the equator, where trade winds from the northern and southern hemispheres converge. Much deeper lows form in the extratropics and migrate toward the poleward flank of the extratropical westeries to form the *subpolar low pressure belts*.

In the real world, surface winds tend to be stronger over the oceans than over land because they are not slowed as much by surface friction. Over the Atlantic and Pacific Oceans, the surface winds mirror many of the features in Fig. 1.15, but a longitudinally dependent structure is apparent as well. The subtropical high-pressure belt, rather than being continuous, manifests itself as distinct high-pressure centers, referred to as *subtropical anticyclones*, centered over the mid-oceans, as shown in Fig. 1.16.

In accord with the relationships between wind and pressure described in the previous subsection, surface winds at lower latitudes exhibit an equatorward

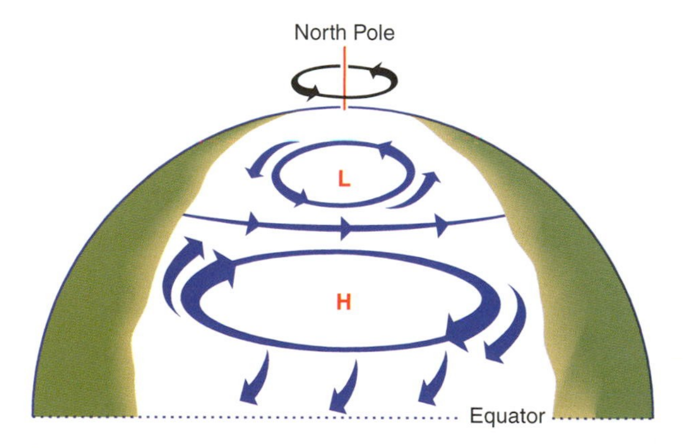

Fig. 1.16 Schematic of the surface winds and sea-level pressure maxima and minima over the Atlantic and Pacific Oceans showing subtropical anticyclones, subpolar lows, the midlatitude westerly belt, and trade winds.

component on the eastern sides of the oceans and a poleward component on the western sides. The equatorward surface winds along the eastern sides of the oceans carry (or *advect*) cool, dry air from higher latitudes into the subtropics; they drive coastal ocean currents that advect cool water equatorward; and they induce coastal upwelling of cool, nutrient-rich ocean water, as explained in the next chapter. On the western sides of the Atlantic and Pacific Oceans, poleward winds advect warm, humid, tropical air into middle latitudes.

In an analogous manner, the subpolar low-pressure belt manifests itself as mid-ocean cyclones referred to, respectively, as the *Icelandic low* and the *Aleutian low*. The poleward flow on the eastern flanks of these semipermanent, subpolar cyclones moderates the winter climates of northern Europe and the Pacific coastal zone poleward of ~40 °N. The subtropical anticyclones are most pronounced during summer, whereas the subpolar lows are most pronounced during winter.

The idealized tropical circulation depicted in Fig. 1.15, with the northeasterly and southeasterly trade winds converging along the equator, is not realized in the real atmosphere. Over the Atlantic and Pacific Oceans, the trade winds converge, not along the equator, but along ~7 °N, as depicted schematically in the upper panel of Fig. 1.17. The belt in which the convergence takes place is referred to as the *intertropical convergence zone (ITCZ)*. The asymmetry with respect to the equator is a consequence of the land–sea geometry, specifically the northwest–southeast orientation of the west coastlines of the Americas and Africa.

Surface winds over the tropical Indian Ocean are dominated by the seasonally reversing *monsoon circulation*,[13] consisting of a broad arc originating as a westward flow in the winter hemisphere, crossing the equator, and curving eastward to form a belt of moisture-laden westerly winds in the summer hemisphere, as depicted [for the northern hemisphere (i.e., *boreal*) summer] in the lower panel of Fig. 1.17. The monsoon is driven by the presence of India and southeast Asia in the northern hemisphere subtropics versus the southern hemisphere subtropics. Surface temperatures over land respond much more strongly to the seasonal variations in solar heating than those over ocean. Hence, during July the

[13] From *mausin*, the Arabic word for season.

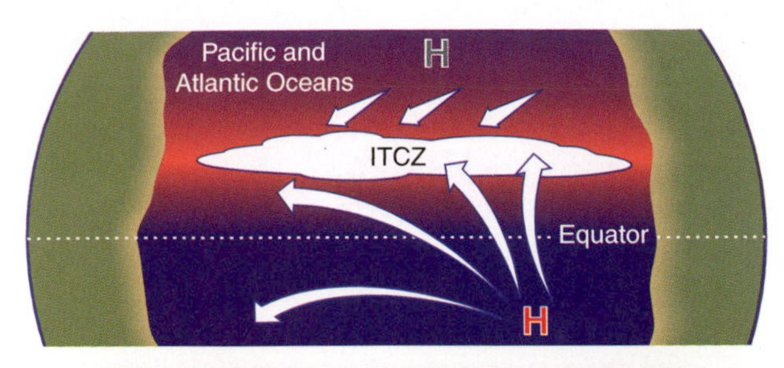

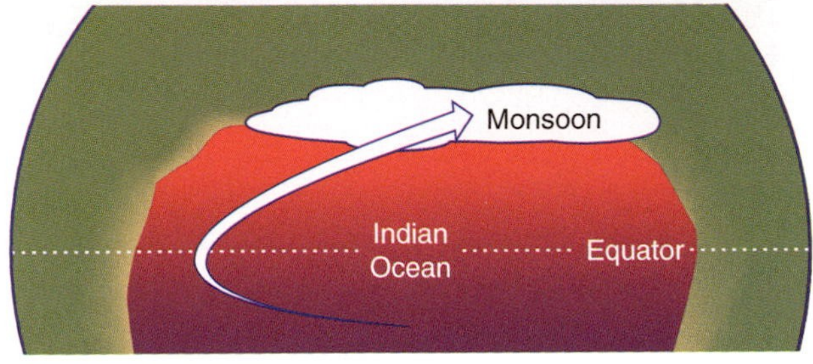

Fig. 1.17 Schematic depicting surface winds (arrows), rainfall (cloud masses), and sea surface temperature over the tropical oceans between ~30 °N and 30 °S. Pink shading denotes warmer, blue cooler sea surface temperature, and khaki shading denotes land. (Top) Atlantic and Pacific sectors where the patterns are dominated by the intertropical convergence zone (ITCZ) and the equatorial dry zone to the south of it. (Bottom) Indian Ocean sector during the northern (boreal) summer monsoon, with the Indian subcontinent to the north and open ocean to the south. During the austral summer (not shown) the flow over the Indian Ocean is in the reverse direction and the rain belt lies just to the south of the equator.

subtropical continents of the northern hemisphere are much warmer than the sea surface temperature over the tropical Indian Ocean. It is this temperature contrast that drives the monsoon flow depicted in the lower panel of Fig. 1.17. In January, when India and southeast Asia are cooler than the sea surface temperature over the tropical Indian Ocean, the monsoon flow is in the reverse sense (not shown).

The reader is invited to compare the observed climatological-mean surface winds for January and July shown in Figs. 1.18 and 1.19 with the idealized flow patterns shown in the two previous figures. In Fig. 1.18, surface winds, based on satellite data, are shown together with the rainfall distribution, indicated by shading, and in Fig. 1.19 a different version of the surface wind field, derived from a blending of many datasets, is superimposed on the climatological-mean sea-level pressure field.

By comparing the surface wind vectors with the shading in Fig. 1.18, it is evident that the major rain belts, which are discussed in the next subsection, tend to be located in regions where the surface wind vectors flow together (i.e., converge). Convergence at low levels in the atmosphere is indicative of ascending motion aloft. Through the processes discussed in Chapter 3, lifting of air leads to condensation of water vapor and ultimately to precipitation. Figure 1.19 provides verification that the surface winds tend to blow parallel to the isobars, except in the equatorial belt. At all latitudes a systematic drift across the isobars from higher toward lower pressure is also clearly apparent.

The observed winds over the southern hemisphere (Figs. 1.18 and 1.19) exhibit well-defined extratropical westerly and tropical trade wind belts reminiscent of those in the idealized aqua-planet simulations (Fig. 1.15). Over the northern hemisphere the surface winds are strongly influenced by the presence of high latitude continents. The subpolar low-pressure belt manifests itself as oceanic pressure minima (the *Icelandic* and *Aleutian lows*) surrounded by cyclonic (counterclockwise) circulations, as discussed in connection with Fig. 1.16. These features and the belts of westerly winds to the south of them are more pronounced during January than during July. In contrast, the northern hemisphere oceanic subtropical anticyclones are more clearly discernible during July.

c. Motions on smaller scales

Over large areas of the globe, the heating of the Earth's surface by solar radiation gives rise to buoyant plumes analogous to those rising in a pan of water heated from below. As the plumes rise, the displaced air subsides slowly, creating a two-way circulation. Plumes of rising air are referred to by glider pilots as *thermals*, and when sufficient moisture is present they are visible as cumulus clouds (Fig. 1.20). When the overturning circulations are confined to the lowest 1 or 2 km of the atmosphere (the so-called *mixed layer* or *atmospheric boundary layer*), as is often the case, they are referred to as *shallow convection*. Somewhat deeper, more vigorous convection gives rise to showery weather in cold air masses flowing over a warmer surface (Fig. 1.21).

Under certain conditions, buoyant plumes originating near the Earth's surface can break through the weak temperature inversion that usually caps the mixed layer, giving rise to towering clouds that extend all the way to the tropopause, as shown in Fig. 1.22. These clouds are the signature of *deep convection*,

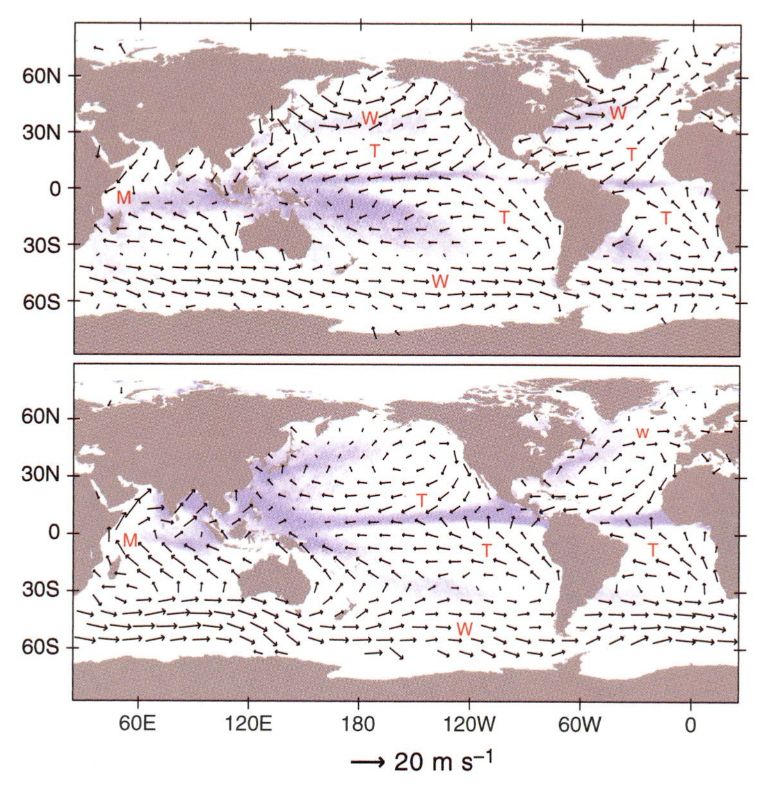

Fig. 1.18 December–January–February (top) and June–July–August (bottom) surface winds over the oceans based on 3 years of satellite observations of capillary waves on the ocean surface. The bands of lighter shading correspond to the major rain belts. **M**'s denote monsoon circulations, **W**'s westerly wind belts, and **T**'s trade winds. The wind scale is at the bottom of the figure. [Based on QuikSCAT data. Courtesy of Todd P. Mitchell.]

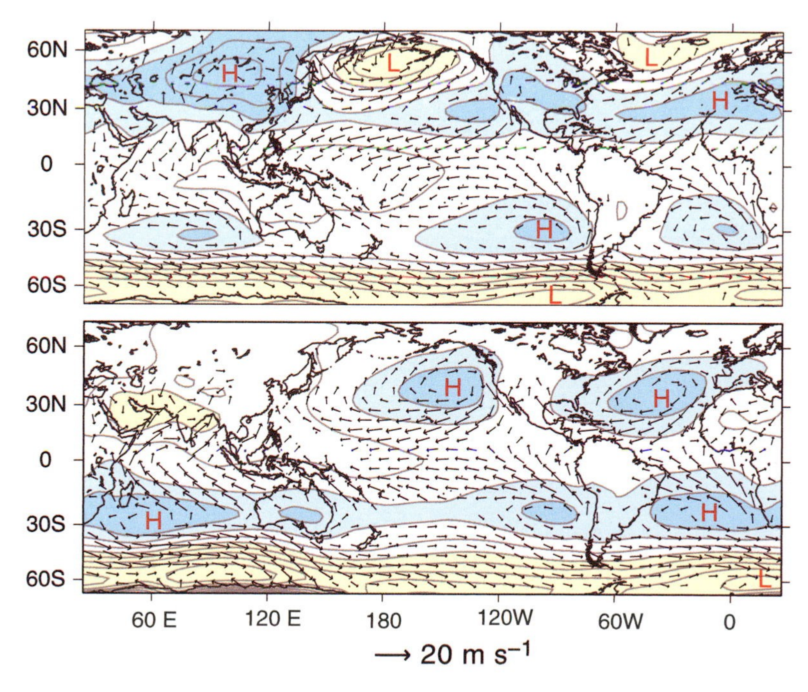

Fig. 1.19 December–January–February (top) and June–July–August (bottom) surface winds, as in Fig. 1.18, but superimposed on the distribution of sea-level pressure. The contour interval for sea-level pressure is 5 hPa. Pressures above 1015 hPa are shaded blue, and pressures below 1000 hPa are shaded yellow. The wind scale is at the bottom of the figure. [Based on the NCEP/NCAR Reanalyses. Courtesy of Todd P. Mitchell.]

Fig. 1.20 Lumpy cumulus clouds reveal the existence of shallow convection that is largely confined to the atmospheric boundary layer. [Photograph courtesy of Bruce S. Richardson.]

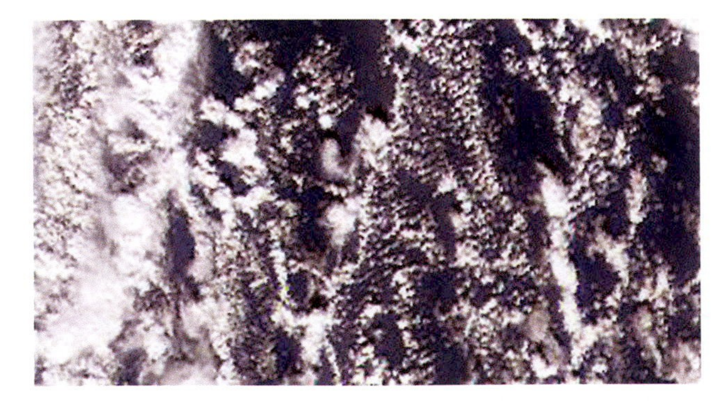

Fig. 1.21 Enlargement of the area enclosed by the red rectangle in Fig. 1.12 showing convection in a cold air mass flowing over warmer water. The centers of the convection cells are cloud free, and the cloudiness is concentrated in narrow bands at the boundaries between cells. The clouds are deep enough to produce rain or snow showers. [NASA MODIS imagery. Photograph courtesy of NASA.]

Fig. 1.22 Clouds over the South China Sea as viewed from a research aircraft flying in the middle troposphere. The foreground is dominated by shallow convective clouds, while deep convection is evident in the background. [Photograph courtesy of Robert A. Houze.]

which occurs intermittently in the tropics and in warm, humid air masses in middle latitudes. Organized deep convection can cause locally heavy rainstorms, often accompanied by lightning and sometimes by hail and strong winds.

Convection is not the only driving mechanism for small-scale atmospheric motions. Large-scale flow over small surface irregularities induces an array of chaotic waves and eddies on scales ranging up to a few kilometers. Such *boundary layer turbulence*, the subject of Chapter 9, is instrumental in causing smoke plumes to widen as they age (Fig. 1.23), in limiting the strength of the winds in the atmosphere, and in mixing momentum, energy, and trace constituents between the atmosphere and the underlying surface.

Turbulence is not exclusively a boundary layer phenomenon: it can also be generated by flow instabilities higher in the atmosphere. The cloud pattern shown in Fig. 1.24 reveals the presence of waves that develop spontaneously in layers with strong *vertical wind shear* (layers in which the wind changes rapidly with height in a vectorial sense). These waves amplify and break, much as ocean waves do when they encounter a beach. *Wave breaking* generates smaller scale waves and eddies, which, in turn, become unstable. Through this succession of instabilities, kinetic energy extracted from the large-scale wind field within the planetary boundary layer and within patches of strong vertical wind shear in the free atmosphere gives rise to a spectrum of small-scale

Fig. 1.23 Exhaust plume from the NASA space shuttle launch on February 7, 2001. The widening of the plume as it ages is due to the presence of small-scale turbulent eddies. The curved shape of the plume is due to the change in horizontal wind speed and direction with height, referred to as *vertical wind shear*. The bright object just above the horizon is the moon and the dark shaft is the shadow of the upper, sunlit part of the smoke plume. [Photograph courtesy of Patrick McCracken, NASA Headquarters.]

motions extending down to the molecular scale, inspiring Richardson's[14] celebrated rhyme:

> Big whirls have smaller whirls that feed on their velocity, and little whirls have lesser whirls, and so on to viscosity. . . . in the molecular sense.

Within localized patches of the atmosphere where wave breaking is particularly intense, eddies on

Fig. 1.24 Billows along the top of this cloud layer reveal the existence of breaking waves in a region of strong vertical wind shear. The right-to-left component of the wind is increasing with height. [Photograph courtesy of Brooks Martner, NOAA.]

scales of tens of meters can be strong enough to cause discomfort to airline passengers and even, in exceptional cases, to pose hazards to aircraft. Turbulence generated by shear instability is referred to as *clear air turbulence (CAT)* to distinguish it from the turbulence that develops within the cloudy air of deep convective storms.

1.3.6 Precipitation

Precipitation tends to be concentrated in space and time. Annual-mean precipitation at different points on Earth ranges over two orders of magnitude, from a few tens of centimeters per year in dry zones to several meters per year in the belts of heaviest rainfall, such as the ITCZ. Over much of the world climatological-mean precipitation exhibits equally dramatic seasonal variations. The global-mean, annual-mean precipitation rate is ~1 m of liquid water per year or ~0.275 cm per day or 1 m per year.

Climatological-mean distributions of precipitation for the months of January and July are shown in Fig. 1.25. The narrow bands of heavy rainfall that dominate the tropical Atlantic and Pacific sectors coincide with the ITCZ in the surface wind field. In the Pacific and Atlantic sectors the ITCZ is flanked by expansive *dry zones* that extend westward from the continental deserts and cover much of the subtropical oceans. These features coincide with the

[14] **Lewis F. Richardson** (1881–1953). English physicist and meteorologist. Youngest of seven children of a Quaker tanner. Served as an ambulance driven in France during World War I. Developed a set of finite differences for solving differential equations for weather prediction, but his formulation was not quite correct and at that time (1922) computations of this kind could not be performed quickly enough to be of practical use. Pioneer in the causes of war, which he described in his books "Arms and Insecurity" and "Statistics of Deadly Quarrels," Boxward Press, Pittsburg, 1960. Sir Ralph Richardson, the actor, was his nephew.

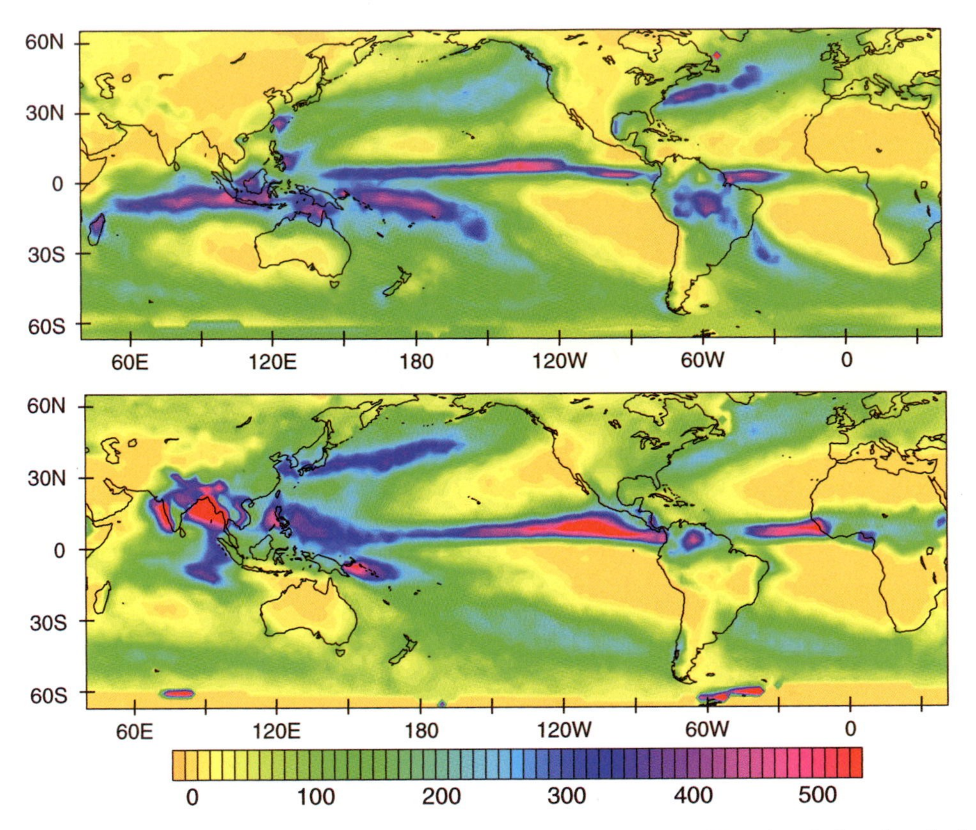

Fig. 1.25 January (top) and July (bottom) climatological-mean precipitation in cm. [Based on infrared and microwave satellite imagery over the oceans and rain gauge data over land, as analyzed by the NOAA National Centers for Environmental Prediction CMAP project. Courtesy of Todd P. Mitchell.]

subtropical anticyclones and, in the Pacific and Atlantic sectors, they encompass equatorial regions as well.

Small seasonal or year-to-year shifts in the position of the ITCZ can cause dramatic local variations in rainfall. For example, at Canton Island (3 °S, 170 °W) near the western edge of the equatorial dry zone, rainfall rates vary from zero in some years to over 30 cm per month (month after month) in other years in response to subtle year-to-year variations in sea surface temperature over the equatorial Pacific that occur in association with El Niño, as discussed in Section 10.2.2.

Over the tropical continents, rainfall is dominated by the monsoons, which migrate northward and southward with the seasons, following the sun. Most equatorial regions receive rainfall year-round, but the belts that lie 10–20° away from the equator experience pronounced dry seasons that correspond to the time of year when the sun is overhead in the opposing hemisphere. The rainy season over India and southeast Asia coincides with the time of year in which the surface winds over the northern Indian Ocean blow from the west (Figs. 1.17 and 1.18). Analogous relationships exist between wind and rainfall in Africa and the Americas. The onset of the rainy season, a cause for celebration in many agricultural regions of the subtropics, is remarkably regular from one year to the next and it is often quite dramatic: for example, in Mumbai (formerly Bombay) on the west coast of India, monthly mean rainfall jumps from less than 2 cm in May to ~50 cm in June.

The flow of warm humid air around the western flanks of the subtropical anticyclones brings copious summer rainfall to eastern China and Japan and the eastern United States. In contrast, Europe and western North America and temperate regions of the southern hemisphere experience dry summers. These regions derive most of their annual precipitation from wintertime extratropical cyclones that form within the belts of westerly surface winds over the oceans and propagate eastward over land. The rainfall maxima extending across the Pacific and Atlantic at latitudes ~45 °N in Fig. 1.25 are manifestations of these oceanic *storm tracks*.

Rainfall data shown in Fig. 1.25, which are averaged over 2.5° latitude × 2.5° longitude grid boxes, do not fully resolve the fine structure of the distribution

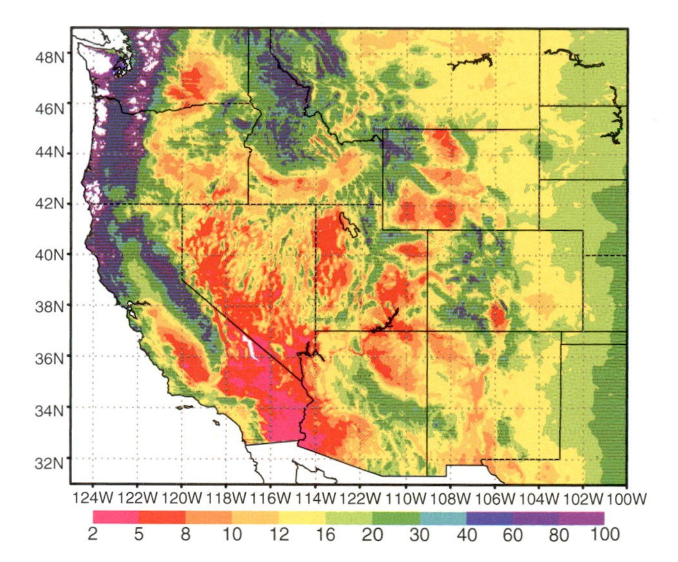

Fig. 1.26 Annual-mean precipitation over the western United States resolved on a 10-km scale making use of a model. The color bar is in units of inches of liquid water (1 in = 2.54 cm). Much of the water supply is derived from winter snow pack, which tends to be concentrated in regions of blue, purple, and white shading. [Map produced by the NOAA Western Regional Climate Center using PRISM data from Oregon State University. Courtesy of Kelly Redmond.]

of precipitation in the presence of *orography* (i.e., terrain). Flow over and around mountain ranges imparts a fractal-like structure to the precipitation distribution, with enhanced precipitation in regions where air tends to be lifted over terrain features and suppressed precipitation in and downstream of regions of descent (i.e., *subsidence*).

The distribution of annual-mean precipitation over the western United States, shown in Fig. 1.26, illustrates the profound influence of orography. Poleward of ~35°, which corresponds to the equatorward limit of the extratropical westerly flow regime that prevails during the wintertime, annual-mean precipitation tends to be enhanced where moisture-laden marine air is lifted as it moves onshore and across successive ranges of mountains. The regions of suppressed precipitation on the lee side of these ranges are referred to as *rain shadows*.

On any given day, the cloud patterns revealed by global satellite imagery exhibit patches of deep convective clouds that can be identified with the ITCZ and the monsoons over the tropical continents of the summer hemisphere; a relative absence of clouds in the subtropical dry zones; and a succession of comma-shaped, frontal cloud bands embedded in the baroclinic waves tracking across the mid-latitude oceans. These features are all present in the example shown in Fig. 1.27.

1.4 What's Next?

The brief survey of the atmosphere presented in this chapter is just a beginning. All the major themes introduced in this survey are developed further in subsequent chapters. The first section of the next chapter provides more condensed surveys of the other components of the Earth system that play a role in climate: the oceans, the crysophere, the terrestrial biosphere, and the Earth's crust and mantle.

Fig. 1.27 Composite satellite image showing sea surface temperature and land surface air temperature and clouds. [Courtesy of the University of Wisconsin Space Science and Engineering Center.]

Exercises[15]

1.6 Explain or interpret the following:

(a) Globally averaged surface pressure is 28 hPa lower than globally averaged sea-level pressure (1013 hPa).

(b) Density decreases exponentially with height in the atmosphere, whereas it is nearly uniform in the oceans.

(c) Pressure in the atmosphere and ocean decreases monotonically with height. The height dependence is almost exponential in the atmosphere and linear in the ocean.

(d) Concentrations of some atmospheric gases, such as N_2, O_2, and CO_2, are nearly uniform below the turbopause, whereas concentrations of other gases such as water vapor and ozone vary by orders of magnitude.

(e) Below ~100 km, radar images of meteor trails become distorted and break up into puffs much like jet aircraft contrails do. In contrast, meteor trails higher in the atmosphere tend to vanish before they have time to become appreciably distorted.

(f) Airline passengers flying at high latitudes are exposed to higher ozone concentrations than those flying in the tropics.

(g) In the tropics, deep convective clouds contain ice crystals, whereas shallow convective clouds do not.

(h) Airliners traveling between Tokyo and Los Angeles often follow a great circle route westbound and a latitude circle eastbound.

(i) Aircraft landings on summer afternoons tend to be bumpier than nighttime landings, especially on clear days.

(j) Cumulus clouds like the ones shown in Fig. 1.20 are often observed during the daytime over land when the sky is otherwise clear.

(k) New York experiences warmer, wetter summers than Lisbon, Portugal, which is located at nearly the same latitude.

1.7 To what feature in Fig. 1.15 does the colloquial term *horse latitudes* refer? What is the origin of this term?

1.8 Prove that exactly half the area of the Earth lies equatorward of 30° latitude.

1.9 How many days would it take a hot air balloon traveling eastward along 40 °N at a mean speed of 15 m s^{-1} to circumnavigate the globe?

1.10 Prove that pressure expressed in cgs units of millibars (1 mb = 10^{-3} bar) is numerically equal to pressure expressed in SI units of hPa (1 hPa = 10^2 Pa).

1.11 How far below the surface of the water does a diver experience a pressure of 2 atmospheres (i.e., a doubling of the ambient atmospheric pressure) due to the weight of the overlying water.

1.12 In a sounding taken on a typical winter day at the South Pole the temperature at the ground is −80 °C and the temperature at the top of a 30-m high tower is −50 °C. Estimate the lapse rate within the lowest 30 m, expressed in °C km^{-1}.

1.13 "Cabin altitude" in typical commercial airliners is around 1.7 km. Estimate the typical pressure and density of the air in the passenger cabin.[16]

1.14 Prove that density and pressure, which decrease more or less exponentially with height, decrease by a factor of 10 over a depth of 2.3 H, where H is the scale height.

1.15 Consider a perfectly elastic ball of mass m bouncing up and down on a horizontal surface under the action of a downward gravitational acceleration g. Prove that in the time average over an integral number of bounces, the

---

[15] A list of constants and conversions that may be useful in working the exercises is printed at the end of the book. Answers and solutions to most of the exercises are provided on the Web site for the book.

[16] Over many generations humans are capable of adapting to living at altitudes as high as 5 km (~550 hPa) and surviving for short intervals at altitudes approaching 9 km (~300 hPa). The first humans to visit such high latitudes may have been British meteorologist James Glaisher and balloonist Henry Coxwell in 1862. Glaisher lost consciousness for several minutes and Coxwell was barely able to arrest the ascent of the balloon after temporarily losing control.

downward force exerted by the ball upon the surface is equal to the weight of the ball. [**Hint**: The downward force is equal to the downward momentum imparted to the surface with each bounce divided by the time interval between successive bounces.] Does this result suggest anything about the "weight" of an atmosphere comprised of gas molecules?

1.16 Estimate the percentage of the mass of the atmosphere that resides in the stratosphere based on the following information. The mean pressure level of tropical tropopause is around 100 hPa and that of the extratropical tropopause is near 300 hPa. The break between the tropical and the extratropical tropopause occurs near 30° latitude so that exactly half the area of the Earth lies in the tropics and half in the extratropics. On the basis of an inspection of Fig. 1.11, verify that the representation of tropopause height in this exercise is reasonably close to observed conditions.

1.17 If the Earth's atmosphere were replaced by an incompressible fluid whose density was everywhere equal to the atmospheric density observed at sea level (1.25 kg m^{-3}), how deep would it have to be to account for the observed mean surface pressure of $\sim 10^5$ Pa?

1.18 The mass of water vapor in the atmosphere ($\sim$100 kg m^{-2}) is equivalent to a layer of liquid water how deep?

1.19 Assuming that the density of air decreases exponentially with height from a value of 1.25 kg m^{-3} at sea level, calculate the scale height that is consistent with the observed global mean surface pressure of $\sim 10^3$ hPa. [**Hint**: Write an expression analogous to (1.8) for density and integrate it from the Earth's surface to infinity to obtain the atmospheric mass per unit area.]

1.20 The equatorward flow in the trade winds averaged around the circumference of the Earth at 15 °N and 15 °S is $\sim$1 m s^{-1}. Assume that this flow extends through a layer extending from sea level up to the 850-hPa pressure surface. Estimate the equatorward mass flux into the equatorial zone. [**Hint**: The equatorward mass flux across the 15 °N, in units of kg s^{-1}, is given by

$$-\oint_{15\,^\circ\mathrm{N}} \int_0^{z_{850}} \rho v dz dx$$

where ρ is the density of the air, v is the meridional (northward) velocity component, the line integral denotes an integration around the 15 °N latitude circle, and the vertical integral is from sea level up to the height of the 850-hPa surface.] Evaluate the integral, making use of the relations

$$\oint_{15\,^\circ\mathrm{N}} dx = 2\pi R_E \cos 15^\circ$$

and

$$\int_0^{z_{850}} \rho dz = \frac{(1000 - 850)\ \mathrm{hPa} \times 100\ \mathrm{Pa/hPa}}{g}.$$

1.21 During September, October, and November the mean surface pressure over the northern hemisphere increases at a rate of $\sim$1 hPa per month. Calculate the mass averaged northward velocity across the equator

$$v_m \equiv \frac{\oint \int_0^\infty \rho v dz dx}{\oint \int_0^\infty \rho dz dx}$$

that is required to account for this pressure rise. [**Hint**: Assume that atmospheric mass is conserved, i.e., that the pressure rise in the northern hemisphere is entirely due to the influx of air from the southern hemisphere.]

1.22 Based on the climatological-mean monthly temperature and rainfall data provided on the Web site, select several stations that fit into the following climate regimes: (a) equatorial belt, wet year-round; (b) monsoon; (c) equatorial dry zone; (d) extratropical, with dry summers; and (e) extratropical, with wet summers. Locate each station with reference to the features in Figs. 1.18, 1.19, and 1.25.

The Earth System

2

Climate depends not only on atmospheric processes, but also on physical, chemical, and biological processes involving other components of the Earth system. This chapter reviews the structure and behavior of those other components. We show how the cycling of water, carbon, and oxygen among the components of the Earth system has affected the evolution of the atmosphere. Drawing on this background, we summarize the history of climate over the lifetime of the Earth, with emphasis on causal mechanisms. The final section discusses why Earth is so much more habitable than its neighbors in the solar system. Chapter 10 revisits some of these same topics in the context of climate dynamics, with a quantitative discussion of feedbacks and climate sensitivity.

2.1 Components of the Earth System

This section introduces the cast of characters and briefly describes their roles and interrelations in the ongoing drama of climate. The atmosphere, which in some sense plays the starring role, has already been introduced in Chapter 1. The interplay between atmospheric radiation and convection regulates the temperature at the Earth's surface, setting the limits for snow and ice cover and for the various life zones in the biosphere. The stratospheric ozone layer protects the biosphere from the lethal effects of solar ultraviolet radiation. Atmospheric wind patterns regulate the patterns of oceanic upwelling that supplies nutrients to the marine biosphere, they determine the distribution of water that sustains the terrestrial (land) biosphere, and they transport trace gases, smoke, dust, insects, seeds, and spores over long distances. Rain, frost, and wind erode the Earth's crust, wearing down mountain ranges, reshaping the landscape, and replenishing the soils and the supply of metallic ions needed to sustain life.

Other components of the Earth system also play important roles in climate. The oceans are notable for their large "thermal inertia" and their central role in the cycling of carbon, which controls atmospheric carbon dioxide concentrations. Extensive snow and ice-covered surfaces render the Earth more reflective, and consequently cooler, than it would be in their absence. By evaporating large quantities of water through their leaves, land plants exert a strong moderating influence on tropical and extratropical summer climate. Living organisms on land and in the sea have been instrumental in liberating oxygen and sequestering of carbon in the Earth's crust, thereby reducing the atmospheric concentration carbon dioxide. On timescales of millions of years or longer, plate tectonics exerts an influence on climate through continental drift, mountain building, and volcanism. This section describes these processes and the media in which they occur.

2.1.1 The Oceans

The oceans cover 72% of the area of the Earth's surface and they reach an extreme depth of nearly 11 km. Their total volume is equivalent to that of a layer 2.6 km deep, covering the entire surface of the Earth. The mass of the oceans is ~250 times as large as that of the atmosphere.

a. Composition and vertical structure

The density of sea water is linearly dependent on the concentration of dissolved salt. On average, sea water in the open oceans contains ~35 g of dissolved salts per kg of fresh water, with values typically ranging from 34 to 36 g kg^{-1} (or parts per thousand by mass, abbreviated as o/oo). Due to the presence of these dissolved salts, sea water is ~2.4% denser than fresh water at the same temperature.

The density σ of sea water (expressed as the departure from 1 in g kg^{-1} or o/oo) typically ranges from 1.02 to 1.03. It is a rather complicated function of temperature T, salinity s, and pressure p; i.e., $\sigma = \sigma(T, s, p)$. The pressure dependence of density in liquids is much weaker than in gases and, for purposes of this qualitative discussion, will be ignored.[1] As in fresh water, $\partial\sigma/\partial T$ is temperature dependent, but the fact that sea water is saline makes the relationship somewhat different: in fresh water, density increases with increasing temperature between 0 and 4 °C, whereas in sea water, density decreases monotonically with increasing temperature.[2] In both fresh water and sea water, $\partial\sigma/\partial T$ is smaller near the freezing point than at higher temperatures. Hence, a salinity change of a prescribed magnitude δs is equivalent, in terms of its effect on density, to a larger temperature change δT in the polar oceans than in the tropical oceans, as illustrated in Fig. 2.1.

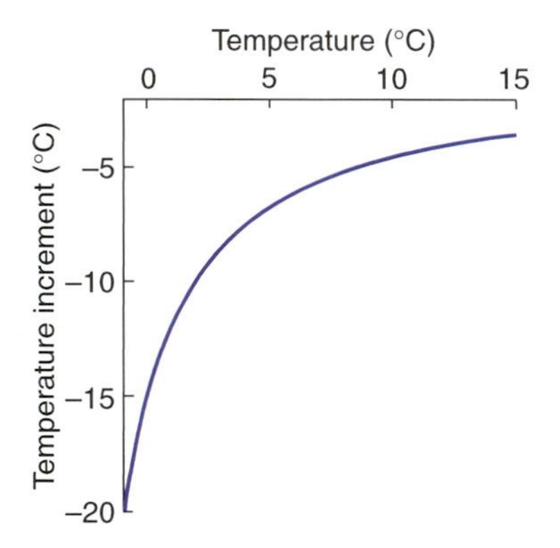

Fig. 2.1 The change in temperature of a water parcel required to raise the density of sea water at sea level as much as a salinity increase of 1 g kg^{-1}, plotted as a function of the temperature of the parcel. For example, for sea water at a temperature of 10 °C, a salinity increase of 1 g kg^{-1} would raise the density as much as a temperature decrease of ~5 °C, whereas for sea water at 0 °C the same salinity increase would be equivalent to a temperature change of ~17 °C. [Adapted from data in M. Winton, Ph.D. thesis, University of Washington, p. 124 (1993).]

Over most of the world's oceans, the density of the water in the wind-stirred, *mixed layer* is smaller, by a few tenths of a percent, than the density of the water below it. Most of the density gradient tends to be concentrated within a layer called the *pycnocline*, which ranges in depth from a few tens of meters to a few hundred meters below the ocean surface. The density gradient within the pycnocline tends to inhibit vertical mixing in the ocean in much the same manner that the increase of temperature with height inhibits vertical mixing in atmospheric temperature inversions and in the stratosphere. In particular, the pycnocline strongly inhibits the exchange of heat and salt between the mixed layer, which is in direct contact with the atmosphere, and the deeper layers of the ocean. At lower latitudes, *pycnocline* is synonymous with the *thermocline* (i.e., the layer in which temperature increases with height), but in polar oceans, *haloclines* (layers with fresher water above and saltier water below) also play an important role in inhibiting vertical mixing. The strength and depth of the thermocline vary with latitude and season, as illustrated in the idealized profiles shown in Fig. 2.2.

Within the oceanic mixed layer, temperature and salinity (and hence density) vary in response to

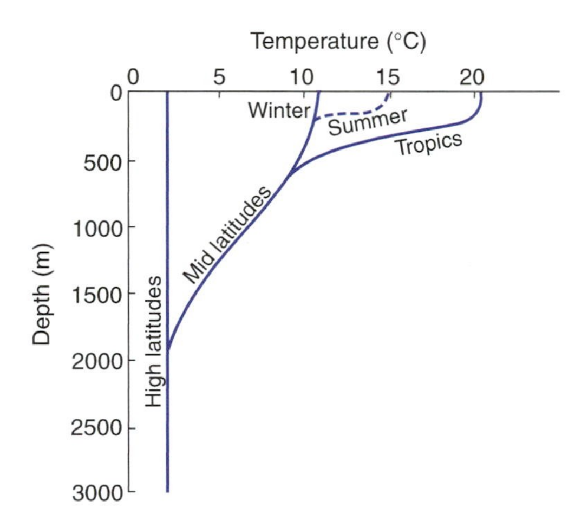

Fig. 2.2 Idealized profiles of the temperature plotted as a function of depth in different regions of the world's oceans. The layer in which the vertical temperature gradient is strongest corresponds to the thermocline. [From J. A. Knauss, *Introduction to Physical Oceanography*, 2nd Edition, p. 2, © 1997. Adapted by permission of Pearson Education, Inc., Upper Saddle River, NJ.]

[1] The small effect of pressure upon density is taken into account through the use of *potential density*, the density that a submerged water parcel would exhibit if it were brought up to sea level, conserving temperature and salinity. (See Exercise 3.54.)

[2] Ice floats on lakes because the density of fresh water decreases with temperature from 0 to 4 °C. In contrast, sea ice floats because water rejects salt as it freezes.

exchanges of heat and water with the atmosphere. Precipitation lowers the salinity by diluting the salts that are present in the oceanic mixed layer, and evaporation raises the salinity by removing fresh water and thereby concentrating the residual salts, as illustrated in the following example.

Exercise 2.1 A heavy tropical storm dumps 20 cm of rainfall in a region of the ocean in which the salinity is 35.00 g kg^{-1} and the mixed layer depth is 50 m. Assuming that the water is well mixed, by how much does the salinity decrease?

Solution: The volume of water in a column extending from the surface of the ocean to the bottom of the mixed layer is increased by a factor

$$\frac{0.2 \text{ m}}{50 \text{ m}} = 4 \times 10^{-3}$$

and (ignoring the small difference between the densities of salt water and fresh water) the mass of the water in the column increases by a corresponding amount. The mass of salt dissolved in the water remains unchanged. Hence, the salinity drops to

$$\frac{35.00 \text{ g of salt}}{1.004 \text{ kg of water}} = 34.86 \text{ g kg}^{-1}. \qquad \blacksquare$$

Water parcels that are not in contact with the ocean surface tend to conserve temperature and salinity as they move over long distances. Hence, *water masses* (layers of water extending over large areas that exhibit nearly uniform temperature and salinity) can be tracked back to the regions of the mixed layer in which they were formed by exchanges of heat and mass with the atmosphere. Among the important water masses in the Atlantic Ocean, in order of increasing density, are:

- *Mediterranean outflow*, which is conspicuously warm and saline due to the excess of evaporation over precipitation in the Mediterranean Sea.
- *North Atlantic deep water (NADW)*, formed by the sinking of water along the ice edge in the Greenland, Iceland, and Norwegian Seas.
- *Antarctic bottom water (AABW)*, formed by sinking along the ice edge in the Weddell Sea.

The NADW and AABW, each marked by its own distinctive range of temperatures and salinities, are

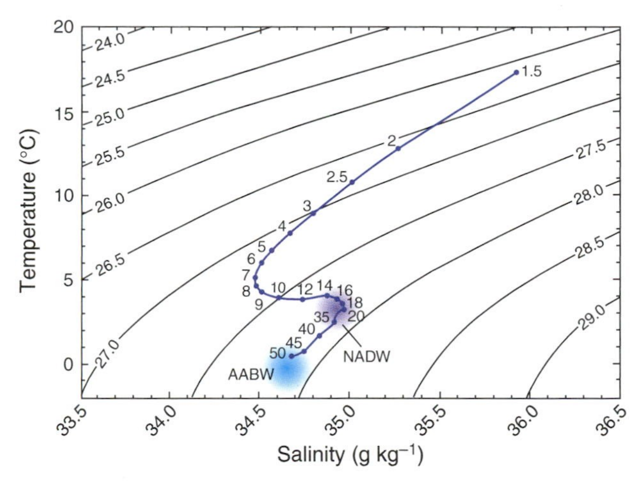

Fig. 2.3 Vertical sounding of water temperature and salinity in a vertical sounding in the subtropical Atlantic Ocean. Numbers along the sounding indicate depths in hundreds of meters. Potential (i.e., pressure-adjusted) density (in *o/oo*) is indicated by the contours. Characteristic temperature and salinity ranges for North Atlantic deep water (NADW) and Antarctic bottom water (AABW) are indicated by shading. [Reprinted from *Seawater: Its Composition, Properties and Behavior*, The Open University in association with Pergamon Press, p. 48 (1989), with permission from Elsevier.]

both clearly evident near the bottom of the tropical sounding shown in Fig. 2.3. The AABW is slightly colder and fresher than the NADW. When both temperature and salinity are taken into account, the AABW is slightly denser than the NADW, consistent with its placement at the bottom of the water column.

b. The ocean circulation

The ocean circulation is composed of a *wind-driven* component and a *thermohaline* component. The wind-driven circulation dominates the surface currents, but it is largely restricted to the topmost few hundred meters. The circulation deeper in the oceans is dominated by the slower thermohaline circulation.

By generating ocean waves, surface winds transfer horizontal momentum from the atmosphere into the ocean. The waves stir the uppermost layer of the ocean, mixing the momentum downward. The momentum, as reflected in the distribution of surface currents shown in Fig. 2.4, mirrors the pattern of surface winds shown in Figs. 1.18 and 1.19, with closed anticyclonic circulations (referred to as *gyres*) at subtropical latitudes and cyclonic gyres at subpolar latitudes. Another notable feature of the wind-driven

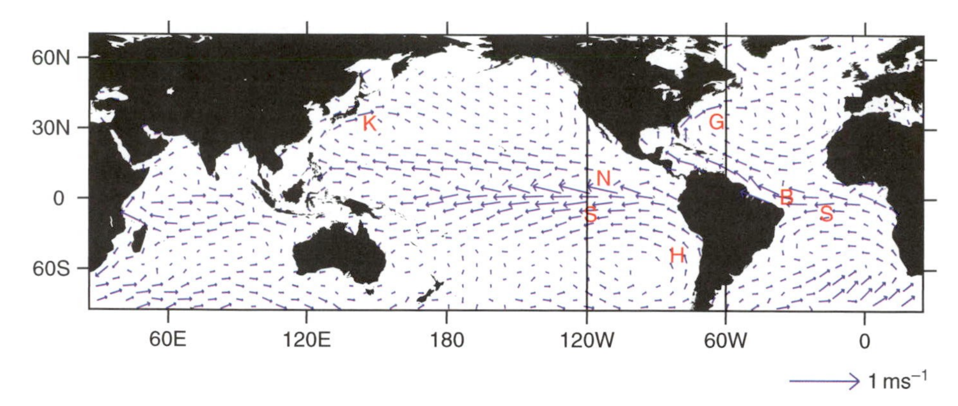

Fig. 2.4 Annual mean ocean surface currents based on the rate of drift of ships. The *Gulf Stream* (G) and the *Kuroshio Current* (K) are warm, *western boundary currents*. The *Humboldt Current* (H) is the most prominent of the cold, equatorward currents driven by the winds along the eastern flanks of the subtropical anticyclones. The westward *South Equatorial Current* (S) is driven by the easterlies along the equator and the weaker eastward *North Equatorial Countercurrent* (N) is a response to the winds in the vicinity of the ITCZ. [Data courtesy of Philip Richardson, WHOI; graphic courtesy of Todd P. Mitchell.]

circulation is the west-to-east *Antarctic circumpolar current* along 55 °S, the latitude of the Drake passage that separates Antarctica and South America. Velocities in these wind-driven currents are typically on the order of 10 cm s^{-1}, a few percent of the speeds of the surface winds that drive them, but in the narrow *western boundary currents* such as the *Gulf Stream* off the east coast of the United States (Figs. 2.4 and 2.5) velocities approach 1 m s^{-1}. The relatively warm water transported poleward by the western boundary currents contributes to moderating winter temperatures over high latitude coastal regions.

Over certain regions of the polar oceans, water in the mixed layer can become sufficiently dense, by virtue of its high salinity, to break through the pycnocline and sink all the way to the ocean floor to become what oceanographers refer to as *deep water* or *bottom water*. In some sense, these negatively buoyant plumes are analogous to the plumes of warm, moist air in low latitudes that succeed in breaking through the top of the atmospheric mixed layer and continue ascending until they encounter the tropopause. The presence of CFCs[3] in NADW and AABW indicates that these water

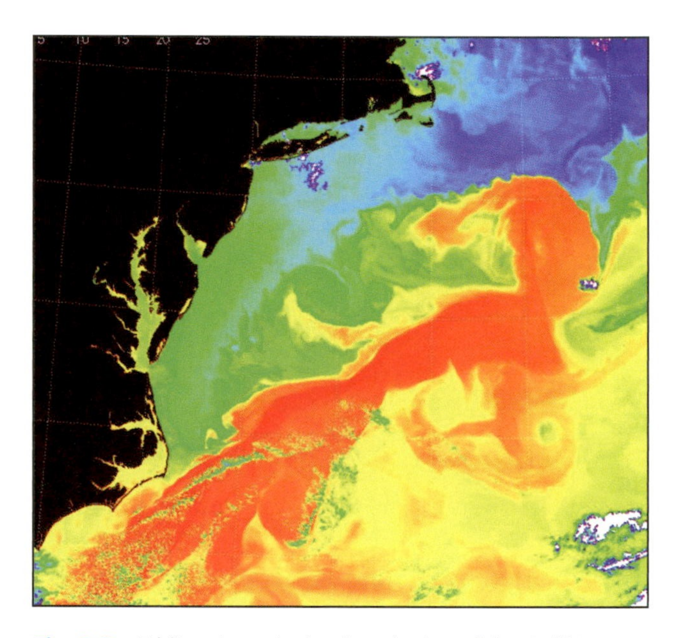

Fig. 2.5 Eddies along the landward edge of the Gulf Stream, as revealed by the pattern of sea surface temperature. Temperatures range from ~20 °C in the orange regions down to ~6 °C in the darkest blue regions. Note the sharpness of the boundary and the indications of turbulent mixing between the waters of the Gulf Stream and the colder Labrador Current to the north of it. [Based on NASA Terra/MODIS imagery. Courtesy of Otis Brown.]

[3] The term *chlorofluorocarbons (CFCs)* refers to a family of gaseous compounds that have no natural sources; first synthesized in 1928. Atmospheric concentrations of CFCs rose rapidly during the 1960s and 1970s as these gases began to be used for a widening range of purposes.

masses were in relatively recent contact with the atmosphere.

By virtue of their distinctive chemical and isotopic signatures, it is possible to track the flow of water masses and to infer how long ago water in various parts of the world's oceans was in contact with the atmosphere. Such chemical analyses indicate the existence of a slow overturning characterized by a spreading of deep water from the high latitude sinking regions, a resurfacing of the deep water, and a return flow of surface waters toward the sinking regions, as illustrated in Fig. 2.6. The timescale in which a parcel completes a circuit of this so-called *thermohaline circulation* is on the order of hundreds of years.

The resurfacing of deep water in the thermohaline circulation requires that it be *ventilated* (i.e., mixed with and ultimately replaced by less dense water that has recently been in contact with the ocean surface). Still at issue is just how this ventilation occurs in the presence of the pycnocline. One school of thought attributes the ventilation to mixing along sloping isopycnal (constant density) surfaces that cut through the pycnocline. Another school of thought attributes it to irreversible mixing produced by tidal motions propagating downward into the deep oceans along the continental shelves, and yet another to vertical mixing in restricted regions characterized by strong winds and steeply sloping isopycnal surfaces, the most important of which coincides with the *Antarctic circumpolar current*, which lies beneath the ring of strong westerly surface winds that encircles Antarctica.

Although most of the deep and bottom water masses are formed in the Atlantic sector, the thermohaline circulation involves the entire world's oceans, as illustrated in Fig. 2.7. Within the Atlantic sector itself, the thermohaline circulation is comprised of two different cells: one involving NADW and the other involving AABW, as illustrated in Fig. 2.8.

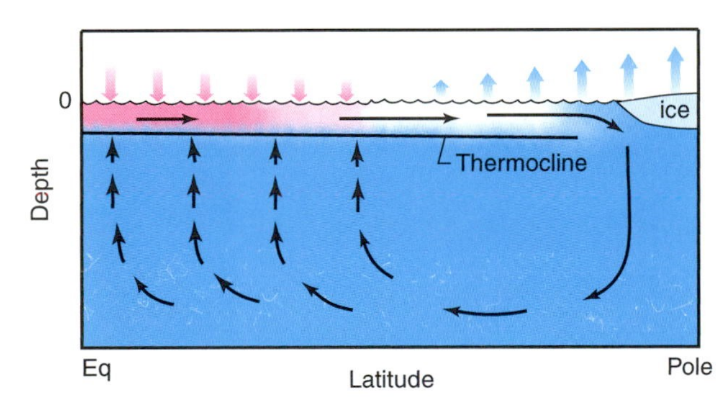

Fig. 2.6 Idealized schematic of the thermohaline circulation in an equatorially symmetric ocean. The domain extends from the sea floor to the ocean surface and from equator to pole. Pink shading indicates warmer water and blue shading indicates colder water. The shaded arrows represent the exchange of energy at the air-sea interface: pink downward arrows indicate a heating of the ocean mixed layer and blue upward arrows indicate a cooling. The role of salinity is not specifically represented in this schematic but it is the rejection of salt when water freezes along the ice edge that makes the water dense enough to enable it to sink to the bottom.

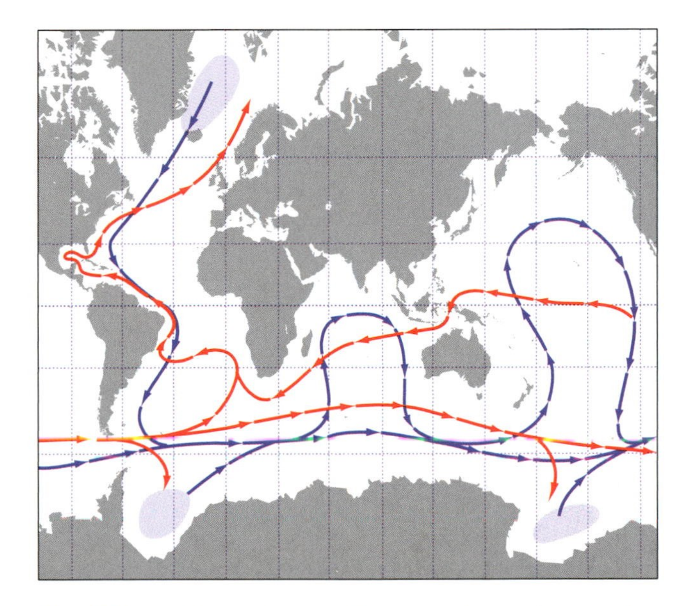

Fig. 2.7 Highly simplified schematic of the thermohaline circulation. Shading denotes regions of downwelling, blue arrows denote transport of bottom water, and red arrows denote the return flow of surface water. [Adapted from W. J. Schmitz, Jr., "On the interbasin-scale thermohaline circulation," *Rev. Geophys.*, **33**, p. 166, Copyright 1995 American Geophysical Union.]

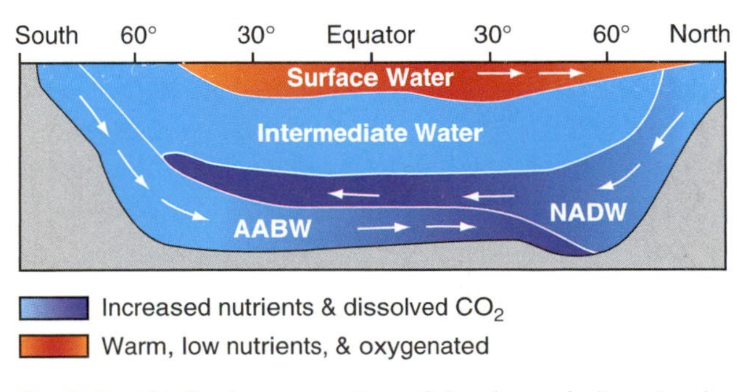

Fig. 2.8 Idealized cross section of the thermohaline circulation in the Atlantic Ocean. In this diagram, *Intermediate Water* comprises several different water masses formed at temperate latitudes. Note the consistency with Fig. 2.3. [Courtesy of Steve Hovan.]

c. The marine biosphere

Virtually all the sunlight that reaches the surface of the ocean is absorbed within the topmost hundred meters. Within this shallow *euphotic zone*,[4] life abounds wherever there are sufficient nutrients, such as phosphorous and iron, to sustain it. In regions of the ocean where the marine biosphere is active, the uppermost layers are enriched in dissolved oxygen (a product of photosynthesis) and depleted in nutrients and dissolved carbon, as illustrated in Fig. 2.9. *Phyto-* (i.e., plant) *plankton* are capable of consuming the nutrients in the euphotic zone within a matter of days. Hence, the maintenance of high *primary productivity* (i.e., photosynthesis) requires a continual supply of nutrients. The most productive regions of the oceans tend to be concentrated in regions of upwelling, where nutrient-rich sea water from below the euphotic zone is first exposed to sunlight.

Nutrients consumed within the euphotic zone by phytoplankton return to the deeper layers of the oceans when marine plants and animals that feed on them die, sink, and decompose. The continual exchange of nutrients between the euphotic zone and the deeper layers of the ocean plays an important role in the carbon cycle, as discussed in Section 2.3. The distribution of upwelling, in turn, is controlled by the pattern of surface winds discussed earlier. The distribution of *ocean color* (Fig. 2.10) shows evidence of high biological productivity and, by inference, upwelling

- beneath cyclonic circulations such as Aleutian and Icelandic lows,
- along the eastern shores of the oceans at subtropical latitudes,
- in a narrow strip along the equator in the equatorial Atlantic and Pacific Oceans.

In contrast, the ocean regions that lie beneath the subtropical anticyclones are biological deserts. The dynamical basis for these relationships is discussed in Section 7.2.5. Through their effect in mediating the geographical distribution of upwelling and the depth of the mixed layer, year-to-year changes in the atmospheric circulation, such as those that occur in association with El Niño, perturb the entire food chain that supports marine mammals, seabirds, and commercial fisheries.

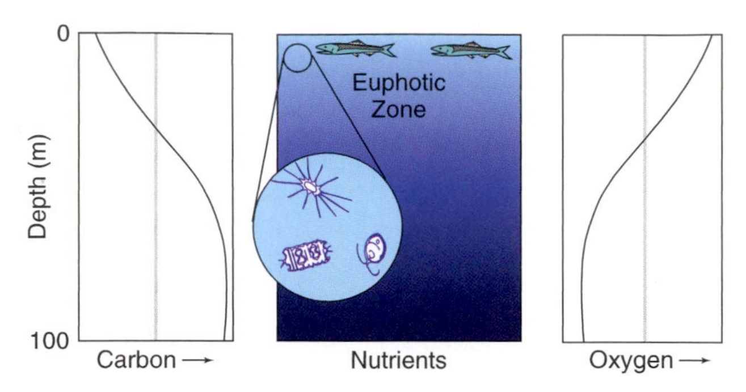

Fig. 2.9 Idealized vertical profiles of dissolved carbon (left) and oxygen (right) in biologically active regions of the oceans. The intensity of sunlight is indicated by the depth of the shading in the middle panel.

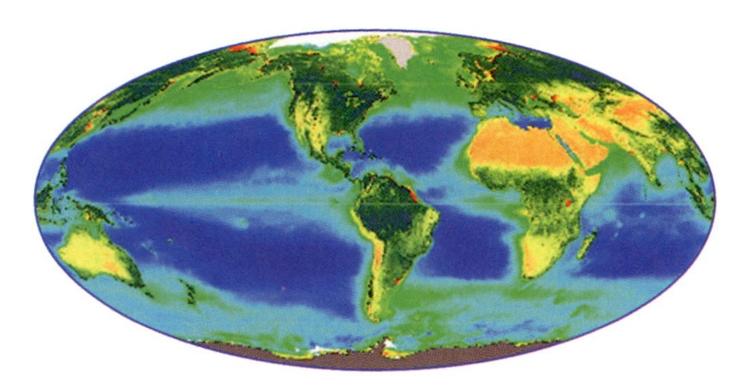

Fig. 2.10 Distribution of primary productivity in the marine and terrestrial biosphere, averaged over a 3-year period. Over the oceans the dark blue areas are indicative of very low productivity and the green and yellow areas are relatively more productive. Over land dark green is indicative of high productivity. [Imagery courtesy of SeaWiFS Project, NASA/GSFC and ORBIMAGE, Inc.]

d. Sea surface temperature

The global distribution of sea surface temperature is shaped by both radiative and dynamical factors relating to the pattern of seasonally varying, climatological-mean surface wind field over the oceans (Fig. 1.18). Radiative heating is the dominant factor. That incident solar radiation is so much stronger in the tropics than in the polar regions gives rise to a strong north–south temperature gradient, which dominates the annual-mean field shown in Fig. 2.11 (top).

The effects of the winds on the sea surface temperature pattern become more clearly apparent when the zonally averaged sea surface temperature at each latitude is removed from the total field, leaving just

[4] Greek: *eu*-good and *photic*-light.

the departures from the zonal-mean, shown in Fig. 2.11 (bottom). The coolness of the eastern oceans relative to the western oceans at subtropical latitudes derives from circulation around the subtropical anticyclones (Fig. 1.16). The equatorward flow of cool air around the eastern flanks of the anticyclones extracts a considerable quantity of heat from the ocean surface, as explained in Section 9.3.4, and drives cool, southward ocean currents (Fig. 2.4). In contrast, the warm, humid poleward flow around their western flanks extracts much less heat and drives warm western boundary currents such as the Gulf Stream. At higher latitudes the winds circulating around the subpolar cyclones have the opposite effect, cooling the western sides of the oceans and warming the eastern sides. The relative warmth of the eastern Atlantic at these higher latitudes is especially striking.

Wind-driven upwelling is responsible for the relative coolness of the equatorial eastern Pacific and Atlantic, where the southeasterly trade winds protrude northward across the equator (Fig. 1.18). Wind-driven upwelling along the coasts of Chile, California, and continents that occupy analogous positions with respect to the subtropical anticyclones, although not well resolved in Fig. 2.11, also contributes to the coolness of the subtropical eastern oceans, as do the highly reflective cloud layers that tend to develop at the top of the atmospheric boundary layer over these regions (Section 9.4.4).

The atmospheric circulation feels the influence of the underlying sea surface temperature pattern, particularly in the tropics. For example, from a comparison of Figs. 1.25 and 2.11 it is evident that the intertropical convergence zones in the Atlantic and Pacific sectors are located over bands of relatively warm sea surface temperature and that the dry zones lie over the *equatorial cold tongues* on the eastern sides of these ocean basins.

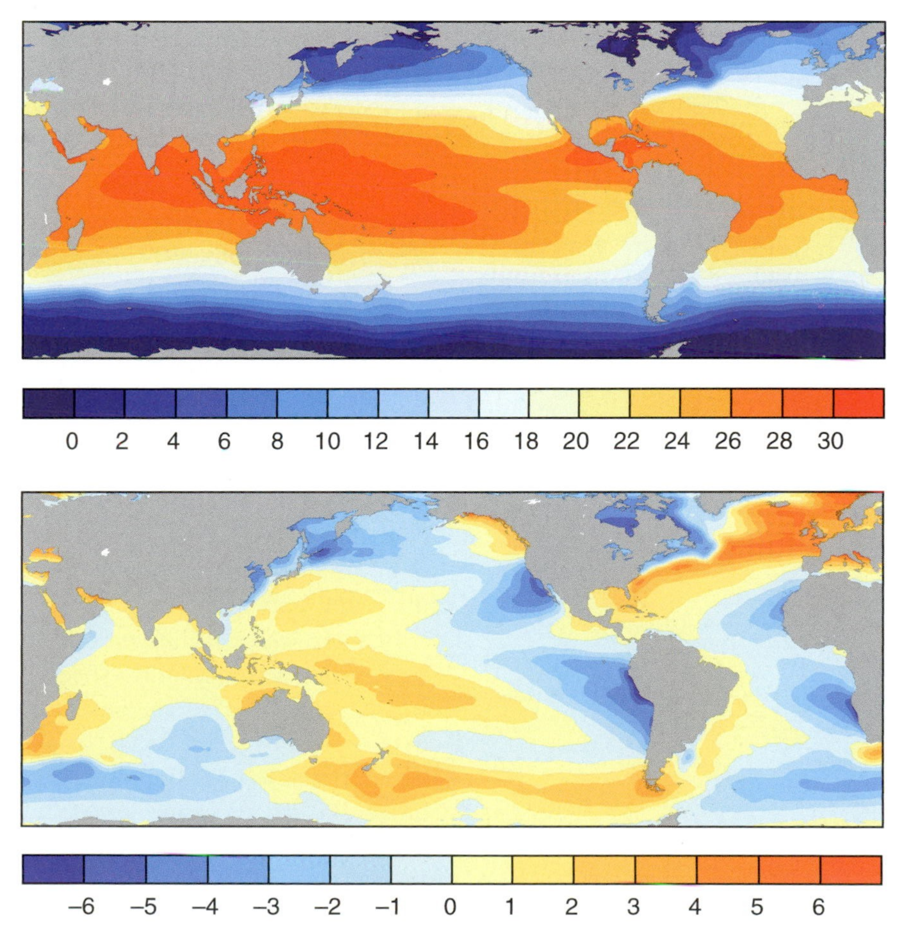

Fig. 2.11 Annual mean sea surface temperature. (Top) The total field. (Bottom) Departure of the local sea surface temperature at each location from the zonally average field. [Based on data from the U.K. Meteorological Office HadISST dataset. Courtesy of Todd P. Mitchell.]

2.1.2 The Cryosphere

The term *cryo-* (frozen) *sphere* refers to components of the Earth system comprised of water in its solid state, or in which frozen water is an essential component. The cryosphere contributes to the thermal inertia of the climate system; it contributes to the reflectivity or *albedo* of the Earth; by taking up and releasing fresh water in the polar regions, it influences oceanic thermohaline circulation; and it stores enough water to significantly the influence global sea level. The elements of the cryosphere are listed in Table 2.1 and all of them, with the exception of alpine glaciers, are represented in Fig. 2.12.

The *continental ice sheets*, dominated by Antarctica and Greenland, are the most massive elements of the cryosphere. The ice sheets are continually replenished by snowfall; they lose mass by sublimation, by the calving of icebergs, and, in summer, by runoff in streams and rivers along their periphery. The net *mass balance* (i.e., the balance between the mass sources and sinks) at any given time determines whether an ice sheet is growing or shrinking.

Over periods of tens of thousands of years and longer, annual layers of snow that fall in the relatively flat interior of the ice sheets are compressed by the accumulation of new snow on top of them. As the pressure increases, snow is transformed into ice. Due to the dome-like shape of the ice sheets and the plasticity of the ice itself, the compressed layers of ice gradually creep downhill toward the periphery of the ice sheet, causing the layer as a whole to spread out horizontally and (in accordance with the conservation of mass) to thin in the vertical dimension. Much of the flow toward the periphery tends to be concentrated in relatively narrow, fast-moving ice streams tens of kilometers in width (Fig. 2.13).

Table 2.1 Surface area and mass of the various components of the cryosphere[a]

Cryospheric component	Area	Mass
Antarctic ice sheet	2.7	53
Greenland ice sheet	0.35	5
Alpine glaciers	0.1	0.2
Arctic sea ice (March)	3	0.04
Antarctic sea ice (September)	4	0.04
Seasonal snow cover	9	<0.01
Permafrost	5	1

[a] Surface area is expressed as percentage of the area of the surface of the Earth. Mass is expressed in units of 10^3 kg m^{-2} (numerically equivalent to meters of liquid water) averaged over the entire surface area of the Earth. For reference, the total surface area of the Earth and the area of the Earth covered by land are 5.12 and 1.45 × 10^{14} m^2, respectively. [Courtesy of S. G. Warren.]

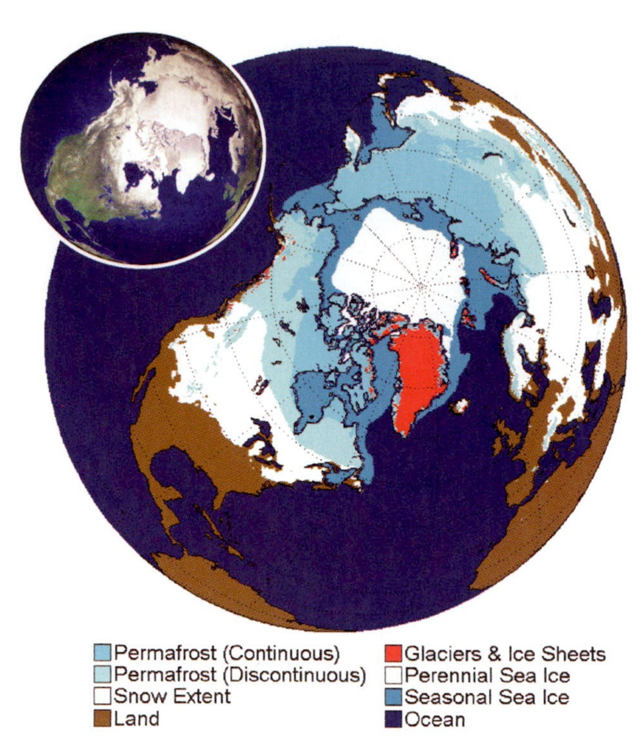

Fig. 2.12 Elements of the northern hemisphere cryosphere. The equatorward edge of the snow cover corresponds to ~50% coverage during the month of maximum snow extent. [Courtesy of Ignatius Rigor.] The inset at the upper left shows a NASA RADARSAT image highlighting these features.

Along the divides of the ice sheets the movement is very slow and the layering of the ice is relatively undisturbed. In *ice cores* extracted from these regions, the age of the ice increases monotonically with depth to ~100,000 years in the Greenland ice sheet and over 500,000 years in the Antarctic ice sheet. Analysis of air bubbles, dust, and chemical and biological tracers embedded within these ice cores is providing a wealth of information on the climate of the past few hundred thousand years, as discussed later in this chapter.

In many respects, *alpine* (i.e., mountain) *glaciers* behave like continental ice sheets, but they are much smaller in areal coverage and mass. Their fate is also determined by their mass balance. Parcels of ice within them flow continually downhill from an upper dome-like region where snow and ice accumulate toward their snouts where mass is lost continually due to melting. Because of their much smaller masses, glaciers respond much more quickly to climate change than continental ice sheets, and ice cycles through them much more rapidly. Some alpine glaciers also exhibit time-dependent behavior that is not climate

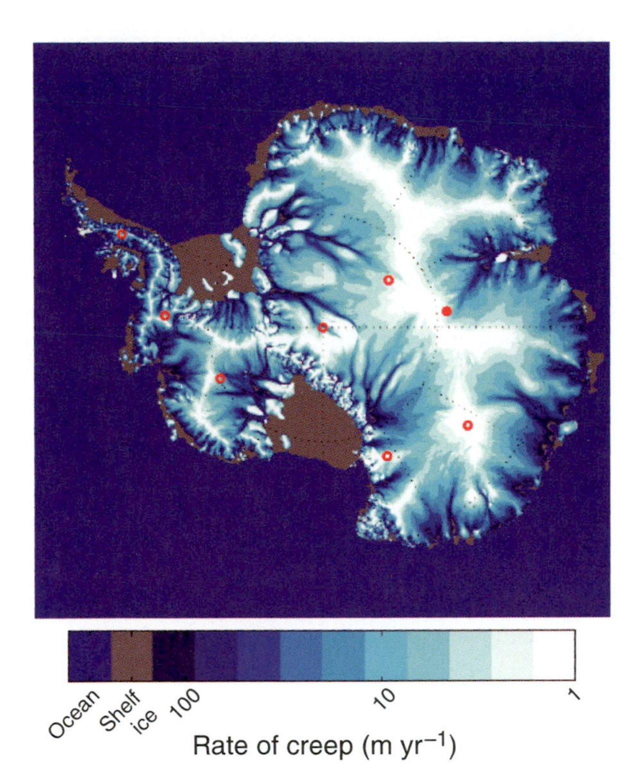

Fig. 2.13 Satellite image of the Antarctic ice sheet showing rate of creep of the ice (in m year^{-1}) on a logarithmic scale. Dots show the locations of ice core sites. Vostok, the site of the ice core shown in Fig. 2.31, is indicated by the solid red dot. [Adapted with permission from Bamber, J. L., D. G. Vaughan and I. Joughin, "Widespread Complex Flow in the Interior of the Antarctic Ice Sheet," *Science*, **287**, 1248–1250. Copyright 2000 AAAS. Courtesy of Ignatius Rigor.]

related: episodic surges of a few months' to a few years' duration interspersed with much longer periods of slow retreat.

Sea ice covers a larger area of the Earth's surface area than the continental ice sheets (Table 2.1) but, with typical thicknesses of only 1–3 m, is orders of magnitude less massive. The ice is not a continuous surface, but a fractal field comprised of ice floes (pieces) of various of shapes and sizes, as shown in Figs. 2.14 and 2.15. The individual floes are separated by patches of open water (called *leads*) that open and close as the ice pack moves, dragged by surface winds.

Seasonal limits of the northern hemisphere pack ice are shown in Fig. 2.12. During winter, ice covers not only the Arctic, but also much of the Bering Sea and the Sea of Okhotsk, but during the brief polar summer the ice retreats dramatically and large leads are sometimes observed, even in the vicinity of the North Pole. Antarctic pack ice also advances and retreats with the seasons.

The annual-mean sea ice motion, shown in Fig. 2.16, is dominated by the clockwise *Beaufort Gyre* to the north of Alaska and the *transpolar drift stream* from Siberia toward Greenland and Spitzbergen.[5] Some ice floes remain in the Arctic for a decade or more, circulating around and around the Beaufort Gyre, whereas others spend just a year or two in the Arctic before they exit either through the Fram Strait between Greenland and Spitzbergen or through the Nares Strait into Baffin Bay along the west side of Greenland. Ice floes exiting the Arctic make a one-way trip into warmer waters, where they are joined by much thicker *icebergs* that break off the Greenland ice sheet.

New pack ice is formed during the cold season by the freezing of water in newly formed leads and in regions where offshore winds drag the pack ice away from the coastline, exposing open water. The new ice thickens rapidly at first and then more gradually as it begins to insulate the water beneath it from the subfreezing air above. Ice thicker than a meter is formed, not by a thickening of newly formed layer of

[5] The existence of a transpolar drift stream was hypothesized by Nansen[6] when he learned that debris from a shipwreck north of the Siberian coast had been recovered, years later, close to the southern tip of Greenland. Motivated by this idea, he resolved to sail a research ship as far east as possible off the coast of Siberia and allow it to be frozen into the pack ice in the expectation that it would be carried across the North Pole along the route suggested by Fig. 2.16. He supervised the design and construction of a research vessel, the *Fram* ("Forward"), with a hull strong enough to withstand the pressure of the ice. The remarkable voyage of the *Fram*, which began in summer of 1893 and lasted for 3 years, confirmed the existence of the transpolar drift stream and provided a wealth of scientific data.

[6] **Fridtjof Nansen** (1861–1930). Norwegian scientist, polar explorer, statesman, and humanitarian. Educated as a zoologist. Led the first traverse of the Greenland ice cap on skis in 1888. The drift of his research vessel the *Fram* across the Arctic (1893–1896) was hailed as a major achievement in polar research and exploration. Midway through this voyage, Nansen turned over command of the *Fram* to Harald Sverdrup and set out with a companion on what proved to be a 132-day trek across the pack ice with dog-drawn sledges and kayaks, reaching 86 °N before adverse conditions forced them to turn southward.

Sacrificed his subsequent aspirations for Antarctic exploration to serve the needs of his country and to pursue humanitarian concerns. Was instrumental in peacefully resolving a political dispute between Norway and Sweden in 1905–1906 and negotiating a relaxation of an American trade embargo that threatened Norwegian food security during World War I. Awarded the Nobel Peace Prize in 1922 in recognition of his extensive efforts on behalf of war refugees and famine victims.

Fig. 2.14 Ice floes and leads in Antarctic pack ice. The lead in the foreground is 4–5 m across. The floe behind it consists of multi-year ice that may have originated as an iceberg; it is unusually thick, extending from ~15 m below to ~1 m above sea level. Most of the portion of the floe that extends above sea level is snow. At the time this picture was taken, the pack ice in the vicinity was under lateral pressure, as evidenced by the fact that a *pressure ridge* had recently developed less than 100 m away. [Photograph courtesy of Miles McPhee.]

ice, but by mechanical processes involving collisions of ice floes. *Pressure ridges* up to 5 m in thickness are created when floes collide, and thickening occurs when part of one floe is pushed or *rafted* on top of another.

When sea water freezes, the ice that forms is composed entirely of fresh water. The concentrated salt water known as *brine* that is left behind mixes with the surrounding water, increasing its salinity. Brine rejection is instrumental in imparting enough negative buoyancy to parcels of water to enable them to break through the pycnocline and sink to the bottom. Hence, it is no accident that the sinking regions in the oceanic thermohaline circulation are in high latitudes, where sea water freezes.

Land snow cover occupies an even larger area of the northern hemisphere than sea ice and it varies

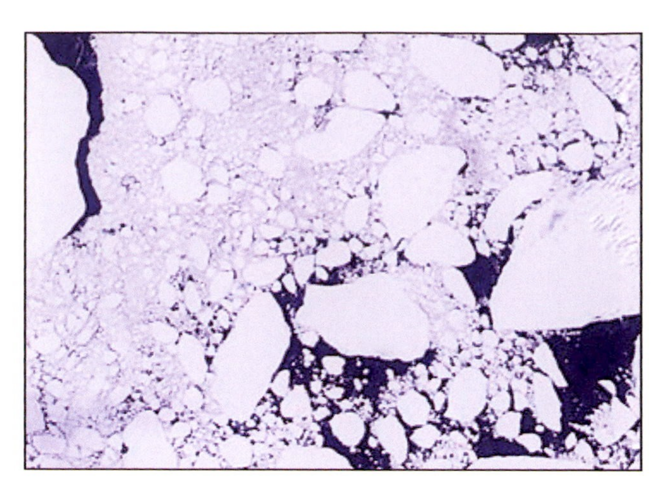

Fig. 2.15 Floes in pack ice streaming southward off the east coast of Greenland. The white area at the upper left is *landfast ice* that is attached to the coast, and the black channel adjacent to it is open water, where the mobile pack ice has become detached from the landfast ice. [NASA MODIS imagery.]

much more widely from week to week and month to month than does sea ice. With the warming of the land surface during spring, the snow virtually disappears, except in the higher mountain ranges.

Permafrost embedded in soils profoundly influences terrestrial ecology and human activities over large areas of Siberia, Alaska, and northern Canada. If the atmosphere and the underlying land surface

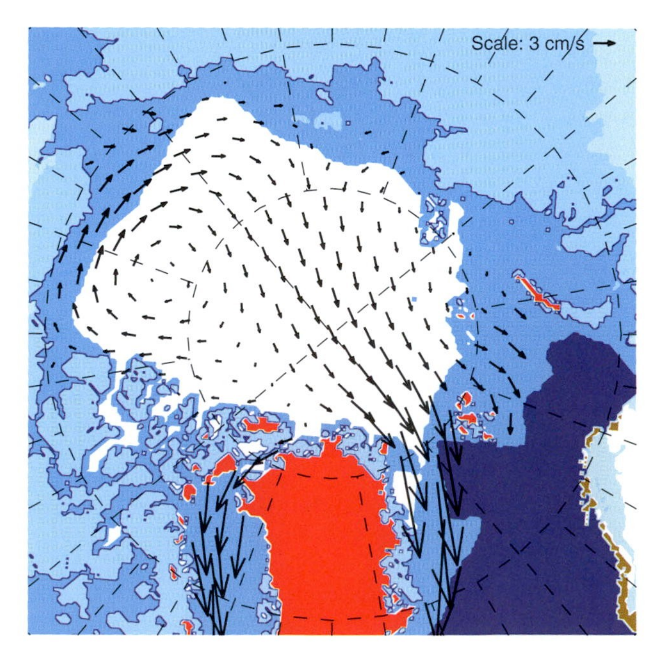

Fig. 2.16 Wintertime Arctic sea ice motion as inferred from the tracks of an array of buoys dropped on ice floes by aircraft. [Courtesy of Ignatius Rigor.]

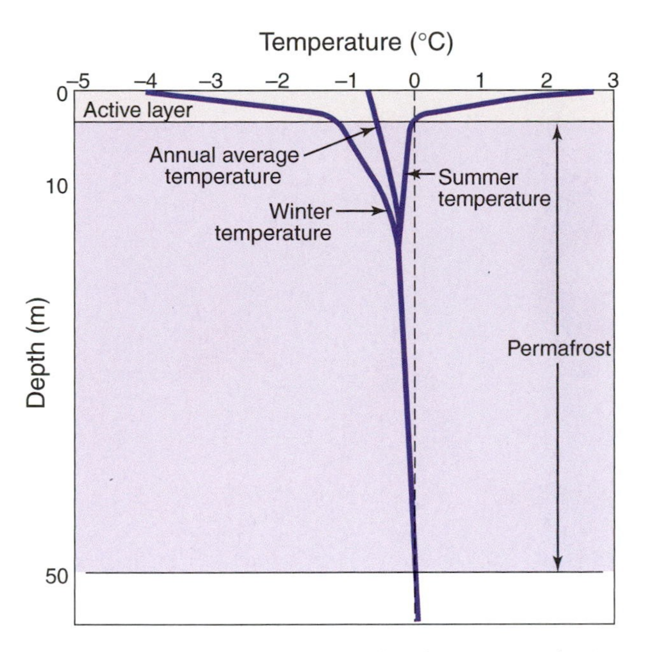

Fig. 2.17 Schematic vertical profile of summer and winter soil temperatures in a region of permafrost. The depth of the permafrost layer varies from as little as a few meters in zones of intermittent permafrost to as much as 1 km over the coldest regions of Siberia.

were in thermal equilibrium, the zones of continuous and intermittent permafrost in Fig. 2.12 would straddle the 0 °C isotherm in annual-mean surface air temperature. There is, in fact, a close correspondence between annual-mean surface air temperature and the limit of continuous permafrost, but the critical value of surface air temperature tends to be slightly above 0 °C due to the presence of snow cover, which insulates the land surface during the cold season, when it is losing heat.

Even in the zone of continuous permafrost, the topmost few meters of the soil thaw during summer in response to the downward diffusion of heat from the surface, as shown in Fig. 2.17. The upward diffusion of heat from the Earth's interior limits the vertical extent of the permafrost layer. Because the molecular diffusion of heat in soil is not an efficient heat transfer mechanism, hundreds of years are required for the permafrost layer to adjust to changes in the temperature of the overlying air.

2.1.3 The Terrestrial Biosphere

Much of the impact of climate upon animals and humans is through its role in regulating the condition and geographical distribution of forests, grasslands, tundra, and deserts, elements of the *terrestrial* (land) *biosphere*. A simple conceptual framework for relating climate (as represented by annual-mean temperature and precipitation) and vegetation type is shown in Fig. 2.18. The boundary between tundra and forest corresponds closely to the limit of the permafrost zone, which, as noted earlier, is determined by annual-mean temperature. The other boundaries in Fig. 2.18 are determined largely by the water requirements of plants. Plants utilize water both as raw material in producing chlorophyll and to keep cool on hot summer days, as described later. Forests require more water than grasslands, and grasslands, in turn, require more water than desert vegetation. The water demands of any specified type of vegetation increase with temperature.

Biomes are geographical regions with climates that favor distinctive combinations of plant and animal species. For example, tundra is the dominant form of vegetation in regions in which the mean temperature of the warmest month is ≤10 °C, and sparse, desert vegetation prevails in regions in which potential evaporation (proportional to the quantity of solar radiation reaching the ground) exceeds precipitation. The global distribution of biomes is determined by the *insolation* (i.e., the incident solar radiation) at the top of the atmosphere and by the climatic variables:

- annual-mean temperature,
- the annual and diurnal temperature ranges,
- annual-mean precipitation, and
- the seasonal distributions of precipitation and cloudiness.

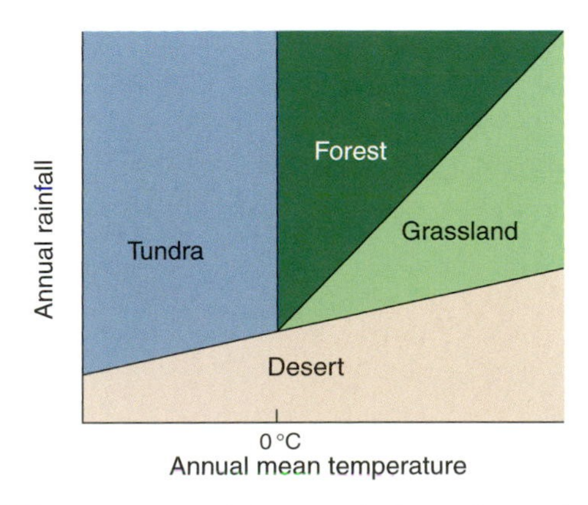

Fig. 2.18 A conceptual framework for understanding how the preferred types of land vegetation over various parts of the globe depend on annual-mean temperature and precipitation.

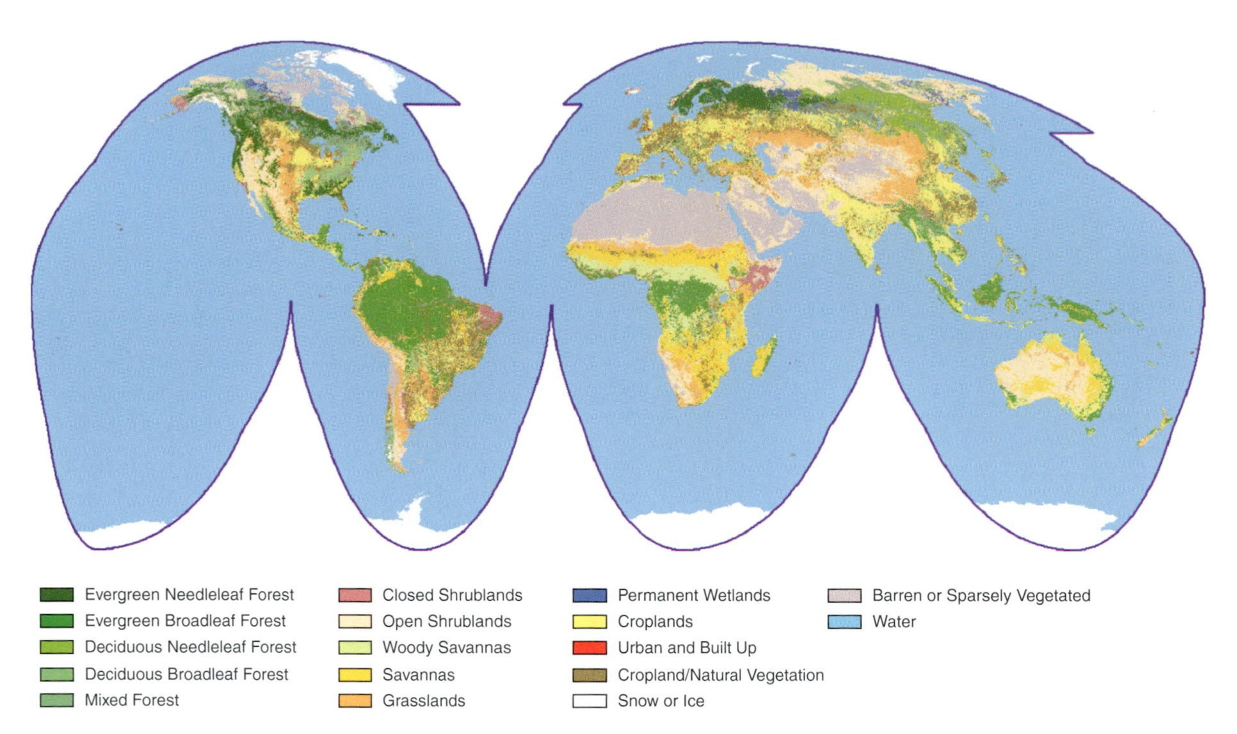

Fig. 2.19 Global land cover characterization, as inferred from NASA AVHRR NDVI satellite imagery and ground-based data relating to ecological regions, soils, vegetation, land use, and land cover. [From USGS Land Processes DAAC.]

Insolation and climate at a given location, in turn, are determined by latitude, altitude, and position with reference to the land–sea configuration and terrain. The combined influence of altitude upon temperature (Fig. 1.9), terrain upon precipitation (Fig. 1.25), and local terrain slope upon the incident solar radiation (Exercise 4.16) gives rise to a variegated distribution of biomes in mountainous regions.

Several different systems exist for assigning biomes, each of which consists of a comprehensive set of criteria that are applied to the climate statistics for each geographical location.[7] The "ground truth" for such classification schemes is the observed distribution of land cover, as inferred from ground-based measurements and high-resolution satellite imagery. An example is shown in Fig. 2.19.

The state of the terrestrial biosphere feeds back upon the climate through its effects on

- the hydrologic cycle: for example, during intervals of hot weather, plants control their temperatures by *evapo-transpiration* (i.e., by giving off water vapor through their leaves or needles). Energy derived from absorbed solar radiation that would otherwise contribute to heating the land surface is used instead to evaporate liquid water extracted from the soil by the roots of the plants. In this manner, the solar energy is transferred to the atmosphere without warming the land surface. Hence, on hot summer days, grass-covered surfaces tend to be cooler than paved surfaces and vegetated regions do not experience as high daily maximum temperatures as deserts and urban areas.
- the local albedo (the fraction of the incident solar radiation that is reflected, without being absorbed): for example, snow-covered tundra is more reflective, and therefore cooler during the daytime, than a snow-covered forest.
- the roughness of the land surface: wind speeds in the lowest few tens of meters above the ground tend to be higher over bare soil and tundra than over forested surfaces.

[7] These systems are elaborations of a scheme developed by Köppen[8] a century ago.

[8] **Wladimir Peter Köppen** (1846–1940) German meteorologist, climatologist, and amateur botanist. His Ph.D. thesis (1870) explored the effect of temperature on plant growth. His climate classification scheme, which introduced the concept of biomes, was published in 1900. For many years, Köppen's work was better known to physical geographers than to atmospheric scientists, but in recent years it is becoming more widely appreciated as a conceptual basis for describing and modeling the interactions between the atmosphere and the terrestrial biosphere.

2.1.4 The Earth's Crust and Mantle

The current configuration of continents, oceans, and mountain ranges is a consequence of plate tectonics and continental drift.[9] The Earth's crust and mantle also take part in chemical transformations that mediate the composition of the atmosphere on timescales of tens to hundreds of millions of years.

The Earth's crust is broken up into plates that float upon the denser and much thicker layer of porous but viscous material that makes up the Earth's mantle. Slow convection within the mantle moves the plates at speeds ranging up to a few centimeters per year (tens of kilometers per million years). Plates that lie above regions of upwelling in the mantle are spreading, whereas plates that lie above regions of downwelling in the mantle are being pushed together. Earthquakes tend to be concentrated along plate boundaries.

Oceanic plates are thinner, but slightly denser than continental plates so that when the two collide, the ocean plate is *subducted* (i.e., drawn under the continental plate) and incorporated into the Earth's mantle, as shown schematically in Fig. 2.20. Rocks in the subducted oceanic crust are subjected to increasingly higher temperatures and pressures as they descend, giving rise to physical and chemical transformations.

Collisions between plate boundaries are often associated with volcanic activity and with the uplift of mountain ranges. The highest of the Earth's mountain ranges, the Himalayas, was created by folding of the Earth's crust following the collision of the Indian and Asian plates, and it is still going on today. The Rockies, Cascades, and Sierra ranges in western North America have been created in a similar manner by the collision of the Pacific and North American plates. These features have all appeared within the past 100 million years.

Oceanic plates are continually being recycled. The Pacific plate is being subducted along much of the extent of its boundaries, while new oceanic crust is being formed along the mid-Atlantic ridge as magma upwelling within the mantle rises to the surface, cools, and solidifies. As this newly formed crust diverges away from the mid-Atlantic ridge, the floor of the Atlantic Ocean is spreading, pushing other parts of the crust into the spaces formerly occupied by the subducted portions of the Pacific plate. As the Atlantic widens and the Pacific shrinks, the continents may be viewed as drifting away from the Atlantic sector on trajectories that will, in 100–200 million years, converge over what is now the mid-Pacific. A similar congregation of the continental plates is believed to have occurred about 200 million years ago, when they were clustered around the current position of Africa, forming a supercontinent called *Pangaea* (all Earth).

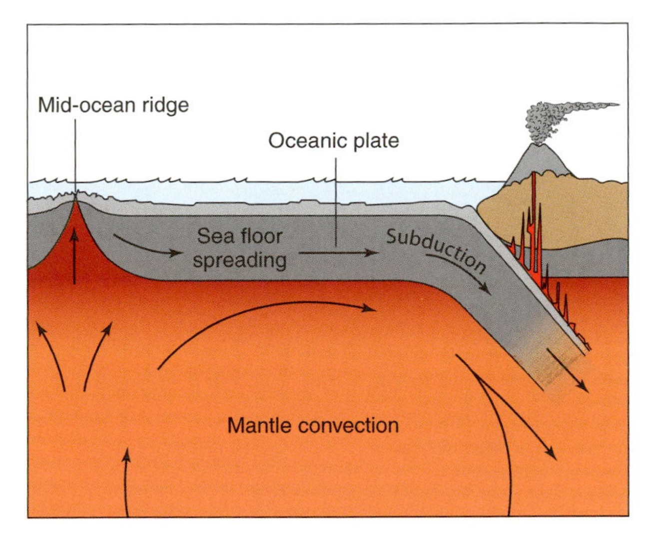

Fig. 2.20 Schematic showing subduction, sea floor spreading, and mountain building. [Adapted by permission of Pearson Education, Inc., Upper Saddle River, NJ. Edward J. Tarbuck, Frederick K. Lutgens and Dennis Tasa, Earth: An Introduction to Physical Geology, 8th Edition, © 2005, p. 426, Fig. 14.9.]

Some of the material incorporated into the mantle when plates are subducted contains *volatile substances* (i.e., substances that can exist in a gaseous form, such as water in hydrated minerals). As the temperature of these materials rises, pressure builds

[9] The theory of continental drift was first proposed by Alfred Wegener[10] in 1912 on the basis of the similarity between the shapes of coastlines, rock formations, and fossils on the two sides of the Atlantic. Wegener's radical reinterpretation of the processes that shaped the Earth was largely rejected by the geological community and did not become widely accepted until the 1960s, with the advent of geomagnetic evidence of sea-floor spreading.

[10] **Alfred Wegener** (1880–1930). German meteorologist, professor at University of Graz. Began his career at the small University of Marburg. First to propose that ice particles play an important role in the growth of cloud droplets. Set endurance record for time a aloft in a hot air balloon (52 h) in 1906. Played a prominent role in the first expeditions to the interior of Greenland. Died on a relief mission on the Greenland icecap. The Alfred Wegener Institute in Bremerhaven is named in his honor. Son-in-law of Vladimir Köppen and co-authored a book with him.

up beneath the Earth's crust, leading to volcanic eruptions. As will be explained later in this chapter, gases expelled in volcanic eruptions are the source of the Earth's present atmosphere, and they are continually renewing it.

2.1.5 Roles of Various Components of the Earth System in Climate

Atmospheric processes play the lead role in determining such fundamental properties of climate as the disposition of incoming solar radiation, temperatures at the Earth's surface, the spatial distribution of water in the terrestrial biosphere, and the distribution of nutrients in the euphotic zone of the ocean. However, other components of the Earth system are also influential. Were it not for the large storage of heat in the ocean mixed layer and cryosphere during summer, and the extraction of that same heat during the following winter, seasonal variations in temperature over the middle and high latitude continents would be much larger than observed and, were it not for the existence of widespread vegetation, summertime daily maximum temperatures in excess of 40 °C would be commonplace over the continents. The oceanic thermohaline circulation warms the Arctic and coastal regions of Europe by several degrees, while wind-driven upwelling keeps the equatorial eastern Pacific cool enough to render the Galapagos Islands a suitable habitat for penguins!

Plate tectonics shaped the current configuration of continents and topography, which, in turn, shapes many of the distinctive regional features of today's climate. The associated recycling of minerals through the Earth's upper mantle is believed to have played a role in regulating the concentration of atmospheric carbon dioxide, which exerts a strong influence upon the Earth's surface temperature.

These are but a few examples of how climate depends not only on atmospheric processes, but on processes involving other components of the Earth system. As explained in Section 10.3, interactions between the atmosphere and other components of the Earth system give rise to feedbacks that can either amplify or dampen the climatic response to an imposed external forcing of the climate system, such as a change in the luminosity of the sun or human-induced changes in atmospheric composition.

The next three sections of this chapter describe the exchanges and cycling of water, carbon, and oxygen among the various components of the Earth system.

2.2 The Hydrologic Cycle

Life on Earth is critically dependent on the cycling of water back and forth among the various *reservoirs* in the Earth system listed in Table 2.2, which are collectively known as the *hydrosphere*. In discussing the exchanges between the smaller reservoirs, we make use of the concept of *residence time* of a substance within a specified reservoir, defined as the mass in the reservoir divided by the *efflux* (the rate at which the substance exits from the reservoir).[11] Residence time provides an indication of amount of time that a typical molecule spends in the reservoir between visits to other reservoirs. Long residence times are indicative of large reservoirs and/or slow rates of exchange with other reservoirs, and vice versa.

Based on current estimates, the largest reservoir of water in the Earth system is the mantle. The rate at which water is expelled from the mantle in volcanic emissions is estimated to be $\sim 2 \times 10^{-4}$ kg m^{-2} year^{-1} averaged over the Earth's surface, which is the basis for the 10^{11} year residence time in Table 2.2. At this

Table 2.2 Masses of the various reservoirs of water in the Earth system (in 10^3 kg m^{-2}) averaged over the surface of the Earth, and corresponding residence times

Reservoirs of water	Mass	Residence time
Atmosphere	0.01	Days
Fresh water (lakes and rivers)	0.6	Days to years
Fresh water (underground)	15	Up to hundreds of years
Alpine glaciers	0.2	Up to hundreds of years[a]
Greenland ice sheet	5	10,000 years[b]
Antarctic ice sheet	53	100,000 years
Oceans	2,700	
Crust and mantle	20,000	10^{11} years

[a] Estimated by dividing typical ice thicknesses of a large alpine glacier (~300 m) by the annual rate of ice accumulation (~1 m).
[b] Estimated by dividing typical ice thicknesses in the interior of the Greenland ice sheet (2000 m) by the annual rate of ice accumulation (~0.2 m).

[11] The concept of residence time is developed more fully in Chapter 5.1.

rate of exchange, only roughly 5% of the water estimated to reside in the mantle would be expelled over the $\sim 4.5 \times 10^9$-year lifetime of the Earth—not even enough to fill the oceans.

After the mantle and oceans, the next largest reservoir of water in the Earth system is the continental ice sheets, the volumes of which have varied widely on timescales of tens of thousands of years and longer, causing large variations in global sea level.

Exercise 2.2 Based on data provided in Table 2.1, estimate how much the sea level would rise if the entire Greenland ice sheet were to melt.

Solution: The mass of the Greenland ice sheet is equal to its mass per unit area averaged over the surface of the Earth (as listed in Table 2.1) times the area of the Earth or

$$(5 \times 10^3 \text{ kg m}^{-2}) \times (5.10 \times 10^{14} \text{ m}^2) = 2.55 \times 10^{18} \text{ kg}$$

If the ice cap were to melt, this mass would be distributed uniformly over the ocean-covered area of the Earth's surface. Hence, if x is the sea level rise, we can write

$$(\text{Area of oceans}) \times (\text{density of water})\, x = \text{mass of ice sheet}$$

$$((5.10 - 1.45) \times 10^{14} \text{ m}^2) \times (10^3 \text{ kg m}^{-2})\, x = 2.55 \times 10^{18} \text{ kg}$$

Solving, we obtain $x = 7$ m.

Because the masses given in Table 2.2 are expressed in units numerically equivalent to the depth (in m) of a layer covering the entire surface of the Earth, we could have written simply

$$(5.10 - 1.45)\, x = 5.10 \times 5 \text{ m}$$
$$x = 7 \text{ m}$$

■

Of the reservoirs listed in Table 2.2, the atmosphere is by far the smallest and it is the one with the largest rates of exchange with the other components of the Earth system. The residence time of water in the atmosphere, estimated by dividing the mass of water residing in the atmosphere ($\sim$30 kg m^{-2}, equivalent to a layer of liquid water $\sim$3 cm deep) by the mean rainfall rate averaged over the Earth's surface (roughly 1 m per year or 0.3 cm day^{-1}), is $\sim$10 days. By virtue of the large exchange rate and the large latent heat of vaporization of water, the cycling of water vapor through the atmospheric branch of the hydrologic cycle is effective in transferring energy from the Earth's surface to the atmosphere.

Averaged over the globe, the rate of precipitation P equals the rate of evaporation E: any appreciable imbalance between these terms would result in a rapid accumulation or depletion of atmospheric water vapor, which is not observed. However, in analyzing the water balance for a limited region, the horizontal transport of water vapor by winds must also be considered. For example, within the region of the ITCZ, $P >> E$: the excess precipitation is derived from an influx of water vapor carried by the converging trade winds shown in Fig. 1.18. Conversely, in the region of the relatively dry, cloud-free subtropical anticyclones, $E >> P$: the excess water vapor is carried away, toward the ITCZ on the equatorward side and toward the midlatitude storm tracks on the poleward side, by the diverging low-level winds. For the continents as a whole, $P > E$: the excess precipitation returns to the sea in rivers. Local evapotranspiration E, as described in Section 2.1.3, accounts for an appreciable fraction of the moisture in summer rainfall P over the continents.

Under steady-state conditions, the mass balance for water vapor over in a column of area A, extending from the Earth's surface to the top of the atmosphere, can be written in the form

$$\overline{E} - \overline{P} = \overline{Tr} \tag{2.1}$$

where overbars denote averages over the area of the column and $\overline{Tr}$ denotes the horizontal transport (or flux) of water vapor out of the column by the winds, as discussed in the previous paragraph. Figure 2.21 shows the distributions of the export of water vapor (i.e., divergence of water vapor transport) by the winds over the low latitude oceans together with the observed distribution of $E - P$. Two aspects of Fig. 2.21 are worthy of note.

1. Apart from the sign reversal, the distribution of $E - P$ in the lower panel resembles the rainfall distribution in Fig. 1.25. That $-P$ and $E - P$ exhibit similar distributions indicates that the horizontal gradients of P must be much stronger than those in E. It follows that the strong observed gradients in climatological-mean rainfall are due to wind patterns rather than to gradients in local evaporation.

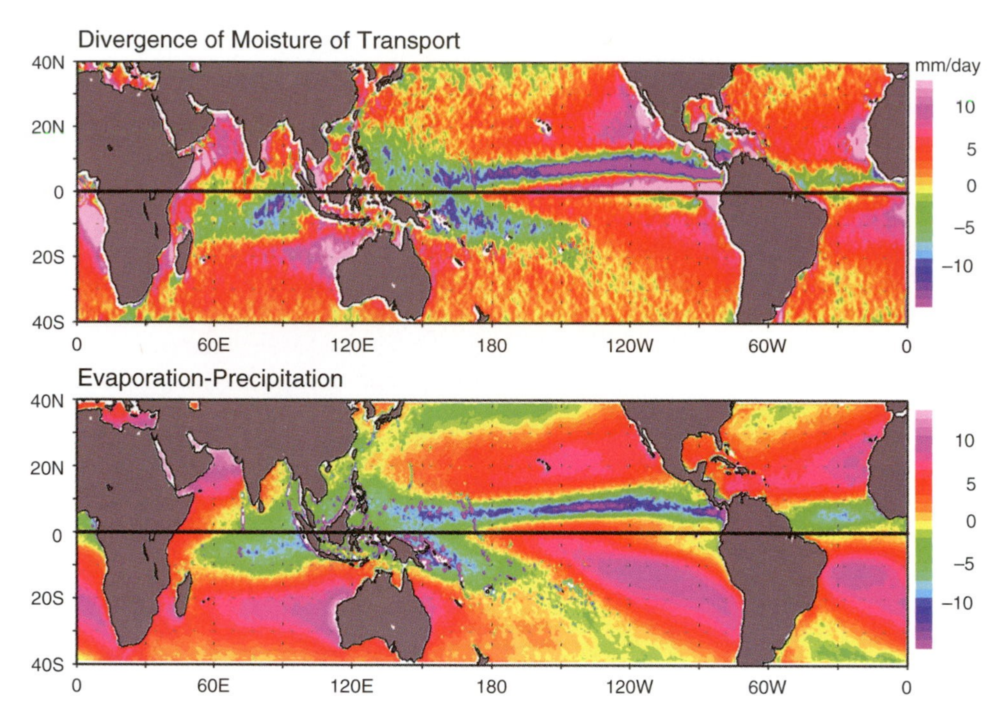

Fig. 2.21 Terms in the annual mean mass balance of atmospheric water vapor in units of mm day^{-1} of liquid water. (Top) The local rate of change of vertically integrated water vapor due to horizontal transport by the winds. (Bottom) Difference between local evaporation and local precipitation. If the estimates were perfect, the maps would be identical. [Based on data from NASA's QuikSCAT and Tropical Rain Measuring Mission (TRMM). Courtesy of W. Timothy Liu and Xiaosu Xie.]

2. In accordance with (2.1), the geographical distributions of $E - P$ and Tr in Fig. 2.21 are similar. The agreement is noteworthy because the measurements used in constructing these two maps are entirely different. The distribution of Tr is constructed from data on winds and atmospheric water vapor concentrations, without reference to evaporation and precipitation.

Also of interest is the time-dependent hydrologic mass balance over land for a layer extending from the land surface downward to the base of the deepest aquifers. In this case

$$\frac{d\overline{St}}{dt} = \overline{P} - \overline{E} - \overline{T} \tag{2.2}$$

where $\overline{St}$ is the area averaged storage of water within some prescribed region and the transport term involves the inflow or outflow of water in rivers and subsurface aquifers. For the special case of a land-locked basin from which there is no inflow or outflow of surface water, the transport term vanishes, and

$$\frac{d\overline{St}}{dt} = \overline{P} - \overline{E} \tag{2.3}$$

Hence the storage of water within the basin, which is reflected in the level of the lake into which the rivers within the basin drain, increases and decreases in response to time variations in $\overline{P} - \overline{E}$.

Figure 2.22 shows how the level of the Great Salt Lake in the reat Basin of the western United States has varied in response to variations in precipitation. From the time of its historic low[12] in 1963 to the time of its high in 1987, the level of the Great Salt Lake rose by 6.65 m, the area of the lake increased by a factor of 3.5, and the volume increased by a factor of 4. The average precipitation during this 14-year interval was heavier than the long-term average, but there were large, year-to-year ups and downs. It is notable that the lake level rose smoothly and

[12] The historical time series dates back to 1847.

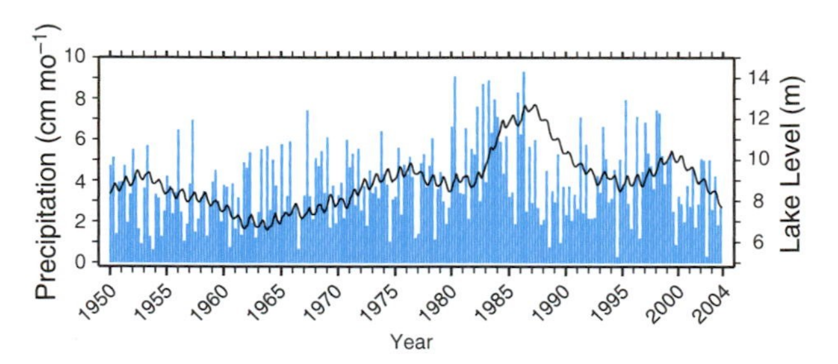

Fig. 2.22 The black curve shows variations in the depth of the Great Salt Lake based on a reference level of 4170 feet above sea level (in m). Depth scale (in m) at right. Blue bars indicate seasonal-mean precipitation at nearby Logan, Utah (in cm month^{-1}). [Lake level data from the U.S. Geological Survey. Courtesy of John D. Horel and Todd P. Mitchell.]

monotonically despite the large year-to-year variations in the precipitation time series. Exercises 2.11–2.13 at the end of this chapter are designed to provide some insight into this behavior.

2.3 The Carbon Cycle

Most of the exchanges between reservoirs in the hydrologic cycle considered in the previous section involve phase changes and transports of a single chemical species, H_2O. In contrast, the cycling of carbon involves chemical transformations. The carbon cycle is of interest from the point of view of climate because it regulates the concentrations of two of the atmosphere's two most important greenhouse gases: carbon dioxide (CO_2) and methane (CH_4).

The important carbon reservoirs in the Earth system are listed in Table 2.3 together with their masses and the residence times, in the same units as in Table 2.2. The atmospheric CO_2 reservoir is intermediate in size between the active biospheric reservoir (green plants, plankton, and the entire food web) and the gigantic reservoirs in the Earth's crust. The exchange rates into and out of the small reservoirs are many orders of magnitude faster than those that involve the large reservoirs. The carbon reservoirs in the Earth's crust have residence times many orders of magnitude longer than the atmospheric reservoirs, reflecting not only their larger sizes, but also the much slower rates at which they exchange carbon with the other components of the Earth system. Figure 2.23 provides an overview of the cycling of carbon between the various carbon reservoirs.

Table 2.3 Major carbon reservoirs in the Earth system and their present capacities in units of kg m^{-2} averaged over the Earth's surface and their residence times[a]

Reservoir	Capacity	Residence time
Atmospheric CO_2	1.6	10 years
Atmospheric CH_4	0.02	9 years
Green part of the biosphere	0.2	Days to seasons
Tree trunks and roots	1.2	Up to centuries
Soils and sediments	3	Decades to millennia
Fossil fuels	10	–
Organic C in sedimentary rocks	20,000	2×10^8 years
Ocean: dissolved CO_2	1.5	12 years
Ocean CO_3^{2-}	2.5	6,500 years
Ocean HCO^-	70	200,000 years
Inorganic C in sedimentary rocks	80,000	10^8 years

[a] Capacities based on data in Fig. 8.3 (p. 150) of Kump, Lee R.; Kasting, James F.; Crane, Robert G., The Earth System, 2nd Edition, © 2004. Adapted by permission of Pearson Education, Inc., Upper Saddle River, NJ.

Exercise 2.3 Carbon inventories are often expressed in terms of gigatons of carbon (Gt C), where the prefix *giga* indicates 10^9 and *t* indicates a metric ton or 10^3 kg. (Gt is equivalent to Pg in cgs units, where the prefix *peta* denotes 10^{15}.) What is the conversion factor between these units and the units used in Table 2.3?

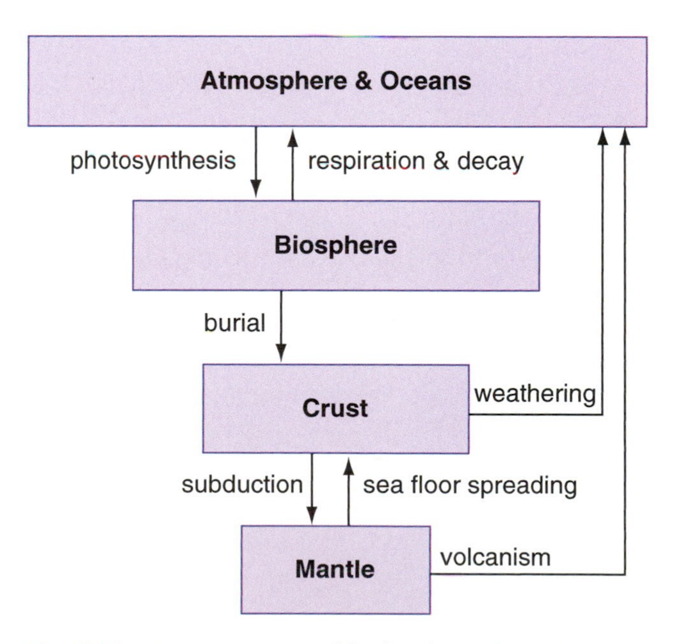

Fig. 2.23 Processes responsible for the cycling of carbon between the various reservoirs in the Earth system.

Solution: The surface area of the Earth is $4\pi R_E^2$, where $R_E = 6.37 \times 10^6$ m is the mean radius of the Earth, or 5.10×10^{14} m^2. To convert mass per unit area in units of kg m^{-2} averaged over the surface of the Earth to Gt C or Pg C, we multiply by 5.10×10^{14} m^2 to obtain the mass in kg and divide by 10^{12} to get the mass in units of Gt (or Pg). Hence, the conversion factor is 510 Gt (or Pg) C per kg m^{-2}. ■

2.3.1 Carbon in the Atmosphere

Most of the carbon in the atmospheric reservoir is in the form of CO_2. Because of its chemical inertness, CO_2 is relatively well mixed within the atmosphere: away from forest canopies and other sites in close contact with vegetation, CO_2 concentrations vary by only ~1% over the surface of the Earth (e.g., compare the concentrations at Mauna Loa and the South Pole in Fig. 1.3).

Methane (CH_4) is present only in trace concentrations in the Earth's atmosphere, but it contributes to the greenhouse effect and is chemically active. It enters the atmosphere mainly through the escape of natural gas in mining operations and pipelines and through the anaerobic breakdown of organic matter, much of which is also human induced through activities such as the production of rice and livestock.[13] Methane has a ~9-year residence time in the atmosphere: it is removed by the oxidation reaction

$$CH_4 + 2O_2 \rightarrow CO_2 + 2H_2O \qquad (2.4)$$

and by the oxidation reactions described in Section 5.3.[14]

Exercise 2.4 Reconcile the present atmospheric CO_2 concentration of ~380 ppmv with the mass concentration of elemental carbon in CO_2 given in Table 2.3.

Solution: Making use of (1.7) with the volume concentration for the ith constituent $c_i = n_i/n$, where n is the total number of molecules, we can write

$$m\,(CO_2) = m_a \times \frac{c\,(CO_2)\,M_C}{\Sigma c_i M_i}$$

where the i subscripts refer to the major constituents of air, m_a is the mass of the atmosphere per m^2, and M_C is the molecular weight of the carbon atom. In carrying out this calculation, we find that it is sufficient to take into account the three major constituents: N_2, O_2, and A. Substituting data from Table 1.1 into this expression (taking care to use the molecular weight of just the carbon atom rather than that of the CO_2 molecule), we obtain

$$m\,(CO_2) = m_a \times \frac{(380 \times 10^{-6}) \times 12}{[(0.7808 \times 28.016) + (0.2095 \times 32.00) + (0.0093 \times 39.94)]}$$

which yields

$$m\,(CO_2) = m_a \times \frac{4.56 \times 10^{-3}}{(21.87 + 6.70 + 0.37)} = m_a \times \frac{4.56 \times 10^{-3}}{28.94}$$

Substituting $m_a = 1.004 \times 10^4$ kg m^{-2} from Exercise 1.1, we obtain $m\,(CO_2) = 1.58$ kg m^{-2}, in agreement with Table 2.3. Multiplying by the area of the Earth's surface (5.10×10^{14} m^2), we obtain an atmospheric mass of 8.06×10^{14} kg, or 806 GtC. ■

2.3.2 Carbon in the Biosphere

On short timescales, large quantities of carbon pass back and forth between the atmosphere and the biosphere. These exchanges involve the *photosynthesis* reaction:

$$CO_2 + H_2O \rightarrow CH_2O + O_2 \qquad (2.5)$$

which removes carbon from the atmosphere and stores it in organic molecules in phytoplankton and leafy plants, and the *respiration and decay* reaction:

$$CH_2O + O_2 \rightarrow CO_2 + H_2O \qquad (2.6)$$

which oxidizes organic matter and returns the CO_2 to the atmosphere. Photosynthesis involves the absorption of energy in the form of visible light at

[13] Ruminants, such as cows, release (burp) methane as they digest the cellulose in grass.

[14] The oxidation of methane is an important source of stratospheric water vapor.

wavelengths near 0.43 (blue) and 0.66 μm (orange), and the respiration and decay reaction releases an equivalent amount of energy in the form of heat. By comparing the intensity of reflected radiation at various wavelengths in the visible part of the spectrum, it is possible to estimate the rate of photosynthesis in (2.5) by phytoplankton and land plants, which is referred to as *net primary productivity (NPP)*.

Figure 2.24 shows the global distribution of net primary productivity for June 2002. Enhanced marine productivity is clearly evident in the bands of equatorial and coastal upwelling, but NPP is generally higher over land areas with growing vegetation than anywhere over the oceans. Rates are particularly high in the boreal forests. The "greening" of the large northern hemisphere continents in spring and summer draws a substantial amount of CO_2 out of the atmosphere and stores it in plant biomass, which is subject to decay at a more uniform rate throughout the year. These exchanges are responsible for the pronounced annual cycle in the Mauna Loa CO_2 time series seen in Fig. 1.3. The annual cycle is even more pronounced at high-latitude northern hemisphere stations. In contrast, CO_2 concentrations at the South Pole exhibit a much weaker annual cycle (Fig. 1.3).

The rate of exchange of carbon between the atmosphere and the biosphere is estimated to be ~0.1–0.2 kg C m^{-2} $year^{-1}$. Hence, the time that a typical molecule of CO_2 resides in the atmosphere is 1.6 kg C m^{-2} (from Table 2.3) divided by ~0.15 kg C m^{-2} $year^{-1}$, or ~10 years. The green part of the biosphere responsible for this large exchange rate is capable of storing only ~10% of the atmospheric carbon at any given time. Hence, if a large quantity of CO_2 were injected into the atmosphere instantaneously, the concentration would remain elevated for a time interval much longer than 10 years. The relaxation time would be determined by the rate of exchange of carbon between the atmosphere and the larger reservoirs in the Earth system listed in Table 2.3. The timescale for the growth of tree trunks and root systems is on the order of decades, and the corresponding timescale for the burial of organic matter is much longer than that, because only ~0.1% of the plant biomass that is photosynthesized each year is eventually buried and incorporated into sedimentary rocks within the Earth's crust (the organic carbon reservoir in Table 2.3). Most of the organic carbon generated by photosynthesis undergoes oxidation by (2.6) when plants decay, when soils weather, or when forests and peat deposits burn. In anoxic (i.e., oxygen deficient) environments, the carbon in decaying organic matter is returned to the atmosphere in the form of methane.

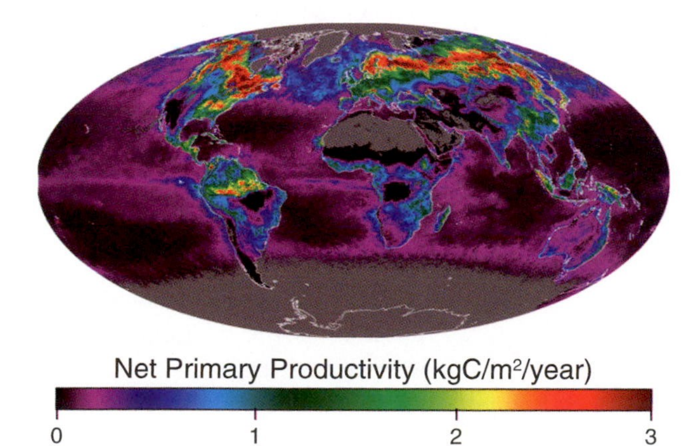

Fig. 2.24 Rate of carbon uptake by photosynthesis, commonly referred to as *net primary productivity*, averaged over June 2002, in units of kg m^{-2} $year^{-1}$. Values are low over high latitudes of the southern hemisphere because of the lack of sunlight. [Based on NASA Sea WiFS imagery.]

The marine biosphere absorbs dissolved CO_2 within the euphotic zone and releases it throughout the deeper layer of the ocean in which plants, animals, and detritus decay as they sink toward the ocean floor. The sinking of organic matter has the effect of transporting CO_2 downward, reducing its concentration in the topmost few tens of meters of the ocean. Were it not for the action of this gravity-driven *biological pump*, atmospheric CO_2 concentrations, which are in equilibrium with concentrations in water at the ocean surface, would be ~1000 ppmv, roughly 2.6 times greater than observed, and the acidity of the water in the euphotic zone would be high enough to quickly dissolve the world's coral reefs.

Within anoxic regions of the oceans [i.e., the regions in which the ventilation of dissolved O_2 into the waters below the euphotic zone is insufficient to keep pace with the rate of oxygen consumption in the decay reaction (2.6)], the organic debris that settles out of the euphotic zone reaches the ocean floor and forms layers of sediment, some of which are eventually incorporated into the *organic carbon reservoir* in the Earth's crust. Cores from sediments containing organic carbon are among the principal sources of proxy data on the climate of the past few million years. Shells and skeletons of sea animals that settle to the ocean floor are converted into limestone ($CaCO_3$) rocks. This *inorganic carbon reservoir* of the Earth's crust is the largest of the carbon reservoirs in the Earth system.

2.3.3 Carbon in the Oceans

The carbon in the oceanic reservoir exists in three forms: (1) dissolved CO_2 or H_2CO_3, also known as carbonic acid, (2) carbonate (CO_3^{2-}) ions paired with Ca^{2+} and Mg^{2+} and other metallic cations, and (3) bicarbonate (HCO_3^-) ions. The third form is by far the largest of the oceanic carbon reservoirs (Table 2.3). The dissolved carbon dioxide concentration equilibrates with atmospheric concentrations through the reaction

$$CO_2 + H_2O \leftrightarrows H_2CO_3 \qquad (2.7)$$

An increase in the atmospheric concentration of CO_2 thus tends to raise the equilibrium concentration of dissolved CO_2. Carbonic acid, in turn, dissociates to form bicarbonate ions and hydrogen ions

$$H_2CO_3 \leftrightarrows H^+ + HCO_3^- \qquad (2.8)$$

thereby causing the water to become more acidic. The increasing concentration of H^+ ions shifts the equilibrium between carbonate and bicarbonate ions

$$HCO_3^- \leftrightarrows H^+ + CO_3^{2-} \qquad (2.9)$$

toward the left. The net effect, obtained by adding (2.7), (2.8), and the reverse of (2.9) is

$$CO_2 + CO_3^{2-} + H_2O \leftrightarrows 2HCO_3^- \qquad (2.10)$$

which incorporates the added carbon into the bicarbonate reservoir without any net increase in the acidity of the ocean. The ability of the ocean to take up and buffer CO_2 in this manner is limited by the availability of ions in the carbonate reservoir.

Marine organisms incorporate bicarbonate ions into their shells and skeletons through the reaction

$$Ca^{2+} + 2HCO_3^- \rightarrow CaCO_3 + H_2CO_3 \qquad (2.11)$$

A fraction of the calcium carbonate created in Eq. (2.11) settles on the sea floor and forms limestone deposits, while the remainder dissolves through the reverse reaction.

$$CaCO_3 + H_2CO_3 \rightarrow Ca^{2+} + 2HCO_3^- \qquad (2.12)$$

Limestone deposits tend to be concentrated in continental shelves beneath shallow tropical seas that provide an environment hospitable to coral. At these levels in the ocean, the acidity of the water is low enough that shells and skeletons deposited on the ocean floor do not dissolve.

The Ca^{2+} ions that marine organisms incorporate into their shells enter the ocean by way of the weathering of rocks in rain water that is carried to the oceans in rivers. Some of these ions are derived from the weathering of calcium-silicate rocks in the reaction

$$CaSiO_3 + H_2CO_3 \rightarrow Ca^{2+} + 2HCO_3^- + SiO_2 + H_2O \qquad (2.13)$$

The net effect of Eqs. (2.11) and (2.13), in combination with Eq. (2.7), is

$$CaSiO_3 + CO_2 \rightarrow CaCO_3 + SiO_2 \qquad (2.14)$$

which has the effect of taking up CO_2 from the atmospheric and oceanic reservoirs and incorporating it into the much larger reservoir of inorganic carbon sedimentary rocks in the Earth's crust.

From a climate perspective, the chemical reactions (2.7)–(2.14) are virtually instantaneous. In contrast, the timescale over which the oceanic reservoirs adjust to changes in atmospheric CO_2 is governed by the ventilation time for the deeper layers of the ocean, which is on the order of centuries. Calcium carbonate formation is limited by the availability of calcium ions, which is determined by the rate of weathering of calcium silicate rocks, as described in the following subsection.

2.3.4 Carbon in the Earth's Crust

The organic and inorganic carbon reservoirs in the Earth's crust are both very large, the exchange rates in and out of them (apart from the burning of fossil fuels) are very slow, and residence times are on the order of many millions of years. Carbon enters both of these reservoirs by way of the biosphere, as described earlier. Most deposits of the organic carbon in natural gas, oil, coal, and shales and other sedimentary rocks were formed in anoxic ocean basins. The even larger inorganic carbon reservoir, consisting mostly of calcium carbonate

($CaCO_3$), is almost exclusively a product of the marine biosphere.

Weathering exposes organic carbon in sedimentary rock to the atmosphere, allowing it to be oxidized, thereby completing the loop in what is sometimes referred to as the *long term inorganic carbon cycle.* Currently the burning of fossil fuels is returning as much carbon to the atmosphere in a single year as weathering would return in hundreds of thousands of years! The mass of carbon that exists in a form concentrated enough to be classified as "fossil fuels" represents only a small fraction of the organic carbon stored in the Earth's crust, but it is nearly an order of magnitude larger than the mass of carbon currently residing in the atmosphere.

On timescales of tens to hundreds of millions of years, plate tectonics and volcanism play an essential role in renewing atmospheric CO_2. This "inorganic carbon cycle," summarized in Fig. 2.25, involves *subduction, metamorphism*, and *weathering*. Limestone sediments on the sea floor are subducted into the Earth's mantle along plate boundaries where continental plates are overriding denser oceanic plates. At the high temperatures within the mantle, limestone is transformed into metamorphic rocks by the reaction

$$CaCO_3 + SiO_2 \rightarrow CaSiO_3 + CO_2 \quad (2.15)$$

The CO_2 released in this reaction eventually returns to the atmosphere by way of volcanic eruptions. The metamorphic rocks containing calcium in chemical combination with silicate are recycled in the form of newly formed crust that emerges in the mid-ocean ridges. The metamorphism reaction (2.15), in combination with weathering, and the carbonate formation reaction (2.14) form a closed loop in which carbon atoms cycle back and forth between the atmospheric

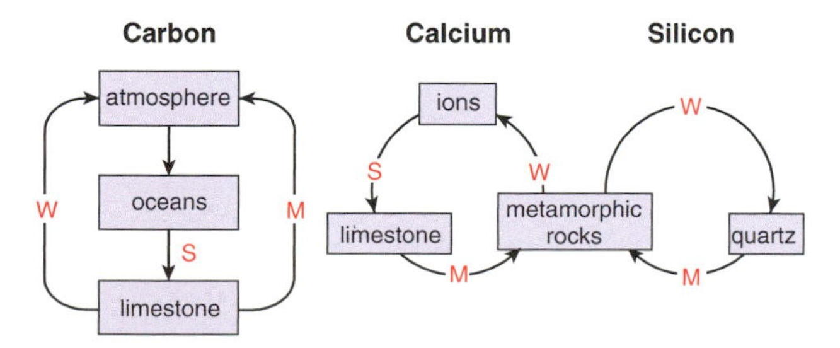

Fig. 2.25 Schematic of the long-term inorganic carbon cycle, also referred to as the carbonate-silicate cycle. The symbol **S** denotes sedimentation, **M** denotes metamorphosis, and **W** denotes weathering.

CO_2 reservoir and the inorganic carbon reservoir in the Earth's crust on a timescale of tens to hundreds of millions of years.

At times when the rate at which CO_2 is injected into the atmosphere by volcanic eruptions exceeds the rate at which calcium ions are made available by weathering, atmospheric CO_2 concentrations increase and vice versa. The injection rate is determined by rate of metamorphism of carbonate rocks, which, in turn, depends on the rate of plate movement along convergent boundaries where subduction is occurring. The rate of weathering, however, is proportional to the rate of cycling of water in the atmospheric branch of the hydrologic cycle, which increases with increasing temperature. The fact that weathering involves the chemical reaction (2.13) makes the temperature dependence even stronger. Hence, high ambient temperatures and slow plate movements are conducive to a draw-down of atmospheric CO_2 and vice versa. The changes in atmospheric CO_2 in response to imbalances between (2.14) and (2.15) on timescales of tens of millions of years are believed to have been quite substantial.

2.4 Oxygen in the Earth System

Earth is unique among the planets of the solar system with respect to the abundance of atmospheric oxygen O_2 and the presence of an ozone (O_3) layer. Atmospheric oxygen accounts for only a very small fraction of the "free" oxygen (i.e., oxygen not bound to hydrogen atoms in water molecules) in the Earth system. Much larger quantities of free oxygen are present in the form of oxidized minerals in sediments and in the crust and upper mantle. The current level of oxidation of the Earth system as a whole is much higher than it was at the time when the planets first formed.

The formation of the Earth's molten metallic iron had the effect of enriching oxygen concentrations in the mantle, the source of volcanic emissions. Yet geological evidence suggests that oxygen was only a trace atmospheric constituent early in the Earth's history. Iron in sedimentary rock formations that date back more than 2.2 billion years is almost exclusively in the partially oxidized ferrous (FeO) form. Had substantial amounts of oxygen been present in the atmosphere and oceans at the time when these sediments formed, the iron in them would have been fully oxidized to ferric oxide (Fe_2O_3). To account for the large concentrations of Fe_2O_3 that currently reside in the Earth's

Table 2.4 Mass of oxygen (in units of 10^3 kg m^{-2}) averaged over the surface of the Earth required to raise the level of oxidation of the various reservoirs in the Earth system to their present state, starting from their state at the time of formation of the Earth[a]

Reservoir	Mass
Atmospheric O_2	2.353
Oceans and sediments	31
Crust Fe^{3+}	>100
Crust CO_3	~100
Crust (other)	>100
Mantle Fe^{3+}	>100

[a] The estimates for the crust and mantle represent conservative lower limits. [Based on data adapted with permission from Catling, D. C., K. J. Zahnle and C. P. McKay, "Biogenic Methane, Hydrogen Escape, and the Irreversible Oxidation of Early Earth," *Science*, **293**, p. 841. Copyright 2001 AAAS.]

crust and upper mantle, it is necessary to invoke the existence of a source of free oxygen that came into play later in the history of the Earth system. Free oxygen is also required to account for the formation of carbonates in the Earth's crust by marine organisms, as illustrated in the following exercise.

Exercise 2.5 Estimate the mass of oxygen required to form the carbonate deposits in the Earth's crust.

Solution: From Table 2.3, the mass of carbon in the carbonate reservoir in the Earth's crust is 80,000 kg m^{-2}. Combining (2.7), (2.8), (2.9), and (2.11), we obtain

$$H_2O + CO_2 + Ca^{2+} \rightarrow CaCO_3 + 2H^+$$

The two hydrogen ions generated as a by-product of this reaction combine with a single oxygen ion to form water. Hence, the final result is

$$CO_2 + Ca^{2+} + O^{2-} \rightarrow CaCO_3 \qquad (2.16)$$

from which it is clear that one "free" oxygen atom is paired with each carbon atom that resides in the Earth's crust in the form of carbonates. Hence, the mass of free oxygen required to account for the carbonates is

$$80{,}000 \text{ kg m}^{-2} \times \frac{\text{molecular weight of oxygen}}{\text{molecular weight of carbon}}$$
$$= 80{,}000 \text{ kg m}^{-2} \times (16/12)$$

or ~100,000 kg m^{-2}, which is the basis for the estimate in Table 2.4. ■

2.4.1 Sources of Free Oxygen

The photosynthesis reaction (2.5) does not affect the overall state of oxidation of the Earth system because for every liberated O_2 molecule, an atom of organic carbon is stored in sedimentary rocks in the Earth's crust. That carbon storage can be used to estimate the mass of carbon liberated by photosynthesis, as illustrated in the following exercise.

Exercise 2.6 On the basis of data in Table 2.3, estimate the oxygen liberated by photosynthesis over the lifetime of the Earth.

Solution: From Table 2.3, the mass of organic carbon in sedimentary rocks in the Earth's crust is 20,000 kg m^{-2}. From (2.5) it is evident that the burial of each carbon atom in the Earth's crust has liberated one O_2 molecule. Hence, the mass of oxygen liberated by photosynthesis is

$$20{,}000 \text{ kg m}^{-2} \times \frac{\text{molecular weight of } O_2}{\text{molecular weight of C}}$$

or ~50,000 kg m^{-2}. ■

Until quite recently, photosynthesis was believed to have been the major source of free oxygen on Earth. However, it is evident from Exercise 2.6 that a much larger source is needed to account for the current degree of oxidation of the Earth system. The only other possible candidate is redox reactions that ultimately lead to the escape of the hydrogen atoms to space, raising the state of oxidation of the Earth system. The rate of escape of hydrogen atoms is proportional to their number concentration in the upper atmosphere. The principal source of free hydrogen is a four-step process.

1. Subduction or deep burial of rocks containing hydrated minerals (i.e., minerals with chemically bound H_2O molecules or interstitial H_2O molecules in their lattices).
2. Breakdown of these hydrated minerals at the high temperatures in the Earth's mantle, releasing the water in the form of steam.
3. The reaction

$$2FeO + H_2O \rightarrow Fe_2O_3 + H_2 \qquad (2.17)$$

in which the steam oxidizes ferrous oxide in the crust or mantle to ferric oxide.

4. The ejection of the hydrogen in volcanism or metamorphism.

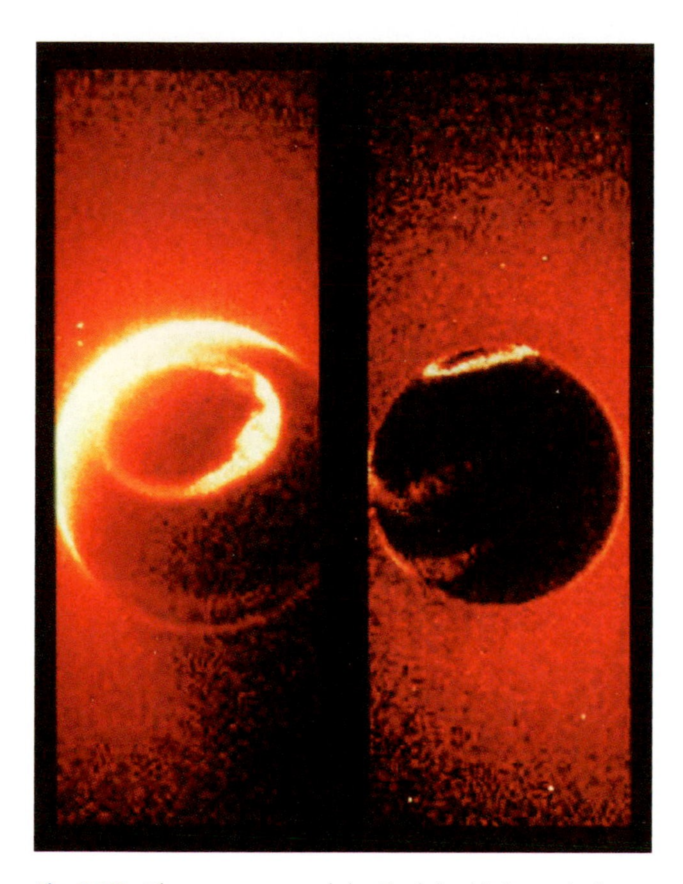

Fig. 2.26 The corona around the Earth in this image is due to the scattering of solar radiation by hydrogen atoms that are escaping from the atmosphere. [Image from NASA/Dynamics Explorer/Spin-Scan Auroral Imaging. Courtesy of David Catling.]

The escape of hydrogen from the Earth's atmosphere is detectable (Fig. 2.26), but the rate at which it is occurring is far too small to account for the degree to which the minerals in the Earth's crust and mantle have become oxidized over the lifetime of the Earth. The rate of escape is slow because the two gases that supply hydrogen atoms to the upper atmosphere (i.e., CH_4 and H_2O) are present only in concentrations of a few parts per million by volume. Air entering the upper atmosphere from below loses most of its water vapor as it passes through the cold equatorial tropopause (see Sections 1.3.3 and 3.5). There are indications that atmospheric methane concentrations may have been much higher in the past, as discussed in the next section.

Photosynthesis and the escape of hydrogen operate independently to liberate oxygen. The crust of a lifeless planet could become highly oxidized due solely to the action of plate tectonics. Conversely, photosynthesis could occur on a planet on which the minerals in the crust and the gases in the atmosphere were in a highly reduced state. However, of the two mechanisms, only photosynthesis is capable of producing atmospheric oxygen, and only the escape of hydrogen from the Earth system is capable of liberating oxygen in the quantities required to account for the present degree of oxidation of the minerals in the crust and mantle.

2.1 Isotope Abundances: Proxies for Climate Data

Isotopes of a given element are atoms with different numbers of neutrons in their nuclei. *Unstable isotopes*, such as ^{14}C, which spontaneously change form by radioactive decay with a known "half-life," are used for dating ice, tree and sediment cores, fossils, and rock samples. By comparing the abundance of the isotope in the sample with the current atmospheric abundance, it is possible to infer how long the sample has been out of contact with the atmosphere (provided, of course, that the atmospheric abundance has not changed since the sample was deposited). Relative abundances of *stable isotopes*, such as ^{13}C, vary in accordance with local (and in some cases regional or even global) environmental conditions that prevailed at the time when the sample was deposited. Some of the more widely used isotopes are the following.

- The relative abundance of deuterium (^2H, or D)[15] in the snow samples recovered from ice cores depends on (and hence can be used as a *proxy* for) the temperature of the surface from which the water vapor that condensed to form the snow was evaporated. The greater the relative abundance of HDO in the core sample, the warmer that evaporating surface

Continued on next page

[15] The relative abundance of D is given by

$$\delta D(o/oo) = \frac{D/H - R}{R}$$

where R is a reference value. Positive (negative) values of δD are indicative of an enrichment (depletion) of D relative to the reference value, expressed in parts per thousand (o/oo). The same formalism applies to the isotopes in subsequent bullets.

2.1 Continued

must have been at the time the sample was deposited.

- Oxygen-18 (^{18}O) abundance in marine sediment cores containing carbonates reflects the temperature of the water in the euphotic zone where the carbonate formation took place. The lower the temperature of the water, the greater the relative abundance of ^{18}C that was incorporated into the shells and skeletons of the marine organisms from which the carbonates formed.
- Worldwide ^{18}O abundances also depend on the volume of the continental ice sheets. ^{16}O evaporates more readily than ^{18}O, so a disproportionately large fraction of ^{16}O tends to be incorporated into the snow that falls on, and becomes incorporated into, the ice sheets. When the ice sheets grow, ocean waters throughout the world are enriched in ^{18}O. Hence ^{18}O abundance in marine sediment cores and ice cores can be used as a proxy for ice volume.
- Carbon-13 (^{13}C) abundances in deposits of organic carbon reflect the ambient CO_2 concentrations at the time that photosynthesis occurred. Plants prefer the lighter isotope ^{12}C, and the higher the ambient CO_2 level, the more strongly they exert this preference. Hence, low relative abundances of ^{13}C in organic carbon deposits are indicative of high ambient CO_2 concentrations, and vice versa.
- Carbon-13 is also an indicator of the sources and sinks of atmospheric CO_2. Emissions from the decay of plants, forest fires and agricultural burning, and the consumption of fossil fuels tend to be low in ^{13}C, whereas CO_2 outgassed from the oceans has the same ^{13}C abundance as atmospheric CO_2. In a similar manner, the presence of a biospheric CO_2 sink should tend to raise atmospheric ^{13}C levels, whereas the presence of an oceanic sink should not.

2.5 A Brief History of Climate and the Earth System

This section describes evolution of the Earth system on a logarithmically telescoping time line, as depicted in the bottom panel of Fig. 2.27, with subsections focusing on (1) the lifetime of the Earth, (2) the past 100 million years, (3) the past million years, and (4) the past 20,000 years. If the life (to date) of a 20-year-old student were viewed on an analogous and proportional time line, the respective subsections would focus on his/her entire 20-year life span, the past 6 months, the past 2 days, and the past hour.

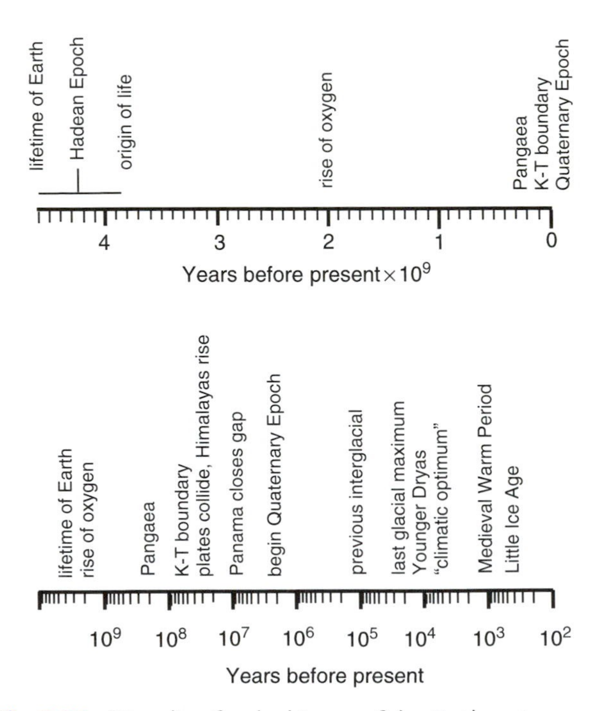

Fig. 2.27 Time line for the history of the Earth system on a linear scale (top) and on a logarithmic scale (bottom).

2.5.1 Formation and Evolution of the Earth System

The sun and the planets are believed to have formed 4.5 billion years ago from the gravitational collapse of a cold cloud of interstellar gas and dust.[16] The absence of the noble gases neon, xenon, and krypton[17] in the

[16] The age of the of the Earth is inferred from a comparison of the ratios of radiogenic (formed by decay of uranium) and nonradiogenic isotopes of lead in meteorites and rocks of various ages.

[17] Argon in the Earth's atmosphere is a product of the radioactive decay of the radioactive isotope ^{40}K in the crust.

atmospheres of the Earth and the other planets, relative to their cosmic abundance, is evidence that the planets formed from the coalescence of the dust into chunks of solid materials called *planetesimals* that were drawn together by gravitation. Present within the condensing cloud were *volatile compounds* (i.e., water, methane, ammonia, and other substances with low boiling points), mainly in the form of ices. When the sun formed, the inner part of the cloud should have warmed, driving out most of the volatiles: hence the relatively low concentrations of these substances in the atmospheres of the inner planets.

During the first 700–800 million years of its history, referred to by geologists as the *Hadean Epoch*, Earth was still under continual bombardment by smaller planetesimals. The heating and degassing resulting from impacts of these collisions should have liberated water vapor and other volatile substances, forming a primordial atmosphere and the oceans. The energy released by the larger objects may well have been sufficient to entirely vaporize the oceans from time to time. The formation of the moon has been attributed to one of these impacts. The bombardment gradually subsided and, by ~3.8 billion years ago, conditions on Earth had become stable enough to allow early microbial life forms to develop in the oceans. Cataclysmic collisions still occur occasionally, as evidenced by the *K–T meteorite impact*[18] that took place only 65 million years ago. Earth is still being bombarded by vast numbers of much smaller objects, as illustrated in Fig. 2.28, but their impact on the Earth system is minimal.

The emission of radiation from stars increases gradually over their lifetime due to the increasing rate of fusion within their cores, which contract and heat up as progressively more of the hydrogen is fused into helium. The luminosity of the sun is believed to have increased by 30% over the lifetime of the solar system. Geological evidence indicates that, with the exception of a few relatively brief intervals, the oceans have been largely free of ice throughout the Earth's history. That the Earth's surface was not perpetually frozen during its early history, when the sun was relatively faint, suggests that its early atmosphere must have contained substantially higher concentrations of greenhouse gases at that time than it does now.

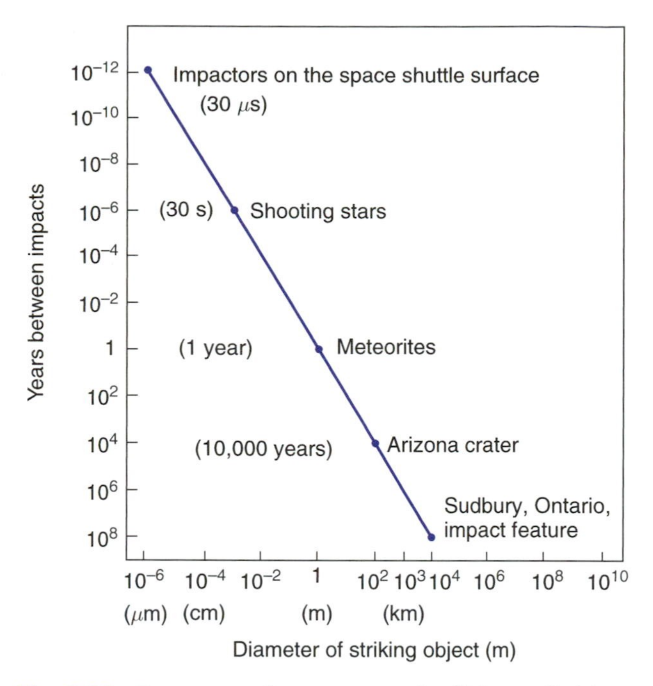

Fig. 2.28 Frequency of occurrence of collisions of objects with the Earth as a function of the size of the objects. [Reprinted with permission from L. W. Alvarez, "Mass extinctions caused by large bolide impacts," *Physics Today*, **40**, p. 27. Copyright 1987, American Institute of Physics.]

Over the lifetime of the Earth, its atmosphere has continuously been recycled and renewed by volcanism and plate tectonics. The makeup of present-day volcanic emissions is 80–90% steam, 6–12% CO_2, 1–2% SO_2, and traces of H_2, CO, H_2S, CH_4, and N_2. The relative concentrations of the reduced gases H_2, CO, and CH_4 could have been much higher earlier in the Earth's history when the mantle was less oxidized than it is today.

An important milestone in the evolution of the Earth system was the rise of atmospheric oxygen. Cyano-(blue-green) bacteria capable of liberating oxygen are believed to have been present in the oceans for at least 3.0 and perhaps as long as 3.8 billion years, yet geological evidence indicates that oxygen did not begin to accumulate in the atmosphere until about 2.4–2.2 billion years ago. Oxygen liberated by photosynthesis early in the Earth's history would have been consumed quickly by H_2 and other reduced gases emanating from the Earth's crust, formed in reactions such as (2.17) or by the oxidation of minerals exposed to the atmosphere by weathering. Only after the

[18] The distinctive marker of this event is the distinctive iridium-enriched layer produced by the explosion that attended the impact. This layer, which is evident in sediments worldwide, occurs at the boundary between *Cretaceous* and *Tertiary* sediments in the Earth's crust (hence the name *K–T*). It has been hypothesized that this event was responsible for the extinction of many species of life forms, including dinosaurs.

minerals in the crust became more highly oxidized due to the gradual escape of hydrogen from the Earth system and reactions such as (2.17) slowed down, would it have it possible for oxygen liberated by photosynthesis to begin to accumulate in the atmosphere. In this sense, atmospheric O_2 can be viewed as "surplus" oxygen in the Earth system.

An array of geological evidence indicates that the rise of atmospheric oxygen, once it began, was quite rapid, with concentrations rising from less than 0.01% of the present concentration 2.4 billion years ago to at least 1–3% of the present concentration 1.9 billion years ago. Concurrent with the rise of oxygen came the formation of ozone layer. Photochemical models based on the equations in Section 5.7.1 indicate that atmospheric oxygen concentrations of even a few percent of today's values should have been capable of supporting an ozone layer thick enough to protect life on Earth from the harmful effects of solar ultraviolet radiation.

The conditions that existed on Earth prior to the rise of oxygen have been the subject of considerable speculation. Given today's rates of production of methane, it is estimated that, in the absence of atmospheric oxygen, methane concentrations could have been two or three orders of magnitude greater than their present concentration of ~1.7 ppmv, in which case, methane might well have been the dominant greenhouse gas. With higher methane concentrations, the number densities of hydrogen atoms in the upper

2.2 "Snowball Earth"

Geological evidence suggests that major glaciations, extending all the way into the tropics, occurred three times in Earth's history: the first around 2.2–2.4 billion years ago, concurrent with the rise of oxygen,[19] the second between 600 and 750 million years ago, and the latest so-called *Permian Glaciation*, ~280 million years ago. Until quite recently, most climate researchers discounted this evidence on the grounds that had the oceans ever been in a completely ice-covered state, they would have remained so. These skeptics argued that the albedo of an ice-covered planet would be so low that very little of the incident solar radiation would be absorbed. Hence, the surface of the Earth would have been so cold that the ice could not have melted. Recently, this argument has been called into question by proponents of the following "freeze-fry scenario."

i. During an unusually cold period, a sufficiently large fraction of the Earth's surface becomes ice covered so that the ice-albedo feedback mechanism described in Section 10.3 renders the climate unstable: the expanding ice cover cools the Earth's surface, the cooling causes the ice to expand still farther, and the process continues until the entire Earth becomes ice covered, as depicted schematically in Fig. 2.29.
ii. During the ensuing *snowball Earth* phase, the carbonate formation reaction (2.11) cannot occur because the oceans are frozen.

Fig. 2.29 Artist's conception of the onset of a world-wide glaciation. [Courtesy of Richard Peltier.]

Continued on next page

[19] It has been hypothesized that the decline in atmospheric methane concentrations brought about by the rise of oxygen might have precipitated a sudden global cooling at this time.

2.2 Continued

However, continental drift continues and, with it, metamorphism and volcanic emissions of CO_2 into the atmosphere. In the absence of a carbon sink, atmospheric CO_2 concentrations increase.

iii. Eventually, the combination of the increasing greenhouse effect and the blackening of parts of the ice sheets by windblown dust raise the surface temperature up to the threshold value at which ice in the tropical oceans begins to melt. Once this process begins, the ice-albedo feedback exerts a powerful warming influence, which abruptly flips the Earth system into an ice-free state.

iv. With the thawing of the oceans, carbonate production resumes, but, compared to the reaction time of the cryosphere in (iii), this is a slow process, limited by the rate at which weathering supplies Ca^{2+} ions. Hence, for several million years, the Earth system resides in the *hothouse* phase, with global-mean temperatures initially as high as 50 °C.

The occurrence of an extended hothouse phase is supported by the existence of rock formations (i.e., banded iron formations and cap carbonates) suggestive of a period of very high temperatures around the time of the Permian glaciation.

atmosphere could have been orders of magnitude higher than their current values. Under those conditions, large numbers of hydrogen atoms would have escaped to space, gradually raising the level of oxidation of the Earth system. The abrupt transition of the atmosphere from an anoxic state, with relatively high concentrations of CH_4, CO, H_2S, and other reduced gases, to a more oxidized state was marked by the extinction of anaerobic life forms that are intolerant of O_2. The rise of O_2 set the stage for the evolution of more complex life forms.

2.5.2 The Past 100 Million Years

During the Cretaceous epoch, which ended ~65 million years ago, surface air temperatures were substantially higher than they are today, especially at the higher latitudes. This view is supported by the discovery of remains of dinosaurs and lush tropical plants dating back to that time in Siberia, Canada, and other subarctic sites. Geological evidence indicates that atmospheric concentrations of CO_2 were about an order of magnitude higher at that time than they are today. The Cretacous epoch was followed by an extended interval of cooling and declining CO_2 concentrations, culminating in *Pleistocene glaciation*,[20] which began around 2.5 million years ago.

The cooling that set the stage for the Pleistocene glaciation is widely attributed to the role of plate tectonics in regulating the reactions (2.14) and (2.15) in the carbonate–silicate cycle. Geological evidence indicates that the rate of movement of the plates has slowed down over the past 100 million years. A reduced rate of ingestion of limestone sediments into the mantle implies reduced metamorphism, which would favor reduced volcanic emissions of CO_2. Meanwhile, the rise of the Himalayas following the collision of the Indian and Asian plates (Fig. 2.30) is believed to have increased the rate of weathering of $CaSiO_3$ rocks, making more Ca^{2+} ions available for

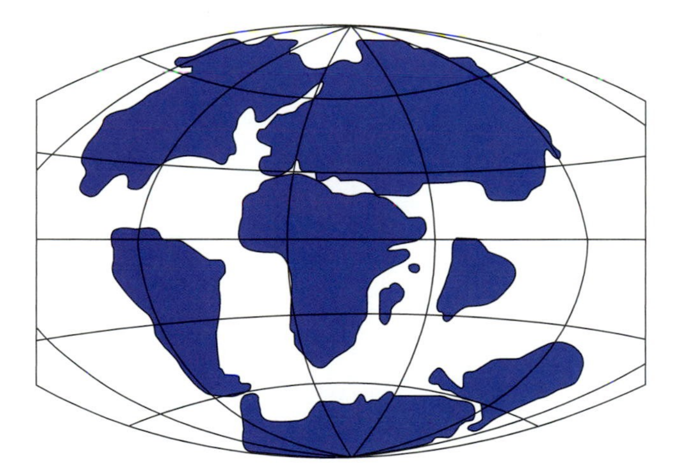

Fig. 2.30 Continental configuration 65 million years ago, at the end of the Cretaceous epoch. Note the separation between the Indian and Eurasian plates. [Courtesy of the U.S. Geological Survey.]

[20] *Pleistocene* (from the Greek: *pleistos*: most, and *ceno*: new) in reference to geological sediments. The *Pleistocene epoch* and the subsequent *Holocene* (all new) *epoch* comprise the *Quaternary period* in the Earth's history.

limestone formation in the oceans, thereby accelerating the removal of CO_2 from the atmosphere and oceans. A decreasing source and increasing sink of atmospheric CO_2 would account for the apparent decline in atmospheric CO_2 levels, and the consequent weakening of the greenhouse effect is consistent with the observed cooling.

Another factor that contributed to the cooling was glaciation of the Antarctic continent 15–30 million years ago as it drifted into higher latitudes, which would have increased the fraction of the incident solar radiation reflected back to space. Other important milestones in the drift of the continents toward their present configuration were the opening of the Drake passage in the southern hemisphere 15–30 million years ago, which gave rise to the formation of the Antarctic circumpolar current, and the joining of the North and South American continents at the isthmus of Panama ~3 million years ago. It has been suggested that these events could have caused major reorganizations of the oceanic thermohaline circulation, resulting in a reduction of the poleward heat flux in the North Atlantic, thereby accelerating the cooling of the Arctic.

2.5.3 The Past Million Years

The past 2.5 million years have been marked by climatic swings back and forth between extended *glacial epochs* in which thick ice sheets covered large areas of North America, northern Europe, and Siberia and shorter *interglacial epochs* such as the present one, in which only Antarctica (and sometimes Greenland) remain ice covered. Carbon and oxygen isotopes (see Box 2.1) and the remains of living organisms buried in marine sediments in anoxic ocean basins around the world reveal a great deal about the history of the pronounced climatic swings during this so-called *Quaternary period* of Earth's history.

A more accurate and detailed history of the climate swings of the past few hundred thousand years, as revealed by an ice core extracted from the dome of the Antarctic ice sheet, is shown in Fig. 2.31. The temperature variations shown in Fig. 2.31 are inferred from changes in concentrations of deuterium in the ice, as explained in Box 2.1. Atmospheric concentrations of CO_2 and CH_4 are inferred from microscopic bubbles of air that became trapped in the ice as the snow from which it was formed was compressed and consolidated. Because these gases tend to be well mixed, their concentrations in the cores are indicative of global conditions. The chemical signature of the dust in the cores provides a basis for identifying it with one or more specific source regions such as the Gobi Desert, where fine particles of soil are exposed directly to strong winds.

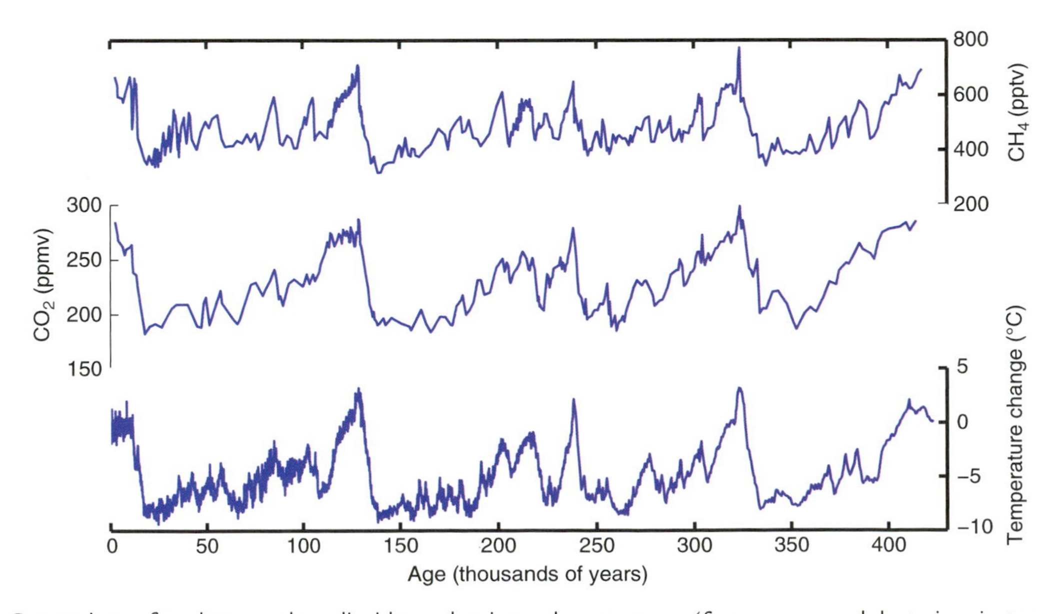

Fig. 2.31 Comparison of methane, carbon dioxide, and estimated temperature (from oxygen and deuterium isotope ratios) from the Vostok ice core, Antarctica, over the last 440 thousand years. The location of Vostok is indicated by the red dot in Fig. 2.13. Note that the time axis runs from right to left. [Adapted from J. R. Petit et al., "Climate and atmospheric history of the past 420,000 years from the Vostok ice core, Antarctica." *Nature*, **399**, p. 431, 1999. Courtesy of Eric Steig.]

Based on the analysis of ice core records like the ones shown in Fig. 2.31 and comparisons with numerous (but less highly resolved and accurately dated) marine sediment cores, it is clear that on timescales of tens of thousands of years or longer, temperature and a number of other climate-related parameters vary coherently with one another and that these variations are global in extent. Global temperatures cooled at an irregular rate during the extended glacial epochs and rose much more rapidly at the beginning of the interglacials, which have been recurring at intervals of roughly 100,000 years. The last glacial maximum occurred around 20,000 years ago. Atmospheric CO_2 and CH_4 concentrations have risen and fallen synchronously with temperature, and the colder epochs have been dustier, perhaps because of higher wind speeds or drier conditions over the source regions.

Figure 2.32 contrasts the extent of the northern hemisphere continental ice sheets at the time of the last glacial maximum with their current extent. Parts of Canada were covered by ice as thick as 3 km. As a consequence of the large amount of water sequestered in the ice sheets, the global sea level was ~125 m lower than it is today. The concentration of atmospheric CO_2 was ~180 ppm, as compared with a mean value of ~260 ppm during the current interglacial prior to the industrial revolution. Hence, the Earth's albedo must have been higher at the time of the last glacial maximum than it is today and the greenhouse effect must have been weaker, both of which would have favored lower surface temperatures. Temperatures in Greenland were ~10 °C lower than they are today and tropical temperatures are estimated to have been ~4 °C lower.

There is evidence of a small time lag between fluctuations in atmospheric CO_2 concentrations and fluctuations in the volume of ice stored in the continental ice sheets, with ice volume leading CO_2. Hence, it would appear that the cause of these fluctuations is intimately related to the growth and shrinkage of the continental ice sheets. The CO_2 fluctuations represent a positive feedback that amplifies the temperature contrasts between glacial and interglacial epochs, as discussed further in Section 10.3.2.

The pronounced climatic swings during the Quaternary period are believed to be driven by subtle variations in the Earth's orbit that affect the summer *insolation* (i.e., the average intensity of incident solar radiation) at high latitudes of the northern hemisphere. During intervals when the summer insolation is relatively weak, snow deposited during winter does not completely melt, leaving a residual, which, over a time span of thousands of years, accumulates to form thick ice sheets. The high reflectivity of the growing ice sheets exacerbates the coolness of the summers, amplifying the orbital forcing. Based on the same reasoning, it is believed that the continental ice sheets are most prone to melting during periods

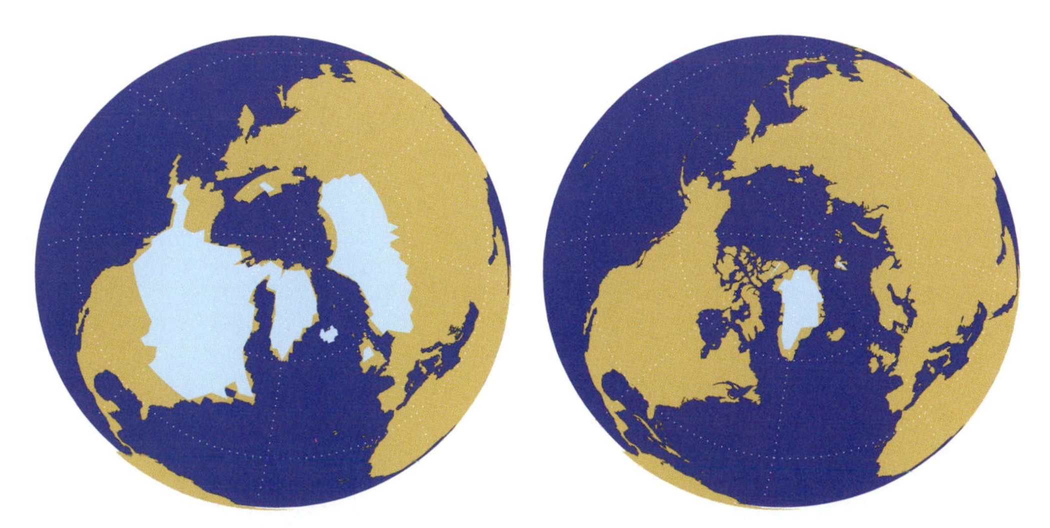

Fig. 2.32 The extent of the northern hemisphere continental ice sheets at the time of the last glacial maximum 20,000 years ago (left) as compared with their current extent (right). [Courtesy of Camille Li.]

when summer insolation at high northern latitudes is strong.[21]

The orbital variations believed to be responsible for these pronounced climatic swings involve:

i. 100,000-year cycle in *eccentricity* (the degree of ellipticity, defined as the distance from the center to either focus of the ellipse divided by the length of the major axis), which ranges from 0 to 0.06 and is currently 0.017,
ii. 41,000-year cycle in the *obliquity* (i.e., the tilt of the Earth's axis of rotation relative to the plane of the Earth's orbit) which ranges from 22.0° to 24.5° and is currently 23.5°, and
iii. 23,000- and 19,000-year cycles in the *precession* of the Earth's orbit. As a result of the precession cycle, the day of the year on which the Earth is closest to the sun (currently January 3) progresses through the year at a rate of ~1.7 calendar day per century.[24]

Figure 2.33 shows a schematic visual representation of these three types of orbital perturbations. When the eccentricity and obliquity are both near the peaks of their respective cycles, summertime insolation at 65 °N varies by up to ~20% between the extremes of the precession cycle (see Exercise 4.19).

In Fig. 2.34 a time series of the rate of growth of the continental ice sheets, as inferred from oxygen-18 concentrations in marine sediment cores, is compared with a time series of summertime insolation over high latitudes of the northern hemisphere, as inferred from orbital calculations. The degree of correspondence between the series is

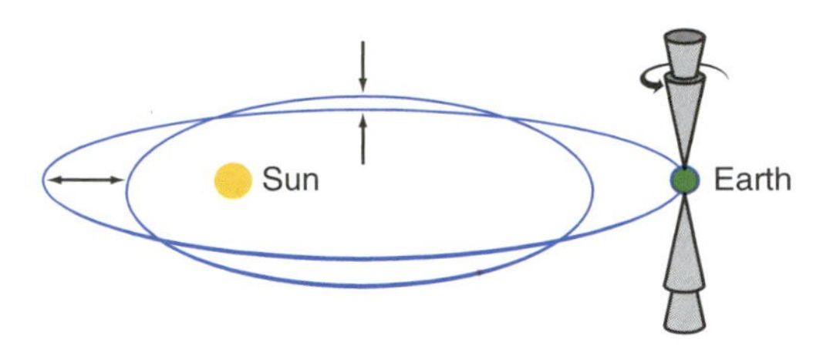

Fig. 2.33 Schematic of the Earth's orbital variations. The precession cycle in the tilt of the Earth's axis is represented by a single cone; the cycle in the obliquity of the axis is represented by the presence of two concentric cones, and the extrema in the ellipticity of the orbit are represented by the pair of ellipses. The figure is not drawn to scale. [Adapted from J. T. Houghton, *Global Warming: The Complete Briefing*, 2nd Edition, Cambridge University Press, p. 55 (1997).]

quite striking. The fit can be made even better by synchronizing the maxima and minima in the oxygen-18 record with nearby features in the insolation time series.[25]

2.5.4 The Past 20,000 Years

The transition from the last glacial to the current interglacial epoch was dramatic. The ice sheets started shrinking around 15,000 years ago. By 12,000 years ago the Laurentide ice sheet was pouring huge volumes of melt water into newly formed lakes and rivers, setting the stage for a series of flood events that shaped many of the features of today's landscape. Around this time the emergence from the ice age was interrupted by an ~800 year relapse into ice age conditions, an event referred to by geologists as the *Younger Dryas*.[26] The signature of the Younger

[21] Many of the elements of the orbital theory of the ice ages are embodied in works of James Croll,[22] published in 1864 and 1875. In 1920, Milutin Milankovitch[23] published a more accurate time series of insolation over high latitudes of the northern hemisphere based on newly available calculations of the variations in the Earth's orbit. Wladimir Köppen and Alfred Wegener included several of Milankovitch's time series in their book *Climates of the Geologic Past* (1924). The idea that summer is the critical season in determining the fate of the continental ice sheets is widely attributed to Köppen. Analysis of extended sediment core records, which did not become widely available until the 1970s, has provided increasingly strong support for orbital theory.

[22] **James Croll** (1821–1890). Largely self-educated Scottish intellectual. Variously employed as a tea merchant, manager of a temperance hotel, insurance agent, and janitor at a museum before his achievements earned him an appointment in the Geological Survey of Scotland and substantial scientific recognition.

[23] **Milutin Milankovitch** (1879–1958). Serbian mathematician. Professor, University of Belgrade.

[24] Evidence of the precession cycle dates back to the Greek astronomer, **Hipparchus**, who inferred, from observations made more than a century apart, that the axis around which the heavens rotate was slowly shifting.

[25] The relationship between depth within the core and time depends on the rate of sedimentation, which varies from one site to another and is not guaranteed to be linear. Hence, in assigning dates on the features in the cores, it is necessary to rely on supplementary information. The reversal in the polarity of the Earth's magnetic field, which is known to have occurred 780,000 years ago and is detectable in the cores, provides a critical "anchor point" in dating the sediment core time series.

[26] *Dryas* is a plant that currently grows only in Arctic and alpine tundra, fossil remains of which are found in a layer of sediments from northern Europe deposited during this interval. *Younger* signifies the topmost (i.e., most recent layer in which dryas is present).

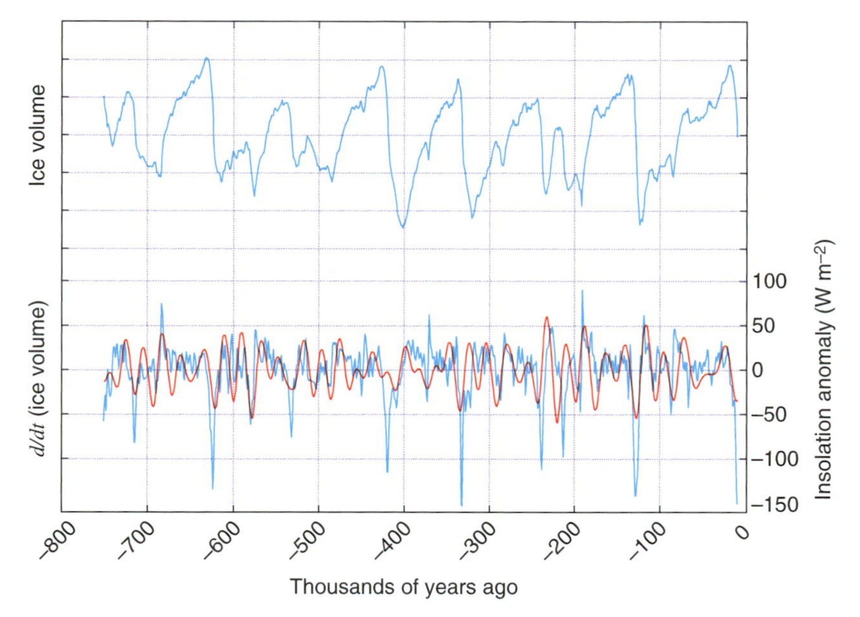

Fig. 2.34 Top time axis: global ice volume (arbitrary scale) as inferred from ^{18}O concentrations in a large collection of marine sediment cores. Bottom time axis: the time rate of change of ice volume (arbitrary scale), obtained by differentiating the curve in the top panel (blue-green curve) and summer insolation at the top of the atmosphere due to variations in the Earth's orbit (red curve). The dating of the sediment core has been carried out without reference to the insolation time series. [The dating of the sediment core time series is based on estimates of Peter Huybers; the insolation time series is based on orbital calculations using the algorithm developed by Jonathan Levine. Figure courtesy of Gerard Roe.]

Dryas shows up clearly in the Greenland ice core (Fig. 2.35) and in European proxy climate records, but whether it was truly global in extent, such as the climatic swings on timescales of 10,000 years and longer, is still under debate.

The timescale of the Younger Dryas event is much shorter than that of the orbital cycles discussed in the previous section, so it is not clear what caused it. One hypothesis is that the sudden freshening of the surface waters over higher latitudes of the North Atlantic, caused by a surge of glacial melt water out of the St. Lawrence River, precipitated a shutdown of the oceanic thermohaline circulation, which reduced the poleward heat transport by the Gulf Stream that warms Greenland and northern Europe.

High-resolution analysis of the Greenland ice core indicates that the Younger Dryas event ended abruptly ~11,700 years ago. The current interglacial, referred to in the geology literature as the *Holocene epoch*, has not witnessed the large temperature swings that characterized the previous glacial epoch. However, even the minor swings appear to have had important societal impacts. For example, during the relatively cold interval from the 14th through most of the 19th century, popularly referred to as the *Little*

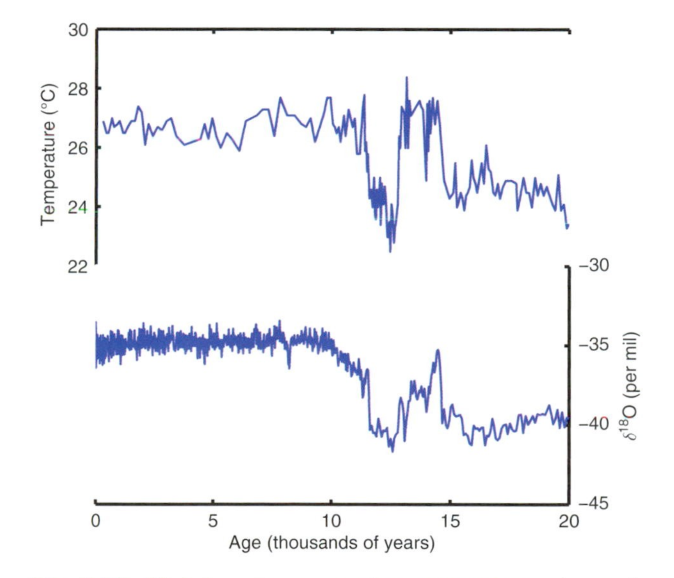

Fig. 2.35 Variations in sea surface temperature change in Cariaco Basin (Venezuela) sediment (top) and oxygen-18 in Greenland ice cores (bottom) during the last 20,000 years. Sea surface temperature is inferred from Mg/Ca ratios. ^{18}O values from the ice cores are indicative of air temperature over and around Greenland. Note that the time axis runs from right to left. [Based on data presented by Lea, D. W., D. K. Pak, L. C. Peterson and K. A. Hughen, "Synchroneity of tropical and high-latitude Atlantic temperatures over the last glacial termination," *Science*, **301**, p. 1364 (2003) and Grootes, P. M., M. Stuiver, J. W. C. White, S. Johnsen and J. Jouzel, "Comparison of oxygen isotope records from the GISP2 and GRIP Greenland ice cores," *Nature*, **366**, p. 552 (1993). Courtesy of Eric Steig.]

Ice Age, the Viking colony in Greenland failed, the population of Iceland declined substantially, and farms were abandoned in parts of Norway and the Alps.

The global distribution of rainfall has varied significantly during the Holocene epoch. Fossils collected from lake beds indicate that areas of what is now the Sahara Desert were vegetated and, in some cases, swampy 6000 years ago. Areas of the Middle East where grains were first cultivated have subsequently become too dry to support extensive agriculture, and aqueducts that the Romans built in north Africa seem out of place in the context of today's climate. Superimposed upon such long-term trends are variations on timescales of decades to century, which have influenced the course of human events. For example, collapses of the Akkadian Empire in the Middle East and the Mayan and Anasazi civilizations in the New World have been attributed to the failure of those societies to adapt to trends toward drier climates.

2.6 Earth: The Habitable Planet

Photographs of Earth and its neighbors in the solar system are shown in Fig. 2.36 and pertinent astronomical and atmospheric data are shown in Table 2.5. That Earth is the only planet in the solar system on which advanced life forms have evolved is due to a very special combination of circumstances.

i. The range of surface temperatures on Earth has allowed for the possibility of oceans that have remained unfrozen throughout most of Earth's history. The oceans have provided (a) an essential pathway in the carbonate– silicate cycle that sequesters large amounts of carbon in the Earth system,

Fig. 2.36 Venus, Earth, Mars, and Jupiter from space. Venus and Jupiter are cloud covered. Not shown to scale. [Photographs courtesy of NASA.]

Table 2.5 Astronomical and atmospheric data for Earth and neighboring planets[a]

Parameter	Venus	Earth	Mars	Jupiter
Radius (km × 10^3)	6,051	6,371	3390	66,911
Gravity (m s^{-2})	8.87	9.80	3.71	24.79
Distance from sun (AU)	0.72	1.000	1.524	5.20
Length of year (Earth years)	0.615	1.000	1.88	11.86
Length of day (Earth days)	117	1.000	1.027	0.41
Orbital eccentricity	0.0067	0.0167	0.093	0.049
Orbital obliquity	2.36	23.45	25.19	3.13
Dominant constituent (% by volume)	CO_2 (96.5)	N_2 (78.1)	CO_2 (95.3)	H_2 (90)
Secondary constituent (% by volume)	N_2 (3.5)	O_2 (21)	N_2 (2.7)	He (10)
Surface pressure (hPa)	92,000	997	8[b]	$>>10^6$
Surface temperature (K)	737	288	210	
Diurnal temperature range (K)	~0	10	40	

[a] Based on *Planetary Fact Sheets* on NASA Web site; Mars surface data based on records at the Viking 1 Lander site.
[b] Varies seasonally from 7.0 hPa during the austral winter, when Mars is farthest from the sun, to 9.0 hPa during the austral summer.

(b) the medium in which simple life forms capable of photosynthesis were able to evolve, and (c) a thermal and chemical buffer that has served to reduce the amplitude of short-term climate variations.

ii. Earth's distance from the the sun and its planetary gravity are in the range that has allowed some, but not all, of its hydrogen to escape to space. The escape of hydrogen has led to oxidation of the minerals in the crust and upper mantle, a necessary condition for the accumulation of O_2 in the atmosphere. It also liberated the oxygen required for the removal of CO_2 from the atmosphere through the formation of carbonates. Yet it is also important that enough hydrogen remains in the Earth system to provide for an abundance of water.

iii. Active plate tectonics has served to continually renew the atmosphere by injecting gases expelled from the mantle in volcanic eruptions.

iv. An active hydrologic cycle sustains life on land.

v. The massive outer planets (Jupiter, in particular) have tended to deflect comets away from Earth's orbit, reducing the frequency of catastrophic collisions.

vi. The strong gravitational pull of the moon has tended to limit the range of obliquity of the Earth's axis. Had there been no such limit, harsh seasonal temperature contrasts would have occurred from time to time in the history of the Earth.

vii. A rotation rate sufficiently rapid to prevent the occurrence of extreme daytime and nighttime temperatures.

The surface of Venus is far too hot to allow for the possibility of oceans. Furthermore, it appears that nearly all of the hydrogen atoms that presumably once resided in the atmosphere and crust of Venus in the form of water have escaped to space, leaving the planet virtually devoid of life-sustaining water. Without oceans and an active biosphere, there is no photosynthesis, the primary source of atmospheric oxygen on Earth, and there is no formation, of carbonates, the primary long-term sink for carbon on Earth. Hence, nearly all the oxygen in "the Venus system" is in the crust, and a large fraction of the carbon is in the atmosphere. As a consequence of the accumulation of CO_2 expelled in volcanic eruptions, the mass of Venus' atmosphere is nearly 100 times that of the Earth's atmosphere. The correspondingly stronger greenhouse effect accounts for the high temperatures on the surface of Venus. Whether the climate of Venus might have once been more habitable is a matter of speculation.

The surface of Mars is far too cold to allow water to exist in the liquid phase. The presence of geological features that resemble river valleys, but are pockmarked by crater impacts, suggests that oceans might have existed on the surface of Mars early in its history but were not sustainable. As the interior of the planet cooled down quickly following the early period of bombardment, the rate of fission of radioactive substances within its core may have been insufficient to maintain the plasticity of the mantle. In the absence of plate tectonics, the crust would have solidified and volcanism would have ceased—and indeed there is no evidence of present-day volcanic activity on Mars. In the absence of volcanism, CO_2 would have accumulated in the crust rather than being recycled back through the atmosphere. Diminishing concentrations of atmospheric CO_2 would have caused the surface of the planet to cool, and any liquid water that was present to freeze. An alternative hypothesis, supported by the lack of spectroscopic evidence of carbonate minerals on the surface of Mars, is that the low Martian gravity allowed the atmosphere to be blasted away by comet or asteroid impacts. In both these scenarios, it is the small size of Mars that ultimately leads to the demise of most of its atmosphere.

Jupiter formed quickly and was large enough to accrete H_2 and He directly from the solar nebula. Hence, these gases are much more abundant in Jupiter's atmosphere than in the atmospheres of the inner planets. It is also believed that the planetesimals that accreted onto Jupiter contained much higher concentrations of volatile compounds with low freezing points, giving rise to the relatively high concentrations of ammonia (NH_3) and CH_4, which served as additional sources of hydrogen. In contrast to the inner planets, Jupiter is so massive that the geothermal energy released by the gravitational collapse of its core is roughly comparable in magnitude to the solar energy that it absorbs (see Exercise 4.27).

Earth is the only planet with an ozone layer, the heating of which produces the temperature maximum that defines the stratopause. Hence the vertical temperature profiles of the other planets consist of only three layers: a troposphere, an isothermal mesosphere, and a thermosphere.[27]

The long-range outlook for the habitability of Earth is not good. If stellar evolution follows its normal course, by a billion years from now the luminosity of the sun will have become strong enough to evaporate the oceans. The impacts of human activities upon the Earth system, while arguably not as disastrous, are a million times more imminent.

Fig. 2.37 Abandoned equipment on the Antarctic ice sheet (top) and on Arctic pack ice (bottom). [Photographs courtesy of Norbert Untersteiner.]

Exercises

2.7 Explain or interpret the following

(a) In the atmosphere, most of the deep convection occurs at low latitudes, whereas in the oceans convection occurs at high latitudes.

(b) The salinity of the oceanic mixed layer is relatively low beneath the areas of heavy rainfall such as the ITCZ.

(c) The outflow from the Mediterranean Sea does not rise to the surface of the North Atlantic, even though it is warmer than the surface water.

(d) Variations in sea ice extent do not affect global sea level.

(e) Industrially produced CFCs are entirely absent in some regions of the oceans.

(f) North Atlantic Deep water and Antarctic Bottom Water become progressively depleted in oxygen and enriched in nutrients and CO_2 as they drift away from their respective high latitude source regions. (Note the gradations in coloring of these water masses in Fig. 2.8.)

(g) The oceanic thermohaline circulation slows the buildup of atmospheric CO_2 concentrations in response to the burning of fossil fuels.

(h) Equipment abandoned on a continental ice sheet is eventually buried by snow, whereas equipment buried on an ice floe remains accessible as long as the floe remains intact (Fig. 2.37).

(i) Regions of permafrost tend to be swampy during summer.

(j) An increase in wintertime snow cover over the continental regions surrounding the Arctic would cause the zone of continuous permafrost to retreat.

(k) In regions of mixed forests and grasslands, forests tend to grow on the poleward-facing slopes.

(l) Many summertime high temperature records have been set during periods of extended drought.

(m) Ice samples retrieved from near the bottoms of Greenland ice cores are much older than the residence time listed in Table 2.2.

(n) Seas in closed drainage basins such as the Great Salt Lake and the Caspian Sea are subject to much larger year-to-year variations in level than those that have outlets to the ocean.

(o) Under certain conditions, wet hay can undergo spontaneous combustion.

[27] The atmosphere of Titan, Saturn's largest moon, exhibits an intermediate temperature maximum analogous to Earth's stratopause. This feature is due to the absorption of sunlight by aerosols: organic compounds formed by the decomposition of CH_4 by ultraviolet radiation.

(p) Atmospheric CO_2 exhibits a pronounced late summer minimum at northern hemisphere stations but not at the South Pole (Fig. 1.3).

(q) The concentration of atmospheric O_2 exhibits a pronounced late summer maximum at northern hemisphere stations.

(r) The relative abundance of ^{13}C atoms in atmospheric CO_2 exhibits a pronounced late summer maximum at northern hemisphere stations.

(s) Weathering of $CaSiO_3$ rocks leads to decreasing atmospheric CO_2 concentrations, whereas the weathering of limestone rocks leads to increasing CO_2 concentrations.

(t) Weathering of ferrous oxide and organic carbon sediments is a sink of atmospheric O_2.

(u) Fossil fuel deposits are concentrated within limited geographical regions of the globe.

(v) Limestone deposits are found in sedimentary rock formations located thousands of kilometers away from the nearest ocean.

(w) Carbon-14 concentrations in fossil fuels are negligible.

(x) Photosynthesis is the primary source of atmospheric oxygen, yet the escape of hydrogen to space may well have been the primary mechanism for oxidizing the Earth system over its lifetime.

(y) CO_2 emissions per unit mass of fuel are different for different kinds of fuels.

(z) Earth's climate exhibits distinct seasons that are out of phase in the northern and southern hemispheres. Can you conceive of a planet whose orbital geometry is such that the seasons in the two hemispheres would be in phase?

(aa) The northern hemisphere continents are warmest at the time of year when Earth is farthest from the sun.

(bb) Scientists are convinced that the major ice ages were global in extent.

(cc) The atmospheres of Jupiter and Saturn consist largely of H and He, gases that are found only in trace amounts in Earth's atmosphere.

(dd) Venus is in a more oxidized state than Earth, yet its atmosphere contains only trace amounts of atmospheric O_2.

(ee) In contrast to the other planets listed in Table 2.5, Mars exhibits a seasonally varying surface pressure.

(ff) There are very few, if any, Earth-like planets in the universe.[28]

2.8 In the oceanographic literature, the unit of mass transport is the sverdrup (Sv)[29] (millions of cubic meters per second). The transport of ocean water by the Gulf Stream is estimated to be on the order of 150 Sv. Compare the transport by the Gulf Stream with the transport by the trade winds estimated in the Exercise 1.20.

2.9 If the air flowing equatorward in the trade winds in Exercise 1.20 contains 20 g of water vapor per kg of air, estimate the rainfall rate averaged over the equatorial zone (15 °N–15 °S) attributable to this transport. [**Hint**: For purposes of this problem, the latitude belt between 15 °N and 15 °S can be treated as a cylinder.]

2.10 By how much would global sea level rise if both the Greenland and the Antarctic ice sheets were to entirely melt?

2.11 Consider the water balance over a closed basin of area A over which precipitation falls at a

[28] For further discussion of this topic, see *Rare Earth: Why Complex Life is Uncommon in the Universe*, by Peter Ward and Donald Brownlee, Springer, 2000, 333 p., ISBN: 0-387-98701-0.

[29] **Harald Ulrik Sverdrup** (1888–1957) Norwegian oceanographer and meteorologist. Began career with V. F. K. Bjerknes in Oslo and Leipzig (1911–1917). Scientific director of Roald Amundsen's polar expedition on Maud (1918–1925). Became head of the Department of Meteorology in Bergen in 1926 and director of the Scripps Institute of Oceanography in 1936. Collaborated on sea, surf, and swell forecasting for the Allied landings in North Africa, Europe, and the Pacific in WWII. Helped initiate large-scale modeling of ocean circulations. Director of the Norwegian–British–Swedish Scientific Expedition to Antarctica (1949–1952).

time varying rate $P(t)$ (averaged over the basin) and drains instantaneously into a lake of area a. The evaporation rate E is assumed to be constant E_0 over the lake and zero elsewhere. P and E are expressed in units of meters per year.

(a) For steady-state conditions, show that

$$P_0/E_0 = a/A$$

(b) For a lake with a flat bottom and vertical sides (i.e., area a independent of depth z) show that

$$\frac{dz}{dt} = \frac{PA}{a} - E_0$$

(c) Based on this expression, describe in general how the level of the lake varies in response to random time variations in P. Can you think of an analogy in physics?

2.12 (a) Describe how the level of the lake in the previous exercise would vary in response to time variations in precipitation of the form

$$P = P_0 + P' cos(2\pi t/T)$$

where $P_0 = E_0$ does not vary with time. Show that the amplitude of the response is directly proportional to the period of the forcing. Does this help explain the prevalence of decade-to-decade variability in Fig. 2.22, as opposed to year-to-year variability?

(b) Taking into account the finite depth of the lake, describe qualitatively the character of the response if P_0 were to gradually (i.e., on a timescale much longer than T) drop below the equilibrium value $P_0 = E_0$.

2.13 Consider a lake that drains an enclosed basin as in Exercise 2.11, but in this case assume that the lake bottom is shaped like an inverted cone or pyramid.

(a) Show that

$$\frac{dz}{dt} = \frac{PA}{a_1 z^2} - E_0$$

where a_1 is the area of the lake when it is 1 m deep, expressed in dimensionless units. [**Hint**: Note that the area of the lake is $a_1 z^2$ and the volume is $\frac{1}{3} a_1 z^3$.]

(b) If precipitation falls over the basin at the steady rate P_0, show that the equilibrium lake level is

$$z_0 = \sqrt{\frac{P_0 A}{E_0 a_1}}$$

(c) Describe in physical terms how and why the response to a time varying forcing is different from that in Exercise 2.11. Why do residents of Salt Lake City and Astrakhan have reason to be grateful for this difference?

2.14 Reconcile the mass of oxygen in the atmosphere in Table 2.4 with the volume concentration given in Table 1.1.

2.15 The current rate of consumption of fossil fuels is 7 Gt(C) per year. Based on data in Table 2.3, how long would it take to deplete the entire fossil fuel reservoir of fossil fuels

(a) if consumption continues at the present rate and

(b) if consumption increases at a rate of 1% per year over the next century and remains constant thereafter.

2.16 If all the carbon in the fossil fuel reservoir in Table 2.3 were consumed and if half of it remained in the atmosphere in the form of atmospheric CO_2, by what proportion would the concentration of CO_2 increase relative to current values? By what proportions would the atmospheric O_2 concentration decrease?

2.17 Using the tables presented in this chapter, compare the mass of water lost from the hydrosphere due to the escape of hydrogen to space over the lifetime of the Earth with the mass of water currently residing in the oceans.

2.18 The half-life of ^{14}C is 5730 years. If c_0 is the ambient concentration of atmospheric ^{14}C, estimate the abundance of ^{14}C remaining in a 50,000-year-old sample.

2.19 (a) If all the carbon in the inorganic and organic sedimentary rock reservoirs in Table 2.3 were in the atmosphere instead, in the form of CO_2, together with the atmosphere's present constituents, what would be the mean pressure on the surface of the Earth? What fraction of the atmospheric mass would be N_2? What would be the fraction of the atmosphere (by volume) that was composed of N_2?

(b) Does this exercise lend insight as to why N_2 is not the dominant constituent of the atmosphere of Venus?

2.20 Averaged over the atmosphere as a whole, the drawdown of atmospheric carbon dioxide due to photosynthesis in the terrestrial biosphere during the growing season in the northern hemisphere is ~4 ppmv, or about 1% of the annual mean atmospheric concentration. Estimate the area-averaged mass per unit area of carbon incorporated into leafy plants in the extratropical northern hemisphere continents. Assume that these plants occupy roughly 15% of the area of the Earth's surface.

2.21 At the time of the last glacial maximum (LGM), the global sea level was ~125 m lower than it is today. Assuming that the lower sea level was due to the larger storage of water in the northern hemisphere continental ice sheets, compare the mass of the northern hemisphere ice sheets at the time of the LGM with the current mass of the Antarctic ice sheet. Ignore the change in the fractional area of the Earth covered by oceans due to the change in sea level.

Atmospheric Thermodynamics

3

The theory of thermodynamics is one of the cornerstones and crowning glories of classical physics. It has applications not only in physics, chemistry, and the Earth sciences, but in subjects as diverse as biology and economics. Thermodynamics plays an important role in our quantitative understanding of atmospheric phenomena ranging from the smallest cloud microphysical processes to the general circulation of the atmosphere. The purpose of this chapter is to introduce some fundamental ideas and relationships in thermodynamics and to apply them to a number of simple, but important, atmospheric situations. Further applications of the concepts developed in this chapter occur throughout this book.

The first section considers the *ideal gas equation* and its application to dry air, water vapor, and moist air. In Section 3.2 an important meteorological relationship, known as the *hydrostatic equation*, is derived and interpreted. The next section is concerned with the relationship between the mechanical work done by a system and the heat the system receives, as expressed in the *first law of thermodynamics*. There follow several sections concerned with applications of the foregoing to the atmosphere. Finally, in Section 3.7, the *second law of thermodynamics* and the concept of *entropy* are introduced and used to derive some important relationships for atmospheric science.

3.1 Gas Laws

Laboratory experiments show that the pressure, volume, and temperature of any material can be related by an *equation of state* over a wide range of conditions. All gases are found to follow approximately the same equation of state, which is referred to as the *ideal gas equation*. For most purposes we may assume that atmospheric gases, whether considered individually or as a mixture, obey the ideal gas equation exactly. This section considers various forms of the ideal gas equation and its application to dry and moist air.

The ideal gas equation may be written as

$$pV = mRT \tag{3.1}$$

where p, V, m, and T are the pressure (Pa), volume (m^3), mass (kg), and absolute temperature (in kelvin, K, where K = °C + 273.15) of the gas, respectively, and R is a constant (called the *gas constant*) for 1 kg of a gas. The value of R depends on the particular gas under consideration. Because $m/V = \rho$, where ρ is the density of the gas, the ideal gas equation may also be written in the form

$$p = \rho RT \tag{3.2}$$

For a unit mass (1 kg) of gas $m = 1$ and we may write (3.1) as

$$p\alpha = RT \tag{3.3}$$

where $\alpha = 1/\rho$ is the *specific volume* of the gas, i.e., the volume occupied by 1 kg of the gas at pressure p and temperature T.

If the temperature is constant (3.1) reduces to *Boyle's law*,[1] which states *if the temperature of a fixed mass of gas is held constant, the volume of the*

[1] The Hon. **Sir Robert Boyle** (1627–1691) Fourteenth child of the first Earl of Cork. Physicist and chemist, often called the "father of modern chemistry." Discovered the law named after him in 1662. Responsible for the first sealed thermometer made in England. One of the founders of the Royal Society of London, Boyle declared: "The Royal Society values no knowledge but as it has a tendency to use it!"

gas is inversely proportional to its pressure. Changes in the physical state of a body that occur at constant temperature are termed *isothermal*. Also implicit in (3.1) are *Charles' two laws*.[2] The first of these laws states *for a fixed mass of gas at constant pressure, the volume of the gas is directly proportional to its absolute temperature*. The second of Charles' laws states *for a fixed mass of gas held within a fixed volume, the pressure of the gas is proportional to its absolute temperature*.

3.1 Gas Laws and the Kinetic Theory of Gases: Handball Anyone?

The kinetic theory of gases pictures a gas as an assemblage of numerous identical particles (atoms or molecules)[3] that move in random directions with a variety of speeds. The particles are assumed to be very small compared to their average separation and are perfectly elastic (i.e., if one of the particles hits another, or a fixed wall, it rebounds, on average, with the same speed that it possessed just prior to the collision). It is shown in the kinetic theory of gases that the mean kinetic energy of the particles is proportional to the temperature in degrees kelvin of the gas.

Imagine now a handball court in a zero-gravity world in which the molecules of a gas are both the balls and the players. A countless (but fixed) number of elastic balls, each of mass m and with mean velocity v, are moving randomly in all directions as they bounce back and forth between the walls.[7] The force exerted on a wall of the court by the bouncing of balls is equal to the momentum exchanged in a typical collision (which is proportional to mv) multiplied by the frequency with which the balls impact the wall. Consider the following thought experiments.

i. Let the volume of the court increase while holding v (and therefore the temperature of the gas) constant. The frequency of collisions will decrease in inverse proportion to the change in volume of the court, and the force (and therefore the pressure) on a wall will decrease similarly. This is Boyle's law.
ii. Let v increase while holding the volume of the court constant. Both the frequency of collisions with a wall and the momentum exchanged in each collision of a ball with a wall will increase in linear proportion to v. Therefore, the pressure on a wall will increase as mv^2, which is proportional to the mean kinetic energy of the molecules and therefore to their temperature in degrees kelvin. This is the second of Charles' laws. It is left as an exercise for the reader to prove Charles' first law, using the same analogy.

[2] **Jacques A. C. Charles** (1746–1823) French physical chemist and inventor. Pioneer in the use of hydrogen in man-carrying balloons. When Benjamin Franklin's experiments with lightning became known, Charles repeated them with his own innovations. Franklin visited Charles and congratulated him on his work.

[3] The idea that a gas consists of atoms in random motion was first proposed by Lucretius.[4] This idea was revived by Bernouilli[5] in 1738 and was treated in mathematical detail by Maxwell.[6]

[4] **Titus Lucretius Carus** (*ca*. 94–51 B.C.) Latin poet and philosopher. Building on the speculations of the Greek philosophers Leucippus and Democritus, Lucretius, in his poem *On the Nature of Things*, propounds an atomic theory of matter. Lucretius' basic theorem is "nothing exists but atoms and voids." He assumed that the quantity of matter and motion in the world never changes, thereby anticipating by nearly 2000 years the statements of the conservation of mass and energy.

[5] **Daniel Bernouilli** (1700–1782) Member of a famous family of Swiss mathematicians and physicists. Professor of botany, anatomy, and natural philosophy (i.e., physics) at University of Basel. His most famous work, *Hydrodynamics* (1738), deals with the behavior of fluids.

[6] **James Clark Maxwell** (1831–1879) Scottish physicist. Made fundamental contributions to the theories of electricity and magnetism (showed that light is an electromagnetic wave), color vision (produced one of the first color photographs), and the kinetic theory of gases. First Cavendish Professor of Physics at Cambridge University; designed the Cavendish Laboratory.

[7] In the kinetic theory of gases, the appropriate velocity of the molecules is their root mean square velocity, which is a little less than the arithmetic mean of the molecular velocities.

We define now a *gram-molecular weight* or *a mole* (abbreviated to *mol*) of any substance as the molecular weight, M, of the substance expressed in grams.[8] For example, the molecular weight of water is 18.015; therefore, 1 mol of water is 18.015 g of water. The number of moles n in mass m (*in grams*) of a substance is given by

$$n = \frac{m}{M} \tag{3.4}$$

Because the masses contained in 1 mol of different substances bear the same ratios to each other as the molecular weights of the substances, 1 mol of any substance must contain the same number of molecules as 1 mol of any other substance. Therefore, the number of molecules in 1 mol of any substance is a universal constant, called *Avogadro's*[9] *number*, N_A. The value of N_A is 6.022×10^{23} per mole.

According to *Avogadro's hypothesis*, gases containing the same number of molecules occupy the same volumes at the same temperature and pressure. It follows from this hypothesis that provided we take the same number of molecules of any gas, the constant R in (3.1) will be the same. However, 1 mol of any gas contains the same number of molecules as 1 mol of any other gas. Therefore, the constant R in (3.1) for 1 mol is the same for all gases; it is called the *universal gas constant* (R^*). The magnitude of R^* is 8.3145 J K^{-1} mol^{-1}. The ideal gas equation for 1 mol of any gas can be written as

$$pV = R^*T \tag{3.5}$$

and for n moles of any gas as

$$pV = nR^*T \tag{3.6}$$

The gas constant for one molecule of any gas is also a universal constant, known as *Boltzmann's*[10] *constant*, k. Because the gas constant for N_A molecules is R^*, we have

$$k = \frac{R^*}{N_A} \tag{3.7}$$

Hence, for a gas containing n_0 molecules per unit volume, the ideal gas equation is

$$p = n_0 kT \tag{3.8}$$

If the pressure and specific volume of dry air (i.e., the mixture of gases in air, excluding water vapor) are p_d and α_d, respectively, the ideal gas equation in the form of (3.3) becomes

$$p_d \alpha_d = R_d T \tag{3.9}$$

where R_d is the gas constant for 1 kg of dry air. By analogy with (3.4), we can define the *apparent molecular weight* M_d of dry air as the total mass (in grams) of the constituent gases in dry air divided by the total number of moles of the constituent gases; that is,

$$M_d = \frac{\sum_i m_i}{\sum_i \frac{m_i}{M_i}} \tag{3.10}$$

where m_i and M_i represent the mass (in grams) and molecular weight, respectively, of the ith constituent in the mixture. The apparent molecular weight of dry air is 28.97. Because R^* is the gas constant for 1 mol of any substance, or for M_d (= 28.97) grams of dry air, the gas constant for 1 g of dry air is R^*/M_d, and for 1 kg of dry air it is

$$R_d = 1000\frac{R^*}{M_d} = 1000\frac{8.3145}{28.97} = 287.0 \text{ J K}^{-1}\text{kg}^{-1} \tag{3.11}$$

[8] In the first edition of this book we defined a *kilogram*-molecular weight (or *kmole*), which is 1000 moles. Although the kmole is more consistent with the SI system of units than the mole, it has not become widely used. For example, the mole is used almost universally in chemistry. One consequence of the use of the mole, rather than kmole, is that a factor of 1000, which serves to convert kmoles to moles, appears in some relationships [e.g. (3.11) and (3.13) shown later].

[9] **Amedeo Avogadro, Count of Quaregna** (1776–1856) Practiced law before turning to science at age 23. Later in life became a professor of physics at the University of Turin. His famous hypothesis was published in 1811, but it was not generally accepted until a half century later. Introduced the term "molecule."

[10] **Ludwig Boltzmann** (1844–1906) Austrian physicist. Made fundamental contributions to the kinetic theory of gases. Adhered to the view that atoms and molecules are real at a time when these concepts were in dispute. Committed suicide.

The ideal gas equation may be applied to the individual gaseous components of air. For example, for water vapor (3.3) becomes

$$e\alpha_v = R_v T \quad (3.12)$$

where e and α_v are, respectively, the pressure and specific volume of water vapor and R_v is the gas constant for 1 kg of water vapor. Because the molecular weight of water is M_w (= 18.016) and the gas constant for M_w grams of water vapor is R^*, we have

$$R_v = 1000\frac{R^*}{M_w} = 1000\frac{8.3145}{18.016} = 461.51 \text{ J K}^{-1}\text{kg}^{-1} \quad (3.13)$$

From (3.11) and (3.13),

$$\frac{R_d}{R_v} = \frac{M_w}{M_d} \equiv \varepsilon = 0.622 \quad (3.14)$$

Because air is a mixture of gases, it obeys *Dalton's*[11] *law of partial pressures*, which states *the total pressure exerted by a mixture of gases that do not interact chemically is equal to the sum of the partial pressures of the gases*. The *partial pressure* of a gas is the pressure it would exert at the same temperature as the mixture if it alone occupied all of the volume that the mixture occupies.

Exercise 3.1 If at 0 °C the density of dry air alone is 1.275 kg m^{-3} and the density of water vapor alone is 4.770 × 10^{-3} kg m^{-3}, what is the total pressure exerted by a mixture of the dry air and water vapor at 0 °C?

Solution: From Dalton's law of partial pressures, the total pressure exerted by the mixture of dry air and water vapor is equal to the sum of their partial pressures. The partial pressure exerted by the dry air is, from (3.9),

$$p_d = \frac{1}{\alpha_d} R_d T = \rho_d R_d T$$

where ρ_d is the density of the dry air (1.275 kg m^{-3} at 273 K), R_d is the gas constant for 1 kg of dry air (287.0 J K^{-1} kg^{-1}), and T is 273.2 K. Therefore,

$$p_d = 9.997 \times 10^4 \text{Pa} = 999.7 \text{ hPa}$$

Similarly, the partial pressure exerted by the water vapor is, from (3.12),

$$e = \frac{1}{\alpha_v} R_v T = \rho_v R_v T$$

where ρ_v is the density of the water vapor (4.770 × 10^{-3} kg m^{-3} at 273 K), R_v is the gas constant for 1 kg of water vapor (461.5 J K^{-1} kg^{-1}), and T is 273.2 K. Therefore,

$$e = 601.4 \; Pa = 6.014 \text{ hPa}$$

Hence, the total pressure exerted by the mixture of dry air and water vapor is (999.7 + 6.014) hPa or 1006 hPa. ■

3.1.1 Virtual Temperature

Moist air has a smaller apparent molecular weight than dry air. Therefore, it follows from (3.11) that the gas constant for 1 kg of moist air is larger than that for 1 kg of dry air. However, rather than use a gas constant for moist air, the exact value of which would depend on the amount of water vapor in the air (which varies considerably), it is convenient to retain the gas constant for dry air and use a fictitious temperature (called the *virtual temperature*) in the ideal gas equation. We can derive an expression for the virtual temperature in the following way.

Consider a volume V of moist air at temperature T and total pressure p that contains mass m_d of dry air and mass m_v of water vapor. The density ρ of the moist air is given by

$$\rho = \frac{m_d + m_v}{V} = \rho'_d + \rho'_v$$

[11] **John Dalton** (1766–1844) English chemist. Initiated modern atomic theory. In 1787 he commenced a meteorological diary that he continued all his life, recording 200,000 observations. Showed that the rain and dew deposited in England are equivalent to the quantity of water carried off by evaporation and by the rivers. This was an important contribution to the idea of a hydrological cycle. First to describe color blindness. He "never found time to marry!" His funeral in Manchester was attended by 40,000 mourners.

where ρ'_d is the density that the same mass of dry air would have if it alone occupied all of the volume V and ρ'_v is the density that the same mass of water vapor would have if it alone occupied all of the volume V. We may call these *partial densities*. Because $\rho = \rho'_d + \rho'_v$, it might appear that the density of moist air is greater than that of dry air. However, this is not the case because the partial density ρ'_v is less than the true density of dry air.[12] Applying the ideal gas equation in the form of (3.2) to the water vapor and dry air in turn, we have

$$e = \rho'_v R_v T$$

and

$$p'_d = \rho'_d R_d T$$

where e and p'_d are the partial pressures exerted by the water vapor and the dry air, respectively. Also, from Dalton's law of partial pressures,

$$p = p'_d + e$$

Combining the last four equations

$$\rho = \frac{p - e}{R_d T} + \frac{e}{R_v T}$$

or

$$\rho = \frac{p}{R_d T}\left[1 - \frac{e}{p}(1 - \varepsilon)\right]$$

where ε is defined by (3.14). The last equation may be written as

$$p = \rho R_d T_v \tag{3.15}$$

where

$$T_v \equiv \frac{T}{1 - \frac{e}{p}(1 - \varepsilon)} \tag{3.16}$$

T_v is called the *virtual temperature*. If this fictitious temperature, rather than the actual temperature, is used for moist air, the total pressure p and density ρ of the moist air are related by a form of the ideal gas equation [namely, (3.15)], but with the gas constant the same as that for a unit mass of *dry air* (R_d) and the actual temperature T replaced by the virtual temperature T_v. It follows that the virtual temperature is the temperature that dry air would need to attain in order to have the same density as the moist air at the same pressure. Because moist air is less dense than dry air at the same temperature and pressure, the virtual temperature is always greater than the actual temperature. However, even for very warm and moist air, the virtual temperature exceeds the actual temperature by only a few degrees (e.g., see Exercise 3.7 in Section 3.5).

3.2 The Hydrostatic Equation

Air pressure at any height in the atmosphere is due to the force per unit area exerted by the weight of all of the air lying above that height. Consequently, atmospheric pressure decreases with increasing height above the ground (in the same way that the pressure at any level in a stack of foam mattresses depends on how many mattresses lie above that level). The net upward force acting on a thin horizontal slab of air, due to the decrease in atmospheric pressure with height, is generally very closely in balance with the downward force due to gravitational attraction that acts on the slab. If the net upward force on the slab is equal to the downward force on the slab, the atmosphere is said to be in *hydrostatic balance*. We will now derive an important equation for the atmosphere in hydrostatic balance.

Consider a vertical column of air with unit horizontal cross-sectional area (Fig. 3.1). The mass of air between heights z and $z + \delta z$ in the column is $\rho \delta z$, where ρ is the density of the air at height z. The downward force acting on this slab of air due to the weight of the air is $g\rho\delta z$, where g is the acceleration due to gravity at height z. Now let us consider the net

[12] The fact that moist air is less dense than dry air was first clearly stated by Sir Isaac Newton[13] in his "*Opticks*" (1717). However, the basis for this relationship was not generally understood until the latter half of the 18th century.

[13] **Sir Isaac Newton** (1642–1727) Renowned English mathematician, physicist, and astronomer. A posthumous, premature ("I could have been fitted into a quart mug at birth"), and only child. Discovered the laws of motion, the universal law of gravitation, calculus, the colored spectrum of white light, and constructed the first reflecting telescope. He said of himself: "I do not know what I may appear to the world, but to myself I seem to have been only like a boy playing on the seashore, and diverting myself in now and then finding a smoother pebble or a prettier shell than ordinary, while the great ocean of truth lay all undiscovered before me."

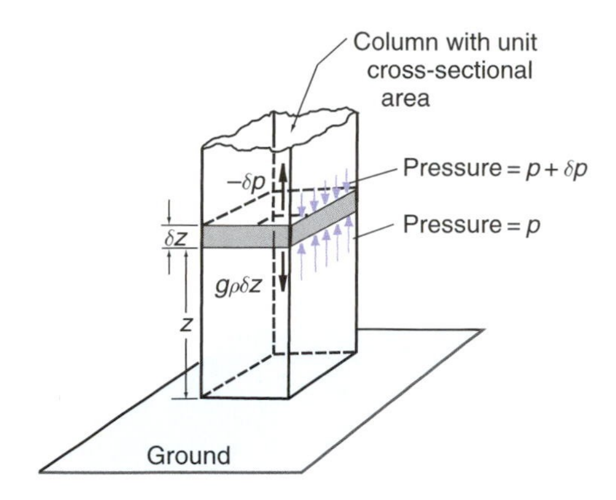

Fig. 3.1 Balance of vertical forces in an atmosphere in which there are no vertical accelerations (i.e., an atmosphere in hydrostatic balance). Small blue arrows indicate the downward force exerted on the air in the shaded slab due to the pressure of the air above the slab; longer blue arrows indicate the upward force exerted on the shaded slab due to the pressure of the air below the slab. Because the slab has a unit cross-sectional area, these two pressures have the same numerical values as forces. The net upward force due to these pressures ($-\delta p$) is indicated by the upward-pointing thick black arrow. Because the incremental pressure change δp is a negative quantity, $-\delta p$ is positive. The downward-pointing thick black arrow is the force acting on the shaded slab due to the mass of the air in this slab.

vertical force that acts on the slab of air between z and $z + \delta z$ due to the pressure of the surrounding air. Let the change in pressure in going from height z to height $z + \delta z$ be δp, as indicated in Fig. 3.1. Because we know that pressure decreases with height, δp must be a negative quantity, and the upward pressure on the lower face of the shaded block must be slightly greater than the downward pressure on the upper face of the block. Therefore, the net vertical force on the block due to the vertical gradient of pressure is upward and given by the positive quantity $-\delta p$, as indicated in Fig. 3.1. For an atmosphere in hydrostatic balance, the balance of forces in the vertical requires that

$$-\delta p = g\rho\delta z$$

or, in the limit as $\delta z \rightarrow 0$,

$$\frac{\partial p}{\partial z} = -g\rho \tag{3.17}$$

Equation (3.17) is the *hydrostatic equation*.[14] It should be noted that the negative sign in (3.17) ensures that the pressure decreases with increasing height. Because $\rho = 1/\alpha$ (3.17) can be rearranged to give

$$gdz = -\alpha dp \tag{3.18}$$

If the pressure at height z is $p(z)$, we have, from (3.17), above a fixed point on the Earth

$$-\int_{p(z)}^{p(\infty)} dp = \int_z^{\infty} g\rho dz$$

or, because $p(\infty) = 0$,

$$p(z) = \int_z^{\infty} g\rho dz \tag{3.19}$$

That is, the pressure at height z is equal to the weight of the air in the vertical column of unit cross-sectional area lying above that level. If the mass of the Earth's atmosphere were distributed uniformly over the globe, retaining the Earth's topography in its present form, the pressure at sea level would be 1.013×10^5 Pa, or 1013 hPa, which is referred to as *1 atmosphere* (or *1 atm*).

3.2.1 Geopotential

The *geopotential* Φ at any point in the Earth's atmosphere is defined as the work that must be done against the Earth's gravitational field to raise a mass of 1 kg from sea level to that point. In other words, Φ is the gravitational potential per unit mass. The units of geopotential are J kg^{-1} or m^2 s^{-2}. The force (in newtons) acting on 1 kg at height z above sea level is numerically equal to g. The work (in joules) in raising 1 kg from z to $z + dz$ is gdz; therefore

$$d\Phi \equiv gdz$$

or, using (3.18),

$$d\Phi \equiv gdz = -\alpha dp \tag{3.20}$$

[14] In accordance with Eq. (1.3), the left-hand side of (3.17) is written in partial differential notation, i.e., $\partial p/\partial z$, because the variation of pressure with height is taken with other independent variables held constant.

The geopotential $\Phi(z)$ at height z is thus given by

$$\Phi(z) = \int_0^z g dz \tag{3.21}$$

where the geopotential $\Phi(0)$ at sea level ($z = 0$) has, by convention, been taken as zero. The geopotential at a particular point in the atmosphere depends only on the height of that point and not on the path through which the unit mass is taken in reaching that point. The work done in taking a mass of 1 kg from point A with geopotential Φ_A to point B with geopotential Φ_B is $\Phi_B - \Phi_A$.

We can also define a quantity called the *geopotential height* Z as

$$Z \equiv \frac{\Phi(z)}{g_0} = \frac{1}{g_0}\int_0^z g dz \tag{3.22}$$

where g_0 is the globally averaged acceleration due to gravity at the Earth's surface (taken as 9.81 m s^{-2}). Geopotential height is used as the vertical coordinate in most atmospheric applications in which energy plays an important role (e.g., in large-scale atmospheric motions). It can be seen from Table 3.1 that the values of z and Z are almost the same in the lower atmosphere where $g_0 \simeq g$.

In meteorological practice it is not convenient to deal with the density of a gas, ρ, the value of which is generally not measured. By making use of (3.2) or (3.15) to eliminate ρ in (3.17), we obtain

$$\frac{\partial p}{\partial z} = -\frac{pg}{RT} = -\frac{pg}{R_d T_v}$$

Rearranging the last expression and using (3.20) yields

$$d\Phi = g\,dz = -RT\frac{dp}{p} = -R_d T_v \frac{dp}{p} \tag{3.23}$$

Table 3.1 Values of geopotential height (Z) and acceleration due to gravity (g) at 40° latitude for geometric height (z)

z (km)	Z (km)	g (m s^{-2})
0	0	9.81
1	1.00	9.80
10	9.99	9.77
100	98.47	9.50
500	463.6	8.43

If we now integrate between pressure levels p_1 and p_2, with geopotentials Φ_1 and Φ_2, respectively,

$$\int_{\Phi_1}^{\Phi_2} d\Phi = -\int_{p_1}^{p_2} R_d T_v \frac{dp}{p}$$

or

$$\Phi_2 - \Phi_1 = -R_d \int_{p_1}^{p_2} T_v \frac{dp}{p}$$

Dividing both sides of the last equation by g_0 and reversing the limits of integration yields

$$Z_2 - Z_1 = \frac{R_d}{g_0}\int_{p_2}^{p_1} T_v \frac{dp}{p} \tag{3.24}$$

This difference $Z_2 - Z_1$ is referred to as the (geopotential) *thickness* of the layer between pressure levels p_1 and p_2.

3.2.2 Scale Height and the Hypsometric Equation

For an *isothermal* atmosphere (i.e., temperature constant with height), if the virtual temperature correction is neglected, (3.24) becomes

$$Z_2 - Z_1 = H \ln(p_1/p_2) \tag{3.25}$$

or

$$p_2 = p_1 \exp\left[-\frac{(Z_2 - Z_1)}{H}\right] \tag{3.26}$$

where

$$H \equiv \frac{RT}{g_0} = 29.3T \tag{3.27}$$

H is the scale height as discussed in Section 1.3.4.

Because the atmosphere is well mixed below the turbopause (about 105 km), the pressures and densities of the individual gases decrease with altitude at the same rate and with a scale height proportional to the gas constant R (and therefore inversely proportional to the apparent molecular weight of the mixture). If we take a value for T_v of 255 K (the approximate mean value for the troposphere and stratosphere), the scale height H for air in the atmosphere is found from (3.27) to be about 7.5 km.

Above the turbopause the vertical distribution of gases is largely controlled by molecular diffusion and a scale height may then be defined for each of the individual gases in air. Because for each gas the scale height is proportional to the gas constant for a unit mass of the gas, which varies inversely as the molecular weight of the gas [see, for example (3.13)], the pressures (and densities) of heavier gases fall off more rapidly with height above the turbopause than those of lighter gases.

Exercise 3.2 If the ratio of the number density of oxygen atoms to the number density of hydrogen atoms at a geopotential height of 200 km above the Earth's surface is 10^5, calculate the ratio of the number densities of these two constituents at a geopotential height of 1400 km. Assume an isothermal atmosphere between 200 and 1400 km with a temperature of 2000 K.

Solution: At these altitudes, the distribution of the individual gases is determined by diffusion and therefore by (3.26). Also, at constant temperature, the ratio of the number densities of two gases is equal to the ratio of their pressures. From (3.26)

$$\frac{(p_{1400\ \mathrm{km}})_{\mathrm{oxy}}}{(p_{1400\ \mathrm{km}})_{\mathrm{hyd}}} = \frac{(p_{200\ \mathrm{km}})_{\mathrm{oxy}} \exp[-1200\ \mathrm{km}/H_{\mathrm{oxy}}\ (\mathrm{km})]}{(p_{200\ \mathrm{km}})_{\mathrm{hyd}} \exp[-1200\ \mathrm{km}/H_{\mathrm{hyd}}\ (\mathrm{km})]}$$

$$= 10^5 \exp\left[-1200\ \mathrm{km}\left(\frac{1}{H_{\mathrm{oxy}}} - \frac{1}{H_{\mathrm{hyd}}}\right)\right]$$

From the definition of scale height (3.27) and analogous expressions to (3.11) for oxygen and hydrogen atoms and the fact that the atomic weights of oxygen and hydrogen are 16 and 1, respectively, we have at 2000 K

$$H_{\mathrm{oxy}} = \frac{1000R^*}{16}\frac{2000}{9.81}\ \mathrm{m} = \frac{8.3145}{16}\frac{2 \times 10^6}{9.81}\ \mathrm{m}$$

$$= 0.106 \times 10^6\ \mathrm{m}$$

and

$$H_{\mathrm{hyd}} = \frac{1000R^*}{1}\frac{2000}{9.81}\ \mathrm{m} = 8.3145\ \frac{2 \times 10^6}{9.81}\ \mathrm{m}$$

$$= 1.695 \times 10^6\ \mathrm{m}$$

Therefore,

$$\frac{1}{H_{\mathrm{oxy}}} - \frac{1}{H_{\mathrm{hyd}}} = 8.84 \times 10^{-6}\ \mathrm{m}^{-1}$$

$$= 8.84 \times 10^{-3}\ \mathrm{km}^{-1}$$

and

$$\frac{(p_{1400\ \mathrm{km}})_{\mathrm{oxy}}}{(p_{1400\ \mathrm{km}})_{\mathrm{hyd}}} = 10^5 \exp(-10.6) = 2.5$$

Hence, the ratio of the number densities of oxygen to hydrogen atoms at a geopotential height of 1400 km is 2.5. ■

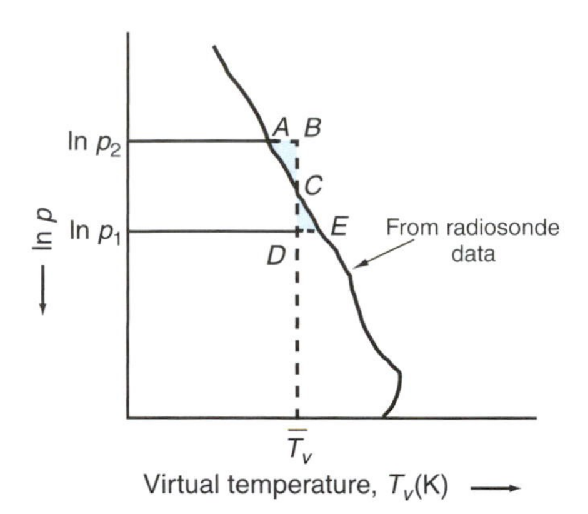

Fig. 3.2 Vertical profile, or sounding, of virtual temperature. If area ABC = area CDE, $\overline{T}_v$ is the mean virtual temperature with respect to ln p between the pressure levels p_1 and p_2.

The temperature of the atmosphere generally varies with height and the virtual temeprature correction cannot always be neglected. In this more general case (3.24) may be integrated if we define a mean virtual temperature $\overline{T}_v$ with respect to p as shown in Fig. 3.2. That is,

$$\overline{T}_v \equiv \frac{\int_{p_2}^{p_1} T_v\, d(\ln p)}{\int_{p_2}^{p_1} d(\ln p)} = \frac{\int_{p_2}^{p_1} T_v \frac{dp}{p}}{\ln\left(\frac{p_1}{p_2}\right)} \qquad (3.28)$$

Then, from (3.24) and (3.28),

$$Z_2 - Z_1 = \overline{H} \ln\left(\frac{p_1}{p_2}\right) = \frac{R_d \overline{T}_v}{g_0} \ln\left(\frac{p_1}{p_2}\right) \qquad (3.29)$$

Equation (3.29) is called the *hypsometric equation*.

Exercise 3.3 Calculate the geopotential height of the 1000-hPa pressure surface when the pressure at sea level is 1014 hPa. The scale height of the atmosphere may be taken as 8 km.

Solution: From the hypsometric equation (3.29)

$$Z_{1000\,\text{hPa}} - Z_{\text{sea level}} = \overline{H} \ln \left(\frac{p_0}{1000} \right)$$
$$= \overline{H} \ln \left(1 + \frac{p_0 - 1000}{1000} \right) \simeq \overline{H} \left(\frac{p_0 - 1000}{1000} \right)$$

where p_0 is the sea-level pressure and the relationship $\ln(1 + x) \simeq x$ for $x \ll 1$ has been used. Substituting $\overline{H} \simeq 8000$ into this expression, and recalling that $Z_{\text{sea level}} = 0$ (Table 3.1), gives

$$Z_{1000\,\text{hPa}} \simeq 8\,(p_0 - 1000)$$

Therefore, with $p_0 = 1014$ hPa, the geopotential height $Z_{1000\,\text{hPa}}$ of the 1000-hPa pressure surface is found to be 112 m above sea level. ■

3.2.3 Thickness and Heights of Constant Pressure Surfaces

Because pressure decreases monotonically with height, pressure surfaces (i.e., imaginary surfaces on which pressure is constant) never intersect. It can be seen from (3.29) that the thickness of the layer between any two pressure surfaces p_2 and p_1 is proportional to the mean virtual temperature of the layer, $\overline{T}_v$. We can visualize that as $\overline{T}_v$ increases, the air between the two pressure levels expands and the layer becomes thicker.

Exercise 3.4 Calculate the thickness of the layer between the 1000- and 500-hPa pressure surfaces (a) at a point in the tropics where the mean virtual temperature of the layer is 15 °C and (b) at a point in the polar regions where the corresponding mean virtual temperature is −40 °C.

Solution: From (3.29)

$$\Delta Z = Z_{500\,\text{hPa}} - Z_{1000\,\text{hPa}} = \frac{R_d \overline{T}_v}{g_0} \ln \left(\frac{1000}{500} \right) = 20.3\,\overline{T}_v \text{ m}$$

Therefore, for the tropics with $\overline{T}_v = 288$ K, $\Delta Z = 5846$ m. For polar regions with $\overline{T}_v = 233$ K, $\Delta Z = 4730$ m. In operational practice, thickness is rounded to the nearest 10 m and is expressed in decameters (dam). Hence, answers for this exercise would normally be expressed as 585 and 473 dam, respectively. ■

Before the advent of remote sensing of the atmosphere by satellite-borne radiometers, thickness was evaluated almost exclusively from radiosonde data, which provide measurements of the pressure, temperature, and humidity at various levels in the atmosphere. The virtual temperature T_v at each level was calculated and mean values for various layers were estimated using the graphical method illustrated in Fig. 3.2. Using soundings from a network of stations, it was possible to construct topographical maps of the distribution of geopotential height on selected pressure surfaces. These calculations, which were first performed by observers working on site, are now incorporated into sophisticated data assimilation protocols, as described in the Appendix of Chapter 8 on the book Web site.

In moving from a given pressure surface to another pressure surface located above or below it, the change in the geopotential height is related geometrically to the thickness of the intervening layer, which, in turn, is directly proportional to the mean virtual temperature of the layer. Therefore, if the three-dimensional distribution of virtual temperature is known, together with the distribution of geopotential height on one pressure surface, it is possible to infer the distribution of geopotential height of any other pressure surface. The same hypsometric relationship between the three-dimensional temperature field and the shape of pressure surface can be used in a qualitative way to gain some useful insights into the three-dimensional structure of atmospheric disturbances, as illustrated by the following examples.

i. The air near the center of a hurricane is warmer than its surroundings. Consequently, the intensity of the storm (as measured by the depression of the isobaric surfaces) must decrease with height (Fig. 3.3a). The winds in such *warm core lows*

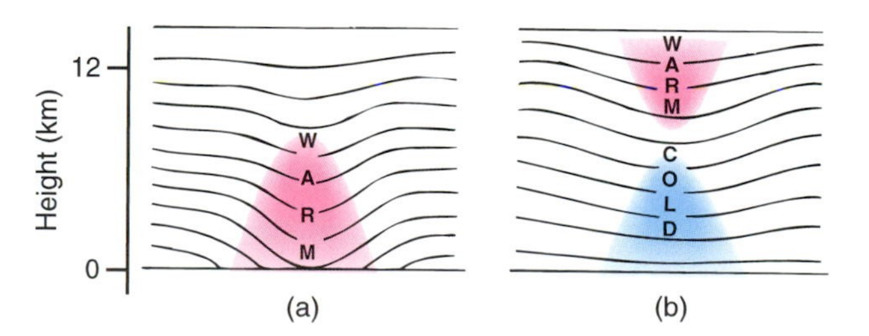

Fig. 3.3 Cross sections in the longitude–height plane. The solid lines indicate various constant pressure surfaces. The sections are drawn such that the thickness between adjacent pressure surfaces is smaller in the cold (blue) regions and larger in the warm (red) regions.

always exhibit their greatest intensity near the ground and diminish with increasing height above the ground.

ii. Some upper level lows do not extend downward to the ground, as indicated in Fig. 3.3b. It follows from the hypsometric equation that these lows must be *cold core* below the level at which they achieve their greatest intensity and *warm core* above that level, as shown in Fig. 3.3b.

3.2.4 Reduction of Pressure to Sea Level

In mountainous regions the difference in surface pressure from one observing station to another is largely due to differences in elevation. To isolate that part of the pressure field that is due to the passage of weather systems, it is necessary to reduce the pressures to a common reference level. For this purpose, sea level is normally used.

Let the subscripts g and 0 refer to conditions at the ground and at sea level ($Z = 0$), respectively. Then, for the layer between the Earth's surface and sea level, the hypsometric equation (3.29) assumes the form

$$Z_g = \overline{H} \ln \frac{p_0}{p_g} \tag{3.30}$$

which can be solved to obtain the sea-level pressure

$$p_0 = p_g \exp\left(\frac{Z_g}{\overline{H}}\right) = p_g \exp\left(\frac{g_0 Z_g}{R_d \overline{T}_v}\right) \tag{3.31}$$

If Z_g is small, the scale height $\overline{H}$ can be evaluated from the ground temperature. Also, if $Z_g / \overline{H} \ll 1$, the exponential in (3.31) can be approximated by $1 + Z_g / \overline{H}$, in which case (3.31) becomes

$$p_0 - p_g \simeq p_g \frac{Z_g}{\overline{H}} = p_g \left(\frac{g_0 Z_g}{R_d \overline{T}_v}\right) \tag{3.32}$$

Because $p_g \simeq 1000$ hpa and $\overline{H} \simeq 8000$ m, the pressure correction (in hPa) is roughly equal to Z_g (in meters) divided by 8. In other words, for altitudes up to a few hundred meters above (or below) sea level, the pressure decreases by about 1 hPa for every 8 m of vertical ascent.

3.3 The First Law of Thermodynamics[15]

In addition to the macroscopic kinetic and potential energy that a system as a whole may possess, it also contains *internal energy* due to the kinetic and potential energy of its molecules or atoms. Increases in internal kinetic energy in the form of molecular motions are manifested as increases in the temperature of the system, whereas changes in the potential energy of the molecules are caused by changes in their relative positions by virtue of any forces that act between the molecules.

Let us suppose that a closed system[16] of unit mass takes in a certain quantity of thermal energy q (measured in joules), which it can receive by thermal conduction and/or radiation. As a result the system may do a certain amount of *external work* w (also measured in joules). The excess of the energy supplied to the body over and above the external work done by the body is $q - w$. Therefore, if there is no change in the macroscopic kinetic and potential energy of the body, it follows from the principle of conservation of energy that the internal energy of the system must increase by $q - w$. That is,

$$q - w = u_2 - u_1 \tag{3.33}$$

where u_1 and u_2 are the internal energies of the system before and after the change. In differential form (3.33) becomes

$$dq - dw = du \tag{3.34}$$

where dq is the differential increment of heat added to the system, dw is the differential element

---

[15] The first law of thermodynamics is a statement of the conservation of energy, taking into account the conversions between the various forms that it can assume and the exchanges of energy between a system and its environment that can take place through the transfer of heat and the performance of mechanical work. A general formulation of the first law of thermodynamics is beyond the scope of this text because it requires consideration of conservation laws, not only for energy, but also for momentum and mass. This section presents a simplified formulation that ignores the macroscopic kinetic and potential energy (i.e., the energy that air molecules possess by virtue of their height above sea level and their organized fluid motions). As it turns out, the expression for the first law of thermodynamics that emerges in this simplified treatment is identical to the one recovered from a more complete treatment of the conservation laws, as is done in J. R. Holton, *Introduction to Dynamic Meteorology*, 4th Edition, Academic Press, New York, 2004, pp. 146–149.

[16] A closed system is one in which the total amount of matter, which may be in the form of gas, liquid, solid or a mixture of these phases, is kept constant.

of work done by the system, and du is the differential increase in internal energy of the system. Equations (3.33) and (3.34) are statements of the *first law of thermodynamics*. In fact (3.34) provides a definition of du. The change in internal energy du depends only on the initial and final states of the system and is therefore independent of the manner by which the system is transferred between these two states. Such parameters are referred to as *functions of state*.[17]

To visualize the work term dw in (3.34) in a simple case, consider a substance, often called the *working substance*, contained in a cylinder of fixed cross-sectional area that is fitted with a movable, frictionless piston (Fig. 3.4). The volume of the substance is proportional to the distance from the base of the cylinder to the face of the piston and can be represented on the horizontal axis of the graph shown in Fig. 3.4. The pressure of the substance in the cylinder can be represented on the vertical axis of this graph. Therefore, every state of the substance, corresponding to a given position of the piston, is represented by a point on this pressure–volume (p–V) diagram. When the substance is in equilibrium at a state represented by point P on the graph, its pressure is p and its volume is V (Fig. 3.4). If the piston moves outward through an incremental distance dx while its pressure remains essentially constant at p, the work dW done by the substance in pushing the external force F through a distance dx is

$$dW = Fdx$$

or, because $F = pA$ where A is the cross-sectional area of the face of the piston,

$$dW = pA\,dx = pdV \tag{3.35}$$

In other words, the work done by the substance when its volume increases by a small increment dV is equal to the pressure of the substance multiplied by its increase in volume, which is equal to the blue-shaded area in the graph shown in Fig. 3.4; that is, it is equal to the area under the curve PQ. When the substance passes from state A with volume V_1 to state B with volume V_2 (Fig. 3.4), during which its pressure p changes, the work W done by the material is equal to the area under the curve AB. That is,

$$W = \int_{V_1}^{V_2} pdV \tag{3.36}$$

Equations (3.35) and (3.36) are quite general and represent work done by any substance (or system) due to a change in its volume. If $V_2 > V_1$, W is positive, indicating that the *substance does work on its environment*. If $V_2 < V_1$, W is negative, which indicates that the *environment does work on the substance*.

The $p-V$ diagram shown in Fig. 3.4 is an example of a *thermodynamic diagram* in which the physical state of a substance is represented by two thermodynamic variables. Such diagrams are very useful in meteorology; we will discuss other examples later in this chapter.

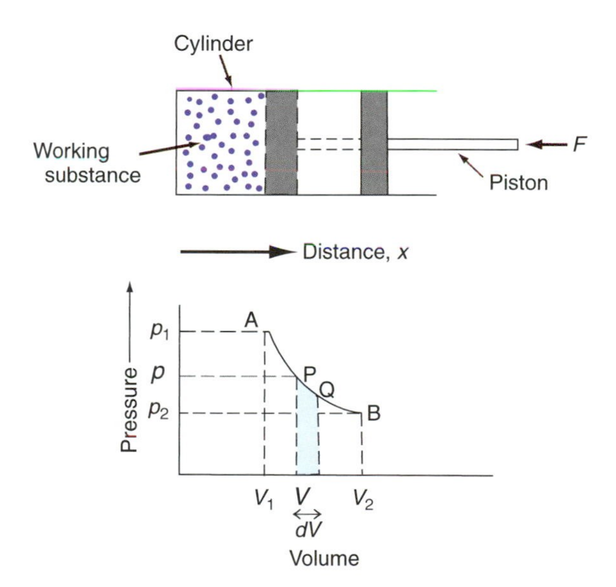

Fig. 3.4 Representation of the state of a working substance in a cylinder on a p–V diagram. The work done by the working substance in passing from P to Q is $p\,dV$, which is equal to the blue-shaded area. [Reprinted from *Atmospheric Science: An Introductory Survey*, 1st Edition, J. M. Wallace and P. V. Hobbs, p. 62, Copyright 1977, with permission from Elsevier.]

[17] Neither the heat q nor the work w are functions of state, since their values depend on *how* a system is transformed from one state to another. For example, a system may or may not receive heat and it may or may not do external work as it undergoes transitions between different states.

If we are dealing with a unit mass of a substance, the volume V is replaced by the specific volume α. Therefore, the work dw that is done when the specific volume increases by $d\alpha$ is

$$dw = pd\alpha \tag{3.37}$$

Combination of (3.34) and (3.37) yields

$$dq = du + pd\alpha \tag{3.38}$$

which is an alternative statement of the first law of thermodynamics.[18]

3.3.1 Joule's Law

Following a series of laboratory experiments on air, Joule[19] concluded in 1848 that when a gas expands without doing external work, by expanding into a chamber that has been evacuated, and without taking in or giving out heat, the temperature of the gas does not change. This statement, which is known as *Joule's law*, is strictly true only for an ideal gas, but air (and many other gases) behaves very similarly to an ideal gas over a wide range of conditions.

Joule's law leads to an important conclusion concerning the internal energy of an ideal gas. If a gas neither does external work nor takes in or gives out heat, $dw = 0$ and $dq = 0$ in (3.38), so that $du = 0$. Also, according to Joule's law, under these conditions the temperature of the gas does not change, which implies that the kinetic energy of the molecules remains constant. Therefore, because the total internal energy of the gas is constant, that part of the internal energy due to the potential energy must also remain unchanged, even though the volume of the gas changes. In other words, the internal energy of an ideal gas is independent of its volume if the temperature is kept constant. This can be the case only if the molecules of an ideal gas do not exert forces on each other. In this case, the internal energy of an ideal gas will depend only on its temperature.[20]

3.2 More Handball?

Box 3.1. showed that the gas laws can be illustrated by picturing the molecules of a gas as elastic balls bouncing around randomly in a handball court. Suppose now that the walls of the court are permitted to move outward when subjected to a force. The force on the walls is supplied by the impact of the balls, and the work required to move the walls outward comes from a decrease in the kinetic energy of the balls that rebound from the walls with lower velocities than they struck them. This decrease in kinetic energy is in accordance with the first law of thermodynamics under adiabatic conditions. The work done by the system by pushing the walls outward is equal to the decrease in the internal energy of the system [see (3.38)]. Of course, if the outside of the walls of the court are bombarded by balls in a similar manner to the inside walls, there will be no net force on the walls and no work will be done.

[18] We have assumed here that the only work done by or on a system is due to a change in the volume of the system. However, there are other ways in which a system may do work, e.g., by the creation of new surface area between two phases (such as between liquid and air when a soap film is formed). Unless stated otherwise, we will assume that the work done by or on a system is due entirely to changes in the volume of the system.

[19] **James Prescott Joule** (1818–1889) Son of a wealthy English brewer; one of the great experimentalists of the 19th century. He started his scientific work (carried out in laboratories in his home and at his own expense) at age 19. He measured the mechanical equivalent of heat, recognized the dynamical nature of heat, and developed the principle of conservation of energy.

[20] Subsequent experiments carried out by Lord Kelvin[21] revealed the existence of small forces between the molecules of a gas.

[21] **Lord Kelvin 1st Baron (William Thomson)** (1824–1907) Scottish mathematician and physicist. Entered Glasgow University at age 11. At 22 became Professor of Natural Philosophy at the same university. Carried out incomparable work in thermodynamics, electricity, and hydrodynamics.

3.3.2 Specific Heats

Suppose a small quantity of heat dq is given to a unit mass of a material and, as a consequence, the temperature of the material increases from T to $T + dT$ without any changes in phase occurring within the material. The ratio dq/dT is called the *specific heat* of the material. The specific heat defined in this way could have any number of values, depending on how the material changes as it receives the heat. If the volume of the material is kept constant, a *specific heat at constant volume* c_v is defined

$$c_v = \left(\frac{dq}{dT}\right)_{v\text{ const}} \tag{3.39}$$

However, if the volume of the material is constant (3.38) becomes $dq = du$. Therefore

$$c_v = \left(\frac{du}{dT}\right)_{v\text{ const}}$$

For an ideal gas, Joule's law applies and therefore u depends only on temperature. Therefore, regardless of whether the volume of a gas changes, we may write

$$c_v = \left(\frac{du}{dT}\right) \tag{3.40}$$

From (3.38) and (3.40), the first law of thermodynamics for an ideal gas can be written in the form[22]

$$dq = c_v dT + p d\alpha \tag{3.41}$$

Because u is a function of state, no matter how the material changes from state 1 to state 2, the change in its internal energy is, from (3.40),

$$u_2 - u_1 = \int_{T_1}^{T_2} c_v dT$$

We can also define a specific heat at constant pressure c_p

$$c_p = \left(\frac{dq}{dT}\right)_{p\text{ const}} \tag{3.42}$$

where the material is allowed to expand as heat is added to it and its temperature rises, but its pressure remains constant. In this case, a certain amount of the heat added to the material will have to be expended to do work as the system expands against the constant pressure of its environment. Therefore, a larger quantity of heat must be added to the material to raise its temperature by a given amount than if the volume of the material were kept constant. For the case of an ideal gas, this inequality can be seen mathematically as follows. Equation (3.41) can be rewritten in the form

$$dq = c_v dT + d(p\alpha) - \alpha\, dp \tag{3.43}$$

From the ideal gas equation (3.3), $d(p\alpha) = RdT$. Therefore (3.43) becomes

$$dq = (c_v + R)dT - \alpha\, dp \tag{3.44}$$

At constant pressure, the last term in (3.44) vanishes; therefore, from (3.42) and (3.44),

$$c_p = c_v + R \tag{3.45}$$

The specific heats at constant volume and at constant pressure for dry air are 717 and 1004 J K^{-1} kg^{-1}, respectively, and the difference between them is 287 J K^{-1} kg^{-1}, which is the gas constant for dry air. It can be shown that for ideal monatomic gases $c_p:c_v:R = 5:3:2$, and for ideal diatomic gases $c_p:c_v:R = 7:5:2$.

By combining (3.44) and (3.45) we obtain an alternate form of the first law of thermodynamics:

$$dq = c_p dT - \alpha\, dp \tag{3.46}$$

3.3.3 Enthalpy

If heat is added to a material at constant pressure so that the specific volume of the material increases from α_1 to α_2, the work done by a unit mass of the material is $p(\alpha_2 - \alpha_1)$. Therefore, from (3.38), the finite quantity of heat Δq added to

[22] The term dq is sometimes called the *diabatic* (or nonadiabatic) *heating* or *cooling*, where "diabatic" means involving the transfer of heat. The term "diabatic" would be redundant if "heating" and "cooling" were always taken to mean "the addition or removal of heat." However, "heating" and "cooling" are often used in the sense of "to raise or lower the temperature of," in which case it is meaningful to distinguish between that part of the temperature change dT due to diabatic effects (dq) and that part due to adiabatic effects ($pd\alpha$).

a unit mass of the material at constant pressure is given by

$$\Delta q = (u_2 - u_1) + p(\alpha_2 - \alpha_1)$$
$$= (u_2 + p\alpha_2) - (u_1 + p\alpha_1)$$

where u_1 and u_2 are, respectively, the initial and final internal energies for a unit mass of the material. Therefore, at constant pressure,

$$\Delta q = h_2 - h_1$$

where h is the *enthalpy* of a unit mass of the material, which is defined by

$$h \equiv u + p\alpha \qquad (3.47)$$

Because u, p, and α are functions of state, h is a function of state. Differentiating (3.47), we obtain

$$dh = du + d(p\alpha)$$

Substituting for du from (3.40) and combining with (3.43), we obtain

$$dq = dh - \alpha dp \qquad (3.48)$$

which is yet another form of the first law of thermodynamics.

By comparing (3.46) and (3.48) we see that

$$dh = c_p dT \qquad (3.49)$$

or, in integrated form,

$$h = c_p T \qquad (3.50)$$

where h is taken as zero when $T = 0$. In view of (3.50), h corresponds to the heat required to raise the temperature of a material from 0 to T K at constant pressure.

When a layer of air that is at rest and in hydrostatic balance is heated, for example, by radiative transfer, the weight of the overlying air pressing down on it remains constant. Hence, the heating is at constant pressure. The energy added to the air is realized in the form of an increase in enthalpy (or *sensible heat*, as atmospheric scientists commonly refer to it) and

$$dq = dh = c_p dT$$

The air within the layer expands as it warms, doing work on the overlying air by lifting it against the Earth's gravitational attraction. Of the energy per unit mass imparted to the air by the heating, we see from (3.40) and (3.41) that $du = c_v dT$ is reflected in an increase in internal energy and $pd\alpha = RdT$ is expended doing work on the overlying air. Because the Earth's atmosphere is made up mainly of the diatomic gases N_2 and O_2, the energy added by the heating dq is partitioned between the increase in internal energy du and the expansion work $pd\alpha$ in the ratio 5:2.

We can write a more general expression that is applicable to a moving air parcel, the pressure of which changes as it rises or sinks relative to the surrounding air. By combining (3.20), (3.48), and (3.50) we obtain

$$dq = d(h + \Phi) = d(c_p T + \Phi) \qquad (3.51)$$

Hence, if the material is a parcel of air with a fixed mass that is moving about in an hydrostatic atmosphere, the quantity $(h + \Phi)$, which is called the *dry static energy*, is constant provided the parcel neither gains nor loses heat (i.e., $dq = 0$).[23]

3.4 Adiabatic Processes

If a material undergoes a change in its physical state (e.g., its pressure, volume, or temperature) without any heat being added to it or withdrawn from it, the change is said to be *adiabatic*.

Suppose that the initial state of a material is represented by the point A on the p–V diagram in Fig. 3.5 and that when the material undergoes an isothermal transformation it moves along the line AB. If the same material underwent a similar change in volume but under adiabatic conditions, the transformation would

[23] Strictly speaking, Eq. (3.51) holds only for an atmosphere in which there are no fluid motions. However, it is correct to within a few percent for the Earth's atmosphere where the kinetic energy of fluid motions represents only a very small fraction of the total energy. An exact relationship can be obtained by using Newton's second law of motion and the continuity equation in place of Eq. (3.20) in the derivation. See J. R. Holton, *An Introduction to Dynamic Meteorology*, 4th ed., Academic Press, pp. 46–49 (2004).

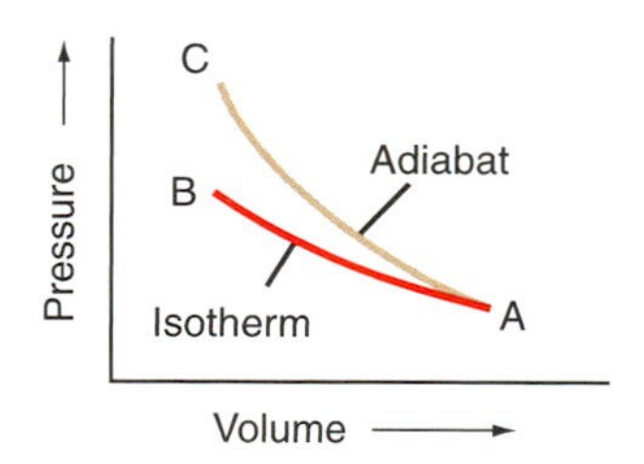

Fig. 3.5 An isotherm and an adiabat on a p–V diagram.

be represented by a curve such as AC, which is called an *adiabat*. The reason why the adiabat AC is steeper than the isotherm AB on a p–V diagram can be seen as follows. During adiabatic compression, the internal energy increases [because $dq = 0$ and $pd\alpha$ is negative in (3.38)] and therefore the temperature of the system rises. However, for isothermal compression, the temperature remains constant. Hence, $T_C > T_B$ and therefore $p_C > p_B$.

3.4.1 Concept of an Air Parcel

In many fluid mechanics problems, mixing is viewed as a result of the random motions of individual molecules. In the atmosphere, molecular mixing is important only within a centimeter of the Earth's surface and at levels above the turbopause (~105 km). At intermediate levels, virtually all mixing in the vertical is accomplished by the exchange of macroscale "air parcels" with horizontal dimensions ranging from millimeters to the scale of the Earth itself.

To gain some insights into the nature of vertical mixing in the atmosphere, it is useful to consider the behavior of an air parcel of infinitesimal dimensions that is assumed to be

i. thermally insulated from its environment so that its temperature changes adiabatically as it rises or sinks, always remaining at exactly the same pressure as the environmental air at the same level,[24] which is assumed to be in hydrostatic equilibrium; and
ii. moving slowly enough that the macroscopic kinetic energy of the air parcel is a negligible fraction of its total energy.

Although in the case of real air parcels one or more of these assumptions is nearly always violated to some extent, this simple, idealized model is helpful in understanding some of the physical processes that influence the distribution of vertical motions and vertical mixing in the atmosphere.

3.4.2 The Dry Adiabatic Lapse Rate

We will now derive an expression for the rate of change of temperature with height of a parcel of dry air that moves about in the Earth's atmosphere while always satisfying the conditions listed at the end of Section 3.4.1. Because the air parcel undergoes only adiabatic transformations ($dq = 0$) and the atmosphere is in hydrostatic equilibrium, for a unit mass of air in the parcel we have, from (3.51),

$$d(c_p T + \Phi) = 0 \qquad (3.52)$$

Dividing through by dz and making use of (3.20) we obtain

$$-\left(\frac{dT}{dz}\right)_{\text{dry parcel}} = \frac{g}{c_p} \equiv \Gamma_d \qquad (3.53)$$

where Γ_d is called the *dry adiabatic lapse rate*. Because an air parcel expands as it rises in the atmosphere, its temperature will decrease with height so that Γ_d defined by (3.53) is a positive quantity. Substituting $g = 9.81$ m s^{-2} and $c_p = 1004$ J K^{-1} kg^{-1} into (3.53) gives $\Gamma_d = 0.0098$ K m^{-1} or 9.8 K km^{-1}, which is the numerical value of the dry adiabatic lapse rate.

It should be emphasized again that Γ_d is the rate of change of temperature following a parcel of dry air that is being raised or lowered adiabatically in the atmosphere. The actual lapse rate of temperature in a column of air, which we will indicate by $\Gamma = \partial T/\partial z$, as measured, for example, by a radiosonde, averages 6–7 K km^{-1} in the troposphere, but it takes on a wide range of values at individual locations.

3.4.3 Potential Temperature

The *potential temperature* θ of an air parcel is defined as the temperature that the parcel of air would have if it were expanded or compressed adiabatically from its existing pressure and temperature to a standard pressure p_0 (generally taken as 1000 hPa).

[24] Any pressure differences between the parcel and its environment give rise to sound waves that produce an almost instantaneous adjustment. Temperature differences, however, are eliminated by much slower processes.

We can derive an expression for the potential temperature of an air parcel in terms of its pressure p, temperature T, and the standard pressure p_0 as follows. For an adiabatic transformation ($dq = 0$) (3.46) becomes

$$c_p dT - \alpha dp = 0$$

Substituting α from (3.3) into this expression yields

$$\frac{c_p}{R}\frac{dT}{T} - \frac{dp}{p} = 0$$

Integrating upward from p_0 (where, by definition, $T = \theta$) to p, we obtain

$$\frac{c_p}{R}\int_{\theta}^{T}\frac{dT}{T} = \int_{p_0}^{p}\frac{dp}{p}$$

or

$$\frac{c_p}{R}\ln\frac{T}{\theta} = \ln\frac{p}{p_0}$$

Taking the antilog of both sides

$$\left(\frac{T}{\theta}\right)^{c_p/R} = \frac{p}{p_0}$$

or

$$\theta = T\left(\frac{p_0}{p}\right)^{R/c_p} \qquad (3.54)$$

Equation (3.54) is called *Poisson's*[25] *equation*. It is usually assumed that $R \simeq R_d = 287$ J K^{-1} kg^{-1} and $c_p \simeq c_{pd} = 1004$ J K^{-1} kg^{-1}; therefore, $R/c_p \simeq 0.286$.

Parameters that remain constant during certain transformations are said to be *conserved*. Potential temperature is a conserved quantity for an air parcel that moves around in the atmosphere under adiabatic conditions (see Exercise 3.36). Potential temperature is an extremely useful parameter in atmospheric thermodynamics, since atmospheric processes are often close to adiabatic, and therefore θ remains essentially constant, like density in an incompressible fluid.

3.4.4 Thermodynamic Diagrams

Poisson's equation may be conveniently solved in graphical form. If pressure is plotted on the ordinate on a distorted scale, in which the distance from the origin is proportional to p^{R_d/c_p}, or $p^{0.286}$ is used, regardless of whether air is dry or moist, and temperature (in K) is plotted on the abscissa, then (3.54) becomes

$$p^{0.286} = \left(\frac{p_0^{0.286}}{\theta}\right)T \qquad (3.55)$$

For a constant value of θ, Eq. (3.55) is of the form $y \propto x$ where $y = p^{0.286}$, $x = T$, and the constant of proportionality is $p_0^{0.286}/\theta$. Each constant value of θ represents a dry adiabat, which is defined by a straight line with a particular slope that passes through the point $p = 0$, $T = 0$. If the pressure scale is inverted so that p increases downward, the relation takes the form shown in Fig. 3.6, which is the basis for the *pseudoadiabatic chart* that used to be widely used for meteorological computations. The region of the chart of greatest interest in the atmosphere is the portion shown within the dotted lines in Fig. 3.6, and this is generally the only portion of the chart that is printed.

In the pseudoadiabatic chart, isotherms are vertical and dry adiabats (constant θ) are oriented at an acute angle relative to isotherms (Fig. 3.6). Because changes in temperature with height in the atmosphere generally lie between isothermal and dry adiabatic, most temperature soundings lie within a narrow range of angles when plotted on a pseudoadiabatic chart. This restriction is overcome in the so-called *skew* $T - \ln p$ *chart*, in which the ordinate (y) is $-\ln p$ (the minus sign ensures that lower pressure levels are located above higher pressure levels on the chart) and the abscissa (x) is

$$x = T + (\text{constant})y = T - (\text{constant})\ln p \qquad (3.56)$$

Since, from (3.56),

$$y = \frac{x - T}{(\text{constant})}$$

and for an isotherm T is constant, the relationship between y and x for an isotherm is of the form

[25] **Simeon Denis Poisson** (1781–1840) French mathematician. Studied medicine but turned to applied mathematics and became the first professor of mechanics at the Sorbonne in Paris.

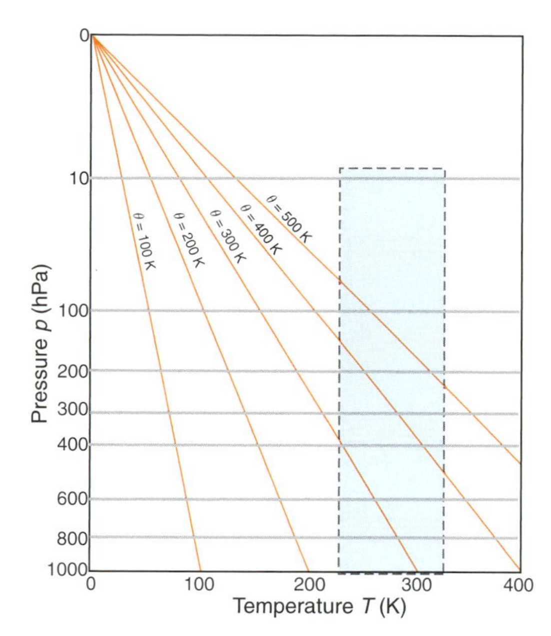

Fig. 3.6 The complete pseudoadiabatic chart. Note that p increases downward and is plotted on a distorted scale (representing $p^{0.286}$). Only the blue-shaded area is generally printed for use in meteorological computations. The sloping lines, each labeled with a value of the potential temperature θ, are dry adiabats. As required by the definition of θ, the actual temperature of the air (given on the abscissa) at 1000 hPa is equal to its potential temperature.

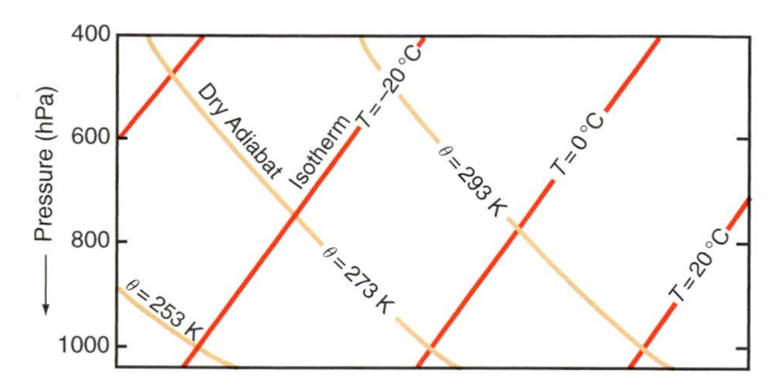

Fig. 3.7 Schematic of a portion of the skew $T - \ln p$ chart. (An accurate reproduction of a larger portion of the chart is available on the book web site that accompanies this book, from which it can be printed and used for solving exercises.)

$y = mx + c$, where m is the same for all isotherms and c is a different constant for each isotherm. Therefore, on the skew $T - \ln p$ chart, isotherms are straight parallel lines that slope upward from left to right. The scale for the x axis is generally chosen to make the angle between the isotherms and the isobars about 45°, as depicted schematically in Fig. 3.7. Note that the isotherms on a skew $T - \ln p$ chart are intentionally "skewed" by about 45° from their vertical orientation in the pseudoadiabatic chart (hence the name *skew* $T - \ln p$ *chart*). From (3.55), the equation for a dry adiabat (θ constant) is

$$-\ln p = (\text{constant}) \ln T + \text{constant}$$

Hence, on a $-\ln p$ versus $\ln T$ chart, dry adiabats would be straight lines. Since $-\ln p$ is the ordinate on the skew $T - \ln p$ chart, but the abscissa is not $\ln T$, dry adiabats on this chart are slightly curved lines that run from the lower right to the upper left. The angle between the isotherms and the dry adiabats on a skew $T - \ln p$ chart is approximately 90° (Fig. 3.7). Therefore, when atmospheric temperature soundings are plotted on this chart, small differences in slope are more apparent than they are on the pseudoadiabatic chart.

Exercise 3.5 A parcel of air has a temperature of −51 °C at the 250-hPa level. What is its potential temperature? What temperature will the parcel have if it is brought into the cabin of a jet aircraft and compressed adiabatically to a cabin pressure of 850 hPa?

Solution: This exercise can be solved using the skew $T - \ln p$ chart. Locate the original state of the air parcel on the chart at pressure 250 hPa and temperature −51 °C. The label on the dry adiabat that passes through this point is 60 °C, which is therefore the potential temperature of the air.

The temperature acquired by the ambient air if it is compressed adiabatically to a pressure of 850 hPa can be found from the chart by following the dry adiabat that passes through the point located by 250 hPa and −51 °C down to a pressure of 850 hPa and reading off the temperature at that point. It is 44.5 °C. (Note that this suggests that ambient air brought into the cabin of a jet aircraft at cruise altitude has to be *cooled* by about 20 °C to provide a comfortable environment.) ■

3.5 Water Vapor in Air

So far we have indicated the presence of water vapor in the air through the vapor pressure e that it exerts, and we have quantified its effect on the density of air by introducing the concept of virtual temperature. However, the amount of water vapor present in a certain quantity of air may be expressed in many different ways, some of the more important of which are presented later. We must also discuss what happens when water vapor condenses in air.

3.5.1 Moisture Parameters

a. *Mixing ratio and specific humidity*

The amount of water vapor in a certain volume of air may be defined as the ratio of the mass m_v of water vapor to the mass of dry air; this is called the *mixing ratio* w. That is

$$w \equiv \frac{m_v}{m_d} \tag{3.57}$$

The mixing ratio is usually expressed in grams of water vapor per kilogram of dry air (but in solving numerical exercises w must be expressed as a dimensionless number, e.g., as kg of water vapor per kg of dry air). In the atmosphere, the magnitude of w typically ranges from a few grams per kilogram in middle latitudes to values of around 20 g kg^{-1} in the tropics. If neither condensation nor evaporation takes place, the mixing ratio of an air parcel is constant (i.e., it is a conserved quantity).

The mass of water vapor m_v in a unit mass of air (dry air plus water vapor) is called the *specific humidity* q, that is

$$q \equiv \frac{m_v}{m_v + m_d} = \frac{w}{1 + w}$$

Because the magnitude of w is only a few percent, it follows that the numerical values of w and q are nearly equivalent.

Exercise 3.6 If air contains water vapor with a mixing ratio of 5.5 g kg^{-1} and the total pressure is 1026.8 hPa, calculate the vapor pressure e.

Solution: The partial pressure exerted by any constituent in a mixture of gases is proportional to the number of moles of the constituent in the mixture. Therefore, the pressure e due to water vapor in air is given by

$$e = \frac{n_v}{n_d + n_v} p = \frac{\dfrac{m_v}{M_w}}{\dfrac{m_d}{M_d} + \dfrac{m_v}{M_w}} p \tag{3.58}$$

n_v and n_d are the number of moles of water vapor and dry air in the mixture, respectively, M_w is the molecular weight of water, M_d is the apparent molecular weight of dry air, and p is the total pressure of the moist air. From (3.57) and (3.58) we obtain

$$e = \frac{w}{w + \varepsilon} p \tag{3.59}$$

where $\varepsilon = 0.622$ is defined by (3.14). Substituting $p = 1026.8$ hPa and $w = 5.5 \times 10^{-3}$ kg kg^{-1} into (3.59), we obtain $e = 9.0$ hPa. ■

Exercise 3.7 Calculate the virtual temperature correction for moist air at 30 °C that has a mixing ratio of 20 g kg^{-1}.

Solution: Substituting e/p from (3.59) into (3.16) and simplifying

$$T_v = T \frac{w + \varepsilon}{\varepsilon (1 + w)}$$

Dividing the denominator into the numerator in this expression and neglecting terms in w^2 and higher orders of w, we obtain

$$T_v - T \simeq \frac{1 - \varepsilon}{\varepsilon} wT$$

or, substituting $\varepsilon = 0.622$ and rearranging,

$$T_v \simeq T(1 + 0.61w) \tag{3.60}$$

With $T = 303$ K and $w = 20 \times 10^{-3}$ kg kg^{-1}, Eq. (3.60) gives $T_v = 306.7$ K. Therefore, the virtual temperature correction is $T_v - T = 3.7$ degrees (K or °C). Note that (3.60) is a useful expression for obtaining T_v from T and the moisture parameter w. ■

b. *Saturation vapor pressures*

Consider a small closed box, the floor of which is covered with pure water at temperature T. Initially assume that the air is completely dry. Water will begin to evaporate and, as it does, the number of water molecules in the box, and therefore the water vapor pressure, will increase. As the water vapor pressure increases, so will the rate at which the water molecules condense from the vapor phase back to the liquid phase. If the rate of condensation is less than the rate of evaporation, the box is said to be *unsaturated* at temperature T (Fig. 3.8a). When the water vapor pressure in the box increases to the point that the rate of condensation is equal to the rate of evaporation (Fig. 3.8b), the air is said to be *saturated with respect*

3.3 Can Air Be Saturated with Water Vapor?[26]

It is common to use phrases such as "the air is saturated with water vapor," "the air can hold no more water vapor," and "warm air can hold more water vapor than cold air." These phrases, which suggest that air absorbs water vapor, rather like a sponge, are misleading. We have seen that the total pressure exerted by a mixture of gases is equal to the sum of the pressures that each gas would exert if it alone occupied the total volume of the mixture of gases (Dalton's law of partial pressures). Hence, the exchange of water molecules between its liquid and vapor phases is (essentially) independent of the presence of air. Strictly speaking, the pressure exerted by water vapor that is in equilibrium with water at a given temperature is referred more appropriately to as *equilibrium vapor pressure* rather than saturation vapor pressure at that temperature. However, the latter term, and the terms "unsaturated air" and "saturated air," provide a convenient shorthand and are so deeply rooted that they will appear in this book.

to a plane surface of pure water at temperature T, and the pressure e_s that is then exerted by the water vapor is called the *saturation vapor pressure over a plane surface of pure water at temperature T*.

Similarly, if the water in Fig. 3.8 were replaced by a plane surface of pure ice at temperature T and the rate of condensation of water vapor were equal to the rate of evaporation of the ice, the pressure e_{si} exerted by the water vapor would be the saturation vapor pressure over a plane surface of pure ice at T. Because, at any given temperature, the rate of evaporation from ice is less than from water, $e_s(T) > e_{si}(T)$.

The rate at which water molecules evaporate from either water or ice increases with increasing temperature.[27] Consequently, both e_s and e_{si} increase with increasing temperature, and their magnitudes depend only on temperature. The variations with temperature of e_s and $e_s - e_{si}$ are shown in Fig. 3.9, where it can be seen that the magnitude of $e_s - e_{si}$ reaches a peak value at about -12 °C. It follows that if an ice particle is in water-saturated air it will grow due to the deposition of water vapor upon it. In Section 6.5.3 it is shown that this phenomenon

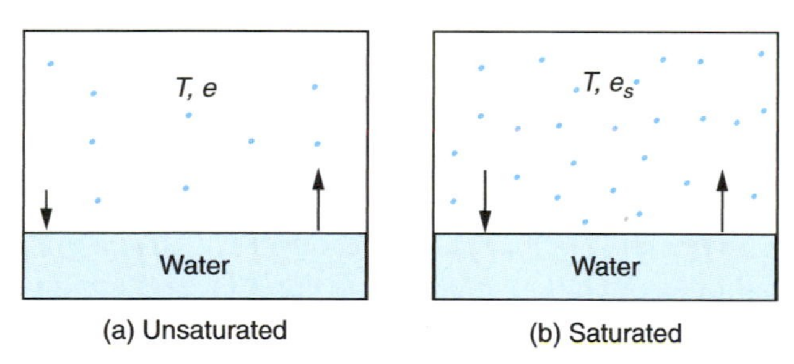

Fig. 3.8 A box (a) unsaturated and (b) saturated with respect to a plane surface of pure water at temperature T. Dots represent water molecules. Lengths of the arrows represent the relative rates of evaporation and condensation. The saturated (i.e., equilibrium) vapor pressure over a plane surface of pure water at temperature T is e_s as indicated in (b).

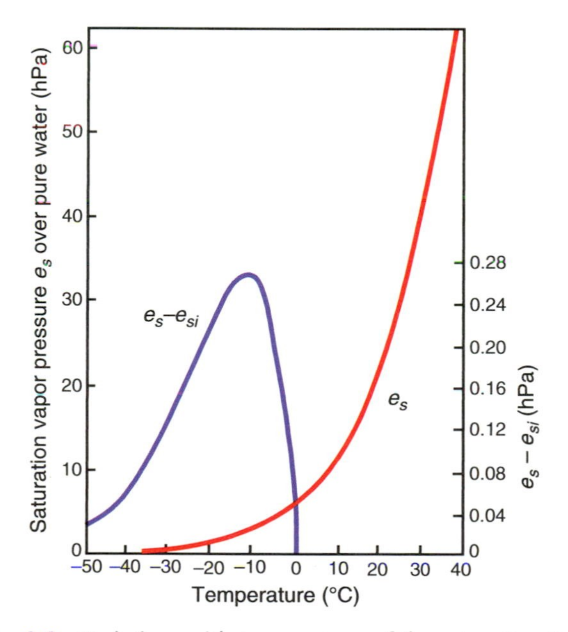

Fig. 3.9 Variations with temperature of the saturation (i.e., equilibrium) vapor pressure e_s over a plane surface of pure water (red line, scale at left) and the difference between e_s and the saturation vapor pressure over a plane surface of ice e_{si} (blue line, scale at right).

[26] For further discussion of this and some other common misconceptions related to meteorology see C. F. Bohren's *Clouds in a Glass of Beer*, Wiley and Sons, New York, 1987.

[27] As a rough rule of thumb, it is useful to bear in mind that the saturation vapor pressure roughly doubles for a 10 °C increase in temperature.

plays a role in the initial growth of precipitable particles in some clouds.

c. Saturation mixing ratios

The *saturation mixing ratio* w_s with respect to water is defined as the ratio of the mass m_{vs} of water vapor in a given volume of air that is saturated with respect to a plane surface of pure water to the mass m_d of the dry air. That is

$$w_s \equiv \frac{m_{vs}}{m_d} \tag{3.61}$$

Because water vapor and dry air both obey the ideal gas equation

$$w_s = \frac{\rho'_{vs}}{\rho'_d} = \frac{e_s}{(R_v T)} \Big/ \frac{(p - e_s)}{(R_d T)} \tag{3.62}$$

where ρ'_{vs} is the partial density of water vapor required to saturate air with respect to water at temperature T, ρ'_d is the partial density of the dry air (see Section 3.1.1), and p is the total pressure. Combining (3.62) with (3.14), we obtain

$$w_s = 0.622 \frac{e_s}{p - e_s}$$

For the range of temperatures observed in the Earth's atmosphere, $p \gg e_s$; therefore

$$w_s \simeq 0.622 \frac{e_s}{p} \tag{3.63}$$

Hence, at a given temperature, the saturation mixing ratio is inversely proportional to the total pressure.

Because e_s depends only on temperature, it follows from (3.63) that w_s is a function of temperature and pressure. Lines of constant saturation mixing ratio are printed as dashed green lines on the skew $T - \ln p$ chart and are labeled with the value of w_s in grams of water vapor per kilogram of dry air. It is apparent from the slope of these lines that at constant pressure w_s increases with increasing temperature, and at constant temperature w_s increases with decreasing pressure.

d. Relative humidity; dew point and frost point

The relative humidity (RH) with respect to water is the ratio (expressed as a percentage) of the actual mixing ratio w of the air to the saturation mixing ratio w_s with respect to a plane surface of pure water at the same temperature and pressure. That is

$$RH \equiv 100 \frac{w}{w_s} \simeq 100 \frac{e}{e_s} \tag{3.64}$$

The *dew point* T_d is the temperature to which air must be cooled at constant pressure for it to become saturated with respect to a plane surface of pure water. In other words, the dew point is the temperature at which the saturation mixing ratio w_s with respect to liquid water becomes equal to the actual mixing ratio w. It follows that the relative humidity at temperature T and pressure p is given by

$$RH = 100 \frac{w_s \text{ (at temperature } T_d \text{ and pressure } p)}{w_s \text{ (at temperature } T \text{ and pressure } p)} \tag{3.65}$$

A simple rule of thumb for converting RH to a dew point depression ($T - T_d$) for moist air ($RH >$ 50%) is that T_d decreases by ~1 °C for every 5% decrease in RH (starting at T_d = dry bulb temperature (T), where $RH = 100\%$). For example, if the RH is 85%, $T_d = T - \left(\frac{100 - 85}{5}\right)$ and the dew point depression is $T - T_d = 3$ °C.

The *frost point* is defined as the temperature to which air must be cooled at constant pressure to saturate it with respect to a plane surface of pure ice. Saturation mixing ratios and relative humidities with respect to ice may be defined in analogous ways to their definitions with respect to liquid water. When the terms mixing ratio and relative humidity are used without qualification they are with respect to liquid water.

Exercise 3.8 Air at 1000 hPa and 18 °C has a mixing ratio of 6 g kg^{-1}. What are the relative humidity and dew point of the air?

Solution: This exercise may be solved using a skew $T - \ln p$ chart. The students should duplicate the following steps. First locate the point with pressure 1000 hPa and temperature 18 °C. We see from the chart that the saturation mixing ratio for this state is ~13 g kg^{-1}. Since the air specified in the problem has a mixing ratio of only 6 g kg^{-1}, it is unsaturated and its relative humidity is, from (3.64), $100 \times 6/13 =$ 46%. To find the dew point we move from right to left along the 1000-hPa ordinate until we intercept the saturation mixing ratio line of magnitude 6 g kg^{-1}; this occurs at a temperature of about 6.5 °C. Therefore, if the air is cooled at constant pressure, the water vapor it contains will just saturate the air with respect to water at a temperature of 6.5 °C. Therefore, by definition, the dew point of the air is 6.5 °C. ■

At the Earth's surface, the pressure typically varies by only a few percent from place to place and from time to time. Therefore, the dew point is a good indicator of the moisture content of the air. In warm, humid weather the dew point is also a convenient indicator of the level of human discomfort. For example, most people begin to feel uncomfortable when the dew point rises above 20 °C, and air with a dew point above about 22 °C is generally regarded as extremely humid or "sticky." Fortunately, dew points much above this temperature are rarely observed even in the tropics. In contrast to the dew point, relative humidity depends as much upon the temperature of the air as upon its moisture content. On a sunny day the relative humidity may drop by as much as 50% from morning to afternoon, just because of a rise in air temperature. Neither is relative humidity a good indicator of the level of human discomfort. For example, a relative humidity of 70% may feel quite comfortable at a temperature of 20 °C, but it would cause considerable discomfort to most people at a temperature of 30 °C.

The highest dew points occur over warm bodies of water or vegetated surfaces from which water is evaporating. In the absence of vertical mixing, the air just above these surfaces would become saturated with water vapor, at which point the dew point would be the same as the temperature of the underlying surface. Complete saturation is rarely achieved over hot surfaces, but dew points in excess of 25 °C are sometimes observed over the warmest regions of the oceans.

e. *Lifting condensation level*

The *lifting condensation level* (LCL) is defined as the level to which an unsaturated (but moist) parcel of air can be lifted adiabatically before it becomes saturated with respect to a plane surface of pure water. During lifting the mixing ratio w and potential temperature θ of the air parcel remain constant, but the saturation mixing ratio w_s decreases until it becomes equal to w at the LCL. Therefore, the LCL is located at the intersection of the potential temperature line passing through the temperature T and pressure p of the air parcel, and the w_s line that passes through the pressure p and dew point T_d of the parcel (Fig. 3.10). Since the dew point and LCL are related in the manner indicated in Fig. 3.10, knowledge of either one is sufficient to determine the other. Similarly, a knowledge of T, p, and any one moisture parameter is sufficient to determine all the other moisture parameters we have defined.

f. *Wet-bulb temperature*

The wet-bulb temperature is measured with a thermometer, the glass bulb of which is covered with a moist cloth over which ambient air is drawn. The heat required to evaporate water from the moist cloth to saturate the ambient air is supplied by the air as it comes into contact with the cloth. When the difference between the temperatures of the bulb and the ambient air is steady and sufficient to supply the heat needed to evaporate the water, the thermometer will read a steady temperature, which is called the *wet-bulb temperature*. If a raindrop falls through a layer of air that has a constant wet-bulb temperature, the raindrop will eventually reach a temperature equal to the wet-bulb temperature of the air.

The definition of wet-bulb temperature and dew point both involve cooling a hypothetical air parcel to saturation, but there is a distinct difference. If the unsaturated air approaching the wet bulb has a mixing ratio w, the dew point T_d is the temperature to which the air must be cooled at constant pressure

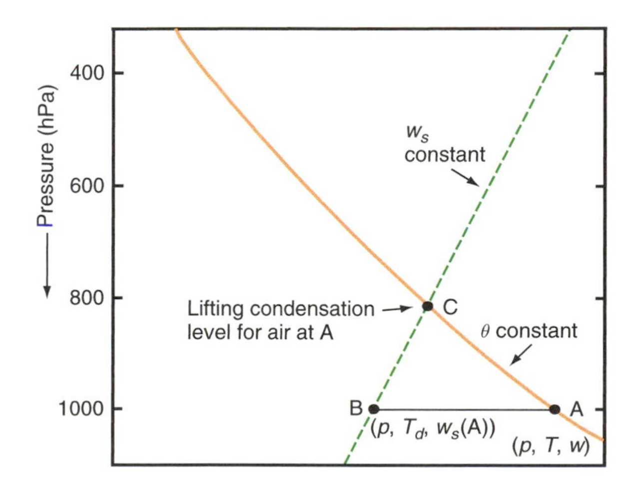

Fig. 3.10 The lifting condensation level of a parcel of air at A, with pressure p, temperature T, and dew point T_d, is at C on the skew $T - \ln p$ chart.

to become saturated. The air that leaves the wet bulb has a mixing ratio w' that saturates it at temperature T_w. If the air approaching the wet bulb is unsaturated, w' is greater than w; therefore, $T_d \leq T_w \leq T$, where the equality signs apply only to air saturated with respect to a plane surface of pure water. Usually T_w is close to the arithmetic mean of T and T_d.

3.5.2 Latent Heats

If heat is supplied to a system under certain conditions it may produce a change in phase rather than a change in temperature. In this case, the increase in internal energy is associated entirely with a change in molecular configurations in the presence of intermolecular forces rather than an increase in the kinetic energy of the molecules (and therefore the temperature of the system). For example, if heat is supplied to ice at 1 atm and 0 °C, the temperature remains constant until all of the ice has melted. The *latent heat of melting* (L_m) is defined as the heat that has to be given to a unit mass of a material to convert it from the solid to the liquid phase without a change in temperature. The temperature at which this phase change occurs is called the *melting point*. At 1 atm and 0 °C the latent heat of melting of the water substance is 3.34×10^5 J kg^{-1}. The *latent heat of freezing* has the same numerical value as the latent heat of melting, but heat is released as a result of the change in phase from liquid to solid.

Similarly, the *latent heat of vaporization* or *evaporation* (L_v) is the heat that has to be given to a unit mass of material to convert it from the liquid to the vapor phase without a change in temperature. For the water substance at 1 atm and 100 °C (the *boiling point* of water at 1 atm), the latent heat of vaporization is 2.25×10^6 J kg^{-1}. The *latent heat of condensation* has the same value as the latent heat of vaporization, but heat is released in the change in phase from vapor to liquid.[28]

As will be shown in Section 3.7.3, the melting point (and boiling point) of a material depends on pressure.

3.5.3 Saturated Adiabatic and Pseudoadiabatic Processes

When an air parcel rises in the atmosphere its temperature decreases with altitude at the dry adiabatic lapse rate (see Section 3.4.2) until it becomes saturated with water vapor. Further lifting results in the condensation of liquid water (or the deposition of ice), which releases latent heat. Consequently, the rate of decrease in the temperature of the rising parcel is reduced. If all of the condensation products remain in the rising parcel, the process may still be considered to be adiabatic (and reversible), even though latent heat is released in the system, provided that heat does not pass through the boundaries of the parcel. The air parcel is then said to undergo a *saturated adiabatic process*. However, if all of the condensation products immediately fall out of the air parcel, the process is irreversible, and not strictly adiabatic, because the condensation products carry some heat. The air parcel is then said to undergo a *pseudoadiabatic process*. As the reader is invited to verify in Exercise 3.44, the amount of heat carried by condensation products is small compared to that carried by the air itself. Therefore, the saturated-adiabatic and the pseudoadiabatic lapse rates are virtually identical.

3.5.4 The Saturated Adiabatic Lapse Rate

In contrast to the dry adiabatic lapse rate Γ_d, which is constant, the numerical value of the saturated adiabatic lapse rate Γ_s varies with pressure and temperature. (The reader is invited to derive an expression for Γ_s in Exercise 3.50; see the book Web site.) Because water vapor condenses when a saturated air parcel rises, it follows that $\Gamma_s < \Gamma_d$. Actual values of Γ_s range from about 4 K km^{-1} near the ground in warm, humid air masses to typical values of 6–7 K km^{-1} in the middle troposphere. For typical temperatures near the tropopause, Γ_s is only slightly less than Γ_d because the saturation vapor pressure of the air is so small that the effect of condensation is negligible.[29] Lines that show the rate of decrease in

[28] Normally, when heat is given to a substance, the temperature of the substance increases. This is called *sensible heat*. However, when heat is given to a substance that is melting or boiling, the temperature of the substance does not change until all of the substance is melted or vaporized. In this case, the heat appears to be *latent* (i.e., hidden). Hence the terms *latent heat of melting* and *latent heat of vaporization*.

[29] **William Thomson** (later Lord Kelvin) was the first (in 1862) to derive quantitative estimates of the dry and saturated adiabatic lapse rates based on theoretical arguments. For an interesting account of the contributions of other 19th-century scientists to the realization of the importance of latent heat in the atmosphere, see W. E. K. Middleton, *A History of the Theories of Rain*, Franklin Watts, Inc., New York, 1965, Chapter 8.

temperature with height of a parcel of air that is rising or sinking in the atmosphere under saturated adiabatic (or pseudoadiabatic) conditions are called *saturated adiabats* (or *pseudoadiabats*). On the skew $T - \ln p$ chart these are the curved green lines that diverge upward and tend to become parallel to the dry adiabats.

Exercise 3.9 A parcel of air with an initial temperature of 15 °C and dew point 2 °C is lifted adiabatically from the 1000-hPa level. Determine its LCL and temperature at that level. If the air parcel is lifted a further 200 hPa above its LCL, what is its final temperature and how much liquid water is condensed during this rise?

Solution: The student should duplicate the following steps on the skew $T - \ln p$ chart (see the book Web site). First locate the initial state of the air on the chart at the intersection of the 15 °C isotherm with the 1000-hPa isobar. Because the dew point of the air is 2 °C, the magnitude of the saturation mixing ratio line that passes through the 1000-hPa pressure level at 2 °C is the actual mixing ratio of the air at 15 °C and 1000 hPa. From the chart this is found to be about 4.4 g kg^{-1}. Because the saturation mixing ratio at 1000 hPa and 15 °C is about 10.7 g kg^{-1}, the air is initially unsaturated. Therefore, when it is lifted it will follow a dry adiabat (i.e., a line of constant potential temperature) until it intercepts the saturation mixing ratio line of magnitude 4.4 g kg^{-1}. Following upward along the dry adiabat ($\theta = 288$ K) that passes through 1000 hPa and 15 °C isotherm, the saturation mixing ratio line of 4.4 g kg^{-1} is intercepted at about the 820-hPa level. This is the LCL of the air parcel. The temperature of the air at this point is about -0.7 °C. For lifting above this level the air parcel will follow a saturated adiabat. Following the saturated adiabat that passes through 820 hPa and -0.7 °C up to the 620-hPa level, the final temperature of the air is found to be about -15 °C. The saturation mixing ratio at 620 hPa and -15 °C is $\sim$1.9 g kg^{-1}. Therefore, about $\sim$4.4 − 1.9 = 2.5 g of water must have condensed out of each kilogram of air during the rise from 820 to 620 hPa. ■

3.5.5 Equivalent Potential Temperature and Wet-Bulb Potential Temperature

We will now derive an equation that describes how temperature varies with pressure under conditions of saturated adiabatic ascent or descent. Substituting (3.3) into (3.46) gives

$$\frac{dq}{T} = c_p \frac{dT}{T} - R\frac{dp}{p} \tag{3.66}$$

From (3.54) the potential temperature θ is given by

$$\ln\theta = \ln T - \frac{R}{c_p}\ln p + \text{constant}$$

or, differentiating,

$$c_p\frac{d\theta}{\theta} = c_p\frac{dT}{T} - R\frac{dp}{p} \tag{3.67}$$

Combining (3.66) and (3.67) and substituting $dq = -L_v\,dw_s$, we obtain

$$-\frac{L_v}{c_p T}dw_s = \frac{d\theta}{\theta} \tag{3.68}$$

In Exercise 3.52 we show that

$$\frac{L_v}{c_p T}\,dw_s \simeq d\left(\frac{L_v w_s}{c_p T}\right) \tag{3.69}$$

From (3.68) and (3.69)

$$-d\left(\frac{L_v w_s}{c_p T}\right) \simeq \frac{d\theta}{\theta}$$

This last expression can be integrated to give

$$-\frac{L_v w_s}{c_p T} \simeq \ln\theta + \text{constant} \tag{3.70}$$

We will define the constant of integration in (3.70) by requiring that at low temperatures, as $w_s/T \to 0$, $\theta \to \theta_e$. Then

$$-\frac{L_v w_s}{c_p T} \simeq \ln\left(\frac{\theta}{\theta_e}\right)$$

or

$$\theta_e \simeq \theta\exp\left(\frac{L_v w_s}{c_p T}\right) \tag{3.71}$$

The quantity θ_e given by (3.71) is called the *equivalent potential temperature*. It can be seen that θ_e is the potential temperature θ of a parcel of air when all the water vapor has condensed so that its saturation mixing ratio w_s is zero. Hence, recalling the definition of θ, the equivalent potential temperature of an air parcel may be found as follows. The air is expanded (i.e., lifted) pseudoadiabatically until all the vapor has condensed, released its latent heat, and fallen out. The air is then compressed dry adiabatically to the standard pressure of 1000 hPa, at which point it will attain the temperature θ_e. (If the air is initially unsaturated, w_s and T are the saturation mixing ratio and temperature at the point where the air first becomes saturated after being lifted dry adiabatically.) We have seen in Section 3.4.3 that potential temperature is a conserved quantity for adiabatic transformations. The equivalent potential temperature is conserved during both dry and saturated adiabatic processes.

If the line of constant equivalent potential temperature (i.e., the pseudoadiabat) that passes through the wet-bulb temperature of a parcel of air is traced back on a skew $T - \ln p$ chart to the point where it intersects the 1000-hPa isobar, the temperature at this intersection is called the *wet-bulb potential temperature* θ_w of the air parcel. Like the equivalent potential temperature, the wet-bulb potential temperature is conserved during both dry and saturated adiabatic processes. On skew $T - \ln p$ charts, pseudoadiabats are labeled (along the 200-hPa isobar) with the wet-bulb potential temperature θ_w (in °C) and the equivalent potential temperature θ_e (in K) of air that rises or sinks along that pseudoadiabat. Both θ_w and θ_e provide equivalent information and are valuable as tracers of air parcels.

When height, rather than pressure, is used as the independent variable, the conserved quantity during adiabatic or pseudoadiabatic ascent or descent with water undergoing transitions between liquid and vapor phases is the *moist static energy* (MSE)[30]

$$MSE = c_p T + \Phi + L_v q \tag{3.72}$$

where T is the temperature of the air parcel, Φ is the geopotential, and q_v is the specific humidity (nearly the same as w). The first term on the right side of (3.72) is the enthalpy per unit mass of air. The second term is the potential energy, and the third term is the latent heat content. The first two terms, which also appear in (3.51), are the *dry static energy*. When air is lifted dry adiabatically, enthalpy is converted into potential energy and the latent heat content remains unchanged. In saturated adiabatic ascent, energy is exchanged among all three terms on the right side of (3.72): potential energy increases, while the enthalpy and latent heat content both decrease. However, the sum of the three terms remains constant.

3.5.6 Normand's Rule

Many of the relationships discussed in this section are embodied in the following theorem, known as *Normand's*[31] *rule*, which is extremely helpful in many computations involving the skew $T - \ln p$ chart. Normand's rule states that on a skew $T - \ln p$ chart the lifting condensation level of an air parcel is located at the intersection of the potential temperature line that passes through the point located by the temperature and pressure of the air parcel, the equivalent potential temperature line (i.e., the pseudoadiabat) that passes through the point located by the wet-bulb temperature and pressure of the air parcel, and the saturation mixing ratio line that passes through the point determined by the dew point and pressure of the air. This rule is illustrated in Fig. 3.11 for the case of an air parcel with temperature T, pressure p, dew point T_d, and wet-bulb temperature T_w. It can be seen that if T, p, and T_d are known, T_w may be readily determined using Normand's rule. Also, by extrapolating the θ_e line that passes through T_w to the 1000-hPa level, the wet-bulb potential temperature θ_w may be found (Fig. 3.11).

3.5.7 Net Effects of Ascent Followed by Descent

When a parcel of air is lifted above its LCL so that condensation occurs and if the products of the condensation fall out as precipitation, the latent heat gained by the air during this process will be retained by the air if the parcel returns to its original level.

[30] The word *static* derives from the fact that the kinetic energy associated with macroscale fluid motions is not included. The reader is invited to show that the kinetic energy per unit mass is much smaller than the other terms on the right side of (3.72), provided that the wind speed is small in comparison to the speed of sound.

[31] **Sir Charles William Blyth Normand** (1889–1982) British meteorologist. Director-General of Indian Meteorological Service, 1927–1944. A founding member of the National Science Academy of India. Improved methods for measuring atmospheric ozone.

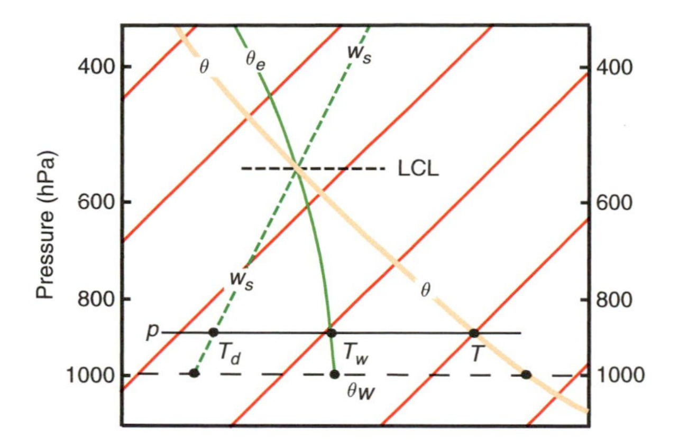

Fig. 3.11 Illustration of Normand's rule on the skew $T - \ln p$ chart. The orange lines are isotherms. The method for determining the wet-bulb temperature (T_w) and the wet-bulb potential temperature (θ_w) of an air parcel with temperature T and dew point T_d at pressure p is illustrated. LCL denotes the lifting condensation level of this air parcel.

The effects of the saturated ascent coupled with the adiabatic descent are:

i. net increases in the temperature and potential temperature of the parcel;
ii. a decrease in moisture content (as indicated by changes in the mixing ratio, relative humidity, dew point, or wet-bulb temperature); and,
iii. no change in the equivalent potential temperature or wet-bulb potential temperature, which are conserved quantities for air parcels undergoing both dry and saturated processes.

The following exercise illustrates these points.

Exercise 3.10 An air parcel at 950 hPa has a temperature of 14 °C and a mixing ratio of 8 g kg^{-1}. What is the wet-bulb potential temperature of the air? The air parcel is lifted to the 700-hPa level by passing over a mountain, and 70% of the water vapor that is condensed out by the ascent is removed by precipitation. Determine the temperature, potential temperature, mixing ratio, and wet-bulb potential temperature of the air parcel after it has descended to the 950-hPa level on the other side of the mountain.

Solution: On a skew $T - \ln p$ chart (see the book Web site), locate the initial state of the air at 950 hPa and 14 °C. The saturation mixing ratio for an air parcel with temperature and pressure is found from the chart to be 10.6 g kg^{-1}. Therefore, because the air has a mixing ratio of only 8 g kg^{-1}, it is unsaturated. The wet-bulb potential temperature (θ_w) can be determined using the method indicated schematically in Fig. 3.11, which is as follows. Trace the constant potential temperature line that passes through the initial state of the air parcel up to the point where it intersects the saturation mixing ratio line with value 8 g kg^{-1}. This occurs at a pressure of about 890 hPa, which is the LCL of the air parcel. Now follow the equivalent potential temperature line that passes through this point back down to the 1000-hPa level and read off the temperature on the abscissa—it is 14 °C. This is in the wet-bulb potential temperature of the air.

When the air is lifted over the mountain, its temperature and pressure up to the LCL at 890 hPa are given by points on the potential temperature line that passes through the point 950 hPa and 14 °C. With further ascent of the air parcel to the 700-hPa level, the air follows the saturated adiabat that passes through the LCL. This saturated adiabat intersects the 700-hPa level at a point where the saturation mixing ratio is 4.7 g kg^{-1}. Therefore, $8 - 4.7 = 3.3$ g kg^{-1} of water vapor has to condense out between the LCL and the 700-hPa level, and 70% of this, or 2.3 g kg^{-1}, is precipitated out. Therefore, at the 700-hPa level 1 g kg^{-1} of liquid water remains in the air. The air parcel descends on the other side of the mountain at the saturated adiabatic lapse rate until it evaporates all of its liquid water, at which point the saturation mixing ratio will have risen to $4.7 + 1 = 5.7$ g kg^{-1}. The air parcel is now at a pressure of 760 hPa and a temperature of 1.8 °C. Thereafter, the air parcel descends along a dry adiabat to the 950-hPa level, where its temperature is 20 °C and the mixing ratio is still 5.7 g kg^{-1}. If the method indicated in Fig. 3.11 is applied again, the wet-bulb potential temperature of the air parcel will be found to be unchanged at 14 °C. (The heating of air during its passage over a mountain, 6 °C in this example, is responsible for the remarkable warmth of *Föhn* or *Chinook* winds, which often blow downward along the lee side of mountain ranges.[32]) ■

[32] The person who first explained the Föhn wind in this way appears to have been J. von Hann[33] in his classic book *Lehrbuch der Meteorologie*, Willibald Keller, Leipzig, 1901.

[33] **Julius F. von Hann** (1839–1921) Austrian meteorologist. Introduced thermodynamic principles into meteorology. Developed theories for mountain and valley winds. Published the first comprehensive treatise on climatology (1883).

3.6 Static Stability

3.6.1 Unsaturated Air

Consider a layer of the atmosphere in which the actual temperature lapse rate Γ (as measured, for example, by a radiosonde) is less than the dry adiabatic lapse rate Γ_d (Fig. 3.12a). If a parcel of unsaturated air originally located at level O is raised to the height defined by points A and B, its temperature will fall to T_A, which is lower than the ambient temperature T_B at this level. Because the parcel immediately adjusts to the pressure of the ambient air, it is clear from the ideal gas equation that the colder parcel of air must be denser than the warmer ambient air. Therefore, if left to itself, the parcel will tend to return to its original level. If the parcel is displaced downward from O it becomes warmer than the ambient air and, if left to itself, the parcel will tend to rise back to its original level. In both cases, the parcel of air encounters a restoring force after being displaced, which inhibits vertical mixing. Thus, the condition $\Gamma < \Gamma_d$ corresponds to a *stable stratification* (or positive static stability) for unsaturated air parcels. In general, the larger the difference $\Gamma_d - \Gamma$, the greater the restoring force for a given displacement and the greater the static stability.[34]

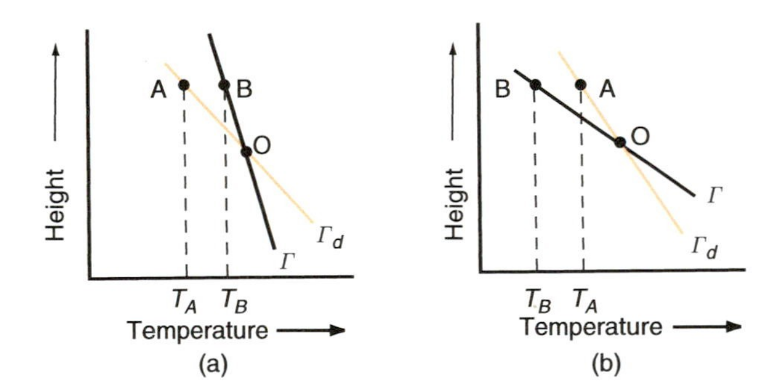

Fig. 3.12 Conditions for (a) positive static stability ($\Gamma < \Gamma_d$) and (b) negative static instability ($\Gamma > \Gamma_d$) for the displacement of unsaturated air parcels.

Exercise 3.11 An unsaturated parcel of air has density ρ' and temperature T', and the density and temperature of the ambient air are ρ and T. Derive an expression for the upward acceleration of the air parcel in terms of T, T', and g.

Solution: The situation is depicted in Fig. 3.13. If we consider a unit volume of the air parcel, its mass is ρ'. Therefore, the downward force acting on unit volume of the parcel is $\rho' g$. From the Archimedes[35] principle we know that the upward force acting on the parcel is equal in magnitude to the gravitational force that acts on the ambient air that is displaced by the air parcel. Because a unit volume of ambient air of density ρ is displaced by the air parcel, the magnitude of the upward force acting on the air parcel is ρg. Therefore, the net upward force (F) acting on a unit volume of the parcel is

$$F = (\rho - \rho')\, g$$

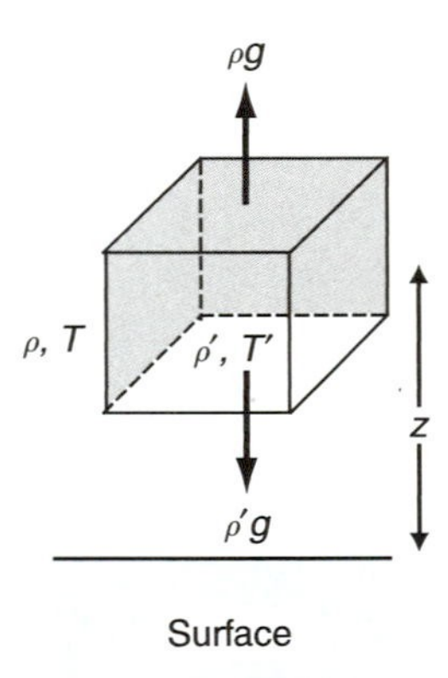

Fig. 3.13 The box represents an air parcel of unit volume with its center of mass at height z above the Earth's surface. The density and temperature of the air parcel are ρ' and T', respectively, and the density and temperature of the ambient air are ρ and T. The vertical forces acting on the air parcel are indicated by the thicker arrows.

[34] A more general method for determing static stability is given in Section 9.3.4.

[35] **Archimedes** (287–212 B.C.) The greatest of Greek scientists. He invented engines of war and the water screw and he derived the principle of buoyancy named after him. When Syracuse was sacked by Rome, a soldier came upon the aged Archimedes absorbed in studying figures he had traced in the sand: "Do not disturb my circles" said Archimedes, but was killed instantly by the soldier. Unfortunately, right does not always conquer over might.

Because the mass of a unit volume of the air parcel is ρ', the upward acceleration of the parcel is

$$\frac{d^2z}{dt^2} = \frac{F}{\rho'} = \left(\frac{\rho - \rho'}{\rho'}\right) g$$

where z is the height of the air parcel. The pressure of the air parcel is the same as that of the ambient air, since they are at the same height in the atmosphere. Therefore, from the gas equation in the form of (3.2), the densities of the air parcel and the ambient air are inversely proportional to their temperatures. Hence,

$$\frac{d^2z}{dt^2} = \frac{\frac{1}{T} - \frac{1}{T'}}{\frac{1}{T'}} g$$

or

$$\frac{d^2z}{dt^2} = g\left(\frac{T' - T}{T}\right) \tag{3.73}$$

■

Strictly speaking, virtual temperature T_v should be used in place of T in all expressions relating to static stability. However, the virtual temperature correction is usually neglected except in certain calculations relating to the boundary layer.

Exercise 3.12 The air parcel in Fig. 3.12a is displaced upward from its equilibrium level at $z' = 0$ by a distance z' to a new level where the ambient temperature is T. The air parcel is then released. Derive an expression that describes the subsequent vertical displacement of the air parcel as a function of time in terms of T, the lapse rate of the ambient air (Γ), and the dry adiabatic lapse rate (Γ_d).

Solution: Let $z = z_0$ be the equilibrium level of the air parcel and $z' = z - z_0$ be the vertical dispalcement of the air parcel from its equilibrium level. Let T_0 be the environmental air temperature at $z = z_0$. If the air parcel is lifted dry adiabatically through a distance z' from its equilibrium level, its temperature will be

$$T' = T_0 - (\Gamma_d)\, z'$$

Therefore

$$T' - T = -(\Gamma_d - \Gamma)\, z'$$

Substituting this last expression into (3.73), we obtain

$$\frac{d^2z'}{dt^2} = -\frac{g}{T}(\Gamma_d - \Gamma)\, z'$$

which may be written in the form

$$\frac{d^2z'}{dt^2} + N^2 z' = 0 \tag{3.74}$$

where

$$N = \left[\frac{g}{T}(\Gamma_d - \Gamma)\right]^{1/2} \tag{3.75}$$

N is referred to as the *Brunt*[36]*–Väisälä*[37] *frequency*. Equation (3.74) is a second order ordinary differential equation. If the layer in question is stably stratified (that is to say, if $\Gamma_d > \Gamma$), then we can be assured that N is real, N^2 is positive, and the solution of (3.74) is

$$z' = A \cos Nt + B \sin Nt$$

Making use of the conditions at the point of maximum displacement at time $t = 0$, namely that $z' = z'(0)$ and $dz'/dt = 0$ at $t = 0$, it follows that

$$z'(t) = z'(0) \cos Nt$$

That is to say, the parcel executes a *buoyancy oscillation* about its equilibrium level z with amplitude equal to its initial displacement $z'(0)$, and frequency N (in units of radians per second). The Brunt–Väisälä frequency is thus a measure of the static stability: the higher the frequency, the greater the ambient stability. ■

Air parcels undergo buoyancy oscillations in association with *gravity waves*, a widespread phenomenon in planetary atmospheres, as illustrated in Fig. 3.14. Gravity waves may be excited by flow over

[36] **Sir David Brunt** (1886–1995) English meteorologist. First full-time professor of meteorology at Imperial College (1934–1952). His textbook *Physical and Dynamical Meteorology*, published in the 1930s, was one of the first modern unifying accounts of meteorology.

[37] **Vilho Väisälä** (1899–1969) Finnish meteorologist. Developed a number of meteorological instruments, including a version of the radiosonde in which readings of temperature, pressure, and moisture are telemetered in terms of radio frequencies. The modern counterpart of this instrument is one of Finland's successful exports.

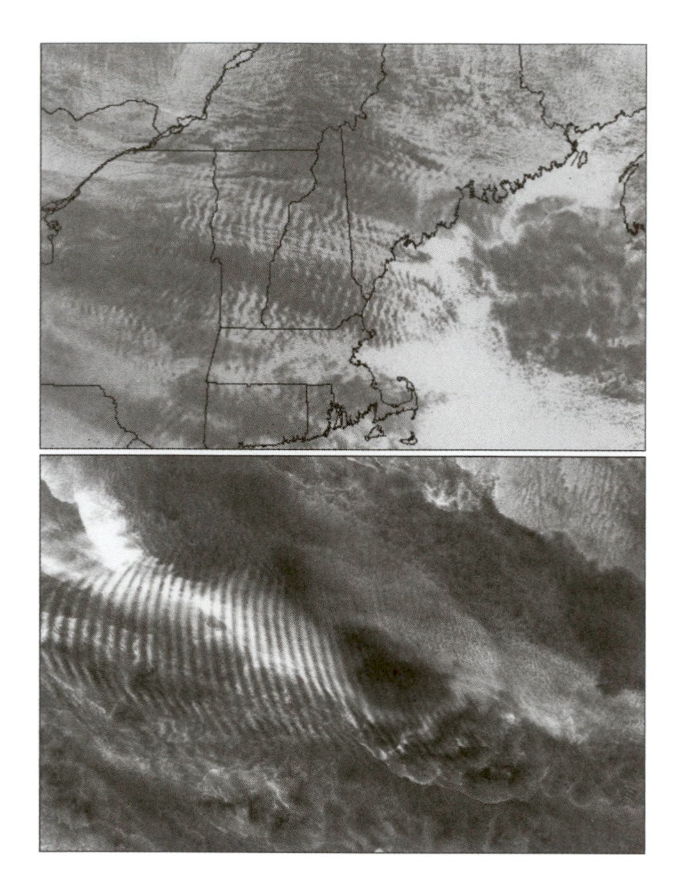

Fig. 3.14 Gravity waves, as revealed by cloud patterns. The upper photograph, based on NOAA GOES 8 visible satellite imagery, shows a wave pattern in west to east (right to left) airflow over the north–south-oriented mountain ranges of the Appalachians in the northeastern United States. The waves are transverse to the flow and their horizontal wavelength is ~20 km. The atmospheric wave pattern is more regular and widespread than the undulations in the terrain. The bottom photograph, based on imagery from NASA's multiangle imaging spectro-radiometer (MISR), shows an even more regular wave pattern in a thin layer of clouds over the Indian Ocean.

mountainous terrain, as shown in the top photograph in Fig. 3.14 or by an intense local disturbance, as shown in the bottom photograph. The following exercise illustrates how buoyancy oscillations can be excited by flow over a mountain range.

Exercise 3.13 A layer of unsaturated air flows over mountainous terrain in which the ridges are 10 km apart in the direction of the flow. The lapse rate is 5 °C km^{-1} and the temperature is 20 °C. For what value of the wind speed U will the period of the orographic (i.e., terrain-induced) forcing match the period of a buoyancy oscillation?

Solution: For the period τ of the orographic forcing to match the period of the buoyancy oscillation, it is required that

$$\tau = \frac{L}{U} = \frac{2\pi}{N}$$

where L is the spacing between the ridges. Hence, from this last expression and (3.75),

$$U = \frac{LN}{2\pi} = \frac{L}{2\pi}\left[\frac{g}{T}(\Gamma_d - \Gamma)\right]^{1/2}$$

or, in SI units,

$$U = \frac{10^4}{2\pi}\left[\frac{9.8}{293}\left((9.8 - 5.0) \times 10^{-3}\right)\right]^{1/2}$$

$$\simeq 20 \text{ m s}^{-1}$$

■

Layers of air with negative lapse rates (i.e., temperatures increasing with height) are called *inversions*. It is clear from the aforementioned discussion that these layers are marked by very strong static stability. A low-level inversion can act as a "lid" that traps pollution-laden air beneath it (Fig. 3.15). The layered structure of the stratosphere derives from the fact that it represents an inversion in the vertical temperature profile.

If $\Gamma > \Gamma_d$ (Fig. 3.12b), a parcel of unsaturated air displaced upward from O will arrive at A with a temperature greater than that of its environment. Therefore, it will be less dense than the ambient air

Fig. 3.15 Looking down onto widespread haze over southern Africa during the biomass-burning season. The haze is confined below a temperature inversion. Above the inversion, the air is remarkably clean and the visibility is excellent. (Photo: P. V. Hobbs.)

and, if left to itself, will continue to rise. Similarly, if the parcel is displaced downward it will be cooler than the ambient air, and it will continue to sink if left to itself. Such *unstable* situations generally do not persist in the free atmosphere, because the instability is eliminated by strong vertical mixing as fast as it forms. The only exception is in the layer just above the ground under conditions of very strong heating from below. ■

Exercise 3.14 Show that if the potential temperature θ increases with increasing altitude the atmosphere is stable with respect to the displacement of unsaturated air parcels.

Solution: Combining (3.1), (3.18), and (3.67), we obtain for a unit mass of air

$$c_p T \frac{d\theta}{\theta} = c_p \, dT + g dz$$

Letting $d\theta = (\partial\theta/\partial z)dz$ and $dT = (\partial T/\partial z)dz$ and dividing through by $c_p T dz$ yields

$$\frac{1}{\theta}\frac{\partial\theta}{\partial z} = \frac{1}{T}\left(\frac{\partial T}{\partial z} + \frac{g}{c_p}\right) \tag{3.76}$$

Noting that $-dT/dz$ is the actual lapse rate Γ of the air and the dry adiabatic lapse rate Γ_d is g/c_p (3.76) may be written as

$$\frac{1}{\theta}\frac{\partial\theta}{\partial z} = \frac{1}{T}(\Gamma_d - \Gamma) \tag{3.77}$$

However, it has been shown earlier that when $\Gamma < \Gamma_d$ the air is characterized by positive static stability. It follows that under these same conditions $\partial\theta/\partial z$ must be positive; that is, the potential temperature must increase with height. ■

3.6.2 Saturated Air

If a parcel of air is saturated, its temperature will decrease with height at the saturated adiabatic lapse rate Γ_s. It follows from arguments similar to those given in Section 3.6.1 that if Γ is the actual lapse rate of temperature in the atmosphere, saturated air parcels will be stable, neutral, or unstable with respect to vertical displacements, depending on whether $\Gamma < \Gamma_s$, $\Gamma = \Gamma_s$, or $\Gamma > \Gamma_s$, respectively. When an environmental temperature sounding is plotted on a skew $T - \ln p$ chart the distinctions between Γ, Γ_d, and Γ_s are clearly discernible (see Exercise 3.53).

3.6.3 Conditional and Convective Instability

If the actual lapse rate Γ of the atmosphere lies between the saturated adiabatic lapse rate Γ_s and the dry adiabatic lapse rate Γ_d, a parcel of air that is lifted sufficiently far above its equilibrium level will become warmer than the ambient air. This situation is illustrated in Fig. 3.16, where an air parcel lifted from its equilibrium level at O cools dry adiabatically until it reaches its lifting condensation level at A. At this level the air parcel is colder than the ambient air. Further lifting produces cooling at the moist adiabatic lapse rate so the temperature of the parcel of air follows the moist adiabat ABC. If the air parcel is sufficiently moist, the moist adiabat through A will cross the ambient temperature sounding; the point of intersection is shown as B in Fig. 3.16. Up to this point the parcel was colder and denser than the ambient air, and an expenditure of energy was required to lift it. If forced lifting had stopped prior to this point, the parcel would have returned to its equilibrium level at point O. However, once above point B, the parcel develops a positive buoyancy that carries it upward even in the absence of further forced lifting. For this reason, B is referred to as the *level of free convection* (LFC). The level of free convection depends on the amount of moisture in the rising parcel of air, as well as the magnitude of the lapse rate Γ.

From the aforementioned discussion it is clear that for a layer in which $\Gamma_s < \Gamma < \Gamma_d$, vigorous convective overturning will occur if forced vertical motions are

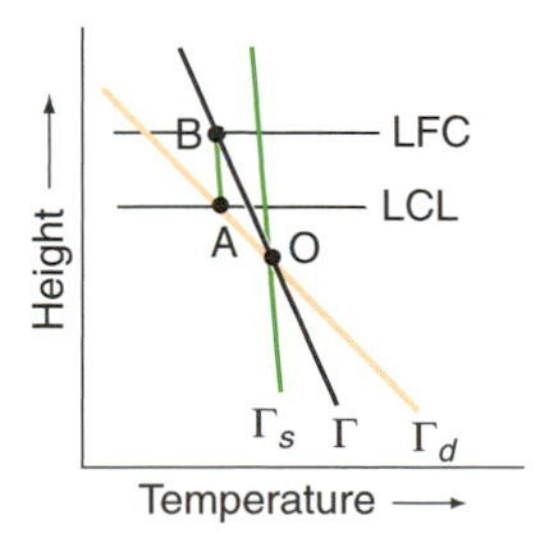

Fig. 3.16 Conditions for conditional instability ($\Gamma_s < \Gamma < \Gamma_d$). Γ_s and Γ_d are the saturated and dry adiabatic lapse rates, and Γ is the lapse rate of temperature of the ambient air. LCL and LFC denote the *lifting condensation level* and the *level of free convection*, respectively.

3.4 Analogs for Static Stability, Instability, Neutral Stability, and Conditional Instability

Sections 3.6.1 and 3.6.2 discussed the conditions for parcels of unsaturated air and saturated air to be stable, unstable, or neutral when displaced vertically in the atmosphere. Under stable conditions, if an air parcel is displaced either upward or downward and is then left to itself (i.e., the force causing the original displacement is removed), the parcel will return to its original position. An analogous situation is shown in Fig. 3.17a where a ball is originally located at the lowest point in a valley. If the ball is displaced in any direction and is then left to itself, it will return to its original location at the base of the valley.

Under unstable conditions in the atmosphere, an air parcel that is displaced either upward or downward, and then left to itself, will continue to move upward or downward, respectively. An analog is shown in Fig. 3.17b, where a ball is initially on top of a hill. If the ball is displaced in any direction, and is then left to itself, it will roll down the hill.

If an air parcel is displaced in a neutral atmosphere, and then left to itself, it will remain in the displaced location. An analog of this condition is a ball on a flat surface (Fig. 3.17c). If the ball is displaced, and then left to itself, it will not move.

If an air parcel is conditionally unstable, it can be lifted up to a certain height and, if left to itself, it will return to its original location. However, if the air parcel is lifted beyond a certain height (i.e., the level of free convection), and is then left to itself, it will continue rising (Section 3.6.3). An analog of this situation is shown in Fig. 3.17d, where a displacement of a ball to a point A, which lies to the left of the hillock, will result in the ball rolling back to its original position. However, if the displacement takes the ball to a point B on the other side of the hillock, the ball will not return to its original position but will roll down the right-hand side of the hillock.

It should be noted that in the analogs shown in Fig. 3.17 the only force acting on the ball after it is displaced is that due to gravity, which is always downward. In contrast, an air parcel is acted on by both a gravitational force and a buoyancy force. The gravitational force is always downward. The buoyancy force may be either upward or downward, depending on whether the air parcel is less dense or more dense than the ambient air.

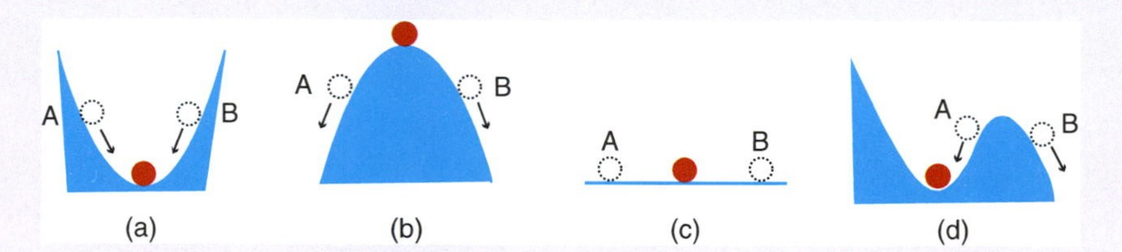

Fig. 3.17 Analogs for (a) stable, (b) unstable, (c) neutral, and (d) conditional instability. The red circle is the original position of the ball, and the white circles are displaced positions. Arrows indicate the direction the ball will move from a displaced position if the force that produced the displacement is removed.

large enough to lift air parcels beyond their level of free convection. Such an atmosphere is said to be *conditionally unstable* with respect to convection. If vertical motions are weak, this type of stratification can be maintained indefinitely.

The potential for instability of air parcels is also related to the vertical stratification of water vapor. In the profiles shown in Fig. 3.18, the dew point decreases rapidly with height within the inversion layer AB that marks the top of a moist layer. Now, suppose that this layer is lifted. An air parcel at A will reach its LCL quickly, and beyond that point it will cool moist adiabatically. In contrast, an air parcel starting at point B will cool dry adiabatically through a deep layer before it reaches its LCL. Therefore, as the inversion layer is lifted, the top part of it cools much more rapidly than the bottom part, and the lapse rate quickly becomes destabilized. Sufficient lifting may cause the layer to become conditionally unstable, even if the entire sounding is absolutely

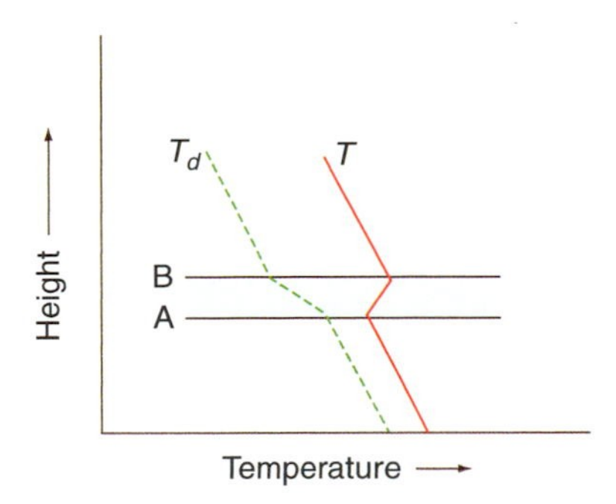

Fig. 3.18 Conditions for convective instability. T and T_d are the temperature and dew point of the air, respectively. The blue-shaded region is a dry inversion layer.

stable to begin with. It may be shown that the criterion for this so-called *convective (or potential) instability* is that $\partial\theta_e/\partial z$ be negative (i.e., θ_e decrease with increasing height) within the layer.

Throughout large areas of the tropics, θ_e decreases markedly with height from the mixed layer to the much drier air above. Yet deep convection breaks out only within a few percent of the area where there is sufficient lifting to release the instability.

3.7 The Second Law of Thermodynamics and Entropy

The first law of thermodynamics (Section 3.3) is a statement of the principle of conservation of energy. The *second law of thermodynamics*, which was deduced in various forms by Carnot,[38] Clausius,[39] and Lord Kelvin, is concerned with the maximum fraction of a quantity of heat that can be converted into work. The fact that for any given system there is a theoretical limit to this conversion was first clearly demonstrated by Carnot, who also introduced the important concepts of cyclic and reversible processes.

3.7.1 The Carnot Cycle

A *cyclic process* is a series of operations by which the state of a substance (called the *working substance*) changes but the substance is finally returned to its original state in all respects. If the volume of the working substance changes, the working substance may do external work, or work may be done on the working substance, during a cyclic process. Since the initial and final states of the working substance are the same in a cyclic process, and internal energy is a function of state, the internal energy of the working substance is unchanged in a cyclic process. Therefore, from (3.33), the *net* heat absorbed by the working substance is equal to the external work that it does in the cycle. A working substance is said to undergo a *reversible* transformation if each state of the system is in equilibrium so that a reversal in the direction of an infinitesimal change returns the working substance and the environment to their original states. A *heat engine* (or *engine* for short) is a device that does work through the agency of heat.

If during one cycle of an engine a quantity of heat Q_1 is absorbed and heat Q_2 is rejected, the amount of work done by the engine is $Q_1 - Q_2$ and its *efficiency* η is defined as

$$\eta = \frac{\text{Work done by the engine}}{\text{Heat absorbed by the working substance}} = \frac{Q_1 - Q_2}{Q_1} \tag{3.78}$$

Carnot was concerned with the important practical problem of the efficiency with which heat engines can do useful mechanical work. He envisaged an ideal heat engine (Fig. 3.19) consisting of a working substance contained in a cylinder (Y) with insulating walls and a conducting base (B) that is fitted with an insulated, frictionless piston (P) to which a variable force can be applied, a nonconducting stand (S) on which the cylinder may be placed to insulate its base, an infinite warm reservoir of heat (H) at constant temperature T_1, and an infinite cold reservoir for heat (C) at constant temperature T_2 (where $T_1 > T_2$). Heat can be supplied from the warm reservoir to the working substance contained in the cylinder, and heat can be extracted from the working substance by the cold reservoir. As the working substance expands (or contracts), the piston moves outward (or inward) and external work is done by (or on) the working substance.

[38] **Nicholas Leonard Sadi Carnot** (1796–1832) Born in Luxenbourg. Admitted to the École Polytechnique, Paris, at age 16. Became a captain in the Corps of Engineers. Founded the science of thermodynamics.

[39] **Rudolf Clausius** (1822–1888) German physicist. Contributed to the sciences of thermodynamics, optics, and electricity.

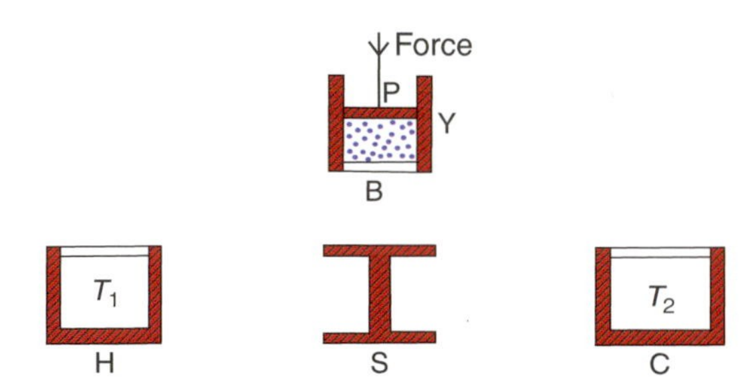

Fig. 3.19 The components of Carnot's ideal heat engine. Red-shaded areas indicate insulating material, and white areas represent thermally conducting material. The working substance is indicated by the blue dots inside the cylinder.

Carnot's cycle consists of taking the working substance in the cylinder through the following four operations that together constitute a reversible, cyclic transformation:

i. The substance starts with temperature T_2 at a condition represented by A on the p–V diagram in Fig. 3.20. The cylinder is placed on the stand S and the working substance is compressed by increasing the downward force applied to the piston. Because heat can neither enter nor leave the working substance in the cylinder when it is on the stand, the working substance undergoes an adiabatic compression to the state represented by B in Fig. 3.20 in which its temperature has risen to T_1.
ii. The cylinder is now placed on the warm reservoir H, from which it extracts a quantity of heat Q_1. During this process the working substance expands isothermally at temperature T_1 to point C in Fig. 3.20. During this process the working substance does work by expanding against the force applied to the piston.
iii. The cylinder is returned to the nonconducting stand and the working substance undergoes an adiabatic expansion along book web site in Fig. 3.20 until its temperature falls to T_2. Again the working substance does work against the force applied to the piston.
iv. Finally, the cylinder is placed on the cold reservoir and, by increasing the force applied to the piston, the working substance is compressed isothermally along DA back to its original state A. In this transformation the working substance gives up a quantity of heat Q_2 to the cold reservoir.

It follows from (3.36) that the net amount of work done by the working substance during the Carnot cycle is equal to the area contained within the figure ABCD in Fig. 3.20. Also, because the working substance is returned to its original state, the net work done is equal to $Q_1 - Q_2$ and the efficiency of the engine is given by (3.78). In this cyclic operation the engine has done work by transferring a certain quantity of heat from a warmer (H) to a cooler (C) body. One way of stating the second law of thermodynamics is "only by transferring heat from a warmer to a colder body can heat be converted into work in a cyclic process." In Exercise 3.56 we prove that *no engine can be more efficient than a reversible engine working between the same limits of temperature, and that all reversible engines working between the same temperature limits have the same efficiency*. The validity of these two statements, which are known as *Carnot's theorems*, depends on the truth of the second law of thermodynamics.

Exercise 3.15 Show that in a Carnot cycle the ratio of the heat Q_1 absorbed from the warm reservoir at temperature T_1 K to the heat Q_2 rejected to the cold reservoir at temperature T_2 K is equal to T_1/T_2.

Solution: To prove this important relationship we let the substance in the Carnot engine be 1 mol of an ideal gas and we take it through the Carnot cycle ABCD shown in Fig. 3.20.

For the adiabatic transformation of the ideal gas from A to B we have (using the adiabatic equation that the reader is invited to prove in Exercise 3.33)

$$p_A V_A^\gamma = p_B V_B^\gamma$$

where γ is the ratio of the specific heat at constant pressure to the specific heat at constant volume. For

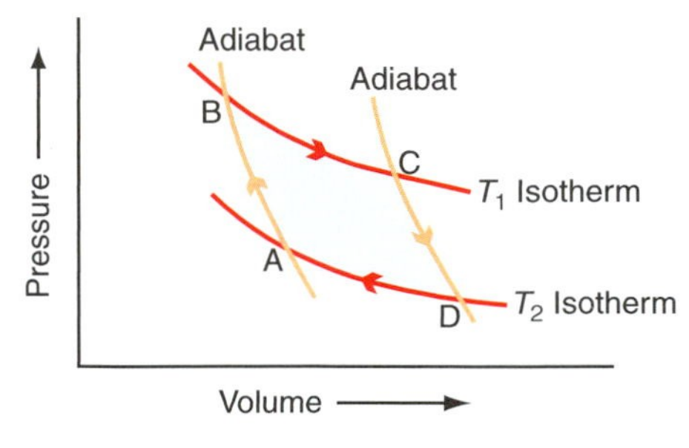

Fig. 3.20 Representations of a Carnot cycle on a p–V diagram. Red lines are isotherms, and orange lines are adiabats.

the isothermal transformation from B to C, we have from Boyle's law

$$p_B V_B = p_C V_C$$

The transformation from C to D is adiabatic. Therefore, from the adiabatic equation,

$$p_C V_C^\gamma = p_D V_D^\gamma$$

For the isothermal change from D to A

$$p_D V_D = p_A V_A$$

Combining the last four equations gives

$$\frac{V_C}{V_B} = \frac{V_D}{V_A} \tag{3.79}$$

Consider now the heats absorbed and rejected by the ideal gas. In passing from state B to C, heat Q_1 is absorbed from the warm reservoir. Since the internal energy of an ideal gas depends only on temperature, and the temperature of the gas does not change from B to C, it follows from (3.33) that the heat Q_1 given to the gas goes solely to do work. Therefore, from (3.36),

$$Q_1 = \int_{V_B}^{V_C} p\,dV$$

or, using (3.6) applied to 1 mol of an ideal gas,

$$Q_1 = \int_{V_B}^{V_C} \frac{R^* T_1}{V} dV = R^* T_1 \int_{V_B}^{V_C} \frac{dV}{V}$$

Therefore

$$Q_1 = R^* T_1 \ln\left(\frac{V_C}{V_B}\right) \tag{3.80}$$

Similarly, the heat Q_2 rejected to the cold reservoir in the isothermal transformation from D to A is given by

$$Q_2 = R^* T_2 \ln\left(\frac{V_D}{V_A}\right) \tag{3.81}$$

From (3.80) and (3.81)

$$\frac{Q_1}{Q_2} = \frac{T_1 \ln(V_C/V_B)}{T_2 \ln(V_D/V_A)} \tag{3.82}$$

Therefore, from (3.79) and (3.82),

$$\frac{Q_1}{Q_2} = \frac{T_1}{T_2} \tag{3.83}$$ ■

Examples of real heat engines are the steam engine and a nuclear power plant. The warm and cold reservoirs for a steam engine are the boiler and the condenser, respectively. The warm and cold reservoirs for a nuclear power plant are the nuclear reactor and the cooling tower, respectively. In both cases, water (in liquid and vapor forms) is the working substance that expands when it absorbs heat and thereby does work by pushing a piston or turning a turbine blade. Section 7.4.2 discusses how differential heating within the Earth's atmosphere maintains the winds against frictional dissipation through the action of a global heat engine.

Carnot's cycle can be reversed in the following way. Starting from point A in Fig. 3.20, the material in the cylinder may be expanded at constant temperature until the state represented by point D is reached. During this process a quantity of heat Q_2 is *taken from* the cold reservoir. An adiabatic expansion takes the substance from state D to C. The substance is then compressed from state C to state B, during which a quantity of heat Q_1 is *given up* to the warm reservoir. Finally, the substance is expanded adiabatically from state B to state A.

In this reverse cycle, Carnot's ideal engine serves as a *refrigerator* or *air conditioner*, for a quantity of heat Q_2 is taken from a cold body (the cold reservoir) and heat Q_1 ($Q_1 > Q_2$) is given to a hot body (the warm reservoir). To accomplish this transfer of heat, a quantity of mechanical work equivalent to $Q_1 - Q_2$ must be expended by some outside agency (e.g., an electric motor) to drive the refrigerator. This leads to another statement of the second law of thermodynamics, namely "heat cannot of itself (i.e., without the performance of work by some external agency) pass from a colder to a warmer body in a cyclic process."

3.7.2 Entropy

We have seen that isotherms are distinguished from each other by differences in temperature and that dry adiabats can be distinguished by their potential temperature. Here we describe another way of characterizing the differences between adiabats. Consider the three adiabats labeled by their potential temperatures θ_1, θ_2, and θ_3 on the p–V diagram shown in Fig. 3.21.

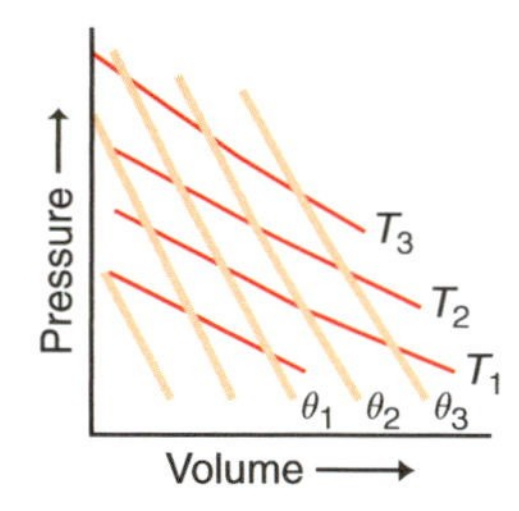

Fig. 3.21 Isotherms (red curves labeled by temperature T) and adiabats (tan curves labeled by potential temperature θ) on a p–V diagram.

In passing reversibly from one adiabat to another along an isotherm (e.g., in one operation of a Carnot cycle) heat is absorbed or rejected, where the amount of heat Q_{rev} (the subscript "rev" indicates that the heat is exchanged reversibly) depends on the temperature T of the isotherm. Moreover, it follows from (3.83) that the ratio Q_{rev}/T is the same no matter which isotherm is chosen in passing from one adiabat to another. Therefore, the ratio Q_{rev}/T could be used as a measure of the difference between the two adiabats; Q_{rev}/T is called the *difference in entropy (S)* between the two adiabats. More precisely, we may define the increase in the entropy dS of a system as

$$dS \equiv \frac{dQ_{rev}}{T} \tag{3.84}$$

where dQ_{rev} is the quantity of heat that is added reversibly to the system at temperature T. For a unit mass of the substance

$$ds \equiv \frac{dq_{rev}}{T} \tag{3.85}$$

Entropy is a function of the state of a system and not the path by which the system is brought to that state. We see from (3.38) and (3.85) that the first law of thermodynamics for a reversible transformation may be written as

$$Tds = du + pd\alpha \tag{3.86}$$

In this form the first law contains functions of state only.

When a system passes from state 1 to state 2, the change in entropy of a unit mass of the system is

$$s_2 - s_1 = \int_1^2 \frac{dq_{rev}}{T} \tag{3.87}$$

Combining (3.66) and (3.67) we obtain

$$\frac{dq}{T} = c_p \frac{d\theta}{\theta} \tag{3.88}$$

Therefore, because the processes leading to (3.66) and (3.67) are reversible, we have from (3.85) and (3.88)

$$ds = c_p \frac{d\theta}{\theta} \tag{3.89}$$

Integrating (3.89) we obtain the relationship between entropy and potential temperature

$$s = c_p \ln \theta + \text{constant} \tag{3.90}$$

Transformations in which entropy (and therefore potential temperature) is constant are called *isentropic*. Therefore, *adiabats* are often referred to as *isentropies* in atmospheric science. We see from (3.90) that the potential temperature can be used as a surrogate for entropy, as is generally done in atmospheric science.

Let us consider now the change in entropy in the Carnot cycle shown in Fig. 3.20. The transformations from A to B and from C to D are both adiabatic and reversible; therefore, in these two transformations there can be no changes in entropy. In passing from state B to state C, the working substance takes in a quantity of heat Q_1 reversibly from the source at temperature T_1; therefore, the entropy of the *source* decreases by an amount Q_1/T_1. In passing from state D to state A, a quantity of heat Q_2 is rejected reversibly from the working substance to the sink at temperature T_2; therefore, the entropy of the *sink* increases by Q_2/T_2. Since the working substance itself is taken in a cycle, and is therefore returned to its original state, it does not undergo any net change in entropy. Therefore, the net increase in entropy in the complete Carnot cycle is $Q_2/T_2 - Q_1/T_1$. However, we have shown in Exercise 3.15 that $Q_1/T_1 = Q_2/T_2$. Hence, there is no change in entropy in a Carnot cycle.

It is interesting to note that if, in a graph (called a *temperature–entropy* diagram[40]), temperature (in kelvin) is taken as the ordinate and entropy as the abscissa, the Carnot cycle assumes a rectangular shape, as shown in Fig. 3.22 where the letters A, B, C, and D correspond to the state points in the previous discussion. Adiabatic processes (AB and CD) are represented by vertical lines (i.e., lines of constant entropy) and isothermal processes (BC and DA) by horizontal lines. From (3.84) it is evident that in a cyclic transformation ABCDA, the heat Q_1 taken in reversibly by the working substance from the warm reservoir is given by the area XBCY, and the heat Q_2 rejected by the working substance to the cold reservoir is given by the area XADY. Therefore, the work $Q_1 - Q_2$ done in the cycle is given by the difference between the two areas, which is equivalent to the shaded area ABCD in Fig. 3.22. Any reversible heat engine can be represented by a closed loop on a temperature–entropy diagram, and the area of the loop is proportional to the net work done by or on (depending on whether the loop is traversed clockwise or counterclockwise, respectively) the engine in one cycle.

Thermodynamic charts on which equal areas represent equal net work done by or on the working substance are particularly useful. The skew $T - \ln p$ chart has this property.

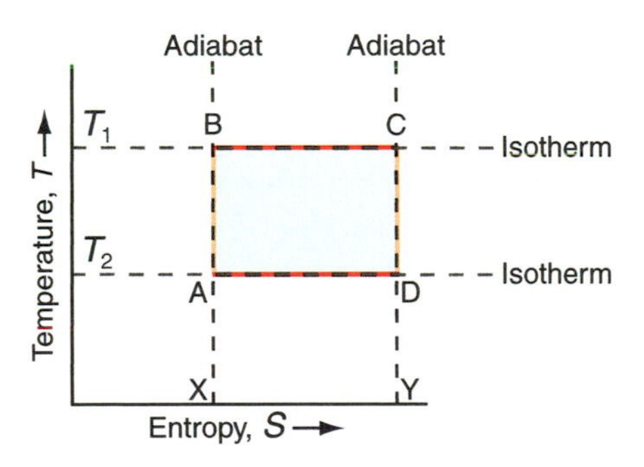

Fig. 3.22 Representation of the Carnot cycle on a temperature (T)-entropy (S) diagram. AB and CD are adiabats, and BC and DA are isotherms.

3.7.3 The Clausius–Clapeyron Equation

We will now utilize the Carnot cycle to derive an important relationship, known as the *Clausius–Clapeyron*[42] *equation* (sometimes referred to by physicists as the *first latent heat equation*). The Clausius–Clapeyron equation describes how the saturated vapor pressure above a liquid changes with temperature and also how the melting point of a solid changes with pressure.

Let the working substance in the cylinder of a Carnot ideal heat engine be a liquid in equilibrium with its saturated vapor and let the initial state of the substance be represented by point A in Fig. 3.23 in which the saturated vapor pressure is $e_s - de_s$ at temperature $T - dT$. The adiabatic compression from state A to state B, where the saturated vapor pressure is e_s at temperature T, is achieved by placing the cylinder on the nonconducting stand and compressing the piston infinitesimally (Fig. 3.24a). Now let the cylinder be placed on the source of heat at temperature T and let the substance expand isothermally until a unit mass of the liquid evaporates (Fig. 3.24b).

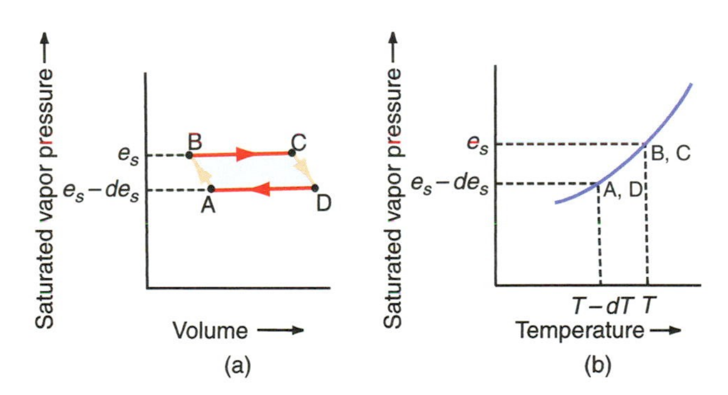

Fig. 3.23 Representation on (a) a saturated vapor pressure versus volume diagram and on (b) a saturated vapor pressure versus temperature diagram of the states of a mixture of a liquid and its saturated vapor taken through a Carnot cycle. Because the saturated vapor pressure is constant if temperature is constant, the isothermal transformations BC and DA are horizontal lines.

[40] The temperature–entropy diagram was introduced into meteorology by Shaw.[41] Because entropy is sometimes represented by the symbol ϕ (rather than S), the temperature–entropy diagram is sometimes referred to as a *tephigram*.

[41] **Sir (William) Napier Shaw** (1854–1945) English meteorologist. Lecturer in Experimental Physics, Cambridge University, 1877–1899. Director of the British Meteorological Office, 1905–1920. Professor of Meteorology, Imperial College, University of London, 1920–1924. Shaw did much to establish the scientific basis of meteorology. His interests ranged from the atmospheric general circulation and forecasting to air pollution.

[42] **Benoit Paul Emile Clapeyron** (1799–1864) French engineer and scientist. Carnot's theory of heat engines was virtually unknown until Clapeyron expressed it in analytical terms. This brought Carnot's ideas to the attention of William Thomson (Lord Kelvin) and Clausius, who utilized them in formulating the second law of thermodynamics.

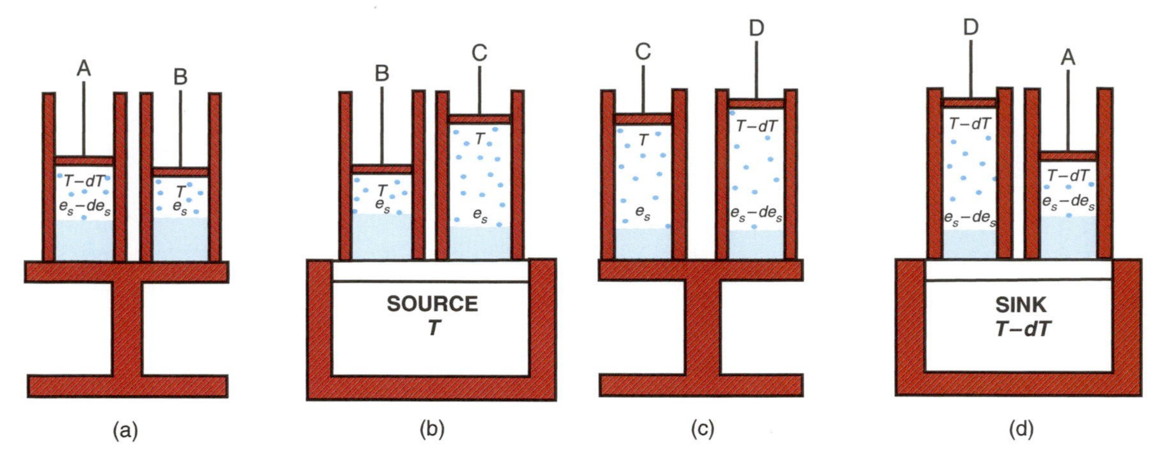

Fig. 3.24 Transformations of a liquid (solid blue) and its saturated vapor (blue dots) in a Carnot cycle. The letters A, B, C, D indicate the states of the mixture shown in Fig. 3.23. Red-shaded areas are thermally insulating materials.

In this transformation the pressure remains constant at e_s and the substance passes from state B to state C (Fig. 3.24b). If the specific volumes of liquid and vapor at temperature T are α_1 and α_2, respectively, the increase in the volume of the system in passing from state B to C is $(\alpha_2 - \alpha_1)$. Also the heat absorbed from the source is L_v where L_v is the latent heat of vaporization. The cylinder is now placed again on the nonconducting stand and a small adiabatic expansion is carried out from state C to state D in which the temperature falls from T to $T - dT$ and the pressure from e_s to $e_s - de_s$ (Fig. 3.24c). Finally, the cylinder is placed on the heat sink at temperature $T - dT$ and an isothermal and isobaric compression is carried out from state D to state A during which vapor is condensed (Fig. 3.24d). All of the aforementioned transformations are reversible.

From (3.83)

$$\frac{Q_1}{T_1} = \frac{Q_2}{T_2} = \frac{Q_1 - Q_2}{T_1 - T_2} \tag{3.91}$$

where $Q_1 - Q_2$ is the net heat absorbed by the working substance in the cylinder during one cycle, which is also equal to the work done by the working substance in the cycle. However, as shown in Section 3.3, the work done during a cycle is equal to the area of the enclosed loop on a p–V diagram. Therefore, from Fig. 3.23, $Q_1 - Q_2 = BC \times de_s = (\alpha_2 - \alpha_1)de_s$. Also, $Q_1 = L_v$, $T_1 = T$, and $T_1 - T_2 = dT$. Therefore, substituting into (3.91),

$$\frac{L_v}{T} = \frac{(\alpha_2 - \alpha_1)de_s}{dT}$$

or

$$\frac{de_s}{dT} = \frac{L_v}{T(\alpha_2 - \alpha_1)} \tag{3.92}$$

which is the *Clausius–Clapeyron equation* for the variation of the equilibrium vapor pressure e_s with temperature T.

Since the volume of a unit mass of vapor is very much greater than the volume of a unit mass of liquid ($\alpha_2 \gg \alpha_1$), Eq. (3.92) can be written to close approximation as

$$\frac{de_s}{dT} \simeq \frac{L_v}{T\alpha_2} \tag{3.93}$$

Because α_2 is the specific volume of water vapor that is in equilibrium with liquid water at temperature T, the pressure it exerts at T is e_s. Therefore, from the ideal gas equation for water vapor,

$$e_s\,\alpha_2 = R_vT \tag{3.94}$$

combining (3.93) and (3.94), and then substituting $R_v = 1000\,R^*/M_w$ from (3.13), we get

$$\frac{1}{e_s}\frac{de_s}{dT} \simeq \frac{L_v}{R_vT^2} = \frac{L_vM_w}{1000\,R^*T^2} \tag{3.95}$$

which is a convenient form of the Clausius–Clapeyron equation. Over the relatively small range of temperatures of interest in the atmosphere, to good approximation (3.95) can be applied in incremental form, that is

$$\frac{1}{e_s}\frac{\Delta e_s}{\Delta T} \simeq \frac{L_vM_w}{1000\,R^*T^2} \tag{3.96}$$

Applying (3.95) to the water substance, and integrating from 273 K to T K,

$$\int_{e_s(273\text{ K})}^{e_s(T\text{ K})} \frac{de_s}{e_s} = \frac{L_v M_w}{1000\, R^*} \int_{273}^{T} \frac{dT}{T^2}$$

Alternatively, because e_s at 273 K = 6.11 hPa (Fig. 3.9), $L_v = 2.500 \times 10^6$ J kg^{-1}, the molecular weight of water (M_w) is 18.016, and $R^* = 8.3145$ J K^{-1} mol^{-1}, the saturated vapor pressure of water e_s (in hPa) at temperature T K is given by

$$\ln \frac{e_s\,(\text{in hPa})}{6.11} = \frac{L_v M_w}{1000\, R^*}\left(\frac{1}{273} - \frac{1}{T}\right) \simeq 5.42 \times 10^3 \left(\frac{1}{273} - \frac{1}{T}\right) \tag{3.97}$$

3.5 Effect of Ambient Pressure on the Boiling Point of a Liquid

A liquid is said to boil when it is heated to a temperature that is sufficient to produce copious small bubbles within the liquid. Why do bubbles form at a certain temperature (the *boiling point*) for each liquid? The key to the answer to this question is to realize that if a bubble forms in a liquid, the interior of the bubble contains only the vapor of the liquid. Therefore, the pressure inside the bubble is the saturation vapor pressure at the temperature of the liquid. If the saturation vapor pressure is less than the ambient pressure that acts on the liquid (and therefore on a bubble just below the surface of a liquid), bubbles cannot form. As the temperature increases, the saturation vapor pressure increases (see Fig. 3.9) and, when the saturation vapor pressure is equal to the ambient pressure, bubbles can form at the surface of the liquid and the liquid boils (Fig. 3.25).

Water boils at a temperature T_B such that the saturation vapor pressure at T_B is equal to the atmospheric (or ambient) pressure (p_{atmos})[43]

$$e_s(T_B) = p_{\text{atmos}} \tag{3.98}$$

From (3.92) expressed in incremental form, and (3.98)

$$\frac{\Delta p_{\text{atmos}}}{\Delta T_B} = \frac{L_v}{T_B(\alpha_2 - \alpha_1)}$$

or

$$\frac{\Delta T_B}{\Delta p_{\text{atmos}}} = \frac{T_B(\alpha_2 - \alpha_1)}{L_v} \tag{3.99}$$

Equation (3.99) gives the change in the boiling point of water with atmospheric pressure (or ambient pressure in general). Because $\alpha_2 > \alpha_1$, T_B increases with increasing p_{atmos}. If the atmospheric pressure is significantly lower than 1 atm, the boiling point of water will be significantly lower than 100 °C. This is why it is difficult to brew a good cup of hot tea on top of a high mountain (see Exercise 3.64)!

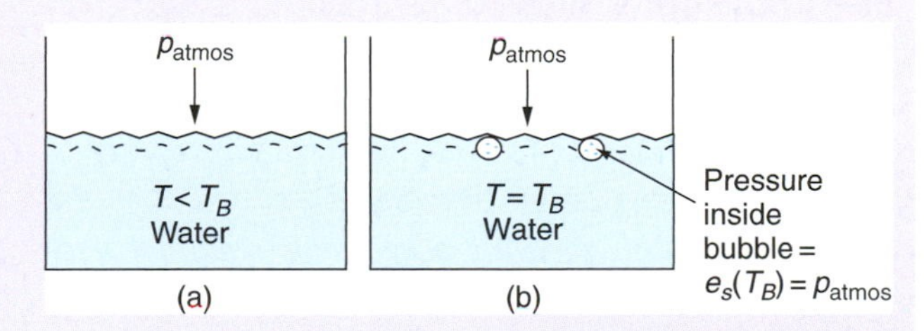

Fig. 3.25 (a) Water below its boiling point (T_B): bubbles cannot form because $e_s(T) < p_{\text{atmos}}$. (b) Water at its boiling point: bubbles can form because the pressure inside them, $e_s(T_B)$, is equal to the atmospheric pressure (p_{atmos}) acting on them.

[43] If, as is generally the case, the water is in a vessel that is heated from below, the pressure where the bubbles originate is slightly greater than atmospheric pressure due to the extra pressure exerted by the water above the bubble. Therefore, when the water is boiling steadily, the temperature at the bottom of the vessel will be slightly in excess of T_B. When water is heated in a transparent vessel, the first visible sign of bubbling occurs well below T_B as trains of small bubbles of *dissolved air* rise to the surface. (Note: the solubility of a gas in a liquid decreases with increasing temperature.) The "singing" that precedes boiling is due to the collapse of bubbles of water vapor in the upper part of the vessel. Those vapor bubbles probably form around air bubbles that act as nuclei, which originate in the slightly hotter water nearer the source of the heat. Nuclei of some sort appear to be necessary for continuous steady boiling at T_B. Without nuclei the water will not begin to boil until it is superheated with respect to the boiling point and "bumping" (i.e., delayed boiling) occurs. When bubbles finally form, the vapor pressure in the bubbles is much greater than the ambient pressure, and the bubbles expand explosively as they rise. Chapter 6 discusses the formation of water drops from the vapor phase, and ice particles from the vapor and liquid phases both of which require nucleation.

3.7.4 Generalized Statement of the Second Law of Thermodynamics

So far we have discussed the second law of thermodynamics and entropy in a fairly informal manner, and only with respect to ideal reversible transformations. The second law of thermodynamics states (in part) that *for a reversible transformation there is no change in the entropy of the universe* (where "universe" refers to a system and its surroundings). In other words, if a system receives heat reversibly, the increase in its entropy is exactly equal in magnitude to the decrease in the entropy of its surroundings.

The concept of reversibility is an abstraction. A reversible transformation moves a system through a series of equilibrium states so that the direction of the transformation can be reversed at any point by making an infinitesimal change in the surroundings. All natural transformations are irreversible to some extent. In an *irreversible* (sometimes called a *spontaneous*) *transformation*, a system undergoes finite transformations at finite rates, and these transformations cannot be reversed simply by changing the surroundings of the system by infinitesimal amounts. Examples of irreversible transformations are the flow of heat from a warmer to a colder body, and the mixing of two gases.

If a system receives heat dq_{irrev} at temperature T during an irreversible transformation, the change in the entropy of the system is *not* equal to dq_{irrev}/T. In fact, for an irreversible transformation there is no simple relationship between the change in the entropy of the system and the change in the entropy of its surroundings. However, the remaining part of the second law of thermodynamics states that *the entropy of the universe increases as a result of irreversible transformations*.

The two parts of the second law of thermodynamics stated earlier can be summarized as follows

$$\Delta S_{\text{universe}} = \Delta S_{\text{system}} + \Delta S_{\text{surroundings}} \tag{3.100a}$$

$$\Delta S_{\text{universe}} = 0 \text{ for reversible (equilibrium) transformations} \tag{3.100b}$$

$$\Delta S_{\text{universe}} > 0 \text{ for irreversible (spontaneous) transformations} \tag{3.100c}$$

The second law of thermodynamics cannot be proved. It is believed to be valid because it leads to deductions that are in accord with observations and experience. The following exercise provides an example of such a deduction.

Exercise 3.16 Assuming the truth of the second law of thermodynamics, prove that an isolated ideal gas can expand spontaneously (e.g., into a vacuum) but it cannot contract spontaneously.

Solution: Consider a unit mass of the gas. If the gas is isolated it has no contact with its surroundings, hence $\Delta S_{\text{surroundings}} = 0$. Therefore, from (3.100a)

$$\Delta S_{\text{universe}} = \Delta S_{\text{gas}} \tag{3.101}$$

Because entropy is a function of state, we can obtain an expression for ΔS_{gas} by taking any reversible and isothermal path from state 1 to state 2 and evaluating the integral

$$\Delta S_{\text{gas}} = \int_1^2 \frac{dq_{\text{rev}}}{T}$$

Combining (3.46) with (3.3), we have for a reversible transformation of a unit mass of an ideal gas

$$\frac{dq_{\text{rev}}}{T} = c_p \frac{dT}{T} - R\frac{dp}{p}$$

Therefore

$$\Delta S_{\text{gas}} = c_p \int_{T_1}^{T_2} \frac{dT}{T} - R \int_{p_1}^{p_2} \frac{dp}{p}$$

or

$$\Delta S_{\text{gas}} = c_p \ln \frac{T_2}{T_1} - R \ln \frac{p_2}{p_1}$$

Because the gas is isolated, $\Delta q = \Delta w = 0$; therefore, from (3.34), $\Delta u = 0$. If $\Delta u = 0$, it follows from Joule's law for an ideal gas that $\Delta T = 0$. Hence, the gas must pass from its initial state (1) to its final state (2) isothermally.

For an isothermal process, the ideal gas equation reduces to Boyle's law, which can be written as $p_1\alpha_1 = p_2\alpha_2$, where the α's are specific volumes. Therefore, the last expression becomes

$$\begin{aligned} \Delta S_{\text{gas}} &= c_p \ln 1 - R \ln \frac{p_2}{p_1} \\ &= -R \ln \frac{\alpha_1}{\alpha_2} = R \ln \frac{\alpha_2}{\alpha_1} \end{aligned} \tag{3.102}$$

From (3.101) and (3.102)

$$\Delta S_{\text{universe}} = R \ln \frac{\alpha_2}{\alpha_1} \tag{3.103}$$

Hence, if the second law of thermodynamics is valid, it follows from (3.100c) and (3.103) that

$$R \ln \frac{\alpha_2}{\alpha_1} > 0$$

or

$$\alpha_2 > \alpha_1$$

That is, the gas expands spontaneously. However, if the gas contracted spontaneously, $\alpha_2 < \alpha_1$ and $\Delta S_{\text{universe}} < 0$, which would violate the second law of thermodynamics.

When a gas expands, the disorder of its molecules increases and, as shown in this exercise, the entropy of the gas increases. This illustrates what is, in fact, a general result, namely that entropy is a measure of the degree of disorder (or randomness) of a system. ■

Section 3.7.2 showed that there is no change in entropy in a Carnot cycle. Because *any* reversible cycle can be divided up into an infinite number of adiabatic and isothermal transformations, and therefore into an infinite number of Carnot cycles, it follows that in any reversible cycle the total change in entropy is zero. This result is yet another way of stating the second law of thermodynamics.

In the real world (as opposed to the world of reversible cycles), systems left to themselves tend to become more disordered with time, and therefore their entropy increases. Consequently, a parallel way of stating the two laws of thermodynamics is (1) "the energy of the universe is constant" and (2) "the entropy of the universe tends to a maximum."

Exercise 3.17 One kilogram of ice at 0 °C is placed in an isolated container with 1 kg of water at 10 °C and 1 atm. (a) How much of the ice melts? (b) What change is there in the entropy of the universe due to the melting of the ice?

Solution: (a) The ice will melt until the ice-water system reaches a temperature of 0 °C. Let mass m kg of ice melt to bring the temperature of the ice-water system to 0 °C. Then, the latent heat required to melt m kg of ice is equal to the heat released when the temperature of 1 kg of water decreases from 10 to 0 °C. Therefore,

$$mL_M = c\Delta T$$

where L_M is the latent heat of melting of ice (3.34×10^5 J kg^{-1}), c is the specific heat of water (4218 J K^{-1} kg^{-1}), and ΔT is 10 K. Hence, the mass of ice that melts (m) is 0.126 kg. (Note: Because $m < 1$ kg, it follows that when the system reaches thermal equilibrium some ice remains in the water, and therefore the final temperature of the ice-water system must be 0 °C.)

(b) Because the container is isolated, there is no change in the entropy of its surroundings. Therefore, (3.100a) becomes

$$\Delta S_{\text{universe}} = \Delta S_{\text{system}}$$

Because the ice-water system undergoes an irreversible transformation, it follows from (3.100c) that its entropy increases. (We could also have deduced that the entropy of the ice-water system increases when some of the ice melts, because melting increases the disorder of the system.)

There are two contributions to ΔS_{system}: the melting of 0.126 kg of ice (ΔS_{ice}) and the cooling of 1 kg of water from 10 to 0 °C (ΔS_{water}). The change in entropy when 0.126 kg of ice is melted at 0 °C is $\Delta S_{\text{ice}} = \Delta Q/T = mL_M/T = (0.126)(3.34 \times 10^5)/273 = 154$ J K^{-1}. The change in entropy associated with cooling the 1 kg of water from 10 to 0 °C is

$$\Delta S_{\text{water}} = \int_{283\,K}^{273\,K} \frac{dQ}{T} = \int_{283\,K}^{273\,K} \frac{c\,dT}{T}$$
$$= c\int_{283\,K}^{273\,K} \frac{dT}{T} = c \ln \frac{273}{283}$$

Because $c = 4218$ J K^{-1} kg^{-1}

$$\Delta S_{\text{water}} = 4218 \ln \frac{273}{283}$$
$$= 4218\,(-0.036)$$
$$= -152 \text{ J K}^{-1}.$$

Hence

$$\Delta S_{\text{universe}} = \Delta S_{\text{system}} = \Delta S_{\text{ice}} + \Delta S_{\text{water}}$$
$$= 154 - 152$$
$$= 2 \text{ J K}^{-1}$$

Exercises

3.18 Answer or explain the following in light of the principles discussed in this chapter.

(a) To carry a given payload, a hot air balloon cruising at a high altitude needs to be bigger or hotter than a balloon cruising at a lower altitude.

(b) More fuel is required to lift a hot air balloon through an inversion than to lift it through a layer of the same depth that exhibits a steep temperature lapse rate. Other conditions being the same, more fuel is required to operate a hot air balloon on a hot day than on a cold day.

(c) Runways are longer at high altitude airports such as Denver and stricter weight limits are imposed on aircraft taking off on hot summer days.

(d) The gas constant for moist air is greater than that for dry air.

(e) Pressure in the atmosphere increases approximately exponentially with depth, whereas the pressure in the ocean increases approximately linearly with depth.

(f) Describe a procedure for converting station pressure to sea-level pressure.

(g) Under what condition(s) does the hypsometric equation predict an exponential decrease of pressure with height?

(h) If a low pressure system is colder than its surroundings, the amplitude of the depression in the geopotential height field increases with height.

(i) On some occasions low surface temperatures are recorded when the 1000- to 500-hPa thickness is well above normal. Explain this apparent paradox.

(j) Air released from a tire is cooler than its surroundings.

(k) Under what conditions can an ideal gas undergo a change of state without doing external work?

(l) A parcel of air cools when it is lifted. Dry parcels cool more rapidly than moist parcels.

(m) If a layer of the atmosphere is well mixed in the vertical, how would you expect the potential temperature within it to change with height?

(n) In cold climates the air indoors tends to be extremely dry.

(o) Summertime dew points tend to be higher over eastern Asia and the eastern United States than over Europe and the western United States.

(p) If someone claims to have experienced hot, humid weather with a temperature in excess of 90 °F and a relative humidity of 90%, it is likely that he/she is exaggerating or inadvertently juxtaposing an afternoon temperature with an early morning relative humidity.

(q) Hot weather causes more human discomfort when the air is humid than when it is dry.

(r) Which of the following pairs of quantities are conserved when unsaturated air is lifted: potential temperature and mixing ratio, potential temperature and saturation mixing ratio, equivalent potential temperature and saturation mixing ratio?

(s) Which of the following quantities are conserved during the lifting of saturated air: potential temperature, equivalent potential temperature, mixing ratio, saturation mixing ratio?

(t) The frost point temperature is higher than the dew point temperature.

(u) You are climbing in the mountains and come across a very cold spring of water. If you had a glass tumbler and a thermometer, how might you determine the dew point of the air?

(v) Leaving the door of a refrigerator open warms the kitchen. (How would the refrigerator need to be reconfigured to make it have the reverse effect?)

(w) A liquid boils when its saturation vapor pressure is equal to the atmospheric pressure.

3.19 Determine the apparent molecular weight of the Venusian atmosphere, assuming that it consists of 95% of CO_2 and 5% N_2 by volume. What is the gas constant for 1 kg of such an atmosphere? (Atomic weights of C, O, and N are 12, 16, and 14, respectively.)

3.20 If water vapor comprises 1% of the volume of the air (i.e., if it accounts for 1% of the molecules in air), what is the virtual temperature correction?

3.21 Archimedes' buoyancy principle asserts that an object placed in a fluid (liquid or gas) will be lighter by an amount equal to the weight of the fluid it displaces. Provide a proof of this principle. [**Hint**: Consider the vertical forces that act on a stationary element of fluid prior to the element being displaced by an object.]

3.22 Typical hot air balloons used on sightseeing flights attain volumes of 3000 m^3. A typical gross weight (balloon, basket, fuel and passengers, but not the air in the balloon) on such a balloon flight is 600 kg. If the ground temperature is 20 °C, the lapse rate is zero, and the balloon is in hydrostatic equilibrium at a cruising altitude of 900 hPa, determine the temperature of the air inside the balloon.

3.23 The gross weight (balloon, basket, fuel and passengers but not the gas in the balloon) of two balloons is the same. The two balloons are cruising together at the same altitude, where the temperature is 0 °C and the ambient air is dry. One balloon is filled with helium and the other balloon with hot air. The volume of the helium balloon is 1000 m^3. If the temperature of the hot air balloon is 90 °C, what is the volume of the hot air balloon?

3.24 Using Eq. (3.29) show that pressure decreases with increasing height at about 1 hPa per 15 m at the 500-hPa level.

3.25 A cheap aneroid barometer aboard a radiosonde is calibrated to the correct surface air pressure when the balloon leaves the ground, but it experiences a systematic drift toward erroneously low pressure readings. By the time the radiosonde reaches the 500-hPa level, the reading is low by the 5-hPa level (i.e., it reads 495 hPa when it should read 500 hPa). Estimate the resulting error in the 500-hPa height. Assume a surface temperature of 10 °C and an average temperature lapse rate of 7 °C km^{-1}. Assume the radiosonde is released from sea level and that the error in the pressure reading is proportional to the height of the radiosonde above sea level (which, from Eq. (3.29), makes it nearly proportional to $\ln p$). Also, assume that the average decrease of pressure with height is 1 hPa per 11 m of rise between sea level and 500 hPa.

3.26 A hurricane with a central pressure of 940 hPa is surrounded by a region with a pressure of 1010 hPa. The storm is located over an ocean region. At 200 hPa the depression in the pressure field vanishes (i.e., the 200-hPa surface is perfectly flat). Estimate the average temperature difference between the center of the hurricane and its surroundings in the layer between the surface and 200 hPa. Assume that the mean temperature of this layer outside the hurricane is −3 °C and ignore the virtual temperature correction.

3.27 A meteorological station is located 50 m below sea level. If the surface pressure at this station is 1020 hPa, the virtual temperature at the surface is 15 °C, and the mean virtual temperature for the 1000- to 500-hPa layer is 0 °C, compute the height of the 500-hPa pressure level above sea level at this station.

3.28 The 1000- to 500-hPa layer is subjected to a heat source having a magnitude of 5.0×10^6 J m^{-2}. Assuming that the atmosphere is at rest (apart from the slight vertical motions associated with the expansion of the layer) calculate the resulting increase in the mean temperature and in the thickness of the layer. [**Hint**: Remember that pressure is force per unit area.]

3.29 The 1000- to 500-hPa thickness is predicted to increase from 5280 to 5460 m at a given station. Assuming that the lapse rate remains constant, what change in surface temperature would you predict?

3.30 Derive a relationship for the height of a given pressure surface (p) in terms of the pressure p_0 and temperature T_0 at sea level assuming that the temperature decreases uniformly with height at a rate Γ K km^{-1}.

Solution: Let the height of the pressure surface be z; then its temperature T is given by

$$T = T_0 - \Gamma_z \qquad (3.104)$$

combining the hydrostatic equation (3.17) with the ideal gas equation (3.2) yields

$$\frac{dp}{p} = -\frac{g}{RT}dz \quad (3.105)$$

From (3.104) and (3.105)

$$\frac{dp}{p} = -\frac{g}{R(T_0 - \Gamma_z)}dz$$

Integrating this equation beween pressure levels p_0 and p and corresponding heights 0 and z and neglecting the variation of g with z, we obtain

$$\int_{p_0}^{p} \frac{dp}{p} = -\frac{g}{R}\int_0^z \frac{dz}{(T_0 - \Gamma_z)}$$

or

$$\ln\frac{p}{p_0} = \frac{g}{R\Gamma}\ln\left(\frac{T_0 - \Gamma_z}{T_0}\right)$$

Therefore,

$$z = \frac{T_0}{\Gamma}\left[1 - \left(\frac{p}{p_0}\right)^{R\Gamma/g}\right] \quad (3.106)$$

This equation forms the basis for the calibration of aircraft altimeters. An altimeter is simply an aneroid barometer that measure ambient air pressure p. However, the scale of the altimeter is expressed at the height z of the aircraft, where z is related to p by (3.106) with values of T_0, p_0 and Γ appropriate to the U.S. Standard Atmosphere, namely, $T_0 = 288$ K, $p_0 = 1013.25$ hPa, and $\Gamma = 6.50$ K km^{-1}. ■

3.31 A hiker sets his pressure altimeter to the correct reading at the beginning of a hike during which he climbs from near sea level to an altitude of 1 km in 3 h. During this same time interval the sea-level pressure drops by 8 hPa due to the approach of a storm. Estimate the altimeter reading at the end of the hike.

3.32 Calculate the work done in compressing isothermally 2 kg of dry air to one-tenth of its volume at 15 °C.

3.33 (a) Prove that when an ideal gas undergoes an adiabatic transformation pV^{γ} = constant, where γ is the ratio of the specific heat at constant pressure (c_p) to the specific heat at constant volume (c_v). [**Hint**: By combining (3.3) and (3.41) show that for an adiabatic transformation of a unit mass of gas $c_v(pd\alpha + \alpha dp) + Rpd\alpha = 0$. Then combine this last expression with (3.45) and proceed to answer.] (b) 7.50 cm^3 of air at 17 °C and 1000 hPa is compressed isothermally to 2.50 cm^3. The air is then allowed to expand adiabatically to its original volume. Calculate the final temperature and final pressure of the gas.

3.34 If the balloon in Exercise 3.22 is filled with air at the ambient temperature of 20 °C at ground level where the pressure is 1013 hPa, estimate how much fuel will need to be burned to lift the balloon to its cruising altitude of 900 hPa. Assume that the balloon is perfectly insulated and that the fuel releases energy at a rate of 5×10^7 J kg^{-1}.

3.35 Calculate the change in enthalpy when 3 kg of ice at 0 °C is heated to liquid water at 40 °C. [The specific heat at constant pressure of liquid water (in J K^{-1} kg^{-1}) at T K is given by $c_{pw} = 4183.9 + 0.1250\ T$.]

3.36 Prove that the potential temperature of an air parcel does not change when the parcel moves around under adiabatic and reversible conditions in the atmosphere. [**Hint**: Use Eq. (3.1) and the adiabatic equation pV^{γ} = constant (see Exercise 3.33) to show that $T\,(p_0/p)^{R/c_p}$ = constant, and hence from Eq. (3.54) that θ = constant.]

3.37 The pressure and temperature at the levels at which jet aircraft normally cruise are typically 200 hPa and −60 °C. Use a skew $T - \ln p$ chart to estimate the temperature of this air if it were compressed adiabatically to 1000 hPa. Compare your answer with an accurate computation.

3.38 Consider a parcel of dry air moving with the speed of sound (c_s), where

$$c_s = (\gamma R_d T)^{\frac{1}{2}}$$

$\gamma = c_p/c_v = 1.40$, R_d is the gas constant for a unit mass of dry air, and T is the temperature of the air in degrees kelvin.

(a) Derive a relationship between the macroscopic kinetic energy of the air parcel K_m and its enthalpy H.
(b) Derive an expression for the fractional change in the speed of sound per degree kelvin change in temperature in terms of c_v, R_d, and T.

3.39 A person perspires. How much liquid water (as a percentage of the mass of the person) must evaporate to lower the temperature of the person by 5 °C? (Assume that the latent heat of evaporation of water is 2.5×10^6 J kg^{-1}, and the specific heat of the human body is 4.2×10^3 J K^{-1} kg^{-1}.)

3.40 Twenty liters of air at 20 °C and a relative humidity of 60% are compressed isothermally to a volume of 4 liters. Calculate the mass of water condensed. The saturation vapor pressure of water at 20 °C is 23 hPa. (Density of air at 0 °C and 1000 hPa is 1.28 kg m^{-3}.)

3.41 If the specific humidity of a sample of air is 0.0196 at 30 °C, find its virtual temperature. If the total pressure of the moist air is 1014 hPa, what is its density?

3.42 A parcel of moist air has a total pressure of 975 hPa and a temperature of 15 °C. If the mixing ratio is 1.80 g kg^{-1}, what are the water vapor pressure and the virtual temperature?

3.43 An isolated raindrop that is evaporating into air at a temperature of 18 °C has a temperature of 12 °C. Calculate the mixing ratio of the air. (Saturation mixing ratio of air at 12 °C is 8.7 g kg^{-1}. Take the latent heat of evaporation of water to be 2.25×10^6 J kg^{-1}.)

3.44 Four grams of liquid water condense out of 1 kg of air during a moist-adiabatic expansion. Show that the internal energy associated with this amount of liquid water is only 2.4% of the internal energy of the air.

3.45 The current mean air temperature at 1000 hPa in the tropics is about 25 °C and the lapse rate is close to saturated adiabatic. Assuming that the lapse rate remains close to saturated adiabatic, by how much would the temperature change at 250 hPa if the temperature in the tropics at 1000 hPa were to increase by 1 °C. [**Hint**: Use a skew $T - \ln p$ chart.]

3.46 An air parcel at 1000 hPa has an initial temperature of 15 °C and a dew point of 4 °C. Using a skew $T - \ln p$ chart,

(a) Find the mixing ratio, relative humidity, wet-bulb temperature, potential temperature, and wet-bulb potential temperature of the air.

(b) Determine the magnitudes of the parameters in (a) if the parcel rises to 900 hPa.

(c) Determine the magnitudes of the parameters in (a) if the parcel rises to 800 hPa.

(d) Where is the lifting condensation level?

3.47 Air at 1000 hPa and 25 °C has a wet-bulb temperature of 20 °C.

(a) Find the dew point.

(b) If this air were expanded until all the moisture condensed and fell out and it were then compressed to 1000 hPa, what would be the resulting temperature?

(c) What is this temperature called?

3.48 Air at a temperature of 20 °C and a mixing ratio of 10 g kg^{-1} is lifted from 1000 to 700 hPa by moving over a mountain. What is the initial dew point of the air? Determine the temperature of the air after it has descended to 900 hPa on the other side of the mountain if 80% of the condensed water vapor is removed by precipitation during the ascent. (**Hint**: Use the skew $T - \ln p$ chart.)

3.49 (a) Show that when a parcel of dry air at temperature T' moves adiabatically in ambient air with temperature T, the temperature lapse rate following the parcel is given by

$$-\frac{dT'}{dz} = \frac{T'}{T}\frac{g}{c_p}$$

(b) Explain why the lapse rate of the air parcel in this case differs from the dry adiabatic lapse rate (g/c_p). [**Hint**: Start with Eq. (3.54) with $T = T'$. Take the natural logarithm of both sides of this equation and then differentiate with respect to height z.]

Solution:

(a) From (3.54) with $T = T'$ we have for the air parcel

$$\theta = T'\left(\frac{p_0}{p}\right)^{R/c_p}$$

Therefore,

$$\ln \theta = \ln T' + \frac{R}{c_p}(\ln p_0 - \ln p)$$

Differentiating this last expression with respect to height z

$$\frac{1}{\theta}\frac{d\theta}{dz} = \frac{1}{T'}\frac{dT'}{dz} - \frac{R}{c_p}\frac{1}{p}\frac{dp}{dz} \qquad (3.110)^{44}$$

However, for the ambient air we have, from the hydrostatic equation,

$$\frac{dp}{dz} = -g\rho \qquad (3.111)$$

From (3.110) and (3.111):

$$\frac{1}{\theta}\frac{d\theta}{dz} = \frac{1}{T'}\frac{dT'}{dz} - \frac{R}{c_p}\frac{1}{p}(-g\rho)$$

For an adiabatic process θ is conserved (i.e., $\frac{d\theta}{dz} = 0$. Therefore,

$$0 = \frac{1}{T'}\frac{dT'}{dz} + \frac{Rg\rho}{pc_p}$$

or

$$\frac{dT'}{dz} = -\frac{R\rho T'g}{pc_p} \qquad (3.112)$$

However, the ideal gas equation for the ambient air is

$$p = R\rho T \qquad (3.113)$$

From (3.112) and (3.113),

$$\frac{dT'}{dz} = -\frac{T'}{T}\frac{g}{c_p} \qquad (3.114)$$

(b) The derivation of an expression for the dry adiabatic lapse rate, namely $\Gamma_d \equiv -\left(\frac{dT}{dz}\right)_{\text{dry parcel}} = \frac{g}{c_p}$, was based on the assumption that the macroscopic kinetic energy of the air parcel was negligible compared to its total energy (see Sections 3.4.1 and 3.4.2). However, in the present exercise the temperature of the air parcel (T') differs from the temperature of the ambient air (T). Therefore, the air parcel is acted upon by a buoyancy force, which accelerates the air parcel in the vertical and gives it macroscopic kinetic energy. Note that if $T' = T$, Eq. (3.114) reduces to

$$-\frac{dT}{dz} = \frac{g}{c_p} = \Gamma_d$$ ■

3.50 Derive an expression for the rate of change in temperature with height (Γ_s) of a parcel of air undergoing a saturated adiabatic process. Assume that $\rho L_v \left(\frac{dw_s}{dp}\right)_T$ is small compared to 1.

Solution: Substituting (3.20) into (3.51) yields

$$dq = c_p dT + g dz \qquad (3.115)$$

If the saturation ratio of the air with respect to water is w_s, the quantity of heat dq released into (or absorbed from) a unit mass of dry air due to condensation (or evaporation) of liquid water is $-L_v\, dw_s$, when L_v is the latent heat of condensation. Therefore,

$$-L_v dw_s = c_p dT + g dz \qquad (3.116)$$

If we neglect the small amounts of water vapor associated with a unit mass of dry air, which are also warmed (or cooled) by the release (or absorption) of the latent heat, then c_p in (3.116) is the specific heat at constant pressure of dry air. Dividing both sides of (3.116) by $c_p\, dz$ and rearranging terms, we obtain

$$\frac{dT}{dz} = -\frac{L_v}{c_p}\frac{dw_s}{dz} - \frac{g}{c_p}$$

$$= -\frac{L_v}{c_p dz}\left[\left(\frac{dw_s}{dp}\right)_T dp + \left(\frac{dw_s}{dT}\right)_p dT\right] - \frac{g}{c_p}$$

Therefore,

$$\frac{dT}{dz}\left[1 + \frac{L_v}{c_p}\left(\frac{dw_s}{dT}\right)_p\right] = -\frac{g}{c_p}\left[1 + \frac{L_v}{g}\left(\frac{dw_s}{dp}\right)_T \frac{dp}{dz}\right] \qquad (3.117)$$

[44] Eqs. (3.107)–(3.110) appear in Exercise solutions provided on the book web site.

Alternatively, using the hydrostatic equation on the last term on the right side of (3.117)

$$\Gamma_s \equiv -\frac{dT}{dz} = \frac{\frac{g}{c_p}\left[1 - \rho L_v \left(\frac{dw_s}{dp}\right)_T\right]}{1 + \frac{L}{c_p}\left(\frac{dw_s}{dT}\right)_p}$$

or

$$\Gamma_s \equiv -\frac{dT}{dz} = \Gamma_d \frac{\left[1 - \rho L_v \left(\frac{dw_s}{dp}\right)_T\right]}{\left[1 + \frac{L_v}{c_p}\left(\frac{dw_s}{dT}\right)_p\right]} \quad (3.118)$$

In Exercise (3.51) we show that

$$-\rho L_v \left(\frac{dw_s}{dp}\right)_T \simeq 0.12$$

If we neglect this small term in (3.118) we obtain

$$\Gamma_s \equiv -\frac{dT}{dz} \simeq \frac{\Gamma_d}{1 + \frac{L_v}{c_p}\left(\frac{dw_s}{dT}\right)_p}$$ ■

3.51 In deriving the expression for the saturated adiabatic lapse rate in the previous exercise, it is assumed that $\rho L_v\, (dw_s/dp)_T$ is small compared to 1. Estimate the magnitude of $\rho L_v\, (dw_s/dp)_T$. Show that this last expression is dimensionless. [**Hint**: Use the skew $T - \ln p$ chart given in the book web site enclosed with this book to estimate the magnitude of $(dw_s/dp)_T$ for a pressure change of, say, 1000 to 950 hPa at 0 °C.]

Solution: Estimation of magnitude of

$$\rho L_v \left(\frac{dw_s}{dp}\right)_T$$

Take $\rho \simeq 1.275$ kg m^{-3} and $L_v = 2.5 \times 10^6$ J kg^{-1}. Suppose pressure changes from 1000 to 950 hPa so that $dp = -50$ hPa $= -5000$ Pa. Then, from the skew $T - \ln p$ chart, we find that

$$dw_s \simeq (4 - 3.75) = 0.25 \text{ g/kg}$$
$$\simeq 0.25 \times 10^{-3} \text{ kg/kg}$$

Hence,

$$\rho L_v \left(\frac{dw_s}{dp}\right)_T \simeq (1.275 \text{ kg m}^{-3})(2.5 \times 10^6 \text{ J kg}^{-1}) \left(\frac{0.25 \times 10^{-3} \text{ kg kg}^{-1}}{-5000 \text{ Pa}}\right) \simeq -0.12$$

The units of $\rho L_v \left(\frac{dw_s}{dp}\right)_T$ are

$$(\text{kg m}^{-3})\ (\text{J kg}^{-1})(\text{kg kg}^{-1})\left(\frac{1}{\text{Pa}}\right)$$

which is dimensionless. ■

3.52 In deriving Eq. (3.71) for equivalent potential temperature it was assumed that

$$\frac{L_v}{c_p T} dw_s \simeq d\left(\frac{L_v w_s}{c_p T}\right) \quad (3.119)$$

Justify this assumption. [**Hint**: Differentiate the right-hand side of the aforementioned expression and, assuming L_v/c_p is independent of temperature, show that the aforementioned approximation holds provided

$$\frac{dT}{T} \ll \frac{dw_s}{w_s}$$

Verify this inequality by noting the relative changes in T and w_s for small incremental displacements along saturated adiabats on a skew $T - \ln p$ chart.]

Solution: Differentiating the right side of (3.119) and assuming L_v/c_p is a constant,

$$\frac{L_v}{c_p}\left[\frac{1}{T} dw_s - w_s \frac{dT}{T^2}\right] = \frac{L_v}{Tc_p}\left[dw_s - w_s \frac{dT}{T}\right]$$
$$= \frac{L_v\, dT}{Tc_p}\left[\frac{dw_s}{dT} - \frac{w_s}{T}\right] \quad (3.120)$$

If $\frac{dT}{T} \ll \frac{dw_s}{w_s}$ (which can be verified from skew $T - \ln p$ chart) then

$$\frac{dw_s}{dT} \gg \frac{w_s}{T} \quad (3.121)$$

Therefore, from (3.120) and (3.121):

Right side of (3.119) $= \frac{L_v}{Tc_p} dw_s =$ Left side of (3.119) ■

3.53 Plot the following sounding on a skew $T - \ln p$ chart:

	Pressure level (hPa)	Air temperature (°C)	Dew point (°C)
A	1000	30.0	21.5
B	970	25.0	21.0
C	900	18.5	18.5
D	850	16.5	16.5
E	800	20.0	5.0
F	700	11.0	−4.0
G	500	−13.0	−20.0

(a) Are layers AB, BC, CD, etc. in stable, unstable, or neutral equilibrium?
(b) Which layers are convectively unstable?[45]

3.54 *Potential density* D is defined as the density that dry air would attain if it were transformed reversibly and adiabatically from its existing conditions to a standard pressure p_0 (usually 1000 hPa).

(a) If the density and pressure of a parcel of the air are ρ and p, respectively, show that

$$D = \rho \left(\frac{p_0}{p} \right)^{c_v/c_p}$$

where c_p and c_v are the specific heats of air at constant pressure and constant volume, respectively.

(b) Calculate the potential density of a quantity of air at a pressure of 600 hPa and a temperature of −15 °C.

(c) Show that

$$\frac{1}{D} \frac{dD}{dz} = -\frac{1}{T}(\Gamma_d - \Gamma)$$

where Γ_d is the dry adiabatic lapse rate, Γ the actual lapse rate of the atmosphere, and T the temperature at height z. [**Hint**: Take the natural logarithms of both sides of the expression given in (a) and then differentiate with respect to height z.]

(d) Show that the criteria for stable, neutral, and unstable conditions in the atmosphere are that the potential density decreases with increasing height, is constant with height, and increases with increasing height, respectively. [**Hint**: Use the expression given in (c).]

(e) Compare the criteria given in (d) with those for stable, neutral, and unstable conditions for a liquid.

3.55 A necessary condition for the formation of a mirage is that the density of the air increases with increasing height. Show that this condition is realized if the decrease of atmospheric temperature with height exceeds 3.5 Γ_d, where Γ_d is the dry adiabatic lapse rate. [**Hint**: Take the natural logarithm of both sides of the expression for D given in Exercise 3.54a and then differentiate with respect to height z. Follow the same two steps for the gas equation in the form $p = \rho R_d T$. Combine the two expressions so derived with the hydrostatic equation to show that $\frac{1}{\rho}\frac{d\rho}{dt} = -\frac{1}{T}(dT/dz + g/R_d)$. Hence, proceed to the solution.]

3.56 Assuming the truth of the second law of thermodynamics, prove the following two statements (known as *Carnot's theorems*):

(a) No engine can be more efficient than a reversible engine working between the same limits of temperature. [**Hint**: The efficiency of any engine is given by Eq. (3.78); the distinction between a reversible (**R**) and an irreversible (**I**) engine is that **R** can be driven backward but **I** cannot. Consider a reversible and an irreversible engine working between the same limits of

[45] For a more realistic treatment of the stability of a layer, see Chapter 9.3.5.

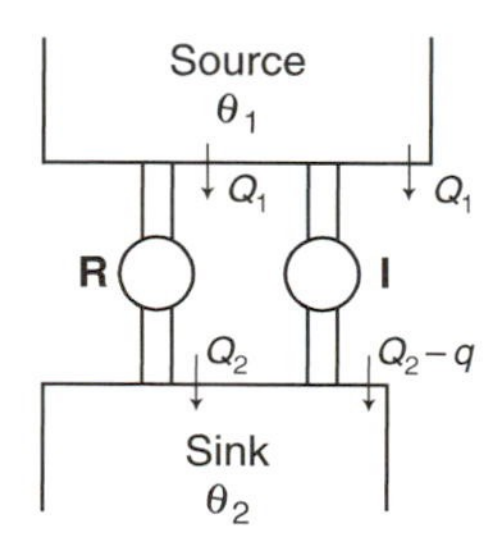

Fig. 3.26

temperature. Suppose initially that **I** is more efficient than **R** and use **I** to drive **R** backward. Show that this leads to a violation of the second law of thermodynamics, and hence prove that **I** cannot be more efficient than **R**.]

(b) All reversible engines working between the same limits of temperature have the same efficiency. [**Hint**: Proof is similar to that for part (a).]

Solution:

(a) To prove that no engine can be more efficient than a reversible engine working between the same limits of temperature, consider a reversible (**R**) and irreversible (**I**) engine working between θ_1 and θ_2. Assume **I** is more efficient than **R** and that **R** takes heat Q_1 from source and yields heat Q_2 to sink (Fig. 3.26). Therefore, if **I** takes Q_1 from source it must yield heat $Q_2 - q$ (q positive) to sink. Now let us use **I** to drive **R** backward. This will require **I** to do work $Q_1 - Q_2$ on **R**. However, in one cycle, **I** develops work $Q_1 - (Q_2 - q) = (Q_1 - Q_2) + q$. Hence, even when **I** is drawing **R** backward, mechanical work q is still available. However, in one cycle of the combined system, heat $Q_2 - (Q_2 - q) = q$ is taken from a colder body. Because this violates the second law of thermodynamics, **I** cannot be more efficient than **R**.

(b) Take two reversible engines operating between θ_1 and θ_2 and assume one engine is more efficient than the other. Then, following same procedure as in (a), it can be shown that if one reversible engine is more efficient than another the second law is violated. ■

3.57 Lord Kelvin introduced the concept of *available energy*, which he defined as the maximum amount of heat that can be converted into work by using the coldest available body in a system as the sink for an ideal heat engine. By considering an ideal heat engine that uses the coldest available body as a sink, show that the available energy of the universe is tending to zero and that

$$\text{loss of available energy} = T_0\,(\text{increase in entropy})$$

where T_0 is the temperature of the coldest available body.

Solution: For an ideal reversible engine

$$\frac{Q_1 - Q_2}{Q_1} = \frac{T_1 - T_2}{T_1}$$

$$\text{Work done in 1 cycle} = Q_1 - Q_2 = \frac{T_1 - T_2}{T_1} Q_1.$$

If an engine operates with sink at T_0 ($= T_2$):

$$\text{Available energy} = \frac{T_1 - T_0}{T_1} Q_1$$

Let Q pass from T_1 to T_2 ($T_1 > T_2$) by, say, conduction or radiation. Then,

Loss of available energy

$$= \left(\frac{T_1 - T_0}{T_1}\right)Q - \left(\frac{T_2 - T_0}{T_2}\right)Q$$

$$= QT_0\left(\frac{T_1 - T_2}{T_1 T_2}\right)$$

Because $T_1 > T_2$, there is a loss of available energy for natural processes

Loss of available energy

$$= T_0\left(\frac{Q}{T_2} - \frac{Q}{T_1}\right)$$

$$= T_0\,(\text{increase in entropy})$$ ■

3.58 An ideal reversible engine has a source and sink at temperatures of 100 and 0 °C, respectively. If the engine receives 20 J of heat from the source in every cycle, calculate the work done by the engine in 10 cycles. How much heat does the engine reject to the sink in 10 cycles?

3.59 A refrigerator has an internal temperature of 0 °C and is situated in a room with a steady temperature of 17 °C. If the refrigerator is driven by an electric motor 1 kW in power, calculate the time required to freeze 20 kg of water already cooled to 0 °C when the water is placed in the refrigerator. The refrigerator may be considered to act as an ideal heat engine in reverse.

3.60 A Carnot engine operating in reverse (i.e., as an air conditioner) is used to cool a house. The indoor temperature of the house is maintained at T_i and the outdoor temperature is T_o ($T_o > T_i$). Because the walls of the house are not perfectly insulating, heat is transferred into the house at a constant rate given by

$$\left(\frac{dq}{dt}\right)_{\text{leakage}} = K(T_o - T_i)$$

where K (>0) is a constant.

(a) Derive an expression for the power (i.e., energy used per second) required to drive the Carnot engine in reverse in terms of T_o, T_i, and K.

(b) During the afternoon, the outdoor temperature increases from 27 to 30 °C. What percentage increase in power is required to drive the Carnot engine in reverse to maintain the interior temperature of the house at 21 °C?

3.61 Calculate the change in entropy of 2 g of ice initially at −10 °C that is converted to steam at 100 °C due to heating.

3.62 Calculate the change in entropy when 1 mol of an ideal diatomic gas initially at 13 °C and 1 atm changes to a temperature of 100 °C and a pressure of 2 atm.

3.63 Show that the expression numbered (3.118) in the solution to Exercise 3.50 can be written as

$$\Gamma_s = \Gamma_d \frac{(1 + w_s L_v / R_d T)}{(1 + w_s L_v^2 / c_p R_v T^2)}$$

3.64 The pressure at the top of Mt. Rainier is about 600 hPa. Estimate the temperature at which water will boil at this pressure. Take the specific volumes of water vapor and liquid water to be 1.66 and 1.00×10^{-3} m^3 kg^{-1}, respectively.

3.65 Calculate the change in the melting point of ice if the pressure is increased from 1 to 2 atm. (The specific volumes of ice and water at 0 °C are 1.0908×10^{-3} and 1.0010×10^{-3} m^3 kg^{-1}, respectively.) [**Hint**: Use (3.112).]

3.66 By differentiating the enthalpy function, defined by Eq. (3.47), show that

$$\left(\frac{dp}{dT}\right)_s = \left(\frac{ds}{d\alpha}\right)_p$$

where s is entropy. Show that this relation is equivalent to the Clausius–Clapeyron equation.

Solution: From Eq. (3.47) in the text,

$$h = u + p\alpha$$

Therefore,

$$dh = du + pd\alpha + \alpha dp$$

or, using Eq. (3.38) in the text,

$$dh = (dq - pd\alpha) + pd\alpha + \alpha dp$$

$$= dq + \alpha dp$$

Using Eq. (3.85) in the text,

$$dh = Tds + \alpha dp \qquad (3.151)$$

We see from (3.151) that h is a function of two variables, namely s and p. Hence, we can write

$$dh = \left(\frac{dh}{ds}\right)_p ds + \left(\frac{dh}{dp}\right)_s dp = \alpha \qquad (3.152)$$

From (3.151) and (3.152),

$$\left(\frac{dh}{ds}\right)_p = T \text{ and } \left(\frac{dh}{dp}\right)_s = \alpha \qquad (3.153)$$

Because the order of differentiating does not matter,

$$\frac{\partial}{\partial p}\left(\frac{dh}{ds}\right)_p = \frac{\partial}{\partial s}\left(\frac{dh}{dp}\right)_s \qquad (3.154)$$

From (3.153) and (3.154),

$$\left(\frac{dT}{dp}\right)_s = \left(\frac{d\alpha}{ds}\right)_p$$

or

$$\left(\frac{dp}{dT}\right)_s = \left(\frac{ds}{d\alpha}\right)_p \tag{3.155}$$

Because, from Eq. (3.85) in the text,

$$ds = \frac{dq}{T}$$

for a phase change from liquid to vapor at temperature T

$$ds = \frac{L_v}{T} \tag{3.156}$$

where L_v is the latent heat of evaporation. If the vapor is saturated so that $p = e_s$ and $d\alpha = \alpha_2 - \alpha_1$, where α_2 and α_1 are the specific volume of the vapor and liquid, respectively, we have from (3.155) and (3.156),

$$\left(\frac{de_s}{dT}\right)_s = \frac{L_v}{T(\alpha_2 - \alpha_1)}$$

which is the Clausius–Clapeyron equation [see Eq. (3.92) in the text]. Equation (3.155) is one of **Maxwell's four thermodynamic equations**. The others are

$$\left(\frac{ds}{d\alpha}\right)_T = \left(\frac{dp}{dT}\right)_\alpha \tag{3.157}$$

$$\left(\frac{d\alpha}{dT}\right)_p = -\left(\frac{ds}{dp}\right)_T \tag{3.158}$$

and

$$\left(\frac{dT}{d\alpha}\right)_s = -\left(\frac{dp}{ds}\right)_\alpha \tag{3.159}$$

Equations (3.157)–(3.159) can be proven in analogous ways to the proof of (3.155) given earlier but, in place of (3.151), starting instead with the state functions $f = u - Ts$, $g = u - Ts + p\alpha$, and $du = Tds - pd\alpha$, respectively. The state functions f and g are called *Helmholtz free energy* and *Gibbs function*, respectively. ■

Radiative Transfer[1]

4

With Qiang Fu
Department of Atmospheric Sciences
University of Washington

Radiative transfer is a branch of atmospheric physics. Like a nation composed of many different ethnic groups, radiative transfer has a rich, but sometimes confusing language that reflects its diverse heritage, which derives from quantum physics, astronomy, climatology, and electrical engineering. Solving radiative transfer problems requires consideration of the geometry and spectral distribution of radiation, both of which are straightforward in principle, but can be quite involved in real world situations.

This chapter introduces the fundamentals of radiative transfer in planetary atmospheres. The first two sections describe the electromagnetic spectrum and define the terms that are used to quantify the field of radiation. The third section reviews physical laws relating to blackbody radiation and concludes with a qualitative discussion of the so-called "greenhouse effect." The fourth section describes the processes by which gases and particles absorb and scatter radiation. The fifth section presents an elementary quantitative treatment of radiative transfer in planetary atmospheres, with subsections on radiative heating rates and remote sensing. The final section describes the radiation balance at the top of the atmosphere as determined from measurements made by satellite-borne sensors.

4.1 The Spectrum of Radiation

Electromagnetic radiation may be viewed as an ensemble of waves propagating at the speed of light ($c^* = 2.998 \times 10^8$ m s^{-1} through a vacuum). As for any wave with a known speed of propagation, frequency $\widetilde{\nu}$, wavelength λ, and wave number ν (i.e., the number of waves per unit length in the direction of propagation) are interdependent. Wave number is the reciprocal of wavelength

$$\nu = 1/\lambda \tag{4.1}$$

and

$$\widetilde{\nu} = c^*\nu = c^*/\lambda \tag{4.2}$$

Small variations in the speed of light within air give rise to mirages, as well as the annoying distortions that limit the resolution of ground-based telescopes, and the difference between the speed of light in air and water produces a number of interesting optical phenomena such as rainbows.

Wavelength, frequency, and wave number are used alternatively in characterizing radiation. Wavelength, which is in some sense the easiest of these measures to visualize, is used most widely in elementary texts on the subject and in communicating between radiation specialists and scientists in other fields; wave number and frequency tend to be preferred by those working in the field of radiative transfer because they are proportional to the quantity of energy carried by photons, as discussed in Section 4.4. Wavelength is used exclusively in other chapters of this book, but in this chapter λ, ν, and $\widetilde{\nu}$ are used in accordance with current practice in the field. Here we express wavelength exclusively in units of micrometers (μm), but in the literature it is also often expressed in nanometers (nm).

Because radiative transfer in planetary atmospheres involves an ensemble of waves with a continuum of

[1] We thank Qiang Fu for his guidance in preparing this chapter.

wavelengths and frequencies, the energy that it carries can be partitioned into the contributions from various wavelength (or frequency or wave number) bands. For example, in atmospheric science the term *shortwave*[2] ($\lambda < 4\ \mu$m) refers to the wavelength band that carries most of the energy associated with solar radiation and *longwave* ($\lambda > 4\ \mu$m) refers to the band that encompasses most of the terrestrial (Earth-emitted) radiation.

In the radiative transfer literature, the spectrum is typically divided into the regions shown in Fig. 4.1. The relatively narrow *visible* region, which extends from wavelengths of 0.39 to 0.76 μm, is defined by the range of wavelengths that the human eye is capable of sensing. Subranges of the visible region are discernible as colors: violet on the short wavelength end and red on the long wavelength end. The term *monochromatic* denotes a single color (i.e., one specific frequency or wavelength).

The visible region of the spectrum is flanked by *ultraviolet* (above violet in terms of frequency) and *infrared* (below red) regions. The *near infrared* region, which extends from the boundary of the visible up to ~4 μm, is dominated by solar radiation, whereas the remainder of the infrared region is dominated by *terrestrial* (i.e., Earth emitted) radiation: hence, the near infrared region is included in the term *shortwave radiation*. Microwave radiation is not important in the Earth's energy balance but it is widely used in remote sensing because it is capable of penetrating through clouds.

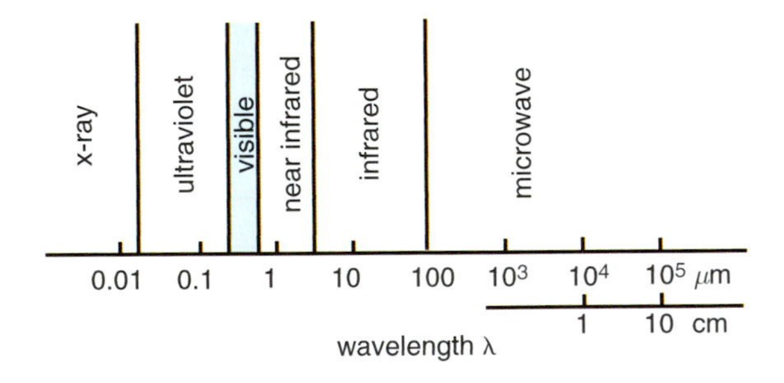

Fig. 4.1 The electromagnetic spectrum.

4.2 Quantitative Description of Radiation

The energy transferred by electromagnetic radiation in a specific direction passing through a unit area (normal to the direction considered) per unit time at a specific wavelength (or wave number) is called *monochromatic intensity* (or *spectral intensity* or *monochromatic radiance*) and is denoted by the symbol I_λ (or I_ν). Monochromatic intensity is expressed in units of watts per square meter per unit arc of solid angle,[3] per unit wavelength (or per unit wave number or frequency) in the electromagnetic spectrum.

The integral of the monochromatic intensity over some finite range of the electromagnetic spectrum is called the *intensity* (or *radiance*) I, which has units of W m^{-2} sr^{-1}

$$I = \int_{\lambda_1}^{\lambda_2} I_\lambda d\lambda = \int_{\nu_1}^{\nu_2} I_\nu d\nu \tag{4.3}$$

For quantifying the energy emitted by a laser, the interval from λ_1 to λ_2 (or ν_1 to ν_2) is very narrow, whereas for describing the Earth's energy balance, it encompasses the entire electromagnetic spectrum. Separate integrations are often carried out for the shortwave and longwave parts of the spectrum corresponding, respectively, to the wavelength ranges of incoming solar radiation and outgoing terrestrial radiation. Hence, the intensity is the area under some finite segment of the the spectrum of monochromatic intensity (i.e., the plot of I_λ as a function of λ, or I_ν as a function of ν, as illustrated in Fig. 4.2).

Although I_λ and I_ν both bear the name *monochromatic intensity*, they are expressed in different units. The shapes of the associated spectra tend to be somewhat different in appearance, as will be apparent in several of the figures later in this chapter. In Exercise 4.13, the student is invited to prove that

$$I_\nu = \lambda^2 I_\lambda \tag{4.4}$$

[2] The term *shortwave* as used in this book is not to be confused with the region of the electromagnetic spectrum exploited in shortwave radio reception, which involves wavelengths on the order of 100 m, well beyond the range of Fig. 4.1.

[3] The unit of solid angle is the dimensionless *steradian* (denoted by the symbol Ω) defined as the area σ subtended by the solid angle on the unit sphere. Alternatively, on a sphere of radius r, $\Omega = \sigma/r^2$. Exercise 4.1 shows that a hemisphere corresponds to a solid angle of 2π steradians.

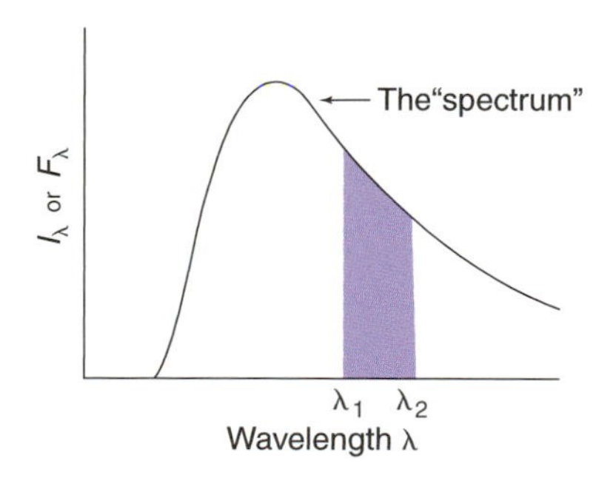

Fig. 4.2 The curve represents a hypothetical spectrum of monochromatic intensity I_λ or monochromatic flux density F_λ as a function of wavelength λ. The shaded area represents the intensity I or flux density F of radiation with wavelengths ranging from λ_1 to λ_2.

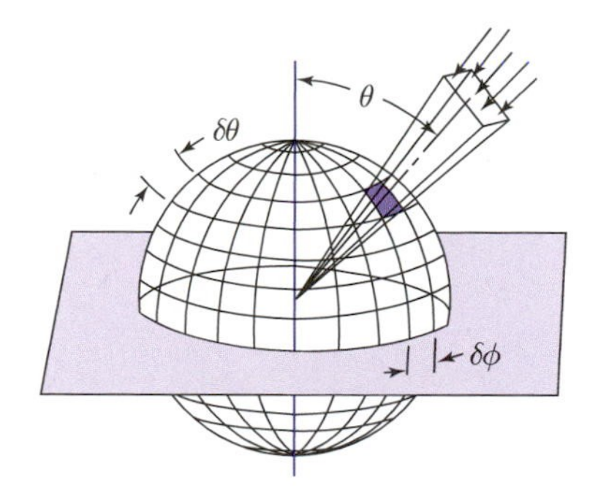

Fig. 4.3 Relationship between intensity and flux density. θ is the angle between the incident radiation and the normal to the surface. For the case of radiation incident upon a horizontal surface from above, θ is called the *zenith angle*. ϕ is referred to as the *azimuth angle*.

The *monochromatic flux density* (or *monochromatic irradiance*) F_λ is a measure of the rate of energy transfer per unit area by radiation with a given wavelength through a plane surface with a specified orientation in three-dimensional space. If the radiation impinges on a plane surface from one direction (e.g., upon the horizontal plane from above) the flux density is said to be *incident* upon that surface, in which case

$$F_\lambda = \int_{2\pi} I_\lambda \cos\theta \, d\omega \tag{4.5}$$

The limit on the bottom of the integral operator indicates that the integration extends over the entire hemisphere of solid angles lying above the plane, $d\omega$ represents an elemental arc of solid angle, and θ is the angle between the incident radiation and the direction normal to dA. The factor $\cos\theta$ represents the spreading and resulting dilution of radiation with a slanted orientation relative to the surface. Monochromatic flux density F_λ has units of W m^{-2} μm^{-1}. Analogous quantities can be defined for the wave number and frequency spectra.

Exercise 4.1 By means of a formal integration over solid angle, calculate the arc of solid angle subtended by the sky when viewed from a point on a horizontal surface.

Solution: The required integration is performed using a spherical coordinate system centered on a point on the surface, with the pole pointing straight upward toward the zenith, where θ is called the *zenith angle* and ϕ the *azimuth angle*, as defined in Fig. 4.3. The required arc of solid angle is given by

$$\int_{2\pi} d\omega = \int_{\phi=0}^{2\pi}\int_{\theta=0}^{\pi/2} \sin\theta d\theta d\phi = 2\pi\int_{\theta=0}^{\pi/2} \sin\theta d\theta = 2\pi$$

■

Combining (4.3) and (4.5), we obtain an expression for the *flux density* (or *irradiance*) of radiation incident upon a plane surface

$$F = \int_{2\pi} I \cos\theta \, d\omega = \int_{\lambda_1}^{\lambda_2}\int_{2\pi} I_\lambda \cos\theta \, d\omega \, d\lambda \tag{4.6}$$

Flux density, the rate at which radiant energy passes through a unit area on a plane surface, is expressed in units of watts per square meter. The following two exercises illustrate the relation between intensity and flux density.

Exercise 4.2 The flux density F_s of solar radiation incident upon a horizontal surface at the top of the Earth's atmosphere at zero zenith angle is 1368 W m^{-2}. Estimate the intensity of solar radiation. Assume that solar radiation is *isotropic* (i.e., that every point on the "surface" of the sun emits radiation with the same intensity in all directions, as indicated in Fig. 4.4). For reference, the radius of the sun R_s is 7.00×10^8 m and the Earth–sun distance d is 1.50×10^{11} m.

Solution: Let I_s be the intensity of solar radiation. If the solar radiation is isotropic, then from (4.5) the

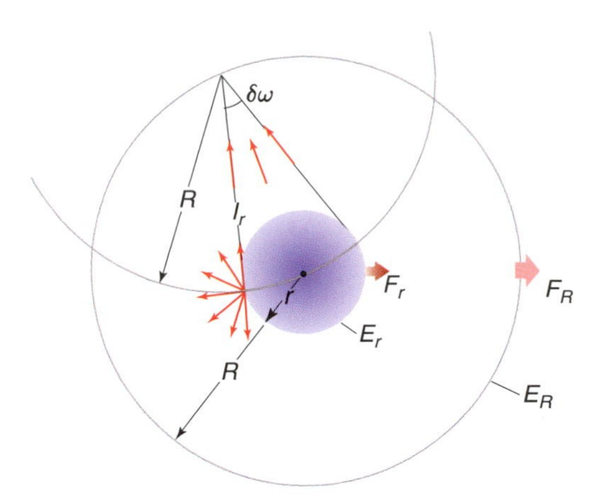

Fig. 4.4 Relationships involving intensity I, flux density F, and flux E of isotropic radiation emitted from a spherical source with radius r, indicated by the blue shading, and incident upon a much larger sphere of radius R, concentric with the source. Thin arrows denote intensity and thick arrows denote flux density. Fluxes $E_R = E_r$ and intensities $I_R = I_r$. Flux density F decreases with the square of the distance from the source.

flux density of solar radiation at the top of the Earth's atmosphere is

$$F_s = \int_{\delta\omega} I_s \cos\theta\, d\omega$$

where $\delta\omega$ is the arc of solid angle subtended by the sun in the sky. Because $\delta\omega$ is very small, we can ignore the variations in $\cos\theta$ in the integration. With this so-called *parallel beam approximation*, the integral reduces to

$$F_s = I_s \times \cos\theta \times \delta\omega$$

and because the zenith angle, in this case, is zero,

$$F_s = I_s \times \delta\omega$$

The fraction of the hemisphere of solid angle (i.e., "the sky") that is occupied by the sun is the same as the fraction of the area of the hemisphere of radius d, centered on the Earth, i.e., occupied by the sun, i.e.,

$$\frac{\delta\omega}{2\pi} = \frac{\pi R_s^2}{2\pi d^2}$$

from which

$$\delta\omega = \pi\left(\frac{R_s}{d}\right)^2 = \pi\left(\frac{7.00 \times 10^8}{1.50 \times 10^{11}}\right)^2 = 6.84 \times 10^{-5}\ \text{sr}$$

and

$$I_s = \frac{F_s}{\delta\omega} = \frac{1368\ \text{W m}^{-2}}{6.84 \times 10^{-5}\ \text{sr}} = 2.00 \times 10^7\ \text{W m}^{-2}\ \text{sr}^{-1}$$

■

The intensity of radiation is constant along ray paths through space and is thus independent of distance from its source, in this case, the sun. The corresponding flux density is directly proportional to the arc solid angle subtended by the sun, which is inversely proportional to the square of the distance from the sun. It follows that flux density varies inversely with the square of the distance from the sun, i.e.,

$$F \propto d^{-2} \tag{4.7}$$

This so-called *inverse square law* also follows from the fact that the *flux* of solar radiation E_s (i.e., the flux density F_s multiplied by the area of spheres, concentric with the sun, through which it passes as it radiates outward) is independent of distance from the Sun, i.e.,

$$E_s = F_s \times 4\pi d^2 = const.$$

Exercise 4.3 Radiation is emitted from a plane surface with a uniform intensity in all directions. What is the flux density of the emitted radiation?

Solution:

$$F = \int_{2\pi} I\cos\theta\, d\omega = \int_{\phi=0}^{2\pi}\int_{\theta=0}^{\pi/2} I\cos\theta\sin\theta\, d\theta d\phi \tag{4.8}$$

$$= 2\pi I\int_0^{\pi/2}\cos\theta\sin\theta\, d\theta$$

$$= \pi I\,[(\sin^2(\pi/2) - \sin^2(0)]$$
$$= \pi I$$

Although the geometrical setting of this exercise is quite specific, the result applies generally to isotropic radiation, as illustrated, for example, in Exercise 4.31. ■

Performing the integrations over wavelength and solid angle in reverse order (with solid angle first) yields the monochromatic flux density F_λ as an intermediate by-product. The relationships discussed in this section can be summarized in terms of an expression for the *flux* of radiation emitted by, incident upon, or passing through a surface δA

$$E = \int_{\delta A} \int_{2\pi} \int_{\lambda_1}^{\lambda_2} I_\lambda \, (\phi, \theta) \, d\lambda \cos\theta \, d\omega dA \quad (4.9)$$

which is expressed in units of watts (W).

4.3 Blackbody Radiation

A *blackbody*[4] is a surface that completely absorbs all incident radiation. Examples include certain substances such as coal and a small aperture of a much larger cavity. The entrances of most caves appear nearly black, even though the interior walls may be quite reflective, because only a very small fraction of the sunlight that enters is reflected back through the entrance: most of the light that enters the cave is absorbed in multiple reflections off the walls. The narrower the entrance and the more complex the interior geometry of the cave, the smaller the fraction of the incident light that is returned back through it, and the blacker the appearance of the cave when viewed from outside (Fig. 4.5).

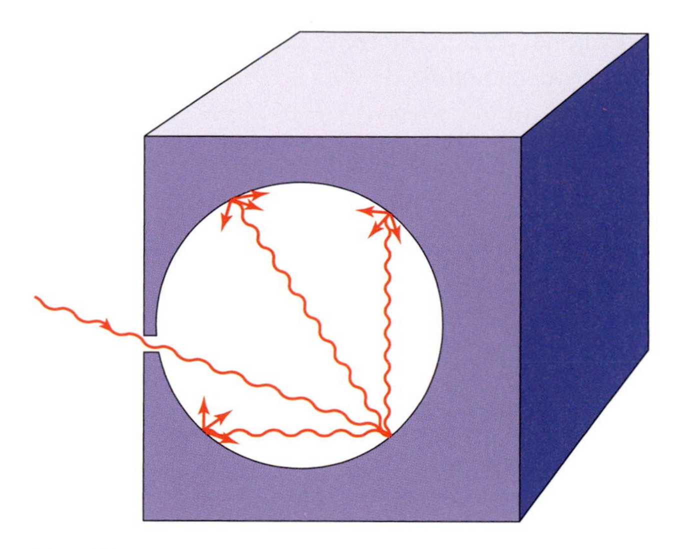

Fig. 4.5 Radiation entering a cavity with a very small aperture and reflecting off the interior walls. [Adapted from K. N. Liou, *An Introduction to Atmospheric Radiation*, Academic Press, p. 10, Copyright (2002), with permission from Elsevier.]

4.3.1 The Planck[5] Function

It has been determined experimentally that the intensity of radiation emitted by a blackbody is given by

$$B_\lambda(T) = \frac{c_1 \lambda^{-5}}{\pi \, (e^{c_2/\lambda T} - 1)} \quad (4.10)$$

where $c_1 = 3.74 \times 10^{-16}$ W m^2 and $c_2 = 1.45 \times 10^{-2}$ m K. The quest for a theoretical justification of this empirical relationship led to the development of the theory of quantum physics. It is observed and has been verified theoretically that blackbody radiation is isotropic. When $B_\lambda(T)$ is plotted as a function of wavelength on a linear scale, the resulting spectrum of monochromatic intensity exhibits the shape shown in Fig. 4.6, with a sharp short wavelength cutoff, a steep rise to a

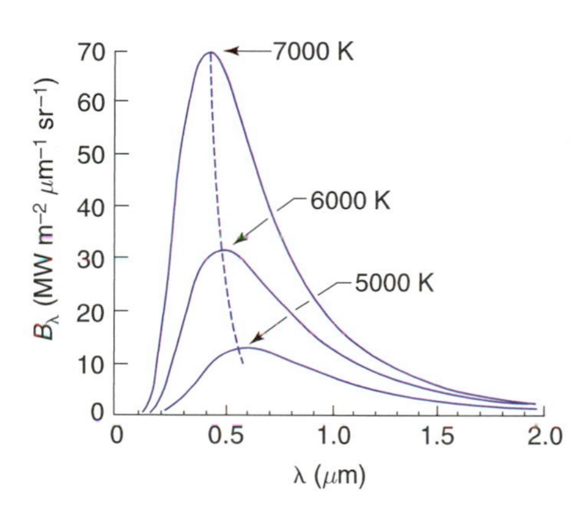

Fig. 4.6 Emission spectra for blackbodies with absolute temperatures as indicated, plotted as a function of wavelength on a linear scale. The three-dimensional surface formed by the ensemble of such spectra is the *Planck function*. [Adapted from R. G. Fleagle and J. A. Businger, *An Introduction to Atmospheric Physics*, Academic Press, p. 137, Copyright (1963) with permission from Elsevier.]

[4] The term *body* in this context refers to a coherent mass of material with a uniform temperature and composition. A body may be a gaseous medium, as long as it has well-defined interfaces with the surrounding objects, media, or vacuum, across which the intensity of the incident and emitted radiation can be defined. For example, it could be a layer of gas of a specified thickness or the surface of a mass of solid material.

[5] **Max Planck** (1858–1947) German physicist. Professor of physics at the University of Kiel and University of Berlin. Studied under Helmholtz and Kirchhoff. Played an important role in the development of quantum theory. Awarded the Nobel Prize in 1918.

maximum, and a more gentle drop off toward longer wavelengths.

4.3.2 Wien's Displacement Law

Differentiating (4.10) and setting the derivative equal to zero (Exercise 4.24) yield the wavelength of peak emission for a blackbody at temperature T

$$\lambda_m = \frac{2897}{T} \tag{4.11}$$

where λ_m is expressed in micrometers and T in degrees kelvin. On the basis of (4.11), which is known as *Wien's*[6] *displacement law*, it is possible to estimate the temperature of a radiation source if we know its emission spectrum, as illustrated in the following example.

Exercise 4.4 Use Wien's displacement law to compute the "color temperature" of the sun, for which the wavelength of maximum solar emission is observed to be ~0.475 μm.

Solution:

$$T = \frac{2897}{\lambda_m} = \frac{2897}{0.475} = 6100 \text{ K.}$$ ■

Wien's displacement law explains why solar radiation is concentrated in the *visible* (0.4–0.7 μm) and *near infrared* (0.7–4 μm) regions of the spectrum, whereas radiation emitted by planets and their atmospheres is largely confined to the infrared (>4 μm), as shown in the top panel of Fig. 4.7.

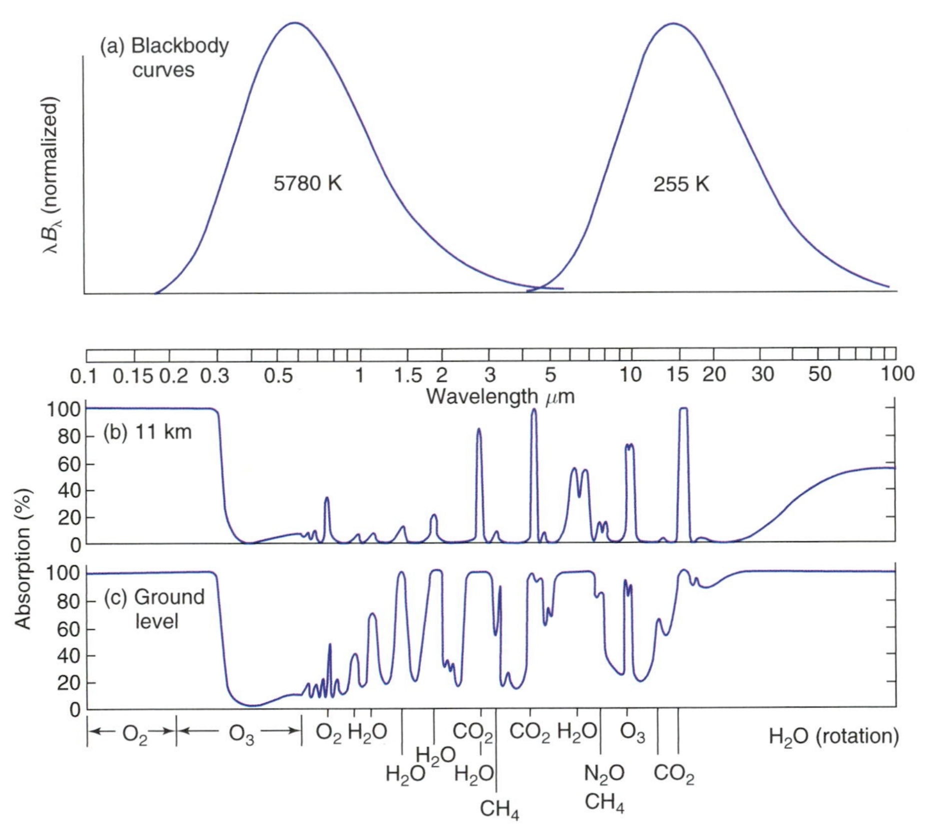

Fig. 4.7 (a) Blackbody spectra representative of the sun (left) and the Earth (right). The wavelength scale is logarithmic rather than linear as in Fig. 4.6, and the ordinate has been multiplied by wavelength in order to retain the proportionality between areas under the curve and intensity. In addition, the intensity scales for the two curves have been scaled to make the areas under the two curves the same; (b) Spectrum of monochromatic absorptivity of the part of the atmosphere that lies above the 11-km level; (c) spectrum of monochromatic absorptivity of the entire atmosphere. [From R. M. Goody and Y. L. Yung, *Atmospheric Radiation: Theoretical Basis*, 2nd ed., Oxford University Press (1995), p. 4. By permission of Oxford University Press, Inc.]

[6] **Wilhelm Wien** (1864–1925) German physicist. Received the Nobel Prize in 1911 for the discovery (in 1893) of the displacement law named after him. Also made the first rough determination of the wavelength of x-rays.

4.3.3 The Stefan–Boltzmann Law

The blackbody flux density (or irradiance) obtained by integrating the Planck function πB_λ over all wavelengths is given by the *Stefan*[7]*–Boltzmann law*

$$F = \sigma T^4 \tag{4.12}$$

where σ, the Stefan–Boltzmann constant, is equal to 5.67×10^{-8} W m^{-2} K^{-4}. Given measurements of the flux density F from any black or nonblack body, (4.12) can be solved for the *equivalent blackbody temperature* (also referred to as the *effective emission temperature*) T_E; that is, the temperature T_E that a blackbody would need to be at in order to emit radiation at the measured rate F. If the body in question emits as a blackbody, its actual temperature and its equivalent blackbody temperature will be the same. Applications of the Stefan–Boltzmann law and the concept of equivalent blackbody temperature are illustrated in the following exercises.

Exercise 4.5 Calculate the equivalent blackbody temperature T_E of the solar *photosphere* (i.e., the outermost visible layer of the sun) based on the following information. The flux density of solar radiation reaching the Earth, F_s, is 1368 W m^{-2}. The Earth–sun distance d is 1.50×10^{11} m and the radius of the solar photosphere R_s is 7.00×10^8 m.

Solution: We first calculate the flux density at the top of the layer, making use of the inverse square law (4.7).

$$F_{\text{photosphere}} = F_s \left(\frac{R_s}{d}\right)^{-2}$$

$$F_{\text{photosphere}} = 1{,}368 \times \left(\frac{1.50 \times 10^{11}}{7.00 \times 10^8}\right)^2$$
$$= 6.28 \times 10^7 \text{ W m}^{-2}$$

From the Stefan–Boltzmann law

$$\sigma T_E^4 = 6.28 \times 10^7 \text{ W m}^{-2}$$

Therefore, the equivalent blackbody temperature is

$$\begin{aligned} T_E &= \left(\frac{6.28 \times 10^7}{5.67 \times 10^{-8}}\right)^{1/4} \\ &= (1108 \times 10^{12})^{1/4} \\ &= 5.77 \times 10^3 \\ &= 5770 \text{ K} \end{aligned}$$

■

That this value is slightly lower than the sun's color temperature estimated in the previous exercise is evidence that the spectrum of the sun's emission differs slightly from the blackbody spectrum prescribed by Planck's law (4.10).

Exercise 4.6 Calculate the equivalent blackbody temperature of the Earth as depicted in Fig. 4.8, assuming a *planetary albedo* (i.e., the fraction of the incident solar radiation that is reflected back into space without absorption) of 0.30. Assume that the Earth is in *radiative equilibrium*; i.e., that it experiences no net energy gain or loss due to radiative transfer.

Solution: Let F_s be the flux density of solar radiation incident upon the Earth (1368 W m^{-2}); F_E the flux density of longwave radiation emitted by the Earth, R_E the radius of the Earth, as shown in Fig. 4.8; A the planetary albedo of the Earth (0.30);

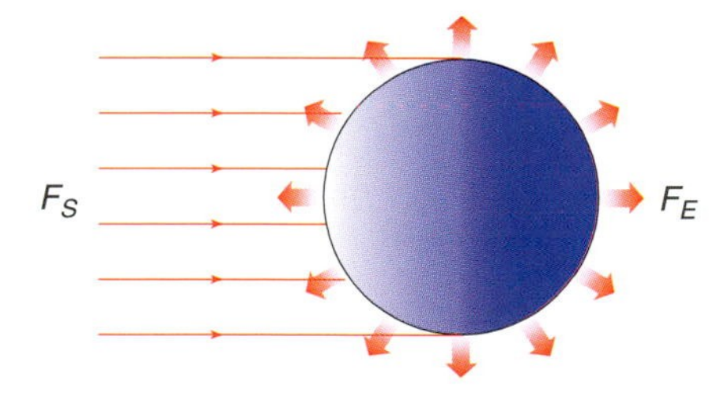

Fig. 4.8 Radiation balance of the Earth. Parallel beam solar radiation incident on the Earth's orbit, indicated by the thin red arrows, is intercepted over an area πR_E^2 and outgoing (blackbody) terrestrial radiation, indicated by the wide red arrows, is emitted over the area $4\pi R_E^2$.

[7] **Joseph Stefan** (1835–1893) Austrian physicist. Professor of physics at the University of Vienna. Originated the theory of the diffusion of gases as well as carrying out fundamental work on the theory of radiation.

and T_E the Earth's equivalent blackbody temperature. From the Stefan–Boltzmann law (4.12)

$$F_E = \sigma T_E^4 = \frac{(1-A)F_s}{4} = \frac{(1-0.30)\times 1368}{4} = 239.4 \text{ W m}^{-2}$$

Solving for T_E, we obtain

$$T_E = \sqrt[4]{\frac{F_E}{\sigma}} = \left(\frac{239.4}{5.67\times 10^{-8}}\right)^{1/4} = 255 \text{ K} \quad ■$$

Analogous estimates for some of the other planets in the solar system are shown in Table 4.1.

Exercise 4.7 A small blackbody satellite is orbiting the Earth at a distance far enough away so that the flux density of Earth radiation is negligible, compared to that of solar radiation. Suppose that the satellite suddenly passes into the Earth's shadow. At what rate will it initially cool? The satellite has a mass $m = 10^3$ kg and a specific heat $c = 10^3$ J (kg K)$^{-1}$: it is spherical with a radius $r = 1$ m, and temperature is uniform over its surface.

Solution: Consideration of the energy balance, as in the previous problem, but with an albedo of zero yields an emission $F = 342$ W m^{-2} from the satellite, which implies an equivalent blackbody temperature $T_E = 279$ K. When the satellite passes into the Earth's shadow, it will no longer be in radiative equilibrium. The flux density of solar radiation abruptly drops to zero while the emitted radiation, which is determined by the temperature of the satellite, drops gradually as the satellite cools by emitting radiation. The instant that the satellite passes into the shadow, its temperature is still equal to T_E, so we can write

$$mc\frac{dT}{dt} = 4\pi r^2 \sigma T_E^4$$

Solving, we obtain $dT/dt = 4.30\times 10^{-3}$ K s^{-1} or 15.5 K hr^{-1}. ■

Table 4.1 Flux density of solar radiation F_s, planetary albedo A, and equivalent blackbody temperature T_E of some of the planets based on the assumption that they are in radiative equilibrium. Astronomical units are multiples of Earth–sun distance

Planet	Distance from sun[a]	F_s (W m^{-2})	A	T_E (K)
Mercury	0.39	8994	0.06	439
Venus	0.72	2639	0.78	225
Earth	1.00	1368	0.30	255
Mars	1.52	592	0.17	216
Jupiter	5.18	51	0.45	105

[a] Astronomical units are multiples of Earth–Sun distance.

4.3.4 Radiative Properties of Nonblack Materials

Unlike blackbodies, which absorb all incident radiation, nonblack bodies such as gaseous media can also reflect and transmit radiation. However, their behavior can nonetheless be understood by applying the radiation laws derived for blackbodies. For this purpose it is useful to define the (monochromatic) *emissivity* ε_λ, i.e., the ratio of the monochromatic intensity of the radiation emitted by the body to the corresponding blackbody radiation

$$\varepsilon_\lambda = \frac{I_\lambda(\text{emitted})}{B_\lambda(T)} \tag{4.13}$$

and the (monochromatic) *absorptivity, reflectivity, and transmissivity*[8] the fractions of the incident monochromatic intensity that a body absorbs, reflects, and transmits, i.e.,

$$\alpha_\lambda = \frac{I_\lambda(\text{absorbed})}{I_\lambda(\text{incident})}, \quad R_\lambda = \frac{I_\lambda(\text{reflected})}{I_\lambda(\text{incident})}$$

$$\text{and } T_\lambda = \frac{I_\lambda(\text{transmitted})}{I_\lambda(\text{incident})} \tag{4.14}$$

[8] Absorptivity, emissivity, reflectivity, and transmissivity may refer either to intensities or to flux densities, depending on the context in which they are used. In this subsection they refer to intensities. Also widely used in the literature is the term *transmittance*, which is synonymous with *transmissivity*. In contrast, *emittance* and *reflectance* sometimes refer to actual intensities or flux densities. Here we use only the terms ending in ..*ivity*, which always refer to the dimensionless ratios defined in (4.13) and (4.14).

4.3.5 Kirchhoff's Law

Kirchhoff's[9] *law* states that

$$\varepsilon_\lambda = \alpha_\lambda \tag{4.15}$$

To understand the basis for this relationship, consider an object (or medium) fully enclosed within a cavity with interior walls that radiates as a blackbody. Every point on the surface of the object is exchanging radiation with the walls of the cavity. The object and the walls of the cavity are in radiative equilibrium. Hence the second law of thermodynamics requires that they be at the same temperature. The radiation incident on the object at a particular wavelength λ coming from a particular direction is given by the Planck function (4.10). The fraction α_λ of that incident radiation is absorbed. Because the system is in radiative equilibrium, the same monochromatic intensity must be returned from the body at each wavelength and along each ray path. Because the object and the walls of the cavity are at the same temperature, it follows that at all wavelengths its emissivity must be the same as its absorptivity.

Because absorptivity and emissivity are intrinsic properties of matter, Kirchhoff postulated that the equality $\varepsilon_\lambda = \alpha_\lambda$ should hold even when the object is removed from the cavity and placed in a field of radiation that is not isotropic, and with which it is not necessarily in radiative equilibrium. Additional rationalizations of Kirchhoff's law are provided in Exercises 4.35–4.38.

Kirchhoff's law is applicable to gases, provided that the frequency of molecular collisions is much larger than the frequency with which molecules absorb and emit radiation in the vicinity of the electromagnetic spectrum near the wavelength of interest. When this condition is satisfied, the gas is said to be in *local thermodynamic equilibrium (LTE)*. In the Earth's atmosphere LTE prevails below altitudes of ~60 km.

4.3.6 The Greenhouse Effect

Solar and terrestrial radiation occupy different ranges of the electromagnetic spectrum that we have been referring to as *shortwave* and *longwave*. For reasons explained in the next section, water vapor, carbon dioxide, and other gases whose molecules have electric dipole moments absorb radiation more strongly in the longwave part of the spectrum occupied by outgoing terrestrial radiation than the shortwave part occupied by incoming solar radiation. This distinction is reflected in the transmissivity spectra of the atmosphere shown in the lower part of Fig. 4.7. Hence, incoming solar radiation passes through the atmosphere quite freely, whereas terrestrial radiation emitted from the Earth's surface is absorbed and reemitted in its upward passage through the atmosphere. The following highly simplified exercise shows how the presence of such greenhouse gases in a planetary atmosphere tends to warm the surface of the planet.

Exercise 4.8 The Earth-like planet considered in Exercise 4.6 has an atmosphere consisting of multiple isothermal layers, each of which is transparent to shortwave radiation and completely opaque to longwave radiation. The layers and the surface of the planet are in radiative equilibrium. How is the surface temperature of the planet affected by the presence of this atmosphere?

Solution: Begin by considering an atmosphere composed of a single isothermal layer. Because the layer is opaque in the longwave part of the spectrum, the equivalent blackbody temperature of the planet corresponds to the temperature of the atmosphere. Hence, the atmosphere must emit F units radiation to space as a blackbody to balance the F units of incoming solar radiation transmitted downward through the top of the atmosphere. Because the layer is isothermal, it also emits F units of radiation in the downward direction. Hence, the downward radiation at the surface of the planet is F units of incident solar radiation plus F units of longwave radiation emitted from the atmosphere, a total of $2F$ units, which must be balanced by an upward emission of $2F$ units of longwave radiation from the surface. Hence, from the Stefan–Boltzmann law (4.12) the temperature of the surface of the planet is 303 K, i.e., 48 K higher than it would be in the absence of an atmosphere, and the temperature of the atmosphere is the same as the temperature of the surface of the planet calculated in Exercise 4.6.

If a second isothermal, opaque layer is added, as illustrated in Fig. 4.9, the flux density of downward radiation incident upon the lower layer will be $2F$

[9] **Gustav Kirchhoff** (1824–1887) German physicist; professor at the University of Breslau. In addition to his work in radiation, he made fundamental discoveries in electricity and spectroscopy. Discovered cesium and iridium.

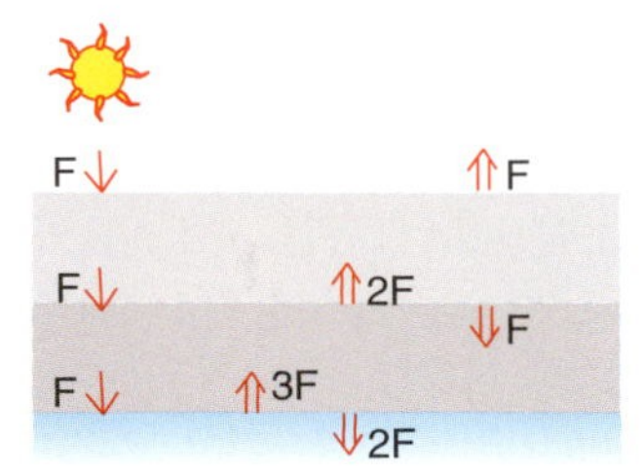

Fig. 4.9 Radiation balance for a planetary atmosphere that is transparent to solar radiation and consists of two isothermal layers that are opaque to planetary radiation. Thin downward arrows represent the flux of *F* units of shortwave solar radiation transmitted downward through the atmosphere. Thicker arrows represent the emission of longwave radiation from the surface of the planet and from each of the layers. For radiative equilibrium the net radiation passing through the Earth's surface and the top of each of the layers must be equal to zero.

(F units of solar radiation plus F units of longwave radiation emitted by the upper layer). To balance the incident radiation, the lower layer must emit $2F$ units of longwave radiation. Because the layer is isothermal, it also emits $2F$ units of radiation in the downward radiation. Hence, the downward radiation at the surface of the planet is F units of incident solar radiation plus $2F$ units of longwave radiation emitted from the atmosphere, a total of $3F$ units, which must be balanced by an upward emission of $3F$ units of longwave radiation from the surface. ■

By induction, the aforementioned analysis can be extended to an N-layer atmosphere. The emissions from the atmospheric layers, working downward from the top, are F, $2F$, $3F \ldots NF$ and the corresponding radiative equilibrium temperatures are 303, 335 .... $[(N+1)F/\sigma]^{1/4}$ K. The geometric thickness of opaque layers decreases approximately exponentially as one descends through the atmosphere due to the increasing density of the absorbing medium with depth. Hence, the radiative equilibrium lapse rate steepens with increasing depth. In effect, radiative transfer becomes less and less efficient at removing the energy absorbed at the surface of the planet due to the increasing blocking effect of greenhouse gases. Once the radiative equilibrium lapse rate exceeds the adiabatic lapse rate (Eq. 3.53), convection becomes the primary mode of energy transfer.

That the global mean surface temperature of the Earth is 289 K rather than the equivalent blackbody temperature 255 K, as calculated in Exercise 4.6, is attributable to the greenhouse effect. Were it not for the upward transfer of latent and sensible heat by fluid motions within the Earth's atmosphere, the disparity would be even larger.

To perform more realistic radiative transfer calculations, it will be necessary to consider the dependence of absorptivity upon the wavelength of the radiation. It is evident from the bottom part of Fig. 4.7 that the wavelength dependence is quite pronounced, with well-defined *absorption bands* identified with specific gaseous constituents, interspersed with *windows* in which the atmosphere is relatively transparent. As shown in the next section, the wavelength dependence of the absorptivity is even more complicated than the low-resolution absorption spectra in Fig. 4.7 would lead us to believe.

4.4 Physics of Scattering and Absorption and Emission

The scattering and absorption of radiation by gas molecules and aerosols all contribute to the extinction of the solar and terrestrial radiation passing through the atmosphere. Each of these contributions is linearly proportional to (1) the intensity of the radiation at that point along the ray path, (2) the local concentration of the gases and/or particles that are responsible for the absorption and scattering, and (3) the effectiveness of the absorbers or scatterers.

Let us consider the fate of a beam of radiation passing through an arbitrarily thin layer of the atmosphere along a specific path, as depicted in Fig. 4.10. For each kind of gas molecule and particle that the beam encounters, its monochromatic intensity is decreased by the increment

$$dI_\lambda = -I_\lambda K_\lambda N \sigma ds \tag{4.16}$$

where N is the number of particles per unit volume of air, σ is the areal cross section of each particle, K_λ is the (dimensionless) *scattering* or *absorption efficiency*, and ds is the differential path length along the ray path of the incident radiation. An *extinction efficiency*, which represents the combined effects of scattering and absorption in depleting the intensity of radiation passing through the layer, can be defined in a similar manner. The product KnNs is called the scattering, absorption, or extinction cross section. In the case of a gaseous atmospheric constituent, it is sometimes convenient to express the rate of scattering or absorption in the form

$$dI_\lambda = -I_\lambda \rho r k_\lambda ds \tag{4.17}$$

where ρ is the density of the air, r is the mass of the absorbing gas per unit mass of air, and k_λ is the *mass absorption coefficient*, which has units of $m^2\ kg^{-1}$.

In the aforementioned expressions the products $N\sigma K_\lambda$ and $\rho r k_\lambda$ are *volume scattering, absorption, or extinction coefficients*, depending on the context, and have units of m^{-1}. The contributions of the various species of gases and particles are additive (i.e., $K_\lambda N\sigma = (K_\lambda)_1 N_1\sigma_1 + (K_\lambda)_2 N_2\sigma_2 + \ldots$), as are the contributions of scattering and absorption to the extinction of the incident beam of radiation; i.e.,

$$K_\lambda(\text{extinction}) = K_\lambda(\text{scattering}) + K_\lambda(\text{absorption}) \tag{4.18}$$

4.4.1 Scattering by Air Molecules and Particles

At any given place and time, particles including aerosols with a wide variety of shapes and sizes, as well as cloud droplets and ice crystals, may be present. Nonetheless it is instructive to consider the case of scattering by a spherical particle of radius r, for which the scattering, absorption, or extinction efficiency K_λ in (4.16) can be prescribed on the

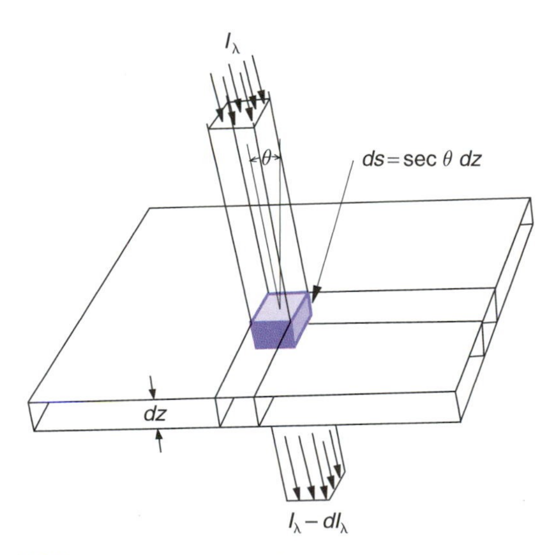

Fig. 4.10 Extinction of incident parallel beam solar radiation as it passes through an infinitesimally thin atmospheric layer containing absorbing gases and/or aerosols.

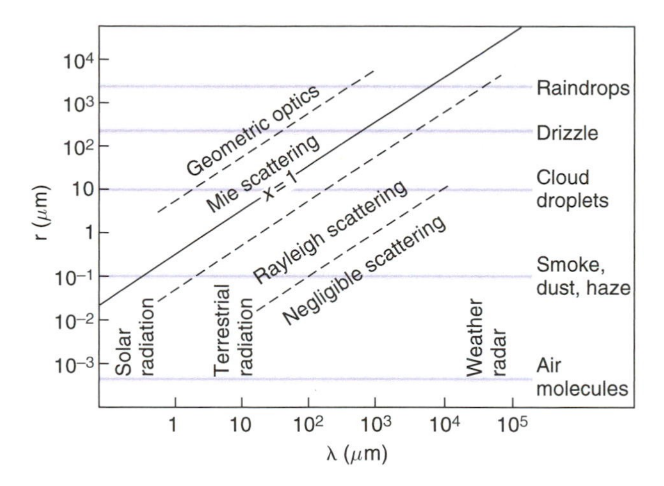

Fig. 4.11 Size parameter x as a function of wavelength (λ) of the incident radiation and particle radius r.

basis of theory, as a function of a dimensionless *size parameter*

$$x = \frac{2\pi r}{\lambda} \tag{4.19}$$

and a complex *index of refraction* of the particles ($\mathbf{m} = m_r + im_i$), whose real part m_r is the ratio of the speed of light in a vacuum to the speed at which light travels when it is passing through the particle. Figure 4.11 shows the range of size parameters for various kinds of particles in the atmosphere and radiation in various wavelength ranges. For the scattering of radiation in the visible part of the spectrum, x ranges from much less than 1 for air molecules to ~1 for haze and smoke particles to >>1 for raindrops.

Particles with $x << 1$ are relatively ineffective at scattering radiation. Within this so-called *Rayleigh scattering* regime the expression for the scattering efficiency is of the form

$$K_\lambda \propto \lambda^{-4} \tag{4.20}$$

and the scattering is divided evenly between the forward and backward hemispheres, as indicated in Fig. 4.12a. For values of the size parameter comparable to or greater than 1 the scattered radiation is directed mainly into the forward hemisphere, as indicated in subsequent panels.

Figure 4.13 shows K_λ as a function of size parameter for particles with $m_r = 1.5$ and a range of values of m_i. Consider just the top curve that corresponds

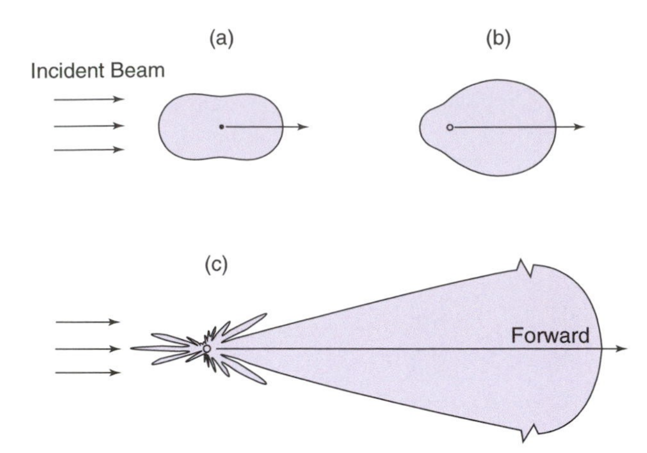

Fig. 4.12 Schematic showing the angular distribution of the radiation at visible (0.5 μm) wavelength scattered by spherical particles with radii of (a) 10^{-4} μm, (b) 0.1 μm, and (c) 1 μm. The forward scattering for the 1-μm aerosol is extremely large and is scaled for presentation purposes. [Adapted from K. N. Liou, *An Introduction to Atmospheric Radiation*, Academic Press, p. 7, Copyright (2002), with permission from Elsevier.]

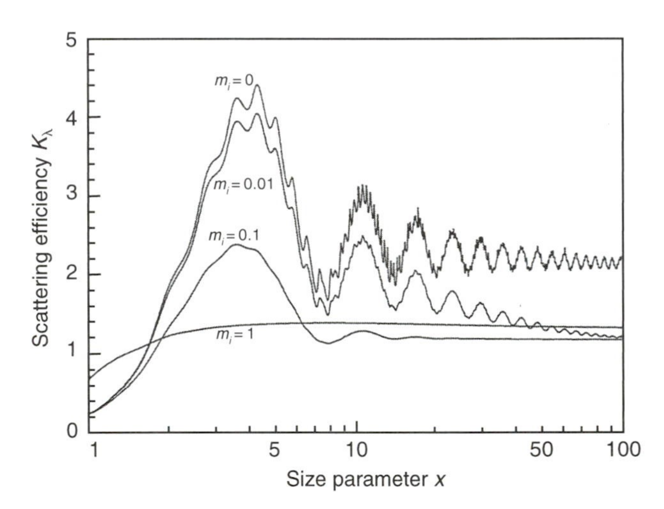

Fig. 4.13 Scattering efficiency K_λ as a function of size parameter x, plotted on a logarithmic scale, for four different refractive indices with $m_r = 1.5$ and m_i ranging from 0 to 1, as indicated. [From K. N. Liou, *An Introduction to Atmospheric Radiation*, Academic Press, p. 191, Copyright (2002), with permission from Elsevier.]

to $m_i = 0$ (no absorption). For $0.1 \leqslant x \lesssim 50$, referred to as the *Mie*[10] *scattering regime,* K_λ exhibits a damped oscillatory behavior, with a mean around a value of 2, and for $x \gtrsim 50$, the range referred to as the *geometric optics regime*, the oscillatory behavior is less prominent and $K_\lambda \approx 2$.

Exercise 4.9 Estimate the relative efficiencies with which red light ($\lambda \simeq 0.64\mu$m) and blue light ($\lambda \simeq 0.47\mu$m) are scattered by air molecules.

Solution: From (4.20)

$$\frac{K(\text{blue})}{K(\text{red})} = \left(\frac{0.64}{0.47}\right)^4 = 3.45$$

■

Hence, the preponderance of blue in light scattered by air molecules, as evidenced by the blueness of the sky on days when the air is relatively free from aerosols.

Figure 4.14 shows an example of the coloring of the sky and sunlit objects imparted by Rayleigh scattering. The photograph was taken just after sunrise. Blue sky is visible overhead, while objects in the foreground, including the aerosol layer, are illuminated by sunlight in which the shorter wavelengths (bluer colors) have been depleted by scattering along its long, oblique path through the atmosphere.

Ground-based weather radars and remote sensing of rainfall from instruments carried aboard satellites exploit the size strong dependence of scattering efficiency K upon size parameter x for microwave radiation in the 1- to 10-cm wavelength range incident upon clouds with droplet radii on the order of millimeters. In contrast to infrared radiation, which

Fig. 4.14 Photograph of the Great Wall of China, taken just after sunrise.

[10] **Gustav Mie** (1868–1957) German physicist. Carried out fundamental studies on the theory of electromagnetic scattering and kinetic theory.

is strongly absorbed by clouds, microwave radiation passes through clouds with droplets ranging up to hundreds of micrometers in radius with little or no scattering. Radiation backscattered from the pulsed radar signal by the larger drops reveals the regions of heavy precipitation.

Scattering of parallel beam solar radiation and moonlight gives rise to a number of distinctive optical effects, that include

- *rainbows* (Figs. 4.15 and 4.16) formed by the refraction and internal reflection of sunlight in water drops (usually raindrops). The bow appears as a circle, rarely complete, centered on the *antisolar point*, which lies on the line from the sun passing through the eye of the observer. The colors of the rainbow from its outer to its inner circumference are red, orange, yellow, green, blue, indigo, and violet. Rainbows are usually seen in falling rain with drop diameters of a few millimeters. Because the drops are much larger than the wavelength of the light, the classical theory of geometric optics provides a reasonable description for the primary bow.
- *bright haloes* produced by the refraction of light by hexagonal, prismshaped ice crystals in high, thin cirrostratus cloud decks, the most common of which are at angles of 22° and 46°, as shown in Fig. 4.17 and the schematic Fig. 4.18.[11]

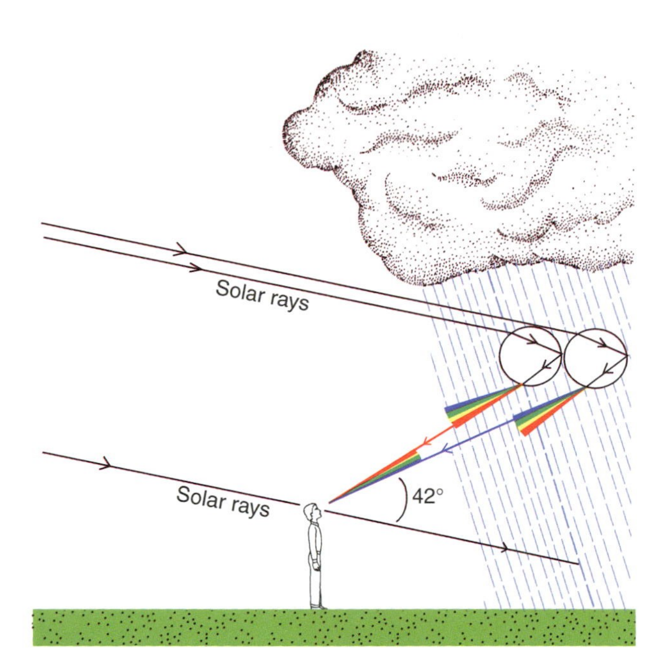

Fig. 4.16 A solar ray is refracted and internally reflected (once) and refracted again by a raindrop to form the (primary) rainbow. The rainbow ray is the brightest and has the smallest angle of deviation of all the rays that encounter raindrop undergo these optical processes. Like a prism, refraction by the raindrop disperses visible light into its component colors forming the rainbow color band. The secondary rainbow is produced by double reflection within raindrops, which appears about 8 degrees above the primary rainbow, with the order of the colors reversed. The rainbow is a mosaic produced by passage of light through the circular cross section of myriad raindrops.

Fig. 4.15 Primary rainbow with a weaker secondary rainbow above it and supernumerary bows below it. [Photograph courtesy of Joanna Gurstelle.]

Fig. 4.17 Haloes of 22° and 46° (faint) formed in a thin cloud consisting of ice crystals. [Photograph courtesy of Alistair Fraser.]

[11] The angular radius of the 22° halo is roughly the same as that subtended at the eye by the distance between the top of the thumb and the little finger when the fingers are spread wide apart and held at arms length. (The reader is warned not to stare at the sun since this can cause eye damage.)

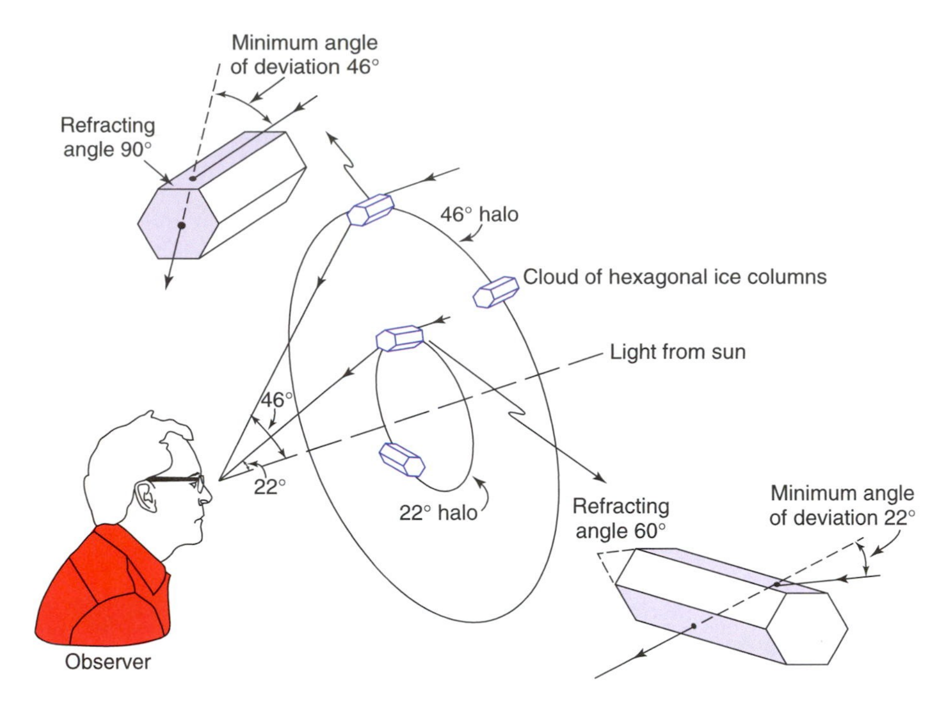

Fig. 4.18 Refraction of light in hexagonal ice crystals to produce the 22° and 46° haloes.

- *coronas* produced by the diffraction of light in water droplets in low or (sometimes) middle cloud decks. They consist of colored rings at an angular radius of less than 15° from the light sources, e.g., the sun or moon. If the cloud droplets are fairly uniform in size, several sequences of rings may be seen: the spacing of the rings depends upon droplet size. In each sequence the inside ring is violet or blue and the outside ring is red. An example is shown in Fig. 4.19.

Fig. 4.19 A corona around the sun produced by the diffraction of light in cloud droplets. [Photograph courtesy of Harald Edens.]

4.4.2 Absorption by Particles

The absorption of radiation by particles is of interest in its own right, and affects scattering as well. The theoretical framework discussed in the previous subsection (commonly referred to as *Mie theory*) predicts both the scattering and the absorption of radiation by homogeneous spherical particles, where the real part of the refractive index relates to the scattering and the imaginary part relates to the absorption. This brief section mentions just a few of the consequences of absorption. In the plot of K versus x (Fig. 4.13) the presence of absorption tends to damp the oscillatory behavior in the Mie regime. In the limit of $x >> 1$, the extinction coefficient always approaches 2 but, in accordance with (4.13) the scattering coefficient, ranges from as low as 1 in the case of strong absorption to as high as 2 for negligible absorption. In the longwave part of the electromagnetic spectrum, cloud droplets absorb radiation so strongly that even a relatively thin cloud layer of clouds behaves as a blackbody, absorbing virtually all the radiation incident from above and below.

4.4.3 Absorption and Emission by Gas Molecules

Whenever radiation interacts with matter it is absorbed, scattered, or emitted in discrete packets called *photons*. Each photon contains energy

$$E = h\tilde{\nu} \tag{4.21}$$

where h is *Planck's constant* (6.626×10^{-34} J s). Hence, the energy carried by a photon is inversely proportional to the wavelength of the radiation.

a. Absorption continua

Extreme ultraviolet radiation with wavelengths $\lesssim 0.1$ μm, emitted by hot, rarefied gases in the sun's outer atmosphere, is sufficiently energetic to strip electrons from atoms, a process referred to as *photoionization*. Solar radiation in this wavelength range, which accounts for only around 3 millionths of the sun's total output, is absorbed in the *ionosphere*, at altitudes of 90 km and above, giving rise to sufficient numbers of free electrons to affect the propagation of radio waves.

Radiation at wavelengths up to 0.24 μm is sufficiently energetic to break O_2 molecules apart into oxygen atoms, a process referred to as *photodissociation*. The oxygen atoms liberated in this reaction are instrumental in the production of ozone (O_3), as explained in Section 5.7.1. Ozone, in turn, is dissociated by solar radiation with wavelengths extending up to 0.31 μm, almost to the threshold of visible wavelengths. This reaction absorbs virtually all of the ~2% of the sun's potentially lethal ultraviolet radiation. The ranges of heights and wavelengths of the primary photoionization and photodissociation reactions in the Earth's atmosphere are shown in Fig. 4.20.

Photons that carry sufficient energy to produce these reactions are absorbed and any excess energy is imparted to the kinetic energy of the molecules, raising the temperature of the gas. Since the energy required to liberate electrons and/or break molecular bonds is very large, the so-called *absorption continua* associated with these reactions are confined to the x-ray and ultraviolet regions of the spectrum. Most of the solar radiation with wavelengths longer than 0.31 μm penetrates to the Earth's surface.

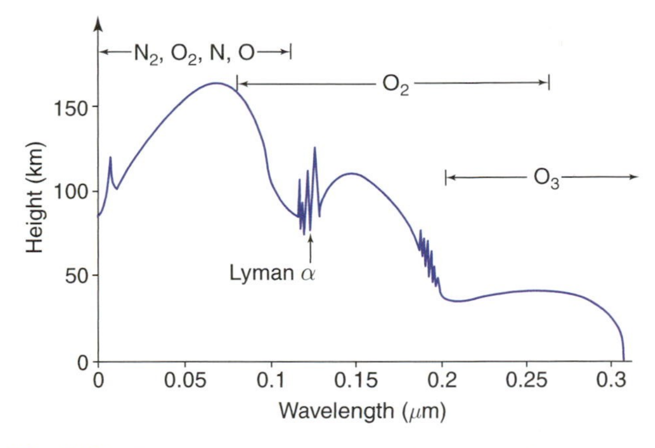

Fig. 4.20 Depth of penetration of solar ultraviolet radiation in the Earth's atmosphere for overhead sun and an average ozone profile. [Adapted from K. N. Liou, *An Introduction to Atmospheric Radiation*, Academic Press, p. 78, Copyright (2002), with permission from Elsevier.]

b. Absorption lines

Radiation at visible and infrared wavelengths does not possess sufficient energy to produce photoionization or photodissociation, but under certain conditions appreciable absorption can nonetheless occur. To understand the processes that are responsible for absorption at these longer wavelengths, it is necessary to consider other kinds of changes in the state of a gas molecule. The internal energy of a gas molecule can be written in the form

$$E = E_o + E_v + E_r + E_t \tag{4.22}$$

where E_o is the energy level of the orbits of the electrons in the atoms, E_v and E_r refer to the energy levels corresponding to the *vibrational* and *rotational* state of the molecule, and E_t is the *translational* energy associated with the random molecular motions. In discussing the first law of thermodynamics in Chapter 3, we considered only changes E_t, but in dealing with radiative transfer it is necessary to consider changes in the other components of the internal energy as well.

Quantum mechanics predicts that only certain configurations of electron orbits are permitted within each atom, and only certain vibrational frequencies and amplitudes and only certain rotation rates are permitted for a given molecular species. Each possible combination of electron orbits, vibration, and rotation is characterized by its own energy level, which represents the sum of the three kinds of energy. (The translational component of the energy is not quantized in this manner.) A molecule may undergo a transition to a higher energy level by absorbing electromagnetic radiation and it may drop to a lower level by emitting radiation. Absorption and emission can occur only in association with discrete changes in energy level ΔE. The frequency of the absorbed or emitted radiation is related to the change in energy level through the relation

$$\Delta E = h\widetilde{\nu} \tag{4.23}$$

Absorptivity at visible and longer wavelengths can thus be described in terms of a *line spectrum* consisting of extremely narrow *absorption lines* separated by much wider gaps in which the gas is virtually transparent to incident radiation.

The changes in state of molecules that give rise to these absorption lines may involve *orbital*, *vibrational*, or *rotational* transitions or combinations thereof. Orbital transitions are associated with absorption lines in the ultraviolet and visible part of the spectrum; vibrational changes with near-infrared and infrared wavelengths; and rotational lines, which involve the smallest changes in energy, with infrared and microwave radiation. The absorption spectra of the dominant species O_2 and N_2 exhibit a sparse population of absorption lines because these molecular species do not possess an electric dipole, even when they are vibrating. In contrast, so-called "greenhouse gases" (notably H_2O, CO_2, O_3, and trace species such as such as CH_4, N_2O, CO, and the chlorofluorocarbons) exhibit myriads of closely spaced absorption lines in the infrared region of the spectrum that are due to pure rotational or simultaneous vibrational–rotational transitions.

c. Broadening of absorption lines

The absorption lines of molecules are of finite width due to the inherent uncertainty in quantizing their energy levels, but this "natural broadening" is inconsequential in comparison to the broadening attributable to the motions and collisions of the gas molecules, i.e.,

- *Doppler broadening*: the Doppler shifting of frequencies at which the gas molecules experience the incident radiation by virtue of their random motions toward or away from the source of the radiation, and
- *pressure broadening*: (also referred to as *collision broadening*) associated with molecular collisions.

The absorption spectra in the vicinity of pressure- and Doppler-broadened absorption lines can be represented by

$$k_\nu = Sf(\nu - \nu_0) \tag{4.24}$$

where

$$S = \int_0^\infty k_\nu d\nu \tag{4.25}$$

is the line intensity, ν_0 is the wave number on which the line is centered, and f is the so-called *shape factor* or *line profile*. The shape factor for Doppler broadening is inferred from the Maxwell[12]–Boltzmann distribution of the velocity of the molecules in a gas, which has the shape of the familiar Gaussian probability distribution. It is of the form

$$f = \frac{1}{\alpha_D\sqrt{\pi}} \exp\left[-\left(\frac{\nu - \nu_0}{\alpha_D}\right)^2\right] \tag{4.26}$$

where

$$\alpha_D = \frac{\nu_0}{c^*}\left(\frac{2kT}{m}\right)^{1/2} \tag{4.27}$$

In this expression the so-called *half-width* of the line (i.e., the distance between the center of the line

[12] **James Clerk Maxwell** (1831–1879) Scottish physicist. Often rated as second only to Newton in terms of his contributions to physics. First Cavendish Professor of physics at Cambridge University. He showed that light is an electromagnetic wave, made one of the first color photographs, and made major contributions to thermodynamics and the kinetic theory of gases.

and the points at which the amplitude is equal to half the peak amplitude) is $\alpha_D\sqrt{\ln 2}$, m is the mass of the molecule and k is the Boltzmann's constant (1.381×10^{-23} J K^{-1} molecule^{-1}).

The shape factor for pressure broadening, commonly referred to as the *Lorentz*[13] *line shape*, is given by

$$f = \frac{\alpha}{\pi\left[(\nu - \nu_0)^2 + \alpha^2\right]} \tag{4.28}$$

In this expression the half-width of the line is determined by

$$\alpha \propto \frac{p}{T^N} \tag{4.29}$$

which is proportional to the frequency of molecular collisions. The exponent N ranges from 1/2 to 1 depending on the molecular species.

Shapes of absorption lines of the same strength and half-width, but broadened by these two distinctly different processes, are contrasted in Fig. 4.21. The "wings" of the absorption lines shaped by pressure broadening extend out farther from the center of the line than those shaped by Doppler broadening. For a water vapor line at 400 cm^{-1} and a temperature of 300 K, the Doppler line width is 7×10^{-4} cm^{-1}. A typical water vapor line width for air at the same temperature at the Earth's surface is ~100 times wider due to the presence of pressure broadening.[14] Below ~20 km, pressure broadening is the dominant factor in determining the width of absorption lines, whereas above 50 km, where molecular collisions are much less frequent, Doppler broadening is the dominant factor. In the intermediate layer between 20 and 50 km, the line shape is a convolution of the Doppler and Lorentz shapes.

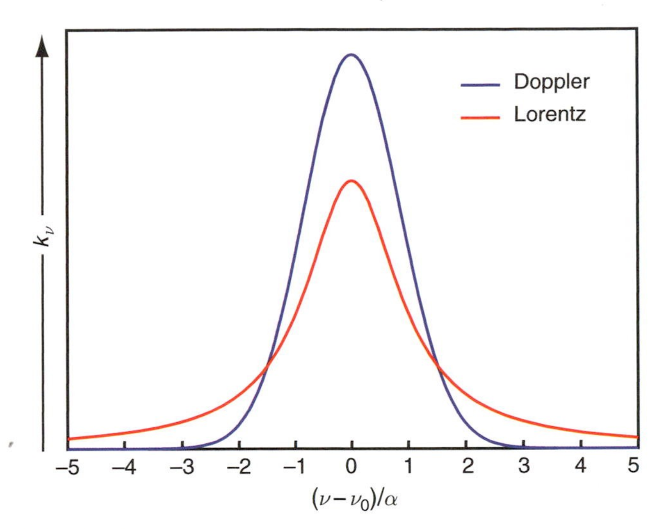

Fig. 4.21 Contrasting absorption line shapes associated with Doppler broadening and pressure broadening. Areas under the two profiles, indicative of the line intensity S, are the same. [Courtesy of Qiang Fu.]

Laboratory measurements of absorption spectra exist for only a very limited sampling of pressures and temperatures. However, through the use of theoretically derived absorption line information, adjusted empirically to improve the fit with existing measurements, atmospheric physicists and climate modelers are able to calculate the absorption spectra for each of the radiatively important atmospheric gases for any specified thermodynamic conditions.[15] An example showing the excellent agreement between observed and theoretically derived absorption spectra is shown in Fig. 4.22. Note the narrowness of the lines, even when the effects of Doppler and pressure broadening are taken into account. The greatest uncertainties in theoretically derived absorption spectra are in the so-called "continua," where the superposition of the outermost parts of the wings of many different lines in nearby line clusters produces weak but in some cases significant absorption.

[13] **Hendrick Antoon Lorentz** (1853–1928) Dutch physicist. Won Nobel prize for physics in 1902 for his theory of electromagnetic radiation, which gave rise to the special theory of relativity. Refined Maxwell's theory of electromagnetic radiation so it better explained the reflection and refraction of light.

[14] The dominance of pressure broadening is mainly due to the fact that, for typical temperatures and pressures in the lower atmosphere, $\alpha >> \alpha_D$. The difference in line shape also contributes, but it is of secondary importance.

[15] The most comprehensive archive of these theoretically derived absorption line information, the *high-resolution transmission molecular absorption (HITRAN)* data base, contains absorption lines for many different gases. This data base contains line intensities at reference temperature, the wave numbers at which the lines are centered, the pressure half-widths at reference temperature and pressure, and the lower energy levels for over a million absorption lines.

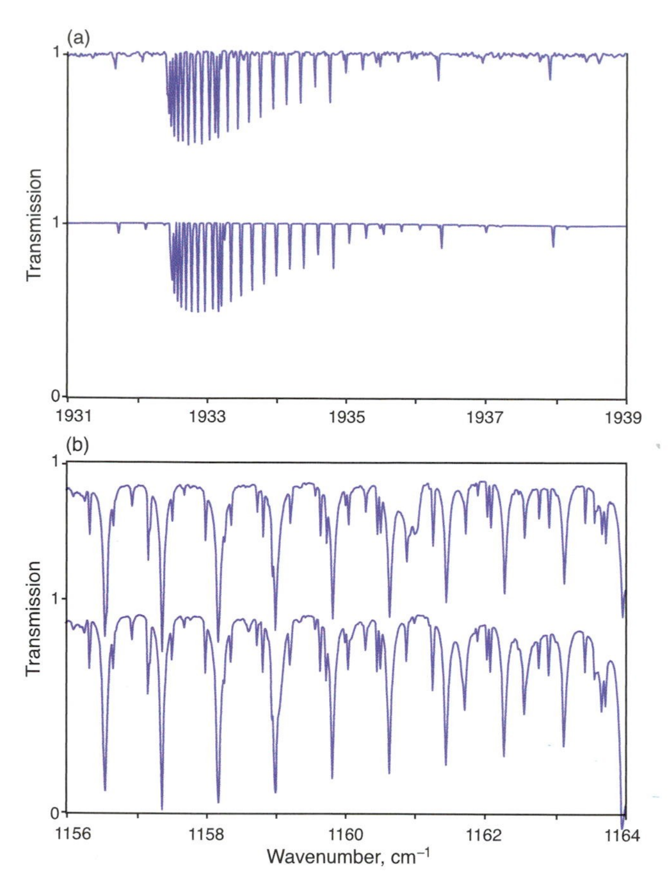

Fig. 4.22 Comparisons of observed and calculated transmissivity spectra. (a) Spectral range 1931 to 1939 cm^{-1} for carbon dioxide and (b) Spectral range 1156 to 1164 cm^{-1} for ozone and nitrous oxide. The upper plot in each panel is the observed spectrum and the lower plot is the calculated spectrum. [From R. M. Goody, R. M. and Y. L. Yung, *Atmospheric Radiation*, 2nd ed., Oxford University Press (1995), p.120. By permission of Oxford University press, Inc.]

4.5 Radiative Transfer in Planetary Atmospheres

4.5.1 Beer's Law

Equations (4.17) and/or (4.16) may be integrated from the top of the atmosphere ($z = \infty$) down to any level (z) to determine what fraction of the incident beam or "pencil" of radiation has been attenuated due to absorption and/or scattering and how much remains undepleted. Integrating (4.17) with $ds = \sec\theta\, dz$ yields

$$\ln I_{\lambda\infty} - \ln I_\lambda = \sec\theta \int_z^\infty k_\lambda \rho r\, dz \qquad (4.30)$$

Taking the antilog of both sides we obtain

$$I_\lambda = I_{\lambda\infty} e^{-\tau_\lambda \sec\theta} = I_{\lambda\infty} T_\lambda \qquad (4.31)$$

where

$$\tau_\lambda = \int_z^\infty k_\lambda \rho r\, dz \qquad (4.32)$$

and

$$T_\lambda = e^{-\tau_\lambda \sec\theta} \qquad (4.33)$$

is the *transmissivity* of the layer.

This set of relationships and definitions, collectively referred to here as *Beer's law*, but also known as *Bouguer's law*, and *Lambert's law*,[16,17,18] states that the monochromatic intensity I_λ decreases monotonically with path length as the radiation passes through the layer. The dimensionless quantity τ_λ, referred to as the *normal optical depth* or *optical thickness* depending on the context in which it is used, is a measure of the cumulative depletion that a beam of radiation directed straight downward (zenith angle $\theta = 0$) would experience in passing through the layer. It follows that, in the absence of scattering, the monochromatic absorptivity

$$\alpha_\lambda = 1 - T_\lambda = 1 - e^{-\tau_\lambda \sec\theta} \qquad (4.34)$$

approaches unity exponentially with increasing optical depth. Optical depths for the scattering and extinction of radiation passing through a medium containing aerosols or cloud droplets can be defined in a similar manner.

[16] **August Beer** (1825–1863). German physicist, noted for his work on optics.

[17] **Pierre Bouguer** (1698–1758). Taught by his father. Awarded the Grand Prix of the Academie des Sciences for studies of naval architecture, observing the stars at sea, and for observations of the magnetic declination at sea. First to attempt to measure the density of the Earth, using the deflection of a plumb line due to the attraction of a mountain. Sometimes known as the "father of photometry." He compared the brightness of the moon to that of a "standard" candle flame (1725). Bouguer's law was published in 1729.

[18] **Johann Heinrich Lambert** (1728–1777) Swiss–German mathematician, astronomer, physicist, and philosopher. Son of a tailor, Lambert was largely self-educated. Proved that π is an irrational number (1728). Made the first systematic investigation of hyperbolic functions. Carried out many investigations of heat and light.

Exercise 4.10 Parallel beam radiation is passing through a layer 100 m thick, containing an absorbing gas with an average density of 0.1 kg m^{-3}. The beam is directed at an angle of 60° relative to the normal to the layer. Calculate the optical thickness, transmissivity, and absorptivity of the layer at wavelengths λ_1, λ_2, and λ_3, for which the mass absorption coefficients are 10^{-3}, 10^{-1}, and 1 m^2 kg^{-1}.

Solution: The mass of the absorbing gas that the beam of radiation encounters along its slant path length is given by

$$u \equiv \sec\theta \int_{z_B}^{z_T} \rho r\, dz \tag{4.35}$$

where z_B and z_T are the heights of the bottom and top of the layer. Substituting, $\sec\theta = 2$, $\rho = 0.1$ kg m^{-3}, $r = 1$, and a layer thickness of 100 m, we obtain

$$u = 2 \times 0.1 \text{ kg m}^{-3} \times 100 \text{ m}$$
$$= 20 \text{ kg m}^{-2}$$

Since k_λ can be assumed to be uniform within through the layer, Eq. (4.33) can be rewritten as

$$T_\lambda = e^{-\tau_\lambda} = e^{-k_\lambda u}$$

and (4.34) as

$$\alpha_\lambda = 1 - T_\lambda = 1 - e^{-k_\lambda u}$$

where

$$\tau_\lambda = k_\lambda \sec\theta \int_{z_B}^{z_T} \rho r\, dz = k_\lambda u \tag{4.36}$$

is the *slant path optical thickness*. Substituting for k_λ and u in the aforementioned equation yields

	$\lambda = \lambda_1$	$\lambda = \lambda_2$	$\lambda = \lambda_3$
τ_λ	0.02	2	20
T_λ	0.98	0.135	2×10^{-9}
α_λ	0.02	0.865	1.00

■

I_λ and T_λ decrease monotonically with increasing geometric depth in the atmosphere. For downward directed radiation ($\sec\theta = 1$), it is shown in the Exercise 4.44 at the end of this chapter that they decrease most rapidly around the level where $\tau_\lambda = 1$, commonly referred to as the *level of unit optical depth*. This result can be understood by considering the shape of the vertical profile of the absorption rate dI_λ/dz, which is shown in Fig. 4.23 together with profiles of I_λ and ρ. We recall from (4.17) that if r, the mixing ratio of the absorbing gas, and k_λ, the mass absorption coefficient, are both independent of height,

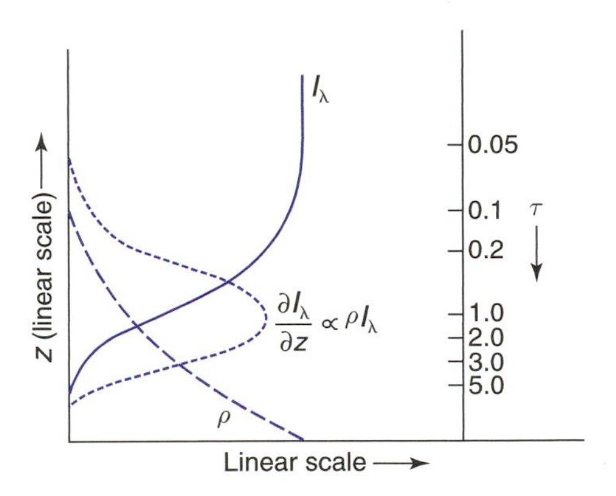

Fig. 4.23 Vertical profiles of the monochromatic intensity of incident radiation, the rate of absorption of incident radiation per unit height, air density and optical depth, for k_λ and r independent of height.

$$\frac{dI_\lambda}{dz} \propto (I_\lambda \times \rho)$$

The scale for optical depth is shown at the right-hand side of Fig. 4.23. Well above the level of unit optical depth, the incoming beam is virtually undepleted, but the density is so low that there are too few molecules to produce appreciable amounts of absorption per unit path length. Well below the level of unit optical depth, there is no shortage of molecules, but there is very little radiation left to absorb.

The larger the value of the absorption coefficient k_λ and the larger the secant of the zenith angle, the smaller the density required to produce significant amounts of absorption and the higher the level of unit optical depth. For small values of k_λ, the radiation may reach the bottom of the atmosphere long before it reaches the level of unit optical depth. It is shown in Exercise 4.47 that for overhead parallel beam radiation incident upon an optically thick atmosphere, 80% of the energy is absorbed at levels between $\tau_\lambda = 0.2$ and $\tau_\lambda = 4.0$, which corresponds to a geometric depth of three scale heights.

The level of slant path unit optical depth is strongly dependent on the solar zenith angle. It is lowest in the atmosphere when the sun is directly overhead and it rises sharply as the sun drops close to the horizon. This dependence is exploited in remote sensing, as discussed in Box 4.1.

4.1 Indirect Determination of the Solar Spectrum

Before the advent of satellites, Beer's law in the form (4.30) was used to infer the emission spectrum of the sun on the basis of ground based measurements. Over the course of a single day with clear sky and good visibility, the incident solar radiation was measured at different time of day, yielding data like that shown in Fig 4.24. Over the relatively short interval in which the measurements were taken, $I_{\lambda\infty}$ and the normal optical depth are assumed to be constant so that $\ln I_\lambda$ is proportional to $\sec\theta$. This assumption is borne out by the almost perfect alignment of the data points in Fig. 4.24.

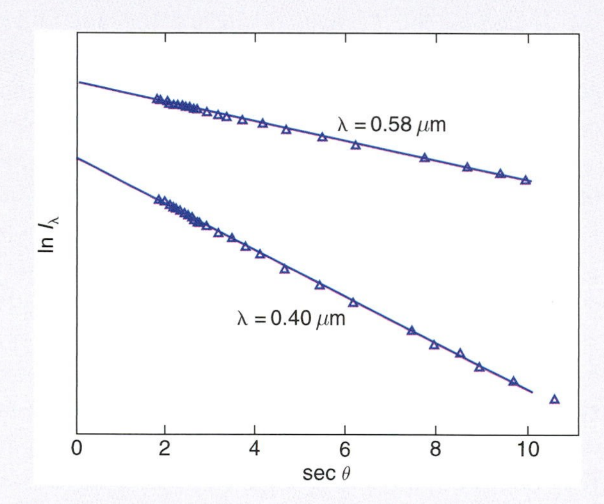

Fig. 4.24 Monochromatic intensity of solar radiation measured at the ground as a function of solar zenith angle under clear, stable conditions at Tucson, Arizona on 12 December 1970. [From *J. Appl. Meteor.*, **12**, 376 (1973).]

Measurements are inherently limited to the range $\sec\theta > 1$, but the linear fit to the data can be extrapolated back to $\sec\theta = 0$ to estimate $\ln I_{\lambda\infty}$.

An instrument called the *sunphotometer* is used to make instantaneous measurements of the optical thickness due to scattering and absorption by aerosols (called the *aerosol optical depth*) of the radiation from the sun (or moon) in its transit through the atmosphere to the location of the measurement.[19] For example, if the sunphotometer is located on the ground, it measures the total column aerosol optical depth. A sunphotometer uses photodiodes and appropriate narrow-band interference filters to measure I_{λ_1} at zenith angle θ_1 and I_{λ_2} at zenith angle θ_2. It follows from (4.30) that

$$\ln \frac{I_{\lambda_1}}{I_{\lambda_2}} = \tau_\lambda (\sec\theta_2 - \sec\theta_1) \qquad (4.37)$$

from which τ_λ can be derived. To isolate attenuation due to aerosols alone, λ_1 and λ_2 must be chosen so as not to coincide with molecular atmospheric absorption lines or bands and the influence of Rayleigh scattering by air molecules must be taken into account. Because particles of different sizes attenuate light differently at different wavelengths, variations of τ_λ with λ can be used to infer particle size spectra.

Sunphotometers can be calibrated using so-called *Langley*[20] *plots* like the one in Fig. 4.24,

Continued on next page

[19] Bouguer made visual estimates of the diminution of moonlight passing through the atmosphere. His measurements, made in 1725 in Britanny, showed much cleaner air than now.

[20] **Samuel Pierpont Langley** (1834–1906) American astronomer, physicist, and aeronautics pioneer. Built the first successful heavier-than-air, flying machine. This unmanned machine, weighing 9.7 kg and propelled by a steam engine, flew 1280 m over the Potomac River in 1896. His chief scientific interest was solar activity and its effects on the weather. Invented the bolometer to study radiation, into the infrared, from the sun. First to provide a clear explanation of how birds soar and glide without moving their wings. His first manned aircraft, catapulted off a houseboat in 1903, never flew but crashed "like a handful of wet mortar" into the Potomac. Nine days later the Wright brothers successfully flew the first manned aircraft. Langley died 3 years later, some say broken by the ridicule that the press treated his flying attempts. The NASA Langley Research Center is named after him.

4.1 Continued

from which both $I_{\lambda\infty}$ and τ_λ can be determined. The derived values of $I_{\lambda\infty}$ can then be compared with tabulated values of solar radiation at the top of the atmosphere at each filter wavelength to estimate the attenuation. Such calibrations are best done in an atmosphere that is temporally invariant and horizontally homogeneous (within ~50 km of the observer), because the Langley method assumes these conditions hold during measurements at different zenith angles. A favorite location for calibrating sunphotometers is atop Mauna Loa, Hawaii.

Networks of sunphotometers are deployed worldwide for monitoring atmospheric aerosol. Sunphotometers may also be mounted on aircraft, as was done in acquiring data for Fig. 4.25; in this case, the aerosol optical depths of layers of the atmosphere can be determined by subtracting the column optical depth at the top of a layer from that at the base of the layer. Small handheld sunphotometers are also available.

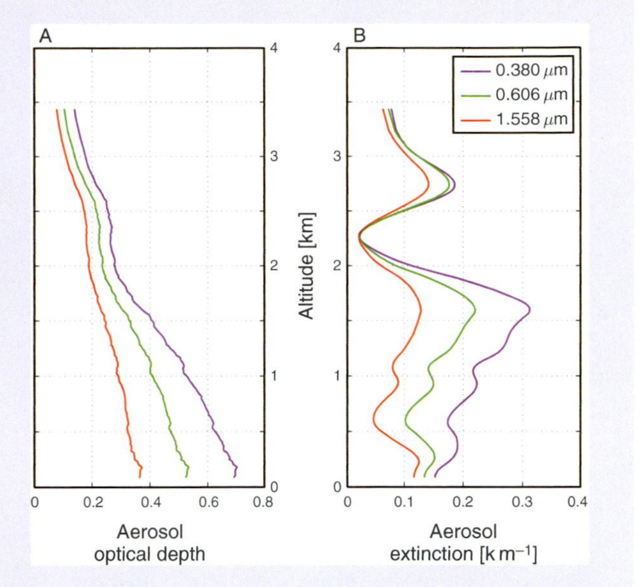

Fig. 4.25 Aerosol optical depth and aerosol extinction coefficient measured with an airborne sunphotometer south of Korea. The enhanced extinction coefficients between ~2.5–3 km were due to dust from Asia. [Courtesy of Gregory Schmidt, NASA-Ames Research Center.]

In this subsection, Beer's law and the concept of optical depth were derived from equations in which height is used as a vertical coordinate. If atmospheric pressure or some function of pressure had been used as a vertical coordinate, the profiles in Fig. 4.23 would be quite different in appearance. For example, in Exercise 4.45 the reader is invited to show that in an isothermal atmosphere, the rate of absorption per unit increment of pressure (i.e., *per unit mass*) is largest, not at the level of unit depth, but at the top of the atmosphere, where the optical depth is much less than 1 and the incident radiation is virtually undepleted. Hence, Beer's law should not be interpreted as indicating that layers of the atmosphere that are far removed from the level of unit optical depth are unaffected by the incident radiation.

4.5.2 Reflection and Absorption by a Layer of the Atmosphere

The conservation of energy requires that for radiation incident on a layer of aerosols or clouds

$$\alpha_\lambda^f + R_\lambda^f + T_\lambda^f = 1 \tag{4.38}$$

where α_λ^f, R_λ^f, and T_λ^f are the flux absorptivity, flux reflectivity, and flux transmissivity of the layer, i.e., the fractions of the incident flux density of solar radiation that are absorbed, transmitted, and reflected.

The concept of scattering and absorption efficiencies was introduced in the previous section. The combined effects of scattering and absorption in reducing the intensity are referred to as *extinction*, as defined by (4.18). The incident radiation may be scattered more than once in its passage through a layer, with each successive scattering event increasing the diversity of ray paths. In the absence of absorption, what started out as parallel beam radiation would (after a sufficient number of scattering events) be converted to isotropic radiation. So-called *multiple scattering* also greatly increases the path length of the incident radiation in its passage through the layer.

Three basic parameters are used to characterize the optical properties of aerosols, cloud droplets, and ice crystals:

- The volume extinction coefficient $N\sigma K_\lambda$ (extinction), a measure of the overall importance of the particles in removing radiation of the incident beam.

- The single scattering albedo, defined as

$$\omega_0(\lambda) = \frac{K_\lambda(\text{scattering})}{K_\lambda(\text{scattering}) + K_\lambda(\text{absorption})} \quad (4.39)$$

a measure of the relative importance of scattering and absorption. Values of single scattering albedo range from 1.0 for nonabsorbing particles to below 0.5 for strongly absorbing particles.
- The asymmetry parameter

$$g(\lambda) = \frac{1}{2}\int_{-1}^{1} P(\cos\theta')\cos\theta' d\cos\theta' \quad (4.40)$$

where $P(\cos\theta')$ is the normalized angular distribution of the scattered radiation (the so-called *scattering phase function*) and θ' is the angle between the incident radiation and the scattered radiation. The asymmetry factor is zero for isotropic radiation and ranges from -1 to $+1$, with positive values indicative of a predominance of forward scattering. It is evident from Fig. 4.12 that forward scattering tends to predominate: typical values of g range from 0.5 for aerosols to 0.80 for ice crystals and 0.85 for cloud droplets. Because of the predominance of forward scattering, a cloud of a given optical depth consisting of spherical cloud droplets reflects much less solar radiation back to space than a "cloud" of isotropic scatterers having the same optical depth.

Whether the presence of a particular kind of aerosol serves to increase or decrease the planetary albedo depends on the interplay among all three parameters just discussed. If the other parameters are held fixed, the higher the value of the single scattering albedo, the higher the fraction of the incident radiation that will be backscattered to space. However, even if the single scattering albedo is close to unity, the enhancement of the Earth's albedo due to the presence of the aerosol layer may be quite small if the sun is nearly overhead and the scattering is strongly biased toward the forward direction. However, if the optical thickness of the aerosol layer is sufficiently large, the incident radiation will be subject to multiple scattering events, each of which increases the fraction of the incident radiation that is reflected back to space. Whether a layer of aerosols contributes to or detracts from the planetary albedo also depends on the albedo of the underlying surface or cloud layer. If the underlying surface were completely black, then any backscattering whatsoever would contribute to increasing the planetary albedo, and if the underlying surface were white, any absorption whatsoever would detract from the albedo.

The optical properties of cloud layers are determined by similar considerations, but the geometry of the radiation is quite different because of the much larger volume scattering coefficients. Even though the scattering of radiation by spherical cloud droplets is predominantly in the forward direction, multiple scattering of parallel beam solar radiation yields a distribution of intensity that is much more isotropic than the distribution in cloud-free air. This is why clouds are the dominant contributor to the planetary albedo. Due to the multiplicity of scattering events within clouds, even minute concentrations of black carbon and other absorbing substances in the droplets can produce appreciable amounts of absorption, reducing the albedo of the cloud layer and serving as a heat source within the cloud layer, which may cause the cloud to evaporate. The penetration of sunlight into a deep cloud layer decreases with zenith angle and the cloud albedo increases with zenith angle.

4.5.3 Absorption and Emission of Infrared Radiation

The two previous subsections dealt with the scattering and absorption of radiation in planetary atmospheres in the absence of emission. The relationships discussed in those subsections are applicable to the transfer of solar radiation through planetary atmospheres. The remainder of this section is concerned with the absorption and emission of infrared radiation in the absence of scattering. This simplified treatment is justified by the fact that the wavelength of infrared radiation is very long in comparison to the circumference of the air molecules so the scattering efficiency is negligible.

a. Schwarzschild's equation

First we will derive the equation that governs the transfer of infrared radiation through a gaseous medium. Rewriting (4.17), the rate of change of the monochromatic intensity of outgoing terrestrial radiation along the path length ds due to the absorption within the layer is

$$dI_\lambda(\text{absorption}) = -I_\lambda k_\lambda \rho r ds = -I_\lambda \alpha_\lambda$$

where α_λ is the absorptivity of the layer. The corresponding rate of change due to the emission of radiation is

$$dI_\lambda(\text{emission}) = B_\lambda(T)\varepsilon_\lambda$$

Invoking Kirchhoff's law (4.15) and summing the two expressions, we obtain *Schwarzschild's*[21] *equation*

$$dI_\lambda = -(I_\lambda - B_\lambda(T))k_\lambda \rho r ds \qquad (4.41)$$

By analogy with Beer's law [Eqs. (4.30)–(4.33)], it follows that as radiation passes through an isothermal layer, its monochromatic intensity exponentially approaches that of blackbody radiation corresponding to the temperature of the layer. By the time the radiation has passed through an optical thickness of 1, the departure $|B_\lambda(T) - I_\lambda|$ has decreased by the factor $1/e$. It can also be argued, on the basis of Exercise (4.44), that the monochromatic intensity of the radiation emitted to space is emitted from levels of the atmosphere near the level of unit optical depth. For $\tau_\lambda \ll 1$, the emissivity is so small that the emission from the atmosphere is negligible. For $\tau_\lambda \gg 1$, the transmissivity of the overlying layer is so small that only a minute fraction of the monochromatic intensity emitted from these depths escapes from the top of the atmosphere.

As in the case of absorbed solar radiation, optical depth is strongly dependent on zenith angle. This dependence is exploited in the design of satellite-borne *limb scanning* radiometers, which monitor radiation emitted along very long, oblique path lengths through the atmosphere for which the level of unit optical depth is encountered at very low air density. With these instruments it is possible to sample the radiation emitted from much higher levels in the atmosphere than is possible with nadir (directly downward) viewing instruments.

The level of unit optical depth of emitted radiation is strongly dependent upon the wavelength of the radiation. At the centers of absorption lines it is encountered much higher in the atmosphere than in the gaps between the lines. In the gaps between the major absorption bands the atmosphere absorbs and emits radiation so weakly that even at sea level τ_λ is less than 1. Within these so-called *windows* of the electromagnetic spectrum, an appreciable fraction of the radiation emitted from the Earth's surface escapes directly to space without absorption during its passage through the atmosphere. The level of unit optical depth is also dependent upon the vertical profiles of the concentrations r of the various greenhouse gases. For example, throughout much of the infrared region of the spectrum, the level of unit optical depth is encountered near ~300 hPa, which corresponds to the top of the layer in which sufficient water vapor is present to render even relatively thin layers of the atmosphere opaque.

The integral form of Schwarzschild's equation is obtained by integrating (4.41) along a ray path from 0 to s_1, depicted in Fig. 4.26, as shown in Exercise 4.51. The monochromatic intensity of the radiation reaching s_1 is

$$I_\lambda(s_1) = I_{\lambda 0} e^{-\tau_\lambda(s_1, 0)} + \int_0^{s_1} k_\lambda \rho r B_\lambda[T(s)] e^{-\tau_\lambda(s_1, s)} ds \qquad (4.42)$$

The first term on the right-hand side represents the monochromatic intensity from $s = 0$ that reaches s_1,

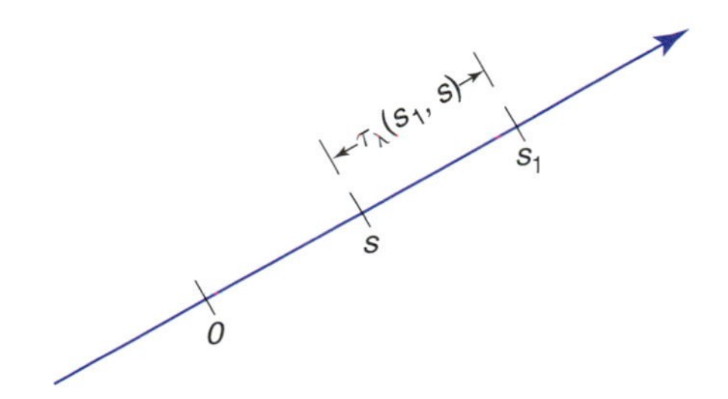

Fig. 4.26 Radiation passing through an absorbing medium along a ray path from 0 to S_1. The ray path may be directed either upward or downward. [From K. N. Liou, *An Introduction to Atmospheric Radiation*, Academic Press, p. 30, Copyright (2002), with permission from Elsevier.]

[21] **Karl Schwarzschild** (1873–1916) German astronomer and physicist. Held positions at the Universites of Göttingen and Potsdam and in the Berlin Academy. Used photogrammetric methods to document emissions from stars and correctly attributed changes in emission of variable stars to changes in their temperature. Formulated the first exact solutions to Einstein's gravitational equations. Although he was not convinced of its physical relevance at the time, his work provided the theoretical basis for understanding black holes.

accounting for the depletion along the way, and the second term represents the monochromatic intensity of the radiation emitted by the gas along the path from $s = 0$ to $s = s_1$ that reaches s_1. The factor $e^{-\tau_\lambda}$ in both terms represents the transmissivity of the radiation along the path from its source to its destination: in the first term the path extends from 0 to s_1, whereas in the second term it extends from s to s_1.

b. The plane-parallel approximation

Many atmospheric radiative transfer calculations can be simplified through use of the *plane-parallel approximation* in which temperature and the densities of the various atmospheric constituents are assumed to be functions of height (or pressure) only. With this simplification, the flux density passing through a given atmospheric level is given by

$$F_\nu^{\downarrow\uparrow}(\tau_\nu) = \int_{2\pi} I_\nu^{\downarrow\uparrow}(\tau_\nu, \cos\theta)\cos\theta\, d\omega \qquad (4.43)$$

where τ_ν, the normal optical depth as defined in (4.32), is used as a vertical coordinate and the arrows ($\downarrow\uparrow$) denote that both upward and downward fluxes are taken into account. Since this section focuses on infrared radiative transfer, we use wave number (ν) notation to be consistent with the literature in this subfield. Integrating over azimuth angle and denoting $\cos\theta$ by μ, we obtain

$$F_\nu^{\downarrow\uparrow}(\tau_\nu) = 2\pi\int_0^1 I_\nu^{\downarrow\uparrow}(\tau_\nu, \mu)\mu\, d\mu \qquad (4.44)$$

The monochromatic intensity in this expression may be broken down into three components:

- Upward emission from the Earth's surface that reaches level τ_ν without being absorbed.
- Upward emission from the underlying atmospheric layer.
- Downward emission from the overlying layer.

The expressions used in evaluating these terms are analogous to Beer's law, with the intensity transmissivity T_ν replaced by the *flux transmissivity*

$$T_\nu^f = 2\int_0^1 e^{-\tau_\nu/\mu}\mu\, d\mu \qquad (4.45)$$

For many purposes it is sufficient to estimate the flux transmissivity from the approximate formula

$$T_\nu^f \simeq e^{-\tau_\nu/\overline{\mu}} \qquad (4.46)$$

where the "average" or "effective zenith angle" is

$$\frac{1}{\overline{\mu}} \equiv \sec 53° = 1.66 \qquad (4.47)$$

In other words, the flux transmissivity of a layer is equivalent to the intensity transmissivity of parallel beam radiation passing through it with a zenith angle $\theta = 53°$. The factor $1/\overline{\mu}$ is widely used in radiative transfer calculations and is referred to as the *diffusivity factor*. Applying the concept of a diffusivity factor (4.44) can be integrated over μ to obtain

$$F_\nu^{\downarrow\uparrow}(\tau_\nu) = \pi I_\nu^{\downarrow\uparrow}(\tau_\nu, \overline{\mu}) \qquad (4.48)$$

c. Integration over wave number[22]

To calculate the flux density F, it is necessary to integrate the monochromatic flux density F_ν over wave number, where the integrand involves the product of the slowly varying Planck function $B_\nu(T)$ and a rapidly varying flux transmissivity T_ν^f. For this purpose it is convenient to separate the smooth B_ν dependence from the much more complex and T_ν^f dependences by breaking up the wave number spectrum into intervals $\Delta\nu$ narrow enough so that $B_\nu(T)$ within each interval can be regarded as a function of T only. Hence, B may be taken outside the integral so that the integrands reduce to expressions for the flux transmissivities T_ν^f of various layers of the atmosphere. These wave number-integrated flux transmissivities $T_{\overline{\nu}}$ for the respective wave number intervals $\Delta\nu$ are the basic building blocks in infrared radiative transfer calculations.

Integration of equations such as (4.44) over wave number, when performed in the conventional manner, requires resolution of myriads of absorption lines with half-widths ranging from 10^{-1} cm^{-1} near the Earth's surface to 10^{-3} cm^{-1} in the upper atmosphere. A single line-by-line integration over the entire infrared

[22] This subsection consists of more advanced material that can be skipped without loss of continuity.

region of the spectrum requires a summation over roughly a million points. The transmissivity T_ν needs to be calculated for each infinitesimal wavelength interval $d\nu$, and these values then need to be summed.

Here we describe an alternative method of performing the integration in which this computationally intensive summation over wave number is greatly simplified. In effect, the same summation is performed, but instead of adding the individual values of T_ν in order of increasing wave number, they are ordered in terms of increasing value of k_ν, i.e.,

$$T_{\bar{\nu}} \equiv \frac{1}{\Delta\nu}\int_{\Delta\nu} e^{-k_\nu u} d\nu = \int_0^1 e^{-k(g)u} dg \quad (4.49)$$

In this expression $g(k)$ is the cumulative probability density function; i.e., the fraction of the individual values of k_ν that lie within the broad frequency interval $\Delta\nu$ with numerical values smaller than k.[23] Unlike k_ν, $k(g)$ is a monotonic, smooth function, which can be inverted, as illustrated in Fig. 4.27.

The integration is performed by grouping the ranked data into M bins, each spanning width Δg. The transmissivity is estimated from the summation

$$T_{\bar{\nu}} = \sum_{j=1}^{M} e^{-k(g_j)u} \Delta g_j \quad (4.50)$$

Because the function $g(k)$ is monotonic and very smooth, it can be closely approximated by the discrete function $g_j(k)$ with only ~10 bins, thereby enabling a reduction of several orders of magnitude in the number of calculations required to evaluate the transmissivity.

As noted in Section 4.4.3, k_ν is a function of temperature and pressure, both of which vary along the path of the radiation. This dependence can be accounted for by assuming that within each spectral interval $\Delta\nu$

$$\frac{1}{\Delta\nu}\int_{\Delta\nu} e^{-\int k_\nu \rho dz} d\nu = \int_0^1 e^{-\int k(g,p,T)\rho dz} dg \quad (4.51)$$

that is, g may replace ν as an independent variable in the transmittance calculations. If the wave number spectrum k_ν is independent of height within the layer for which the integration is performed, then (4.51) is exact to within the degree of approximation inherent in estimating the integral, and the layer is said to be homogeneous. If k_ν varies significantly with height in

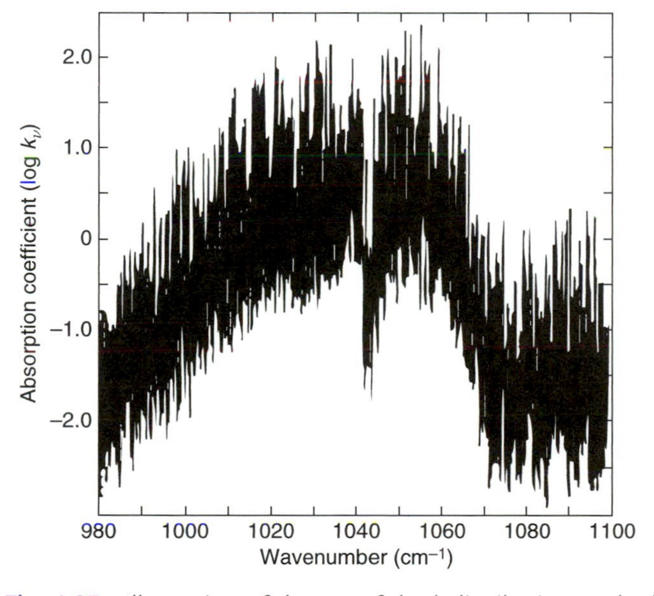

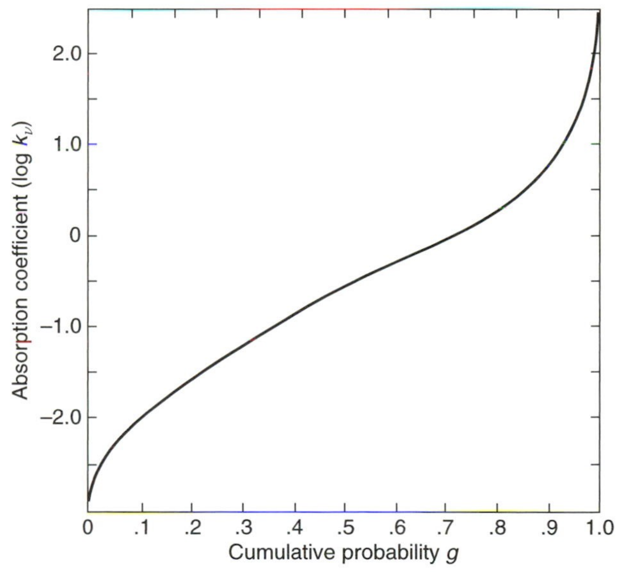

Fig. 4.27 Illustration of the use of the k distribution method for integrating over wave number. (Left) Log k_ν [in units of (cm atm)$^{-1}$] plotted as a function of wave number. (The log scale is used to capture the wide range of values of k_ν.) (Right) g on the x axis plotted as a function of log k on the y axis. For example, within the frequency range shown in the left panel, k_ν is less than 1 (i.e., log 0) over 70% of the interval $\Delta\nu$. Hence (reading from the scale in the right-hand panel) $g(k = 1) = 0.7$. [Courtesy of Qiang Fu.]

[23] The derivative of $g(k)$ with respect to k (represented by the slope of g plotted as a function of k) is the corresponding *probability density function $h(k)$* within the broad frequency interval $\Delta\nu$.

response to vertical gradients of temperature and pressure within the layer, then (4.51) is only an approximation, but it has proven to be a very good one because the dependence of k_ν on T and p tends to be quite similar in different wavelength bands $d\nu$. When applied to such a vertically inhomogeneous layer, this method of estimating the transmissivity is thus referred to as the *correlated k method*.

4.5.4 Vertical Profiles of Radiative Heating Rate

The radiatively induced time rate of change of temperature due to the absorption or emission of radiation within an atmospheric layer is given by

$$\rho c_p \frac{dT}{dt} = -\frac{dF(z)}{dz} \tag{4.52}$$

where $F = F\uparrow - F\downarrow$ is the net flux and ρ is the total density of the air, including the radiatively inactive constituents. The contribution per unit wave number interval at wave number ν is

$$\begin{aligned}\left(\frac{dT}{dt}\right)_\nu &= -\frac{1}{\rho c_p}\frac{dF_\nu(z)}{dz} \\ &= -\frac{1}{\rho c_p}\frac{d}{dz}\left[\int_{4\pi} I_\nu \mu d\omega\right] \\ &= -\frac{1}{\rho c_p}\int_{4\pi}\frac{dI_\nu}{ds}d\omega \\ &= -\frac{2\pi}{\rho c_p}\int_{-1}^{1}\frac{dI_\nu}{ds}d\mu\end{aligned} \tag{4.53}$$

where $\mu = \cos\theta$ and $ds = dz/\mu$. Substituting for dI_ν/ds from Schwarzschild's equation (4.41), we obtain

$$\left(\frac{dT}{dt}\right)_\nu = \frac{2\pi}{c_p}\int_{-1}^{1} k_\nu r(I_\nu - B_\nu)d\mu \tag{4.54}$$

Here we describe an approximation of (4.54) that is widely used in estimating infrared radiative heating rates.

A thin, isothermal layer of air cools by emitting infrared radiation to space, from which it receives no (infrared) radiation in return. Such a layer also exchanges radiation with the layers above and below it, and it may exchange radiation with the Earth's surface as well, as illustrated in Fig. 4.28. If the lapse rate were isothermal, these two-way exchanges would exactly cancel and the only net effect of infrared radiative transfer would be cooling to space. In the real atmosphere there often exists a sufficient degree of cancellation in these exchanges that they can be ignored in calculating the flux of radiation emitted from the layer. With this assumption (4.54) reduces to

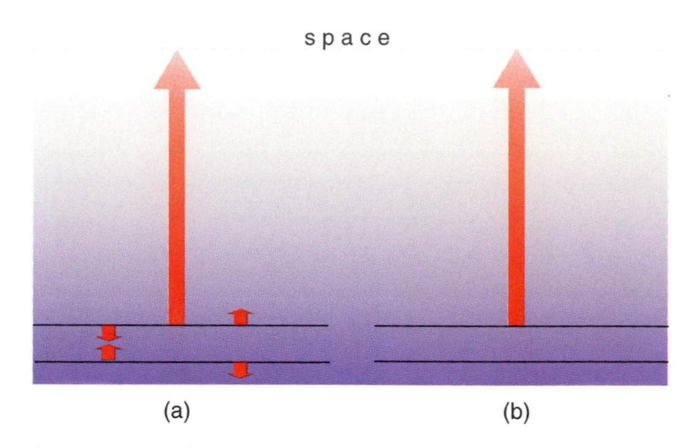

Fig. 4.28 Radiation geometry in cooling to space. (a) The complete longwave radiative balance for the layer in question (not drawn to scale) and (b) the simplified balance based on the assumption of cooling to space.

$$\left(\frac{dT}{dt}\right)_\nu = -\frac{2\pi}{c_p}\int_0^1 k_\nu r B_\nu(z) e^{-\tau_\nu/\mu} d\mu \tag{4.55}$$

which can be integrated over solid angle (as shown in Exercise 4.53 at the end of the chapter) to obtain

$$\left(\frac{dT}{dt}\right)_\nu = -\frac{\pi}{c_p} k_\nu r B_\nu(z)\frac{e^{-\tau_\nu/\overline{\mu}}}{\overline{\mu}} \tag{4.56}$$

where $\overline{\mu} = 1.66$ is the diffusivity factor, as defined in Eqs. (4.46) and (4.47). Based on the form of (4.56), it can be inferred that $\rho c_p(dT/dt)_\nu$ is greatest at and near the level where $\tau_\nu = \overline{\mu}$, which corresponds to the level of unit optical depth for flux density. Using this so-called *cooling to space* approximation, it is possible to estimate the radiative heating rates for water vapor and carbon dioxide surprisingly well, without the need for more detailed radiative transfer calculations.

Vertical profiles of radiative heating rates for the atmosphere's three most important radiatively active greenhouse gases, H_2O, CO_2, and O_3, are shown in Fig. 4.29. In the troposphere, all three constituents produce radiative cooling in the longwave part of the spectrum.[24] Water vapor is the dominant contributor,

[24] The effects of tropospheric ozone are not included in this plot.

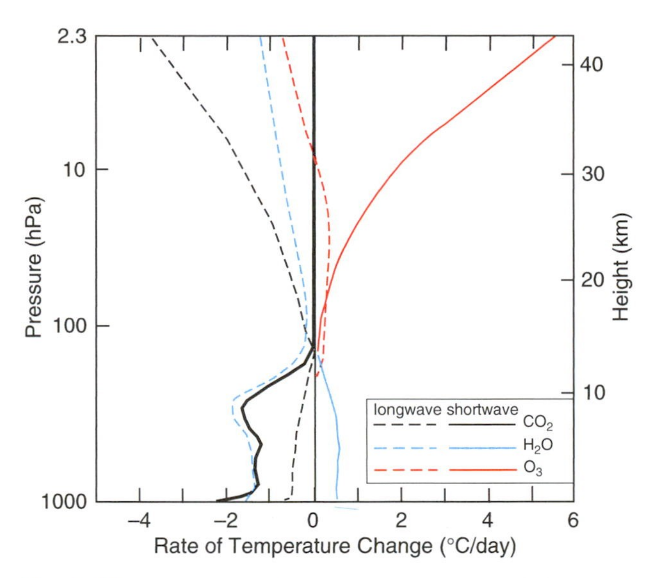

Fig. 4.29 Vertical profiles of the time rate of change of temperature due to the absorption of solar radiation (solid curves) and the transfer of infrared radiation (dashed curves) by water vapor (blue), carbon dioxide (black), and ozone (red). The heavy black solid curve represents the combined effects of the three gases. [Adapted from S. Manabe and R. F. Strickler, *J. Atmos. Sci.*, **21**, p. 373 (1964).]

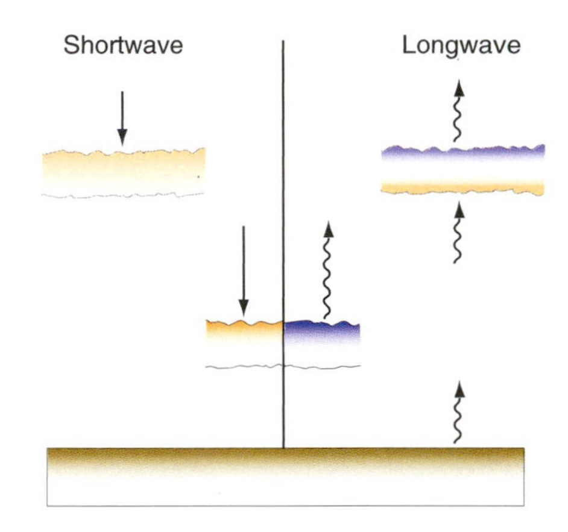

Fig. 4.30 Schematic of vertical profiles of heating in cloud layers at various heights in the atmosphere as indicated. Orange shading indicates warming and blue shading indicates cooling. Effects of shortwave radiation are represented on the left, and effects of longwave radiation on the right.

but its influence decreases with height in parallel with the decline in water vapor mixing ratio. The cooling by tropospheric water vapor is partially offset by the absorption of solar radiation at near-infrared wavelengths by water vapor molecules. However, the troposphere experiences net radiative cooling due to the presence of greenhouse gases.

In contrast to the troposphere, the stratosphere is very close to radiative equilibrium (i.e., a net radiative heating rate of zero). Longwave cooling to space by CO_2, H_2O, and O_3 almost exactly balances the radiative heating due to the absorption of solar radiation in the ultraviolet region of the spectrum by ozone molecules. The most important contributor to the longwave cooling at stratospheric levels is CO_2.

The vertical distribution of heating in cloud layers is depicted schematically in Fig. 4.30. During the daytime, heating rates due to the absorption of solar radiation by ice crystals and cloud droplets range from a few °C day^{-1} in high cirrostratus cloud layers up to a few tens of °C day^{-1} near the tops of dense stratus cloud layers. The emission of infrared radiation from the tops of low and middle cloud decks results in cooling rates ranging up to 50 °C day^{-1} averaged over the 24-h day. If the base of the cloud layer is much colder than the Earth's surface (e.g., as in middle and high cloud decks in the tropics), the infrared radiation emitted from the Earth's surface and absorbed near the base of the layer can produce substantial heating as well. Hence, the overall effect of infrared radiative transfer is to increase the lapse rates within cloud layers, promoting convection. During the daytime this tendency is counteracted by the shortwave heating near the top of the cloud layer.

4.5.5 Passive Remote Sensing by Satellites

Monitoring of radiation emitted by and reflected from the Earth system by satellite-borne radiometers provides a wealth of information on weather and climate. Fields that are currently routinely monitored from space include temperature, cloud cover, cloud droplet concentrations and sizes, rainfall rates, humidity, surface wind speed and direction, concentrations of trace constituents and aerosols, and lightning. Many of these applications are illustrated in the figures that appear in the various chapters of this book. This section focuses on just a few of the many applications of remote sensing in atmospheric science.

a. Satellite imagery

Many of the figures in this book are based on high (horizontal)-resolution satellite imagery in one of three discrete wavelength bands (or *channels*) identified in the bottom part of Fig. 4.6.

- *Visible imagery* is based on reflected solar radiation and is therefore available only during the daylight hours. Because the atmosphere's gaseous constituents are nearly transparent at these wavelengths, visible imagery is used mainly for mapping the distributions of clouds and aerosols.
- *Infrared imagery* corresponds to a spectral window at a wavelength of 10.7 μm, in which radiation emitted from the Earth's surface and cloud tops penetrates through cloud-free air with little absorption. Intensities in this channel (or *radiances*, as they are more commonly referred to in the remote sensing literature) are indicative of the temperatures of the surfaces from which the radiation is emitted as inferred from the Planck function (4.10). In contrast to visible imagery, high clouds are usually clearly distinguishable from low clouds in infrared imagery by virtue of their lower temperatures.[25]
- *Water vapor imagery* exploits a region of the spectrum that encompasses a dense complex of absorption lines associated with vibrational–rotational transitions of water vapor molecules near 6.7 μm. In regions that are free of high clouds, the emission in this channel is largely determined by the vertical profile of humidity in the middle or upper troposphere, at the level where the infrared cooling rate in Fig. 4.29 drops off sharply with height. The more moist the air at these levels, the higher (and colder) the level of unit optical depth and the weaker the emitted radiation.[26]

b. Remote temperature sensing

By comparing the monochromatic radiances in a number of different channels in which the atmosphere exhibits different absorptivities in the infrared or microwave regions of the spectrum, it is possible to make inferences about the distribution of temperature along the path of the emitted radiation. With the assumption of plane parallel radiation, the radiance emitted by gas molecules along a slanting path through the atmosphere can be interpreted in terms of a local vertical temperature profile or "sounding." The physical basis for remote temperature sensing is that (1) most of the radiation reaching the satellite in any given channel is emitted from near the level of unit optical depth for that channel and (2) wave number ranges with higher absorptivities are generally associated with higher levels of unit optical depth, and vice versa.

Figure 4.31 gives a qualitative impression of how remote temperature sensing works. The smooth curves are radiance (or intensity) spectra as inferred from the Planck function (4.10) for various blackbody temperatures, plotted as a function of wave number. The jagged curve is an example of the emission spectrum along a cloud-free path through the atmosphere, as sensed by a satellite-borne instrument called an infrared interferometer spectrometer. The monochromatic radiance at any given wave number can be identified with an equivalent blackbody temperature for that wave number simply by noting which of the Planck function spectra $B(\nu, T)$ passes through that point on the plot. These wavelength-dependent equivalent blackbody temperatures are referred to as *brightness temperatures*.

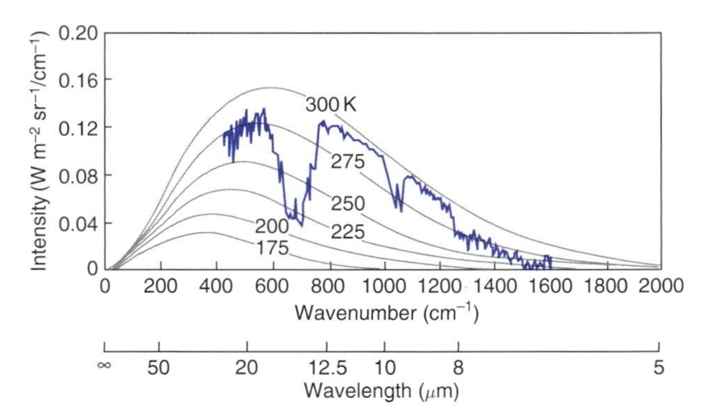

Fig. 4.31 Monochromatic intensity of radiation emitted from a point on the Earth measured by an infrared interferometer spectrometer carried aboard a spacecraft. The gray curves are blackbody spectra computed from Eq. (4.10). The monochromatic intensity of the radiation at any given wavelength is indicative of the temperature of the layer of the atmosphere from which that radiation is emitted (i.e., the level of unit optical depth for that wavelength), which can be inferred from the blackbody spectra. [Adapted from K. N. Liou, *An Introduction to Atmospheric Radiation*, Academic Press, p. 117, Copyright (2002), with permission from Elsevier.]

[25] Higher radiances, indicative of higher equivalent blackbody temperatures, are often rendered in darker shades of gray so that high clouds appear brighter than low clouds.

[26] As in infrared imagery, higher radiances in the water vapor channel are rendered in darker shades of gray, which, in this case, should be interpreted as a drier (more transparent) upper troposphere.

The highest brightness temperature for the spectrum in Fig. 4.31 is just above 290 K, which corresponds closely to the temperature of the Earth's surface at the particular place and time that the sounding was taken. This maximum value is realized within a broad "window" covering most of the range from 800 to 1200 cm^{-1}; the same window in the infrared region of the spectrum that was mentioned in the discussion of Fig. 4.7. This feature is revealed more clearly by the more finely resolved emission spectrum shown in Fig. 4.32, in which the ordinate is brightness temperature. Values as low as 220 K in Fig. 4.31 and 207 K in Fig. 4.32 are realized near the center of a broad CO_2 absorption band extending from around 570 to 770 cm^{-1}. The radiation in much of this range is emitted from around the tropopause level. Intermediate values of brightness temperature, observable at various points along the spectrum, tend to be associated with intermediate absorptivities for which the levels of unit optical depth are encountered in the troposphere.

Throughout most of the emission spectrum shown in Figs. 4.31 and 4.32 absorption bands appear inverted; i.e., higher atmospheric absorptivity is associated with lower brightness temperature, indicative of temperatures decreasing with height. A notable exception is the spike at 667 cm^{-1}, which corresponds to a band made up of particularly strong CO_2 absorption lines. This feature is more clearly evident near the left end of the more finely resolved spectrum in Fig. 4.32. Within this narrow range of the spectrum the radiation reaching the satellite is emitted from levels high in the stratosphere, where the air is substantially warmer than at the tropopause level. Such upward pointing spikes in emission spectra of upwelling radiation are indicative of temperature inversions.

If the radiation reaching the satellite in each channel were emitted from a discrete level of the atmosphere, precisely at the level of unit optical depth for that channel, remote temperature sensing would be straightforward. One could simply monitor the monochromatic intensities in a set of channels whose optical depths match the levels at which temperatures are desired and use the measurements in the respective channels to infer the temperatures. Vertical resolution would be limited only by the number of available channels. In the real world, retrieval of the temperatures from the radiances is not quite so simple. The vertical resolution that is practically achievable is inherently limited by the fact that the radiation in each channel is emitted, not from a single level in the atmosphere, but from a deep layer on the order of three scale heights in depth (Exercise 4.47).

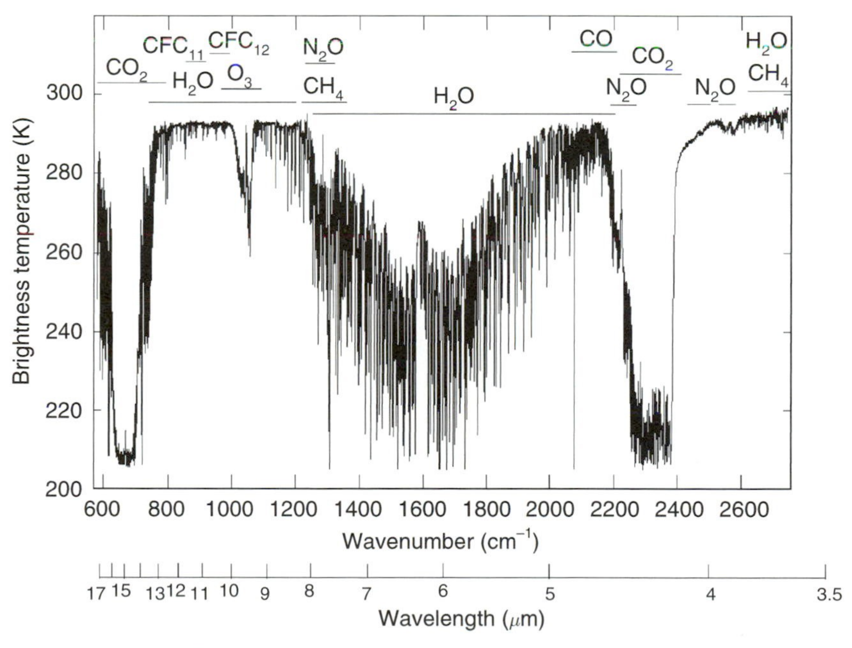

Fig. 4.32 As in Fig. 4.31, but a more finely resolved top of atmosphere spectrum sensed by an instrument aboard the NASA ER-2 aircraft flying over the Gulf of Mexico at the 20-km level. Note that in this plot the ordinate is brightness temperature rather than monochromatic intensity. [From K. N. Liou, *An Introduction to Atmospheric Radiation*, Academic Press, p. 122 (2002). Courtesy of H. -L. Allen Huang and David Tobin.]

Applying (4.42) to radiation reaching the satellite from directly below, we can write

$$I_{\nu\infty} = B_\nu(T_s)e^{-\tau_{\nu *}} + \int_0^\infty B_\nu[T(z)]e^{-\tau_\nu(z)}k_\nu \rho r dz \quad (4.57)$$

where $I_{\nu\infty}$ is the monochromatic radiance sensed by the satellite at wave number ν, T_s is the temperature of the underlying surface, $\tau_{\nu *}$ is the optical thickness of the entire atmosphere, and $\tau_\nu(z)$ is the optical thickness of the layer extending from level z to the top of the atmosphere. Integrating over the narrow range of wave numbers encompassed by the ith channel on the satellite sensor, we obtain

$$I_i = B_i(T_s)e^{-\tau_{i*}} + \int_0^\infty w_i B_i[T(z)]dz \quad (4.58)$$

where I_i is the radiance and

$$w_i = e^{-\tau_i(z)}k_i \rho r \quad (4.59)$$

called the *weighting function*, represents the contribution from a layer of unit thickness located at level z to the radiance sensed by the satellite. The weighting function can also be expressed as the vertical derivative of the transmittance of the overlying layer (see Exercise 4.55).

Figure 4.33 shows transmittances and weighting functions for six channels that were used in one of

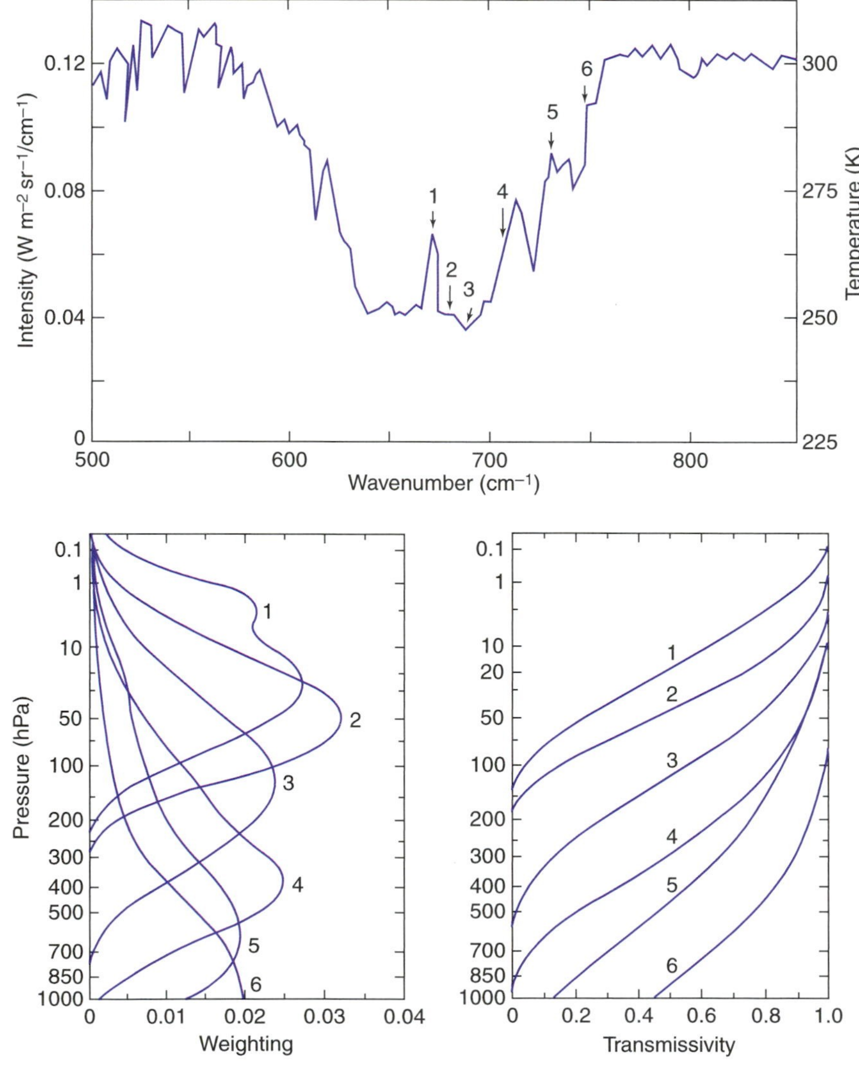

Fig. 4.33 (Top) Intensities or radiances (scale at left) and equivalent blackbody temperatures (scale at right) in the vicinity of the 15-μm CO_2 absorption band, as observed from the satellite. Arrows denote the spectral bands or "channels" sampled by the satellite-borne vertical temperature profiling radiometer (VTPR). The weighting functions and transmittances for each of the channels are shown below. [Adapted from K. N. Liou, *An Introduction to atmospheric radiation*, Academic Press, pp. 389–390, Copyright (2002), with permission from Elsevier.]

the early instruments used for operational remote temperatures sensing. The locations of the channels with respect to the emission spectrum, shown in the inset, span a range of absorptances within the broad CO_2 absorption band centered near $\lambda = 15\ \mu m$ ($\nu = 600–750\ cm^{-1}$). The channels are numbered in order of decreasing absorptance. The weighting functions for channels 1 and 2 peak in the stratosphere. The weighting function for channel 3, which exhibits the lowest radiance in the spectrum, straddles the tropopause level, but because of the substantial depth of its weighting function, its brightness temperature is substantially higher than the tropopause temperature. The weighting function for channel 4 peaks in the middle troposphere and those for channels 5 and 6 peak at or near the Earth's surface. Note the high degree of overlap (or redundancy) between the weighting functions, especially those for channels 5 and 6.

Well-mixed trace gases are well suited for remote temperature sensing because it can be assured that the variations in the radiances from one sounding to another are mainly due to differences in the vertical profile of B_ν rather than to differences in the vertical profiles of the concentrations of the absorbing constituents. In the foregoing example, the differences in absorptivity among the six channels are dominated by features in the absorption spectrum of CO_2.

Over much of the globe the interpretation of satellite-sensed radiation in the infrared region of the spectrum is complicated by the presence of cloud layers. To circumvent this problem, several operational instruments have been designed that make use of radiation emitted by oxygen molecules in the microwave region of the spectrum at frequencies around 55 GHz, in which clouds are nearly transparent. In interpreting microwave radiances it is necessary to take into account variations in the emissivity of the Earth's surface, which are much larger than those for infrared radiation. Emissivities over land range from close to 1 for vegetated or dry surfaces to below 0.8 for wet surfaces. Emissivities over water average ~ 0.5 and vary with surface roughness. Remote sensing in the microwave region of the spectrum is also used to infer the distributions of precipitation and surface winds over the oceans.

c. Retrieval of temperatures from radiances

There are two fundamentally different approaches to retrieving a temperature sounding from the radiances in a suite of channels on a satellite instrument: one involves solving the radiative transfer equation and the other is based on a statistical analysis of the relationships between the radiances in the various channels and the temperatures at various atmospheric levels along the path of the radiation. Both approaches are subject to the inherent limitations in the vertical resolution that is achievable with remote temperature soundings.

The appeal of the radiative transfer approach for retrieving temperatures from radiances is that it requires no prior data on the relationship between the radiances measured by the instrument and *in situ* temperature measurements. It is therefore the only method applicable to atmospheric soundings on planets lacking *in situ* measurements. An approximate solution to the radiative transfer equation can be obtained by representing the atmosphere in terms of n isothermal layers with its own temperature T_n. Rewriting (4.58) as

$$I_i = W_s B_i(T_s) + \sum_n W_{i,n} B_i(T_n) \tag{4.60}$$

where W_s is the weighting function for radiation emitted from the Earth's surface, $W_{i,n}$ is the effective weighting function for the atmospheric emission in the nth atmospheric layer, and $B_i(T_n)$ is the radiance in the ith channel emitted by a blackbody at temperature T_n. The resulting set of equations, one for each channel, in which the I_i are measured and the $W_{i,n}$ are estimated from (4.59) can be solved by a nonlinear iterative scheme[27] to obtain the temperatures T_n that yield the appropriate blackbody emissions. Because of measurement errors, a constraint needs to be applied in the retrieval process.

The statistical approach is less computationally intensive than the radiative transfer approach. Other advantages are that it takes into account the systematic biases and random error characteristics of the radiance measurements and it offers the option of combining radiance measurements with *in situ* temperature measurements and other pertinent observational data. The radiances in the channels measured by a satellite-borne instrument are used to estimate

[27] See K. N. Liou, *An Introduction to Atmospheric Radiation*, Academic Press, p. 391 (2002).

the departures of the temperatures from some prescribed "reference sounding" defined at a prescribed set of levels using standard linear regression techniques. The reference sounding incorporates *in situ* data capable of resolving features such as the tropopause that tend to be smoothed out in the retrievals based on solutions of the radiative transfer equations. Statistically based temperature and moisture retrievals are carried out routinely as an integral part of the multivariate data assimilation protocols, as explained in the supplement to Chapter 8 on the book Web site.

4.6 Radiation Balance at the Top of the Atmosphere

This chapter concludes with a brief survey of some of the global fields relating to the energy balance at the top of the atmosphere, defined on the basis of the analysis of 1 year of satellite observations. The map projection used in this section is area preserving, so that the maps presented here can be interpreted as depicting local contributions to global-mean quantities. The top image in Fig. 4.34 shows the annual mean net (downward) shortwave radiation, taking into account the geographical variations in solar declination angle and local albedo. Values are $\sim$300 W m^{-2} in the tropics, where the sun is nearly directly overhead at midday throughout the year. Within the tropics the highest values are observed over cloud-free regions of the oceans, where annual-mean local albedos range as low as 0.1, and the lowest values are observed over the deserts where albedos are $\sim$0.2 and locally range as high as 0.35. Net incoming solar radiation drops below 100 W m^{-2} in the polar regions where winters are dark and the continuous summer daylight is offset by the high solar zenith angles, widespread cloudiness, and the high albedo of ice-covered surfaces.

The corresponding distribution of outgoing longwave radiation (OLR) at the top of the atmosphere, shown in the bottom image in Fig. 4.34, exhibits a gentler equator-to-pole gradient and more regional variability within the tropics. As shown in Exercise 4.56, the observed equator-to-pole contrast in surface air temperature is sufficient to produce a 2:1 difference in outgoing OLR between the equator and the polar regions, but this is partially offset by the fact that cloud tops and the top of the moist layer are higher in the tropics than over high latitudes. The regions of conspicuously low OLR over Indonesia

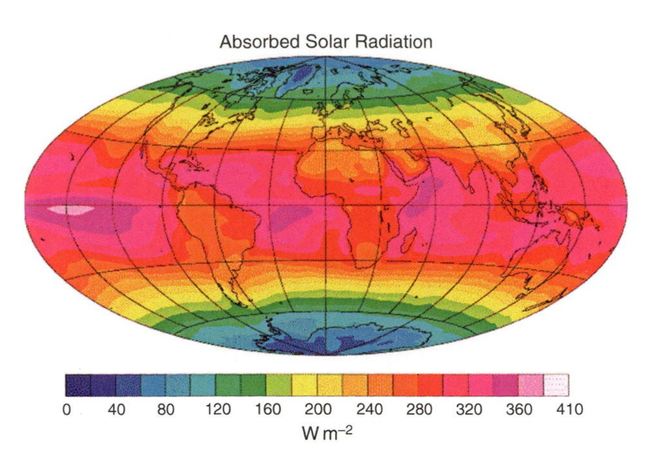

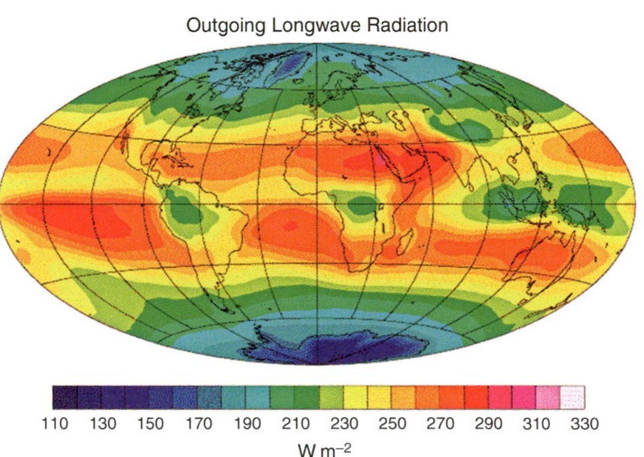

Fig. 4.34 Global distributions of the annual-mean radiation at the top of the atmosphere. [Based on data from the NASA Earth Radiation Budget Experiment. Courtesy of Dennis L. Hartmann.]

and parts of the tropical continents reflect the prevalence of deep convective clouds with high, cold tops: the intertropical convergence zone is also evident as a local OLR minimum, but it is not as pronounced because the cloud tops are not as high as those associated with convection over the continents and the extreme western Pacific and Indonesia. Areas with the highest annual mean OLR are the deserts and the equatorial dry zones over the tropical Pacific, where the atmosphere is relatively dry and cloud free, allowing more of the radiation emitted by the Earth's surface to escape unimpeded.

The net downward radiation at the top of the atmosphere (i.e., the imbalance between net solar and outgoing longwave radiation at the top of the atmosphere) obtained by taking the difference between the two panels of Fig. 4.34 is shown in Fig. 4.35. The surplus of incoming solar radiation over outgoing longwave radiation in low latitudes and the deficit in high latitudes has important impli-

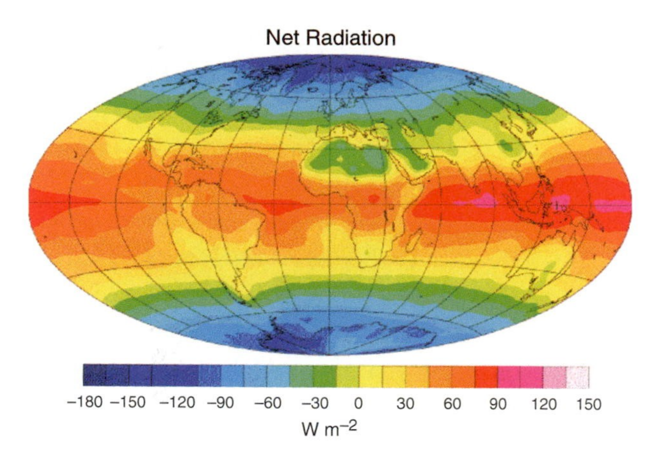

Fig. 4.35 Global distribution of the net imbalance between the annual-mean net incoming solar radiation and the outgoing longwave radiation. Positive values indicate a downward flux. [Based on data from the NASA Earth Radiation Budget Experiment. Courtesy of Dennis L. Hartmann.]

cations for the global energy balance, as discussed in Section 10.1. It is notable that over some of the world's hottest desert regions, the outgoing longwave radiation exceeds absorbed solar radiation.

The exclusive focus on radiative fluxes in this section is justified by the fact that radiative transfer is the only process capable of exchanging energy between the Earth and the rest of the universe. The energy balance at the Earth's surface is more complicated because conduction of latent and sensible heat across the Earth's surface also plays important roles, as discussed in Chapter 9.

Exercises

4.11 Explain or interpret the following in terms of the principles discussed in this chapter.

(a) Species of plants that require a relatively cool, moist environment tend to grow on poleward-facing slopes.

(b) On a clear day a snow surface is much brighter when the sun is nearly overhead than when it is just above the horizon.

(c) Light-colored clothing is often worn in hot climates.

(d) The stars twinkle, whereas the planets do not.

(e) The radiation emitted by the sun is isotropic, yet solar radiation incident on the Earth's atmosphere may be regarded as parallel beam.

(f) The colors of the stars are related to their temperatures, whereas the colors of the planets are not.

(g) The color temperature of the sun is slightly different from its equivalent blackbody temperature.

(h) The equivalent blackbody temperature of Venus is lower than that of Earth even though Venus is closer to the sun.

(i) The equivalent blackbody temperature of the Earth is lower than the global-mean surface temperature by 34 °C.

(j) Frost may form on the ground when temperatures just above the ground are above freezing.

(k) At night if air temperatures are uniform, frost on highways is likely to form first on bridges and hilltops.

(l) Aerosols in the atmosphere (or on windows) are more clearly visible when viewed looking toward a light source than away from it.

(m) Clouds behave as blackbodies in the infrared region of the spectrum, but are relatively transparent in the microwave region.

(n) Sunlit objects appear reddish around sunrise and sunset on days when the air is relatively free of aerosols.

(o) A given mass of liquid water would produce more scattering if it were distributed among a large number of small cloud droplets than among a smaller number of larger droplets (assume that both sizes of droplets fall within the *geometric optics* regime).

(p) The upper layers of the atmosphere in Fig. 4.36 appear blue, whereas the lower layers appear red.

(q) Smoke particles with a radius of ~ 0.5 mm appear bluish when viewed against a dark background but reddish when viewed against a light background (e.g., the sky).

(r) The disk of the full moon appears uniformly bright all the way out to the edge.

(s) The absorption coefficient of a gas is a function of temperature and pressure.

(t) The absorptivity of greenhouse gases in the troposphere is enhanced by the presence of high concentrations of N_2 and O_2.

Fig. 4.36 In these views of the limb of the Earth from space, the upper atmosphere shows up blue, while the lower atmosphere exhibits an orange hue. The lower photo, taken 2 months after the eruption of Mt. Pinatubo, shows layers of sulfate aerosols in the lower stratosphere. [Photographs courtesy of NASA.]

(u) Low clouds emit more infrared radiation than high clouds of comparable optical thickness.

(v) The presence of cloud cover tends to favor lower daytime surface temperatures and higher nighttime surface temperatures.

(w) Temperature inversions tend to form at night immediately above the tops of cloud layers.

(x) Convection cells are often observed within cloud layers.

(y) At night, when a surface based inversion is present, surface temperatures tend to rise when a deck of low clouds moves overhead.

(z) Low clouds are not visible in satellite imagery in the water vapor channel.

(aa) Ground fog is more clearly evident in visible than in infrared satellite imagery.

(bb) The fraction of the incoming solar radiation that is backscattered to space by clouds is higher when the sun is low in the sky than when it is overhead.

(cc) Under what conditions can the flux density of solar radiation at the Earth's surface (locally) be greater than that at the top of the atmosphere?

4.12 Remote sensing in the microwave part of the spectrum relies on radiation emitted by oxygen molecules at frequencies near 55 GHz. Calculate the wavelength and wave number of this radiation.

4.13 The spectrum of monochromatic intensity can be defined either in terms of wavelength λ or wave number ν such that the area under the spectrum, plotted as a linear function of λ or ν, is proportional to intensity. Show that $I_\nu = \lambda^2 I_\lambda$.

4.14 A body is emitting radiation with the following idealized spectrum of monochromatic flux density.

$$
\begin{array}{ll}
\lambda < 0.35\ \mu\text{m} & F_\lambda = 0 \\
0.35\ \mu\text{m} < \lambda < 0.50\ \mu\text{m} & F_\lambda = 1.0\ \text{W m}^{-2}\mu\text{m}^{-1} \\
0.50\ \mu\text{m} < \lambda < 0.70\ \mu\text{m} & F_\lambda = 0.5\ \text{W m}^{-2}\mu\text{m}^{-1} \\
0.70\ \mu\text{m} < \lambda < 1.00\ \mu\text{m} & F_\lambda = 0.2\ \text{W m}^{-2}\mu\text{m}^{-1} \\
\lambda > 1.00\ \mu\text{m} & F_\lambda = 0
\end{array}
$$

Calculate the flux density of the radiation.

4.15 An opaque surface with the following absorption spectrum is subjected to the radiation described in the previous exercise:

$$
\begin{array}{ll}
\lambda < 0.70\ \mu\text{m} & \alpha_\lambda = 0 \\
\lambda > 0.70\ \mu\text{m} & \alpha_\lambda = 1
\end{array}
$$

How much of the radiation is absorbed? How much is reflected?

4.16 Calculate the ratios of the incident solar radiation at noon on north- and south-facing 5° slopes (relative to the horizon) in seasons in which the solar zenith angle is (a) 30° and (b) 60°.

4.17 Compute the daily insolation at the North Pole at the time of the summer solstice when the Earth–sun distance is 1.52×10^8 km. The tilt of the Earth's axis is 23.5°.

4.18 Compute the daily insolation at the top of the atmosphere at the equator at the time of the equinox (a) by integrating the flux density over a 24-h period and (b) by simple geometric considerations. Compare your result with the value in the previous exercise and with Fig. 10.5.

4.19 Orbitally induced variations in solar flux density incident in the top of the atmosphere during summer at high latitudes of the northern hemisphere play a central role in the orbital theory of the ice ages discussed in Section 2.5.3. By what factor does the flux density at noon, at 55 °N on the day of the summer solstice, vary between the extremes of the orbital cycles?

4.20 What fraction of the flux of energy emitted by the sun does the Earth intercept?

4.21 Show that for small perturbations in the Earth's radiation balance

$$\frac{\delta T_E}{T_E} = \frac{1}{4}\frac{\delta F_E}{F_E}$$

where T_E is the Earth's equivalent blackbody temperature and F_E is the flux density of radiation emitted from the top of its atmosphere. [**Hint**: Take the logarithm of the Stefan–Boltzmann law (4.12) and then take the differential.] Use this relationship to estimate the change in equivalent blackbody temperature that would occur in response to (a) the seasonal variations in the sun–Earth distance due to the eccentricity of the Earth's orbit (presently ~3.5%) and (b) an increase in the Earth's albedo from 0.305 to 0.315.

4.22 Show that the flux density of incident solar radiation on any planet in the solar system is 1368 W $m^{-2} \times r^{-2}$, where r is the planet–sun distance, expressed in astronomical units.

4.23 Estimate the flux density of the radiation emitted from the solar photosphere using two different approaches: (a) starting with the intensity (2.00×10^7 W m^{-2} sr^{-1}) and making use of the results in Exercise 4.3 and (b) making use of the relationship derived in the previous exercise. (c) Estimate the output of the sun in watts.

4.24 By differentiating the Planck function (4.10), derive Wien's displacement law. [**Hint**: in the wavelength range of interest, the exponential term in the denominator of (4.10) is much larger than 1.]

4.25 Show that for radiation with very long wavelengths, the Planck monochromatic intensity $B_\lambda(T)$ is linearly proportional to absolute temperature. This is referred to as the *Rayleigh–Jeans limit.*

4.26 Use the relationship derived in Exercise 4.22 to check the numerical values of T_E in Table 4.1.

4.27 The observed equivalent blackbody temperature of Jupiter is 125 K, 20 K higher than the value in Table 4.1. Assuming that the temperature of Jupiter is in a steady state, estimate the flux density of radiation emitted from the top of its atmosphere that is generated internally by processes on the planet.

4.28 If the moon subtends the same arc of solid angle in the sky that the sun does and it is directly overhead, prove that the flux density of moonlight on a horizontal surface on Earth is given by $F_s a(R_s/d)^2$, where F_s is the flux density of solar radiation intercepted by the Earth, a is the moon's albedo, R_s is the radius of the sun, and d is the Earth–sun distance. Estimate the flux density of moonlight under these conditions, assuming a lunar albedo of 0.07.

4.29 Suppose that the sun's emission or the Earth's albedo were to change abruptly by a small increment. Show that the *radiative relaxation rate* for the atmosphere (i.e., the initial rate at which the Earth's equivalent blackbody temperature would respond to the change, assuming that the atmosphere is thermally isolated from the other components of the Earth system) is given by

$$\frac{dT}{dt} = -\frac{4\sigma T_E^3 \delta T_E}{c_p p_s g^{-1}}$$

where δT_E is the initial departure of the equivalent blackbody temperature from radiative equilibrium, σ is the Stefan–Boltzmann constant, T_E is the equivalent blackbody temperature in K, c_p is the specific heat of air, p_s is the global-mean surface pressure, and g is the gravitational acceleration. The time $\delta T_E(dT/dt)^{-1}$ required for the atmosphere to fully adjust to the change in radiative forcing, if this initial time rate of change of temperature were maintained until the new equilibrium was established, is called the *radiative relaxation time*. Estimate the radiative relaxation time for the Earth's atmosphere.

4.30 A small, perfectly black, spherical satellite is in orbit around the Earth at an altitude of 2000 km as depicted in Fig. 4.37. What angle does the Earth subtend when viewed from the satellite?

4.31 If the Earth radiates as a blackbody at an equivalent blackbody temperature $T_E = 255$ K, calculate the radiative equilibrium temperature of the satellite when it is in the Earth's shadow. [**Hint**: Let dE be the amount of radiation flux imparted to the satellite by the flux density dE received within the infinitesimal element of solid angle $d\omega$.] Then,

$$dE = \pi r^2 I d\omega$$

where r is the radius of the satellite and I is the intensity of the radiation emitted by the Earth, i.e., the flux density of blackbody radiation, as given by (4.12), divided by π. Integrate the above expression over the arc of solid angle subtended by the Earth, as computed in the previous exercise, noting that the radiation is isotropic, to obtain the total energy absorbed by the satellite per unit time

$$Q = 2.21 r^2 \sigma T_E^4$$

Finally, show that the temperature of the satellite is given by

$$T_s = T_E\left(\frac{2.21}{4\pi}\right)^{1/4}$$

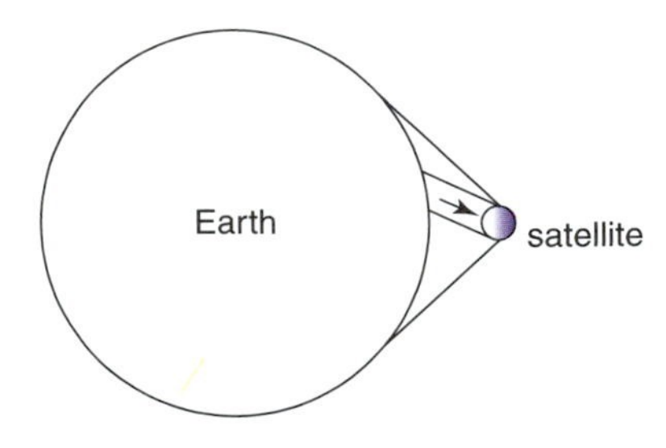

Fig. 4.37 Geometric setting for Exercises 4.30–4.34.

4.32 Show that the approach in Exercise 4.5 in the text, when applied to the previous exercise, yields a temperature of

$$T_s = T_E\left[\frac{1}{4}\left(\frac{6371}{8371}\right)^2\right]^{1/4} = 158 \text{ K}$$

Explain why this approach underestimates the temperature of the satellite. Show that the answer obtained with this approach converges to the exact solution in the previous exercise as the distance between the satellite and the center of the Earth becomes large in comparison to the radius of the Earth R_E. [**Hint**: show that as $d/R_E \rightarrow \infty$, the arc of solid angle subtended by the Earth approaches $\pi R_E^2/d^2$.]

4.33 Calculate the radiative equilibrium temperature of the satellite immediately after it emerges from the Earth's shadow (i.e., when the satellite is sunlit but the Earth, as viewed from the satellite, is still entirely in shadow).

4.34 The satellite has a mass of 100 kg, a radius of 1 m, and a specific heat of 10^3 J kg^{-1} K^{-1}. Calculate the rate at which the satellite heats up immediately after it (instantaneously) emerges from the Earth's shadow.

4.35 Consider two opaque walls facing one another. One of the walls is a blackbody and the other wall is "gray" (i.e., α_λ independent of λ). The walls are initially at the same temperature T and, apart from the exchange of radiation between them, they are thermally insulated from their surroundings. If α and ε are the absorptivity and emissivity of the gray wall, prove that $\varepsilon = \alpha$.

Solution: The flux emitted by the black wall is $F = \sigma T^4$ and the flux absorbed by the gray wall is $\alpha\sigma T^4$. The flux emitted by the gray wall is $\varepsilon\sigma T^4$.

The rate at which the gray wall gains or loses energy from the exchange of radiation with the black wall is

$$\begin{aligned} H &= \alpha\sigma T^4 - \varepsilon\sigma T^4 \\ &= (\alpha - \varepsilon)\sigma T^4 \end{aligned}$$

If H is not zero, then the temperature of the wall must change in response to the energy imbalance, in which case, heat is being transferred from a colder body to a warmer body, in violation of the second law of thermodynamics. It follows that

$$\varepsilon = \alpha$$

■

4.36 (a) Extend the proof in the previous exercise to the case in which absorptivity and emissivity are wavelength dependent. Let one of the walls be black, as in the previous exercise, and let the other wall also be black, except within a very narrow wavelength range of width $\delta\lambda$, centered at $\lambda = \lambda_1$ where $a_{\lambda_1} < 1$. [**Hint**: Because blackbody radiation is isotropic, it follows the blackbody flux in the interval $\delta\lambda$ is $\pi B(\lambda_1, T)$ $\delta\lambda$. Using this relationship, consider the energy balance as in the previous exercise and proceed to show that $\alpha_{\lambda_1} = \varepsilon_{\lambda_1}$.] (b) Indicate how this result could be extended to prove that

$$\varepsilon_\lambda = \alpha_\lambda$$

4.37 Consider a closed spherical cavity in which the walls are opaque and all at the same temperature. The surfaces on the top hemisphere are black and the surfaces on the bottom hemisphere reflect all the incident radiation at all angles. Prove that in all directions $I_\lambda = B_\lambda$.

4.38 (a) Consider the situation described in Exercise 4.35, except the both plates are gray, one with absorptivity α_1 and the other with absorptivity α_2. Prove that

$$\frac{F_1'}{\alpha_1} = \frac{F_2'}{\alpha_2}$$

where F_1' and F_2' are the flux densities of the radiation emitted from the two plates. Make use of the fact that the two plates are in radiative equilibrium at the same temperature but do not make use of Kirchhoff's law. [**Hint**: Consider the total flux densities F_1 from plate 1 to plate 2 and F_2 from plate 2 to plate 1. The problem can be worked without dealing explicitly with the multiple reflections between the plates.]

4.39 Consider the radiation balance of an atmosphere with a large number of layers, each of which is isothermal, transparent to solar radiation, and absorbs the fraction α of the longwave radiation incident on it from above or below. (a) Show that the flux density of the radiation emitted by the topmost layer is $\alpha F/(2-\alpha)$ where F is the flux density of the planetary radiation emitted to space. By applying the Stefan–Boltzmann law (4.12) to an infinitesimally thin topmost layer, show that the radiative equilibrium temperature at the top of the atmosphere, sometimes referred to as the *skin temperature*, is given by

$$T^* = \left(\frac{1}{2}\right)^{1/4} T_E$$

(Were it not for the presence of stratospheric ozone, the temperature of the 20- to 80-km layer in the Earth's atmosphere would be close to the skin temperature.)

4.40 Consider an idealized aerosol consisting of spherical particles of radius r with a refractive index of 1.5. Using Fig. 4.13, estimate the smallest radius for which the particles would impart a bluish cast to transmitted white light, as in the rarely observed "blue moon."

4.41 Consider an idealized cloud consisting of spherical droplets with a uniform radius of 20 μm and concentrations of 1 cm^{-3}. How long a path through such a cloud would be required to deplete a beam of visible radiation by a factor of e due to scattering alone? (Assume that none of the scattered radiation is subsequently scattered back into the path of the beam.)

4.42 Consider solar radiation with a zenith angle of 0° that is incident on a layer of aerosols with a single scattering albedo $\omega_0 = 0.85$, an asymmetry factor $g = 0.7$, and an optical thickness $\tau = 0.1$ averaged over the shortwave part of the spectrum. The albedo of the underlying surface is $R_s = 0.15$.

(a) Estimate the fraction of the incident radiation that is backscattered by the aerosol layer in its downward passage through the atmosphere.

(b) Estimate the fraction of the incident radiation that is absorbed by the aerosol layer in its downward passage through the atmosphere.

(c) Estimate the consequent corresponding impact of the aerosol layer upon the local albedo. Neglect multiple scattering. For simplicity, assume that the radiation back-scattered from the earth's surface and clouds is parallel beam and oriented at 0° or 180° Zenith angle. (In reality it is isotropic.) [**Hint**: Show that the fraction of the radiation that is backscattered in its passage through the layer is

$$b = \omega_0(1 - e^{-\tau})\frac{(1-g)}{2}$$

and the fraction that is transmitted through the layer is

$$t = e^{-\tau} + \omega_0(1 - e^{-\tau})\frac{(1+g)}{2}$$

Then show that the total upward reflection from the top of the atmosphere is

$$b + R_s t^2(1 + bR_s + b^2R_s^2 + \cdots)$$

which can be rewritten in the form

$$b + \frac{R_s t^2}{(1 - bR_s)}\Bigg]$$

4.43 Consider radiation with wavelength λ and zero zenith angle passing through a gas with an absorption coefficient k_λ of 0.01 $m^2\ kg^{-1}$. What fraction of the beam is absorbed in passing through a layer containing 1 $kg\ m^{-2}$ of the gas? What mass of gas would the layer have to contain in order to absorb half the incident radiation?

4.44 Show that for overhead parallel beam radiation incident on an isothermal atmosphere in which the r, the mixing ratio of the absorbing gas, and k_λ, the volume absorption coefficient, are both independent of height, the strongest absorption *per unit volume* (i.e., dI_λ/dz) is strongest at the level of unit optical depth.

Solution: From (4.17) if r and k_λ are both independent of height,

$$\frac{dI_\lambda}{dz} = I_{\lambda\infty} \times T_\lambda \times (k_\lambda r)\rho \qquad (4.61)$$

where $I_{\lambda\infty}$ is the intensity of the radiation incident on the top of the atmosphere, $T_\lambda = I_\lambda/I_{\lambda\infty}$ is the transmissivity of the overlying layer, and ρ is the density of the ambient air. From (4.33)

$$T_\lambda = e^{-\tau_\lambda}$$

and from the hypsometric equation applied to an isothermal layer

$$\rho = \rho_0 e^{-z/H}$$

Substituting for T_λ and ρ in (4.61) we obtain

$$\frac{dI_\lambda}{dz} = I_{\lambda\infty}(k_\lambda r\rho_0)e^{-z/H}e^{-\tau_\lambda}$$

From (4.32)

$$\begin{aligned}\tau_\lambda &= (k_\lambda r\rho_0)\int_z^\infty e^{-z/H}dz \\ &= H(k_\lambda r\rho_0)e^{-z/H} \qquad (4.62)\end{aligned}$$

Using (4.62) to express $e^{-z/H}$ in the aforementioned equation in terms of optical depth, we obtain

$$\frac{dI_\lambda}{dz} = \frac{I_{\lambda\infty}}{H}\tau_\lambda e^{-\tau_\lambda}$$

Now at the level where the absorption is strongest,

$$\frac{d}{dz}\frac{dI_\lambda}{dz} = \frac{I_{\lambda\infty}}{H}\frac{d}{dz}(\tau_\lambda e^{-\tau_\lambda}) = 0$$

Performing the indicated differentiation, we obtain

$$e^{-\tau_\lambda}\frac{d\tau_\lambda}{dz}(1 - \tau_\lambda) = 0$$

from which it follows that $\tau_\lambda = 1$. Although this result is strictly applicable only to an isothermal atmosphere in which k_λ and r are independent of height, it is qualitatively representative of conditions in planetary atmospheres in which the mixing ratios of the principal absorbing constituents do not change rapidly with height. It was originally developed by Chapman[28] for understanding radiative and photochemical processes related to the stratospheric ozone layer.

4.45 For incident parallel beam solar radiation in an isothermal atmosphere in which k_λ is independent of height (a) show that optical depth is linearly proportional to pressure and (b) show that the absorption *per unit mass* (and consequently the heating rate) is strongest, not at the level of unit optical depth but near the top of the atmosphere, where the incident radiation is virtually undepleted.

4.46 Consider a hypothetical planetary atmosphere comprised entirely of the gas in Exercise 4.43. The atmospheric pressure at the surface of the planet is 1000 hPa, the lapse rate is isothermal, the scale height is 10 km, and the gravitational acceleration is 10 m s^{-2}. Estimate the height and pressure of the level of unit normal optical depth.

4.47 (a) What percentage of the incident monochromatic intensity with wavelength λ and zero zenith angle is absorbed in passing through the layer of the atmosphere extending from an optical depth $\tau_\lambda = 0.2$ to $\tau_\lambda = 4.0$?

(b) What percentage of the outgoing monochromatic intensity to space with wavelength λ and zero zenith angle is emitted from the layer of the atmosphere extending from an optical depth $\tau_\lambda = 0.2$ to $\tau_\lambda = 4.0$?

(c) In an isothermal atmosphere, through how many scale heights would the layer in (a) and (b) extend?

4.48 For the atmosphere in Exercise 4.46, estimate the levels and pressures of unit (slant path) optical depth for downward parallel beam radiation with zenith angles of 30° and 60°.

4.49 Prove that the optical thickness of a layer is equal to (-1) times the natural logarithm of the transmissivity of the layer.

4.50 Prove that the fraction of the flux density of overhead solar radiation that is backscattered to space in its first encounter with a particle in the atmosphere is given by

$$b = \frac{1-g}{2}$$

where g is the asymmetry factor defined in (4.35). [**Hint**: The intensity of the scattered radiation must be integrated over zenith angle.]

4.51 Show that Schwarzschild's equation in the form that it appears in (4.41) in the text

$$dI_\lambda = (-I_\lambda + B_\lambda(T))k_\lambda \rho r ds$$

can be integrated along a path extending from 0 to s_1 in Fig. 4.26 to obtain an expression for $I_\lambda(s_1)$.

4.52 Prove that as the optical thickness of a layer approaches zero, the flux transmissivity, as defined in Eq. (4.45), becomes equal to the intensity transmissivity at a zenith angle of 60 degrees.

4.53 Integrate the expression for the heating rate, as approximated by cooling to space that appears as Eq. (4.55) of the text, namely

$$\left(\frac{dT}{dt}\right)_\nu = -\frac{2\pi}{c_p}\int_0^1 k_\nu r B_\nu(z) e^{-\tau_\nu/\mu} d\mu \qquad (4.55)$$

over solid angle.

4.54 A thin, isothermal layer of air in thermal equilibrium at temperature T_0 is perturbed about that equilibrium value (e.g., by absorption of a burst of ultraviolet radiation emitted by the sun during a short-lived solar flare) by the temperature increment δT. Using the cooling to space approximation (4.56) show that

$$\delta\left(\frac{dT}{dt}\right)_\nu = -\alpha_\nu \delta T \qquad (4.63)$$

where

$$\alpha_\nu = \frac{\pi k_\nu r}{c_p}\frac{e^{-\tau_\nu/\overline{\mu}}}{\overline{\mu}}\left(\frac{dB_\nu}{dT}\right)_{T_0} \qquad (4.64)$$

This formulation, in which cooling to space acts to bring the temperature back toward radiative equilibrium, is known as *Newtonian cooling* or *radiative relaxation*. It is widely used in

[28] **Sydney Chapman** (1888–1970) English geophysicist. Made important contributions to a wide range of geophysical problems, including geomagnetism, space physics, photochemistry, and diffusion and convection in the atmosphere.

parameterizing the effects of longwave radiative transfer in the middle atmosphere.

4.55 Prove that weighting function w_i used in remote sensing, as defined in (4.59), can also be expressed as the vertical derivative of the transmittance of the overlying layer.

4.56 The annual mean surface air temperature ranges from roughly 23 °C in the tropics to −25 °C in the polar cap regions. On the basis of the Stefan–Boltzmann law, estimate the ratio of the flux density of the emitted longwave radiation in the tropics to that in the polar cap region.

Atmospheric Chemistry[1]

5

Atmospheric chemistry studies were originally concerned with determining the major gases in the Earth's atmosphere. In the latter half of the last century, as air pollution became an increasing problem in many large cities, attention turned to identifying the sources, properties, and effects of the myriad of chemical species that exist in the natural and polluted atmosphere. Acid deposition, which became recognized as a widespread problem in the 1970s, led to the realization that chemical species emitted into the atmosphere can be transported over large distances and undergo significant transformations as they move along their trajectories. The identification in 1985 of significant depletion of ozone in the Antarctic stratosphere focused attention on stratospheric chemistry and the susceptibility of the stratosphere to modification. More recently, studies of the effects of trace chemical constituents in the atmosphere on the climate of the Earth have moved to center stage.

The focus of this chapter is on some of the basic concepts and principles underlying atmospheric chemistry, as illustrated by the effects of both natural and anthropogenic trace constituents. For the most part, we confine our attention in this chapter to gas-phase chemistry and atmospheric aerosols. The interactions of gases and aerosols with tropospheric clouds are discussed in Chapter 6.

5.1 Composition of Tropospheric Air

The ancient Greeks considered air to be one of the four "elements" (the others being earth, fire and water). Leonardo da Vinci,[2] and later Mayow,[3] suggested that air is a mixture consisting of one component that supports combustion and life ("fire-air") and the other that does not ("foul-air"). "Fire-air" was isolated by Scheele[4] in 1773 and independently by Priestley[5]

[1] This chapter is based in part on P. V. Hobbs' *Introduction to Atmospheric Chemistry*, Cambridge University Press, New York, 2000, to which the reader is referred for more details on atmospheric chemistry. If the reader feels a need for a review of the basic principles of chemistry, this is given in P. V. Hobbs' *Basic Physical Chemistry for the Atmospheric Sciences*, 2nd Edition, Cambridge University Press, New York, 2000. Both of these books are designed for upper-class undergraduate and first-year graduate students and, like the present book, contain numerous worked exercises and exercises for the student.

[2] **Leonardo da Vinci** (1452–1519) Renowned Italian painter, architect, engineer, mathematician, and scientist. Best known for his paintings (e.g., the *Mona Lisa*, the *Last Supper*). His notebooks reveal his knowledge of human anatomy and natural laws, as well as his mechanical inventiveness.

[3] **John Mayow** (1640–1679) English chemist and physiologist. Described the muscular actions around the chest involved in respiration.

[4] **Carl Wilhelm Scheele** (1742–1786) Swedish chemist. Discoverer of many chemicals, including oxygen, nitrogen, chlorine, manganese, hydrogen cyanide, citric acid, hydrogen sulfide, and hydrogen fluoride. Discovered a process similar to pasteurization.

[5] **Joseph Priestley** (1733–1804) English "tinkerer" *par excellence*. Never took a formal science course. Discovered that graphite (i.e., carbon) can conduct electricity (carbon is the main ingredient in modern electrical circuitry), the respiration of plants (i.e., they take in carbon dioxide and release oxygen), and he isolated photosynthesis, and isolated nitrous oxide, N_2O ("laughing gas"—later to become the first surgical anesthetic), ammonia (NH_3), sulfur dioxide (SO_2), hydrogen sulfide (H_2S), and carbon monoxide (CO). Also discovered that India gum can be used to rub out lead pencil marks. Due to his support of both the American and the French Revolution, his home and church in England were burned down by a mob. Emigrated to the United States in 1794.

in 1774. It was named oxygen [from the Greek *oxus* (acid) and *genan* (to beget)] by Lavoisier.[6] The role of oxygen in the Earth system was discussed in Section 2.4.

Oxygen occupies 20.946% by volume of dry air. "Foul air" (now called nitrogen[8]) occupies 78.084%. The next two most abundant gases in air are argon (0.934%) and carbon dioxide[9] (0.03%). Together these four gases account for 99.99% of the volume of air. Many of the remaining minute amounts of the many other gases in air (some of which are listed in Table 5.1) are of prime importance in atmospheric chemistry because of their reactivity.

The most common unit for expressing the quantity of a gas in air is the fraction of the total volume of air that the gas occupies. The volumes occupied by different gases at the same temperature and pressure are proportional to the numbers of molecules of the respective gases [see Eq. (3.6)]. Also, for a mixture of ideal gases (such as air) the partial pressure exerted by a gas is proportional to the mole fraction of the gas in the mixture. For example, if CO_2 occupies 0.04% of the volume of air, the fraction of the total number of molecules in air that are CO_2 (i.e., the mole fraction of CO_2) is 0.04% and, if the total air pressure is 1 atm, the partial pressure exerted by CO_2 is 0.04% of 1 atm.

Exercise 5.1 N_2O occupies 310 ppbv of air, how many N_2O molecules are there in 1 m^3 of air at 1 atm and 0 °C?

Solution: We need to calculate first the number of molecules in 1 m^3 of any gas (or mixture of gases such as air) at 1 atm and 0 °C (called *Loschmidt's*[10] *number*).

From (3.8), $p = n_0 kT$ where n_0 is the number of molecules in 1 m^3. Substituting $p = 1$ atm $= 1.013 \times 10^5$ Pa, $T = 273$ K, and $k = 1.381 \times 10^{-23}$ J K^{-1} molecule^{-1} into this expression yields Loschmidt's number

$$n_0 = \frac{1.013 \times 10^5}{(1.381 \times 10^{-23})\, 273} = 2.687 \times 10^{25} \text{ molecules m}^{-3}$$

Because the volumes occupied by gases at the same temperature and pressure are proportional to the numbers of molecules of the gases,

$$\frac{\text{Volume occupied by } N_2O \text{ molecules}}{\text{Volume occupied by air}} = \frac{\text{Number of } N_2O \text{ molecules per m}^3}{\text{Total number of molecules per m}^3\ (n_0)}$$

The left side of this relation is equal to 310 ppbv $= 310 \times 10^{-9}$. Therefore, the number of N_2O molecules in 1 m^3 of air $= 310 \times 10^{-9}\, n_0 = (310 \times 10^{-9}) \times (2.687 \times 10^{25}) = 8.33 \times 10^{18}$. ■

[6] **Antoine-Laurent Lavoisier** (1743–1794) French chemist. Father of modern chemistry. His early studies included the best means for lighting large cities, analysis of gypsum, thunder, and the aurora. Confirmed that oxygen is absorbed by burning. Confirmed Cavendish's[7] conclusion that water is formed by the combustion of hydrogen and oxygen. Recognized some 30 elements and proposed that compounds be named after the elements they contained. Beheaded in the French Terror. ("La république n'a pas besoin des savants.")

Lavoisier thought that all acids contain oxygen. It is now known that many acids do not contain oxygen (e.g., hydrochloric acid, HCl) but they all contain hydrogen.

[7] **Henry Cavendish** (1731–1810) English chemist and physicist. Perfected the technique of collecting gases above water. Discovered "flammable air," called oxygen by Lavoisier. Used a sensitive torsion balance to measure the gravitational constant (G), from which he calculated the mass of the Earth. Terrified of women, with whom he communicated only by letter.

[8] The Scottish chemist **Daniel Rutherford** (1749–1819) is attributed with the discovery of nitrogen (1772), which he called "phlogisticated air."

[9] Carbon dioxide was discovered in 1750 by the Scottish physicist **Joseph Black** (1728–1799). Black is also known for his work on melting and evaporation, which led to the concept of latent heats and specific heats.

[10] **Joseph Loschmidt** (1821–1895) Czech physicist and chemist. Son of a poor Bohemian farmer. Moved to Vienna at age 20, where he attended lectures in chemistry and physics at the Polytechnic Institute. Considered becoming a settler in the new state of Texas, but instead started a company in Vienna to produce potassium nitrate. In 1856 became a high school teacher in Vienna. It was during this period that he became the first person to use the kinetic theory of gases to obtain an estimate for the diameter of a molecule. Proposed the first structural chemical formulae for many molecules, including markings for double and triple carbon bonds. Appointed to a faculty position at the University of Vienna in 1866. What is called "Avogadro's number" in English textbooks is, in German-speaking countries, called "Loschmidt's number."

Table 5.1 Some gases in dry tropospheric air at a pressure of 1 atm

Gas	Chemical formula	Fraction of volume of air occupied by the species[a]	Residence time (or lifetime)[b]	Major sources
Nitrogen	N_2	78.084%	1.6×10^7 years	Biological
Oxygen	O_2	20.946%	3000–4000 years	Biological
Argon	Ar	0.934%	—	Radiogenic
Carbon dioxide	CO_2	379 ppmv[c]	3–4 years[d]	Biological, oceanic, combustion (concentration increasing)
Neon	Ne	18.18 ppmv	—	Volcanic (?)
Helium	He	5.24 ppmv	—	Radiogenic
Methane[e]	CH_4	1.7 ppmv	9 years	Biological, anthropogenic
Hydrogen	H_2	0.56 ppmv	~2 years	Biological, anthropogenic
Nitrous oxide	N_2O	0.31 ppmv	150 years	Biological, anthropogenic
Carbon monoxide	CO	40–200 ppbv	~60 days	Photochemical, combustion, anthropogenic
Ozone	O_3	10–100 ppbv	Days–weeks	Photochemical
Nonmethane hydrocarbons (NMHC)[e]	—	5–20 ppbv	Variable	Biological, anthropogenic
Halocarbons	—	3.8 ppbv	Variable	Mainly anthropogenic
Hydrogen peroxide	H_2O_2	0.1–10 ppbv	1 day	Photochemical
Formaldehyde	HCHO	0.1–1 ppbv	~1.5 h	Photochemical
Nitrogen species ($NO + NO_2 (= NO_x) + NO_3 + N_2O_5 + HNO_3$ + PAN)	NO_y	10 pptv–1 ppmv	Variable	Soils, anthropogenic, lightning
Ammonia	NH_3	10 pptv–1 ppbv	2–10 days	Biological
Sulfur dioxide	SO_2	10 pptv–1 ppbv	Days	Photochemical, volcanic, anthropogenic
Dimethyl sulfide (DMS)	CH_3SCH_3	10–100 pptv	0.7 days	Biological, oceanic
Hydrogen sulfide	H_2S	5–500 pptv	1–5 days	Biogenic, volcanic
Carbon disulfide	CS_2	1–300 pptv	~120 h	Biological, anthropogenic
Hydroxyl radical[f]	OH	0–0.4 pptv	~1 s	Photochemical
Hydroperoxyl radical[f]	HO_2	0–5 pptv	—	Photochemical

[a] In addition to percentage by volume, the units used are parts per million by volume (ppmv), 10^{-6}; parts per billion by volume (ppbv), 10^{-9}; and parts per trillion by volume (pptv), 10^{-12}.
[b] See Box 5.1.
[c] See Fig. 1.3.
[d] This is the average time a CO_2 molecule is in the atmosphere before it is taken up by plants or dissolved in the ocean. However, the time required for atmospheric CO_2 to adjust to a new equilibrium if its sources or sinks were changed is ~50–200 years (see Sections 2.3 and 10.4.1).
[e] Hydrocarbons other than CH_4 are referred to as *nonmethane hydrocarbons* (NMHC). They originate from fossil fuel combustion, biomass burning, forest vegetation, etc. As is the case for CH_4, the primary sink for most NMHC is oxidation by OH. Because NMHC are more reactive than CH_4, their atmospheric residence times are much shorter (hours to months). Also, unlike CH_4, NMHC contribute significantly to O_3 formation in urban and regional smogs (see Section 5.5.2). For these reasons, it has been traditional to separate CH_4 from NMHC. In addition to CH_4 and NMHC, a variety of organic compounds are important in tropospheric chemistry. These other organics include *volatile organic compounds* (VOCs) such as carbonyls, organic sulfur compounds, and alcohols.
[f] *Radicals* (sometimes called *free radicals*) are chemical species with an unpaired electron in their outer (*valence*) shell. Consequently, a radical has an odd number of total electrons[g] (e.g., OH has 8 + 1 = 9 electrons). The unpaired electrons make radicals more reactive than nonradicals.
[g] Important exceptions are atomic oxygen in its ground state [$O(^3P)$ in spectroscopic notation] and in its excited state [$O(^1D)$, or O^* as we will indicate it], which, despite the fact that each has 8 electrons, is very reactive.

5.1 Residence Time and Spatial Scales of Variation of Chemicals in the Atmosphere

If the globally averaged concentration of a trace constituent in the atmosphere does not change significantly over a given time period, the rate at which the constituent is injected into (and/or produced within) the atmosphere must equal the rate at which it is removed from the atmosphere. Under such steady-state conditions, we can define the *residence time* (or *lifetime*) τ of a trace constituent in the atmosphere as

$$\tau = \frac{M}{F} \tag{5.1}$$

where M is the amount of the constituent in the atmosphere (in kg) and F is the rate of its removal (in kg s^{-1}) from the atmosphere.

The following analogy may be helpful in understanding the concept of residence time. Suppose a tank is full of water and is over-flowing at its top due to water being pumped into the bottom of the tank at a rate F. Then the rate of removal of water from the tank is F. If we assume that the water entering the bottom of the tank steadily displaces the water above it by pushing it upward without any mixing, the time spent by each small element of water that enters the bottom of the tank before it overflows at the top of the tank is M/F, where M is the volume of the tank in analogy with (5.1).

Although each atmospheric constituent can be assigned a residence time in accordance with (5.1), the residence times of individual molecules of that constituent vary widely, especially if the removal processes tend to be locally concentrated. Furthermore, residence time, defined in this manner, does not always give a representative idea of how long it would take for the atmospheric concentration of a species to react to an abrupt change in the source. For example, CO_2 has a residence time of only a few years in the atmosphere, but a much slower adjustment time (see Section 2.3.2).

In the atmosphere, the very stable gas nitrogen has a residence time of $\sim 10^7$ years. In contrast,

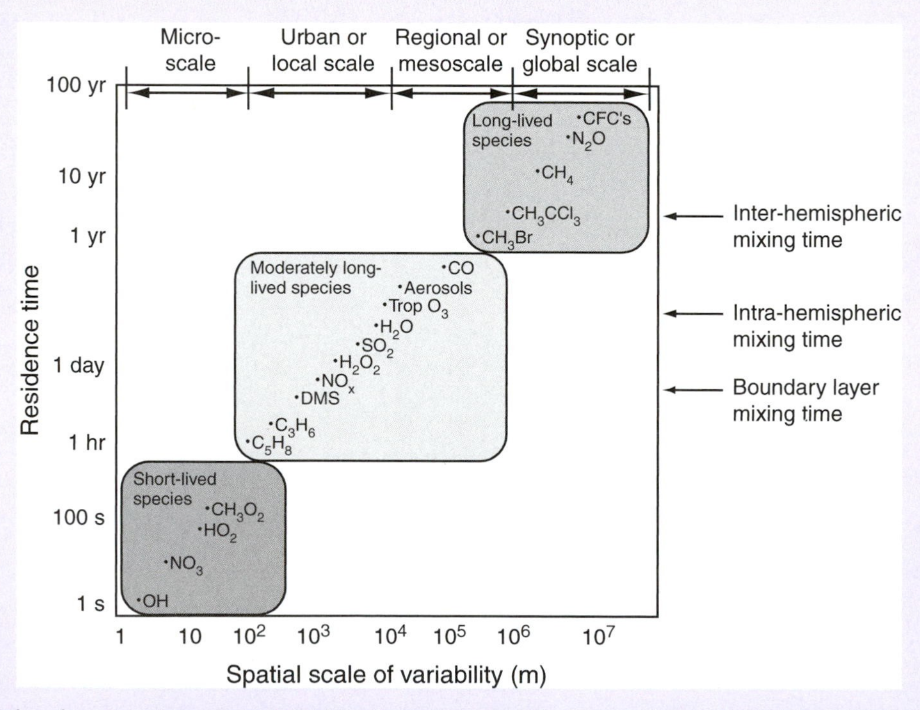

Fig. 5.1 Spatial and temporal scales of variability for some atmospheric constituents. The temporal scale is represented by residence time. [Adapted with permission from *The Atmospheric Sciences Entering the Twenty-First Century*, United States National Academy Press, 1998, p. 137.]

Continued on next page

5.1 Continued

the very reactive hydroxyl radical (OH) has a residence time in the atmosphere of only a second or so. Of course, residence times may be determined by physical removal processes (e.g., scavenging by precipitation) as well as by chemical processes.

If a chemical species has a very short (or very long) residence time in the atmosphere, significant variations in the concentration of the species will generally occur over very short (or very large) spatial scales (Fig. 5.1). Species with short residence times will be present in high concentrations close to localized sources and in low concentrations far removed from their sources. In contrast, chemical species with long residence times exhibit more uniform concentrations.

5.2 Sources, Transport, and Sinks of Trace Gases

5.2.1 Sources

The major natural sources of gases in the troposphere are biogenic, the solid Earth, the oceans, and *in situ* formation. These sources are discussed, in turn, next.

a. Biogenic

As discussed in Sections 2.3 and 2.4, oxygen in the Earth's present atmosphere was liberated by biological activity starting about 3.8 billion years ago, and the Earth's atmosphere is still strongly affected by the biota. Of prime importance is the photosynthesis reaction (2.5), which removes carbon from the atmosphere and stores it in organic matter and releases oxygen to the atmosphere.

Exercise 5.2 If the photosynthesis reaction is represented by the reaction[11]

$$CO_2(g) + H_2O(l) + h\nu \rightarrow CH_2O(s) + O_2(g) \quad (5.2)$$

where $h\nu$ represents a photon. Is the carbon atom reduced or oxidized by this reaction?

Solution: Because the oxidation number of each oxygen atom in CO_2 is -2 and CO_2 has no net electric charge, the oxidation number of the carbon atom in CO_2 is $+4$. In CH_2O, the oxidation numbers of the H and O atoms are $+1$ and -2, respectively. Therefore, the oxidation number of the carbon atom in CH_2O is 0. Therefore, the reaction decreases the oxidation number of the carbon atom from $+4$ to 0 (i.e., the carbon is *reduced*). ■

About 80% of the CH_4 in air derives from the decay of recent organic materials (rather than fossil fuels) through cud-chewing animals (cows, etc.), termites, rice paddies, and wetlands.

Biological processes (often mediated by microbes) convert N_2 into NH_3 (primarily via animal urine and soils), N_2O (through nitrate respiration by bacteria in soils), and NO.

Regions of the ocean with high organic content and biological productivity (e.g., upwelling regions, coastal waters, and salt marshes) are a major source of CS_2 and carbonyl sulfide (COS). Phytoplankton are the major source of atmospheric DMS and dimethyl disulfide (CH_3SSCH_3). DMS is oxidized to SO_2 and then to sulfate aerosols. Microbial degradation of dead organic matter releases H_2S. The most abundant halocarbon in the air, and the major natural source of chlorine (Cl) in the stratosphere, is methyl chloride (CH_3Cl), which derives, in part, from biological activity in seawater, wood molds, and biomass burning. Halogen compounds (e.g., chlorine and bromine species) are also produced by biological activity in the oceans.

Several thousand *volatile organic compounds* (VOCs), emitted by plants and anthropogenic sources, have been identified. In the United States, motor vehicles are the primary source of VOCs, mainly in the form of hydrocarbons produced by the incomplete combustion of fuel and from the vaporization of fuel. The evaporation of solvents is the second

[11] When needed for clarity, the phase of a chemical species is indicated in parenthesis by g for gas, l for liquid, s for solid, and aq for aqueous solution.

largest source of VOCs worldwide. Some of the more important VOCs are isoprene (C_5H_8), ethene (C_2H_4), and monoterpenes. Isoprene accounts for ~50% of the NMHC. The photooxidation of isoprene can produce compounds that have vapor pressures low enough for them to condense onto preexisting particles. This process could account for ~5–20% of the annual secondary organic aerosol from biogenic sources. Terpenes are a class of hydrocarbons that evaporate from leaves. About 80% of these emissions oxidize to organic aerosols in about an hour. Emissions from vegetation are a significant source of hydrocarbons, which can react photochemically with NO and NO_2 to produce O_3, thereby playing a central role in atmospheric chemistry (see Section 5.3.5).

Use of biological materials by humans results in the emissions of many chemicals into the atmosphere, for example, CO_2, CO, NO_x, N_2O, NH_3, SO_2 and hydrogen chloride—HCl (from the combustion of oil, gas, coal, and wood), hydrocarbons (from automobiles, refineries, paints, and solvents), H_2S and DMS (from paper mills, and oil refineries), carbonyl sulfide—COS (from natural gas), and chloroform—$CHCl_3$ (from combustion of petroleum, bleaching of woods, solvents).

Figure 5.2 shows the global distribution of fires during September 2000. September is in the biomass burning season in South America and southern Africa, and hence there are many fires in these two locations. An area of vegetation equal to about half the area of Europe is burned globally each year by natural forest fires (many initiated by lightning) and deliberate deforestation (e.g., in the Amazon Basin), by refertilization of soils and grazing (e.g., in the savannas of southern Africa), and by the use of wood for heating and cooking (e.g., in Africa, India, and southeast Asia). It has been estimated that on an annual basis, biomass burning produces ~38% of the O_3, 32% of the CO, ~39% of the particulate carbon, and more than 20% of the H_2, NMHC, methyl chloride (CH_3Cl), and NO_x in the troposphere. Biomass burning also produces ~40% of the world's annual production of CO_2, but this is largely offset by the uptake of CO_2 by young vegetation that sprouts quickly on burned areas.

As discussed in Section 5.3.5, ozone is produced in the troposphere by photochemical reactions involving oxidation by OH of CO, CH_4, and NMHC in the presence of NO_x. Because all of these precursors are present in smoke from biomass burning, elevated O_3 concentrations are produced in biomass smoke as it disperses in the troposphere. Many of the emissions from biomass burning are carcinogens, they cause significant degradation to air quality on local and regional scales, and they have global effects on atmospheric chemistry and climate.

Biomass smoke can be dispersed over large distances in the atmosphere. For example, under appropriate wind conditions, biomass smoke from Africa is dispersed across the south Atlantic Ocean and even to Australia (Fig. 5.3). Indeed it is likely that even the most remote regions of Earth are not immune from pollution. Biomass smoke can also be lofted into the middle and upper troposphere where it can become a dominant source of HO_x (where $x = 0$, 1, or 2) and NO_x and result in the production of O_3 (see Section 5.3.5).

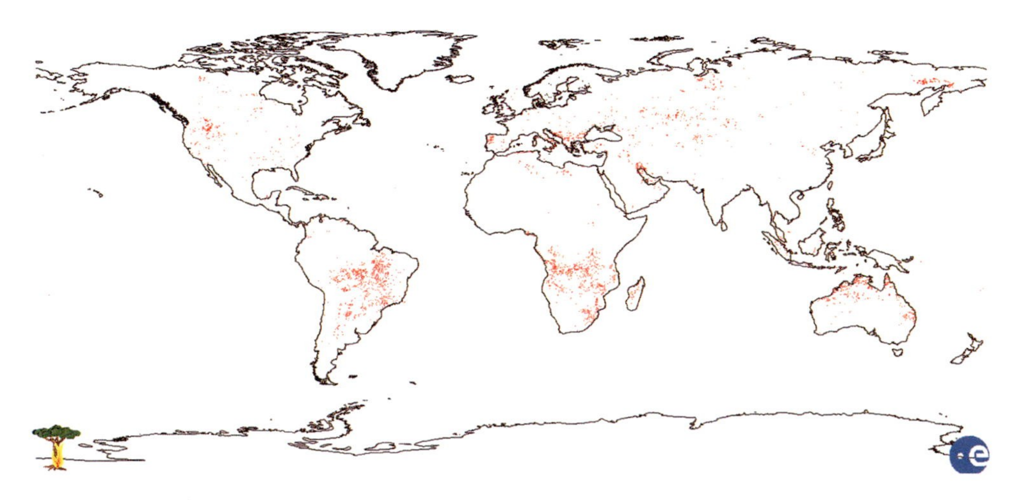

Fig. 5.2 Global distribution of fires detected by satellite in September 2000. [Image courtesy of European Space Agency.]

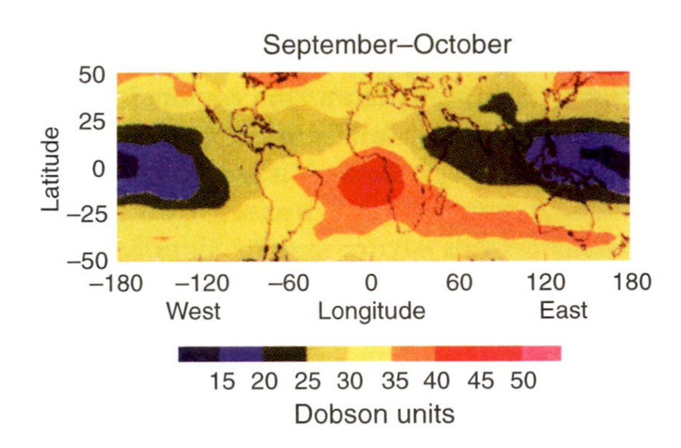

Fig. 5.3 Satellite measurements of tropospheric ozone in September and October for the period 1979–1989. The high column amounts of ozone (indicated by high Dobson[12] units[13]) over tropical and southern Africa are due to smoke from biomass burning. [Excerpted with permission from Fishman et al., *Science* **252**, p. 1694. Copyright 1991 AAAS.]

b. Solid Earth

Various aspects of the role of the solid Earth in the chemistry of the atmosphere have been discussed in Chapter 2. Some additional information is given here.

Volcanoes are the most important source of atmospheric trace gases from the solid Earth. In addition to ash and copious small particles, volcanoes emit H_2O, CO_2, SO_2, H_2S, COS, HCl, hydrogen fluoride (HF), hydrogen bromide (HBr), CH_4, CH_3Cl, H_2, CO, and heavy metals [e.g., mercury (Hg)]. Emissions from violent volcanic eruptions can be blasted into the stratosphere, where constituents with long residence times can be dispersed around the globe.[14]

Rocks are the major sources of He, Ar, and radon (Rn) in the atmosphere. Helium is produced by the radioactive decay of uranium-238 and thorium-232. It does not significantly accumulate in the atmosphere because it is so light that it escapes from the exosphere. Argon has accumulated in air over eons from the radioactive decay of potassium-40 in rocks. Radon-222 is a decay product of uranium in rocks; it has a half-life of only 3.8 days.

Carbonate rocks, such as limestone (e.g., $CaCO_3$), contain about 20,000 times more carbon than the atmosphere (see Table 2.3), but most of this is sequestered. As described in Section 2.3.3 and 2.2.4, carbonate rocks and marine sediments are involved in a long-period cycle with atmospheric CO_2.

c. Oceanic

As mentioned in Section 5.2.1(a), the oceans are a huge reservoir of water-soluble gases (e.g., see the major carbon reservoirs listed in Table 2.3). Thus, the oceans may serve as either a sink or a source for soluble gases.

The oceans are an atmospheric source for many gases produced by biological activity, particularly sulfur-containing gases.

Exercise 5.3 If the Henry's[15] law constant for CO_2 in water is 3.40×10^{-2} mole liter^{-1} atm^{-1}, what is the the solubility of CO_2 in water?

Solution: Henry's law is

$$C_g = k_H p_g \tag{5.3}$$

where C_g is solubility of a gas in a liquid (in mole liter^{-1}), k_H is a temperature-dependent constant of proportionality called the *Henry's law constant*, and p_g is the partial pressure (in atm) of the gas

[12] **G. M. B. Dobson** (1889–1976) English physicist and meteorologist. Made the first measurements of the variation of wind with height using pilot balloons (1913). In 1922 he discovered the presence of a warm layer of air at ~50 km, which he correctly attributed to the absorption of UV radiation by O_3. Built a UV solar spectrograph for measuring the atmospheric O_3 column. Also obtained first measurements of water vapor in the stratosphere.

[13] One Dobson unit (DU) is the thickness, in hundredths of a millimeter, that the total O_3 column would occupy at 0 °C and 1 atm. The Earth's total atmospheric O_3 column is ~300 DU (i.e., if all the O_3 in the atmosphere were brought to 0 °C and 1 atm, it would form a layer just ~3 mm deep).

[14] The violent eruption in 1883 of the Indonesian volcano of Krakatau caused remarkable sunsets and lowered global temperatures at the Earth's surface by ~0.5 °C in the year following the eruption. The largest volcanic eruption in the 20th century, in terms of its atmospheric effects, was Pinatubo in the Philippines in 1991. The emissions from this eruption produced a global average cooling of ~0.5 °C for ~2 years and lowered ozone concentrations in the stratosphere. (See Sections 5.7.3 and 10.2.3.)

[15] **William Henry** (1774–1836) English physician and chemist. First scientific paper was a refutation of a claim that carbon is not an element. Together with his friend John Dalton, his experiments on the dissolution of gases were crucial in the development of the atomic theory of matter. Committed suicide due to pain (from a childhood injury) and sleep deprivation.

above the surface of the liquid. The partial pressure of CO_2 in air is $\sim 3.79 \times 10^{-4}$ atm (see Table 5.1). Therefore,

$$\begin{aligned} C_g &= k_H p_g \\ &= (3.40 \times 10^{-2} \text{ mol liter}^{-1})(3.79 \times 10^{-4} \text{ atm}) \\ &= 1.29 \times 10^{-5} \text{ mol liter}^{-1} \end{aligned}$$

■

d. In situ formation

In situ formation, which refers to the formation of chemical species by chemical reactions in the atmosphere, is a major source of many important atmospheric trace constituents. Most such gaseous reactions are initiated by photolysis involving radicals and occur by uni-, bi- and termolecular reactions.

In situ chemical reactions can be classified as *homogeneous* or *heterogeneous*. A homogeneous reaction is one in which all of the reactants are in the same phase. For example, the reaction

$$NO_2(g) + O_3(g) \rightarrow NO_3(g) + O_2(g) \tag{5.4}$$

which is a major source of the nitrate radical (NO_3) in the atmosphere, is a *homogeneous gas-phase reaction*. A heterogeneous reaction is one involving reactants in two or more phases. The mixing of an inorganic aerosol [e.g., sulfuric acid (H_2SO_4) or nitric acid (HNO_3)] with organic compounds (e.g., aldehydes), which can appreciably increase the rate of aerosol growth, is an example of a heterogeneous reaction.

Trace gases emitted from the biosphere, solid Earth, and oceans are generally in a *reduced* (low) oxidation state (e.g., hydrocarbons, ammonia, hydrogen sulfide), but they are *oxidized* (i.e., raised to a higher oxidation state) by *in situ* reactions in the atmosphere.

e. Anthropogenic sources

We will discuss anthropogenic (i.e., human) sources of gases and particles in Sections 5.5 and 5.6. However, it is important to note here that anthropogenic sources play significant roles in the budgets of many important trace gases in the atmosphere (Table 5.2). As a result of increasing populations, anthropogenic emissions of a number of important trace gases have increased significantly over the past century. As a consequence, the extent of human influences on the atmosphere is one of the main themes of current research in atmospheric chemistry.

5.2.2 Transport

In the atmospheric boundary layer (ABL) the atmosphere interacts directly with the Earth's surface through turbulent mixing. Consequently, during the day over land, chemicals in the ABL are generally well mixed up to a height of $\sim$1–2 km. The dilution of chemical compounds by turbulent mixing is less efficient at night when the ABL depth is usually a few hundred meters or less. Over the oceans, the diurnal cycle is much less apparent.

If a chemical that originates from the Earth's surface is not returned to the surface or transformed by *in situ* reactions in the ABL, it will eventually pass into the free troposphere. Once in the free troposphere, chemicals with long residence times are carried along with the global circulation pattern. For example, in midlatitudes, where the winds are generally from west to east and have speeds of $\sim$10–30 m s^{-1}, a chemical injected into the atmosphere from a "point source," such as a volcano, will become distributed fairly uniformly longitudinally around the latitude belt within a few weeks. Since the transport of tropospheric air across the tropics is relatively restricted, so is the transport of chemicals. It follows that the chemistry of the troposphere in the northern hemisphere is more strongly affected by emissions from the use of fossil fuels than the chemistry of the southern hemisphere; the latter reflects more the effects of emissions from the oceans and from biomass burning. Transport is also restricted between the free troposphere and the stratosphere; most of the upward transport is in the tropics, and most of the downward transport is in higher latitudes. Nevertheless, as shown in Section 5.7.2, certain long-lived chemicals of anthropogenic origin can accumulate in the stratosphere, where they can have major effects.

Satellite observations provide strong evidence for the transport of tropospheric gases and particles. For example, satellite observations reveal large plumes of particles off the east coasts of the United States and Asia, enormous dust plumes carried westward from the Sahara Desert over the Atlantic Ocean, and large smoke plumes from regions of biomass burning. During the winter monsoon (December through April), a plume of pollutants extends from the southwest coast of India over the Indian Ocean. In spring and summer, dust and pollutants are transported from sources in Asia across the north Pacific Ocean.

Table 5.2 Estimates of natural and anthropogenic sources of a number of atmospheric trace gases in 2000[a]

Sources		SO_2 [Tg(S) year^{-1}]	NH_3 [Tg(N) year^{-1}]	N_2O [Tg(N) year^{-1}]	CH_4 [Tg(CH_4) year^{-1}]	CO [Tg(CO) year^{-1}]	NO_x [Tg(N) year^{-1}]	NMHC [Tg(C) year^{-1}]
Natural								
Vegetation			5.1			100 (60–160)		400 (230–1150)
Wetlands					115 (55–150)			
Wild animals			2.5					
Termites					20 (10–50)			
Oceans		25[b]	7.0	3 (1–5)	10 (5–50)	50 (20–200)		50 (20–150)
Soils				6 (3.3–9.7)			7 (5–12)	
Lightning							5 (2–20)	
Volcanoes		10 (7–10)						
Other		7.5[c]			15 (10–40)		1.5 (0–5.7)[d]	
Total natural		42.5	14.6	9 (4–15)	160 (80–290)	150 (80–360)	13.5 (7–38)	450 (250–1300)
Anthropogenic								
Fossil fuels related		75[e]						
	Natural gas				40 (25–50)			
	Coal mines				30 (15–45)			
	Petroleum industry				15 (5–30)			
	Coal combustion				? (1–30)			
	Energy use					500 (300–900)	22 (20–24)	70 (60–100)
	Aircraft						0.5 (0.2–1)	
Biospheric carbon		3						
	Enteric fermentation				85 (65–100)			
	Rice paddies				60 (20–100)			
	Biomass burning		2	0.5 (0.2–1.0)	40 (20–80)	500 (400–700)	8 (3–13)	40 (30–90)
	Landfills				40 (20–70)			
	Animal waste		22		25 (20–30)			
	Domestic sewage				25 (15–80)			
Fertilizer			6.4					
Cultivated soils				3.5 (1.8–5.3)				
Cattle and feedlots				0.4 (0.2–0.5)				
Industrial sources				1.3 (0.7–1.8)				
Total anthropogenic		78	30.4	5.7 (3–9)	360 (206–615)	1000 (700–1600)	30.5 (23–38)	110 (90–190)

[a] The last four listed species are ozone precursors. Numbers in parentheses are possible ranges of values. Estimates are based on several authoritative sources.
[b] Via oxidation of DMS.
[c] Oxidation of H_2S [7 Tg(S) year^{-1}] and CS_2 [0.5 Tg(S) year^{-1}].
[d] Oxidation of NH_3 [~0.9 Tg(N) year^{-1}] and breakdown of N_2O [~0.6 Tg(N) year^{-1}] produced in the stratosphere and transported to the troposphere.
[e] Including other industrial sources.

5.2.3 Sinks

The final stage in the life history of a chemical in the atmosphere is its removal. Sinks include transformations into other chemical species and gas-to-particle (g-to-p) conversion, which can involve both chemical and physical processes. The other important removal process for both gases and aerosols is deposition onto the Earth's surface and vegetation. Deposition is of two types: wet and dry. *Wet deposition*, which involves the scavenging of gases and particles in the air by clouds and precipitation, is one of the major mechanisms by which the atmosphere is cleansed. *Dry deposition* involves the direct collection of gases and particles in the air by vegetation and the Earth's solid and liquid surfaces. Dry deposition is a much slower process than wet deposition, but it is continuous rather than episodic.

The oceans are important sinks for many trace gases. The flux of a gas to the ocean depends on how undersaturated the ocean is with respect to the gas (see Section 5.2.1c). If the surface layers of the ocean are supersaturated with a gas, then the flux is from the ocean to the atmosphere (e.g., the estimated global flux of DMS from the ocean to the atmosphere is ~25 Tg of sulfur per year).

Exercise 5.4 If SO_2 were confined to a layer of the atmosphere extending from the surface of the Earth to a height of 5 km and the average *deposition velocity* of SO_2 onto the ground were 0.800 cm s^{-1}, how long would it take for all of the SO_2 to be deposited on the ground if all sources of SO_2 were suddenly switched off?

Solution: The deposition velocity of a gas onto a surface is defined by

$$\text{deposition velocity} = \frac{\text{flux of the gas to a surface}}{\text{mean concentration of the gas just above the surface}} \quad (5.5)$$

Since the units of flux are kg m^{-2} s^{-1} and the units of concentration are kg m^{-3}, it follows from (5.5) that the units of deposition velocity are m s^{-1}, which are the usual units of velocity. Hence, we can consider the deposition velocity of a gas to be the counterpart of the terminal fall speed of a particle.

Therefore, the time required to remove all of the SO_2 in a vertical column of height 5 km will be the time taken for a molecule of SO_2 to move through a vertical distance of 5 km toward the Earth's surface, which is

$$\frac{5\ \text{km}}{0.800 \times 10^{-5}\ \text{km s}^{-1}} = 6.25 \times 10^5\ \text{s} = 7.23\ \text{days}.$$

The residence time of 7.23 days for SO_2 derived here is an approximate upper limit since it neglects other removal mechanisms, such as *in situ* chemical reactions. ■

5.3 Some Important Tropospheric Trace Gases

Prior to the 1970s, photochemical reactions, and the oxidation of most trace gases, were thought to take place primarily in the stratosphere where the intensity of UV radiation is much greater than in the troposphere.[16] However, in the 1960–1970s it was recognized that the very reactive hydroxyl radical OH can be produced by photochemistry in the troposphere. At about the same time, studies of photochemical smogs (such as those that occur in Los Angeles) began to reveal the roles of OH, nitrogen oxides, and hydrocarbons in the formation of O_3 and other pollutants (see Section 5.5.2b).

This section considers some of the trace gases that play important roles in tropospheric chemistry, including those mentioned earlier. Our main concern in this section is with the nonurban troposphere. Chemical reactions in heavily polluted air, which can produce smogs, are considered in Section 5.5. Stratospheric chemistry is discussed in Section 5.7.

5.3.1 The Hydroxyl Radical

Because of its high reactivity with both inorganic and organic compounds, OH is one of the most important chemical species in the atmosphere, even though it is present in the troposphere in globally and diurnally averaged concentrations of just a few tenths of a pptv (~10^{12} OH molecules m^{-3}, or about 3 OH molecules per 10^{14} molecules in the air). Reaction with OH

[16] Photochemical reactions involve photons (represented by $h\nu$). To trigger a chemical reaction, the energy of the photons must exceed a critical value. Because the energy of a photon is inversely proportional to the wavelength of the electromagnetic radiation, this means that the wavelength of the radiation must be below a critical value if it is to trigger a specified photochemical reaction. In general, this critical value places the effective radiation for photochemical reactions in the UV region.

is the major sink for most atmospheric trace gases. Because it is so reactive, the average lifetime of an OH molecule in the atmosphere is only ~1 s.

Hydroxyl radicals are produced when solar UV radiation (with $\lambda \leq 0.32\ \mu m$) decomposes O_3 into molecular oxygen and energetically excited oxygen atoms (O*)

$$O_3 + h\nu \rightarrow O_2 + O^* \qquad (5.6a)$$

Most of the O* atoms produced by (5.6a) dissipate their excess energy as heat and eventually recombine with O_2 to form O_3, which, together with (5.6a), is a *null cycle* (i.e., it has no net chemical effect). However, a small fraction (~1%) of the O* atoms reacts with water vapor to form two hydroxyl radicals

$$O^* + H_2O \rightarrow 2OH \qquad (5.6b)$$

The net effect for those O* atoms produced by (5.6a) and removed by (5.6b) is

$$O_3 + H_2O + h\nu \rightarrow O_2 + 2OH \qquad (5.7)$$

Once formed, the OH radical is a powerful oxidant that reacts quickly with almost all trace gases containing H, C, N, O, and S and the halogens [except N_2O and the chlorofluorocarbons (CFC)]. For example, OH reacts with CO to form CO_2, NO_2 to form HNO_3, H_2S to form SO_2, SO_2 to form H_2SO_4, etc. (Fig. 5.4). Because of its role in removing many pollutants, OH has been called the atmosphere's detergent.

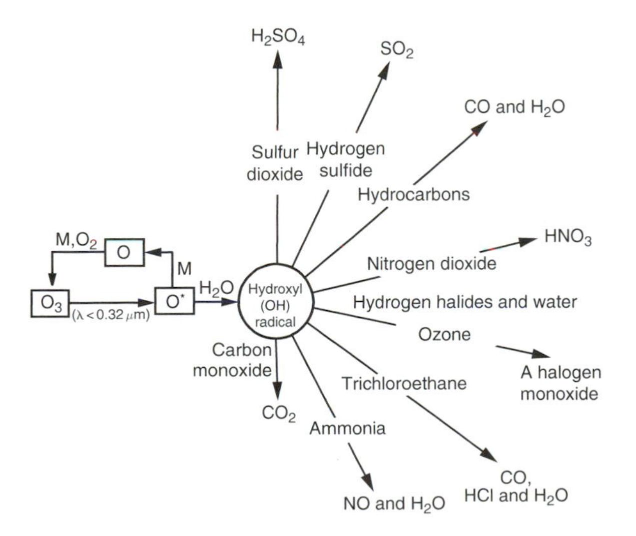

Fig. 5.4 Illustration of the central role of the OH radical in the oxidation of tropospheric trace gases. Little escapes oxidation by OH. [Adapted from *Global Tropospheric Chemistry*, United States National Academy Press, 1984, p. 79.]

The dominant sinks for OH in the global troposphere are the oxidation of CO and CH_4. Over the continents, reactions with NMHC can be strong local sinks for OH. In forests, the dominant reactant with OH is often isoprene (C_5H_8), which is emitted by deciduous trees.

The production of O_3 through reaction (5.7) has probably increased over the past two centuries due to increases in anthropogenic sources of its precursors, and also due to the increasing flux of solar UV radiation into the atmospheric boundary layer in response to the thinning of the stratospheric ozone layer. However, sinks for OH have also increased due to increases in the concentrations of CO and hydrocarbons. Consequently, it is not clear whether the concentration of OH in the troposphere has changed significantly over the past several decades in response to human activity.

5.3.2 Some Reactive Nitrogen Compounds

a. Nitrogen oxides

The oxides of nitrogen, NO (nitric oxide), and NO_2 (nitrogen dioxide), which together are referred to as NO_x, play important roles in atmospheric chemistry. They are produced by fossil fuel combustion, biomass burning, and from soils, lightning, NH_3 oxidation, aircraft emissions, and transport from the stratosphere. NO_x is emitted into the troposphere primarily as NO, but during the day NO rapidly establishes an equilibrium with NO_2 through the following null cycle.

$$NO + O_3 \rightarrow NO_2 + O_2 \qquad (5.8a)$$

$$NO_2 + O_2 + M + h\nu \rightarrow NO + O_3 + M \qquad (5.8b)$$

where M represents an inert molecule that absorbs excess molecular energies. Once NO is converted to NO_2, a number of reaction paths are available. At night NO_x is present only as NO_2 due to reaction (5.8a). The principal sink for NO_x in the daytime is

$$NO_2 + OH + M \rightarrow HNO_3 + M \qquad (5.9)$$

The nitric acid (HNO_3) is removed in about 1 week by dry and wet deposition. At night, NO_2 is oxidized by O_3 to NO_3, the NO_3 then reacts with NO_2 to produce N_2O_5, and the N_2O_5 reacts with water on particles to produce HNO_3. The resulting residence time of NO_2 is ~1 day.

b. The nitrate radical

Because OH is produced primarily by photochemical reactions and has a very short lifetime, it is present in the atmosphere at currently measurable concentrations only during the day. At night, the nitrate radical NO_3 takes over from OH as the major reactive oxidant in the troposphere. The nitrate radical is formed by Eq. (5.4). During the day, NO_3 is photolyzed rapidly by solar radiation via two pathways

$$NO_3 + h\nu\,(\lambda < 0.700\ \mu m) \rightarrow NO + O_2 \quad (5.10)$$

and

$$NO_3 + h\nu\,(\lambda < 0.580\ \mu m) \rightarrow NO_2 + O \quad (5.11)$$

The resulting atmospheric residence time of NO_3 at noon in sunlight is ~5 s. NO_3 also reacts with NO

$$NO_3 + NO \rightarrow 2NO_2 \quad (5.12)$$

Although NO_3 is much less reactive than OH, at night it is present in much higher concentrations than OH is during the day so it can serve as an effective oxidant.

c. Odd nitrogen and other "chemical families"

Odd nitrogen (NO_y), or total reactive nitrogen as it is sometimes called, refers to the sum of NO_x plus all compounds that are the products of the atmospheric oxidation of NO_x, including HNO_3, NO_3, dinitrogen pentoxide (N_2O_5), and peroxyacetyl nitrate (PAN for short). Grouping the nitrogen-containing species in this manner is useful in considering their budgets in the atmosphere.

The main sources of NO_y are anthropogenic emissions (see Section 5.6.1) but, away from pollution sources, soils and lightning can play dominant roles. Interactions between NO_y and NMHC can lead to photochemical smogs in cities (see Section 5.5.2). On regional and global scales, interactions of NO_y with odd hydrogen have a strong influence on OH concentrations.

NO_x and NO_y are examples of "chemical families." Other useful chemical families are *odd oxygen* (e.g., O, O*, O_3 and NO_2), *odd hydrogen* (HO_x, where x = 0, 1 or 2), and *odd chlorine* (ClO_x, where x = 0, 1 or 2). Note that odd hydrogen includes the OH radical and NO_y includes the NO_3 radical.

d. Ammonia

Ammonia (NH_3) originates from soils, animal waste, fertilizers, and industrial emissions. It is the principal basic gas in the atmosphere. Ammonia neutralizes acid species by reactions of the form

$$2NH_3 + H_2SO_4 \rightarrow (NH_4)_2SO_4 \quad (5.13)$$

The primary removal mechanisms for NH_3 involve its conversion to ammonium-containing aerosols by reactions such as (5.13). The aerosols are then transported to the ground by wet and dry deposition. The residence time of NH_3 in the lower troposphere is ~10 days.

5.3.3 Organic Compounds

Organic compounds contain carbon atoms. The four electrons in the outer orbital of the carbon atom can form bonds with up to four other elements: hydrogen, oxygen, nitrogen, sulfur, halogens, etc. *Hydrocarbons* are organics composed of carbon and hydrogen. There are large natural and anthropogenic sources of atmospheric hydrocarbons, gases that play key roles in many aspects of the chemistry of the troposphere.

Methane is the most abundant and ubiquitous hydrocarbon in the atmosphere. The present concentration of CH_4 in the northern hemisphere is ~1.7 ppmv; it has a residence time in the atmosphere of ~9 years. Sources of CH_4 include wetlands, landfills, domestic animals, termites, biomass burning, leakages from natural gas lines, and coal mines. The primary sink for tropospheric CH_4 is its oxidation by OH to form formaldehyde (HCHO); HCHO then photodissociates into CO (see Section 5.3.6) or, in air with sufficient NO_x, OH oxidizes CO to produce O_3 (see Section 5.3.5).

There are numerous nonmethane hydrocarbons (NMHC) in the Earth's atmosphere, and many of them play important roles in tropospheric chemistry. Based on their molecular structures, NMHC can be grouped into several classes. Examples include *alkanes* (C_nH_{2n+2}), which include ethane ($CH_3{-}CH_3$) and propane ($CH_3{-}CH_2{-}CH_3$); *alkenes*, which have a double bond, such as ethene ($CH_2{=}CH_2$) and propene ($CH_3CH{=}CH_2$); and *aromatics*, such as benzene (C_6H_6) and toluene (C_7H_8). *Oxygenated hydrocarbons*, which contain one or more oxygen atoms, such as acetone (CH_3COCH_3), may provide an important source of HO_x in the upper troposphere, thereby influencing O_3 chemistry in this region.

5.3.4 Carbon Monoxide

Carbon monoxide is produced by the oxidation of CH_4 or NMHC, such as isoprene. Other important sources of CO are biomass burning and the combustion of fossil fuels. The dominant sink of CO is oxidation by OH

$$CO + OH \rightarrow H + CO_2 \tag{5.14}$$

Because (5.14) is generally the major sink for OH in nonurban and nonforested locations, the concentration and distribution of OH are often determined by the ambient concentrations of CO, although CH_4, NO_x, H_2O, etc. can also be determining factors. An important feature of CO in extratropical latitudes is its seasonal cycle; it accumulates in the atmosphere during winter when OH concentrations are low, but in spring CO is depleted rapidly due to reaction (5.14).

Figure 5.5 shows satellite measurements of CO. The high concentration of CO emanating from South America and Africa are due to biomass burning. The plumes travel slowly across the southern hemisphere and can be detected in Australia during the dry season. A strong source of CO, due to industrial emissions, is also apparent in southeast Asia; this plume can sometimes reach North America.

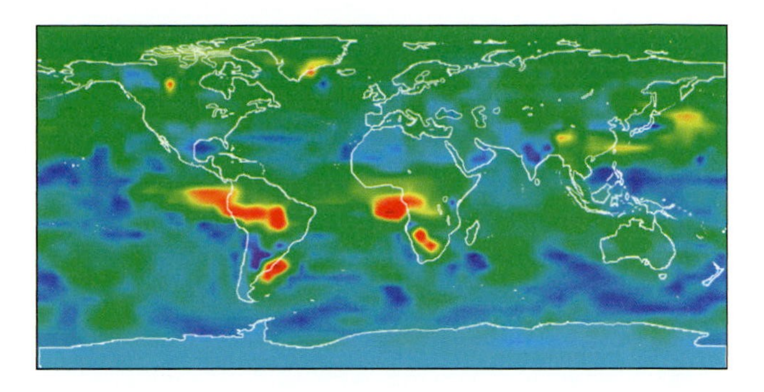

Fig. 5.5 Concentrations of CO at an altitude of ~4.5 km measured from a satellite. Concentrations range from background values ~50 pptv in regions shaded blue to as high as 450 ppbv in the regions shaded red. The CO can be transported upward and also be carried over large horizontal distances. [Courtesy of NASA.]

5.3.5 Ozone

Because about 90% of the O_3 in the Earth's atmosphere is in the stratosphere (see Section 5.7.1), it was suggested in the middle of the 20th century that the stratosphere was a primary source for tropospheric O_3 and that a balance existed between this source and surface sinks. It has subsequently been recognized that trace gases, such as NO, CO, and organic compounds, which are emitted by human activities, lead to the formation of O_3 through photochemical reactions. In addition, various natural sources, such as hydrocarbons from vegetation and NO from lightning, produce O_3 precursors.

Ozone plays a controlling role in the oxidizing capacity of the troposphere. Much of the O_3 in the troposphere is produced by *in situ* homogeneous gas-phase reactions involving the oxidation of CO, CH_4, and NMHC by OH in the presence of NO_x, as outlined later.

5.2 "Bad" and "Good" Ozone

Schönbein[17] discovered ozone by its odor following an electrical discharge (it can sometimes be smelled after a thunderstorm). He called it ozone, after the Greek word *ozein*, meaning to smell.

Ozone is an irritating, pale blue gas that is toxic and explosive. Because of its high reactivity, ozone is an extremely powerful oxidizing agent that damages rubber and plastics and is harmful to humans and plants even at low concentrations (~several tens of ppbv). Haagen-Smit[18] suggested that the formation of O_3 in cities is due to photochemical reactions involving nitrogen oxides and

Continued on next page

[17] **Christian Friedrich Schönbein** (1799–1868) German chemist. Discovered nitrocellulose (an explosive) when he used his wife's cotton apron to wipe up some spilled nitric and sulfuric acid and the apron disintegrated. Taught in England before joining the faculty at the University of Basel.

[18] **Arie Jan Haagen-Smit** (1900–1977) Dutch biologist and chemist. Joined the Biology Department, California Institute of Technology, in 1937, where he worked on terpenes and plant hormones. Carried out pioneering work on analysis of Los Angeles' smogs and the emissions responsible for them.

5.2 *Continued*

hydrocarbons released by cars and oil refineries. Ozone reacts with hydrocarbons from automobile exhausts and evaporated gasoline to form secondary organic pollutants such as aldehydes and ketones. Ozone produced in urban areas can be transported into rural areas far removed from industrial regions. For example, during a summer heat wave in 1988 the Acadia National Park in Maine had dangerously high concentrations of O_3, which probably originated in New York City. Ozone alone, or in combination with SO_2 and NO_2, accounts for ~90% of the annual loss of crops due to air pollution in the United States.

In contrast to the bad effects of O_3 in the troposphere, the much greater concentrations of O_3 in the stratosphere reduce the intensity of dangerous UV radiation from the sun (see Section 5.7.1).

At wavelengths <0.430 μm, NO_2 is photolyzed

$$NO_2 + h\nu \xrightarrow{j} NO + O \qquad (5.15)$$

which is followed quickly by

$$O + O_2 + M \xrightarrow{k_1} O_3 + M \qquad (5.16)$$

where j and k_1 are rate coefficients.[19] However, much of the O_3 produced by (5.16) is removed rapidly by

$$O_3 + NO \xrightarrow{k_2} NO_2 + O_2 \qquad (5.17)$$

In combination, (5.15)–(5.17) constitute a null cycle that neither creates nor destroys ozone.

Exercise 5.5 Derive an expression for the steady-state concentration of O_3 given by Eqs. (5.15)–(5.17) in terms of the concentrations of NO and NO_2, j and k_2.

Solution: All three reactions are fast: NO is formed by (5.15) and NO_2 is formed by (5.17) as rapidly as it is depleted by (5.15). Therefore, from the definition of a rate coefficient,

$$j\,[NO_2] = k_2\,[O_3]\,[NO]$$

where the square brackets indicate species concentration. Therefore,

$$[O_3] = \frac{j}{k_2}\frac{[NO_2]}{[NO]} \qquad (5.18)$$

Expression (5.18) is called the *Leighton*[20] *relationship*. It is an example of a *photostationary state relation*. ■

If typical concentrations of NO_2 and NO are substituted into (5.18), together with values for j and k_2, the concentrations of O_3 obtained are far below observed concentrations even in the free troposphere. Hence, reactions other than (5.15)–(5.17) must be involved in the control of tropospheric O_3. This conclusion led to the suggestion that HO_x, and related radicals derived from organic species, are involved in determining atmospheric concentrations of O_3. The hydroxyl radical can be produced from small amounts of highly reactive atomic oxygen by (5.6b). Then, in unpolluted air, OH is rapidly transformed to HO_2 by

$$OH + CO \rightarrow H + CO_2 \qquad (5.19a)$$

$$H + O_2 + M \rightarrow HO_2 + M \qquad (5.19b)$$

and, in the absence of NO_x, HO_2 is mostly converted back to OH through reactions with NO or O_3, e.g.,

$$HO_2 + O_3 \rightarrow O_2 + OH \qquad (5.19c)$$

In sunlight, a photostationary steady state is established quickly between OH and HO_2. HO_x (=[OH] + [HO_2]) is lost by

$$2HO_2 \rightarrow H_2O_2 + O_2 \qquad (5.20)$$

[19] It is usual to represent chemical rate coefficients for photolytic reactions, such as (5.15), by j and the rate coefficients for nonphotolytic reactions, such as (5.16), by k.

[20] **Philip Albert Leighton** (1897–1983) American chemist. Faculty member at Stanford University. Main research interest was photochemistry. His monograph *The Photochemistry of Air Pollution* (1961) was particularly influential in promoting research on urban air pollution.

December–February

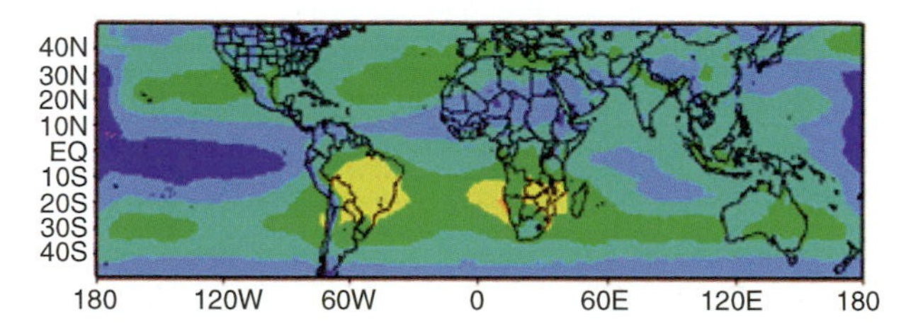

March–May

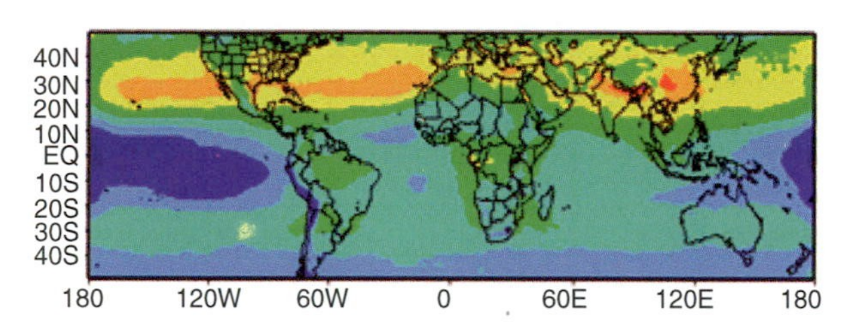

June–August

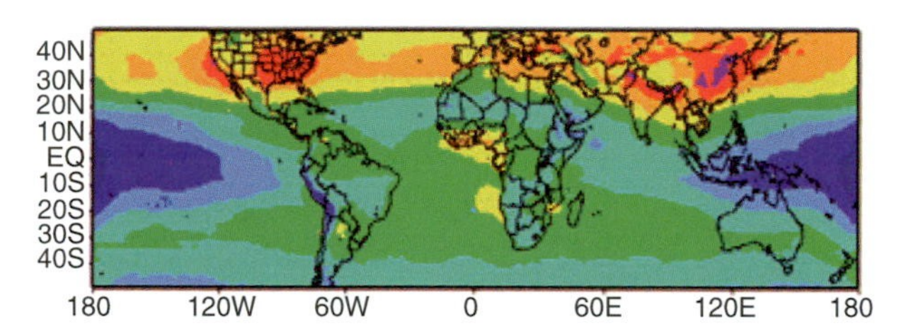

September–November

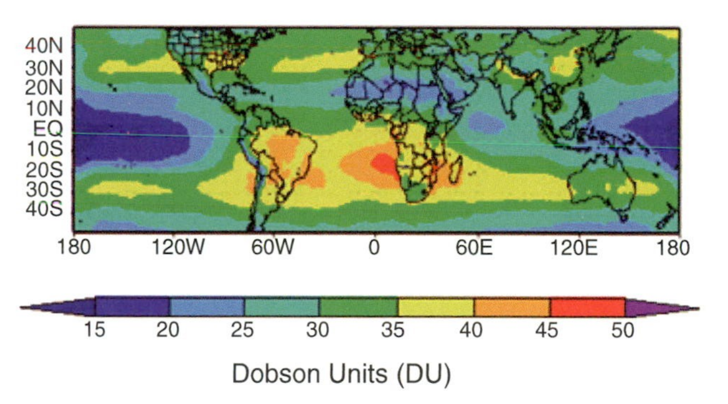

Fig. 5.6 Composite seasonal distribution of the tropospheric ozone column (in Dobson units) determined from satellite measurements from 1979 to 2000. [From *Atmos. Chem. Phys.* **3**, 895 (2003).]

and

$$OH + NO_2 + M \rightarrow HNO_3 + M \qquad (5.21)$$

Because H_2O_2 and HNO_3 are highly soluble, they are removed quickly from the troposphere by wet deposition. The aforementioned reactions, together with values for the appropriate rate coefficients, suggest that OH concentrations near the Earth's surface in sunlight should be $\sim 3 \times 10^{12}$ molecules m^{-3}, which is close to the measured concentration of OH during daylight hours.

In the presence of a sufficiently high mixing ratio of NO ($\gtrsim 10$ pptv), HO_2 formed by the oxidation of VOCs converts NO to NO_2, which then forms O_3 by

$$OH + CO + O_2 \rightarrow HO_2 + CO_2 \qquad (5.22a)$$

$$HO_2 + NO \rightarrow OH + NO_2 \qquad (5.22b)$$

$$NO_2 + h\nu \rightarrow NO + O \qquad (5.22c)$$

$$O + O_2 + M \rightarrow O_3 + M \qquad (5.22d)$$

Note that OH and NO are regenerated in cycle (5.22), and that NO is converted to NO_2 without consuming O_3, providing a pathway for net ozone production. Thus, increases in NO might be expected to produce increases in O_3.

The OH radical also oxidizes methane

$$OH + CH_4 + O_2 \rightarrow H_2O + CH_3O_2 \qquad (5.23)$$

Then, if the concentration of NO_x is low,

$$HO_2 + CH_3O_2 \rightarrow CH_3OOH + O_2 \qquad (5.24)$$

where CH_3O_2 is the methylperoxy radical and CH_3OOH is methyl hydroperoxide. CH_3OOH is removed by wet deposition, resulting in a net loss of HO_x. If NO_x is present, NO_2 from the reaction

$$CH_3O_2 + NO \rightarrow CH_3O + NO_2 \qquad (5.25)$$

leads to O_3 formation by (5.15)–(5.16). These reactions demonstrate the importance of photochemistry and NO_x in determining the concentration of O_3 in the global troposphere, as well as the complexity of tropospheric ozone chemistry.

Figure 5.6 shows the global distribution of O_3 in the troposphere obtained by subtracting satellite measurements of O_3 in the stratosphere from the total column ozone concentration. It is evident that O_3 is generally low throughout the year over the tropical oceans. Over middle latitudes O_3 exhibits pronounced seasonal variations, with a pronounced spring maximum. Over industrialized regions of the northern hemisphere ozone concentrations also tend to be high during summer.

An increase in tropospheric O_3 has occurred globally over the past century, from ~10–15 ppbv in the preindustrial era to ~30–40 ppbv in 2000 in remote regions of the world. The increase is attributable to the increase in NO_x emissions associated with the rapid increase in the use of fossil fuels since the industrial revolution.

5.3.6 Hydrogen Compounds

Hydrogen compounds are the most important oxidants for many chemicals in the atmosphere and are involved in the cycles of many chemical families. Hydrogen compounds include atomic hydrogen (H), which is very short-lived because it combines quickly with O_2 to form the hydroperoxyl radical (HO_2); molecular hydrogen (H_2), which, next to CH_4, is the most abundant reactive trace gas in the troposphere; the hydroxyl radical OH (see Section 5.3.1); the hydroperoxyl radical HO_2; hydrogen peroxide (H_2O_2), which is formed by the reaction of HO_2 radicals and is an important oxidant for SO_2 in cloud droplets; and H_2O, which plays multivarious roles in atmospheric chemistry, including its reaction with excited atomic oxygen to form OH (reaction 5.6b).

The primary source of odd hydrogen species (HO_x), in the form of OH, in the lower troposphere where water vapor is abundant, is (5.7). Another source of HO_x is the photolysis of formaldehyde

$$HCHO + h\nu \rightarrow H + HCO \tag{5.26}$$

followed by (5.19b) and

$$HCO + O_2 \rightarrow HO_2 + CO \tag{5.27}$$

both of which produce HO_2. The simplest loss mechanisms for HO_x are of the form (5.20), and

$$OH + HO_2 \rightarrow H_2O + O_2 \tag{5.28}$$

HO_x and NO_x can react to produce O_3 (see Section 5.3.5). Consequently, there is considerable interest in HO_x chemistry in the upper troposphere where NO_x is emitted in significant quantities by jet aircraft. Cycling between OH and HO_2 occurs on a timescale of a few seconds and is controlled by reactions (5.22). These reactions account for ~80% of the measured concentrations of HO_x in the upper troposphere. NO_x controls the cycling within HO_x that leads to O_3 production. NO_x also regulates the loss of HO_x through reactions of OH with HO_2, NO_2, and HO_2NO_2.

5.3.7 Sulfur Gases

Sulfur is assimilated by living organisms and is then released as an end product of metabolism. The most important sulfur gases in the atmosphere are SO_2, H_2S, dimethyl sulfide (CH_3SCH_3 or DMS for short), COS, and carbon disulfide (CS_2).

Sulfur compounds exist in both reduced and oxidized states, with oxidation numbers from −2 to +6. After being emitted into the Earth's oxidizing atmosphere, the reduced sulfur compounds (most of which are biogenic in origin) are generally oxidized to the +4 oxidation state of SO_2 [i.e., S(IV)] and eventually to the +6 oxidation state of H_2SO_4 [i.e., S(VI)]. The +6 oxidation state is the stable form of sulfur in the presence of oxygen. The oxidation of sulfur compounds illustrates an effect that often applies to other compounds, namely the more oxidized species generally have a high affinity for water (e.g., sulfuric acid). Consequently, the more oxidized species are removed more readily from the atmosphere by wet processes.

The principal natural sources of SO_2 are the oxidation of DMS and H_2S. For example,

$$OH + H_2S \rightarrow H_2O + HS \tag{5.29a}$$

The HS then reacts with O_3 or NO_2 to form HSO

$$HS + O_3 \rightarrow HSO + O_2 \tag{5.29b}$$

$$HS + NO_2 \rightarrow HSO + NO \tag{5.29c}$$

The HSO is then converted rapidly to SO_2 by

$$HSO + O_3 \rightarrow HSO_2 + O_2 \tag{5.29d}$$

$$HSO_2 + O_2 \rightarrow HO_2 + SO_2 \tag{5.29e}$$

Volcanoes and biomass burning are also sources of atmospheric SO_2. However, the largest source of SO_2 is fossil fuel combustion (see Fig. 5.15 for sources and sinks of sulfur compounds).

In the gas phase, SO_2 is oxidized by

$$OH + SO_2 + M \rightarrow HOSO_2 + M \quad (5.30a)$$

$$HOSO_2 + O_2 \rightarrow HO_2 + SO_3 \quad (5.30b)$$

$$SO_3 + H_2O \rightarrow H_2SO_4 \quad (5.30c)$$

SO_2 is also oxidized to H_2SO_4 in cloud water (see Section 6.8.5).

The main sources of H_2S are emissions from soils, marshlands, oceans, and volcanoes. The only significant sink for H_2S is oxidation to SO_2 by (5.29).

Biological reactions in the oceans, involving phytoplankton, emit several sulfur gases, of which DMS has the largest emission rate. DMS is removed from the atmosphere primarily by its reaction with OH to produce SO_2. The sulfur gas with the largest concentration in the unpolluted atmosphere is COS (~0.5 ppbv). The major sources of COS are biogenic and the oxidation of CS_2 by OH; the source of CS_2 is also biogenic. Because COS is very stable in the troposphere, it is eventually transported into the stratosphere where it is the dominant source of sulfate particles during the quiescent periods between large volcanic eruptions (see Section 5.7.3).

5.4 Tropospheric Aerosols

Atmospheric aerosols are suspensions of small solid and/or liquid particles (excluding cloud particles) in air that have negligible terminal fall speeds. Figure 5.7 shows the ranges of particle sizes that play a role in the atmosphere.

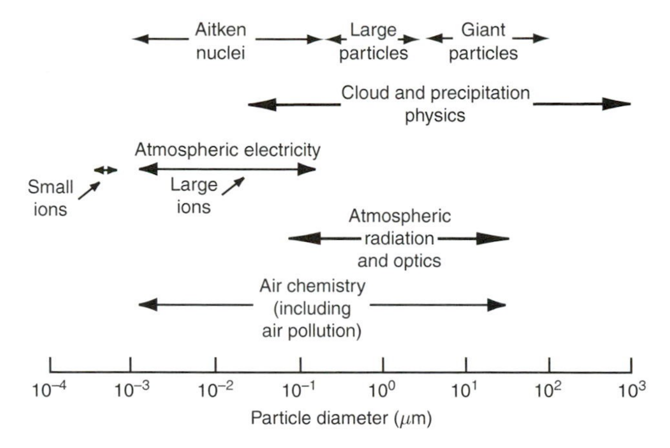

Fig. 5.7 Size range of particles in the atmosphere and their importance.

5.3 Charged Particles

Molecular aggregates that carry an electric charge are called *ions*. The number density and type of ions in the air determine the electrical conductivity of the air, which, in turn, affects the magnitude of the fair weather atmospheric electric field. Ions in the lower atmosphere are produced primarily by cosmic rays, although very close to the Earth's surface ionization due to radioactive materials in the Earth and atmosphere also plays a role. Ions are removed by combining with ions of opposite sign. Small ions, which are not much larger than molecular size, have electrical mobilities (defined as their velocity in a unit electric field) between ~1 and 2×10^{-4} m s^{-1} for an electric field of 1 V m^{-1} at normal temperature and pressure (NTP). Large ions have electrical mobilities in the range from 3×10^{-8} to 8×10^{-7} m s^{-1} in a field of 1 V m^{-1} at NTP. Concentrations of small ions vary from about 40 to 1500 cm^{-3} at sea level, whereas concentrations of large ions vary from about 200 cm^{-3} in marine air to a maximum value of about 8×10^5 cm^{-3} in some cities. It can be seen from these numbers that the electrical conductivity of the air (which is proportional to the product of the ion mobility and the ion concentration) is generally dominated by small ions. However, when the concentrations of large ions and uncharged aerosols are large, as they are in cities, the concentration of small ions tends to be low due to their capture by both large ions and uncharged aerosols. Consequently, the electrical conductivity of air is a minimum (and the fair weather atmospheric electric field a maximum) when the concentration of large ions, and similarly sized uncharged particles, is a maximum. The observed decrease of at least 20% in the electrical conductivity of the air over the north Atlantic Ocean during the 20th century is attributed to a doubling in the concentration of particles with diameters between 0.02 and 0.2 μm, probably due to pollution.

The effects of aerosols on the scattering and absorption of radiation have been discussed in Section 4.4. The role that aerosols play in the formation of cloud particles is discussed in Chapter 6. Here we are concerned with tropospheric aerosols, in particular with their sources, transport, sinks, properties, and with their roles in tropospheric chemistry.

5.4.1 Sources

a. Biological

Solid and liquid particles are released into the atmosphere from plants and animals. These emissions, which include seeds, pollen, spores, and fragments of animals and plants, are usually 1–250 μm in diameter. Bacteria, algae, protozoa, fungi, and viruses are generally <1 μm in diameter. Some characteristic concentrations are: maximum values of grassy pollens $>200\ \text{m}^{-3}$; fungal spores (in water) $\sim$100–400 m^{-3}; bacteria over remote oceans $\sim 0.5\ \text{m}^{-3}$; bacteria in New York City $\sim$80–800 m^{-3}; and bacteria over sewage treatment plants $\sim 10^4\ \text{m}^{-3}$. Microorganisms live on skin: when you change your clothes, you can propel $\sim 10^4$ bacteria per minute into the air, with diameters from $\sim$1 to 5 μm.

The oceans are one of the most important sources of atmospheric aerosols [$\sim$1000–5000 Tg per year, although this includes giant particles ($\sim$2–20 μm diameter) that are not transported very far]. Just above the ocean surface in the remote marine atmosphere, sea salt generally dominates the mass of both supermicrometer and submicrometer particles.

The major mechanism for ejecting ocean materials into the air is bubble bursting (some materials enter the air in drops torn from windblown spray and foam, but because these drops are relatively large, their residence times in the air are very short). Aerosols composed of sea salt originate from droplets ejected into the air when air bubbles burst at the ocean surface (Fig. 5.8). Many small droplets are produced when the upper portion of an air bubble film bursts; these are called *film droplets* (Fig. 5.8b). Bubbles $\gtrsim$2 mm in diameter each eject $\sim$100–200 film droplets into the air. After evaporation, the film droplets leave behind sea-salt particles with diameters less than $\sim$0.3 μm. From one to five larger drops break away from each jet that forms when a bubble bursts (Fig. 5.8d), and these *jet drops* are thrown about 15 cm up into the air. Some of these drops subsequently evaporate and leave behind sea-salt particles with diameters >2 μm. The average rate of production of sea-salt particles over the oceans is $\sim 100\ \text{cm}^{-2}\,\text{s}^{-1}$.

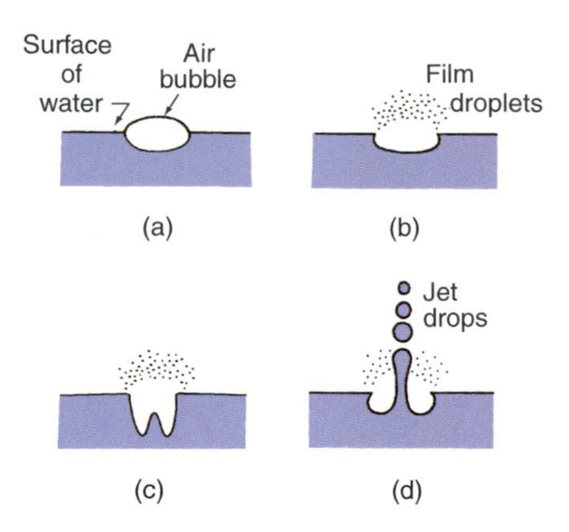

Fig. 5.8 Schematics to illustrate the manner in which film droplets and jet drops are produced when an air bubble bursts at the surface of water. Over the oceans some of the droplets and drops evaporate to leave sea-salt particles and other materials in the air. The time between (a) and (d) is $\sim$2 ms. The film droplets are $\sim$5–30 μm diameter before evaporation. The size of the jet drops are $\sim$15% of the diameter of the air bubble.

Hygroscopic salts [NaCl (85%), KCl, $CaSO_4$, $(NH_4)_2SO_4$] account for $\sim$3.5% of the mass of seawater. These materials are injected into the atmosphere by bubble bursting over the oceans. In addition, organic compounds and bacteria in the surface layers of the ocean are transported to the air by bubble bursting.

Dry sea-salt particles will not form solution droplets until the relative humidity exceeds 75%. Ambient gases (e.g., SO_2 and CO_2) are taken up by these droplets, which changes the ionic composition of the droplets. For example, the reaction of OH(g) with sea-salt particles generates OH^-(aq) in the droplets, which leads to an increase in the production of SO_4^{2-}(aq) by aqueous-phase reactions and a reduction in the concentration of Cl^-(aq). Consequently, the ratio of Cl to Na in sea-salt particles collected from the atmosphere is generally much less than in seawater itself. The excess of SO_4^{2-}(aq) over that of bulk seawater is referred to as *non-sea-salt sulfate* (nss).

The oxidation of Br^-(aq) and Cl^-(aq) in solutions of sea-salt particles can produce BrO_x and ClO_x species. Catalytic reactions involving BrO_x and ClO_x, similar to those that occur in the stratosphere (see Section 5.7.2), destroy O_3. This mechanism has been postulated to explain the depletion of O_3, from ~40 to <0.5 ppbv, that occurs episodically over periods of hours to days in the Arctic boundary layer starting at polar sunrise and continuing through April.

Smoke from forest fires is a major source of atmospheric aerosols. Small smoke particles (primarily organic compounds and elemental carbon) and fly ash are injected directly into the air by forest fires. Several million grams of particles can be released by the burning of 1 hectare (10^4 m^2). It is estimated that about 54 Tg of particles (containing ~6 Tg of elemental carbon) are released into atmosphere each year by biomass burning. The number distribution of particles from forest fires peak at ~0.1 μm diameter, which makes them efficient cloud condensation nuclei. Some biogenic particles (e.g., bacteria from vegetation) may nucleate ice in clouds.

b. Solid Earth

The transfer of particles to the atmosphere from the Earth's surface is caused by winds and atmospheric turbulence. To initiate the motion of particles on the Earth's surface, surface wind speeds must exceed certain threshold values, which depend on the size of the particle and the type of surface. The threshold values are at least ~0.2 m s^{-1} for particles 50–200 μm in diameter (smaller particles adhere better to the surface) and for soils containing 50% clay or tilled soils. To achieve a frictional speed of 0.2 m s^{-1} requires a wind speed of several meters per second a few meters above ground level. A major source of smaller (~10–100 μm diameter) particles is *saltation*, in which larger sand grains become airborne, fly a few meters, and then land on the ground, creating a burst of dust particles.

On the global scale, semiarid regions and deserts (which cover about one-third of the land surface) are the main source of particles from the Earth's surface. They provide ~2000 Tg per year of mineral particles. Dust from these sources can be transported over long distances (see Section 5.4.3).

Volcanoes inject gases and particles into the atmosphere. The large particles have short residence times, but the small particles (produced primarily by gas-to-particle (g-to-p) conversion of SO_2) can be transported globally, particularly if they reach high altitudes. Volcanic emissions play an important role in stratospheric chemistry (see Section 5.7.3).

c. Anthropogenic

The global input of particles into the atmosphere from anthropogenic activities is ~20% (by mass) of that from natural sources. The main anthropogenic sources of aerosols are dust from roads, wind erosion of tilled land, biomass burning, fuel combustion, and industrial processes. For particles with diameters >5 μm, direct emissions from anthropogenic sources dominate over aerosols that form in the atmosphere by g-to-p conversion (referred to as *secondary particles*) of anthropogenic gases. However, the reverse is the case for most of the smaller particles, for which g-to-p conversion is the over-whelming source of the number concentration of anthropogenically derived aerosols.

In 1997 the worldwide direct emission into the atmosphere of particles <10 μm diameter from anthropogenic sources was estimated to be ~350 Tg per year (excluding g-to-p conversion). About 35% of the number concentration of aerosols in the atmosphere was sulfate, produced by the oxidation of SO_2 emissions. Particle emissions worldwide were dominated by fossil fuel combustion (primarily coal) and biomass burning. These emissions are projected to double by the year 2040, due largely to anticipated increases in fossil fuel combustion, with the greatest growth in emissions from China and India.

During the 20th century, the emission of particles into the atmosphere from anthropogenic sources was a small fraction of the mass of particles from natural sources. However, it is projected that by 2040 anthropogenic sources of particles could be comparable to those from natural processes.

d. In situ formation

In situ condensation of gases (i.e., g-to-p conversion) is important in the atmosphere. Gases may condense onto existing particles, thereby increasing the mass (but not the number) of particles, or gases may condense to form new particles. The former path is favored when the surface area of existing particles is high and the supersaturation of the gases is low. If new particles are formed, they are generally <0.01 μm diameter. The quantities of aerosols produced by g-to-p conversion exceed those from direct emissions for anthropogenically derived aerosols and are comparable to direct emission in the case of naturally derived aerosols.

Three major families of chemical species are involved in g-to-p conversion: sulfur, nitrogen, and organic and carbonaceous materials. Various sulfur gases (e.g., H_2S, CS_2, COS, DMS) can be oxidized to SO_2. The SO_2 is then oxidized to sulfate (SO_4^{2-}), the dominant gas phase routes being reactions (5.30). However, on a global scale, heterogeneous reactions of SO_2 in cloud water dominate the conversion of SO_2 to SO_4^{2-} (see Section 6.8.9).

Over the oceans, sulfates derived from DMS contribute to the growth of existing particles. Sulfates are also produced in and around clouds, and nitric acid can form from N_2O_5 in cloud water. Subsequent evaporation of cloud water releases these sulfate and nitrate particles into the air (see Section 6.8).

Organic and carbonaceous aerosols are produced by g-to-p conversion from gases released from the biosphere and from volatile compounds such as crude oil that leak to the Earth's surface. Carbonaceous particles emitted directly into the atmosphere derive mainly from biomass fires.

e. Summary

Table 5.3 summarizes estimates of the magnitudes of the principal sources of direct emission of particles into the atmosphere and *in situ* sources.

Anthropogenic activities emit large numbers of particles into the atmosphere, both directly and through g-to-p conversion (see Sections 5.4.1c and 5.4.1d). For particles ≥5 μm diameter, human activities worldwide are estimated to produce ~15% of natural emissions, with industrial processes, fuel combustion, and g-to-p conversion accounting for ~80% of the anthropogenic emissions. However, in urban areas, anthropogenic sources are much more important. For particles <5 μm diameter, human activities produce ~20% of natural emissions, with g-to-p conversion accounting for ~90% of the human emissions.

Table 5.3 Estimates (in Tg per year) for the year 2000 of (a) direct particle emissions into the atmosphere and (b) *in situ* production

(a) Direct emissions	Northern hemisphere	Southern hemisphere
Carbonaceous aerosols		
Organic matter (0–2 μm)[a]		
Biomass burning	28	26
Fossil fuel	28	0.4
Biogenic (>1 μm)	—	—
Black carbon (0–2 μm)		
Biomass burning	2.9	2.7
Fossil fuel	6.5	0.1
Aircraft	0.005	0.0004
Industrial dust, etc. (>1 μm)		
Sea salt		
<1 μm	23	31
1–16 μm	1,420	1,870
Total	1,440	1,900
Mineral (soil) dust		
<1 μm	90	17
1–2 μm	240	50
2–20 μm	1,470	282
Total	1,800	349

(b) *In situ*	Northern hemisphere	Southern hemisphere
Sulfates (as NH_4HSO_4)	145	55
Anthropogenic	106	15
Biogenic	25	32
Volcanic	14	7
Nitrate (as NO_3^-)		
Anthropogenic	12.4	1.8
Natural	2.2	1.7
Organic compounds		
Anthropogenic	0.15	0.45
Biogenic	8.2	7.4

[a] Sizes refer to diameters. [Adapted from Intergovernmental Panel on Climate Change, *Climate Change 2001*, Cambridge University Press, pp. 297 and 301, 2001.]

5.4.2 Chemical Composition

Except for marine aerosols, the masses of which are dominated by sodium chloride, sulfate is one of the prime contributors to the mass of atmospheric aerosols. The mass fractions of SO_4^{2-} range from ~22–45% for continental aerosols to ~75% for aerosols in the Arctic and Antarctic. Because the sulfate content of the Earth's crust is too low to explain the large percentages of sulfate in atmospheric aerosols, most of the SO_4^{2-} must derive from g-to-p conversion of SO_2. The sulfate is contained mainly in submicrometer particles.

Ammonium (NH_4^+) is the main cation associated with SO_4^{2-} in continental aerosol. It is produced by gaseous ammonia neutralizing sulfuric acid to produce ammonium sulfate [$(NH_4)_2SO_4$]—see reaction

(5.13). The ratio of the molar concentrations of NH_4^+ to SO_4^{2-} ranges from ~1 to 2, corresponding to an aerosol composition intermediate between that for NH_4HSO_4 and $(NH_4)_2SO_4$. The average mass fractions of submicrometer non sea-salt sulfates plus associated NH_4^+ range from ~16 to 54% over large regions of the world.

In marine air the main contributors to the mass of inorganic aerosols are the ions Na^+, Cl^-, Mg^{2+}, SO_4^{2-}, K^+, and Ca^{2+}. Apart from SO_4^{2-}, these compounds are contained primarily in particles a few micrometers in diameter because they originate from sea salt derived from bubble bursting (see Fig. 5.8). Sulfate mass concentrations peak for particles with diameters ~0.1–1 μm. Particles in this size range are effective in scattering light (and therefore reducing visibility) and as cloud condensation nuclei.

Nitrate (NO_3^-) occurs in larger sized particles than sulfate in marine air. Because seawater contains negligible nitrate, the nitrate in these particles must derive from the condensation of gaseous HNO_3, possibly by g-to-p conversion in the liquid phase.

Nitrate is also common in continental aerosols, where it extends over the diameter range ~0.2–20 μm. It derives, in part, from the condensation of HNO_3 onto larger and more alkaline mineral particles.

Organic compounds form an appreciable fraction of the mass of atmospheric aerosols. The most abundant organics in urban aerosols are higher molecular weight alkanes (C_xH_{2x+2}), ~1000–4000 ng m^{-3}, and alkenes (C_xH_{2x}), ~2000 ng m^{-3}. Many of the particles in urban smog are by-products of photochemical reactions involving hydrocarbons and nitrogen oxides, which derive from combustion. In the United States, carbonaceous materials can account for ~50% or more of the total dry aerosol mass.

Elemental carbon (commonly referred to as "soot"), a common component of organic aerosols in the atmosphere, is a strong absorber of solar radiation. For example, in polluted air masses from India, elemental carbon accounts for ~10% of the mass of submicrometer sized particles.

5.4.3 Transport

Aerosols are transported by the airflows they encounter during the time they spend in the atmosphere. The transport can be over intercontinental, even global, scales. Thus, Saharan dust is transported to the Americas, and dust from the Gobi Desert can reach the west coast of North America. If the aerosols are produced by g-to-p conversion, long-range transport is likely because the time required for g-to-p conversion and the relatively small sizes of the particles produced by this process lead to long residence times in the atmosphere. This is the case for sulfates that derive from SO_2 blasted into the stratosphere by large volcanic eruptions. It is also the case for acidic aerosols such as sulfates and nitrates, which contribute to *acid rain*. Thus, SO_2 emitted from power plants in the United Kingdom can be deposited as sulfate far inland in continental Europe.

5.4.4 Sinks

On average, particles are removed from the atmosphere at about the same rate as they enter it. Small particles can be converted into larger particles by coagulation. Because the mobility of a particle decreases rapidly as it increases in size, coagulation is essentially confined to particles less than ~0.2 μm in diameter. Although coagulation does not remove particles from the atmosphere, it modifies their size spectra and shifts small particles into size ranges where they can be removed by other mechanisms.

Exercise 5.6 If the rate of decrease in the number concentration N of a *monodispersed* (i.e., all particles of the same size) aerosol due to coagulation is given by $-dN/dt = KN^2$, where K is a constant (assume $K = 1.40 \times 10^{-15}\ m^3\ s^{-1}$ for 0.10-μm-diameter particles at 20 °C and 1 atm), determine the time required at 20 °C and 1 atm for coagulation to decrease the concentration of a monodispersed atmospheric aerosol with particles of a diameter of 0.100 μm to one-half of its initial concentration of $1.00 \times 10^{11}\ m^{-3}$.

Solution: Because $-dN/dt = KN^2$

$$\int_{N_0}^{N} \frac{dN}{N^2} = -K \int_0^t dt$$

where N_0 and N are the number concentrations of particles at time $t = 0$ and time t, respectively. Therefore,

$$\left[-\frac{1}{N}\right]_{N_0}^{N} = -Kt$$

or

$$\frac{1}{N_0} - \frac{1}{N} = -Kt$$

that is

$$t = \frac{\frac{N_0}{N} - 1}{KN_0}$$

For, $N_0/N = 2, N_0 = 1.00 \times 10^{11}$ m^{-3} and $K = 1.40 \times 10^{-15}$ m^3 s^{-1},

$$t = \frac{2 - 1}{(1.40 \times 10^{-15})10^{11}} = 7140 \text{ s} = 1.98 \text{ h} \quad ■$$

Improvements in visibility that frequently follow periods of precipitation are due, in large part, to the removal (i.e., *scavenging*) of particles by precipitation. It is estimated that, on a global scale, precipitation processes account for about 80–90% of the mass of particles removed from the atmosphere. Prior to the formation of precipitation, some of the particles in the air serve as nuclei upon which cloud particles (water and ice) can form (see Chapter 6). As cloud particles grow, aerosols tend to be driven to their surfaces by the diffusion field associated with the flux of water vapor to the growing cloud particles (called the *diffusiophoretic force*). Aerosol particles less than ~0.1 μm are collected most efficiently by diffusiophoresis. As precipitation particles fall through the air, they collect aerosol particles greater than ~2 μm in diameter with reasonable efficiency by *impaction*. Aerosols are also removed by impaction onto obstacles on the Earth's surface (such as newly washed automobiles).

The terminal fall speeds of particles >1 μm diameter are sufficiently large that gravitational settling (i.e., *dry deposition*) is important as a removal process. For example, the fall speeds of particles 1 and 10 μm in diameter are $\sim 3 \times 10^{-5}$ and $\sim 3 \times 10^{-3}$ m s^{-1}, respectively. It is estimated that ~10–20% of the mass of particles removed from the atmosphere is by dry fallout.

5.4.5 Concentrations and Size Distributions

One of the oldest and most convenient techniques (which in various forms is still in widespread use) for measuring the concentrations of particles in the air is the Aitken[21] nucleus counter. In this instrument, saturated air is expanded rapidly so that it becomes supersaturated by several hundred percent with respect to water (see Exercise 5.15). At these high supersaturations, water condenses onto virtually all of the particles in the air to form a cloud of small water droplets. The concentration of droplets in the cloud (which is close to the concentration of particles) can be determined by allowing the droplets to settle out onto a substrate, where they can be counted under a microscope, or automatically by optical techniques. The concentration of particles measured with an Aitken nucleus counter is referred to as the *Aitken* (or *condensation*) *nucleus* (CN) *count*.

Condensation nucleus counts near the Earth's surface vary widely from one location to another and can also fluctuate by more than an order of magnitude with time at any one site. Generally, they range from average concentrations of $\sim 10^3$ cm^{-3} over the oceans, to $\sim 10^4$ cm^{-3} over rural land areas, to $\sim 10^5$ cm^{-3} or greater in urban polluted air. These observations, together with the fact that CN counts decline with increasing altitude, support the view that land is an important source of atmospheric aerosol, with human and industrial activities being particularly prolific sources.

Atmospheric aerosol particles range in size from $\sim 10^{-4}$ μm to tens of micrometers. The averages of numerous measurements of *particle number distributions* in continental, marine, and urban polluted air are shown in Fig. 5.9. The measurements are plotted in the form of a number distribution in which the ordinate $[dN/d(\log D)]$ and the abscissa (D) are plotted on logarithmic scales, where dN is the number concentration of particles with diameters between D and $D + dD$.[22]

[21] **John Aitken** (1839–1919) Scottish physicist, although originally an apprentice marine engineer. In addition to his pioneering work on atmospheric aerosol, he investigated cyclones, color, and color sensations.

[22] Note that if the ordinate were linear in N (which it is not in Fig. 5.9), the concentration of particles within a diameter interval $d(\log D)$ would be equal to the area under the curve in this diameter interval.

Several conclusions can be drawn from the results shown in Fig. 5.9:

- The concentrations of particles fall off very rapidly as they increase in size. Therefore, the total number concentration (i.e., the Aitken or CN count) is dominated by particles with diameters $<0.2\ \mu m$, which are therefore referred to as *Aitken nuclei* or *condensation nuclei*.
- Those portions of the number distribution curves that are straight lines in Fig. 5.9 can be represented by an expression of the form

$$\log \frac{dN}{d(\log D)} = \text{const} - \beta \log D$$

or, taking antilogs,

$$\frac{dN}{d(\log D)} = CD^{-\beta} \qquad (5.31)$$

where C is a constant related to the concentration of the particles and $-\beta$ is the slope of the number distribution curve. The value of β generally lies between 2 and 4. Continental aerosol particles with diameters larger than $\sim 0.2\ \mu m$ follow (5.31) quite closely with $\beta \simeq 3$. A size distribution with $\beta = 3$ is called a *Junge*[23] *distribution*.
- The number distributions of particles shown in Fig. 5.9 confirm CN measurements, which indicate that the total concentrations of particles are, on average, greatest in urban polluted air and least in marine air.
- The concentrations of particles with diameters $>2\ \mu m$ (*giant particles*) are, on average, rather similar in continental, marine, and urban polluted air.

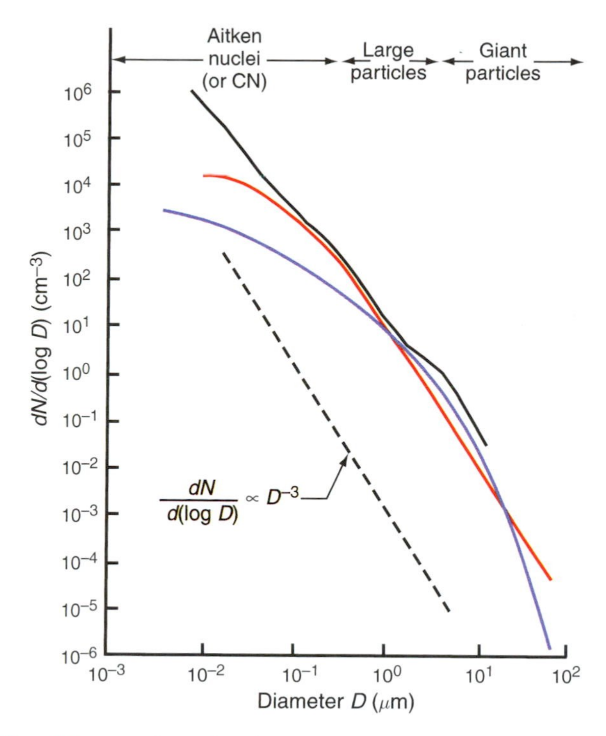

Fig. 5.9 Number distributions of tropospheric particles obtained from averaging many measurements in continental (red), marine (blue), and urban polluted (black) air. Also plotted is Eq. (5.31) with $\beta = 3$, with the line (dashed) displaced from the other curves for the sake of clarity.

Further insights into atmospheric particle size distributions can be obtained by plotting *particle surface area distributions* or *particle volume distributions*. In a surface area distribution the ordinate is $dS/d(\log D)$ and the abscissa is D ploted on a logarithmic scale, where S is the total surface area of particles with diameters between D and $D + dD$. In a volume distribution the ordinate is $dV/d(\log D)$ and the abscissa is D plotted on a logarithmic scale, where V is the total volume of particles with diameters between D and $D + dD$.

If particle densities are independent of their diameter, the mass concentration dM of particles with diameters between D and $D + dD$ is proportional to $D^3\,dN$. If the number distribution follows (5.31) with $\beta = 3$, $D^3\,dN \propto d(\log D)$, therefore, $dM/d(\log D)$ has a fixed value. In this case, the particles in each logarithmic increment of diameter contribute equally to the total mass concentration of the aerosol. It follows from this result that, although particles with diameters from 0.2 to 2 μm (so-called *large particles*) are present in much higher number concentrations than giant particles, the large and giant particles in

[23] **Christian E. Junge** (1912–1996) German meteorologist. Carried out pioneering studies of tropospheric and stratospheric aerosols and trace gases. His book *Atmospheric Chemistry and Radioactivity* (Academic Press, 1963) initiated the modern era of atmospheric chemistry research.

continental air make similar contributions to the total mass of the aerosol. However, despite their relatively large number concentration, the CN contribute only ~10–20% to the total mass of the atmospheric aerosol. This is because, as seen in Fig. 5.9, the CN do not increase in concentration with decreasing size as rapidly as indicated by (5.31).

Small fluctuations in the slope of a particle number distribution about values of −2 and −3 (i.e., $\beta = 2$ and 3, respectively) appear as local maxima and minima in the surface and volume distribution plots, respectively (see Exercise 5.16). Shown in Fig. 5.10 are particle surface area and particle volume distribution plots for continental and urban polluted air in Denver, Colorado. These curves show far more structure, in the form of maxima and minima, than the number distribution plots shown in Fig. 5.9. The maxima and minima in the curves in Fig. 5.10 are associated with sources and sinks, respectively, of the aerosol. The prominent maxima in the surface and volume distribution plots in the particle size range ~0.2–2 μm diameter is due primarily to the growth of the CN by coagulation into this size range, together with particles left behind when cloud droplets evaporate. Since the sinks for particles ~0.2–2 μm diameter are weak, particles in this size range tend to accumulate in the atmosphere: hence, the peak in the particle surface area and volume distribution plots for particles between ~0.2 and 2 μm

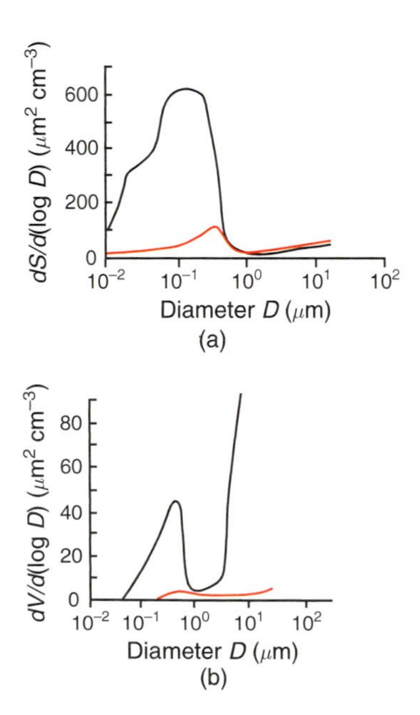

Fig. 5.10 (a) Typical particle surface area and (b) particle volume distribution plots in urban-polluted air (black line) and cleaner continental air (solid red line).

diameter is referred to as the *accumulation mode*. Another peak, called the *coarse particle mode*, occurs in the particle surface and volume distribution plots for particles with D ≳1 μm. This mode is attributable to dusts and industrial processes, which produce fly ash and other large particles, sea salt, and some biological materials.

Another common way of subdividing atmosphere particles by size is to call particles with D <0.01 μm *ultrafine*, those with 0.01 μm ≤ D <2.5 μm *fine particles*, and those with D >2.5 μm *coarse particles*.

5.4.6 Residence Times

Figure 5.11 shows estimates of the residence times of particles in the atmosphere as a function of their size. Particles with diameter <0.01 μm have residence times ≲1 day; the major removal mechanisms for particles in this size range are diffusion to cloud particles and coagulation. Particles ≳20 μm diameter also have residence times ≲1 day, but they are removed by sedimentation, impaction onto surfaces, and precipitation scavenging. In contrast, particles with diameters ~0.2–2 μm have strong sources (from the coagulation of Aitken nuclei and particles left behind by the evaporation of cloud droplets) but weak sinks. Consequently, these particles have relatively long residence times, reaching several hundred days in the upper troposphere, but precipitation scavenging and impaction reduce these residence times to a few tens of days in the middle and lower troposphere. It is for this reason that the accumulation mode is in the size range of these particles (e.g., see Fig. 5.10).

5.5 Air Pollution

In urban and industrialized locations, anthropogenic emissions can become so large that the concentrations of various undesirable chemical species (*pollutants*) cause significant deterioration in air quality and visibility and can pose threats to human health. Severe air pollution episodes, and associated visibility reduction, occur when the rates of emissions or formation of pollutants greatly exceed the rates at which the pollutants are dispersed by winds and vertical transport or are removed from the atmosphere by chemical reactions or deposition. Severe air pollution episodes tend to occur in association with extended intervals of light winds and strong static stability (see Section 3.6 and Box 3.4).

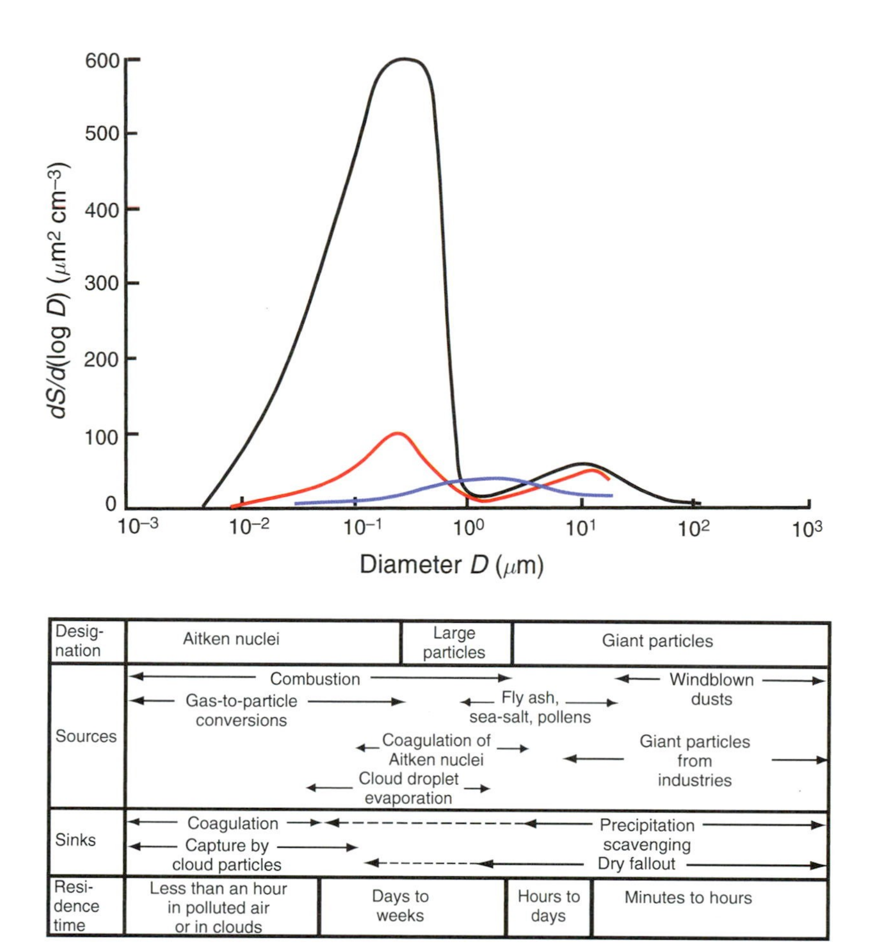

Fig. 5.11 Schematic curves of particle surface area distributions for urban polluted air (black line), continental air (red line), and marine air (blue line). Shown below the curves are the principal sources and sinks of atmospheric particles and estimates of their mean residence times in the troposphere. [Adapted from *Atmos. Environ.* **9**, W. G. N. Slinn, Atmospheric aerosol particles in surface-level air, 763, copyright (1975), with permission from Elsevier.]

5.5.1 Some Sources of Pollutants

Combustion (in power plants, smelters, automobiles, and of wood, vegetation, etc.) is the largest source of air pollutants. On a global scale, fossil-fuel combustion is the major source of CO, CO_2, NO_x, and SO_2. Many other pollutants are released into the air by combustion. For example, about 15% of the total emissions of hydrocarbons are from anthropogenic sources, most notably the burning of hydrocarbon compounds (oil, natural gas, coal, and wood). *Ideal* (or *complete*) *combustion* (or *oxidation*) of a hydrocarbon fuel yields only CO_2 and H_2O. However, for a given quantity of fuel, a precise amount of oxygen is required for complete combustion, and this ideal combination of fuel and oxygen is rarely achieved.

Exercise 5.7 Determine the ratio of the mass of dry air to the mass of isooctane (C_8H_{18})—called the *air–fuel ratio*—for ideal combustion.

Solution: The balanced chemical equation for the ideal or complete combustion of C_8H_{18} can be written as

$$C_8H_{18} + 12.5O_2 \rightarrow 8CO_2 + 9H_2O$$

Therefore, for complete combustion, 1 mol of C_8H_{18} reacts with 12.5 mol of O_2 or, because the molecular weights of C_8H_{18} and O_2 are 114 and 32, respectively, 114 g of C_8H_{18} reacts with 12.5×32 or 400 g of O_2. We now need to calculate what mass of air contains 400 g of oxygen. The amount of oxygen in air *by volume* (or by number of molecules) is 20.95% (see Table 5.1). Because the apparent molecular weight of dry air is 28.97, the amount of oxygen in air *by mass* is $20.95 \times 32/28.97 \simeq 23\%$. Therefore, the mass of air containing 400 g of oxygen is $400/0.23 \simeq 1700$ g. Hence, for complete combustion, 114 g of C_8H_{18} reacts with ~1700 g of air. Therefore, the air–fuel ratio for ideal combustion is $1700/114 \simeq 15$. ■

Since 1981, gasoline-powered internal combustion engines in the United States have used an oxygen sensor in the exhaust system and a computer-controlled fuel flow that keeps the air–fuel ratio to within a few percent of that needed for ideal combustion. This system, together with a catalytic converter that reduces the emissions of particularly undesirable substances, makes the emissions from modern automobiles quite small. Older automobiles tended to run with a fuel-rich mixture (i.e., a mixture that contains less than the required amount of air for ideal combustion) and these emit considerable pollutants. For example, suppose the 12.5 mol of O_2 in Exercise 5.7 is reduced to 11.25 mol. Then, 9 mol of H_2O is produced together with 5.5 mol of CO_2 and 2.5 mol of CO. In this case, more than one-third of the carbon in the emissions is the highly poisonous gas CO. Gas mileage is reduced because the fuel is not completely burned. However, because the engine is pumping less air, peak power can actually increase. Even modern automobiles operate in this "fuel-rich mode" for almost 1 min after the engine is started, and for several seconds each time a strong acceleration is required. Therefore, despite the fact that in most countries automobiles account for only a small percentage of the total fuel burned, they produce a large fraction of the CO.

If an automobile has a broken catalyst or does not have a catalyst, it will emit significant amounts of unburned (and partially burned) hydrocarbons (HC), many of which are toxic and carcinogenic. Measurements of motor vehicle exhausts on highways show that half of the emissions of CO and HC come from less than 10% of the vehicles, namely from those that are old or poorly maintained. Hence the need for regular emission tests of older automobiles.

The high temperatures associated with combustion permit oxidation of atmospheric molecular nitrogen to NO (referred to as *thermal* NO). At temperatures below ~4500 K, the reactions are

$$O_2 + M \rightleftarrows 2O + M \tag{5.32a}$$

$$2O + 2N_2 \rightleftarrows 2NO + 2N \tag{5.32b}$$

$$2N + 2O_2 \rightleftarrows 2NO + O_2 \tag{5.32c}$$

$$\text{Net:} \quad O_2 + N_2 + M \rightleftarrows 2NO + M \tag{5.33}$$

Because of the strong temperature dependence of reactions (5.32a) and (5.32b), thermal formation of NO is highly temperature dependent. Under equilibrium conditions, these reactions produce maximum NO concentrations at ≥3500 K (although equilibrium is not attained in engines). As the combustion gases cool rapidly to ambient temperatures, the rates of the reverse reactions are reduced drastically so that the NO concentration is "frozen in" at the high temperature. An additional strong source of NO during combustion is the oxidation of nitrogen-containing compounds in the fuel (*fuel* NO).

As shown in Section 5.5.2, NO_x emissions from automobiles play a key role in the formation of photochemical smog. However, over large geographic areas, power stations and industries are generally larger sources of NO_x than automobiles.

Exercise 5.8 The balanced chemical equation for the ideal combustion of a general hydrocarbon fuel C_xH_y can be written as

$$C_xH_y + \left(x + \frac{y}{4}\right)O_2 \rightarrow xCO_2 + \frac{y}{2}H_2O \tag{5.34}$$

If 1 mol of C_xH_y is completely burned, show that $3.7(x + \frac{y}{4})$ mol of (unreacting) nitrogen will be contained in the emissions. Hence, derive an expression for the total number of moles of gases in the emissions in terms of x and y.

Solution: From the percentages of O_2 and N_2 in air by volume (namely 20.9 and 78) we have for air

$$\frac{\text{number of moles of } O_2}{\text{number of moles of } N_2} = \frac{20.9}{78} = \frac{1}{3.73} \tag{5.35}$$

Therefore, from (5.35), 1 mol of O_2 is associated with 3.73 mol of N_2 in air. From (5.34), $3.7(x + \frac{y}{4})$ mol of N_2 will be associated with 1 mol of C_xH_y and $3.7(x + \frac{y}{4})$ mol of O_2 burned. Therefore, the total number of moles of gases in the emissions produced by the combustion of 1 mol of C_xH_y is, from (5.34), $x + \frac{y}{2} + 3.7(x + \frac{y}{4})$. ■

Combustion of fuels, particularly coal, dominates the emissions of sulfur oxides, which are mainly SO_2. Heavy-metal smelters (e.g., Ni, Cu, Zn, Pb, and Ag) can be large local sources of SO_2.

There are also lower temperature sources of air pollutants, e.g., leakages of hydrocarbons from natural gas lines, organics from the evaporation of solvents, and nitrogen compounds from fertilizers.

5.5.2 Smogs

The term *smog* derives from *smoke* and *fog*; it was originally coined to refer to heavily polluted air that can form in cities (generally in winter under calm, stable and moist conditions) due to the emissions of sulfur dioxide and aerosols from the burning of fossil fuels (primarily coal and oil). The term is now applied to all forms of severe air pollution, particularly in urban areas, that restrict visibility.

a. London (or classical) smog

Prior to the introduction of air pollution abatement laws in the latter part of the 20th century, many large cities in Europe and North America regularly suffered from severe smogs. The London smogs were sufficiently notorious that such pollution became known as *London smog*.[24] In the London (or *classical*) type of smog, particles swell in size under high relative humidity and some of the particles serve as nuclei on which fog droplets form. Sulfur dioxide gas dissolves in the fog droplets where it is oxidized to form sulfuric acid (see Section 6.8.5).

In December 1952, cold air moved from the English Channel and settled over London, producing a pollution-trapping inversion fog. Over the next 5 days London experienced its worst air pollution episode. The smog was so thick that people had to grope their way along the streets, buses crawled along at a walking pace led by pedestrians with flashlights, and indoor events were canceled because the stage could not be seen.[25] By the time the smog had lifted, 4000 people had died of respiratory problems, and the smog was implicated in an additional 8000 deaths in the months that followed. In the Great Smog, as it was called, SO_2 reached peak mixing ratios of ~0.7 ppmv (compared to typical annual mean mixing ratios of ~0.1 ppmv in polluted cities with large coal usage), and the peak particle concentrations were ~1.7 mg m^{-3} (compared to ~0.1 mg m^{-3} under more typical urban-polluted conditions). Interestingly, there is no direct evidence that even these high concentrations of pollutants would, in themselves, have caused fatalities (see Box 5.4).

After the Great Smog, laws were passed in Britain and elsewhere banning the use of coal on open fires for domestic heating and the emissions of black smoke, and requiring industries to switch to cleaner-burning fuels. Nevertheless, pollution is still a serious problem in many cities in Europe and the United States. Also, many large cities, particularly in developing countries (e.g., China, India), still suffer from London-type smogs due to the burning of coal and wood and to the lack of strict air pollution controls.

Even with stringent pollution controls, increasing numbers of automobiles in large cities can produce high concentrations of NO, which can then be converted to NO_2 by

$$2NO + O_2 \rightarrow 2NO_2 \qquad (5.36)$$

The NO_2 can then be involved in the production of O_3 (see Section 5.5.2b). Because the magnitude of the rate coefficient for (5.36) increases with decreasing temperature, the production of NO_2 by (5.36) can become significant in cities that have cold winters.

5.4 Effects of Particulates on Human Health

Leonardo da Vinci noted, alongside a sketch of the lung, "dust causes damage." Some 500 years later evidence began to mount supporting this conclusion. There appears to be an increasing risk of lung cancer and mortality with increasing concentration of the total mass of particulate matter (PM) in the air, even at concentrations below 0.1 mg m^{-3}. However, because other pollutants, such as SO_2

Continued on next page

[24] In 1661 the English diarist John Evelyn noted the effects of industrial emissions (and, no doubt, domestic wood burning) on the health of plants and people. In the 17th century there were days when a plume of smoke half a mile high and 20 miles wide emanated from London. Evelyn suggested that industries be placed outside of towns and that they be equipped with tall chimneys to disperse the smoke. What are the pros and cons of this suggestion?

[25] Such events were experienced by one of the authors (PVH) of this book.

5.4 Continued

and O_3, generally increase with increasing concentrations of PM, it is difficult to separate the health effects of the different pollutants. Weather also plays a confounding role because of the effects of relative humidity on particles and of temperature on humans.

Most statistical studies of the effects of PM on human health have utilized the total mass of PM with diameters below 10 μm (PM-10). However, there are several reasons for suspecting that particles <1 μm in diameter have a more deleterious effect on human health than the larger particles that dominate PM-10 measurements. Smaller particles can penetrate deeper into the small airways of the lungs. There is also evidence that very small particles (<0.05 μm diameter) have enhanced toxicity. A recent examination of lung tissue samples preserved from the victims of the Great Smog of London (see Section 5.5.2) has revealed high concentrations of small particles and metals such as lead of the type that derive from coal combustion and diesel engines. It might not be chance coincidence that 1952 (the year of the Great Smog) was also the year that London completed a change from electric trams to diesel buses, which emit many small toxic particles.

Despite the introduction of air pollution control strategies, small particles are still prevalent in significant number concentrations in large cities.

b. Photochemical (or Los Angeles) smog

During the second half of the 20th century emissions from automobiles became increasingly important as a source of pollutants in many urban areas. When subjected to sunlight and stagnant meteorological conditions, the combination of chemical species in such strongly polluted urban air can lead to *photochemical* (or *Los Angeles-type*) *smog*. These smogs are characterized by high concentrations of a large variety of pollutants, such as nitrogen oxides, O_3, CO, hydrocarbons, aldehydes (and other materials that are eye irritants), and sometimes sulfuric acid as well. The chemical reactions that lead to photochemical smog are complex and still not completely understood. However, some of the major chemical reactions thought to be involved are described here.

Photochemical smogs are due to the interactions between a variety of organic pollutants (e.g., hydrocarbons such as ethylene and butane) and oxides of nitrogen. The reactions start with (5.15) and (5.16), which form O_3. However, the O_3 is then depleted by the rapid reaction (5.17). As shown in Exercise 5.5, if there were no other reactions (5.15)–(5.17) would lead to a steady-state concentration of O_3 given by (5.18). However, (5.18) predicts O_3 concentrations in urban-polluted air of only ~0.03 ppmv, whereas typical values are well above this concentration and can reach 0.5 ppmv. Therefore, other chemical reactions must be involved in photochemical smogs. Most effective are reactions that oxidize NO to NO_2 without consuming O_3, since this would allow net O_3 production, leading to a buildup of ozone concentrations during the day. The OH radical can initiate a chain reaction that can act in this way by attacking the hydrocarbon pollutants in urban air, e.g.,

$$OH + CH_4 \rightarrow H_2O + CH_3 \tag{5.37}$$

or

$$OH + CO \rightarrow H + CO_2 \tag{5.38}$$

or

$$OH + CH_3CHO \rightarrow H_2O + CH_3CO \tag{5.39}$$

The resulting radicals, CH_3 in (5.37), H in (5.38), and CH_3CO in (5.39), then become involved in reactions that oxidize NO to NO_2 and regenerate OH. For example, CH_3 from (5.37) can initiate the series of reactions

$$CH_3 + O_2 \rightarrow CH_3O_2 \tag{5.40a}$$

$$CH_3O_2 + NO \rightarrow CH_3O + NO_2 \tag{5.40b}$$

$$CH_3O + O_2 \rightarrow HCHO + HO_2 \tag{5.40c}$$

$$HO_2 + NO \rightarrow NO_2 + OH \tag{5.40d}$$

The net results of (5.40), together with (5.37), are

$$CH_4 + 2O_2 + 2NO \rightarrow H_2O + 2NO_2 + HCHO \tag{5.41}$$

Thus, in this case, the oxidation of CH_4 results in the oxidation of NO to NO_2 without consuming O_3. Reaction (5.41) produces formaldehyde (HCHO), which is an eye irritant and a source of HO_x [see (5.26) and (5.27)].

Similarly, the acetyl radical (CH_3CO) from (5.39) is involved in a series of reactions leading to the methyl radical CH_3 and the peroxyacetyl radical (CH_3COO_2). The methyl radical oxidizes NO by reactions (5.40), and the peroxyacetyl radical reacts with nitrogen dioxide

$$CH_3COO_2 + NO_2 \rightarrow CH_3COO_2NO_2 \quad (5.42)$$

The chemical species on the right-hand side of (5.42) is the vapor of a colorless and dangerously explosive liquid called peroxyacetyl nitrate (PAN), which is an important component of photochemical smogs and another major eye irritant. Other alkenes oxidize NO to NO_2 without consuming O_3 and regenerate OH and can do so faster than the aforementioned reactions.

Shown in Fig. 5.12 are typical variations through the course of a day in the concentrations of some of the major components of photochemical smogs in Los Angeles. Ozone precursors (NO_x and hydrocarbons) build up during the morning rush hour, and aldehydes, O_3, and PAN peak in the early afternoon.

The role of *polycyclic aromatic hydrocarbons* (PAHs) in air pollution and public health was first realized in the early 1940s following the discovery that organic extracts of particles from polluted air (e.g., benzo[a]pyrene) produce cancer in laboratory experiments on animals. PAHs are emitted by diesel and gasoline engines, coal-fired electric power-generating plants, biomass burning, and cigarettes. They are present in air as volatile particulates and gases. Reactions initiated by OH in the day and NO_3 at night convert gaseous PAHs to nitro-PAH derivatives, which are responsible for ~50% of the mutagenic activity of respirable airborne particles in southern California.

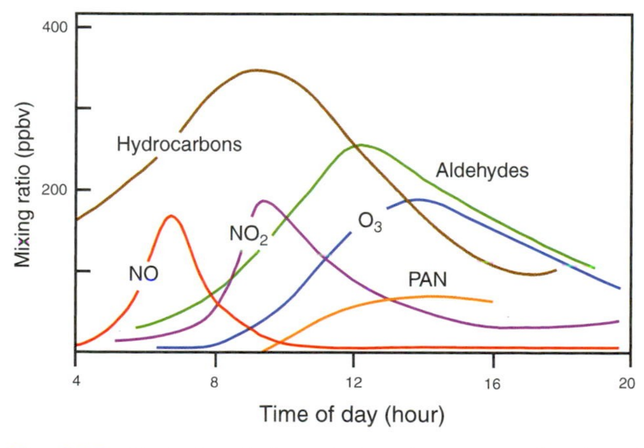

Fig. 5.12 Typical variations during the course of a day of some important pollutants in photochemical smogs in Los Angeles. [Adapted from P. A. Leighton, *Photochemistry of Air Pollution*, Academic Press, New York, 1961, p. 273, with permission of Elsevier.]

5.5.3 Regional and Global Pollution

The effects of anthropogenic pollution now extend to regional and global scales. Europe, Russia, the northeastern United States, India, and large areas of southeastern Asia are regularly covered by enormous palls of polluted air that reduce visibility significantly, produce acid deposition, soil and erode buildings and other materials, and have deleterious effects on human health, animals, and plants.

The fact that pollutants can be transported over large distances is well illustrated by air pollution episodes in the Arctic, known as *arctic haze*, which can be as severe as those in cities. The pollutants originate from fossil-fuel combustion, smelting, and other industrial processes in northern Europe and Russia. The pollutants are transported to the Arctic by synoptic-scale flow patterns, primarily from December to April. Because the arctic atmosphere is generally stably stratified during this time of the year, vertical mixing is limited; also, precipitation is rare so that wet removal processes are weak. Consequently, the pollutants can be transported over large distances with relatively little dilution. A major contributor to arctic haze is SO_2, which is converted to sulfate particles over the long transport distances.

Glacial records show that air pollution in the Arctic has increased markedly since the 1950s, paralleling the increases in SO_2 and NO_x emissions in Europe. Interestingly, ice cores from Greenland show unusually high lead concentrations from ~500 B.C. to 300 A.D. This is attributed to Greek and Roman lead and silver mining and smelting activities, which apparently polluted large regions of the northern hemisphere. However, cumulative lead deposits in the Greenland ice during these eight centuries were only ~15% of those caused by the widespread use of lead additives in gasoline from ~1930 to 1995. Lead additives to gasoline were eliminated in the United States in 1986, more than 60 years after their introduction.

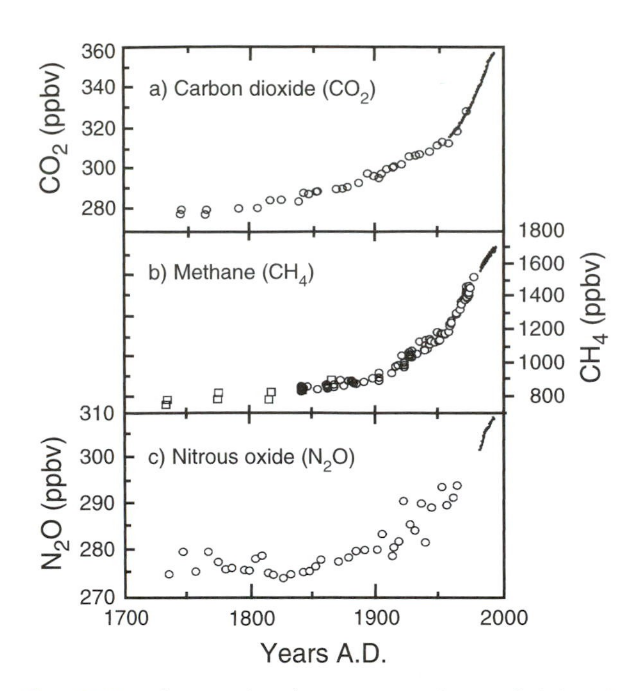

Fig. 5.13 Changes in the concentrations of (a) CO_2, (b) CH_4, and (c) N_2O over 300 years deduced from analysis of Greenland and Antarctic ice cores. [Adapted from D. J. Wuebbles et al., "Changes in the chemical composition of the atmosphere and potential impacts," in Brasseur, Prinn and Pszenny, eds, *Atmospheric Chemistry in a Changing World*, Springer, 2003, p. 12, Copyright 2003 Springer-Verlag.]

Anthropogenic influences are now apparent on a global scale, as illustrated by the worldwide increase in CO_2 concentrations since the industrial revolution (Fig. 5.13a). Other trace gases (e.g., CH_4 and N_2O) also show increasing concentrations worldwide over the past 150 years or so (Figs. 5.13b and 5.13c).

5.6 Tropospheric Chemical Cycles

The reservoirs of chemical species in the Earth system are the solid Earth, the hydrosphere (oceans and fresh water), the cryosphere (ice and snow), the biosphere, and the atmosphere. Chemical species can be transferred between these reservoirs. Because under steady-state conditions, a chemical species cannot accumulate indefinitely in any of the reservoirs, there must be continual cycling of species through the various reservoirs. This is termed *geochemical cycling*.

The geochemical cycling of carbon has been discussed in some detail in Chapter 2. Here we consider the tropospheric portions of the geochemical cycles of two other important chemical species, namely nitrogen and sulfur. We will be concerned with relatively rapid interchanges involving the atmosphere and other reservoirs (generally the oceans and the biosphere). Many aspects of global chemical cycles are not well understood; therefore, in many cases, the magnitudes of the sources and sinks given here are only rough estimates. Also, the relative importance of the various species is not determined exclusively by the magnitude of the emissions: atmospheric residence times must also be taken into account.

5.6.1 The Nitrogen Cycle

Nitrogen gas (N_2) constitutes more than 99.99% of the nitrogen present in the atmosphere, and N_2O (an important greenhouse gas) makes up more than 99% of the rest of the nitrogen. The other nitrogen species in the atmosphere are therefore present in only trace concentrations (see Table 5.1), but are nonetheless of crucial importance in atmospheric chemistry. For example, NH_3 is the only basic gas in the atmosphere. Therefore, it is solely responsible for neutralizing acids produced by the oxidation of SO_2 and NO_2; the ammonium salts of sulfuric and nitric acid so formed become atmospheric aerosols. Nitric oxide (NO) and NO_2 play important roles in both tropospheric and stratospheric chemistry.

All of the nitrogen-containing gases in the air are involved in biological nitrogen fixation and denitrification. *Fixation* refers to the reduction and incorporation of nitrogen from the atmosphere into living biomass. This transformation is accomplished by bacteria that are equipped with a special enzyme system for this task; the usual product is NH_3. The term *fixed nitrogen* refers to nitrogen contained in chemical compounds that can be used by plants and microorganisms. Under *aerobic* (i.e., oxygen-rich) conditions, other specialized bacteria can oxidize ammonia to nitrite and then to nitrate; a process referred to as *nitrification*. Most plants use nitrate taken through their roots to satisfy their nitrogen needs. Some of the nitrate undergoes bacterial reduction to N_2 and N_2O (termed *denitrification*), which returns fixed nitrogen from the biosphere to the atmosphere. In this case, nitrate acts as the oxidizing agent; therefore, denitrification generally occurs under *anaerobic* conditions, where oxygen is unavailable. Fixed nitrogen can also be returned from plants to the atmosphere in the

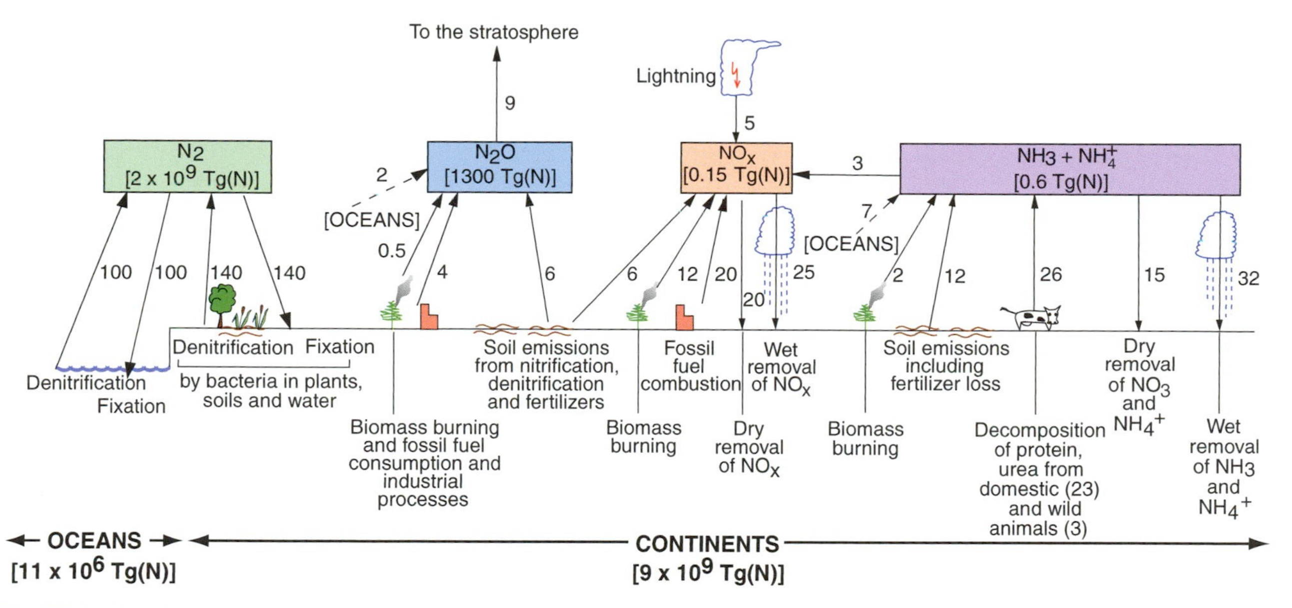

Fig. 5.14 Principal sources and sinks of nitrogen-containing gases in the atmosphere. Numbers alongside the arrows are estimates of average annual fluxes in Tg(N) per year; various degrees of uncertainty, some quite large, are associated with all of the fluxes. Numbers in square brackets are total amounts of the species in the atmosphere. [Adapted from P. V. Hobbs, *Introduction to Atmospheric Chemistry*, Camb. Univ. Press, 2000, p. 148. Reprinted with the permission of Cambridge University Press.]

form of N_2O. Biomass burning returns fixed nitrogen to the atmosphere as N_2, NH_3, N_2O, and NO_x.

The principal sources and sinks of nitrogen-containing species in the atmosphere are shown in Fig. 5.14. Assuming that fixation and denitrification of N_2 are in approximate balance, the main atmospheric sources of nitrogen-containing species are biogenic emissions from the terrestrial and marine biosphere (NH_3, N_2O, and NO_x), decomposition of proteins and urea from animals (NH_3), biomass burning and fossil fuel consumption (NO_x, NH_3, and N_2), and lightning (NO_x). Nitrous oxide (N_2O) is by far the largest reservoir of nitrogen in the atmosphere and, except for N_2, it has a much longer residence time than other nitrogen species.

The main sinks of nitrogen-containing species are wet removal by precipitation (NH_3 and NO_x as NO_3^-), dry deposition (NO_x and NH_3), and the chemical breakdown of N_2O in the stratosphere. Because anthropogenic sources of NH_3, N_2O, and NO_x (from fossil fuel consumption, biomass burning, and agricultural nitrate fertilization) are appreciable, human activities may be causing significant perturbations in the budgets of these species in the atmosphere.

5.6.2 The Sulfur Cycle

Sulfur species enter the atmosphere primarily in a chemically reduced state. The most important reduced sulfur gases are H_2S, DMS, COS, and CS_2. Their main natural sources are biogenic reactions in soils, marshland, and plants. Biogenic reactions in the oceans, due primarily to phytoplankton, are sources of DMS, COS, and CS_2. When reduced sulfur gases are released into the oxygen-rich atmosphere, they are oxidized to SO_2 and then over 65% of the SO_2 is oxidized to SO_4^{2-} (the remainder of the SO_2 is removed by dry deposition). Estimates of the fluxes of these natural emissions of sulfur gases and their transformations to SO_2 and SO_4^{2-} are given in Fig. 5.15.

It can be seen from Fig. 5.15 that DMS dominates the emissions of sulfur gases from the oceans.[26] Most of the sulfides in the air are oxidized rapidly by the OH radical; therefore, their residence times are only

[26] An enormous amount of sulfate is ejected into the air from the oceans in the form of sea salt. However, because these are relatively large particles, they are recycled quickly back to the ocean. Therefore, they do not have a significant effect on the global sulfur cycle.

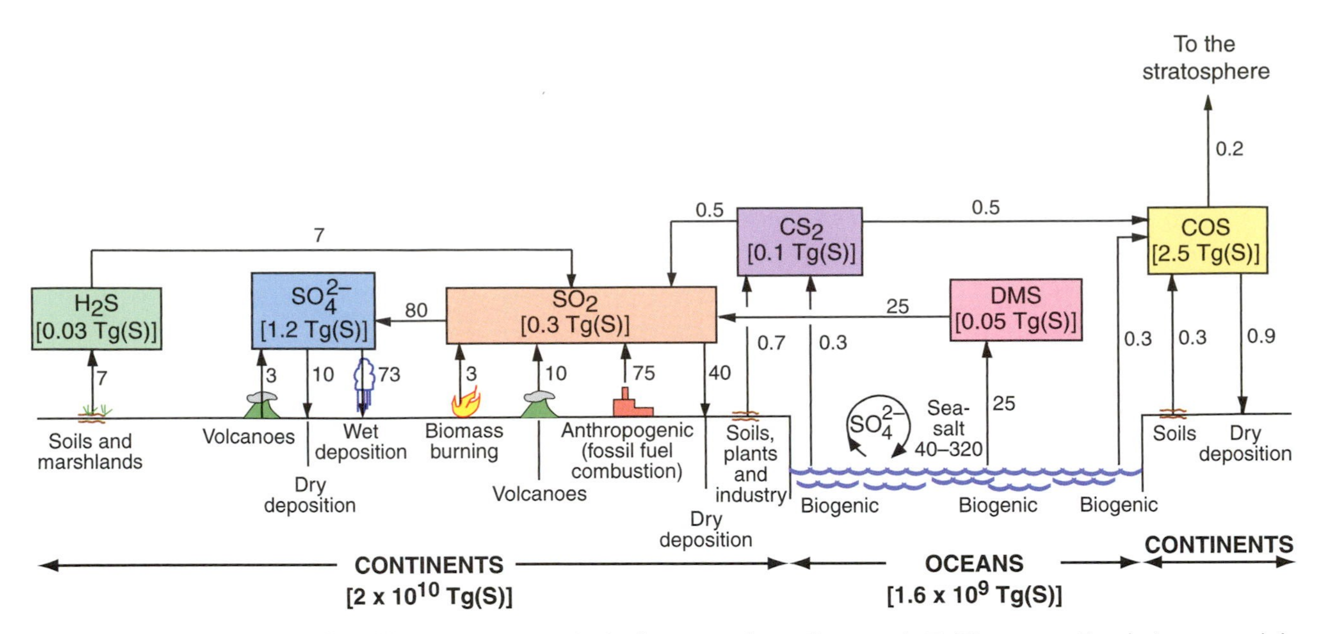

Fig. 5.15 As for Fig. 5.14 but for sulfur-containing species in the troposphere. Fluxes are in Tg(S) per year. For clarity, wet and dry removal are shown only over the continents, although they also occur over the oceans. [Adapted from P. V. Hobbs, *Introduction to Atmospheric Chemistry*, Camb. Univ. Press, 2000, p. 150. Reprinted with the permission of Cambridge University Press.]

days to a week. An important exception is COS, which is very stable in the troposphere and, as a result, has a relatively long residence time (~2 years) and a large and relatively uniform concentration (~0.5 ppbv, compared to ~0.1 ppbv for DMS, ~0.2 ppbv for H_2S and SO_2, and ~0.05 ppbv for CS_2). Consequently, COS is the most abundant sulfur compound in the troposphere. However, because it is relatively unreactive, it is generally ignored in tropospheric chemistry. The relatively long residence time of COS enables it to be mixed into the stratosphere gradually, where, converted by UV radiation, it is the dominant source of the sulfate aerosol during volcanically quiescent periods (see Section 5.7.3).

The flux of anthropogenic emissions of sulfur to the atmosphere is known quite well; it is ~78 Tg(S) per year, which is greater than estimates of the natural emissions of reduced sulfur gases to the atmosphere (Fig. 5.15). Therefore, the global sulfur budget is significantly affected by human activities. Anthropogenic emissions of sulfur are almost entirely in the form of SO_2 and 90% are from the northern hemisphere. The main sources are the burning of coal and the smelting of sulfide ores.

The main mechanisms for removing sulfur from the atmosphere are wet and dry deposition. For example, of the 80 Tg(S) per year of SO_2 that is oxidized to SO_4^{2-}, about 70 Tg(S) per year occurs in clouds, which is subsequently wet deposited; the remainder is oxidized by gas-phase reactions and dry deposited (Fig. 5.15).

Exercise 5.9 Using the information given in Fig. 5.15, estimate the residence time of SO_2 in the troposphere with respect to influxes.

Solution: From Fig. 5.15 we see that the magnitude of the tropospheric reservoir of SO_2 is 0.3 Tg(S). By adding the influxes of SO_2 shown in Fig. 5.15, the total influx is found to be 120.5 Tg(S) per year. Therefore,

$$\begin{array}{c}\text{residence time of } SO_2 \\ \text{in the troposphere}\end{array} = \frac{\begin{array}{c}\text{magnitude of} \\ \text{tropospheric reservoir}\end{array}}{\text{total influx to troposphere}}$$

$$= \frac{0.3}{120.5} \simeq 2 \times 10^{-3} \text{ years} \simeq 1 \text{ day} \quad ■$$

5.7 Stratospheric Chemistry

In passing across the tropopause, from the troposphere to the stratosphere, there is generally an abrupt change in concentrations of several important trace constituents of the atmosphere. For example, within the first few kilometers above the tropopause, water vapor decreases and O_3 concentrations often increase by an order of magnitude. The strong vertical gradients across the tropopause reflect the fact that there is very little vertical mixing between the relatively moist, ozone-poor troposphere and the dry, ozone-rich stratosphere. Within the stratosphere the air is generally neutral or stable with respect to vertical motions. Also, the removal of aerosols and trace gases by precipitation, which is a powerful cleansing mechanism in the troposphere, is generally absent in the stratosphere. Consequently, materials that enter the stratosphere (e.g., volcanic effluents, anthropogenic chemicals that diffuse across the tropopause or are carried across the tropopause by strong updrafts in deep thunderstorms, and effluents from aircraft) can remain there for long periods of time, often as stratified layers.

This section discusses three topics of particular interest in stratospheric chemistry: unperturbed (i.e., natural) stratospheric O_3, anthropogenic perturbations to stratospheric O_3, and sulfur in the stratosphere. The emphasis in this chapter is on chemical processes. However, it should be kept in mind that chemical, physical, and dynamical processes in the atmosphere are often intimately entwined.

5.7.1 Unperturbed Stratospheric Ozone

a. Distribution

Ozone in the stratosphere is of great importance for the following reasons:

- It forms a protective shield that reduces the intensity of UV radiation (with wavelengths between 0.23 and 0.32 μm)[27,28] from the sun that reaches the Earth's surface.
- Because of the absorption of UV radiation by O_3, it determines the vertical profile of temperature in the stratosphere.
- It is involved in many stratospheric chemical reactions.

In 1881 Hartley[29] measured the UV radiation reaching the Earth's surface and found a sharp cutoff at $\lambda = 0.30\ \mu$m; he correctly attributed this to absorption by O_3 in the stratosphere. Subsequent ground-based UV measurements as a function of the sun's elevation and the first measurements of the concentrations of stratospheric O_3 by balloons in the 1930s placed the maximum O_3 concentration in the lower part of the stratosphere.

Figure 5.16 shows results of more recent measurements of O_3. The presence of an *ozone layer* between heights of ~15–30 km is clear. However, the O_3 layer is highly variable: its height and intensity change with latitude, season, and meteorological conditions.

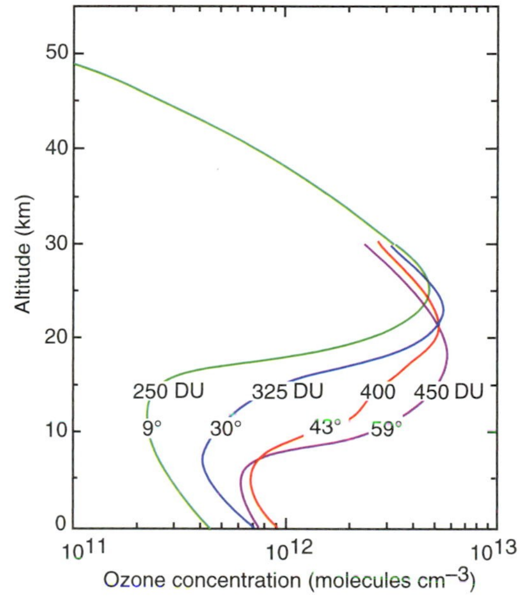

Fig. 5.16 Mean vertical distributions of ozone concentrations based on measurements at different latitudes (given in degrees). Note the increase in the total ozone column abundance (given in DU) with increasing latitude. [Adapted from G. Brasseur and S. Solomon, *Aeronomy of the Middle Atmosphere*, D. Reidel Pub. Co., 1984, Fig. 5.7, p. 215. Copyright 1984 D. Reidel Pub. Co., with kind permission of Springer Science and Business Media.]

[27] Electromagnetic radiation in this wavelength range is dangerous to living cells. Radiation with a wavelength (λ) $\leq 0.29\ \mu$m is lethal to lower organisms and to the cells of higher organisms. Radiation with $\lambda = 0.290$–$0.320\ \mu$m (*UV-B radiation*), which causes sunburn, has adverse effects on human health and on animals and plants. Were it not for O_3 in the stratosphere, radiation from the sun with $\lambda = 0.23$–$0.32\ \mu$m, would reach the Earth's surface unhindered. Ozone strongly absorbs UV radiation in just this wavelength band. For this reason, less than 1 part in 10^{30} of the flux of solar radiation at the top of the atmosphere with $\lambda = 0.25\ \mu$m reaches the Earth's surface. As shown in Chapter 2, the absorption of UV radiation by O_3 was essential for the emergence of life (as we know it) on Earth.

[28] It is common to use nanometers (nm; 1 μm = 10^3 nm) as the unit of wavelength in the ultraviolet and visible regions. However, for consistency with earlier chapters, we use micrometers (μm).

[29] **W. N. Hartley** (1846–1913) Irish spectroscopist. Professor of chemistry at Royal College of Science, Dublin.

Figure 5.16 includes the total *ozone column abundance* in Dobson units (DU)—see Footnote 13 in this chapter for the definition of DU.

The greatest column densities of O_3 in the northern hemisphere occur in polar latitudes in spring; in the southern hemisphere the spring maximum is at midlatitudes. Because O_3 is produced by photochemical reactions, the production is a maximum in the stratosphere over the tropics. The peaks in concentrations at polar and midlatitudes are attributed to meridional and downward transport of O_3 away from the equator, although at any given point in the atmosphere the balance between the production and loss of O_3, and its flux divergence, determines the O_3 concentration at that point. It is clear from Fig. 5.16 that much of the meridional contrast in the total column abundance of O_3 is due to differences in the profiles below 20 km, which are largely determined by transport.

Since 1960 remote sensing measurements from satellites (see Box 5.5) have provided a wealth of information on the global distribution of O_3 and the variations in its vertical profiles and column abundance.

5.5 Techniques for Measuring Atmospheric Ozone

In situ measurements of vertical profiles of O_3 may be obtained from ozonesondes, which can be carried on radiosonde balloons. The O_3 sensor consists of two electrolytic cells, each containing a solution of potassium iodide (KI). The cells are initially in chemical and electrical equilibrium. However, when an air sample containing O_3 is drawn through one of the cells, the equilibrium is perturbed and an electric current flows between the cells. The amount of electric charge, which is proportional to the partial pressure of the O_3 in the ambient air, is continuously transmitted to a ground station along with the ambient pressure and temperature. In this way, a vertical profile of O_3 is obtained, the integration of which provides the O_3 column from ground level up to the height of the balloon.

The O_3 column from the ground to the top of the atmosphere can be measured by passive remote sensing using a *Dobson spectrophotometer*. This is done by measuring the amount of UV sunlight that reaches the ground and deducing from this how much UV absorption occurred due to O_3. Absorption by O_3 occurs in the UV-B region ($\lambda = 0.290$–$0.320\ \mu m$). However, clouds and some aerosol particles also absorb in this wavelength band. Therefore, a region of the electromagnetic spectrum where O_3 absorbs only weakly, but clouds and aerosol absorb similarly to the UV-B region, is also monitored. By ratioing the two measured values, absorption by O_3 in the total vertical column can be obtained.

Ozone can be derived from satellite observations using any of the four *passive* techniques depicted in Fig. 5.17, namely by backscatter UV(BUV), occultation, limb emission, and limb scattering.

For determination of the total O_3 column by the BUV technique (Fig. 5.17a), two pairs of measurements are needed: the incoming UV irradiance and the backscattered UV radiance at a wavelength that is strongly absorbed by O_3 and a similar pair of measurements at a wavelength that is absorbed only weakly by O_3. The difference between these two pairs of measurements can be used to infer the total O_3 column. Vertical profiles of O_3 can be obtained using the BUV technique. Because O_3 absorbs more strongly at shorter wavelengths, solar radiation with progressively shorter wavelengths is absorbed at progressively higher altitudes. Therefore, the radiation at a particular wavelength in the UV can be scattered only above a certain height. By measuring backscattered radiation at a number of wavelengths, the vertical profile of O_3 can be deduced.

Occultation instruments measure radiation through the limb of the atmosphere (see Fig. 5.17b) when the sun, moon, or a star is rising or setting. From measurements of the amounts of radiation at various wavelengths absorbed by the atmosphere, vertical profiles of various trace constituents can be derived. This technique has been used to monitor O_3 since 1984.

In the limb emission technique, the concentration of O_3 is derived from measurements of infrared or microwave radiation emitted by the atmosphere along the line of sight of the instrument (Fig. 5.17c). Limb infrared emissions can be used to derive temperature, O_3, $H_2O(g)$, CH_4, N_2O, NO_2, CFCs, and the location of polar stratospheric clouds in the upper atmosphere. The advantage of utilizing microwave emission is that it passes through clouds and therefore can provide measurements lower in

Continued on next page

5.5 Continued

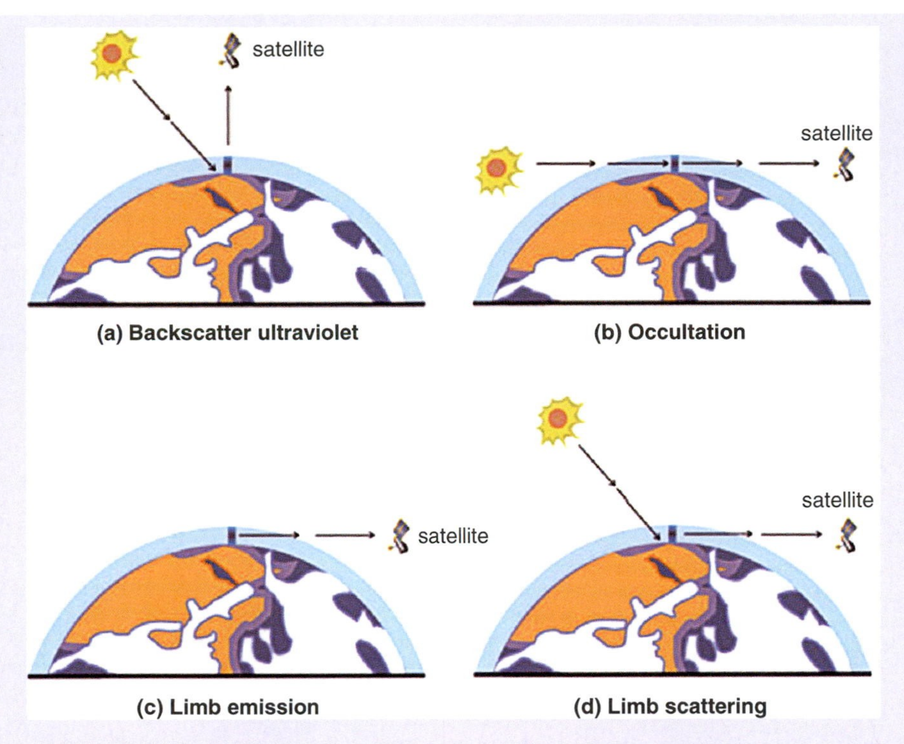

Fig. 5.17 Four passive techniques for measuring ozone from satellites. [Courtesy of P. Newman, NASA Goddard Space Flight Center.]

the atmosphere than infrared emissions. However, microwave detectors are large, heavy, and require more power than infrared detectors.

Limb scattering (Fig. 5.17d) employs aspects of the three passive remote sensing techniques discussed briefly earlier. However, it measures scattered light rather than light directly from the sun or moon. Consequently, if the sun is visible, this technique can provide essentially continuous measurements. Limb scattering works best with O_3, but $H_2O(g)$, NO_2, SO_2, and aerosols can also be measured.

Ozone profiles (as well as those of other trace constituents) can be obtained using differential absorption lidar (DIAL), which is an *active* remote sensing technique. The radiation source is a powerful lidar. A telescope, generally located near the laser, is used to collect the radiation backscattered by the atmosphere along the line of sight of the lidar beam. The time interval between transmission of a pulse and its return can be used to determine the range of the scatterers. The attenuation of the beam provides information on atmospheric trace constituents. Differences between transmitted and returned wavelengths produced by the Doppler effect can be used to infer atmospheric motions. As in the BUV technique, the ratio of two returned signals can be used to deduce the concentration of the species. This technique can provide measurements with high spatial resolution in the vertical. Because DIAL instruments are large and heavy, they have generally been used only from the ground.

The advantage of satellite-born instruments is their extensive spatial coverage over relatively long time periods. Disadvantages are that the measurements of trace constituents are not direct (they have to be derived from radiation measurements) and typically they do not have very high spatial resolution. Remote sensing measurements from the ground can be continuous, but each instrument is generally located at one site. Remote sensing measurements from aircraft provide mobility and better spatial resolution in the vertical. *In situ* measurements (from aircraft and balloons) are generally more direct, they can be related to chemical standards, and they can provide high spatial resolution. However, they are inherently very intermittent. Airborne *in situ* measurements have been used extensively to validate satellite measurements.

b. Chapman's theory

In 1930 Chapman[30] proposed a simple chemical scheme for maintaining steady-state concentrations of O_3 in an "oxygen-only" (i.e., O, O_2, and O_3) stratosphere. The reactions were the dissociation of O_2 by solar UV radiation (at $\lambda < 0.242\ \mu m$)

$$O_2 + h\nu \xrightarrow{j_a} 2O \tag{5.43}$$

the reaction of atomic oxygen and molecular oxygen to form O_3

$$O + O_2 + M \xrightarrow{k_b} O_3 + M \tag{5.44}$$

(where *M* represents N_2 and O_2), the photodissociation of O_3 (which occurs for $\lambda < 0.366\ \mu m$)

$$O_3 + h\nu \xrightarrow{j_c} O_2 + O \tag{5.45}$$

and the combination of atomic oxygen and O_3 to form O_2

$$O + O_3 \xrightarrow{k_d} 2O_2 \tag{5.46}$$

Reactions (5.43)–(5.46) are called the *Chapman reactions*. The rate coefficients (*j*'s and *k*'s) are temperature dependent.

The O atoms produced by (5.43) undergo numerous collisions with N_2 and O_2 molecules, most of which do not result in any permanent products. Occasionally, however, a three-body collision occurs [as represented by the left side of (5.44)] in which M absorbs the excess energy of the collision and a stable O_3 molecule is produced. Note that the excess energy acquired by M, which is in the form of thermal energy that warms the stratosphere, derives primarily from the energy of the photon in (5.43), i.e., from solar energy. Because the concentration of M decreases with increasing altitude, the time constant for converting O atoms and O_2 molecules to O_3 by (5.44) increases with altitude (e.g., from a few seconds at 40 km to ~100 s at 50 km).

After it is formed, the O_3 molecule remains intact until it is split apart by the photolytic reaction (5.45). Reactions (5.44) and (5.45) are fast and they continually cycle oxygen atoms back and forth between O and O_3. Ozone molecules are removed from the system by (5.46). The time constant for this reaction is ~1 day at 40 km and a fraction of a day at 50 km (where the concentration of O atoms is greater). This reaction can take place as given by (5.46) or it can be catalytically accelerated (e.g., by chlorine or bromine), as discussed in Section 5.7.1c. A simplified schematic of the Chapman reactions is shown in Fig. 5.18.

Stratospheric O_3 concentrations exhibit minor diurnal variations with time of day. After sunset, both the source and the sink of O_3 (5.43) and (5.45), respectively, are switched off, and the remaining O atoms are then converted to O_3 within a minute or so by (5.44). When the sun rises, some of the O_3 molecules are destroyed by (5.45) but they are reformed by (5.43) followed by (5.44).

Exercise 5.10 During the daytime in the stratosphere a steady-state concentration of atomic oxygen may be assumed. Use this fact to derive an expression for dn_3/dt in terms of n_1, n_2, n_3, k_d, and j_a, where n_1, n_2, and n_3 are the concentrations of atomic oxygen, molecular oxygen, and ozone, respectively. Assuming that the removal of atomic oxygen by (5.46) is small, and (at 30 km) $j_c n_3 >> j_a n_2$, derive an expression for the steady-state concentration of atomic oxygen in the stratosphere at 30 km during the daytime in terms of n_2, n_3, n_M, k_b, and j_c, where n_M is the concentration of M in (5.44). Hence, derive an expression for dn_3/dt in terms of $n_2, n_3, n_M, j_a, j_c, k_b$, and k_d.

Solution: From (5.43)–(5.46)

$$\frac{dn_1}{dt} = 2j_a n_2 - k_b n_1 n_2 n_M + j_c n_3 - k_d n_1 n_3 \tag{5.47}$$

If atomic oxygen is in steady-state $dn_1/dt = 0$, and

$$2j_a n_2 + j_c n_3 = k_b n_1 n_2 n_M + k_d n_1 n_3 \tag{5.48}$$

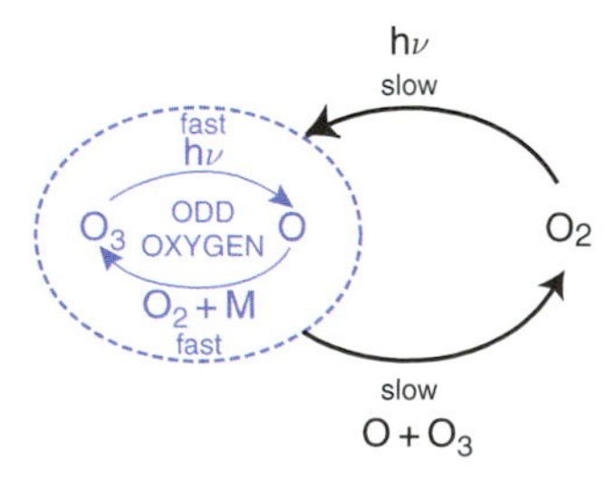

Fig. 5.18 Schematic of the Chapman reactions. [Adapted from *The Chemistry and Physics of Stratospheric Ozone*, A. Dessler, Academic Press, p. 44, copyright (2000), with permission from Elsevier.]

[30] **Sydney Chapman** (1888–1970) English geophysicist. Chapman made important contributions to a wide range of geophysical problems, including geomagnetism, space physics, photochemistry, and diffusion and convection in the atmosphere.

From (5.44)–(5.46)

$$\frac{dn_3}{dt} = k_b n_1 n_2 n_M - j_c n_3 - k_d n_1 n_3 \qquad (5.49)$$

From (5.48) and (5.49)

$$\frac{dn_3}{dt} = 2j_a n_2 - 2k_d n_1 n_3 \qquad (5.50)$$

If the removal of atomic oxygen by (5.46) is small, we have from (5.47)

$$\frac{dn_1}{dt} \simeq 2j_a n_2 - k_b n_1 n_2 n_M + j_c n_3 \qquad (5.51)$$

Therefore, since $dn_1/dt = 0$ and $j_c n_3 \gg 2j_a n_2$,

$$n_1 \simeq \frac{j_c n_3}{k_b n_2 n_M} \qquad (5.52)$$

Substituting (5.52) into (5.50) yields

$$\frac{dn_3}{dt} = 2j_a n_2 - \frac{2k_d j_c n_3^2}{k_b n_2 n_M} \qquad (5.53)$$

■

The Chapman reactions reproduce some of the broad features of the vertical distribution of O_3 in the stratosphere. For example, they predict that O_3 concentrations should reach maximum values at an altitude of ~25 km. Indeed, prior to 1964 it was generally believed that the Chapman theory provided an adequate description of stratospheric ozone chemistry. This is now known not to be the case. For example, although the Chapman reactions predict the correct shape for the vertical profile of O_3, they overpredict the concentrations of stratospheric O_3 by about a factor of two in the tropics and they underpredict O_3 concentration in middle to high latitudes. Also, model calculations based on the Chapman reactions predict that the global rate of production of O_3 in spring is 4.86×10^{31} molecules s^{-1} (of which 0.06×10^{31} molecules s^{-1} are transported to the troposphere). However, the loss of "odd oxygen" is only 0.89×10^{31} molecules s^{-1}. This leaves a net 3.91×10^{31} molecules s^{-1}, which would double atmospheric O_3 concentrations in just 2 weeks.[31] Because O_3 concentrations are not increasing, there must be important sinks of odd oxygen in the stratosphere in addition to reaction (5.46). As shown later, there are catalytic cycles involving nitrogen compounds, H, OH, Cl, and Br that deplete O_3 in the stratosphere. Also, there is an equator-to-pole circulation in the stratosphere, sometimes referred to as the *Brewer–Dobson circulation*, which transports O_3 from its primary source in the tropical stratosphere poleward and downward into the extratropical lower stratosphere.

c. Catalytic chemical cycles

Most of the catalytic reactions that have been proposed for the removal of stratospheric odd oxygen are of the form

$$X + O_3 \rightarrow XO + O_2 \qquad (5.54a)$$

$$XO + O \rightarrow X + O_2 \qquad (5.54b)$$

$$\text{Net:} \quad O + O_3 \rightarrow 2O_2 \qquad (5.55)$$

where X represents the catalyst and XO the intermediate product. Provided that reactions (5.54) are fast, reaction (5.55) can proceed much faster than reaction (5.46). Also, because X is consumed in (5.54a) but regenerated in reaction (5.54b) and provided there is no appreciable sink for X, just a few molecules of X have the potential to eliminate large numbers of O_3 molecules and atomic oxygen.

The first candidate that was suggested for species X in (5.54) in the upper atmosphere (actually in the mesosphere where excited atomic oxygen is common) was OH, which has at least three sources

$$H_2O + O^* \rightarrow 2OH \qquad (5.56)$$

$$CH_4 + O^* \rightarrow CH_3 + OH \qquad (5.57)$$

and

$$H_2 + O^* \rightarrow H + OH \qquad (5.58)$$

The H_2O and CH_4 in (5.56) and (5.57) ascend into the stratosphere in the tropics. During the passage through the cold tropical tropopause, most of the water vapor condenses out, which accounts for the

[31] In the Chapman reactions, odd oxygen is produced only by reaction (5.43) and is lost only in reaction (5.46). Reactions (5.44) and (5.45) intraconvert atomic oxygen and O_3 and determine the ratio $[O]/[O_3]$; both reactions are fast during the day. Therefore, atomic oxygen and O_3 are interconverted rapidly, as shown in Fig. 5.18. Below about 45 km in the stratosphere, O_3 accounts for >99% of the odd oxygen.

dryness of the stratosphere. However, CH_4 is not affected by the low temperatures and is therefore lofted into the lower stratosphere in concentrations similar to those in the upper troposphere.

Below ~40 km the OH from (5.56)–(5.58) acts as catalyst X in (5.54) to destroy odd oxygen

$$OH + O_3 \rightarrow HO_2 + O_2 \quad (5.59a)$$

$$HO_2 + O \rightarrow OH + O_2 \quad (5.59b)$$

$$\text{Net:} \quad O + O_3 \rightarrow 2O_2 \quad (5.60)$$

Below 30 km, where atomic oxygen is very scarce, the following catalytic cycle, which destroys O_3, is increasingly important

$$OH + O_3 \rightarrow HO_2 + O_2 \quad (5.61a)$$

$$HO_2 + O_3 \rightarrow OH + 2O_2 \quad (5.61b)$$

$$\text{Net:} \quad 2O_3 \rightarrow 3O_2 \quad (5.62)$$

A catalytic cycle that is important for the destruction of O_3 in the middle stratosphere is

$$NO + O_3 \rightarrow NO_2 + O_2 \quad (5.63a)$$

$$NO_2 + O \rightarrow NO + O_2 \quad (5.63b)$$

$$\text{Net:} \quad O + O_3 \rightarrow 2O_2 \quad (5.64)$$

where NO has replaced X in (5.54). Nitric oxide (NO) is produced in the stratosphere by

$$N_2O + O^* \rightarrow 2NO \quad (5.65)$$

At a temperature of −53 °C (which is typical of the stratosphere), the rate coefficients for reactions (5.63a) and (5.63b) are 3.5×10^{-21} and 9.3×10^{-18} m^3 molecules^{-1} s^{-1}, respectively, compared to 6.8×10^{-22} m^3 molecules^{-1} s^{-1} for k_d in (5.46). Whether reactions (5.63a) and (5.63b) destroy odd oxygen faster than reaction (5.46) depends on the concentrations of NO_2 and O_3.

In the 1970s it was suggested that catalytic cycles involving reactive chlorine (Cl) and bromine (Br) could be important in the destruction of O_3. The natural sources of Cl and Br in the stratosphere are the destruction of methyl chloride (CH_3Cl) and methyl bromide (CH_3Br), respectively. Chlorine and Br can then serve as X in (5.54), with intermediates ClO and BrO, respectively. The following cycle of reactions, which couple these two intermediates,

$$BrO + ClO \rightarrow Br + Cl + O_2 \quad (5.66a)$$

$$Br + O_3 \rightarrow BrO + O_2 \quad (5.66b)$$

$$Cl + O_3 \rightarrow ClO + O_2 \quad (5.66c)$$

$$\text{Net:} \quad 2O_3 \rightarrow 2O_2 \quad (5.67)$$

is a more powerful mechanism for destroying O_3 in the lower stratosphere than those discussed previously because O_3 is much more abundant than atomic oxygen. Because the stable reservoirs for Br and BrO (i.e., $BrONO_2$ and HOBr) are not as stable as the reservoirs for Cl and ClO (i.e., HCl and $ClONO_2$) on a per molecule basis, Br is more efficient in destroying O_3 than Cl.

The destruction of stratospheric O_3 by the various chemical mechanisms discussed earlier, and many other possible mechanisms not discussed here are not simply additive because, as illustrated by (5.66), the species in one cycle can react with those in another cycle. Therefore, advanced numerical models that consider all of the known suspects and that utilize measurements of species concentrations and temperature-dependent rate coefficients must be used to unravel the relative importance of the various mechanisms for destroying O_3 in the stratosphere. This task is complicated further by the fact that concentrations of many of the prime suspects are changing with time due to anthropogenic emissions (see Section 5.7.2).

5.7.2 Anthropogenic Perturbations to Stratospheric Ozone: The Ozone Hole

If the concentration of catalyst X in (5.54) is increased significantly by human activities, the balance between the sources and sinks of atmospheric O_3 will be disturbed and stratospheric O_3 concentrations can be expected to decrease. One of the first concerns in this respect was raised in response to a proposal in the 1970s to build a fleet of supersonic aircraft flying in the stratosphere. It was noted that aircraft engines emit nitric oxide (NO), which can decrease odd oxygen by the cycle (5.63). The proposal was eventually rejected on both environmental and economic grounds. At the time of writing (2005), there are not sufficient aircraft flying in the stratosphere to significantly perturb stratospheric O_3.

Of much greater concern, with documented impacts, is the catalytic action of chlorine, from industrially manufactured chlorofluorocarbons (CFCs), in depleting stratospheric ozone.[32] CFCs are compounds containing Cl, F, and C; CFC-11 ($CFCl_3$) and CFC-12 (CF_2Cl_2) are the most common.[33] CFCs were first synthesized in 1928 as the result of a search for a nontoxic, nonflammable refrigerant. Over the next half-century they were marketed under the trade name Freon and became widely used, not only as refrigerants, but as propellants in aerosol cans, inflating agents in foam materials, solvents, and cleansing agents. Concern about their effects on the atmosphere began in 1973 when it was found that CFCs were spreading globally and, because of their inertness, were expected to have residence times ranging up to several hundred years in the troposphere. Such long-lived compounds eventually find their way into the stratosphere[34] where, at altitudes $\gtrsim$20 km, they absorb UV radiation in the wavelength interval 0.19–0.22 μm and photodissociate

$$CFCl_3 + h\nu \rightarrow CFCl_2 + Cl \qquad (5.68)$$

and

$$CF_2Cl_2 + h\nu \rightarrow CF_2Cl + Cl \qquad (5.69)$$

The chlorine atom released by these reactions can serve as the catalyst X in cycle (5.54), thereby destroying odd oxygen in the cycle

$$Cl + O_3 \rightarrow ClO + O_2 \qquad (5.70a)$$

$$ClO + O \rightarrow Cl + O_2 \qquad (5.70b)$$

$$\text{Net:} \quad O_3 + O \rightarrow 2O_2 \qquad (5.71)$$

In 1990, ~85% of the chlorine in the stratosphere originated from anthropogenic sources. Because CFCs absorb strongly in the infrared, they are also significant greenhouse gases.

The first evidence of a significant depletion in stratospheric O_3 produced by anthropogenic chemicals in the stratosphere came, surprisingly, from measurements in the Antarctic. In 1985, British scientists, who had been making ground-based, remote sensing measurements of O_3 at Halley Bay (76 °S) in the Antarctic for many years, reported a 30% decrease in springtime (october) total column O_3 since 1977. These observations were subsequently confirmed by remote sensing measurements from satellite and by airborne *in situ* measurements. Shown in Fig. 5.19 are a series of satellite measurements of the total O_3 column in the southern hemisphere. The high O_3 values (red and orange colors) that encircle the Antarctic continent and the lower O_3 values (green) over the continent itself in the 1970s (upper four images) are considered to be close to natural. The *Antarctic ozone hole*, as it came to be called, is apparent in the lower four images in Fig. 5.19 as the anomalously low O_3 values (purple) over the continent in October from 1999 to 2002.

The areal extent of the Antarctic ozone hole, derived from satellite measurements for the period September 7 to October 13 1979–2003, are shown in Fig. 5.20. From 1979 to 2001 the ozone hole grew progressively until it occupied an area (~25 million km^2) similar to that of North America. However, in 2002 the area occupied by the ozone hole decreased dramatically, but returned to its former size in 2003; the reasons for these dramatic changes are discussed later in this chapter.

The presence of an ozone hole over the Antarctic raised several intriguing questions. Why over the Antarctic? Why during spring? Why was the magnitude of the measured decreases in O_3 much greater than predictions based solely on gas-phase chemistry of the

[32] In 1995, Paul Crutzen, Mario Molina, and Sherwood Rowland were awarded the Nobel Prize in chemistry for predicting stratospheric O_3 loss by CFCs and nitrogen-containing gases.

[33] Compounds containing H, F, and C are called *hydrofluorocarbons* (HFCs), and compounds containing H, Cl, F, and C are called *hydrochlorofluorocarbons* (HCFCs). HFCs and HCFCs are considered to be more "environmentally friendly" than CFCs because they are partially destroyed by OH in the troposphere.

[34] Long-lived compounds enter the stratosphere primarily in air that ascends through the tropical tropopause. After entering the lower stratosphere the compounds are carried by the Brewer–Dobson circulation to higher altitudes (where they are broken down photolytically) or to higher latitudes. On average, it takes a year or two for a CFC molecule to move from the tropical tropopause to the upper stratosphere. However, it takes several decades for all of the air in the troposphere to be cycled through the upper stratosphere.

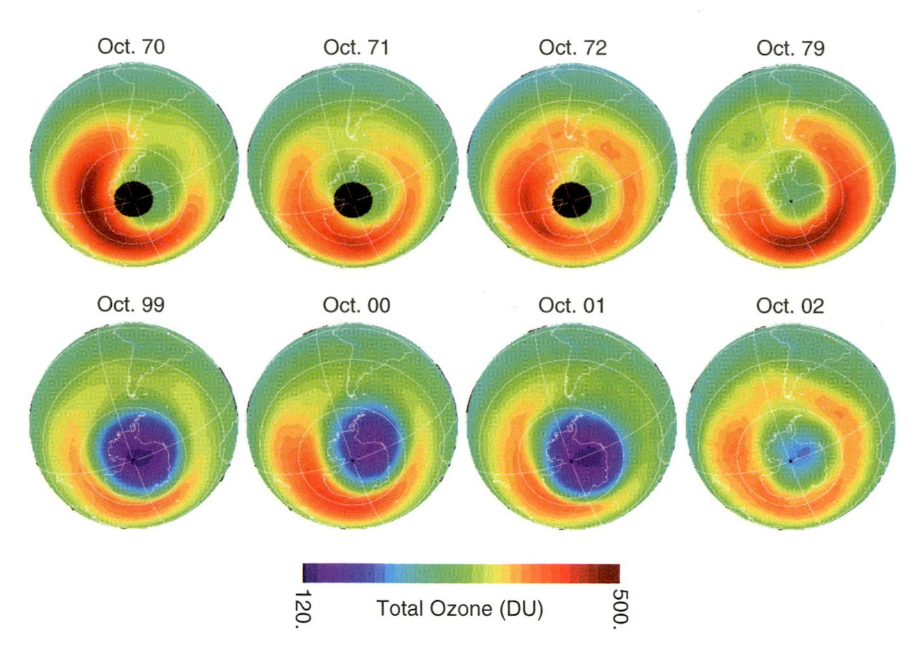

Fig. 5.19 Satellite observations of the total ozone column in the southern hemisphere during October for 8 years from 1970 to 2002. The color scale is in Dobson units (DU). The small black circles over the pole in 1970, 1971, and 1972 are a reflection of missing data. [Courtesy P. Newman, NASA Goddard Space Flight Center.]

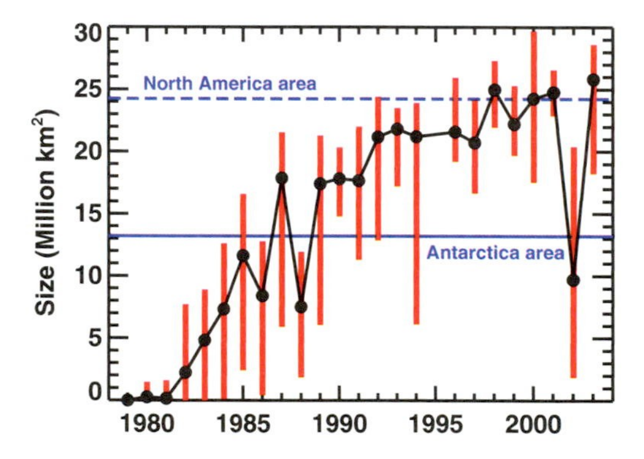

Fig. 5.20 Average areal extent of the ozone hole (defined as less than 220 Dobson units) averaged from September 7-October 13 of each year from 1979 to 2003. Vertical red bars are the range of values for each period. Year-to-year variations are caused by temperature fluctuations near the polar vortex edge. Warmer years have smaller holes (e.g., 2002), whereas colder years have larger ozone holes (e.g., 2003). The blue horizontal line highlights the total area of the Antarctic continent; for comparison, the area of North America (dashed blue line) is about twice the area of Antarctica. (Courtesy of P. Newman, NASA Goddard Space Flight Center.)

type outlined earlier? Why was the size of the ozone hole so much less in 2002 than in other recent years? The research that led to the answers to these questions, which is outlined later, was one of the triumphs of 20th century science.

During the austral winter (June–September), stratospheric air over the Antarctic continent is restricted from interacting with air from lower latitudes by a large-scale vortex circulation, which is bounded at its perimeter (called the *vortex collar*) by strong westerly winds encircling the pole. Because of the lack of solar heating in the austral winter, the air within the vortex is extremely cold and it lies within the sinking branch of the Brewer–Dobson circulation (Fig. 5.21). High-level clouds, called *polar stratospheric clouds* (PSCs), form in the cold core of the vortex, where temperatures can fall below −80 °C.[35] In the austral spring, as temperatures rise, the winds around the vortex weaken, and by late December the vortex disappears. However, during winter, the vortex serves as a giant and relatively isolated chemical reactor in which unique chemistry can occur. For example,

[35] During the drift of the *Deutschland* in the Weddell Sea in 1912, PSCs were observed in the late winter and early spring. Also, the Norwegian–British–Swedish 1949–1952 expedition to the Antarctic frequently observed a thin "cloud-veil" in the lower stratosphere in winter months. However, the widespread nature and periodicity of PSC formation in the Antarctic were not fully appreciated until they were observed by satellites starting in 1979.

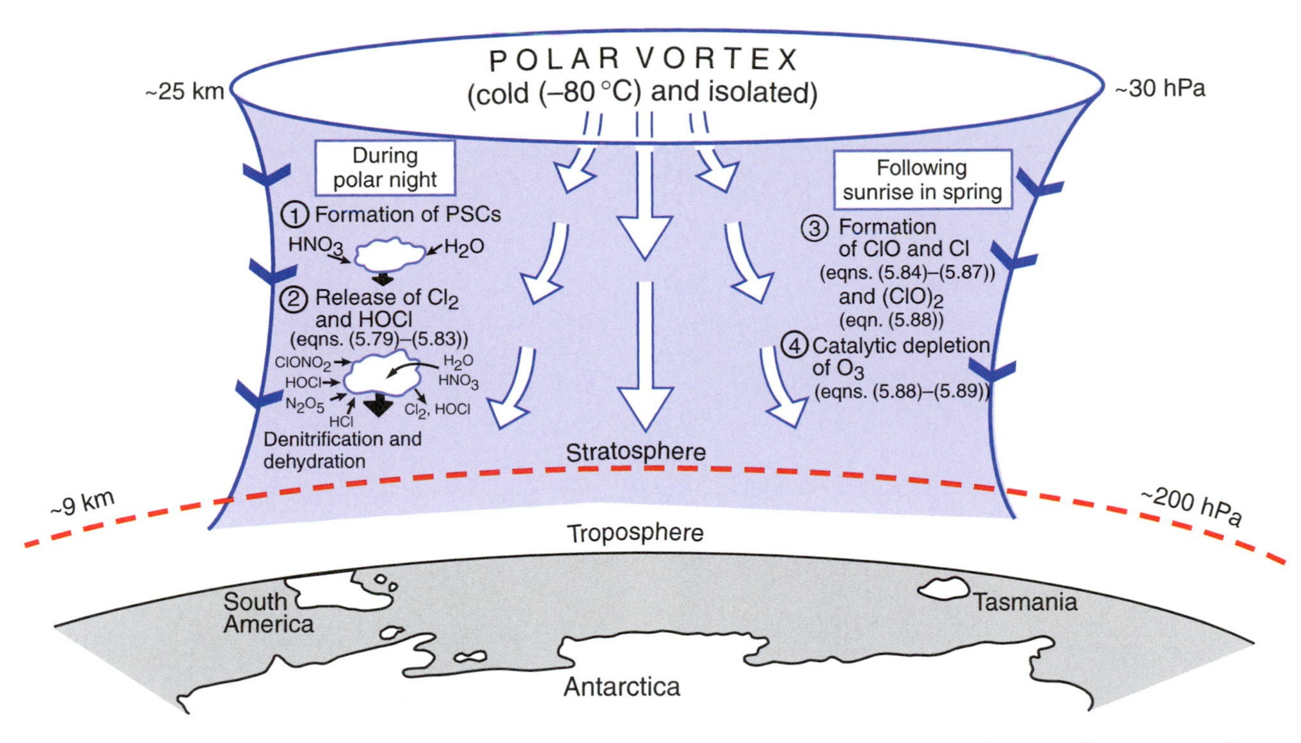

Fig. 5.21 Schematic of the polar vortex (blue) over Antarctica. Large arrows indicate cold descending air. The sequence of events (1 through 4) leading to the Antarctic ozone hole is summarized. For clarity, bromine reactions are not shown. See text for details.

although the concentrations of O_3 in the vortex are normal in August, the concentrations of ClO in the vortex are 10 times greater than just outside the vortex collar. In September, when sunlight returns to the polar cap region, O_3 concentrations within the vortex decrease dramatically. There are also sharp decreases in the concentrations of oxides of nitrogen and water vapor when passing from the outside to the inside of the vortex collar. These decreases are due, respectively, to the formation of nitric acid (HNO_3) and to the condensation of water at the very low temperatures inside the vortex. These two condensates form three types of PSCs. *Type I* PSCs, which condense near −78 °C, probably consist of a mixture of liquid and solid particles of nitric acid trihydrate ($HNO_3(H_2O)_3$—NAT for short), water and sulfuric acid. These particles are ~1 μm in diameter, so they settle out very slowly (~10 m per day). *Type II* PSCs, which form near −85 °C, consist of a mixture of ice and water together with some dissolved HNO_3. Because these particles are >10 μm in diameter, they settle out with appreciable speeds (~1.5 km per day). *Type III* PSCs ("mother-of-pearl" clouds) are produced by the rapid freezing of condensed water in air flow over topography. However, they are of limited extent and duration and do not form over the South Pole.

As the particles in PSCs slowly sink, they remove both water and nitrogen compounds from the stratosphere, processes referred to as *dehydration* and *denitrification*, respectively. As we will see, the chemical reactions involved in the removal of these two species play an important role in depleting O_3 in the Antarctic vortex.

Although (5.70) is important in destroying O_3 in the middle and upper stratosphere, it is less important in the lower stratosphere (accounting for ~5–25% of the total loss of halogens there) because the concentration of atomic oxygen decreases with decreasing altitude in the stratosphere and because most of the Cl and ClO released into the stratosphere by (5.68)–(5.70) are quickly tied up in reservoirs as HCl and chlorine nitrate $ClONO_2$ by

$$Cl + CH_4 \rightarrow HCl + CH_3 \quad (5.72)$$

and

$$ClO + NO_2 + M \rightarrow ClONO_2 + M \quad (5.73)$$

Liberation of the active Cl atoms from the reservoir species HCl and $ClONO_2$ is generally slow. However, on the surface of the ice particles that form PSCs, the following heterogeneous reactions are thought to be important

$$ClONO_2(g) + HCl(s) \rightarrow Cl_2(g) + HNO_3(s) \quad (5.74)$$

$$ClONO_2(g) + H_2O(s) \rightarrow HOCl(g) + HNO_3(s) \quad (5.75)$$

$$HOCl(g) + HCl(s) \rightarrow Cl_2(g) + H_2O(s) \quad (5.76)$$

$$N_2O_5(g) + H_2O(s) \rightarrow 2HNO_3(s) \quad (5.77)$$

$$N_2O_5(g) + HCl(s) \rightarrow ClNO_2(g) + HNO_3(s) \quad (5.78)$$

where the parenthetical s has been included to emphasize those compounds that are on (or in) ice particles, and the parenthetical g indicates species that are released as gases. Note that in addition to catalyzing the reactions (5.74)–(5.78), the ice particles that settle out from PSCs remove $HNO_3(s)$ from the stratosphere (Fig. 5.19). Sedimentation reduces the $ClONO_2(g)$ reservoir that has the potential to tie up Cl and ClO by (5.73). Therefore, on both counts, during the austral winter the ice particles that comprise PSCs in the Antarctic vortex set the stage for the destruction of ozone by enhancing the concentrations of the active species ClO and Cl.

Reactions (5.74), (5.75), (5.76), and (5.78) convert the reservoir species $ClONO_2$ and HCl to Cl_2, HOCl, and $ClNO_2$. When the sun rises in the Antarctic spring, these three species are photolyzed rapidly to produce Cl and ClO

$$Cl_2 + h\nu \rightarrow 2Cl \quad (5.79)$$

$$HOCl + h\nu \rightarrow OH + Cl \quad (5.80)$$

$$ClNO_2 + h\nu \rightarrow Cl + NO_2 \quad (5.81)$$

and

$$Cl + O_3 \rightarrow ClO + O_2 \quad (5.82)$$

Ozone in the Antarctic vortex is then destroyed efficiently by

$$ClO + ClO + M \rightarrow (ClO)_2 + M \quad (5.83a)$$

$$(ClO)_2 + h\nu \rightarrow Cl + ClOO \quad (5.83b)$$

$$ClOO + M \rightarrow Cl + O_2 + M \quad (5.83c)$$

$$2Cl + 2O_3 \rightarrow 2ClO + 2O_2 \quad (5.83d)$$

$$\text{Net:} \quad 2O_3 + h\nu \rightarrow 3O_2 \quad (5.84)$$

The following should be noted about (5.83):

- Because two ClO molecules are regenerated for every two ClO molecules that are destroyed, it is a catalytic cycle in which ClO is the catalyst.
- Unlike (5.54), (5.83) does not depend on atomic oxygen (which is in short supply).
- The Cl atom in the ClO on the left side of (5.83a) derives from Cl released from CFCs via reactions (5.68) and (5.69). However, as we have seen, the Cl atom is then normally quickly tied up as HCl and $ClONO_2$ by reactions (5.72) and (5.73).
- In the presence of PSCs, Cl_2, HOCl, and $ClNO_2$ are released by reactions (5.74)–(5.78) and, as soon as the solar radiation reaches sufficient intensity in early spring, reactions (5.79)–(5.82) release Cl and ClO, which lead to the rapid depletion of O_3 in the Antarctic stratosphere by (5.83).
- The dimer $(ClO)_2$ is formed by reaction (5.83a) only at low temperatures. Low enough temperatures are present in the Antarctic stratosphere, where there are also large concentrations of ClO. Therefore, on both counts, the Antarctic stratosphere in spring is a region in which the reaction cycle (5.83) can destroy large quantities of O_3.

The sequence of events leading to the Antarctic ozone hole is illustrated schematically in Figs. 5.21 and 5.22. Although for illustration we have considered only the role of Cl and its compounds in the formation of the Antarctic ozone hole, Br and its compounds play a similar role. Also, their intermediates ClO and BrO can combine to destroy O_3 through (5.66).

Exercise 5.11 If reaction (5.83a) is the slowest step in the catalytic cycle (5.83), and the pseudo rate coefficient for this reaction is k, derive an expression for the amount of O_3 destroyed over a time period Δt by this cycle.

Solution: Inspection of the reaction cycle (5.83) shows that two O_3 molecules are destroyed in each cycle. Also, the rate of the cycle is determined by the slowest reaction in the cycle, namely reaction (5.83a). Therefore, the rate of destruction of O_3 by this cycle is

$$\frac{d[O_3]}{dt} = -2k[ClO]^2$$

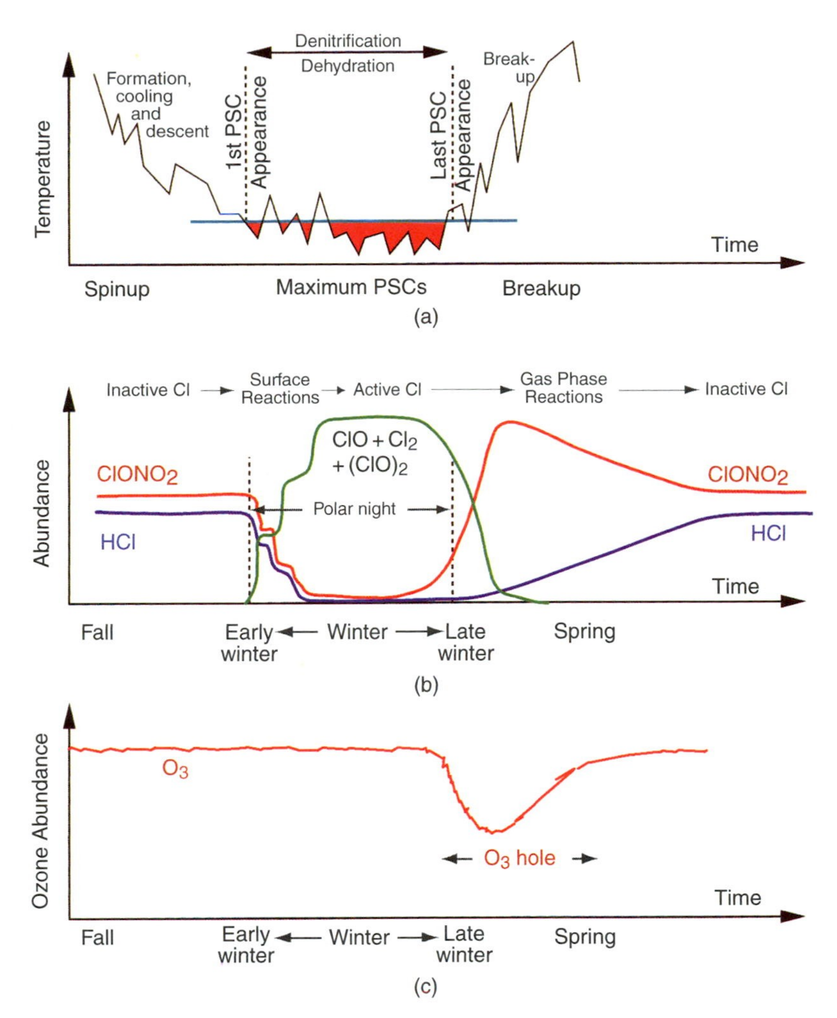

Fig. 5.22 Schematic illustrating time evolution of the main processes associated with the development of the Antarctic ozone hole. (a) The Antarctic vortex. Polar stratospheric clouds (PSCs) form, and reactions occur on the PSC particles when stratospheric temperatures fall into the region colored red. (b) Chlorine reservoirs in the Antarctic vortex. The inactive reservoir species $ClONO_2$ and HCl are converted into the active chlorine species Cl_2, ClO, and $(ClO)_2$ when the temperature falls below the value required for the formation of PSCs. The reservoir species return after the disappearance of the PSCs. (c) Ozone in the Antarctic vortex. Ozone is depleted rapidly by photolysis when the sun rises in September. As the active chlorine species are depleted, and the vortex breaks up, the concentration of O_3 rises. [Figures 5.22a and 5.22c adapted from World Meteorological Organization *Scientific Assessment of Ozone Depletion*, 1994. Figure 5.22b adapted with permission from Webster et al., *Science* **261**, 1131 (1993), Copyright 1993 AAAS.]

(where [M] has been incorporated into the pseudo rate coefficient k). Hence, over a time period Δt, the amount of ozone destroyed, $\Delta[O_3]$, is

$$\int_0^{\Delta[O_3]} d[O_3] = -2k \int_0^{\Delta t} [ClO]^2\, dt$$

or, if [ClO] does not change in the period Δt,

$$\Delta\, [O_3] = -2k[ClO]^2\, \Delta t \qquad \blacksquare$$

The sharp decrease in the area covered by the Antarctic ozone hole in 2002 (Fig. 5.20) and the decrease in the severity of the hole in that year (Fig. 5.19) are attributable to a series of unusual stratospheric warmings that occurred during the winter in the southern hemisphere in 2002. Changes in the circulation weakened and warmed the polar vortex, which cut off the formation of PSCs, turned off O_3 loss in September 2002, and transported O_3-rich air over Antarctica. In 2003, which was a cold year, the ozone hole returned to its earlier extent of ~25 million square kilometers.

Does an ozone hole, similar to that in the Antarctic, develop in the Arctic stratosphere during winter in the northern hemisphere? The first thing to note in this respect is that although a vortex can develop over the Arctic in the northern winter, it is generally not as well developed nor as cold as the Antarctic vortex.[36] Consequently, the conditions that produce severe loss of ozone in the Antarctic are not as common in the Arctic.

The total ozone column in the Arctic spring has, on average, decreased over the past two decades (Fig. 5.23). However, prior to the winter of 1995–1996 there was no evidence of a stratospheric ozone hole in the Arctic comparable to that in the Antarctic. During the northern winter of 1996–1997, the longest lasting polar vortex on record developed over the Arctic, and in March 1997 the average ozone column over the Arctic (354 DU) was the lowest in 20 prior years of observations.

Concerns about the health and environmental hazards of increased UV radiation at the Earth's surface, which accompany depletion in the total column O_3, led to international agreements to eliminate the production and use of compounds known to deplete stratospheric O_3 by the year 2000.[37] Consequently,

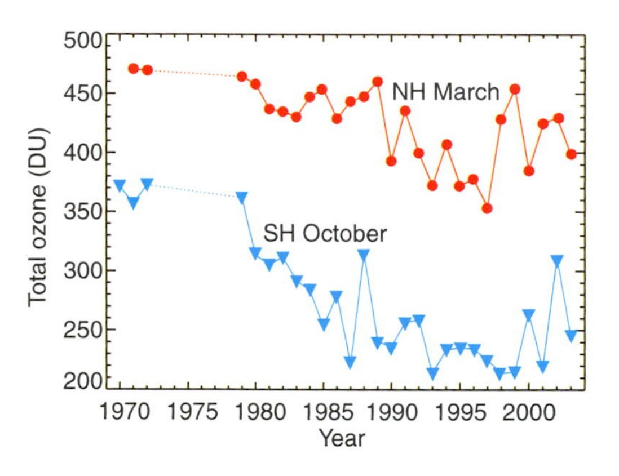

Fig. 5.23 Average ozone columns between latitudes 63°–90° for the northern hemisphere in March (red line and symbols) and the southern hemisphere in October (blue line and symbols). [Adapted with courtesy of P. Newman, NASA Goddard Space Flight Center.]

CFCs in the lower atmosphere are no longer increasing, and their *rate of growth* in the stratosphere is decreasing. An analysis (in 2003) of 20 years of satellite data showed a slowing in the reduction of O_3 at an altitude of ~33 km starting in 1997 and simultaneous slowing in the buildup of harmful Cl. However, due to the long lifetimes of CFCs, the concentrations of Cl in the stratosphere are expected to continue to rise for some time. Therefore, it is predicted that the O_3 layer probably will not recover to the level of the 1970s until the middle or late 21st century.

5.7.3 Stratospheric Aerosols; Sulfur in the Stratosphere

Aitken nucleus concentrations show considerable variations in the lower stratosphere, although they generally decrease slowly with increasing height. In contrast, particles with radii ~0.1–2 μm reach a maximum concentration of ~0.1 cm^{-3} at altitudes of ~17–20 km (Fig. 5.24). Because these particles are composed of about 75% sulfuric acid (H_2SO_4) and ~25% water, the region of maximum sulfate loading in the lower stratosphere is called the *stratospheric sulfate layer*, or sometimes the *Junge layer*, after C. Junge who discovered it in the late 1950s.

Stratospheric sulfate aerosols are produced primarily by the oxidation of SO_2 to SO_3 in the stratosphere

$$SO_2 + OH + M \rightarrow HOSO_2 + M \quad (5.85a)$$

$$HOSO_2 + O_2 \rightarrow HO_2 + SO_3 \quad (5.85b)$$

Also,

$$SO_2 + O + M \rightarrow SO_3 + M \quad (5.86)$$

Followed by

$$SO_3 + H_2O \rightarrow H_2SO_4 \quad (5.87)$$

[36] The northern hemisphere stratospheric polar vortex is weaker than its southern hemisphere counterpart because it is disturbed by planetary-scale waves forced by the temperature contrast between the cold continents (Eurasia and North America) and the warmer oceans.

[37] *The Montreal Protocol on Substances that Deplete the Ozone Layer* was a landmark international agreement originally signed in 1987, with substantial subsequent amendments. The Copenhagen amendment (1992) called for the complete elimination of CFCs by 1996, and the Vienna amendment (1995) called for the complete elimination of HCFCs by 2020. Because it takes several decades for all of the air in the troposphere to cycle through the upper stratosphere, where CFCs are broken down (see Footnote 34), it will take a similar period of time to remove all of the CFCs from the atmosphere from the time their production is brought completely to a halt (see Exercise 5.31).

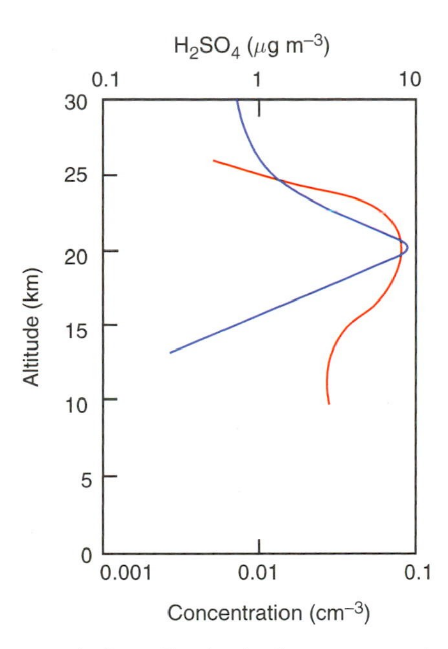

Fig. 5.24 Vertical profiles in the lower stratosphere of the number concentration of particles with radius ~0.1–2 μm (red line and bottom scale) and mass concentration of liquid sulfuric acid at standard temperature and pressure (blue line and upper scale). Measurements were obtained from balloons launched from midlatitudes. [Adapted from P. V. Hobbs, *Introduction to Atmospheric Chemistry*, Camb. Univ. Press, 2000, p. 179. Reprinted with permission of Cambridge University Press.]

The conversion of the H_2SO_4 vapor to liquid H_2SO_4 can occur by two main mechanisms:

- The combination of molecules of H_2SO_4 and H_2O (i.e., *homogeneous, bimolecular nucleation*) and/or the combination of H_2SO_4, H_2O, and HNO_3 to form new (primarily sulfuric acid) droplets (i.e., *homogeneous, heteromolecular nucleation*).
- Vapor condensation of H_2SO_4, H_2O, and HNO_3 onto the surfaces of preexisting particles with radius >0.15 μm (i.e., *heterogeneous, heteromolecular nucleation*).

Model calculations suggest that the second mechanism is the more likely route in the stratosphere. The tropical stratosphere is probably the major region where the nucleation process occurs, and the aerosols are then transported to higher latitudes by large-scale atmospheric motions.

The most significant source of SO_2 in the stratosphere is major volcanic eruptions. The SO_2 from such eruptions is converted to H_2SO_4 with an *e*-folding time of about 1 month in the stratosphere. The effect of major volcanic eruptions on the stratospheric aerosol optical depth (a measure of the aerosol loading) is shown in Fig. 5.25. The period from about 1997 to 2003 represents a stratospheric aerosol "background" state because it followed a period of about 6 years without volcanic eruptions. The 1982 El Chichón volcanic eruption produced the largest perturbation to the stratospheric sulfate layer observed during the 1980s, and the eruption of Mount Pinatubo in June 1991, which was the largest volcanic eruption of the 20th century, had an even larger effect on stratospheric aerosols. The enhancements in aerosol loadings each year in the Antarctic during the local winter and early spring, which can be seen in Fig. 5.25, are due to the formation of PSCs, as discussed in Section 5.7.2. In fact, nitric acid trihydrate particles, which are the major component of type I PSCs, condense onto the particles in the stratospheric sulfate layer.

Enhancement of the sulfate layer by volcanic eruptions can cause depletions in stratospheric O_3 due to the H_2SO_4 droplets acting to modify the distribution of catalytically active free radicals. For example, the eruption of Mount Pinatubo produced

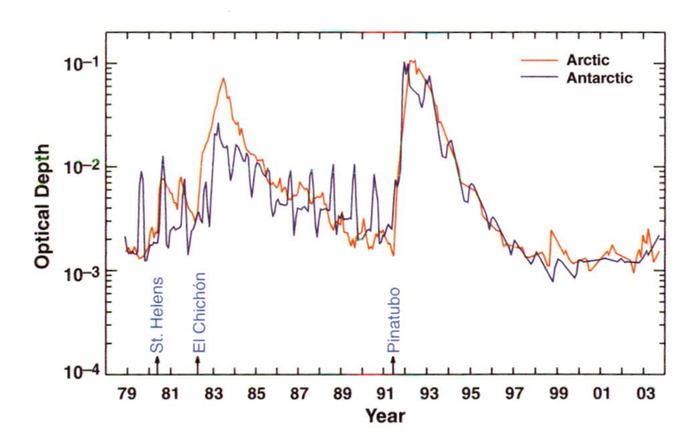

Fig. 5.25 Monthly averaged stratospheric vertical column aerosol optical depths at a wavelength of 1 μm, integrated from 40 km down through 2 km above the tropopause, from 1979 to 2003. "Arctic" data are for measurements north of 65 °N and "Antarctic" data are for measurements south of 65 °S. Some of the major volcanic eruptions during the past 25 years are indicated. The most conspicuous perturbations of optical depth occurred after the eruptions of El Chichón (4 April 1982) and Mt. Pinatubo (15 June 1991). Antarctic data prior to 1991, when measurements from a suitable satellite were available, indicate the presence of polar stratospheric clouds. [Courtesy of M. P. McCormick and NASA.]

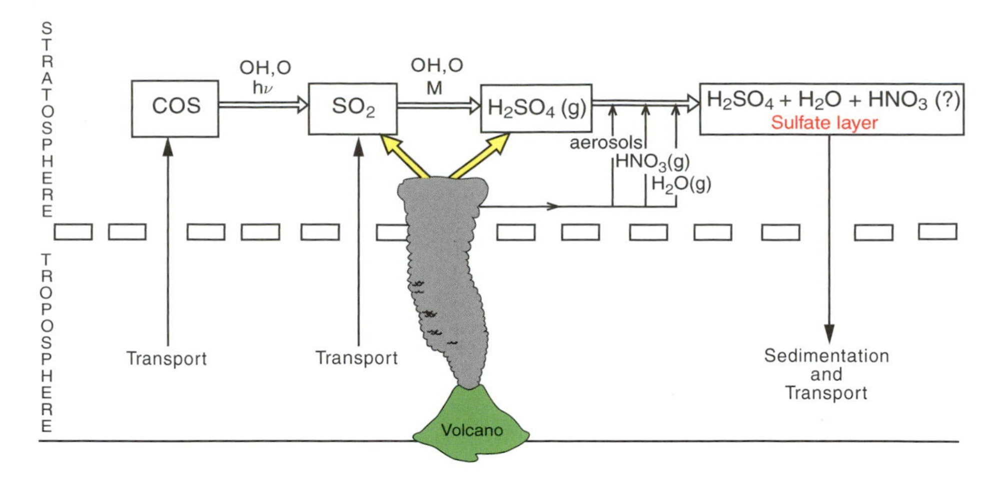

Fig. 5.26 Schematic of the processes responsible for the stratospheric sulfate layer. [Adapted from P. V. Hobbs, *Introduction to Atmospheric Chemistry*, Camb. Univ. Press, 2000, p. 182. Reprinted with the permission of Cambridge University Press.]

dramatic depletions in stratospheric ozone. It is well established that volcanic aerosol reflect shortwave solar radiation and absorb longwave terrestrial radiation. Satellite measurements revealed a 1.4% increase in solar radiation reflected from the atmosphere for several months after the eruption of Mount Pinatubo. The global effects of major volcanic eruptions can persist for up to a few years because stratospheric aerosols are not revomed by wet deposition and their fall speeds are very slow.

When major volcanic activity is low, the primary source of gaseous sulfur compounds that maintain the stratospheric sulfate layer is believed to be the transport of carbonyl sulfide (COS) and SO_2 across the tropopause (Fig. 5.26). COS can be converted to SO_2 by the reaction series

$$COS + h\nu \rightarrow CO + S \tag{5.88a}$$

$$S + O_2 \rightarrow SO + O \tag{5.88b}$$

$$O + COS \rightarrow SO + CO \tag{5.88c}$$

$$SO + O_2 \rightarrow SO_2 + O \tag{5.88d}$$

$$SO + O_3 \rightarrow SO_2 + O_2 \tag{5.88e}$$

$$\text{Net: } 2COS + O_2 + O_3 + h\nu \rightarrow 2CO + 2SO_2 + O \tag{5.89}$$

and also by

$$COS + OH \rightarrow \text{Products} + SO_2 \tag{5.90}$$

The mixing ratio of COS decreases with height in the lower stratosphere (from about 0.4 ppbv at the tropopause to 0.02 ppbv at 30 km), the concentration of SO_2 remains roughly constant (at ~0.05 ppbv), and the concentration of liquid H_2SO_4 peaks at ~20 km. The relationship between these vertical profiles supports the idea that COS is converted to SO_2, which then forms a H_2SO_4 condensate by the mechanisms discussed previously. Numerical modeling results indicate that the direct transfer of SO_2 into the stratosphere from the troposphere via the Brewer–Dobson circulation is also important. The model calculations also show that H_2SO_4 (and O_3) is produced at low latitudes in the stratosphere, with maximum transport toward the poles in winter and spring.

Exercises

5.12 Answer or explain the following in light of the principles discussed in Chapter 5.

(a) Former U.S. President Reagan said "About 93% of our air pollution stems from hydrocarbons released by vegetation, so let's not go overboard in setting and enforcing tough emission standards from man-made sources." Can this statement be justified in any way?

(b) Tropospheric ozone concentrations are generally low over the tropical oceans and high in summer over industrial regions.

(c) The concentration of CO_2 is essentially the same all over the world, but the concentration of H_2S varies considerably from one location to another.

(d) In the shadow of a thick cloud the concentration of OH falls to nearly zero.

(e) Which trace constituents primarily determine the oxidizing capacity of the troposphere in daytime?

(f) Put forward arguments for and against contentions that the atmospheric concentration of OH has increased since the industrial revolution.

(g) When the sun is low in the sky, the sun's rays can be seen when they pass through gaps in a cloud layer. These are called *crepuscular rays* (see Fig. 5.27).

(h) In high altitude landscapes it is difficult to estimate distances accurately.

(i) Particles collected by impaction do not provide an unbiased sample of atmospheric aerosol.

(j) The air above breaking waves along the shoreline is often very hazy and the visibility is low.

(k) Hot objects in a dusty atmosphere are often surrounded by a thin, dust-free space.

(l) The concentrations of aerosol in the plumes from some industrial sources do not decrease as rapidly as predicted by simple diffusion models.

(m) If the rate of removal of a particular toxic chemical from the atmosphere is high, it would be preferable to place a control limit on the amount of the chemical emitted into the atmosphere rather than on the concentration of the chemical in the air.

Fig. 5.27 (a) Crepuscular rays. The rays are parallel to each other (because the sun is so far away) but they appear to converge toward the gap in the cloud. Why? Hint: See (b) below. [Photograph courtesy of UCAR/NCAR.] (b) The parallel lines of tulips appear to converge in the distance. [Photograph courtesy of Art Rangno.]

(n) The residence time of water vapor in middle latitudes is ~5 days but in polar regions it is ~12 days.

(o) Even the cleanest combustion fuels (e.g., hydrogen) are sources of NO_x.

(p) If our eyes detected only UV radiation the world would appear very dark.

(q) Give a qualitative explanation why the Chapman reactions (5.43–5.46) predict a peak concentration of O_3 at some level in the stratosphere.

(r) In the lower stratosphere, the concentration of atomic oxygen decreases rapidly when the sun sets. [**Hint**: Consider the Chapman reactions.]

(s) For catalytic reactions such as (5.54) to be efficient, each reaction comprising the cycle must be exothermic.

(t) Elimination of CFCs, to allow restoration of the stratospheric ozone layer, could lead to higher concentrations of tropospheric pollutants.

(u) Large amounts of chlorine are present in the atmosphere in the form of NaCl, but chlorine from this source is not involved in the destruction of stratospheric ozone.

(v) Very low temperatures develop in the polar vortex in winter.

(w) What chlorine species causes the greatest decrease in O_3 in the Antarctic stratosphere?

(x) How and when is ClO formed in the Antarctic stratosphere?

(y) In the polar vortex, concentrations of HCl are relatively low in the lower stratosphere, but higher up in the stratosphere the concentrations are much greater.

(z) In the stratosphere the concentration of $H_2O(g)$ increases with increasing altitude while the concentration of CH_4 decreases.

(aa) Cooling of the arctic stratosphere over the next decade might cause an "arctic ozone hole," even if Cl levels decrease due to the elimination of CFCs.

5.13 If the burning of wood is represented by the reverse of Eq. (5.2), what change does burning produce in the oxidation number of carbon? Is the carbon atom oxidized or reduced?

5.14 If the concentration of NH_3 in air is 0.456 μg m^{-3} at 0 °C and 1 atm, what is its mixing ratio in ppbv? (Atomic weights of N and H are 14 and 1.01, respectively.)

5.15 Calculate the maximum supersaturation reached in an Aitken nucleus counter if air in the counter, which is initially saturated with respect to water at 15 °C, is suddenly expanded to 1.2 times its initial volume. You may assume that the expansion is adiabatic and use Fig. 3.9 to estimate saturation vapor pressure.

5.16 If the aerosol number distribution is given by (5.31), show that small fluctuations in the value of β about values of 2 and 3 will produce stationary values in the surface and volume distribution plots, respectively. Assume that the aerosol is spherical.

5.17 A particle of mass m and radius r passes horizontally through a small hole in a screen. If the velocity of the particle at the instant ($t = 0$) it passes through the hole is v_o, derive an expression for the horizontal velocity v of the particle at time t. You may assume that the drag force on the particle is given by Eq. (6.23). Use this expression to deduce an expression for the horizontal distance (called the *stop distance*) the particle would travel beyond the hole.

5.18 Ammonia (NH_3), nitrous oxide (N_2O), and methane (CH_4) comprise $1 \times 10^{-8}, 3 \times 10^{-5}$, and 7×10^{-5}% by mass of the Earth's atmosphere, respectively. If the effluxes of these chemicals from the atmosphere are 5×10^{10}, 1×10^{10}, and 4×10^{11} kg per year, respectively, what are the residence times of NH_3, N_2O, and CH_4 in the atmosphere? (Mass of the Earth's atmosphere = 5×10^{18} kg.)

5.19 Assuming that tropospheric O_3 over the continents is confined to a layer of the atmosphere extending from the surface of the Earth up to a height of 5 km and that the average deposition velocity of O_3 onto the ground is 0.40 cm s^{-1}, how long would it take for all the O_3 in the column to be deposited on the ground if all the sources of O_3 were suddenly switched off? How do you reconcile your answer with the residence time for O_3 given in Table 5.1?

5.20 In the troposphere NO reacts with O_3 to produce NO_2 and O_2. Nitric oxide also reacts with the hydroperoxyl (HO_2) radical to produce NO_2 and OH. In turn, NO_2 is photolyzed rapidly to produce NO and atomic oxygen. The atomic oxygen quickly combines with O_2 (when aided by an inert molecule M) to produce O_3.

(a) Write down balanced chemical equations to represent each of these four chemical reactions.

(b) Write down differential equations to represent the time dependencies of the concentrations of NO, O_3, NO_2, HO_2, OH, and O in terms of appropriate constituent concentrations and rate coefficients.

(c) Neglecting hydroxyl-hydroperoxyl chemistry and assuming steady-state conditions, derive an expression for the concentration of O_3 in terms of the concentrations of NO_2 and NO and appropriate rate coefficients.

5.21 Some of the increase in atmospheric CO_2 over, say, the past 50 years may be due to an increase in the average temperature of the oceans, which would cause a decrease in the solubility of CO_2 in the oceans and therefore release CO_2 into the atmosphere. Estimate the percentage change in the CO_2 content of the atmosphere due to an average warming of 0.5 °C in the upper (mixed) layer of the world's oceans over the past 50 years. (Assume that the average temperatures of the mixed layer of all the oceans has increased from 15.0 to 15.5 °C. You may treat the ocean water as pure water.)

Based on your calculation, does it appear likely that the measured increase in atmospheric CO_2 over the past 50 years (~20%) is due to warming of the oceans?

You will need to use the following information. The solubility, C_g, of a gas in a liquid is given by Henry's law:

$$C_g = k_H p_g$$

where k_H is the Henry's law constant and p_g is the partial pressure of the gas over the solution. For CO_2 in pure water, $k_H = 4.5 \times 10^{-2}$ M atm^{-1} at 15 °C. The temperature dependence of k_H is given by

$$\ln \frac{k_H(T_2)}{k_H(T_1)} = \frac{\Delta H}{R^*}\left(\frac{1}{T_1} - \frac{1}{T_2}\right)$$

where for CO_2 in water $\Delta H = -20.4 \times 10^3$ J mol^{-1} and R^* is the universal gas constant (8.31 J K^{-1} mol^{-1}). The total mass of carbon in the form of CO_2 in the mixed layer of the world's oceans is ~6.7×10^5 Tg, which is about the same as the mass of CO_2 in the atmosphere.

Solution:
From Henry's law

$$C_g = k_H p_g$$

For a constant value of p_g

$$\Delta C_g = p_g \Delta k_H \qquad (5.91)$$

Also,

$$\ln \frac{k_H(T_2)}{k_H(T_1)} = \frac{\Delta H}{R^*}\left(\frac{1}{T_1} - \frac{1}{T_2}\right)$$

where,

$$\Delta H = -20.4 \times 10^3 \text{ J mol}^{-1}$$

$$R^* = 8.31 \text{ J deg}^{-1} \text{ mol}^{-1}$$

For $T_1 = 288$ K and $T_2 = 288.5$ K

$$\ln \frac{k_H(288.5)}{k_H(288)} = \frac{-20.4 \times 10^3}{8.31}\left(\frac{1}{288} - \frac{1}{288.5}\right)$$

$$= -2.459\,(3.472 - 3.466)$$

$$= -0.0148$$

Therefore,

$$\frac{k_H(288.5)}{k_H(288)} = 0.985$$

and,

$$k_H(288.5) = 0.985\,(4.5 \times 10^{-2})$$

$$= 4.433 \times 10^{-2} \text{ M atm}^{-1}$$

Therefore,

$$\Delta k_H = (4.433 - 4.5)\,10^{-2}$$

$$= -6.7 \times 10^{-4} \text{ M atm}^{-1} \qquad (5.92)$$

Because CO_2 occupies 373 ppmv of air (see Table 5.1), the partial pressure (p_g) of CO_2 in air is 373×10^{-6} atm. Substituting this value for p_g into (5.89) and using (5.90), we obtain

$$\Delta C_g = -(6.7 \times 10^{-4})\,(380 \times 10^{-6}) \text{ M}$$

$$= -2.5 \times 10^{-7} \text{ M}$$

Therefore, the percentage change in C_g is

$$-100 \frac{2.5 \times 10^{-7} \text{ M}}{(4.5 \times 10^{-2} \text{ M atm}^{-1})\,(373 \times 10^{-6} \text{ atm})}$$

$$= -1.49\%$$

That is, the percentage *decrease* in CO_2 in the mixed layer of the oceans for a change in temperature from 288 to 288.5 K is about 1.5%.

Because the CO_2 capacity of the atmosphere is about the same as for the mixed layer of the oceans, the percentage *increase* in CO_2 in the atmosphere due to 0.5 °C warming of the oceans will also be about 1.5%. Because the measured increase in CO_2 over the past 50 years is ~20%, only about 7.5% of this can be attributed to the warming of the oceans. ■

5.22 This exercise is a follow-on to Exercise 5.8.

(a) In reality, combustion in cars converts most of the hydrogen in the fuel to H_2O and most of the carbon in the fuel to varying amounts of CO_2 and CO depending on the availability of oxygen.

If a fraction f of the C_xH_y fuel is provided in excess of that required for ideal combustion, derive an expression in terms of f, x, and y for the mole fraction of CO in the emissions (i.e., the ratio of the number of moles of CO to the total number of moles in the emissions). Assume that oxygen is made available to the fuel at the rate required for ideal combustion (even though ideal combustion is not achieved) and that the only effect of the excess C_xH_y is to add CO to the emissions and to change the amount of CO_2 emitted.

(b) Assuming that CH_2 is a reasonable approximation for a general hydrocarbon fuel, use the result from (a) to determine the concentrations (in ppmv and percent) of CO in the emissions from an engine for the following values of f: 0.0010, 0.010, and 0.10.

Solution:

(a) If we include the (unreacting) nitrogen in the balanced chemical equation for *complete* combustion, we have from Eq. (5.34) in Chapter 5

$$C_xH_y + \left(x + \frac{y}{4}\right) O_2 + 3.7\left(x + \frac{y}{4}\right) N_2 \rightarrow$$

$$xCO_2 + \frac{y}{2} H_2O + 3.7\left(x + \frac{y}{4}\right) N_2$$

However, if a fraction f of fuel is provided in excess of that needed for complete combustion and, as a consequence, m moles of CO_2 and n moles of CO are contained in the emissions, the chemical equation for combustion becomes

$$(1 + f)\, C_xH_y + \left(x + \frac{y}{4}\right) O_2$$

$$+ 3.7\left(x + \frac{y}{4}\right) N_2 \rightarrow mCO_2 + nCO$$

$$+ (1 + f)\frac{y}{2} H_2O + 3.7\left(x + \frac{y}{4}\right) N_2$$

Balancing the carbon atoms for this reaction yields

$$x(1 + f) = m + n \tag{5.93}$$

and, balancing the oxygen atoms, gives

$$2x + \frac{y}{2} = (1 + f)\frac{y}{2} + 2m + n \tag{5.94}$$

Solving (5.93) and (5.94) for m and n yields

$$m = x - xf - f\frac{y}{2}$$

and,

$$n = f\frac{y}{2} + 2fx$$

Therefore, the mole fraction of CO in the emissions is

$$\frac{f\frac{y}{2} + 2fx}{\underbrace{\left(x - xf - f\frac{y}{2}\right)}_{CO_2} + \underbrace{\left[3.7\left(x + \frac{y}{4}\right)\right]}_{N_2} + \underbrace{\left(f\frac{y}{2} + 2fx\right)}_{CO} + \underbrace{(1 + f)\frac{y}{2}}_{H_2O}}$$

$$= \frac{f\left(2x + \frac{y}{2}\right)}{x\,(4.7 + f) + \frac{y}{2}(2.85 + f)}$$

(b) If the fuel is CH_2, $x = 1$ and $y = 2$. Therefore, from (a) shown earlier, the mole fraction of CO in the emissions is

$$\frac{3f}{7.55 + 2f}$$

Therefore, for $f = 0.001$ the mole fraction of unburned CO is 3.97×10^{-4} or 397 ppmv (=0.0397%). For $f = 0.01$ it is 3.96×10^{-3} or 3960 ppmv (=0.396%). For $f = 0.1$ it is 3.87×10^{-2} or 38700 ppmv (=3.87%). (This last concentration of CO is enough to kill you in a closed garage in about 17 min!) ■

5.23 (a) Write down the rate law for the production of NO_2 by reaction (5.36). Does this rate law explain why the production of NO_2 by (5.36) increases sharply with increasing concentration of NO? (b) Another route for the production of NO_2 from NO is reaction (5.17). If the rate coefficients for the production of NO_2 by (5.36) and (5.17) are $\sim 2 \times 10^{-38}$ cm^6 molecule^{-1} s^{-1} and $\sim 2 \times 10^{-14}$ cm^3 molecule^{-1} s^{-1}, respectively, and the concentrations of NO, O_3, and O_2 are 80 ppbv, 50 ppbv, and 209460 ppmv, respectively, compare the rates of production of NO_2 by these two reactions.

5.24 In the troposphere, the primary sink for CH_4 is

$$CH_4 + OH \rightarrow CH_3 + H_2O$$

The rate coefficient for this reaction at temperatures typical of the troposphere is $\sim 3.5 \times 10^{-15}$ cm^3 molecule^{-1} s^{-1}. If the average 24-h concentration of OH molecules in the atmosphere is 1×10^6 cm^{-3}, what is the residence time of a CH_4 molecule in the atmosphere?

5.25 Propane (C_3H_8) is a nonmethane hydrocarbon (NMHC) that reacts with OH

$$C_3H_8 + OH \rightarrow C_3H_7 + H_2O$$

with a rate coefficient of 6.1×10^{-13} cm^3 molecule^{-1} s^{-1} in the troposphere. (a) Assuming the same OH concentration as in Exercise 5.24, what is the residence time of C_3H_8 in the atmosphere? (b) If the residence times of other NMHC are closer to that of C_3H_8 than to CH_4, would you recommend that more attention be paid to the regulation of CH_4 or to NMHC emissions? (c) Why is CH_4 the only hydrocarbon to enter the stratosphere in appreciable concentrations?

5.26 If the following elementary reactions are responsible for converting ozone into molecular oxygen

$$O_3 \leftrightarrows O_2 + O \qquad \text{(i)}$$

$$O_3 + O \leftrightarrows 2O_2 \qquad \text{(ii)}$$

what is (a) the overall chemical reaction, (b) the intermediate, (c) the rate law for each elementary reaction, (d) the rate-controlling elementary reaction if the rate law for the overall reaction is

$$\text{Rate} = k[O_3]^2\,[O_2]^{-1}$$

where k is a rate coefficient, and (e) on what does [O] depend?

5.27 The rate coefficient for (5.46) is

$$k = 8.0 \times 10^{-12} \exp\left(-\frac{2060}{T}\right) \text{cm}^3 \text{ molecule}^{-1} \text{ s}^{-1}$$

If the temperature decreases from −20 to −30 °C, what would be the percentage change in the rate of removal of O_3 by (5.46)?

5.28 In the middle and upper stratosphere, O_3 concentrations are maintained at roughly steady values by a number of chemical reactions. Assume that at around a temperature of 220 K

$$\frac{dX}{dt} = k_1 - k_2X^2$$

where

$$X = \frac{\text{concentration of } O_3 \text{ molecules}}{\text{concentration of all molecules}}$$

$$k_1 = (\text{constant}) \exp\left(\frac{300}{T}\right) \text{s}^{-1}$$

$$k_2 = 10.0 \exp\left(\frac{-1{,}100}{T}\right) \text{s}^{-1}$$

(a) Doubling the concentration of CO_2 in the atmosphere is predicted to cool the middle stratosphere by about 2 °C. What fractional change in X would you expect from this temperature perturbation?

(b) If X were temporarily raised by 1.0% above its steady-state value of 5.0×10^{-7}, how long would it take for this perturbation to fall to exp(−1) of 1.0% at 220 K? (exp 1 = 2.7)

Solution:

(a)
$$\frac{dX}{dt} = k_1 - k_2X^2 \qquad (5.95)$$

At steady state $\frac{dX_{ss}}{dt} = 0$, where X_{ss} is the fractional concentration of O_3 molecules at steady state. Therefore, from (5.95),

$$X_{ss} = \left(\frac{k_1}{k_2}\right)^{1/2} \qquad (5.96)$$

Since,

$$k_1 = (\text{constant}) \exp\left(\frac{300}{T}\right)$$

and

$$k_2 = 10.0 \exp\left(-\frac{1{,}100}{T}\right)$$

we have

$$X_{ss} = (\text{constant}) \exp\left[\frac{1}{2}\left(\frac{300}{T} + \frac{1100}{T}\right)\right]$$

Therefore,

$$\ln X_{ss} = \text{constant} + \frac{1}{2T}(300 + 1100)$$

Differentiating this last expression

$$\frac{1}{X_{ss}}\frac{dX_{ss}}{dT} = \frac{d}{dT}\left(\frac{700}{T}\right) = -\frac{700}{T^2}$$

or, in incremental form,

$$\frac{\Delta X_{ss}}{X_{ss}} = -\frac{700}{T^2}\Delta T$$

For $T = 220$ K and $\Delta T = -2$ K

$$\frac{\Delta X_{ss}}{X_{ss}} = \frac{1400}{(200)^2} = 0.029 \text{ or } 2.9\%$$

(b) If $Y = X - X_{ss}$ is substituted into Eq. (5.95) in (a) shown earlier,

$$\frac{dY}{dt} = k_1 - k_2(Y + X_{ss})^2 \qquad (5.97)$$

From (5.96) and (5.97)

$$\frac{dY}{dt} = X_{ss}^2\, k_2 - k_2\,(Y + X_{ss})^2$$

$$= -(2k_2X_{ss}Y) + \text{term in } Y^2 \text{ (which is small)}$$

Therefore,

$$\frac{dY}{Y} \simeq -2k_2X_{ss}t$$

and,

$$\int_{Y_o}^{Y}\frac{dY}{Y} \simeq -2k_2X_{ss}\int_o^t dt$$

that is,

$$\ln\frac{Y}{Y_o} \simeq -2k_2X_{ss}t$$

Hence,

$$Y \simeq Y_o \exp(-2k_2X_{ss}t)$$

The time τ required for Y to decline to $\exp(-1)$ of its initial value Y_o is therefore given by

$$\tau \simeq \frac{1}{2k_2X_{ss}}$$

But, at $T = 220$ K,

$$k_2 = 10.0 \exp\left(-\frac{1{,}100}{220}\right) \text{s}^{-1}$$

and for $X_{ss} = 5 \times 10^{-7}$, we have

$$\tau \simeq \frac{1}{2(10 \exp(-5)(5 \times 10^{-7}))} \text{ s}$$

$$\simeq 1.48 \times 10^7 \text{ s}$$

$$\simeq 172 \text{ days}$$

■

5.29 A variation on the catalytic reaction cycle (5.83) is

$$ClO + ClO + M \xrightarrow{k_1} (ClO)_2 + M \qquad \text{(i)}$$

$$(ClO)_2 + h\nu \xrightarrow{j_2} Cl + ClOO \qquad \text{(iia)}$$

$$(ClO)_2 + M \xrightarrow{k_2} 2ClO + M \qquad \text{(iib)}$$

$$ClOO + M \xrightarrow{k_3} Cl + O_2 + M \qquad \text{(iii)}$$

$$2Cl + 2O_3 \xrightarrow{k_4} 2ClO + 2O_2 \qquad \text{(iv)}$$

In this cycle, two possible fates for $(ClO)_2$ are indicated: photolysis to form Cl and ClOO [reaction (iia)], in which case reactions (iii) and (iv) follow, or thermal decomposition to produce ClO [reaction (iib)].

(a) What is the net effect of this cycle if reaction (iia) dominates? What is the net effect if reaction (iib) dominates?

(b) Derive equations for the rate of change of O_3, Cl, $(ClO)_2$, and ClOO, assuming that reaction (iia) dominates and that reaction (iib) can be neglected.

(c) Assume that the concentrations of Cl, $(ClO)_2$, and ClOO are in steady state and, using the results from part (b), derive an expression for the concentration of Cl in terms of the concentrations of ClO, O_3, and M, and the values of k_1 and k_4.

(d) Use the results of parts (b) and (c) to find an expression for the rate of change in the concentration of O_3 in terms of k_1 and the concentrations of ClO and M.

(e) If removal of O_3 by chlorine chemistry becomes a significant part of the stratospheric O_3 budget and total chlorine increases linearly with time, what mathematical form do you expect for the time dependence of the O_3 concentration (e.g., linear in time, square root of time, etc.)?

5.30 In the atmosphere at altitudes near and above 30 km the following reactions significantly affect the chemistry of ozone

$$O_3 + h\nu \xrightarrow{j_1} O_2 + O^* \qquad (j_1 = 1 \times 10^{-4}\ s^{-1})$$

$$O^* + M \xrightarrow{k_1} O + M \qquad (k_1 = 1 \times 10^{-11}\ cm^3\ s^{-1})$$

$$O^* + H_2O \xrightarrow{k_2} OH + OH \qquad (k_2 = 2 \times 10^{-6}\ cm^3\ s^{-1})$$

$$O + O_2 + M \xrightarrow{k_3} O_3 + M \qquad (k_3 = 6 \times 10^{-34}\ cm^6\ s^{-1})$$

$$OH + O_3 \xrightarrow{k_4} HO_2 + O_2 \qquad (k_4 = 2 \times 10^{-14}\ cm^3\ s^{-1})$$

$$HO_2 + O_3 \xrightarrow{k_5} OH + O_2 + O_2 \qquad (k_5 = 3 \times 10^{-16}\ cm^3\ s^{-1})$$

$$OH + HO_2 \xrightarrow{k_6} H_2O + O_2 \qquad (k_6 = 3 \times 10^{-11}\ cm^3\ s^{-1})$$

where O* is an electronically excited metastable state of atomic oxygen. The free radical species OH and HO_2 are collectively labeled "odd hydrogen." At 30 km the molecular density of the atmosphere is about $5 \times 10^{17}\ cm^{-3}$, and the molecular fractions of water vapor and O_3 are each about 2×10^{-6} and that of oxygen is 0.2.

(a) What are the approximate steady-state molecular fractions of O*, HO_2, and OH?

(b) What is the approximate mean residence time of odd hydrogen under steady-state conditions?

(c) For every O_3 molecule that is destroyed under steady-state conditions, how many odd hydrogen species are produced by the reaction associated with k_2?

(**Hint**: The steps associated with k_4 and k_5 occur many times for each formation or loss of odd hydrogen.)

Solution:

(a) From the reactions that are given in this exercise we have

$$\frac{d[O^*]}{dt} = j_1[O_3] - k_1[O^*][M] - k_2[O^*][H_2O] \qquad (5.98)$$

At steady state, $\frac{d[O^*]}{dt} = 0$. Therefore, from (5.98),

$$[O^*] = \frac{j_1[O_3]}{k_1[M] + k_2[H_2O]}$$

since $[M] \simeq 5 \times 10^{17}\ cm^{-3}$ and $[H_2O] = 0.2 \times 5 \times 10^{17}\ cm^{-3}$, $k_1 = 10^{-11}\ cm^3\ s^{-1}$ and $k_2 = 2 \times 10^{-6}\ cm^3\ s^{-1}$, we have

$$\frac{k_1[M]}{k_2[H_2O]} = \frac{(10^{-11})(5 \times 10^{17})}{(2 \times 10^{-6})(2 \times 10^{-6})(5 \times 10^{17})} = 2.5.$$

Therefore,

$$[O^*] \simeq \frac{j_1[O_3]}{k_1[M]}$$

Therefore, the molecular fraction of O* is

$$f(\mathrm{O}^*) \simeq \frac{j_1 f[\mathrm{O}_3]}{k_1 [\mathrm{M}]}$$

where $f(X)$ represents the molecular fraction of species X. Hence,

$$f(\mathrm{O}^*) \simeq \frac{10^{-4}(2\times10^{-6})}{10^{-11}(5\times10^{17})} = 4\times10^{-17}$$

Also,

$$[\text{odd oxygen}] = [\mathrm{HO}] + [\mathrm{HO}_2]$$

Therefore, from the reactions that are given,

$$\frac{d}{dt}[\text{odd oxygen}] = 2k_2[\mathrm{O}^*][\mathrm{H_2O}] - 2k_6[\mathrm{HO}][\mathrm{HO}_2] \quad (5.99)$$

If, as stated, the steps associated with k_4 and k_5 occur frequently compared to the steps associated with k_2 and k_6, they determine the concentrations of HO and HO_2. Therefore, at steady state,

$$k_4[\mathrm{HO}][\mathrm{O}_3] \simeq k_5[\mathrm{HO}_2][\mathrm{O}_3] \quad (5.100)$$

or

$$\frac{[\mathrm{HO}_2]}{[\mathrm{HO}]} \simeq \frac{k_4}{k_5} \simeq \frac{2\times10^{-14}}{3\times10^{-16}} \simeq 70 \quad (5.101)$$

From (5.99) and (5.101), and with

$$\frac{d[\text{odd oxygen}]}{dt} = 0,$$

$$k_2[\mathrm{O}^*][\mathrm{H_2O}] - k_6\frac{k_5}{k_4}[\mathrm{HO}_2]^2 = 0$$

Therefore,

$$\{f(\mathrm{HO}_2)\}^2 = \frac{k_2k_4}{k_5k_6} f(\mathrm{O}^*) f(\mathrm{H_2O})$$

or

$$f(\mathrm{HO}_2) = \left\{\frac{(2\times10^{-6})(2\times10^{-14})}{(3\times10^{-16})(3\times10^{-11})} \times (4\times10^{-17})(2\times10^{-6})\right\}^{1/2} \simeq 2\times10^{-8}$$

Using (5.99),

$$f(\mathrm{HO}) \simeq \frac{f(\mathrm{HO}_2)}{70} \simeq 3\times10^{-10}$$

(b) The mean residence time of odd hydrogen is

$$\tau_{\substack{\text{odd}\\\text{hydrogen}}} = \left(\frac{\text{loss rate of odd hydrogen}}{\text{concentration of odd hydrogen}}\right)^{-1}$$

Since, from (a) shown earlier, $[\mathrm{HO}_2] \simeq 70\,[\mathrm{HO}]$ the concentration of odd hydrogen is approximately equal to the concentration of HO_2. Also, only the reaction associated with $k6$ depletes odd hydrogen, which is given by $2k_6$ [HO] [HO_2]. Therefore,

$$\tau_{\substack{\text{odd}\\\text{hydrogen}}} \simeq \left\{\frac{2k_6[\mathrm{HO}][\mathrm{HO}_2]}{[\mathrm{HO}_2]}\right\}^{-1} \simeq \{2k_6[\mathrm{HO}]\}^{-1} \simeq \{2k_6 f(\mathrm{HO})[\mathrm{M}]\}^{-1}$$

Since, from (a) above, $f(\mathrm{HO}) \simeq 3\times10^{-10}$

$$\tau_{\substack{\text{odd}\\\text{hydrogen}}} \simeq \{2(3\times10^{-11})(3\times10^{-10})(5\times10^{17})\}^{-1} \simeq 111\ \mathrm{s}$$

(c) Let

$$N = \frac{\text{number of odd hydrogen species produced per second by the reaction associated with } k_2}{\text{number of O}_3\text{ molecules destroyed per second}}$$

Because the steps associated with k_4 and k_5 are fast

$$N \simeq \frac{\{[\mathrm{HO}] + [\mathrm{HO}_2]\}/\tau_{\substack{\text{odd}\\\text{hydrogen}}}}{k_4[\mathrm{HO}][\mathrm{O}_3] + k_5[\mathrm{HO}_2][\mathrm{O}_3]}$$

Using (5.100) and (5.101) from (a) given earlier,

$$N \simeq \frac{\{[HO_2]/70 + [HO_2]\}/\tau_{\text{odd hydrogen}}}{2k_5[HO_2][O_3]}$$

or,

$$N \simeq \frac{1}{2k_5[O_3]\,\tau_{\text{odd hydrogen}}}$$

$$\simeq \frac{1}{2k_5\, f(O_3)[M]\,\tau_{\text{odd hydrogen}}}$$

$$\simeq \frac{1}{2(3 \times 10^{-16})(2 \times 10^{-6})(5 \times 10^{17})111}$$

$$N \simeq 15$$ ■

5.31 Suppose the CFC emissions ceased completely in 1996 and chlorine in the stratosphere peaks in 2006 at 5 ppbv. If the CFCs were destroyed by a first-order reaction with a half-life of 35 years, in what year would the chlorine mixing ratio in the stratosphere return to its 1980 mixing ratio of 1.5 ppbv?

Cloud Microphysics

6

Raindrops and snowflakes are among the smallest meteorological entities observable without special equipment. Yet from the perspective of cloud microphysics, the particles commonly encountered in precipitation are quite remarkable precisely because of their *large* sizes. To form raindrops, cloud particles have to increase in mass a million times or more, and these same cloud particles are nucleated by aerosol as small as 0.01 μm. To account for growth through such a wide range of sizes in time periods as short as 10 min or so for some convective clouds, it is necessary to consider a number of physical processes. Scientific investigations of these processes is the domain of *cloud microphysics* studies, which is the main subject of this chapter.

We begin with a discussion of the nucleation of cloud droplets from water vapor and the particles in the air that are involved in this nucleation (Section 6.1). We then consider the microstructures of warm clouds (Sections 6.2 and 6.3) and the mechanisms by which cloud droplets grow to form raindrops (Section 6.4). In Section 6.5 we turn to ice particles in clouds and describe the various ways in which ice particles form and grow into solid precipitation particles.

The artificial modification of clouds and attempts to deliberately modify precipitation are discussed briefly in Section 6.6.

Cloud microphysical processes are thought to be responsible for the electrification of thunderstorms. This subject is discussed in Section 6.7, together with lightning and thunder.

The final section of this chapter is concerned with chemical processes within and around clouds, which play important roles in atmospheric chemistry, including the formation of acid rain.

6.1 Nucleation of Water Vapor Condensation

Clouds form when air becomes *supersaturated*[1] with respect to liquid water (or in some cases with respect to ice). The most common means by which supersaturation is produced in the atmosphere is through the ascent of air parcels, which results in the expansion of the air and adiabatic cooling[2] (see Sections 3.4 and 3.5). Under these conditions, water vapor condenses onto some of the particles in the air to form a cloud of small water droplets or ice particles.[4] This section

[1] If the water vapor pressure in the air is e, the supersaturation (in percent) with respect to liquid water is $(e/e_s - 1)100$, where e_s is the saturation vapor pressure over a plane surface of liquid water at the temperature of the air. A supersaturation with respect to ice may be defined in an analogous way. When the term supersaturation is used without qualification, it refers to supersaturation with respect to liquid water.

[2] The first person to suggest that clouds form by the adiabatic cooling of moist air appears to have been the scientist and poet Erasmus Darwin[3] in 1788.

[3] **Erasmus Darwin** (1731–1802) English freethinker and radical. Anticipated the theory of evolution, expounded by his famous grandson, by suggesting that species modify themselves by adapting to their environment.

[4] It was widely accepted well into the second half of the 19th century that clouds are composed of numerous small bubbles of water! How else can clouds float? Although John Dalton had suggested in 1793 that clouds may consist of water drops that are continually descending relative to the air, it was not until 1850 that James Espy[5] clearly recognized the role of upward-moving air currents in suspending cloud particles.

[5] **James Pollard Espy** (1785–1860) Born in Pennsylvania. Studied law; became a classics teacher at the Franklin Institute. Impressed by the meteorological writings of John Dalton, he gave up teaching to devote his time to meteorology. First to recognize the importance of latent heat release in sustaining cloud and storm circulations. Espy made the first estimates of the dry and saturated adiabatic lapse rates based on experimental data.

is concerned with the formation of water droplets from the condensation of water vapor.

6.1.1 Theory

We consider first the hypothetical problem (as far as the Earth's atmosphere is concerned) of the formation of a pure water droplet by condensation from a supersaturated vapor without the aid of particles in the air (i.e., in perfectly clean air). In this process, which is referred to as *homogeneous nucleation* of condensation, the first stage is the chance collisions of a number of water molecules in the vapor phase to form small embryonic water droplets that are large enough to remain intact.

Let us suppose that a small embryonic water droplet of volume V and surface area A forms from pure supersaturated water vapor at constant temperature and pressure. If μ_l and μ_v are the Gibbs[6] free energies[7] per molecule in the liquid and the vapor phases, respectively, and n is the number of water molecules per unit volume of liquid, the decrease in the Gibbs free energy of the system due to the condensation is $nV(\mu_v - \mu_l)$. Work is done in creating the surface area of the droplet. This work may be written as $A\sigma$, where σ is the work required to create a unit area of vapor–liquid interface (called the *interfacial energy* between the vapor and the liquid, or the *surface energy* of the liquid). The student is invited to show in Exercise 6.9 that the surface energy of a liquid has the same numerical value as its *surface tension*. Let us write

$$\Delta E = A\sigma - nV(\mu_v - \mu_l) \quad (6.1)$$

then ΔE is the net increase in the energy of the system due to the formation of the droplet. It can be shown that[8]

$$\mu_v - \mu_l = kT \ln \frac{e}{e_s} \quad (6.2)$$

where e and T are the vapor pressure and temperature of the system and e_s is the saturation vapor pressure over a plane surface of water at temperature T. Therefore,

$$\Delta E = A\sigma - nVkT \ln \frac{e}{e_s} \quad (6.3)$$

For a droplet of radius R, (6.3) becomes

$$\Delta E = 4\pi R^2 \sigma - \frac{4}{3}\pi R^3 nkT \ln \frac{e}{e_s} \quad (6.4)$$

Under subsaturated conditions, $e < e_s$. In this case, $\ln(e/e_s)$ is negative and ΔE is always positive and increases with increasing R (blue curve in Fig. 6.1). In other words, the larger the embryonic droplet that forms in a subsaturated vapor, the greater the increase in the energy, ΔE, of the system. Because a system approaches an equilibrium state by reducing its energy, the formation of droplets is clearly not favored under subsaturated conditions. Even so, due to random collisions of water molecules, very small embryonic droplets continually form (and evaporate) in a subsaturated vapor, but they do not grow large enough to become visible as a cloud of droplets.

Under supersaturated conditions, $e > e_s$, and $\ln(e/e_s)$ is positive. In this case, ΔE in (6.4) can be either positive or negative depending on the value of R. The variation of ΔE with R for $e > e_s$ is

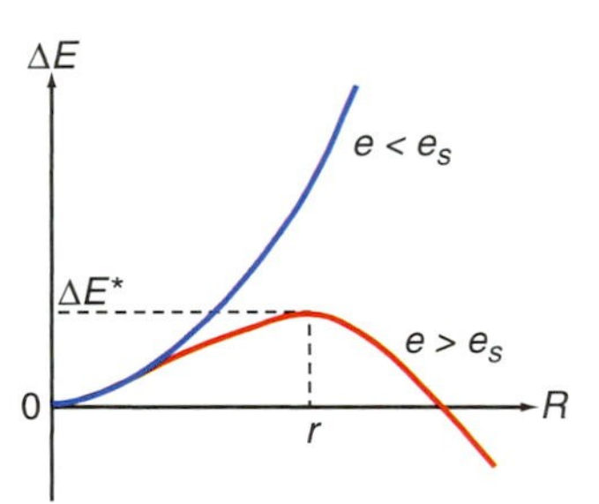

Fig. 6.1 Increase ΔE in the energy of a system due to the formation of a water droplet of radius R from water vapor with pressure e; e_s is the saturation vapor pressure with respect to a plane surface of water at the temperature of the system.

[6] **Josiah Willard Gibbs** (1839–1903) Received his doctorate in engineering from Yale in 1863 for a thesis on gear design. From 1866 to 1869 Gibbs studied mathematics and physics in Europe. Appointed Professor of Mathematical Physics (without salary) at Yale. Subsequently Gibbs used the first and second laws of thermodynamics to deduce the conditions for equilibrium of thermodynamic systems. Also laid the foundations of statistical mechanics and vector analysis.

[7] See Chapter 2 in "Basic Physical Chemistry for the Atmospheric Sciences" by P. V. Hobbs, Camb. Univ. Press, 2000, for a discussion of the Gibbs free energy. For the present purpose, Gibbs free energy can be considered, very loosely, as the microscopic energy of the system.

[8] See Chapter 2 in "Basic Physical Chemistry for the Atmospheric Sciences" *loc. cit.*

also shown in Fig. 6.1 (red curve), where it can be seen that ΔE initially increases with increasing R, reaches a maximum value at $R = r$, and then decreases with increasing R. Hence, under supersaturated conditions, embryonic droplets with $R < r$ tend to evaporate, since by so doing they decrease ΔE. However, droplets that manage to grow by chance collisions to a radius that just exceeds r will continue to grow spontaneously by condensation from the vapor phase, since this will produce a decrease in ΔE. At $R = r$, a droplet can grow or evaporate infinitesimally without any change in ΔE. We can obtain an expression for r in terms of e by setting $d(\Delta E)/dR = 0$ at $R = r$. Hence, from (6.4),

$$r = \frac{2\sigma}{nkT \ln \dfrac{e}{e_s}} \tag{6.5}$$

Equation (6.5) is referred to as *Kelvin's equation*, after Lord Kelvin who first derived it.

We can use (6.5) in two ways. It can be used to calculate the radius r of a droplet that is in (unstable) equilibrium[9] with a given water vapor pressure e. Alternatively, it can be used to determine the saturation vapor pressure e over a droplet of specified radius r. It should be noted that the relative humidity at which a droplet of radius r is in (unstable) equilibrium is $100e/e_s$, where e/e_s is by inverting (6.5). The variation of this relative humidity with droplet radius is shown in Fig. 6.2. It can be seen from Fig. 6.2 that a pure water droplet of radius 0.01 μm requires a relative humidity of ~112% (i.e., a supersaturation of ~12%) to be in (unstable) equilibrium with its environment, while a droplet of radius 1 μm requires a relative humidity of only 100.12% (i.e., a supersaturation of 0.12%).

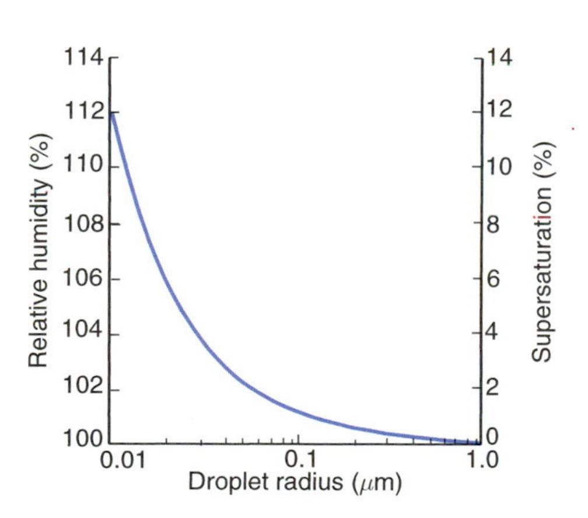

Fig. 6.2 The relative humidity and supersaturation (both with respect to a plane surface of water) at which pure water droplets are in (unstable) equilibrium at 5 °C.

Exercise 6.1 (a) Show that the height of the critical energy barrier ΔE^* in Fig. 6.1 is given by

$$\Delta E^* = \frac{16\pi\sigma^3}{3\left(nkT \ln \dfrac{e}{e_s}\right)^2}$$

(b) Determine the fractional changes in ΔE^* and r if the surface tension, σ, is decreased by 10% by adding sodium laurel sulfate (a common ingredient in soap) to pure water. Neglect the effect of the sodium laurel sulfate on n and e. (c) What effect would the addition of the sodium laurel sulfate have on the homogeneous nucleation of droplets?

Solution: (a) With reference to Fig. 6.1, $\Delta E = \Delta E^*$ when $R = r$. Therefore, from (6.4),

$$\Delta E^* = 4\pi r^2\sigma - \frac{4}{3}\pi r^3 nkT \ln \frac{e}{e_s}$$

Using (6.5)

$$\Delta E^* = 4\pi r^2\sigma - \frac{4}{3}\pi r^2(2\sigma)$$

or

$$\Delta E^* = \frac{4}{3}\pi r^2\sigma$$

Substituting for r from (6.5) into this last expression

$$\Delta E^* = \frac{16\pi\sigma^3}{3\left(nkT \ln \dfrac{e}{e_s}\right)^2}$$

(b) Differentiating the last equation with respect to σ

$$\frac{d(\Delta E^*)}{d\sigma} = \frac{16\pi\sigma^2}{\left(nkT \ln \dfrac{e}{e_s}\right)^2}$$

[9] The equilibrium is unstable in the sense that if the droplet begins to grow by condensation it will continue to do so, and if the droplet begins to evaporate it will continue to evaporate (compare with Fig. B3.4b).

or

$$\frac{d(\Delta E^*)}{\Delta E^*} = 3\frac{d\sigma}{\sigma}$$

If the surface tension of the droplet is decreased by 10%, i.e., if $d\sigma/\sigma = -0.1$, then $d(\Delta E^*)/\Delta E^* = -0.3$ and the critical energy barrier ΔE^* will be decreased by 30%. From (6.5), $dr/r = d\sigma/\sigma$. Therefore, if σ is decreased by 10%, r will also be decreased by 10%.
(c) As shown in Fig. 6.1, r is the critical radius that an embryonic droplet must attain, due to the chance collision of water molecules, if it is to continue to grow spontaneously by condensation. Therefore, for a specified supersaturation of the ambient air, if r is decreased (which it is by the addition of the sodium laurel sulfate), the homogeneous nucleation of droplets will be achieved more readily. ■

Because the supersaturations that develop in natural clouds due to the adiabatic ascent of air rarely exceed a few percent (see Section 6.4.1), it follows from the preceding discussion that even if embryonic droplets of pure water as large as 0.01 μm in radius formed by the chance collision of water molecules, they would be well below the critical radius required for survival in air that is just a few percent supersaturated. Consequently, droplets do not form in natural clouds by the homogeneous nucleation of pure water. Instead they form on atmospheric aerosol[10] by what is known as *heterogeneous nucleation*.[12]

As we have seen in Section 5.4.5, the atmosphere contains many particles that range in size from submicrometer to several tens of micrometers. Those particles that are *wettable*[13] can serve as centers upon which water vapor condenses. Moreover, droplets can form and grow on these particles at much lower supersaturations than those required for homogeneous nucleation. For example, if sufficient water condenses onto a completely wettable particle 0.3 μm in radius to form a thin film of water over the surface of the particle, we see from Fig. 6.2 that the water film will be in (unstable) equilibrium with air that has a supersaturation of 0.4%. If the supersaturation were slightly greater than 0.4%, water would condense onto the film of water and the droplet would increase in size.

Some of the particles in air are soluble in water. Consequently, they dissolve, wholly or in part, when water condenses onto them, so that a solution (rather than a pure water) droplet is formed. Let us now consider the behavior of such a droplet.

The saturation vapor pressure of water adjacent to a solution droplet (i.e., a water droplet containing some dissolved material, such as sodium chloride or ammonium sulfate) is less than that adjacent to a pure water droplet of the same size. The fractional reduction in the water vapor pressure is given by *Raoult's*[14] *law*

$$\frac{e'}{e} = f \tag{6.6}$$

where e' is the saturation vapor pressure of water adjacent to a solution droplet containing a mole fraction f of pure water and e is the saturation vapor pressure of water adjacent to a pure water droplet of the same size and at the same temperature. The *mole fraction of pure water* is defined as the number of moles of pure water in the solution divided by the total number of moles (pure water plus dissolved material) in the solution.

Consider a solution droplet of radius r that contains a mass m (in kg) of a dissolved material of molecular

[10] That aerosol plays a role in the condensation of water was first clearly demonstrated by Coulier[11] in 1875. His results were rediscovered independently by Aitken in 1881 (see Section 5.4.5).

[11] **Paul Coulier** (1824–1890) French physician and chemist. Carried out research on hygiene, nutrition, and the ventilation of buildings.

[12] Cloud physicists use the terms homogeneous and heterogeneous differently than chemists. In chemistry, a homogeneous system is one in which all the species are in the same phase (solid, liquid or gas), whereas a heterogeneous system is one in which species are present in more than one phase. In cloud physics, a homogeneous system is one involving just one species (in one or more phases), whereas a heterogeneous system is one in which there is more than one species. In this chapter these two terms are used in the cloud physicist's sense.

[13] A surface is said to be perfectly wettable (hydrophilic) if it allows water to spread out on it as a horizontal film (detergents are used for this purpose). A surface is completely unwettable (hydrophobic) if water forms spherical drops on its surface (cars are waxed to make them hydrophobic).

[14] **François Marie Raoult** (1830–1901) Leading French experimental physical chemist of the 19th century. Professor of chemistry at Grenoble. His labors were met with ample, though tardy recognition (Commander of de la Legion of d' Honneur; Davy medallist of the Royal Society, etc.).

weight M_s. If each molecule of the material dissociates into i ions in water, the effective number of moles of the material in the droplet is $i(1000\,m)/M_s$. If the density of the solution is ρ' and the molecular weight of water M_w, the number of moles of pure water in the droplet is $1000\,(\frac{4}{3}\pi r^3 \rho' - m)/M_w$. Therefore, the mole fraction of water in the droplet is

$$f = \frac{(\frac{4}{3}\pi r^3 \rho' - m)/M_w}{[(\frac{4}{3}\pi r^3 \rho' - m)/M_w] + im/M_s} = \left[1 + \frac{imM_w}{M_s(\frac{4}{3}\pi r^3 \rho' - m)}\right]^{-1} \tag{6.7}$$

Combining (6.5)–(6.7) (but replacing σ and n by σ' and n' to indicate the surface energy and number concentration of water molecules, respectively, for the solution) we obtain the following expression for the saturation vapor pressure e' adjacent to a solution droplet of radius r

$$\frac{e'}{e_s} = \left[\exp \frac{2\sigma'}{n'kTr}\right]\left[1 + \frac{imM_w}{M_s(\frac{4}{3}\pi r^3 \rho' - m)}\right]^{-1} \tag{6.8}$$

Equation (6.8) may be used to calculate the saturation vapor pressure e' [or relative humidity $100e'/e_s$, or supersaturation $\left(\frac{e'}{e_s} - 1\right) 100$] adjacent to a solution droplet with a specified radius r. If we plot the variation of the relative humidity (or supersaturation) adjacent to a solution droplet as a function of its radius, we obtain what is referred to as a *Köhler*[15] *curve*. Several such curves, derived from (6.8), are shown in Fig. 6.3. Below a certain droplet radius, the relative humidity adjacent to a solution droplet is less than that which is in equilibrium with a plane surface of pure water at the same temperature (i.e., 100%). As the droplet increases in size, the solution becomes weaker, the Kelvin curvature effect becomes the dominant influence, and eventually the relative humidity of the air adjacent to the droplet becomes essentially the same as that adjacent to a pure water droplet of the same size.

To illustrate further the interpretation of the Köhler curves, we reproduce in Fig. 6.4 the Köhler curves for solution droplets containing 10^{-19} kg of NaCl (the red curve from Fig. 6.3) and 10^{-19} kg of $(NH_4)_2SO_4$ (the green curve from Fig. 6.3). Suppose that a particle of NaCl with mass 10^{-19} kg were placed in air with a water supersaturation of 0.4% (indicated by the dashed line in Fig. 6.4). Condensation would occur on this particle to form a solution droplet, and the droplet would grow along the red curve in Fig. 6.4. As it does so, the supersaturation adjacent to the surface of this solution droplet will initially increase, but even at the peak in its Köhler curve the supersaturation adjacent to the droplet is less than the ambient supersaturation. Consequently, the droplet will grow over the peak in

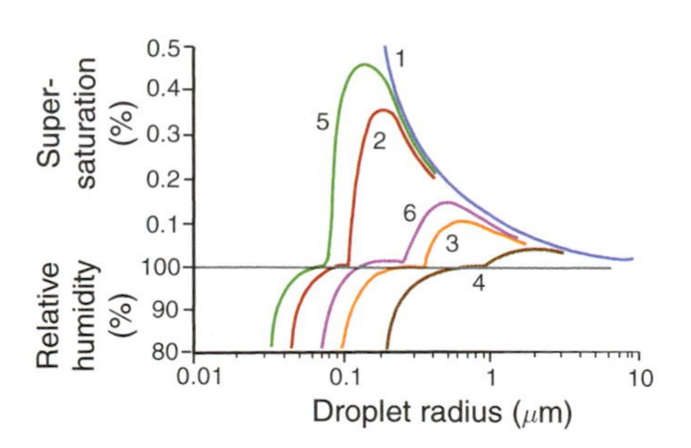

Fig. 6.3 Variations of the relative humidity and supersaturation adjacent to droplets of (1) pure water (blue) and adjacent to solution droplets containing the following fixed masses of salt: (2) 10^{-19} kg of NaCl, (3) 10^{-18} kg of NaCl, (4) 10^{-17} kg of NaCl, (5) 10^{-19} kg of $(NH_4)_2SO_4$, and (6) 10^{-18} kg of $(NH_4)_2SO_4$. Note the discontinuity in the ordinate at 100% relative humidity. [Adapted from H. R. Pruppacher, "The role of natural and anthropogenic pollutants in cloud and precipitation formation," in S. I. Rasool, ed., *Chemistry of the Lower Atmosphere*, Plenum Press, New York, 1973, Fig. 5, p. 16, copyright 1973, with kind permission of Springer Science and Business Media.]

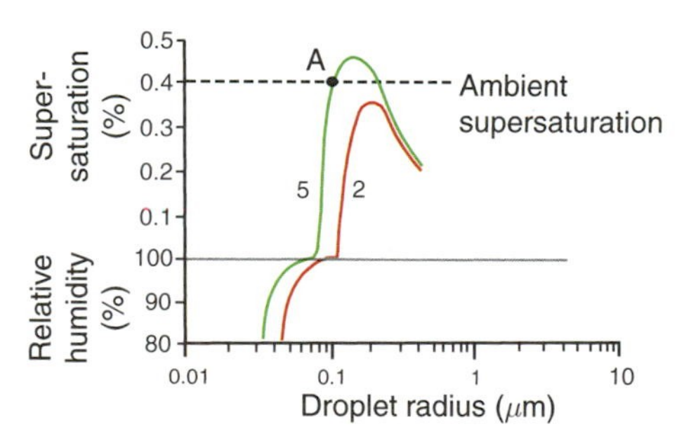

Fig. 6.4 Köhler curves 2 and 5 from Fig. 6.3. Curve 2 is for a solution droplet containing 10^{-19} kg of NaCl, and curve 5 is for a solution droplet containing 10^{-19} kg of $(NH_4)_2SO_4$. The dashed line is an assumed ambient supersaturation discussed in the text.

[15] **Hilding Köhler** (1888–1982) Swedish meteorologist. Former Chair of the Meteorology Department and Director of the Meteorological Observatory, University of Uppsala.

its Köhler curve and down the right-hand side of this curve to form a fog or cloud droplet. A droplet that has passed over the peak in its Köhler curve and continues to grow is said to be *activated*.

Now consider a particle of $(NH_4)_2SO_4$ with mass 10^{-19} kg that is placed in the same ambient supersaturation of 0.4%. In this case, condensation will occur on the particle and it will grow as a solution droplet along its Köhler curve (the green curve in Fig. 6.4) until it reaches point A. At point A the supersaturation adjacent to the droplet is equal to the ambient supersaturation. If the droplet at A should grow slightly, the supersaturation adjacent to it would increase above the ambient supersaturation, and therefore the droplet would evaporate back to point A. If the droplet at A should evaporate slightly, the supersaturation adjacent to it would decrease below the ambient supersaturation, and the droplet would grow by condensation back to A in Fig. 6.4. Hence, in this case, the solution droplet at A is in stable equilibrium with the ambient supersaturation. If the ambient supersaturation were to change a little, the location of A in Fig. 6.4 would shift and the equilibrium size of the droplet would change accordingly. Droplets in this state are said to be *unactivated* or *haze droplets*. Haze droplets in the atmosphere can considerably reduce visibility by scattering light.

6.1.2 Cloud Condensation Nuclei

Section 5.4 discussed atmospheric aerosol. A small subset of the atmospheric aerosol serves as particles upon which water vapor condenses to form droplets that are activated and grow by condensation to form cloud droplets at the supersaturations achieved in clouds (~0.1–1%). These particles are called *cloud condensation nuclei* (CCN). It follows from the discussion in Section 6.1.1 that the larger the size of a particle, the more readily it is wetted by water, and the greater its solubility, the lower will be the supersaturation at which the particle can serve as a CCN. For example, to serve as a CCN at 1% supersaturation, completely wettable but water-insoluble particles need to be at least ~0.1 μm in radius, whereas soluble particles can serve as CCN at 1% supersaturation even if they are as small as ~0.01 μm in radius. Most CCN consist of a mixture of soluble and insoluble components (called *mixed nuclei*).

The concentrations of CCN active at various supersaturations can be measured with a *thermal diffusion chamber*. This device consists of a flat chamber in which the upper and lower horizontal plates are kept wet and maintained at different temperatures, with the lower plate being several degrees colder than the upper plate. By varying the temperature difference between the plates, it is possible to produce maximum supersaturations in the chamber that range from a few tenths of 1% to a few percent (see Exercise 6.14), which are similar to the supersaturations that activate droplets in clouds. Small water droplets form on those particles that act as CCN at the peak supersaturation in the chamber. The concentration of these droplets can be determined by photographing a known volume of the cloud and counting the number of droplets visible in the photograph or by measuring the intensity of light scattered from the droplets. By repeating the aforementioned procedure with different temperature gradients in the chamber, the concentrations of CCN in the air at several supersaturations (called the *CCN supersaturation spectrum*) can be determined.

Worldwide measurements of CCN concentrations have not revealed any systematic latitudinal or seasonal variations. However, near the Earth's surface, continental air masses generally contain larger concentrations of CCN than marine air masses (Fig. 6.5). For example, the concentration of CCN in the continental air mass over the Azores, depicted in Fig. 6.5,

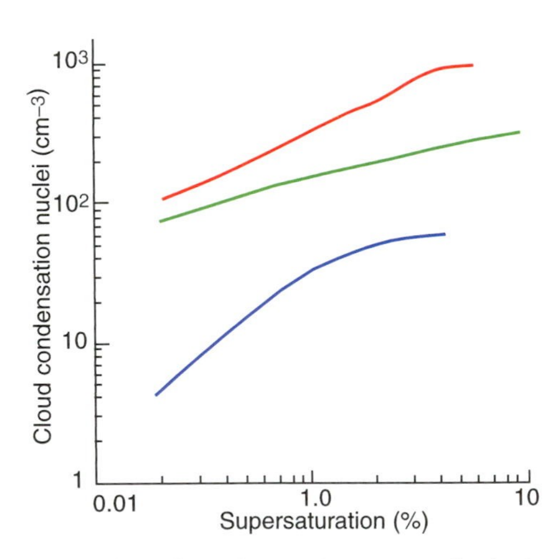

Fig. 6.5 Cloud condensation nucleus spectra in the boundary layer from measurements near the Azores in a polluted continental air mass (brown line), in Florida in a marine air mass (green line), and in clean air in the Arctic (blue line). [Data from J. G. Hudson and S. S. Yun, "Cloud condensation nuclei spectra and polluted and clean clouds over the Indian Ocean," *J. Geophys. Res.* **107**(D19), 8022, doi:10.1029/2001JD000829, 2002. Copyright 2002 American Geophysical Union. Reproduced/modified by permission of American Geophysical Union.]

is about ~300 cm^{-3} at 1% supersaturation, in the marine air mass over Florida it is ~100 cm^{-3}, and in clean Arctic air it is only ~30 cm^{-3}. The ratio of CCN (at 1% supersaturation) to the total number of particles in the air (CN) is ~0.2–0.6 in marine air; in continental air this ratio is generally less than ~0.01 but can rise to ~0.1. The very low ratio of CCN to CN in continental air is attributable to the large number of very small particles, which are not activated at low supersaturations. Concentrations of CCN over land decline by about a factor of five between the planetary boundary layer and the free troposphere. Over the same height interval, concentrations of CCN over the ocean remain fairly constant or may even increase with height, reaching a maximum concentration just above the mean cloud height. Ground-based measurements indicate that there is a diurnal variation in CCN concentrations, with a minimum at about 6 a.m. and a maximum at about 6 p.m.

The observations just described provide clues as to the origins of CCN. First of all it appears that the land acts as a source of CCN because the concentrations of CCN are higher over land and decrease with altitude. Some of the soil particles and dusts that enter the atmosphere probably serve as CCN, but they do not appear to be a dominant source. The rate of production of CCN from burning vegetable matter is on the order of $10^{12}-10^{15}$ per kg of material consumed. Thus, forest fires are a source of CCN. About 80% of the particles emitted by idling diesel engines are CCN at 1% supersaturation. About 70% of the particles emitted by the 1991 Kuwait oil fires were CCN at 1% supersaturation. Although sea-salt particles enter the air over the oceans by the mechanisms discussed in Section 5.4.1, they do not appear to be a dominant source of CCN, even over the oceans.

There appears to be a widespread and probably a fairly uniform source of CCN over both the oceans and the land, the nature of which has not been definitely established. A likely candidate is gas-to-particle conversion, which can produce particles up to a few tenths of a micrometer in diameter that can act as CCN if they are soluble or wettable. Gas-to-particle conversion mechanisms that require solar radiation might be responsible for the observed peak in CCN concentrations at ~6 p.m. Many CCN consist of sulfates. Over the oceans, organic sulfur from the ocean [in the form of the gases dimethyl sulfide (DMS) and methane sulfonic acid (MSA)] provides a source of CCN, with the DMS and MSA being converted to sulfate in the atmosphere. Evaporating clouds also release sulfate particles (see Section 6.8.9).

6.2 Microstructures of Warm Clouds

Clouds that lie completely below the 0 °C isotherm, referred to as *warm clouds*, contain only water droplets. Therefore, in describing the microstructure of warm clouds, we are interested in the amount of liquid water per unit volume of air (called the *liquid water content* (LWC), usually expressed in grams per cubic meter[16]), the total number of water droplets per unit volume of air (called the *cloud droplet concentration*, usually expressed as a number per cubic centimeter), and the size distribution of cloud droplets (called the *droplet size spectrum*, usually displayed as a histogram of the number of droplets per cubic centimeter in various droplet size intervals). These three parameters are not independent; for example, if the droplet spectrum is known, the droplet concentration and LWC can be derived.

In principle, the most direct way of determining the microstructure of a warm cloud is to collect all the droplets in a measured volume of the cloud and then to size and count them under a microscope. In the early days of cloud measurements, oil-coated slides were exposed from an aircraft to cloudy air along a measured path length. Droplets that collided with a slide, and became completely immersed in the oil, were preserved for subsequent analysis. An alternative method was to obtain replicas of the droplets by coating a slide with magnesium oxide powder (obtained by burning a magnesium ribbon near the slide). When water droplets collide with these slides they leave clear imprints, the sizes of which can be related to the actual sizes of the droplets. Direct impaction methods, of the type described earlier, bias against smaller droplets, which tend to follow the streamlines around the slide and thereby avoid

[16] Bearing in mind that the densily of air is approximately 1 kg m^{-3}, a LWC of 1 g m^{-3} expressed as a mixing ratio is approximately the same as 1 g kg^{-3}.

capture. Consequently, corrections have to be made based on theoretical calculations of droplet trajectories around the slide.

Automatic techniques are now available for sizing cloud droplets from an aircraft without collecting the droplets (e.g., by measuring the angular distribution of light scattered from individual cloud drops). These techniques are free from the collection problems described earlier and permit a cloud to be sampled continuously so that variations in cloud microstructures in space and time can be investigated more readily.

Several techniques are available for measuring the LWC of clouds from an aircraft. A common instrument is a device in which an electrically heated wire is exposed to the airstream. When cloud droplets impinge on the wire, they are evaporated and therefore tend to cool and lower the electrical resistance of the wire. The resistance of the wire is used in an electrical feedback loop to maintain the temperature of the wire constant. The power required to do this can be calibrated to give the LWC. Another more recently developed instrument uses light scattering from an ensemble of drops to derive LWC.

The optical thickness (τ_c) and effective particle radius (r_e) of a liquid water or an ice cloud can be derived from satellites or airborne solar spectral reflectance measurements. Such retrievals exploit the spectral variation of bulk water absorption (liquid or ice) in atmospheric window regions. Condensed water is essentially transparent in the visible and near-infrared portions of the spectrum (e.g., 0.4–1.0 μm) and therefore cloud reflectance is dependent only on τ_c and the particle phase function (or the asymmetry parameter, g, a cosine weighting of the phase function that is pertinent to multiple scattering problems). However, water is weakly absorbing in the shortwave and midwave infrared windows (1.6, 2.1, and 3.7 μm bands) with an order of magnitude increase in absorption in each longer wavelength window. Therefore, in these spectral bands, cloud reflectance is also dependent on particle absorption, which is described by the single scattering albedo (ω_o). Specifically, r_e is the relevant radiative measure of the size distribution and is approximately linearly related to ω_o for weak absorption. Therefore, a reflectance measurement in an absorbing band contains information about r_e. Retrieval algorithms use a radiative transfer model to predict the reflectance in transparent and absorbing sensor bands as a function of τ_c and r_e, including specifications of relevant non-cloud parameters (e.g., absorbing atmospheric gases, surface boundary conditions). The unknown cloud optical parameters τ_c and r_e are then adjusted until the differences between predicted and observed reflectances are minimized. The liquid water path, LWP (i.e., the mass of cloud liquid water in a vertical column with a cross-sectional area of 1 m^2), is approximately proportional to the product of τ_c and r_e (see Exercise 6.16) and is often reported as part of the retrieval output.

Shown in Fig. 6.6 are measurements of the vertical velocity of the air, the LWC, and droplet size spectra in a small cumulus cloud. The cloud itself is primarily a region of updrafts, with downdrafts just outside its boundary. Regions of higher LWC correspond quite closely to regions of stronger updrafts, which are, of course, the driving force for the formation of clouds (see Section 3.5). It can be seen from the LWC measurements that the cloud was very inhomogeneous, containing pockets of relatively high LWC interspersed with regions of virtually no liquid water (like Swiss cheese). The droplet spectrum measurements depicted in Fig. 6.6c show droplets ranging from a few micrometers up to about 17 μm in radius.

Cloud LWC typically increases with height above cloud base, reaches a maximum somewhere in the upper half of a cloud, and then decreases toward cloud top.

To demonstrate the profound effects that CCN can have on the concentrations and size distributions of cloud droplets, Fig. 6.7 shows measurements in cumulus clouds in marine and continental air masses. Most of the marine clouds have droplet concentrations less than 100 cm^{-3}, and none has a droplet concentration greater than 200 cm^{-3} (Fig. 6.7a). In contrast, some of the continental cumulus clouds have droplet concentrations in excess of 900 cm^{-3}, and most have concentrations of a few hundred per cubic centimeter (Fig. 6.7c). These differences reflect the much higher concentrations of CCN present in continental air (see Section 6.1.2 and Fig. 6.5). Since the LWC of marine cumulus clouds do not differ significantly from those of continental cumulus, the higher droplet concentrations in the continental cumulus must result in smaller average droplet sizes in continental clouds than in marine clouds. By comparing the results shown in Figs. 6.7b and 6.7d, it can

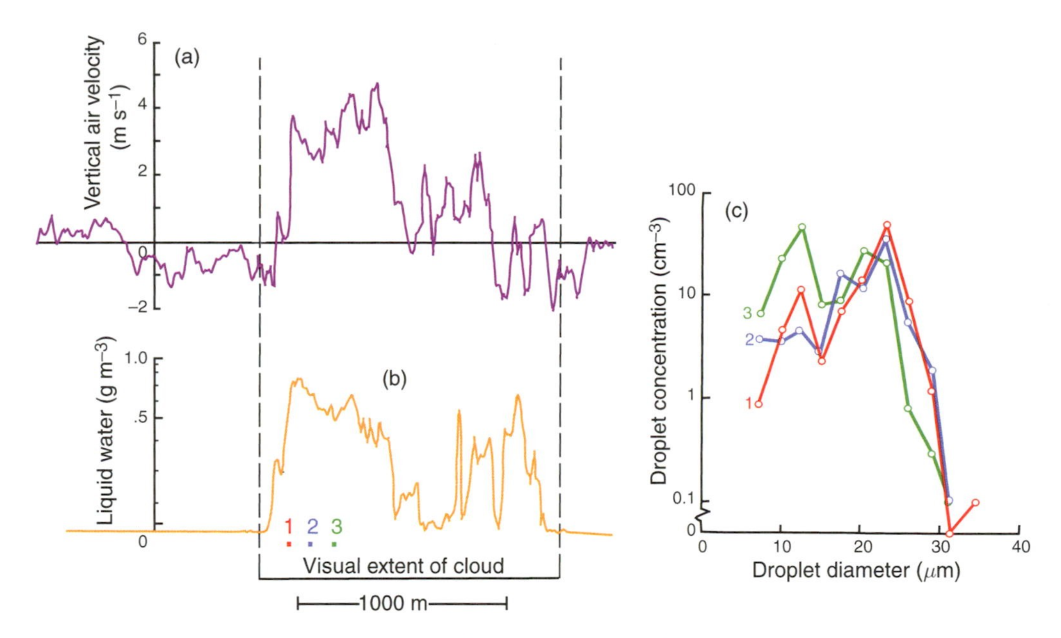

Fig. 6.6 (a) Vertical air velocity (with positive values indicating updrafts and negative values downdrafts), (b) liquid water content, and (c) droplet size spectra at points 1, 2, and 3 in (b), measured from an aircraft as it flew in a horizontal track across the width and about half-way between the cloud base and cloud top in a small, warm, nonraining cumulus cloud. The cloud was about 2 km deep. [Adapted from *J. Atmos. Sci.* **26**, 1053 (1969).]

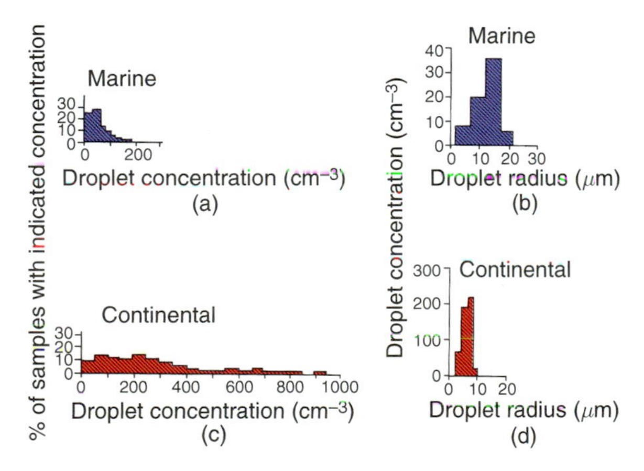

Fig. 6.7 (a) Percentage of marine cumulus clouds with indicated droplet concentrations. (b) Droplet size distributions in a marine cumulus cloud. (c) Percentage of continental cumulus clouds with indicated droplet concentrations. (d) Droplet size distributions in a continental cumulus cloud. Note change in ordinate from (b). [Adapted from P. Squires, "The microstructure and colloidal stability of warm clouds. Part I—The relation between structure and stability," *Tellus* **10**, 258 (1958). Permission from Blackwell Publishing Ltd.]

be seen that not only is the droplet size spectrum for the continental cumulus cloud much narrower than that for the marine cumulus cloud, but the average droplet radius is significantly smaller. To describe this distinction in another way, droplets with a radius of about 20 μm exist in concentrations of a few per cubic centimeter in the marine clouds, whereas in continental clouds the radius has to be lowered to 10 μm before there are droplets in concentrations of a few per cubic centimeter. The generally smaller droplets in continental clouds result in the boundaries of these clouds being well defined because the droplets evaporate quickly in the nonsaturated ambient air. The absence of droplets much beyond the main boundary of continental cumulus clouds gives them a "harder" appearance than maritime clouds.[17] We will see in Section 6.4.2 that the larger droplets in marine clouds lead to the release of precipitation in shallower clouds, and with smaller updrafts, than in continental clouds.

Shown in Fig. 6.8 are retrievals from satellite measurements of cloud optical thickness (τ_c) and cloud droplet effective radius (r_e) for low-level water clouds over the globe. It can be seen that the r_e values are generally smaller over the land than over the oceans, in agreement with the aforementioned discussion.

[17] The formation of ice particles in clouds also affects their appearance (see Section 6.5.3).

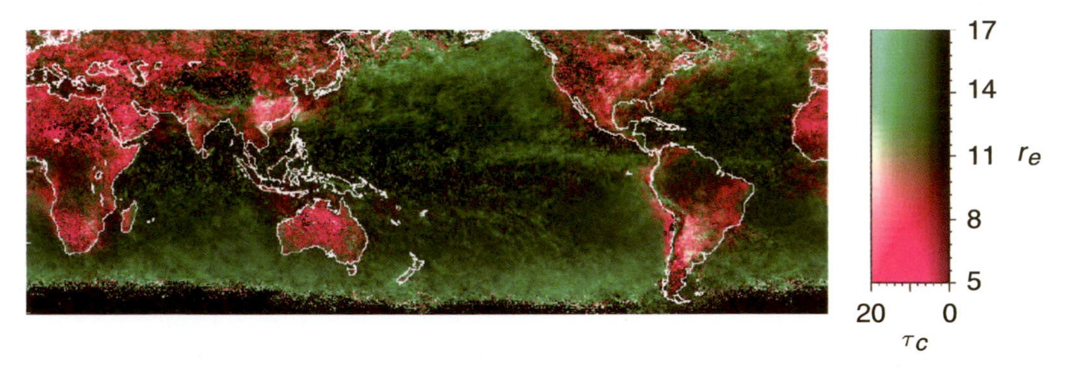

Fig. 6.8 Retrievals from a satellite of cloud optical thickness (τ_c) and cloud particle effective radius (r_e in μm) for low-level water clouds. [From T. Nakajima et al., "A possible correlation between satellite-derived cloud and aerosol microphysical parameters," *Geophys. Res. Lett.* **28**, 1172 (2001). Copyright 2001 American Geophysical Union. Reprinted/modified by permission of American Geophysical Union.]

6.1 Ship Tracks in Clouds

The effects of CCN on increasing the number concentration of cloud droplets is demonstrated dramatically by the longevity of ship tracks, as illustrated in Fig. 6.9. As we have seen, under natural conditions marine air contains relatively few CCN, which is reflected in the low concentrations of small droplets in marine clouds. Ships emit large numbers of CCN ($\sim 10^{15}$ s^{-1} at 0.2% supersaturation), and when these particles are carried up into the bases of marine stratus clouds, they increase the number concentration and decrease the average size of the cloud droplets. The greater concentrations of droplets in these regions cause more sunlight to be reflected back to space so they appear as white lines in the cloud when viewed from a satellite.

Fig. 6.9 Ship tracks (white lines) in marine stratus clouds over the Atlantic Ocean as viewed from the NASA Aqua satellite on January 27, 2003. Brittany and the southwest coast of England can be seen on the upper right side of the image.

6.3 Cloud Liquid Water Content and Entrainment

Exercise 3.10 explained how the skew $T - \ln p$ chart can be used to determine the quantity of liquid water that is condensed when a parcel of air is lifted above its lifting condensation level (LCL). Since the skew $T - \ln p$ chart is based on adiabatic assumptions for air parcels (see Section 3.4.1), the LWC derived in this manner is called the *adiabatic liquid water content.*

Shown in Figs. 6.10 and 6.11 are measurements of LWC in cumulus clouds. The measured LWC are well below the adiabatic LWC because unsaturated ambient air is *entrained* into cumulus clouds. Consequently, some of the cloud water evaporates to saturate the entrained parcels of air, thereby reducing the cloud LWC.

Measurements in and around small cumulus clouds suggest that entrainment occurs primarily at their tops, as shown schematically in Fig. 6.12. Some field measurements have suggested the presence of adiabatic cores deep within cumulus clouds, where the cloud water is undiluted by entrainment. However, more recent measurements, utilizing very fast response

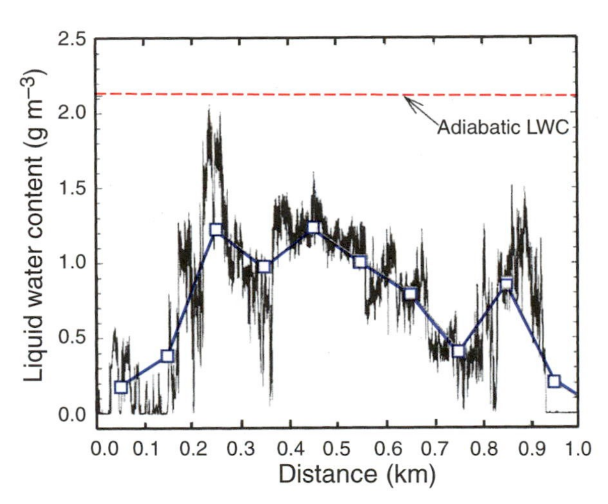

Fig. 6.10 High-resolution liquid water content (LWC) measurements (black line) derived from a horizontal pass through a small cumulus cloud. Note that a small portion of the cumulus cloud had nearly an adiabatic LWC. This feature disappears when the data are smoothed (blue line) to mimic the much lower sampling rates that were prevalent in older measurements. [Adapted from *Proc. 13th Intern. Conf. on Clouds and Precipitation*, Reno, NV, 2000, p. 105.]

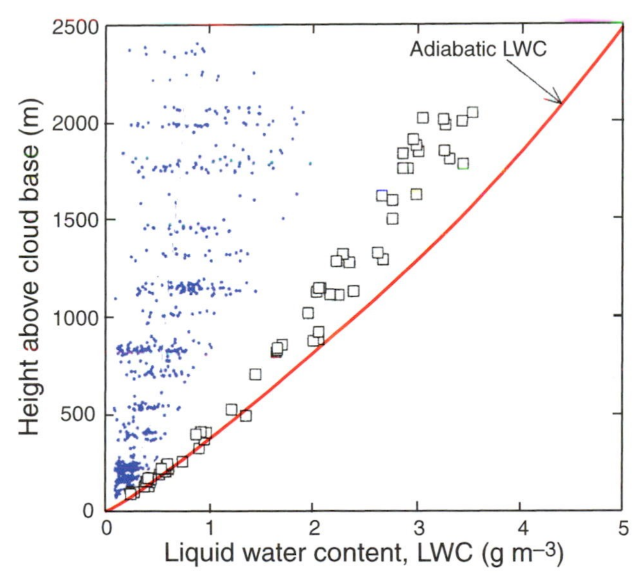

Fig. 6.11 Blue dots are average liquid water contents (LWC) measured in traverses of 802 cumulus clouds. Squares are the largest measured LWC. Note that no adiabatic LWC was measured beyond ~900 m above the cloud base. Cloud base temperatures varied little for all flights, which permitted this summary to be constructed with a cloud base normalized to a height of 0 m. [Adapted from *Proc. 13th Intern. Conf. on Clouds and Precipitation*, Reno, NV, 2000, p. 106.]

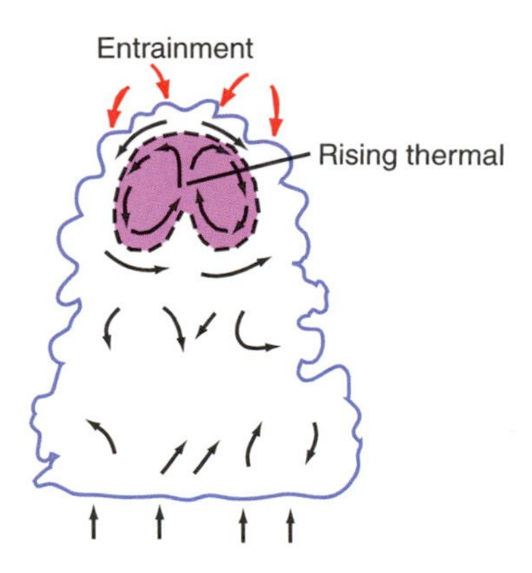

Fig. 6.12 Schematic of entrainment of ambient air into a small cumulus cloud. The thermal (shaded violet region) has ascended from cloud base. [Adapted from *J. Atmos. Sci.* **45**, 3957 (1988).]

instruments that can reveal the fine structures of clouds (Figs. 6.10 and 6.11), indicate that adiabatic cores, if they exist at all, must be quite rare.

Air entrained at the top of a cloud is distributed to lower levels as follows. When cloud water is evaporated to saturate an entrained parcel of air, the parcel is cooled. If sufficient evaporation occurs before the parcel loses its identity by mixing, the parcel will sink, mixing with more cloudy air as it does so. The sinking parcel will descend until it runs out of negative buoyancy or loses its identity. Such parcels can descend several kilometers in a cloud, even in the presence of substantial updrafts, in which case they are referred to as *penetrative downdrafts*. This process is responsible in part for the "Swiss cheese" distribution of LWC in cumulus clouds (see Fig. 6.6). Patchiness in the distribution of LWC in a cloud will tend to broaden the droplet size distribution, since droplets will evaporate partially or completely in downdrafts and grow again when they enter updrafts.

Over large areas of the oceans stratocumulus clouds often form just below a strong temperature inversion at a height of ~0.5–1.5 km, which marks the top of the marine boundary layer. The tops of the stratocumulus clouds are cooled by longwave radiation to space, and their bases are warmed by longwave radiation from the surface. This differential heating drives shallow convection in which cold cloudy air sinks and droplets within it tend to evaporate, while the warm cloudy air rises and the droplets within it tend to grow. These motions are responsible in part for the cellular appearance of stratocumulus clouds (Fig. 6.13).

Fig. 6.13 Looking down on stratocumulus clouds over the Bristol Channel, England. [Photograph courtesy of R. Wood/ The Met Office.]

Entrainment of warm, dry air from the free troposphere into the cool, moist boundary layer air below plays an important role in the marine stratocumulus-topped boundary layer. The rate at which this entrainment occurs increases with the vigor of the boundary layer turbulence, but it is hindered by the stability associated with the temperature inversion. Figure 6.14, based on model simulations, indicates how a parcel of air from the free troposphere might become engulfed into the stratocumulus-topped boundary layer. As in the case of cumulus clouds, following such engulfment, cooling of entrained air parcels by the evaporation of cloud water will tend to drive the parcel downward.

Exercise 6.2 Derive an expression for the fractional change $d\theta'/\theta'$ in the potential temperature θ' of a parcel of cloudy air produced by a fractional change in the mass m of the parcel due to the entrainment of mass dm of unsaturated ambient air.

Solution: Let the parcel of cloudy air have temperature T', pressure p', and volume V', and the ambient air have temperature T and mixing ratio w. The heat dQ_1, needed to warm the entrained air of mass dm, is

$$dQ_1 = c_p(T' - T)\,dm \tag{6.9}$$

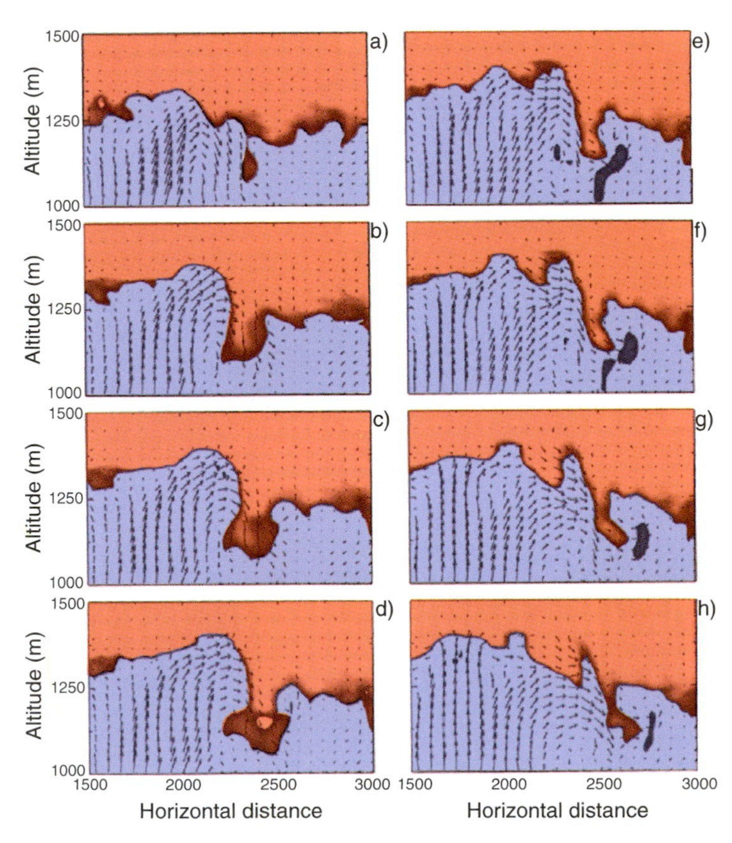

Fig. 6.14 Model simulations showing the entrainment of air (darker orange) from the free troposphere (lighter orange) into the boundary layer (blue) over a period of ~6 min. Arrows show fluid motions. [Adapted from Sullivan et al., *J. Atmos. Sci.* **55**, 3051 (1998).]

where c_p is the specific heat at constant pressure of the entrained air. To evaporate just enough cloud liquid water to saturate the entrained air requires heat

$$dQ_2 = L_v(w_s - w)\,dm \quad (6.10)$$

Suppose the entrainment occurs as the parcel of cloudy air ascends moist adiabatically and cools while its saturation mixing ratio decreases by dw_s. The heat released by the associated condensation of liquid water is

$$dQ_3 = -mL_v\,dw_s \quad (6.11)$$

[Note: Since dw_s is negative, dQ_3 given by (6.11) is positive.] Therefore, the net heat gained by the parcel of cloudy air is $dQ_3 - dQ_1 - dQ_2$.

Applying the first law of thermodynamics (Section 3.3) to the cloudy parcel of air, $dQ_3 - dQ_1 - dQ_2 = dU' + p'dV'$ where $dU' = mc_v dT'$ is the change in the internal energy of the cloudy air parcel. Therefore,

$$dQ_3 - dQ_1 - dQ_2 = mc_v dT' + p'dV' \quad (6.12)$$

If we ignore the difference between the temperature and the virtual temperature of the cloudy air, we have from the ideal gas equation

$$p'V' = mR_d T' \quad (6.13)$$

or

$$p'dV' = mR_d dT' - V'dp' \quad (6.14)$$

Substituting (6.14) into (6.12),

$$dQ_3 - dQ_1 - dQ_2 = mc_v dT' + mR_d dT' - V'dp'$$

or, using (3.48) and (6.13),

$$dQ_3 - dQ_1 - dQ_2 = mc_p dT' - \frac{mR_d T'}{p'}dp' \quad (6.15)$$

From (6.9), (6.10), (6.11), and (6.15)

$$-\frac{L_v}{c_p T'}dw_s - \frac{dm}{m}\left[\frac{T' - T}{T'} + \frac{L_v}{c_p T'}(w_s - w)\right] = \frac{dT'}{T'} - \frac{R_d}{c_p}\frac{dp'}{p'} \quad (6.16)$$

Applying (3.57) to the cloudy air

$$\ln\theta' = \ln T' + \frac{R_d}{c_p}\ln p_o - \frac{R_d}{c_p}\ln p'$$

Differentiating this last expression

$$\frac{d\theta'}{\theta'} = \frac{dT'}{T'} - \frac{R_d}{c_p}\frac{dp'}{p'} \quad (6.17)$$

From (6.16) and (6.17)

$$\frac{d\theta'}{\theta'} = -\frac{L_v}{c_p T'}dw_s - \left[\frac{T' - T}{T'} + \frac{L_v}{c_p T'}(w_s - w)\right]\frac{dm}{m} \quad (6.18)$$ ■

6.4 Growth of Cloud Droplets in Warm Clouds

In warm clouds, droplets can grow by condensation in a supersaturated environment and by colliding and coalescing with other cloud droplets. This section considers these two growth processes and assesses the extent to which they can explain the formation of rain in warm clouds.

6.4.1 Growth by Condensation

In Section 6.1.1 we followed the growth of a droplet by condensation up to a few micrometers in size. We saw that if the supersaturation is large enough to activate a droplet, the droplet will pass over the peak in its Köhler curve (Fig. 6.4) and continue to grow. We will now consider the rate at which such a droplet grows by condensation.

Let us consider first an isolated droplet, with radius r at time t, situated in a supersaturated environment in which the water vapor density at a large distance from the droplet is $\rho_v(\infty)$ and the water vapor density adjacent to the droplet is $\rho_v(r)$. If we assume that the system is in equilibrium (i.e., there is no accumulation of water vapor in the air surrounding the drop), the rate of increase in the mass M of the droplet at time t is equal to the flux of water vapor across any spherical surface of radius x centered on the droplet. Hence, if we define the *diffusion coefficient* D of water vapor in air as the rate of mass flow of water vapor across (and normal to) a unit area in the presence of a unit gradient in water vapor density, the rate of increase in the mass of the droplet is given by

$$\frac{dM}{dt} = 4\pi x^2 D \frac{d\rho_v}{dx}$$

where ρ_v is the water vapor density at distance $x(>r)$ from the droplet. Because, under steady-state conditions, dM/dt is independent of x, the aforementioned equation can be integrated as follows

$$\frac{dM}{dt}\int_{x=r}^{x=\infty}\frac{dx}{x^2} = 4\pi D\int_{\rho_v(r)}^{\rho_v(\infty)} d\rho_v$$

or

$$\frac{dM}{dt} = 4\pi r D[\rho_v(\infty) - \rho_v(r)] \tag{6.19}$$

Substituting $M = \frac{4}{3}\pi r^3 \rho_l$, where ρ_l is the density of liquid water, into this last expression we obtain

$$\frac{dr}{dt} = \frac{D}{r\rho_l}[\rho_v(\infty) - \rho_v(r)]$$

Finally, using the ideal gas equation for the water vapor, and with some algebraic manipulation,

$$\frac{dr}{dt} = \frac{1}{r}\frac{D\rho_v(\infty)}{\rho_l e(\infty)}[e(\infty) - e(r)] \tag{6.20}$$

where $e(\infty)$ is the water vapor pressure in the ambient air well removed from the droplet and $e(r)$ is the vapor pressure adjacent to the droplet.[18]

Strictly speaking, $e(r)$ in (6.20) should be replaced by e', where e' is given by (6.8). However, for droplets in excess of 1 μm or so in radius it can be seen from Fig. 6.3 that the solute effect and the Kelvin curvature effect are not very important so the vapor pressure $e(r)$ is approximately equal to the saturation vapor pressure e_s over a plane surface of pure water (which depends only on temperature). In this case, if $e(\infty)$ is not too different from e_s,

$$\frac{e(\infty) - e(r)}{e(\infty)} \simeq \frac{e(\infty) - e_s}{e_s} = S$$

where S is the supersaturation of the ambient air (expressed as a fraction rather than a percentage). Hence (6.20) becomes

$$r\frac{dr}{dt} = G_\ell S \tag{6.21}$$

where

$$G_\ell = \frac{D\rho_v(\infty)}{\rho_l}$$

which has a constant value for a given environment. It can be seen from (6.21) that, for fixed values of G_ℓ and the supersaturation S, dr/dt is inversely proportional to the radius r of the droplet. Consequently, droplets growing by condensation initially increase in radius very rapidly but their rate of growth diminishes with time, as shown schematically by curve (a) in Fig. 6.15.

In a cloud we are concerned with the growth of a large number of droplets in a rising parcel of air. As the parcel rises it expands, cools adiabatically, and

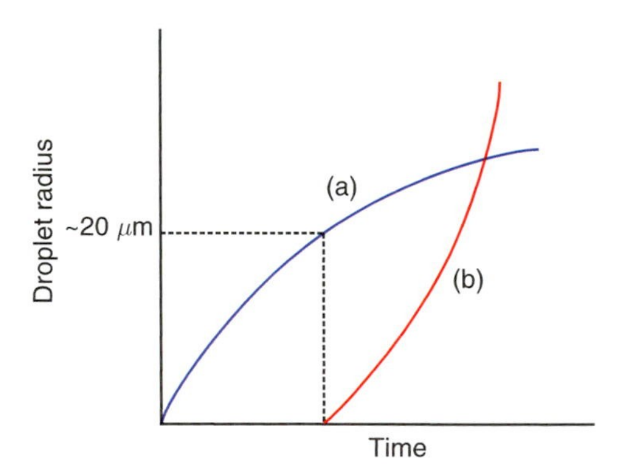

Fig. 6.15 Schematic curves of droplet growth (a) by condensation from the vapor phase (blue curve) and (b) by collection of droplets (red curve).

[18] Several assumptions have been made in the derivation of (6.20). For example, we have assumed that all of the water molecules that land on the droplet remain there and that the vapor adjacent to the droplet is at the same temperature as the environment. Due to the release of latent heat of condensation, the temperature at the surface of the droplet will, in fact, be somewhat higher than the temperature of the air well away from the droplet. We have also assumed that the droplet is at rest; droplets that are falling with appreciable speeds will be ventilated, which will affect both the temperature of the droplet and the flow of vapor to the droplet.

eventually reaches saturation with respect to liquid water. Further uplift produces supersaturations that initially increase at a rate proportional to the updraft velocity. As the supersaturation rises, CCN are activated, starting with the most efficient. When the rate at which water vapor in excess of saturation, made available by the adiabatic cooling, is equal to the rate at which water vapor condenses onto the CCN and droplets, the supersaturation in the cloud reaches a maximum value. The concentration of cloud droplets is determined at this stage (which generally occurs within 100 m or so of cloud base) and is equal to the concentration of CCN activated by the peak supersaturation that has been attained. Subsequently, the growing droplets consume water vapor faster than it is made available by the cooling of the air so the supersaturation begins to decrease. The haze droplets then begin to evaporate while the activated droplets continue to grow by condensation. Because the rate of growth of a droplet by condensation is inversely proportional to its radius [see (6.21)], the smaller activated droplets grow faster than the larger droplets. Consequently, in this simplified model, the sizes of the droplets in the cloud become increasingly uniform with time (i.e., the droplets approach a *monodispersed* distribution). This sequence of events is illustrated by the results of theoretical calculations shown in Fig. 6.16.

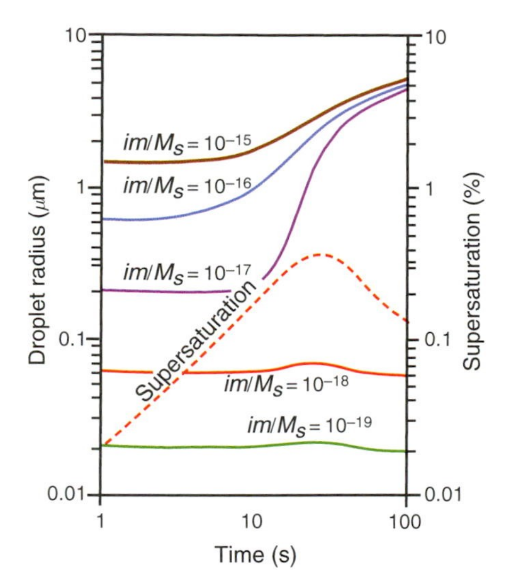

Fig. 6.16 Theoretical computations of the growth of cloud condensation nuclei by condensation in a parcel of air rising with a speed of 60 cm s^{-1}. A total of 500 CCN cm^{-1} was assumed with im/M_s values [see Eq. (6.8)] as indicated. Note how the droplets that have been activated (brown, blue, and purple curves) approach a monodispersed size distribution after just 100 s. The variation with time of the supersaturation of the air parcel is also shown (dashed red line). [Based on data from *J. Meteor.* **6**, 143 (1949).]

Comparisons of cloud droplet size distributions measured a few hundred meters above the bases of nonprecipitating warm cumulus clouds with droplet size distributions computed assuming growth by condensation for about 5 min show good agreement (Fig. 6.17). Note that the droplets produced by condensation during this time period extend up to only about 10 μm in radius. Moreover, as mentioned earlier the rate of increase in the radius of a droplet growing by condensation decreases with time. It is clear, therefore, as first noted by Reynolds[19] in 1877, that growth by condensation alone in warm clouds is much too slow to produce raindrops with radii of several millimeters. Yet rain does form in warm clouds. The enormous increase in size required to transform cloud droplets into raindrops is illustrated by the scaled diagram shown in Fig. 6.18. For a cloud droplet 10 μm in radius to grow to a raindrop 1 mm in radius requires an increases in volume of one millionfold! However, only about one droplet in a million (about

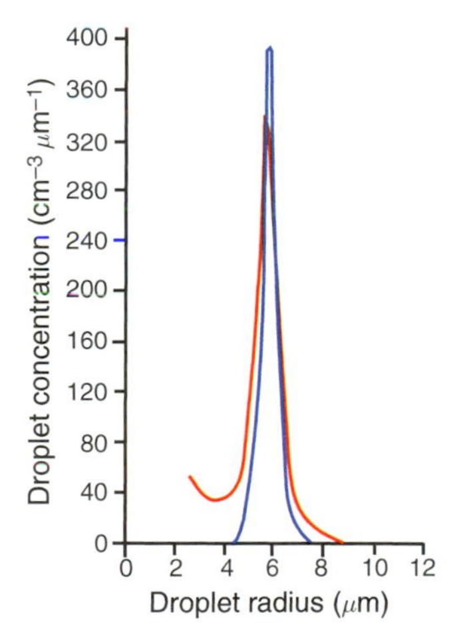

Fig. 6.17 Comparison of the cloud droplet size distribution measured 244 m above the base of a warm cumulus cloud (red line) and the corresponding computed droplet size distribution assuming growth by condensation only (blue line). [Adapted from Tech. Note No. 44, Cloud Physics Lab., Univ. of Chicago.]

[19] **Osborne Reynolds** (1842–1912) Probably the outstanding English theoretical mechanical engineer of the 19th century. Carried out important work on hydrodynamics and the theory of lubrication. Studied atmospheric refraction of sound. The Reynolds number, which he introduced, is named after him.

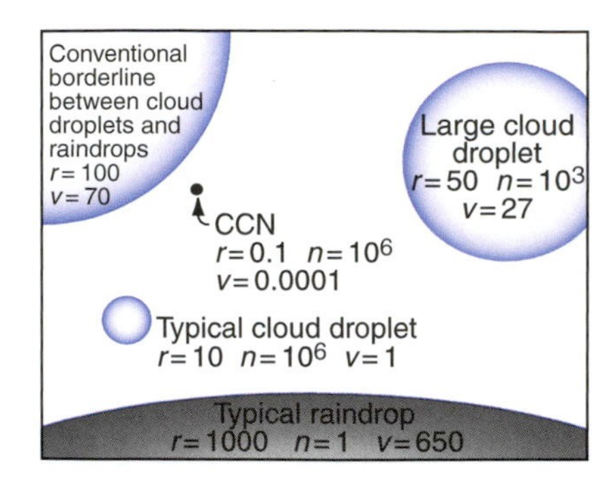

Fig. 6.18 Relative sizes of cloud droplets and raindrops; r is the radius in micrometers, n is the number per liter of air, and v is the terminal fall speed in centimeters per second. The circumferences of the circles are drawn approximately to scale, but the black dot representing a typical CCN is 25 times larger than it should be relative to the other circles. [Adapted from J. E. MacDonald, "The physics of cloud modification," *Adv. Geophys.* **5**, 244 (1958). Copyright 1958, with permission from Elsevier.]

1 liter^{-1}) in a cloud has to grow by this amount for the cloud to rain. The mechanism responsible for the selective growth of a few droplets into raindrops in warm clouds is discussed in the next section.

6.4.2 Growth by Collection

In warm clouds the growth of some droplets from the relatively small sizes achieved by condensation to the sizes of raindrops is achieved by the collision and coalescence of droplets.[20] Because the steady settling velocity of a droplet as it falls under the influence of gravity through still air (called the *terminal fall speed* of the droplet) increases with the size of the droplet (see Box 6.2), those droplets in a cloud that are somewhat larger than average will have a higher than average terminal fall speed and will collide with smaller droplets lying in their paths.

6.2 Was Galileo[22] Correct? Terminal Fall Speeds of Water Droplets in Air

By dropping objects of different masses from the leaning tower of Pisa (so the story goes), Galileo showed that freely falling bodies with different masses fall through a given distance in the same time (i.e., they experience the same acceleration). However, this is true only if the force acting on the body due to gravity is much greater than the frictional drag on the body due to the air and if the density of the body is much greater than the density of air. (Both of these requirements were met by the heavy, dense objects used by Galileo.)

Consider, however, the more general case of a body of density ρ' and volume V' falling through still air of density ρ. The downward force acting on the body due to gravity is $\rho' V' g$, and the (Archimedes') upward force acting on the body due to the mass of air displaced by the body is $\rho V'$. In addition, the air exerts a drag force F_{drag} on the body, which acts upward. The body will attain a steady *terminal fall speed* when these three forces are in balance, that is

$$\rho' V' g = \rho V' g + F_{drag}$$

Continued on next page

[20] As early as the 10th century a secret society of Basra ("The Brethren of Purity") suggested that rain is produced by the collision of cloud drops. In 1715 Barlow[21] also suggested that raindrops form due to larger cloud drops overtaking and colliding with smaller droplets. These ideas, however, were not investigated seriously until the first half of the 20th century.

[21] **Edward Barlow** (1639–1719) English priest. Author of *Meteorological Essays Concerning the Origin of Springs, Generation of Rain, and Production of Wind, with an Account of the Tide*, John Hooke and Thomas Caldecott, London, 1715.

[22] **Galileo Galilei** (1564–1642) Renowned Italian scientist. Carried out fundamental investigations into the motion of falling bodies and projectiles, and the oscillation of pendulums. The thermometer had its origins in Galileo's thermoscope. Invented the microscope. Built a telescope with which he discovered the satellites of Jupiter and observed sunspots. Following the publication of his "Dialogue on the Two Chief Systems of the World," a tribunal of the Catholic Church (the Inquisition) compelled Galileo to renounce his view that the Earth revolved around the sun (he is reputed to have muttered "It's true nevertheless") and committed him to lifelong house arrest. He died the year of Newton's birth. On 31 October 1992, 350 years after Galileo's death, Pope John Paul II admitted that errors had been made by the Church in the case of Galileo and declared the case closed.

6.2 Continued

or, if the body is a sphere of radius r, when

$$\frac{4}{3}\pi r^3 g(\rho' - \rho) = F_{drag} \tag{6.22}$$

For spheres with radius $\leq 20\ \mu m$

$$F_{drag} = 6\pi\eta r v \tag{6.23}$$

where v is the terminal fall speed of the body and η is the viscosity of the air. The expression for F_{drag} given by (6.23) is called the *Stokes' drag force*. From (6.22) and (6.23)

$$v = \frac{2}{9}\frac{g(\rho' - \rho)r^2}{\eta}$$

or, if $\rho' >> \rho$ (which it is for liquid and solid objects),

$$v = \frac{2}{9}\frac{g\rho' r^2}{\eta} \tag{6.24}$$

The terminal fall speeds of 10- and 20-μm-radius water droplets in air at 1013 hPa and 20 °C are 0.3 and 1.2 cm s^{-1}, respectively. The terminal fall speed of a water droplet with radius 40 μm is 4.7 cm s^{-1}, which is about 10% less than given by (6.24). Water drops of radius 100 μm, 1 mm, and 4 mm have terminal fall speeds of 25.6, 403, and 883 cm s^{-1}, respectively, which are very much less than given by (6.24). This is because as a drop increases in size, it becomes increasingly nonspherical and has an increasing wake. This gives rise to a drag force that is much greater than that given by (6.23).

Consider a single drop[23] of radius r_1 (called the *collector drop*) that is overtaking a smaller droplet of radius r_2 (Fig. 6.19). As the collector drop approaches the droplet, the latter will tend to follow the streamlines around the collector drop and thereby might avoid capture. We define an effective collision cross section in terms of the parameter y shown in Fig. 6.19, which represents the critical distance between the center fall line of the collector drop and the center of the droplet (measured at a large distance from the collector drop) that just makes a grazing collision with the collector drop. If the center of a droplet of radius r_2 is any closer than y to the center fall line of a collector drop of radius r_1, it will collide with the collector drop; conversely, if the center of a droplet of radius r_2 is at a greater distance than y from the center fall line, it will not collide with the collector drop. The effective collision cross section of the collector drop for droplets of radius r_2 is then πy^2, whereas the geometrical collision cross section is $\pi(r_1 + r_2)^2$. The *collision efficiency* E of a droplet of radius r_2 with a drop of radius r_1 is therefore defined as

$$E = \frac{y^2}{(r_1 + r_2)^2} \tag{6.25}$$

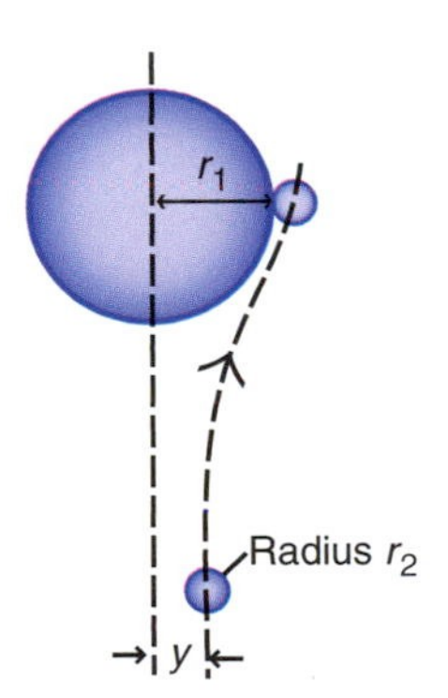

Fig. 6.19 Relative motion of a small droplet with respect to a collector drop. y is the maximum impact parameter for a droplet of radius r_2 with a collector drop of radius r_1.

Determination of the values of the collision efficiency is a difficult mathematical problem, particularly when the drop and droplet are similar in size, in which case they strongly affect each other's motion. Computed values for E are shown in Fig. 6.20, from which it can be seen that the collision efficiency increases markedly as the size of the collector drop increases and that the collision efficiencies for collector drops less than about 20 μm in radius are quite small. When the collector drop is much larger than the droplet, the collision efficiency is small because the droplet tends to follow closely

[23] In this section, "drop" refers to the larger and "droplet" to the smaller body.

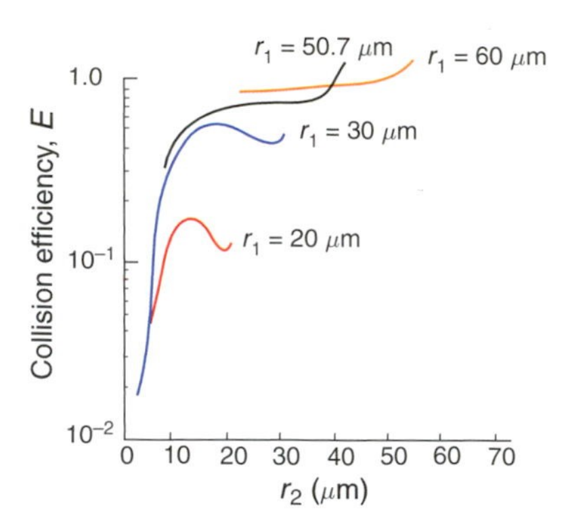

Fig. 6.20 Calculated values of the collision efficiency, E, for collector drops of radius r_1 with droplets of radius r_2. [Adapted from H. R. Pruppacher and J. D. Klett, *Microphysics of Clouds and Precipitation*, Kluwer Academic Pub., 1997, Fig. 14-6, p. 584, Copyright 1997, with kind permission of Springer Science and Business Media. Based on *J. Atmos. Sci.* **30**, 112 (1973).]

the streamlines around the collector drop. As the size of the droplet increases, E increases because the droplet tends to move more nearly in a straight line rather than follow the streamlines around the collector drop. However, as r_2/r_1 increases from about 0.6 to 0.9, E falls off, particularly for smaller collector drops because the terminal fall speeds of the collector drop and the droplets approach one another so the relative velocity between them is very small. Finally, however, as r_2/r_1 approaches unity, E tends to increase again because two nearly equal sized drops interact strongly to produce a closing velocity between them. Indeed, wake effects behind the collector drop can produce values of E greater than unity (Fig. 6.20).

The next issue to be considered is whether a droplet is captured (i.e., does coalescence occur?) when it collides with a larger drop. It is known from laboratory experiments that droplets can bounce off one another or off a plane surface of water, as demonstrated in Fig. 6.21a. This occurs when air becomes trapped between the colliding surfaces so that they deform without actually touching.[24] In effect, the droplet rebounds on a cushion of air. If the cushion of air is squeezed out before rebound occurs, the droplet will make physical contact with the drop and coalescence will occur (Fig. 6.21b).[26] The *coalescence efficiency* E' of a droplet of radius r_2 with a drop of radius r_1 is defined as the fraction of collisions that result in a coalescence. The *collection efficiency* E_c is equal to EE'.

The results of laboratory measurements on coalescence are shown in Fig. 6.22. The coalescence efficiency E' is large for very small droplets colliding with larger drops. E' initially decreases as the size of the droplet being collected increases relative to

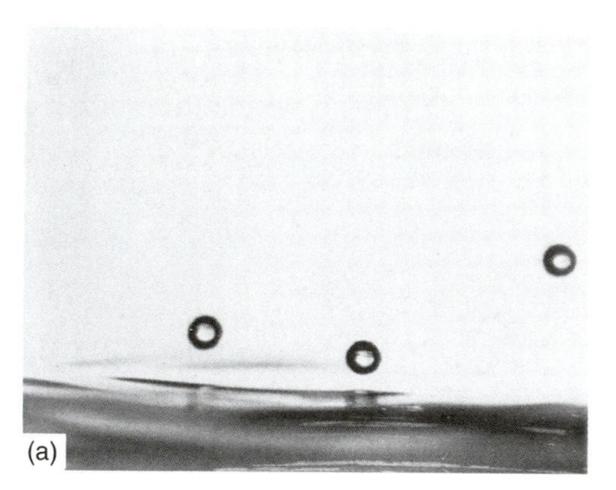

Fig. 6.21 (a) A stream of water droplets (entering from the right), about 100 μm in diameter, rebounding from a plane surface of water. (b) When the angle between the stream of droplets and the surface of the water is increased beyond a critical value, the droplets coalesce with the water. [Photograph courtesy of P. V. Hobbs.]

[24] Lenard[25] pointed out in 1904 that cloud droplets might not always coalesce when they collide, and he suggested that this could be due to a layer of air between the droplets or to electrical charges.

[25] **Phillip Lenard** (1862–1947) Austrian physicist. Studied under Helmholtz and Hertz. Professor of physics at Heidelberg and Kiel. Won the Nobel prize in physics (1905) for work on cathode rays. One of the first to study the charging produced by the disruption of water (e.g., in waterfalls).

[26] Even after two droplets have coalesced, the motions set up in their combined mass may cause subsequent breakup into several droplets (see Box 6.2).

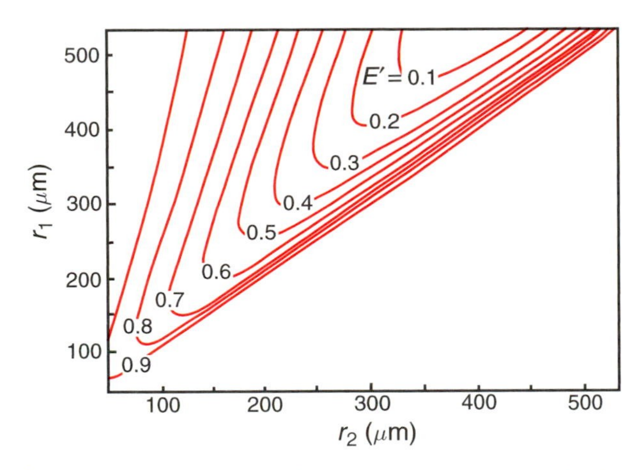

Fig. 6.22 Coalescence efficiencies E' for droplets of radius r_2 with collector drops of radius r_1 based on an empirical fit to laboratory measurements. [Adapted from *J. Atmos. Sci.* **52**, 3985 (1995).]

the collector drop, but as the droplet and drop approach each other in size, E' increases sharply. This behavior can be explained as follows. Whether coalescence occurs depends on the relative magnitude of the impact energy to the surface energy of water. This energy ratio provides a measure of the deformation of the collector drop due to the impact, which, in turn, determines how much air is trapped between the drop and the droplet. The tendency for bouncing is a maximum for intermediate values of the size ratio of the droplet to the drop. At smaller and larger values of the size ratio, the impact energy is relatively smaller and less able to prevent contact and coalescence.

The presence of an electric field enhances coalescence. For example, in the experiment illustrated in Fig. 6.21a, droplets that bounce at a certain angle of incidence can be made to coalesce by applying an electric field of about 10^4 V m^{-1}, which is within the range of measured values in clouds. Similarly, coalescence is aided if the impacting droplet carries an electric charge in excess of about 0.03 pC. The maximum electric charge that a water drop can carry occurs when the surface electrostatic stress equals the surface tension stress. For a droplet 5 μm in radius, the maximum charge is $\sim$0.3 pC; for a drop 0.5 mm in radius, it is $\sim$300 pC. Measured charges on cloud drops are generally several orders of magnitude below the maximum possible charge.

Let us now consider a collector drop of radius r_1 that has a terminal fall speed v_1. Let us suppose that

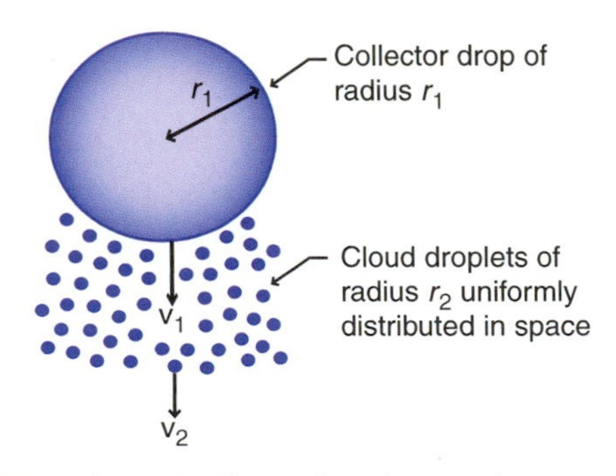

Fig. 6.23 Schematic illustrating the continuous collection model for the growth of a cloud drop by collisions and coalescence.

this drop is falling in still air through a cloud of equal sized droplets of radius r_2 with terminal fall speed v_2. We will assume that the droplets are uniformly distributed in space and that they are collected uniformly at the same rate by all collector drops of a given size. This so-called *continuous collection model* is illustrated in Fig. 6.23. The rate of increase in the mass M of the collector drop due to collisions is given by

$$\frac{dM}{dt} = \pi r_1^2 (v_1 - v_2) w_l E_c \tag{6.26}$$

where w_l is the LWC (in kg m^{-3}) of the cloud droplets of radius r_2. Substituting $M = \frac{4}{3}\pi r_1^3 \rho_l$ into (6.26), where ρ_l is the density of liquid water, we obtain

$$\frac{dr_1}{dt} = \frac{(v_1 - v_2) w_l E_c}{4\rho_l} \tag{6.27}$$

If $v_1 >> v_2$ and we assume that the coalescence efficiency is unity, so that $E_c = E$, (6.27) becomes

$$\frac{dr_1}{dt} = \frac{v_1 w_l E}{4\rho_l} \tag{6.28}$$

Because v_1 increases as r_1 increases (see Box 6.2), and E also increases with r_1 (see Fig. 6.20), it follows from (6.28) that dr_1/dt increases with increasing r_1; that is, the growth of a drop by collection is an accelerating process. This behavior is illustrated by the red curve in Fig. 6.15, which indicates negligible growth by collection until the collector drop has reached a radius of $\sim$20 μm (see Fig. 6.20). It can be seen from Fig. 6.15 that for small cloud droplets, growth by condensation is initially dominant but, beyond a certain radius, growth by collection dominates and accelerates rapidly.

If there is a steady updraft velocity w in the cloud, the velocity of the collector drop with respect to the ground will be $w - v_1$ and the velocity of the cloud droplets will be $w - v_2$. Hence dr_1/dt will still be given by (6.28), but the motion of the collector drops is

$$\frac{dh}{dt} = w - v_1 \tag{6.29}$$

where h is the height above a fixed level (say, cloud base) at time t. Eliminating dt between (6.28) and (6.29), and assuming $v_1 >> v_2$ and $E_c = E$, we obtain

$$\frac{dr_1}{dh} = \frac{v_1 w_l E}{4\rho_l (w - v_1)}$$

or, if the radius of the collector drop at height H above cloud base is r_H and its radius at cloud base is r_0,

$$\int_0^H w_l \, dh = 4\rho_l \int_{r_0}^{r_H} \frac{(w - v_1)}{v_1 E} dr_1$$

Hence, if we assume that w_l is independent of h,

$$H = \frac{4\rho_l}{w_l}\left[\int_{r_0}^{r_H} \frac{w}{v_1 E} dr_1 - \int_{r_0}^{r_H} \frac{dr_1}{E}\right] \tag{6.30}$$

If values of E and v_1 as functions of r_1 are known (6.30) can be used to determine the value of H corresponding to any value of r_H and vice versa. We can also deduce from (6.30) the general behavior of a cloud drop that grows by collection. When the drop is still quite small $w > v_1$, therefore the first integral in (6.30) dominates over the second integral; H then increases as r_H increases, that is, a drop growing by collection is initially carried upward in the cloud. Eventually, as the drop grows, v_1 becomes greater than w and the magnitude of the second integral in (6.30) becomes greater than that of the first integral. H then decreases with increasing r_H, that is, the drop begins to fall through the updraft and will eventually pass through the cloud base and may reach the ground as a raindrop. Some of the larger drops may break up as they fall through the air (see Box 6.3). The resulting fragments may be carried up into the cloud on updrafts, grow by collection, fall out, and perhaps break up again; such a chain reaction would serve to enhance precipitation.

Exercise 6.3 A drop enters the base of a cloud with a radius r_0 and, after growing with a constant collection efficiency while traveling up and down in the cloud, the drop reaches cloud base again with a radius R. Show that R is a function only of r_0 and the updraft velocity w (assumed to be constant).

Solution: Putting $H = 0$ and $r_H = R$ into the equation before (6.30) we obtain

$$0 = 4\rho_l \int_{r_0}^{R} \frac{w - v_1}{v_1 E} dr_1$$

or, because E and w are assumed to be constant,

$$w \int_{r_0}^{R} \frac{dr_1}{v_1} = \int_{r_0}^{R} dr_1 = R - r_0$$

Therefore

$$R = r_0 + w \int_{r_0}^{R} \frac{dr_1}{v_1} \tag{6.31}$$

Because $\int_{r_0}^{R} dr_1/v_1$ is a function only of R and r_0, it follows from (6.31) that R is a function only of r_0 and w. ■

6.4.3 Bridging the Gap between Droplet Growth by Condensation and Collision–Coalescence

Provided a few drops are large enough to be reasonably efficient collectors (i.e., with radius ≥ 20 μm), and a cloud is deep enough and contains sufficient liquid water, the equations derived earlier for growth by continuous collection indicate that raindrops should grow within reasonable time periods (~1 h), see Exercise 6.23, and that deep clouds with strong updrafts should produce rain quicker than shallower clouds with weak updrafts.

We have seen in Section 6.4.1 that growth by condensation slows appreciably as a droplet approaches ~10 μm in radius (e.g., Fig. 6.16). Also, growth by condensation alone in a homogeneous cloud and a uniform updraft tends to produce a monodispersed droplet size distribution (Fig. 6.17), in which the fall speeds of the droplets would be very similar and therefore collisions unlikely. Consequently, there has been considerable interest in the origins of the few ($\sim$1 liter^{-1}) larger drops (with radius $\gtrsim$20 μm) that can become the collectors in warm clouds that go on to produce raindrops, and in the mechanisms responsible for the broad spectrum of droplet sizes measured in clouds (Figs. 6.6 and 6.7). This section describes briefly some of the mechanisms that have been proposed to bridge the gap between droplet growth by condensation and collision–coalescence.

(a) *Role of giant cloud condensation nuclei.* Aerosols containing *giant cloud condensation nuclei* (GCCN) (i.e., wettable particles with a radius greater than $\sim$3 μm) may act as embryos for the formation of collector drops. For example, the addition of 1 liter^{-1} of GCCN (i.e., about 1 particle in 10^6) can account for the formation of precipitation-sized particles even in continental clouds. GCCN concentrations of 10^{-1} to 10 liter^{-1} can transform a nonprecipitating stratocumulus cloud, with CCN concentrations of 50–250 cm^{-3}, into a precipitating cloud. For lower CCN concentrations in marine stratocumulus, drizzle can form anyway and the addition of GCCN (e.g., from sea salt) should have little impact. For polluted convective clouds, model calculations show that with CCN concentrations of $\sim$1700 cm^{-3} and GCCN of 20 liter^{-1}, precipitation can be produced more readily than for a cleaner cloud with $\sim$1000 cm^{-3} of CCN and no GCCN.

(b) *Effects of turbulence on the collision and coalescence of droplets.* Turbulence can influence the growth of droplets by producing fluctuating supersaturations that enhance condensational growth and by enhancing collision efficiencies and collection.

Simple models of homogeneous mixing in the presence of turbulence and associated fluctuations in supersaturation predict only slight broadening of the droplet size distribution. However, if mixing occurs inhomogeneously (i.e., finite blobs of unsaturated air mix with nearly saturated blobs, resulting in complete evaporation of some droplets of all sizes), the overall concentration of droplets is reduced, and the largest drops grow much faster than for homogeneous mixing due to enhanced local supersaturations.

Another view of the role of turbulence in droplet broadening is associated with updrafts and downdrafts in clouds. Downdrafts are formed when saturated air near the cloud top mixes with dry environmental air. The evaporation of drops produces cooling and downdrafts. In downdrafts the air is heated by adiabatic compression, which causes further evaporation of drops. Larger drops may be mixed into the downdrafts from the surrounding undiluted air. When a downdraft is transformed into an updraft, the drops mixed most recently from the undiluted surrounding air will be larger than the other drops and increase in size as they are carried upward. With sufficient entrainment of air and vertical cycling, a broad droplet size spectrum may be produced.

It has also been hypothesized that in turbulent flow the droplets in a cloud are not dispersed randomly. Instead they are concentrated (on the cm scale) in regions of strong deformation and centrifuged away from regions of high vorticity, where the terms *deformation* and *vorticity* are defined in Section 7.1. The high vorticity regions experience high supersaturations, and the high strain regions low supersaturations. Droplets in the regions of low number concentration will experience more rapid condensational growth, whereas those in regions of high number concentration will experience slower condensational growth. This will lead to broadening of the droplet size spectrum.

In turbulent air, droplets will be accelerated and thereby able to cross streamlines more readily than in laminar flow, which will enhance collision efficiencies. Turbulence can also cause fluctuations in droplet fall speeds and horizontal motions, thereby increasing collectional growth. Because little is known about turbulence on small scales (<1 cm) in clouds, it has been difficult to quantify these possible effects of turbulence on drop growth by collection.

(c) *Radiative broadening.* When a droplet is growing by condensation, it is warmer than the environmental air. Therefore, the droplet will lose heat by radiation. Consequently, the saturation vapor pressure above the surface of the droplet will be lower, and the droplet will grow faster, than predicted if radiation is neglected. The loss of heat by radiation will be proportional to the cross-sectional area of a droplet. Therefore, the radiation effect will be greater the larger the drop, which will enhance the growth of potential collector drops. The radiation effect will also be greater for drops that reside near cloud tops, where they can radiate to space.

(d) *Stochastic collection.* In the continuous collection model, it is assumed that the collector drop collides in a continuous and uniform fashion with smaller cloud droplets that are distributed uniformly in space. Consequently, the continuous collection model predicts that collector drops of the same size grow at the same rate if they fall through the same cloud of droplets. The *stochastic* (or *statistical*) *collection model* allows for the fact that collisions are individual events, distributed statistically in time and space. Consider, for example, 100 droplets, initially the same size as shown on line 1 in Fig. 6.24. After a certain interval of time, some of these droplets (let us say 10) will have collided with other droplets so that the distribution will now be as depicted in line 2 of Fig. 6.24. Because of their larger size, these 10 larger droplets are now in a more favored position for making further collisions. The second collisions are similarly statistically distributed, giving a further broadening of the droplet size spectrum, as shown in line 3 of Fig. 6.24 (where it has been assumed that in this time step, nine of the smaller droplets and one of the larger droplets on line 2 each had a collision). Hence, by allowing for a statistical distribution of collisions, three size categories of droplets have developed after just two time steps. This concept is important, because it not only provides a mechanism for developing broad droplet size spectra from the fairly uniform droplet sizes produced by condensation, but it also reveals how a small fraction of the droplets in a cloud can grow much faster than average by statistically distributed collisions.

The growth of drops by collection is also accelerated if they pass through pockets of higher than average LWC. Even if such pockets of high LWC exist for only a few minutes and occupy only a few percent of the cloud volume, they can produce significant concentrations of large drops when averaged over the entire cloud volume. Measurements in clouds reveal such pockets of high LW (e.g., Fig. 6.10).

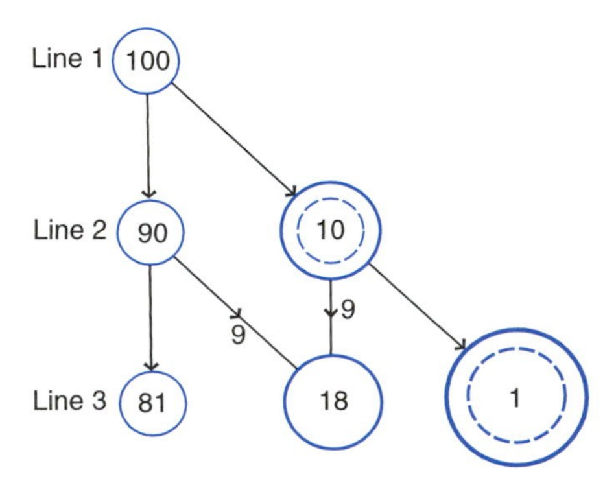

Fig. 6.24 Schematic diagram to illustrate broadening of droplet sizes by statistical collisions. [Adapted from *J. Atmos. Sci.* **24**, 689 (1967).]

Much of what is presently known about both dynamical and microphysical cloud processes can be incorporated into computer models and numerical experiments can be carried out. For example, consider the growth of drops, by condensation and stochastic collisions, in warm cumulus clouds in typical marine and continental air masses. As pointed out in Section 6.2, the average droplet sizes are significantly larger, and the droplet size spectra much broader, in marine than in continental cumulus clouds (Fig. 6.7). We have attributed these differences to the higher concentrations of CCN present in continental air (Fig. 6.5). Figure 6.25 illustrates the effects of these differences in cloud microstructures on the development of larger drops. CCN spectra used as input data to the two clouds were based on measurements, with continental air having much higher concentrations of CCN than marine air (about 200 versus 45 cm^{-3} at 0.2% super-saturation). It can be seen that after 67 min the cumulus cloud in marine air develops some drops between 100 and 1000 μm in radius (i.e., raindrops), whereas the continental cloud does not contain any droplets greater than about 20 μm in radius. These markedly different developments are attributable to the fact that the marine cloud contains a small number of drops that are large enough to grow by collection, whereas the continental cloud does not. These model results support the observation that a marine cumulus cloud is more likely to rain than a continental cumulus cloud with similar updraft velocity, LWC, and depth.

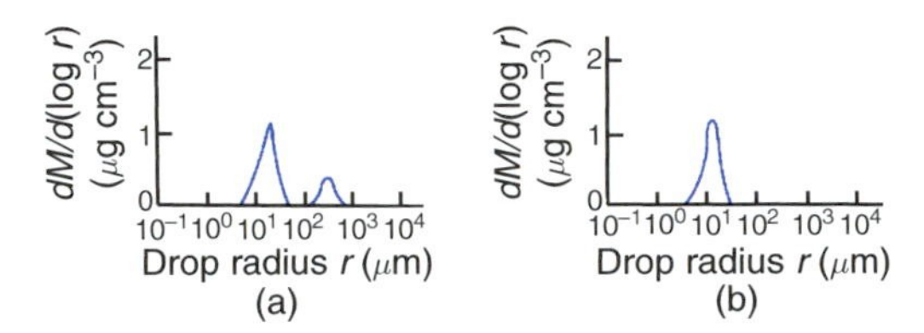

Fig. 6.25 Numerical predictions of the mass spectrum of drops near the middle of (a) a warm marine cumulus cloud and (b) a warm continental cloud after 67 min of growth. (B. C. Scott and P. V. Hobbs, unpublished.)

6.3 Shape, Breakup, and Size Distribution of Raindrops

Raindrops in free fall are often depicted as tear shaped. In fact, as a drop increases in size above about 1 mm in radius it becomes flattened on its underside in free fall and it gradually changes in shape from essentially spherical to increasingly parachute (or jellyfish)-like. If the initial radius of the drop exceeds about 2.5 mm, the parachute becomes a large inverted bag, with a toroidal ring of water around its lower rim. Laboratory and theoretical studies indicate that when the drop bag bursts to produce a fine spray of droplets, the toroidal ring breaks up into a number of large drops (Fig. 6.26). Interestingly, the largest raindrops ever observed at diameters of ~0.8–1 cm must have been on the verge of bag breakup.

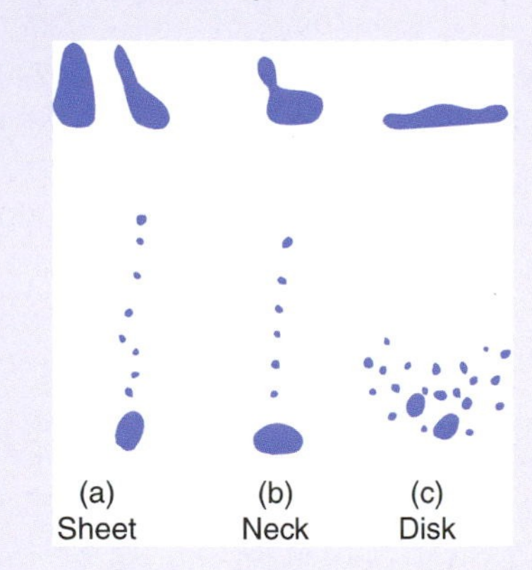

Fig. 6.27 Schematic illustrations of three types of breakup following the collision of two drops. [Adapted from *J. Atmos. Sci.* **32**, 1403 (1975).]

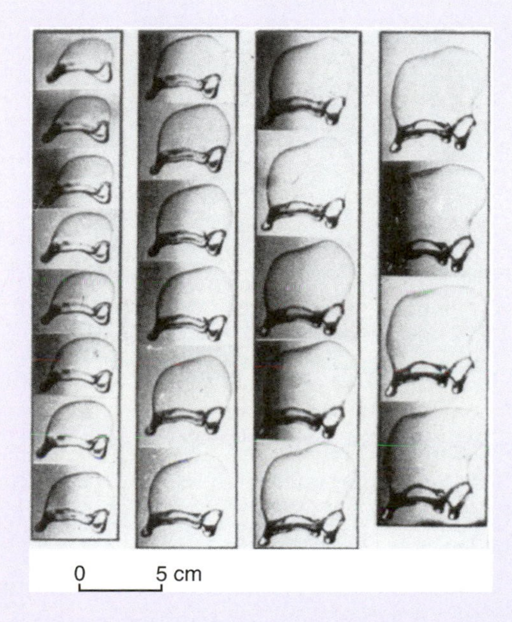

Fig. 6.26 Sequence of high-speed photographs (starting at upper left and moving down and to the right) showing how a large drop in free fall forms a parachute-like shape with a toroidal ring of water around its lower rim. The toroidal ring becomes distorted and develops cusps separated by threads of water. The cusps eventually break away to form large drops, and the thin film of water that forms the upper part of the parachute bursts to produce a spray of small droplets. Time interval between photographs = 1 ms. [Photograph courtesy of B. J. Mason.]

Collisions between raindrops appears to be a more important path to breakup. The three main types of breakup following a collision are shown schematically in Fig. 6.27. The probabilities of the various types of breakup shown in Fig. 2.77 are sheets, 55%; necks, 27%; and disks, 18%. Bag breakup may also occur following the collision of two drops, but it is very rare (<0.5%).

The spectrum of raindrop sizes should reflect the combined influences of the drop growth processes discussed earlier and the breakup of individual drops and of colliding drops. However, as we have seen, individual drops have to reach quite large sizes before they break up. Also, the frequency of collisions between raindrops is quite small, except in very heavy rain. For example, a drop with a radius of ~1.7 mm in rain of 5 mm h^{-1} would experience only one collision as it falls to the ground from a cloud base at 2 km above the ground. Consequently, it cannot be generally assumed that raindrops have had sufficient time to attain an equilibrium size distribution.

Continued on next page

6.3 Continued

Theoretical predictions of the probability of breakup as a function of the sizes of two coalescing drops are shown in Fig. 6.28. For a fixed size of the larger drop, the probability of breakup initially increases as the size of the smaller drop increases, but as the smaller drop approaches the size of the larger, the probability of breakup decreases due to the decrease in the kinetic energy of the impact.

Measurements of the size distributions of raindrops that reach the ground can often be fitted to an expression, known as the *Marshall–Palmer* distribution, which is of the form

$$N(D) = N_o \exp(-\Lambda D) \tag{6.32}$$

where $N(D)dD$ is the number of drops per unit volume with diameters between D and $D + dD$ and N_o and Λ are empirical fitting parameters. The value of N_o tends to be constant, but Λ varies with the rainfall rate.

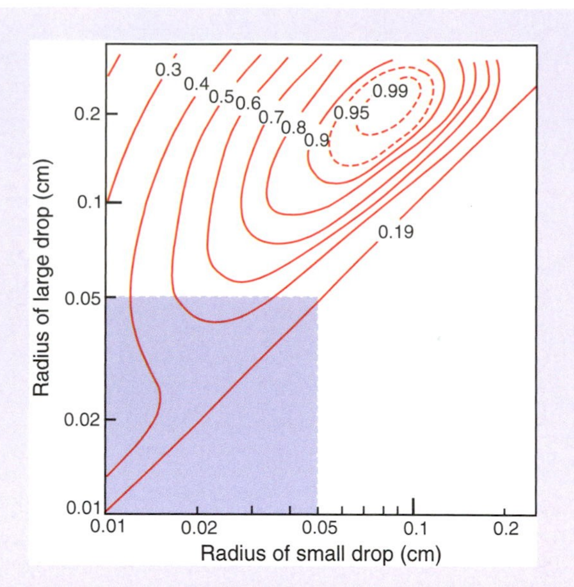

Fig. 6.28 Empirical results for the probability of breakup (expressed as a fraction and shown on the contours) following the collision and initial coalescence of two drops. The shaded region is covered by Fig. 6.21, but note that Fig. 6.21 shows the probability of coalescence rather than breakup. [Based on *J. Atmos. Sci.* **39**, 1600 (1982).]

6.5 Microphysics of Cold Clouds

If a cloud extends above the 0 °C level it is called a *cold cloud*. Even though the temperature may be below 0 °C, water droplets can still exist in clouds, in which case they are referred to as *supercooled droplets*.[27] Cold clouds may also contain ice particles. If a cold cloud contains both ice particles and supercooled droplets, it is said to be a *mixed cloud*; if it consists entirely of ice, it is said to be *glaciated*.

This section is concerned with the origins and concentrations of ice particles in clouds, the various ways in which ice particles can grow, and the formation of precipitation in cold clouds.

6.5.1 Nucleation of Ice Particles; Ice Nuclei

A supercooled droplet is in an unstable state. For freezing to occur, enough water molecules must come together within the droplet to form an embryo of ice large enough to survive and grow. The situation is analogous to the formation of a water droplet from the vapor phase discussed in Section 6.2.1. If an ice embryo within a droplet exceeds a certain critical size, its growth will produce a decrease in the energy of the system. However, any increase in the size of an ice embryo smaller than the critical size causes an increase in total energy. In the latter case, from an energetic point of view, it is preferable for the embryo to break up.

If a water droplet contains no foreign particles, it can freeze only by *homogeneous nucleation*. Because the numbers and sizes of the ice embryos that form by chance aggregations increase with decreasing temperature, below a certain temperature (which depends on the volume of water considered), freezing by homogeneous nucleation becomes a virtual certainty. The results of laboratory experiments on the freezing of very pure water droplets, which were probably nucleated homogeneously, are shown by the blue

[27] Saussure[28] observed, around 1783, that water could remain in the liquid state below 0 °C. A spectacular confirmation of the existence of supercooled clouds was provided by a balloon flight made by Barrel[29] in 1850. He observed water droplets down to −10.5 °C and ice crystals at lower temperatures.

[28] **Horace Bénédict de Saussure** (1740–1799) Swiss geologist, physicist, meteorologist, and naturalist. Traveled extensively, particularly in the Alps, and made the second ascent of Mont Blanc (1787).

[29] **Jean Augustine Barrel** (1819–1884) French chemist and agriculturalist. First to extract nicotine from tobacco leaf.

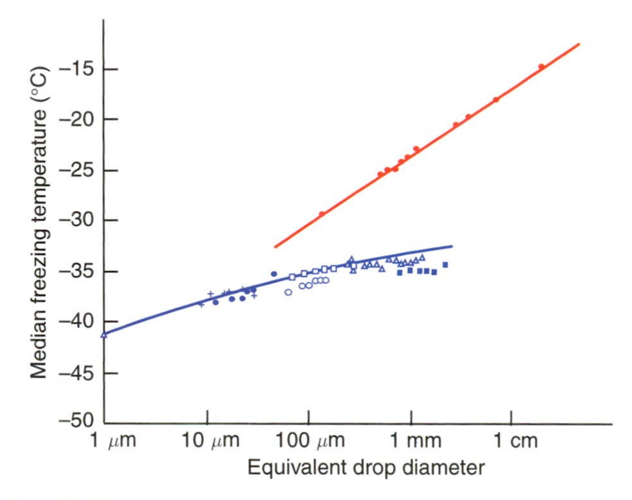

Fig. 6.29 Median freezing temperatures of water samples as a function of their equivalent drop diameter. The different symbols are results from different workers. The red symbols and red line represent heterogeneous freezing, and the blue symbols and line represent homogeneous freezing. [Adapted from B. J. Mason, *The Physics of Clouds*, Oxford Univ. Press, Oxford, 1971, p. 160. By permission of Oxford University Press.]

symbols in Fig. 6.29. It appears from these results that homogeneous nucleation occurs at about −41 °C for droplets about 1 μm in diameter and at about −35 °C for drops 100 μm in diameter. Hence, in the atmosphere, homogeneous nucleation of freezing generally occurs only in high clouds.

If a droplet contains a rather special type of particle, called a *freezing nucleus*, it may freeze by a process known as *heterogeneous nucleation*[30] in which water molecules in the droplet collect onto the surface of the particle to form an ice-like structure that may increase in size and cause the droplet to freeze. Because the formation of the ice structure is aided by the freezing nucleus, and the ice embryo also starts off with the dimensions of the freezing nucleus, heterogeneous nucleation can occur at much higher temperatures than homogeneous nucleation. The red symbols in Fig. 6.29 show results of laboratory experiments on the heterogeneous freezing of water droplets. The droplets consisted of distilled water from which most, but not all, of the foreign particles were removed. A large number of droplets of each of the sizes indicated in Fig. 6.29 were cooled, and the temperature at which half of the droplets had frozen was noted. It can be seen that this median freezing temperature increases as the size of the droplet increases. The size dependence reflects the fact that a larger drop is more likely to contain a freezing nucleus capable of causing heterogeneous nucleation at a given temperature.

We have assumed earlier that the particle that initiates the freezing is contained within the droplet. However, cloud droplets may also be frozen if a suitable particle in the air comes into contact with the droplet, in which case freezing is said to occur by *contact nucleation*, and the particle is referred to as a *contact nucleus*. Laboratory experiments suggest that some particles can cause a drop to freeze by contact nucleation at temperatures several degrees higher than if they were embedded in the drop.

Certain particles in the air also serve as centers upon which ice can form directly from the vapor phase. These particles are referred to as *deposition nuclei*. Ice can form by deposition[32] provided that the air is supersaturated with respect to ice and the temperature is low enough. If the air is supersaturated with respect to water, a suitable particle may serve either as a freezing nucleus (in which case liquid water first condenses onto the particle and subsequently freezes) or as a deposition nucleus (in which case there is no intermediate liquid phase, at least on the macroscopic scale).

If we wish to refer to an ice nucleating particle in general, without specifying its mode of action, we will call it an *ice nucleus*. However, it should be kept in mind that the temperature at which a particle can cause ice to form depends, in general, on the mechanism by which the particle nucleates the ice as well as on the previous history of the particle.

Particles with molecular spacings and crystallographic arrangements similar to those of ice (which has a hexagonal structure) tend to be effective as ice nuclei, although this is neither a necessary nor a sufficient condition for a good ice nucleus. Most effective ice nuclei are virtually insoluble in water. Some inorganic soil particles (mainly clays) can nucleate ice at fairly high temperatures (i.e., above −15 °C) and

[30] Studies of the heterogeneous nucleation of ice date back to 1724 when Fahrenheit[31] slipped on the stairs while carrying a flask of cold (supercooled) water and noticed that the water had become full of flakes of ice.

[31] **Gabriel Daniel Fahrenheit** (1686–1736) German instrument maker and experimental physicist. Lived in Holland from the age of 15 but traveled widely in Europe. Developed the thermometric scale that bears his name. Fahrenheit knew that the boiling point of water varies with atmospheric pressure [see Eq. (3.112)] and he constructed a thermometer from which the atmospheric pressure could be determined by noting the boiling point of water.

[32] The transfer of water vapor to ice is sometimes referred to as *sublimation* by cloud physicists. However, since chemists use this term more appropriately to describe the evaporation of a solid, the term *deposition* is preferable and is used here.

probably play an important role in nucleating ice in clouds. For example, in one study, 87% of the snow crystals collected on the ground had clay mineral particles at their centers and more than half of these were kaolinite. Many organic materials are effective ice nucleators. Decayed plant leaves contain copious ice nuclei, some active as high as −4 °C. Ice nuclei active at −4 °C have also been found in sea water rich in plankton.

The results of laboratory measurements on condensation-freezing and deposition shown in Fig. 6.30 indicate that for a variety of materials the onset of ice nucleation occurs at higher temperatures under water-supersaturated conditions (so that condensation-freezing is possible) than under water-subsaturated conditions (when only ice deposition is possible). For example, kaolinite serves as an ice nucleus at −10.5 °C at water saturation, but at 17% supersaturation with respect to ice (but subsaturation with respect to water), the temperature has to be about −20 °C for kaolinite to act as an ice nucleus.

In some cases, after a particle has served as an ice nucleus and all of the visible ice is then evaporated from it but the particle is not warmed above −5 °C or exposed to a relative humidity with respect to ice of less than 35%, it may subsequently serve as an ice nucleus at a temperature a few degrees higher than it did initially. This is referred to as *preactivation*. Thus, ice crystals from upper level clouds that evaporate before reaching the ground may leave behind preactivated ice nuclei.

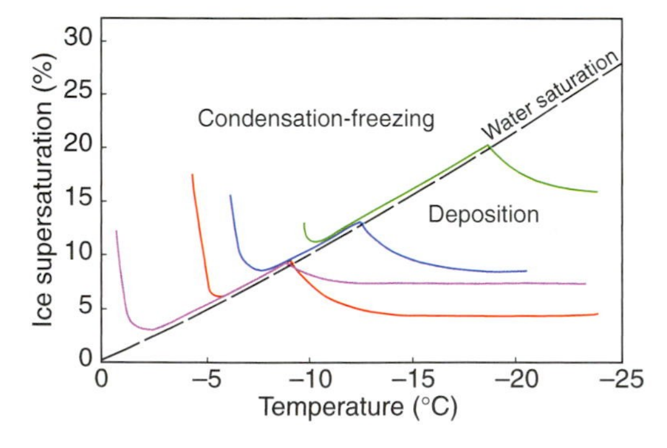

Fig. 6.30 Onset of ice nucleation as a function of temperature and supersaturation for various compounds. Conditions for condensation-freezing and ice deposition are indicated. Ice nucleation starts above the indicated lines. The materials are silver iodide (red), lead iodide (blue), methaldehyde (violet), and kaolinite (green). [Adapted from *J. Atmos. Sci.* **36**, 1797 (1979).]

Several techniques have been used for measuring the concentrations of particles in the air that are active as ice nuclei at a given temperature. A common method is to draw a known volume of air into a container and to cool it until a cloud is formed. The number of ice crystals forming at a particular temperature is then measured. In *expansion chambers*, cooling is produced by compressing the air and then suddenly expanding it; in *mixing chambers*, cooling is produced by refrigeration. In these chambers particles may serve as freezing, contact, or deposition nuclei. The number of ice crystals that appear in the chamber may be determined by illuminating a certain volume of the chamber and estimating visually the number of crystals in the light beam, by letting the ice crystals fall into a dish of supercooled soap or sugar solution where they grow and can be counted, or by allowing the ice crystals to pass through a small capillary tube attached to the chamber where they produce audible clicks that can be counted electronically. In another technique for detecting ice nuclei, a measured volume of air is drawn through a Millipore filter that retains the particles in the air. The number of ice nuclei on the filter is then determined by placing it in a box held at a known supersaturation and temperature and counting the number of ice crystals that grow on the filter. More recently, ice nucleation has been studied using diffusion chambers in which temperature, supersaturation, and pressure can be controlled independently.

Worldwide measurements of ice nucleus concentrations as a function of temperature (Fig. 6.31) indicate that concentrations of ice nuclei tend to be higher in the northern than in the southern hemisphere. It should be noted, however, that ice nucleus concentrations can sometimes vary by several orders of magnitude over several hours. On the average, the number N of ice nuclei per liter of air active at temperature T tends to follow the empirical relationship

$$\ln N = a(T_1 - T) \quad (6.33)$$

where T_1 is the temperature at which one ice nucleus per liter is active (typically about −20 °C) and a varies from about 0.3 to 0.8. For $a = 0.6$, (6.32) predicts that the concentration of ice nuclei increases by about a factor of 10 for every 4 °C decrease in temperature. In urban air, the total concentration of aerosol is on the order of 10^8 liter^{-1} and only about one particle in 10^8 acts as an ice nucleus at −20 °C.

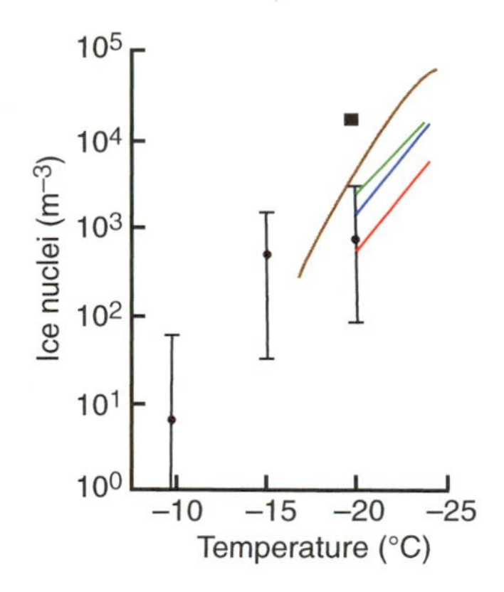

Fig. 6.31 Measurements of average ice nucleus concentrations at close to water saturation in the northern and southern hemispheres. Southern hemisphere, expansion chamber (red); southern hemisphere, mixing chamber (blue); northern hemisphere, expansion chamber (green); northern hemisphere, mixing chamber (black square); Antarctica, mixing chamber (brown). Vertical lines show the range and mean values (dots) of ice nucleus concentrations based on Millipore filter measurements in many locations around the world.

Exercise 6.4 If the concentration of freezing nuclei in a drop that are active at temperature T is given by (6.33), show that the median freezing temperature of a number of drops should vary with their diameter as shown by the red line in Fig. 6.29. [If a drop contains n active freezing nuclei, assume that the probability p that it freezes in a given time interval is given by the Poisson distribution for random events, namely $p = 1 - \exp(-n)$.]

Solution: From (6.33) the number of active freezing nuclei at temperature T in 1 liter of a drop is

$$N = \exp[a(T_1 - T)]$$

where T_1 is the temperature at which $N = 1$ per liter. Therefore, the number of active freezing nuclei n at temperature T in a drop of diameter D (in meters) is equal to the volume of the drop (in m^3) multiplied by the number of active freezing nuclei per m^3 (i.e., $10^3\,N$). Therefore

$$n = \frac{4}{3}\pi\left(\frac{D}{2}\right)^3 10^3 \exp[a(T_1 - T)] \qquad (6.34)$$

The probability p that a drop of diameter D, containing n freezing nuclei, is frozen is given by

$$p = 1 - \exp(-n)$$

When half of the drops are frozen (i.e., $p = 0.5$) n (and therefore $\ln n$) are constants. Hence, from (6.34)

$$\ln n = \text{constant} = \ln\left\{\frac{4}{3}\pi\left(\frac{D}{2}\right)^3 10^3 \exp[a(T_1 - T)]\right\}$$

or

$$3\ln D + a(T_1 - T) = \text{constant}$$

where (since $p = 0.5$) T is now the median freezing temperature. It follows from this last expression that

$$T = (\text{constant})\ln D + (\text{constant})$$

Therefore, $\ln D$ plotted against the median freezing temperature T for drops of diameter D should be a straight line, as shown by the red line in Fig. 6.29. ■

As we have seen, the activity of a particle as a freezing or a deposition nucleus depends not only on the temperature but also on the supersaturation of the ambient air. Supersaturation was not well controlled in many of the measurements shown in Fig. 6.31, on which (6.33) is based. The effect of supersaturation on measurements of ice nucleus concentrations is shown in Fig. 6.32, where it can be seen that at a

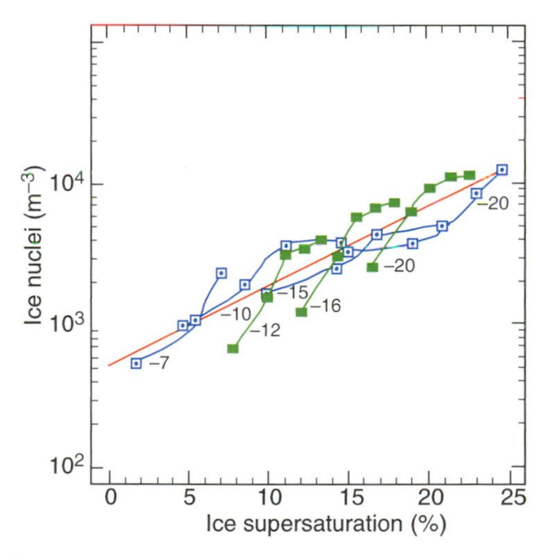

Fig. 6.32 Ice nucleus concentration measurements versus ice supersaturation; temperatures are noted alongside each line. The red line is Eq. (6.35). [Data reprinted from D. C. Rogers, "Measurements of natural ice nuclei with a continuous flow diffusion chamber," *Atmos. Res.* **29**, 209 (1993) with permission from Elsevier—blue squares, and R. Al-Naimi and C. P. R. Saunders, "Measurements of natural deposition and condensation-freezing ice nuclei with a continuous flow chamber," *Atmos. Environ.* **19**, 1872 (1985) with permission from Elsevier—green squares.]

constant temperature the greater the supersaturation with respect to ice, the more particles serve as ice nuclei. The empirical equation to the best-fit line to these measurements (the red line in Fig. 6.32) is

$$N = \exp\{a + b\,[100\,(S_i - 1)]\} \qquad (6.35)$$

where N is the concentration of ice nuclei per liter, S_i is the supersaturation with respect to ice, $a = -0.639$, and $b = 0.1296$.

6.5.2 Concentrations of Ice Particles in Clouds; Ice Multiplication

The probability of ice particles being present in a cloud increases as the temperature decreases below 0 °C. Results shown in Fig. 6.33 indicate that the probability of ice being present is 100% for cloud top temperature below about −13 °C. At higher temperatures the probability of ice being present falls off sharply, but it is greater if the cloud contains drizzle or raindrops. Clouds with top temperatures between about 0 and −8 °C generally contain copious supercooled droplets. It is in clouds such as these that aircraft are most likely to encounter severe icing conditions, since supercooled droplets freeze when they collide with an aircraft.

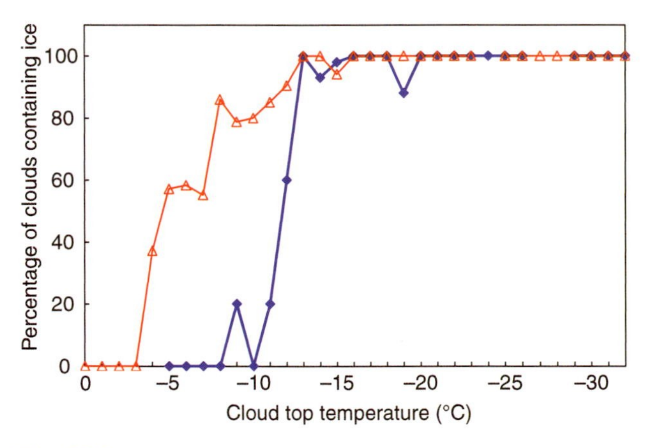

Fig. 6.33 Percentage of clouds containing ice particle concentrations greater than about 1 per liter as a function of cloud top temperature. Note that on the abscissa temperatures decrease to the right. Blue curve: continental cumuliform clouds with base temperatures of 8 to −18 °C containing no drizzle or raindrops prior to the formation of ice. [Data from *Quart. J. Roy. Met. Soc.* **120**, 573 (1994).] Red curve: clean marine cumuliform clouds and clean arctic stratiform clouds with base temperatures from 25 to −3 °C containing drizzle or raindrops prior to the formation of ice. [Based on data from *Quart. J. Roy. Met. Soc.* **117**, 207 (1991); A. L. Rangno and P. V. Hobbs, "Ice particles in stratiform clouds in the Arctic and possible mechanisms for the production of high ice concentrations," *J. Geophys. Res.* **106**, 15,066 (2001) Copyright 2001 American Geophysical Union. Reproduced by permission of American Geophysical Union; Cloud and Aerosol Research Group, University of Washington, unpublished data.]

Shown in Fig. 6.34 are measurements of the concentrations of ice particles in clouds. Also shown are the concentrations of ice nuclei given by (6.33) with $a = 0.6$ and $T_1 = 253$ °K. It can be seen that (6.33) approximates the minimum values of the maximum concentration of ice particles. However, on many occasions, ice particles are present in concentrations several orders of magnitude greater than ice nucleus measurements. At temperatures above about −20 °C, marine clouds show a particular propensity for ice particle concentrations that are many orders of magnitude greater than ice nucleus measurements would suggest.

Several explanations have been proposed to account for the high ice particle concentrations observed in some clouds. First, it is possible that current techniques for measuring ice nuclei do not provide reliable estimates of the concentrations of ice nuclei active in natural clouds under certain conditions. It is also possible that ice particles in clouds

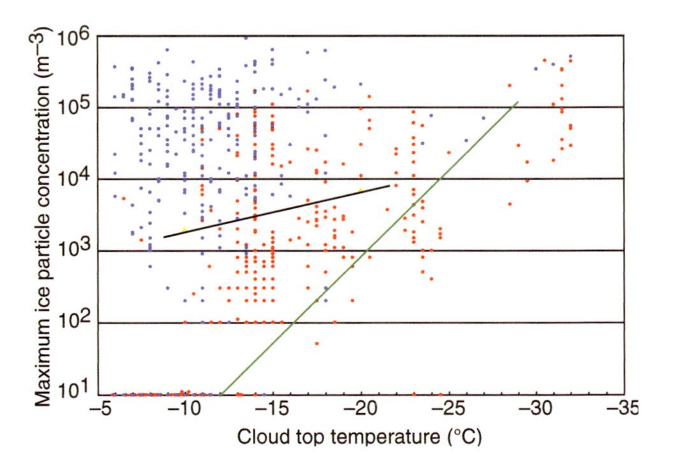

Fig. 6.34 Maximum concentrations of ice particles versus cloud top temperature in mature and aging marine cumuliform clouds (blue dots) and in continental cumuliform clouds (red dots). Note that on the abscissa temperatures decrease to the right. Symbols along the abscissa indicate ice concentrations ≤ 1 liter^{-1}, which was the lower limit of detection. The green line shows ice nucleus concentrations predicted by Eq. (6.33) with $a = 0.6$ and $T_1 = 253$ K. The black line shows ice nucleus concentrations from (6.35) assuming water-saturated conditions. [Data from *J. Atmos. Sci.* **42**, 2528 (1985); and *Quart. J. Roy. Met. Soc.* **117**, 207 (1991) and **120**, 573 (1994). Reproduced by permission of The Royal Meteorological Society.]

increase in number without the action of ice nuclei, by what are termed *ice multiplication* (or *ice enhancement*) processes. For example, some ice crystals are quite fragile and may break up when they collide with other ice particles. However, the strongest contender for an ice enhancement process in clouds is one that involves water droplets freezing. When a supercooled droplet freezes in isolation (e.g., in free fall) or after it collides with an ice particle (the freezing of droplets onto an ice particle is called *riming*), it does so in two distinct stages. In the first stage, which occurs almost instantaneously, a fine mesh of ice shoots through the droplet and freezes just enough water to raise the temperature of the droplet to 0 °C. The second stage of freezing is much slower and involves the transfer of heat from the partially frozen droplet to the colder ambient air. During the second stage of freezing an ice shell forms over the surface of the droplet and then thickens progressively inward. As the ice shell advances inward, water is trapped in the interior of the droplet; as this water freezes it expands and sets up large stresses in the ice shell. These stresses may cause the ice shell to crack and even explode, throwing off numerous small ice splinters.

Exercise 6.5 Determine the fraction of the mass of a supercooled droplet that is frozen in the initial stage of freezing if the original temperature of the droplet is −20 °C. What are the percentage increases in the volume of the droplet due to the first and to the second stages of freezing? (Latent heat of melting = 3.3×10^5 J kg^{-1}; specific heat of liquid water = 4218 J K^{-1} kg^{-1}; specific heat of ice = 2106 J K^{-1} kg^{-1}; density of ice = 0.917×10^3 kg m^{-3}.)

Solution: Let m be the mass (in kg) of the droplet and dm the mass of ice that is frozen in the initial stage of freezing. Then the latent heat (in joules) released due to freezing is $3.3 \times 10^5\, dm$. This heat raises the temperature of the unfrozen water and the ice from −20 to 0 °C (at which temperature the first stage of freezing ceases). Therefore,

$$3.3 \times 10^5\, dm = (2106 \times 20\, dm) + [4218 \times 20(m - dm)]$$

Hence

$$\frac{dm}{m} = \frac{4218}{(3.3 \times 10^5/20) - 2106 + 4218} = 0.23$$

Therefore, 23% of the mass of the droplet is frozen during the initial stage of freezing.

Because the density of water is 10^3 kg m^{-3}, when mass dm of water freezes the increase in volume is $[(1/0.917) - 1]\, dm/10^3$. The fractional increase in volume of the mixture is therefore $[(1/0.917) - 1]\, dm/(10^3 V)$, where V is the volume of mass m of water. However, $m/V = 10^3$ kg m^{-3}. Therefore, the fractional increase in volume produced by the initial stage of freezing is

$$[(1/0.917) - 1]\, dm/m = [(1/0.917) - 1]0.23 = 0.021 \text{ or } 2.1\%$$

If the fraction of the mass of the droplet that is frozen in the initial stage of freezing is 0.23, the fraction that is frozen in the second stage of freezing is 0.77. Therefore, the fractional increase in volume produced by the second stage of freezing is

$$[(1/0.917) - 1]\, dm/m = [(1/0.917) - 1]0.77 = 0.070 \text{ or } 7.0\%$$

■

Because an ice particle falling through a supercooled cloud will be impacted by thousands of droplets, each of which might shed numerous ice splinters as it freezes onto the ice particle, ice splinter production by riming is potentially much more important than ice splinter production during the freezing of isolated droplets. Laboratory experiments indicate that ice splinters are ejected during riming, provided the droplets involved have diameters ≥25 μm, temperatures are between −2.5 and −8.5 °C (with peak ice splinter production from −4 to −5 °C), and the impact speed (determined in a cloud by the fall speed of the ice particle undergoing riming) is between ~0.2 and 5 m s^{-1} with peak splinter production at impact speeds of a few m s^{-1}. For example, laboratory measurements show that for a droplet spectrum characterized by 50 drops cm^{-3} with droplets ranging in diameter from ~5 to 35 μm, a LWC of 0.2 g m^{-3}, a temperature of −4.5 °C, and an impact speed of 3.6 m s^{-1}, ~300 ice splinters are produced for every microgram of rime that is accumulated (for a spherical ice particle 1 mm in radius, 1 μg of accumulated rime corresponds to a layer of ice ~0.1 μm thick).

The high concentrations of ice particles (100 liter^{-1} or more) observed in some clouds (see Fig. 6.34) are associated primarily with older clouds. Young cumulus towers generally consist entirely of water droplets

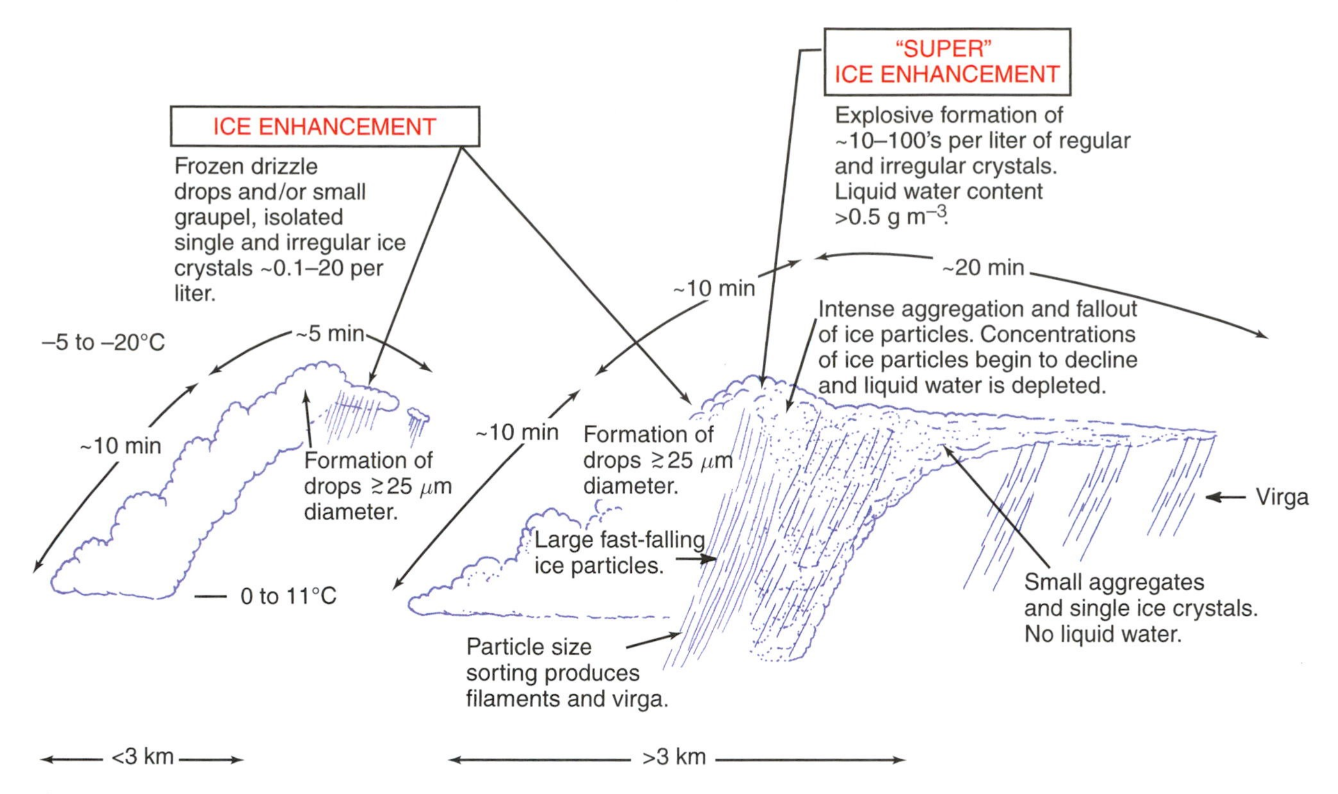

Fig. 6.35 Schematic of ice development in small cumuliform clouds. [Adapted from *Quart. J. Roy. Meteor. Soc.* **117**, 231 (1991). Reproduced by permission of The Royal Meteorological Society.]

and generally require about 10 min before showing signs of plentiful ice particles. It also appears from measurements in clouds that high ice particle concentrations occur after the formation of drops with diameters ≳25 μm and when rimed ice particles appear. These observations are consistent with the hypothesis that the high ice particle concentrations are due to the ejection of ice splinters during riming. However, calculations based on the results of laboratory experiments on ice splinter production during riming suggest that this process is too slow to explain the explosive formation of extremely high concentrations of ice particles observed in some clouds. As indicated schematically in Fig. 6.35, an additional "super" ice enhancement mechanism may sometimes operate, but the exact nature of this mechanism remains a mystery.

6.5.3 Growth of Ice Particles in Clouds

(a) *Growth from the vapor phase.* In a mixed cloud dominated by suprecooled droplets, the air is close to saturated with respect to liquid water and is therefore supersaturation with respect to ice. For example, air saturated with respect to liquid water at −10 °C is supersaturated with respect to ice by 10% and at −20 °C it is supersaturated by 21%. These values are much greater than the supersaturations of cloudy air with respect to liquid water, which rarely exceed 1%. Consequently, in mixed clouds dominated by supercooled water droplets, in which the cloudy air is close to water saturation, ice particles will grow from the vapor phase much more rapidly than droplets. In fact, if a growing ice particle lowers the vapor pressure in its vicinity below water saturation, adjacent droplets will evaporate (Fig. 6.36).

Fig. 6.36 Laboratory demonstration of the growth of an ice crystal at the expense of surrounding supercooled water drops. [Photograph courtesy of Richard L. Pitter.]

Fig. 6.37 The growing cumulus clouds in the foreground with well-defined boundaries contained primarily small droplets. The higher cloud behind with fuzzy boundaries is an older glaciated cloud full of ice crystals. [Photograph courtesy of Art Rangno.]

Fig. 6.38 Fallstreaks of ice crystals from cirrus clouds. The characteristic curved shape of fallstreaks indicates that the wind speed was increasing (from left to right) with increasing altitude. [Photograph courtesy of Art Rangno.]

Cumulus turrets containing relatively large ice particles often have ill-defined, fuzzy boundaries, whereas turrets containing only small droplets have well-defined, sharper boundaries, particularly if the cloud is growing (Fig. 6.37). Another factor that contributes to the difference in appearance of ice and water clouds is the lower equilibrium vapor pressure over ice than over water at the same temperature, which allows ice particles to migrate for greater distances than droplets into the nonsaturated air surrounding a cloud before they evaporate. For the same reason, ice particles that are large enough to fall out of a cloud can survive great distances before evaporating completely, even if the ambient air is subsaturated with respect to ice; ice particles will grow in air that is subsaturated with respect to water, provided that it is supersaturated with respect to ice. The trails of ice crystals so produced are called *fallstreaks* or *virga* (Fig. 6.38).

The factors that control the mass growth rate of an ice crystal by deposition from the vapor phase are similar to those that control the growth of a droplet by condensation (see Section 6.4.1). However, the problem is more complicated because ice crystals are not spherical and therefore points of equal vapor density do not lie on a sphere centered on the crystal (as they do for a droplet). For the special case of a spherical ice particle of radius r, we can write, by analogy with (6.19),

$$\frac{dM}{dt} = 4\pi r D \left[\rho_v(\infty) - \rho_{vc}\right]$$

where ρ_{vc} is the density of the vapor just adjacent to the surface of the crystal and the other symbols were defined in Section 6.4.1. We can derive an expression for the rate of increase in the mass of an ice crystal of arbitrary shape by exploiting the analogy between the vapor field around an ice crystal and the field of electrostatic potential around a charged conductor of the same size and shape.[33] The leakage of charge from the conductor (the analog of the flux of vapor to or from an ice crystal) is proportional to the electrostatic capacity C of the conductor, which is entirely

[33] This analogy was suggested by Harold Jeffreys.[34]

[34] **Harold Jeffreys** (1891–1989) English mathematician and geophysicist. First to suggest that the core of the Earth is liquid. Studied earthquakes and the circulation of the atmosphere. Proposed models for the structure of the outer planets and the origin of the solar system.

determined by the size and shape of the conductor. For a sphere,

$$\frac{C}{\varepsilon_0} = 4\pi r$$

where ε_0 is the permittivity of free space ($8.85 \times 10^{-12}\, C^2\, N^{-1}\, m^{-2}$). Combining the last two expressions, the mass growth rate of a spherical ice crystal is given by

$$\frac{dM}{dt} = \frac{DC}{\varepsilon_0}[\rho_v(\infty) - \rho_{vc}] \tag{6.36}$$

Equation (6.36) is quite general and can be applied to an arbitrarily shaped crystal of capacity C.

Provided that the vapor pressure corresponding to $\rho_v(\infty)$ is not too much greater than the saturation vapor pressure e_{si} over a plane surface of ice, and the ice crystal is not too small, (6.36) can be written as

$$\frac{dM}{dt} = \frac{C}{\varepsilon_0} G_i S_i \tag{6.37}$$

where S_i is the supersaturation (as a fraction) with respect to ice, $[e(\infty) - e_{si}]/e_{si}$, and

$$G_i = D\rho_v(\infty) \tag{6.38}$$

The variation of G_iS_i with temperature for the case of an ice crystal growing in air saturated with water is shown in Fig. 6.39. The product G_iS_i attains a maximum value at about $-14\,°C$, which is due mainly to the fact that the difference between the

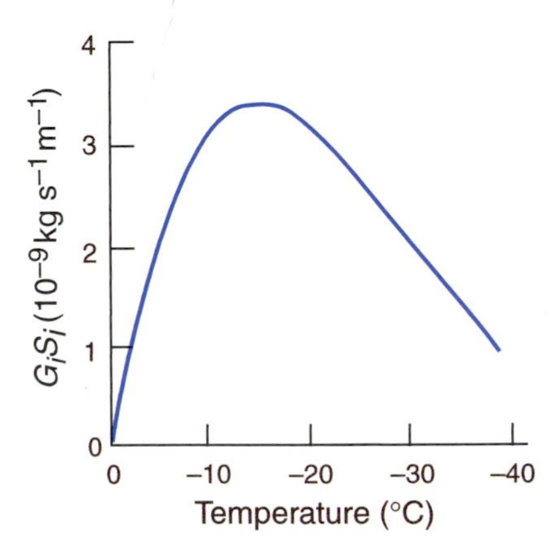

Fig. 6.39 Variation of G_iS_i [see Eq. (6.37)] with temperature for an ice crystal growing in a water-saturated environment at a total pressure of 1000 hPa.

saturated vapor pressures over water and ice is a maximum near this temperature. Consequently, ice crystals growing by vapor deposition in mixed clouds increase in mass most rapidly at temperatures around $-14\,°C$.

The majority of ice particles in clouds are irregular in shape (sometimes referred to as "junk" ice). This may be due, in part, to ice enhancement. However, laboratory studies show that under appropriate conditions ice crystals that grow from the vapor phase can assume a variety of regular shapes (or *habits*) that are either *plate-like* or *column-like*. The simplest plate-like crystals are plane hexagonal plates (Fig. 6.40a), and the simplest column-like

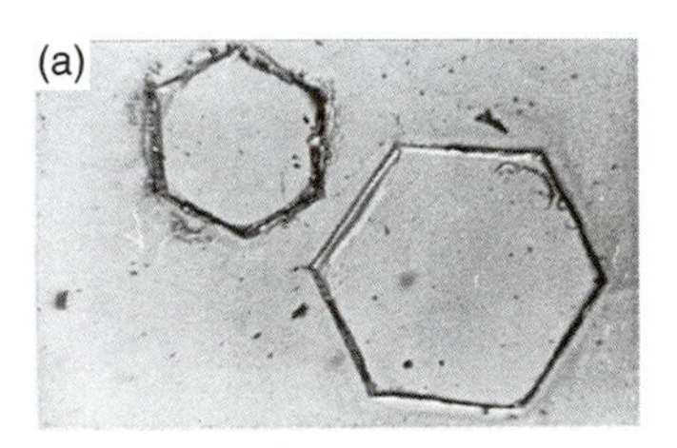

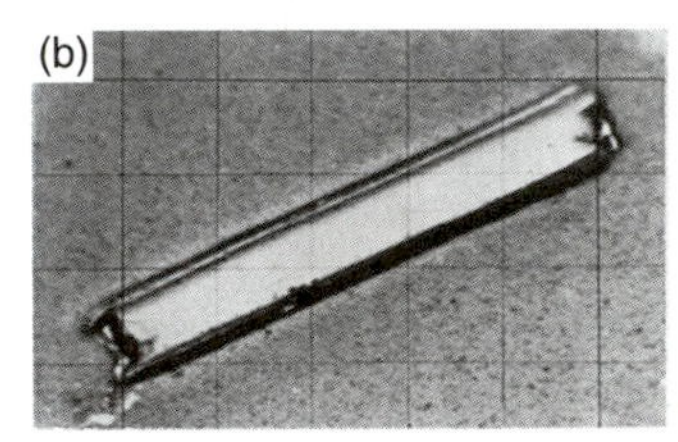

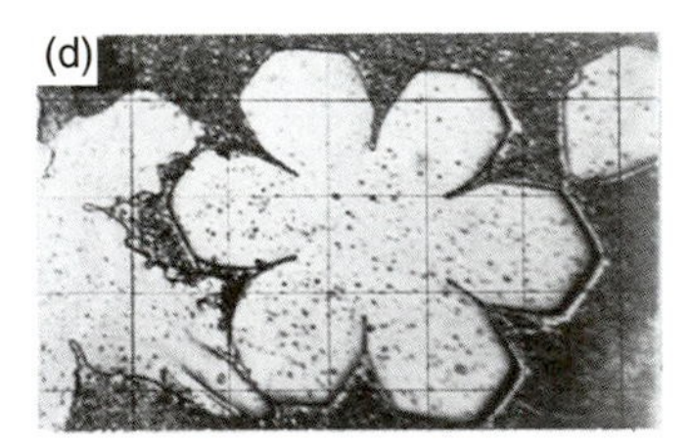

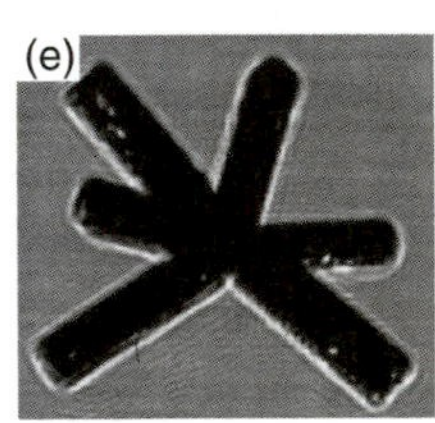

Fig. 6.40 Ice crystals grown from the vapor phase: (a) hexagonal plates, (b) column, (c) dendrite, and (d) sector plate. [Photographs courtesy of Cloud and Aerosol Research Group, University of Washington.] (e) Bullet rosette. [Photograph courtesy of A. Heymsfield.]

Table 6.1 Variations in the basic habits of ice crystals with temperature[a]

Temperature (°C)	Basic habit	Supersaturation[b, c] Between ice and water saturation	Supersaturation[b, c] Near to or greater than water saturation
0 to −2.5	Plate-like	Hexagonal plates	Dendrites −1 to −2 °C
−3	Transition	Equiaxed	Equiaxed
−3.5 to −7.5	Column-like	Columns	Needles −4 to −6 °C Hollow columns −6 to −8 °C
−8.5	Transition	Equiaxed	Equiaxed
−9 to −40	Plate-like	Plates and multiple habits[d]	Scrolls and sector plates −9 to −12 °C Dendrites −12 to −16 °C Sector plates −16 to −20 °C
−40 to −60	Column-like	Solid column rosettes below −41 °C	Hollow column rosettes below −41 °C

[a] From information provided by J. Hallett and M. Bailey.

[b] If the ice crystals are sufficiently large to have significant fall speeds, they will be ventilated by the airflow. Ventilation of an ice crystal has a similar effect on embellishing the crystal habit, as does increasing the supersaturation.

[c] At low supersaturations, crystal growth depends on the presence of molecular defects. As water saturation is approached, surface nucleation occurs near the crystal edges and layers of ice spread toward the crystal interior. Growth at the edges of a crystal is limited by vapor and/or heat transfer and in the interior of a crystal by kinetic processes at the ice-vapor interface.

[d] At lower supersaturations different crystal habits grow under identical ambient conditions depending on the defect structure inherited at nucleation.

crystals are solid columns that are hexagonal in cross section (Fig. 6.40b).

Studies of the growth of ice crystals from the vapor phase under controlled conditions in the laboratory and observations in natural clouds have shown that the basic habit of an ice crystal is determined by the temperature at which it grows (Table 6.1). In the temperature range between 0 and −60 °C the basic habit changes three times. These changes occur near −3, −8, and −40 °C. When the air is saturated or supersaturated with respect to water, the basic habits become embellished. For example, at close to or in excess of water saturation, column-like crystals take the form of long thin needles between −4 and −6 °C, from −12 to −16 °C plate-like crystals appear like ferns, called dendrites (Fig. 6.40c), from −9 to −12 °C and −16 to −20 °C sector plates grow (Fig. 6.40d), and below −40 °C the column-like crystals take the form of column (often called bullet) rosettes (Fig. 6.40e). Because ice crystals are generally exposed to continually changing temperatures and supersaturations as they fall through clouds and to the ground, crystals can assume quite complex shapes.

(b) *Growth by riming; hailstones.* In a mixed cloud, ice particles can increase in mass by colliding with supercooled droplets that then freeze onto them. This process, referred to as growth by riming, leads to the formation of various rimed structures; some examples are shown in Fig. 6.41. Figure 6.41a shows a needle that collected a few droplets on its leading edge as it fell through the air; Fig. 6.41b a uniformly, densely rimed column; Fig. 6.41c a rimed

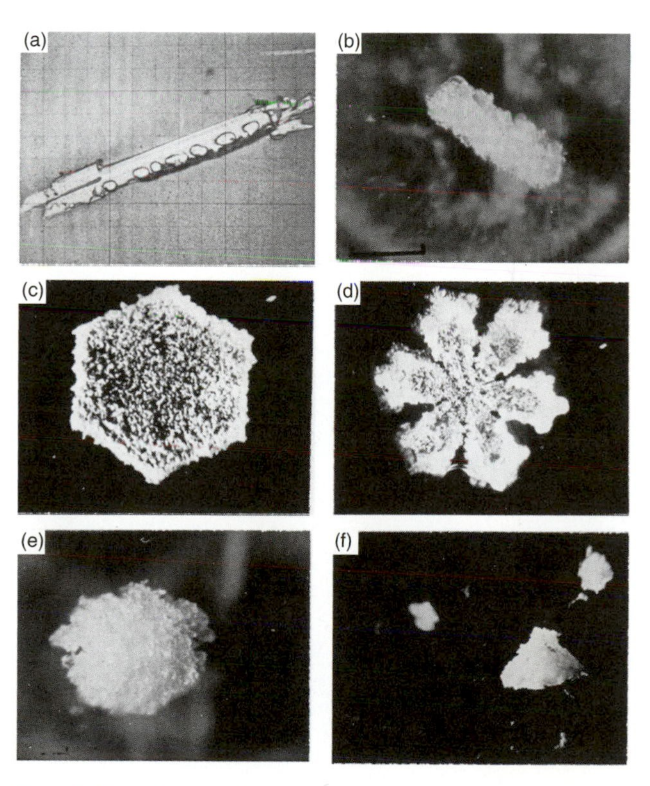

Fig. 6.41 (a) Lightly rimed needle; (b) rimed column; (c) rimed plate; (d) rimed stellar; (e) spherical graupel; and (f) conical graupel. [Photographs courtesy of Cloud and Aerosol Research Group, University of Washington.]

plate; and Fig. 6.41d a rimed stellar crystal. When riming proceeds beyond a certain stage it becomes difficult to discern the original shape of the ice crystal. The rimed particle is then referred to as *graupel*. Examples of spherical and conical graupel are shown in Figs. 6.41e and 6.41f, respectively.

Hailstones represent an extreme case of the growth of ice particles by riming. They form in vigorous convective clouds that have high liquid water contents. The largest hailstone reported in the United States (Nebraska) was 13.8 cm in diameter and weighed about 0.7 kg. However, hailstones about 1 cm in diameter are much more common. If a hailstone collects supercooled droplets at a very high rate, its surface temperature rises to 0 °C and some of the water it collects remains unfrozen. The surface of the hailstone then becomes covered with a layer of water and the hailstone is said to grow *wet*. Under these conditions some of the water may be shed in the wake of the hailstone, but some of the water may be incorporated into a water-ice mesh to form what is known as *spongy hail*.

If a thin section is cut from a hailstone and viewed in transmitted light, it is often seen to consist of alternate dark and light layers (Fig. 6.42). The dark layers are opaque ice containing numerous small air bubbles, and the light layers are clear (bubble-free) ice. Clear ice is more likely to form when the hailstone is growing wet. Detailed examination of the orientation of the individual crystals within a hailstone (which can be seen when the hailstone is viewed between crossed polarizing filters; see inset to Fig. 6.42) can also reveal whether wet growth has occurred. It can be seen from Figs. 6.42 and 6.43 that the surface of a hailstone can contain fairly large lobes. Lobe-like growth appears to be more pronounced when the accreted droplets are small and growth is near the wet limit. The development of lobes may be due to the fact that any small bumps on a hailstone will be areas of enhanced collection efficiencies for droplets.

Fig. 6.43 Artificial hailstone (i.e., grown in the laboratory) showing a lobe structure. Growth was initially dry but tended toward wet growth as the stone grew. [Photograph courtesy of I. H. Bailey and W. C. Macklin.]

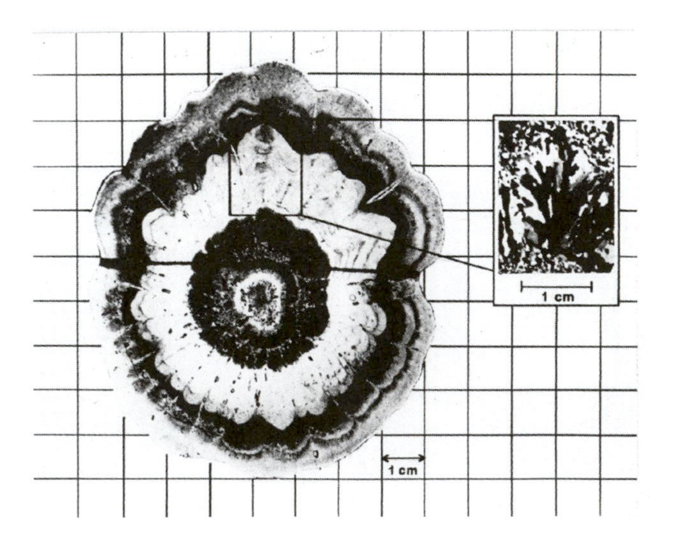

Fig. 6.42 Thin section through the center of a hailstone. [From *Quart. J. Roy. Met. Soc.* **92**, 10 (1966). Reproduced by permission of The Royal Meteorological Society.]

(c) *Growth by aggregation*. The third mechanism by which ice particles grow in clouds is by colliding and aggregating with one another. Ice particles can collide with each other, provided their terminal fall speeds are different. The terminal fall speed of an unrimed column-like ice crystal increases as the length of the crystal increases; for example, the fall speeds of needles 1 and 2 mm in length are about 0.5 and 0.7 m s^{-1}, respectively. In contrast, unrimed plate-like ice crystals have terminal fall speeds that are virtually independent of their diameter for the following reason. The thickness of a plate-like crystal is essentially independent of its diameter, therefore, its mass varies linearly with its cross-sectional area. Because the drag force acting on a plate-like crystal also varies as the cross-sectional area of the crystal, the terminal fall speed, which is determined by a balance between the drag and the gravitational forces acting on a crystal, is independent of the diameter of a plate. Because unrimed plate-like crystals all have similar terminal fall speeds, they are unlikely to collide with each other (unless they come close enough to be influenced by wake effects). The terminal fall speeds of rimed crystals and graupel are strongly dependent on their degrees of riming and their dimensions. For example, graupel particles 1 and 4 mm in diameter have terminal fall speeds of about 1 and 2.5 m s^{-1}, respectively. Consequently, the frequency of collisions of ice particles in clouds is enhanced greatly if some riming has taken place.

The second factor that influences growth by aggregation is whether two ice particles adhere when they collide. The probability of adhesion is determined primarily by two factors: the types of ice particles and the temperature. Intricate crystals, such as dendrites, tend to adhere to one another because they become entwined on collision, whereas two solid plates will tend to rebound. Apart from this dependence upon habit, the probability of two colliding crystals adhering increases with increasing temperature, with adhesion being particularly likely above about −5 °C at which temperatures ice surfaces become quite "sticky." Some examples of ice particle aggregates are shown in Fig. 6.44.

6.5.4 Formation of Precipitation in Cold Clouds

As early as 1789 Franklin[35] suggested that "much of what is rain, when it arrives at the surface of the Earth, might have been snow, when it began its descent . . ." This idea was not developed until the early part of the last century when Wegener, in 1911, stated that ice particles would grow preferentially by deposition from the vapor phase in a mixed cloud. Subsequently, Bergeron, in 1933, and Findeisen[36], in 1938, developed this idea in a more

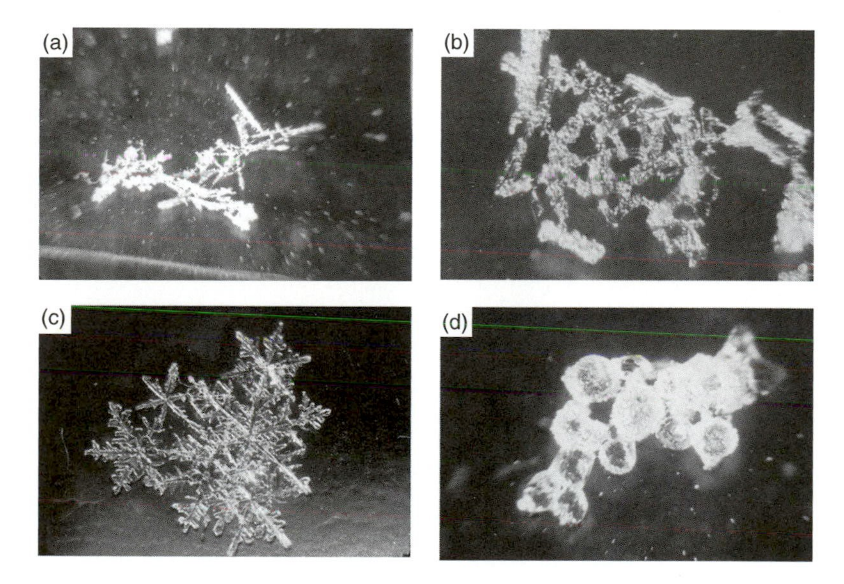

Fig. 6.44 Aggregates of (a) rimed needles; (b) rimed columns; (c) dendrites; and (d) rimed frozen drops. [Photographs courtesy of Cloud and Aerosol Research Group, University of Washington.]

[35] **Benjamin Franklin** (1706–1790) American scientist, inventor, statesman, and philosopher. Largely self-taught, and originally a printer and publisher by trade. First American to win international fame in science. Carried out fundamental work on the nature of electricity (introduced the terms "positive charge," "negative charge," and "battery"). Showed lightning to be an electrical phenomenon (1752). Attempted to deduce paths of storms over North America. Invented the lightning conductor, daylight savings time, bifocals, Franklin stove, and the rocking chair! First to study the Gulf Stream.

[36] **Theodor Robert Walter Findeisen** (1909–1945) German meteorologist. Director of Cloud Research, German Weather Bureau, Prague, Czechoslovakia, from 1940. Laid much of the foundation of modern cloud physics and foresaw the possibility of stimulating rain by introducing artificial ice nuclei. Disappeared in Czechoslovakia at the end of World War II.

quantitative manner and indicated the importance of ice nuclei in the formation of crystals. Because Findeisen carried out his field studies in northwestern Europe, he was led to believe that all rain originates as ice. However, as shown in Section 6.4.2, rain can also form in warm clouds by the collision-coalescence mechanism.

We will now consider the growth of ice particles to precipitation size in a little more detail. Application of (6.36) to the case of a hexagonal plate growing by deposition from the vapor phase in air saturated with respect to water at -5 °C shows that the plate can obtain a mass of $\sim 7\ \mu$g (i.e., a radius of $\sim$0.5 mm) in half an hour (see Exercise 6.27). Thereafter, its mass growth rate decreases significantly. On melting, a 7-μg ice crystal would form a small drizzle drop about 130 μm in radius, which could reach the ground, provided that the updraft velocity of the air were less than the terminal fall speed of the crystal (about 0.3 m s^{-1}) and the drop survived evaporation as it descended through the subcloud layer. Calculations such as this indicate that the growth of ice crystals by deposition of vapor is not sufficiently fast to produce large raindrops.

Unlike growth by deposition, the growth rates of an ice particle by riming and aggregation increase as the ice particle increases in size. A simple calculation shows that a plate-like ice crystal, 1 mm in diameter, falling through a cloud with a liquid content of 0.5 g m^{-3}, could develop into a spherical graupel particle about 0.5 mm in radius in a few minutes (see Exercise 6.28). A graupel particle of this size, with a density of 100 kg m^{-3}, has a terminal fall speed of about 1 m s^{-1} and would melt into a drop about 230 μm in radius. The radius of a snowflake can increase from 0.5 mm to 0.5 cm in $\sim$30 min due to aggregation with ice crystals, provided that the ice content of the cloud is about 1 g m^{-3} (see Exercise 6.29). An aggregated snow crystal with a radius of 0.5 cm has a mass of about 3 mg and a terminal fall speed of about 1 m s^{-1}. Upon melting, a snow crystal of this mass would form a drop about 1 mm in radius. We conclude from these calculations that the growth of ice crystals, first by deposition from the vapor phase in mixed clouds and then by riming and/or aggregation, can produce precipitation-sized particles in reasonable time periods (say about 40 min).

The role of the ice phase in producing precipitation in cold clouds is demonstrated by radar observations. For example, Fig. 6.45 shows a radar screen (on which the intensity of radar echoes reflected from atmospheric targets are displayed) while the radar antenna was pointing vertically upward and clouds drifted over the radar. The horizontal band (in brown) just above a height of 2 km was produced by the melting of ice particles. This is referred to as the "*bright band*." The radar reflectivity is high around the melting level because, while melting, ice particles become coated with a film of water that increases their radar reflectivity greatly. When the crystals have melted completely, they collapse into droplets and their terminal fall speeds increase so that the concentration of

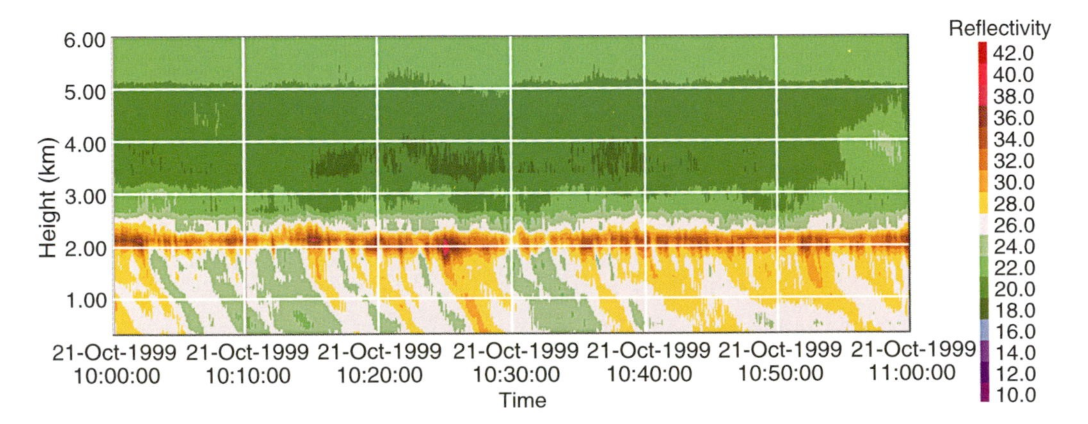

Fig. 6.45 Reflectivity (or "echo") from a vertically pointing radar. The horizontal band of high reflectivity values (in brown), located just above a height of 2 km, is the melting band. The curved trails of relatively high reflectivity (in yellow) emanating from the bright band are *fallstreaks* of precipitation, some of which reach the ground. [Courtesy of Sandra E. Yuter.]

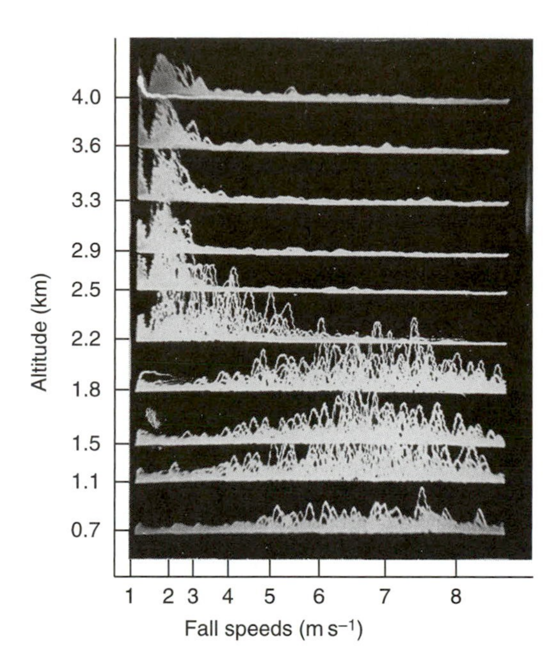

Fig. 6.46 Spectra of Doppler fall speeds for precipitation particles at ten heights in the atmosphere. The melting level is at about 2.2 km. [Courtesy of Cloud and Aerosol Research Group, University of Washington.]

particles is reduced. These changes result in a sharp decrease in radar reflectivity below the melting band.

The sharp increase in particle fall speeds produced by melting is illustrated in Fig. 6.46, which shows the spectrum of fall speeds of precipitation particles measured at various heights with a vertically pointing Doppler radar.[37] At heights above 2.2 km the particles are ice with fall speeds centered around 2 m s^{-1}. At 2.2 km the particles are partially melted, and below 2.2 km there are raindrops with fall speeds centered around 7 m s^{-1}.

6.5.5 Classification of Solid Precipitation

The growth of ice particles by deposition from the vapor phase, riming, and aggregation leads to a very wide variety of solid precipitation particles. A relatively simple classification into 10 main classes is shown in Table 6.2.

6.6 Artificial Modification of Clouds and Precipitation

As shown in Sections 6.2–6.5, the microstructures of clouds are influenced by the concentrations of CCN and ice nuclei, and the growth of precipitation particles is a result of instabilities that exist in the microstructures of clouds. These instabilities are of two main types. First, in warm clouds the larger drops increase in size at the expense of the smaller droplets due to growth by the collision-coalescence mechanism. Second, if ice particles exist in a certain optimum range of concentrations in a mixed cloud, they grow by deposition at the expense of the droplets (and subsequently by riming and aggregation). In light of these ideas, the following techniques have been suggested whereby clouds and precipitation might be modified artificially by so-called *cloud seeding*.

- Introducing large hygroscopic particles or water drops into warm clouds to stimulate the growth of raindrops by the collision-coalescence mechanism.
- Introducing artificial ice nuclei into cold clouds (which may be deficient in ice particles) to stimulate the production of precipitation by the ice crystal mechanism.
- Introducing comparatively high concentrations of artificial ice nuclei into cold clouds to reduce drastically the concentrations of supercooled droplets and thereby inhibit the growth of ice particles by deposition and riming, thereby dissipating the clouds and suppressing the growth of precipitable particles.

6.6.1 Modification of Warm Clouds

Even in principle, the introduction of water drops into the tops of clouds is not a very efficient method for producing rain, since large quantities of water

[37] Doppler radars, unlike conventional meteorological radars, transmit coherent electromagnetic waves. From measurements of the difference in frequencies between returned and transmitted waves, the velocity of the target (e.g., precipitation particles) along the line of sight of the radar can be deduced. Radars used by the police for measuring the speeds of motor vehicles are based on the same principle.

Table 6.2 A classification of solid precipitation[a,b,c]

Typical forms	Symbol	Graphic symbol
	F1	
	F2	
	F3	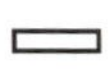
	F4	
	F5	
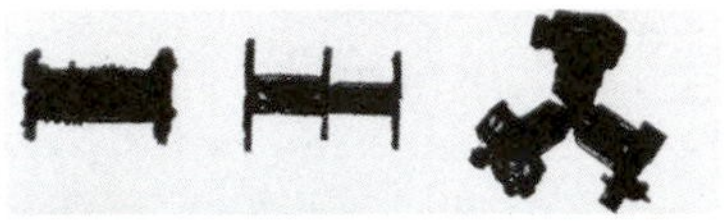	F6	
	F7	
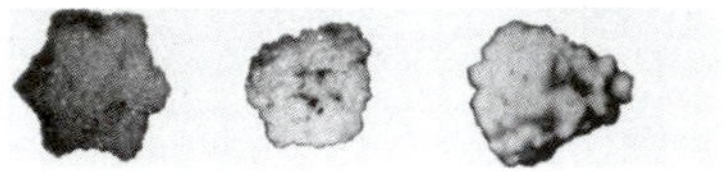	F8	
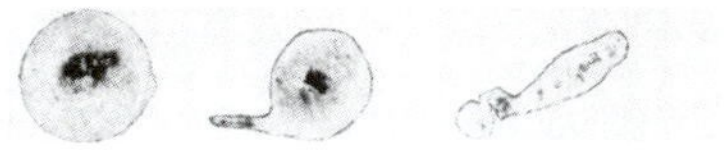	F9	
	F10	

[a] Suggested by the International Association of Hydrology's commission of snow and ice in 1951. [Photograph courtesy of V. Schaefer.]

[b] Additional characteristics: *p*, broken crystals; *r*, rime-coated particles not sufficiently coated to be classed as graupel; *f*, clusters, such as compound snowflakes, composed of several individual snow crystals; *w*, wet or partly melted particles.

[c] Size of particle is indicated by the general symbol D. The size of a crystal or particle is its greatest extension measured in millimeters. When many particles are involved (e.g., a compound snowflake), it refers to the average size of the individual particles.

are required. A more efficient technique might be to introduce small water droplets (radius $\simeq 30\ \mu m$) or hygroscopic particles (e.g., NaCl) into the base of a cloud; these particles might then grow by condensation, and then by collision-coalescence, as they are carried up and subsequently fall through a cloud.

In the second half of the last century, a number of cloud seeding experiments on warm clouds were carried out using water drops and hygroscopic particles. In some cases, rain appeared to be initiated by the seeding, but because neither extensive physical nor rigorous statistical evaluations were carried out, the results were inconclusive. Recently, there has been somewhat of a revival of interest in seeding warm clouds with hygroscopic nuclei to increase precipitation but, as yet, the efficacy of this technique has not been proven.

Seeding with hygroscopic particles has been used in attempts to improve visibility in warm fogs. Because the visibility in a fog is inversely proportional to the number concentration of droplets and

Description
A plate is a thin, plate-like snow crystal the form of which more or less resembles a hexagon or, in rare cases, a triangle. Generally all edges or alternative edges of the plate are similar in pattern and length.
A stellar crystal is a thin, flat snow crystal in the form of a conventional star. It generally has 6 arms but stellar crystals with 3 or 12 arms occur occasionally. The arms may lie in a single plane or in closely spaced parallel planes in which case the arms are interconnected by a very short column.
A column is a relatively short prismatic crystal, either solid or hollow, with plane, pyramidal, truncated, or hollow ends. Pyramids, which may be regarded as a particular case, and combinations of columns are included in this class.
A needle is a very slender, needle-like snow particle of approximately cylindrical form. This class includes hollow bundles of parallel needles, which are very common, and combinations of needles arranged in any of a wide variety of fashions.
A spatial dendrite is a complex snow crystal with fern-like arms that do not lie in a plane or in parallel planes but extend in many directions from a central nucleus. Its general form is roughly spherical.
A capped column is a column with plates of hexagonal or stellar form at its ends and, in many cases, with additional plates at intermediate positions. The plates are arranged normal to the principal axis of the column. Occasionally, only one end of the column is capped in this manner.
An irregular crystal is a snow particle made up of a number of small crystals grown together in a random fashion. Generally the component crystals are so small that the crystalline form of the particle can only be seen with the aid of a magnifying glass or microscope.
Graupel, which includes soft hail, small hail, and snow pellets, is a snow crystal or particle coated with a heavy deposit of rime. It may retain some evidence of the outline of the original crystal, although the most common type has a form that is approximately spherical.
Ice pellets (frequently called sleet in North America) are transparent spheroids of ice and are usually fairly small. Some ice pellets do not have a frozen center, which indicates that, at least in some cases, freezing takes place from the surface inward.
A hailstone[d] is a grain of ice, generally having a laminar structure and characterized by its smooth glazed surface and its translucent or milky-white center. Hail is usually associated with those atmospheric conditions that accompany thunderstorms. Hailstones are sometimes quite large.

[d] Hail, like rain, refers to a number of particles, whereas hailstone, like raindrop, refers to an individual particle.

to their total surface area, visibility can be improved by decreasing either the concentration or the size of the droplets. When hygroscopic particles are dispersed into a warm fog, they grow by condensation (causing partial evaporation of some of the fog droplets) and the droplets so formed fall out of the fog slowly. Fog clearing by this method has not been widely used due to its expense and lack of dependability. At the present time, the most effective methods for dissipating warm fogs are "brute force" approaches, involving evaporating the fog droplets by ground-based heating.

6.6.2 Modification of Cold Clouds

We have seen in Section 6.5.3 that when supercooled droplets and ice particles coexist in a cloud, the ice particles may increase to precipitation size rather rapidly. We also saw in Section 6.5.1 that in some situations the concentrations of ice nuclei may be less than that required for the efficient initiation of the ice crystal mechanism for the formation of precipitation. Under these conditions, it is argued, clouds might be induced to rain by seeding them with artificial ice nuclei or some other material that might increase the concentration of ice particles. This idea was the basis for most of the cloud seeding experiments carried out in the second half of the 20th century.

A material suitable for seeding cold clouds was first discovered in July 1946 in Project Cirrus, which was carried out under the direction of Irving Langmuir.[38] One of Langmuir's assistants, Vincent Schaefer,[39] observed in laboratory experiments that when a small piece of dry ice (i.e., solid carbon dioxide) is dropped into a cloud of supercooled droplets, numerous small ice crystals are produced and the cloud is glaciated quickly. In this transformation, dry ice does not serve as an ice nucleus in the usual sense of this term, but rather, because it is so cold (−78 °C), it causes numerous ice crystals to form in its wake by homogeneous nucleation. For example, a pellet of dry ice 1 cm in diameter falling through air at −10 °C produces about 10^{11} ice crystals.

The first field trials using dry ice were made in Project Cirrus on 13 November 1946, when about 1.5 kg of crushed dry ice was dropped along a line about 5 km long into a layer of a supercooled altocumulus cloud. Snow was observed to fall from the base of the seeded cloud for a distance of about 0.5 km before it evaporated in the dry air.

Because of the large numbers of ice crystals that a small amount of dry ice can produce, it is most suitable for *overseeding* cold clouds rather than producing ice crystals in the optimal concentrations (~1 liter^{-1}) for enhancing precipitation. When a cloud is overseeded it is converted completely into ice crystals (i.e., it is glaciated). The ice crystals in a glaciated cloud are generally quite small and, because there are no supercooled droplets present, supersaturation with respect to ice is either low or nonexistent. Therefore, instead of the ice crystals growing (as they would in a mixed cloud at water saturation) they tend to evaporate. Consequently, seeding with dry ice can dissipate large areas of supercooled cloud or fog (Fig. 6.47). This technique is used for clearing supercooled fogs at several international airports.

Fig. 6.47 A γ-shaped path cut in a layer of supercooled cloud by seeding with dry ice. [Photograph courtesy of General Electric Company, Schenectady, New York.]

Following the demonstration that supercooled clouds can be modified by dry ice, Bernard Vonnegut,[40] who was also working with Langmuir, began searching for artificial ice nuclei. In this search he was guided by the expectation that an effective ice nucleus should have a crystallographic structure similar to that of ice. Examination of crystallographic tables revealed that silver iodide fulfilled this requirement. Subsequent laboratory tests showed that silver iodide could act as an ice nucleus at temperatures as high as −4 °C.

The seeding of natural clouds with silver iodide was first tried as part of Project Cirrus on 21 December 1948. Pieces of burning charcoal impregnated with silver iodide were dropped from an aircraft into about 16 km^2 of supercooled stratus cloud 0.3 km thick at a temperature of −10 °C. The cloud was converted into ice crystals by less than 30 g of silver iodide!

[38] **Irving Langmuir** (1881–1957) American physicist and chemist. Spent most of his working career as an industrial chemist in the GE Research Laboratories in Schenectady, New York. Made major contributions to several areas of physics and chemistry and won the Nobel Prize in chemistry in 1932 for work on surface chemistry. His major preoccupation in later years was cloud seeding. His outspoken advocacy of large-scale effects of cloud seeding involved him in much controversy.

[39] **Vincent Schaefer** (1906–1993) American naturalist and experimentalist. Left school at age 16 to help support the family income. Initially worked as a toolmaker at the GE Research Laboratory, but subsequently became Langmuir's research assistant. Schaefer helped to create the Long Path of New York (a hiking trail from New York City to Whiteface Mt. in the Adirondacks); also an expert on Dutch barns.

[40] **Bernard Vonnegut** (1914–1997) American physical chemist. In addition to his research on cloud seeding, Vonnegut had a lifelong interest in thunderstorms and lightning. His brother, Kurt Vonnegut the novelist, wrote "My longest experience with common decency, surely, has been with my older brother, my only brother, Bernard . . . We were given very different sorts of minds at birth. Bernard could never be a writer and I could never be a scientist." Interestingly, following Project Cirrus, neither Vonnegut nor Schaefer became deeply involved in the quest to increase precipitation by artificial seeding.

Many artificial ice nucleating materials are now known (e.g., lead iodide, cupric sulfide) and some organic materials (e.g., phloroglucinol, metaldehyde) are more effective as ice nuclei than silver iodide. However, silver iodide has been used in most cloud seeding experiments.

Since the first cloud seeding experiments in the 1940s, many more experiments have been carried out all over the world. It is now well established that the concentrations of ice crystals in clouds can be increased by seeding with artificial ice nuclei and that, under certain conditions, precipitation can be artificially initiated in some clouds. However, the important question is: under what conditions (if any) can seeding with artificial ice nuclei be employed to produce significant increases in precipitation on the ground in a predictable manner and over a large area? This question remains unanswered.

So far we have discussed the role of artificial ice nuclei in modifying the microstructures of cold clouds. However, when large volumes of a cloud are glaciated by overseeding, the resulting release of latent heat provides added buoyancy to the cloudy air. If, prior to seeding, the height of a cloud were restricted by a stable layer, the release of the latent heat of fusion caused by artificial seeding might provide enough buoyancy to push the cloud through the inversion and up to its level of free convection. The cloud top might then rise to much greater heights than it would have done naturally. Figure 6.48 shows the explosive growth of a cumulus cloud that may have been produced by overseeding.

Seeding experiments have been carried out in attempts to reduce the damage produced by hailstones. Seeding with artificial nuclei should tend to increase the number of small ice particles competing for the available supercooled droplets. Therefore, seeding should result in a reduction in the average size of the hailstones. It is also possible that, if a hailstorm is overseeded with extremely large numbers of ice nuclei, the majority of the supercooled droplets in the cloud will be nucleated, and the growth of hailstones by riming will be reduced significantly. Although these hypotheses are plausible, the results of experiments on hail suppression have not been encouraging.

Exploratory experiments have been carried out to investigate if orographic snowfall might be redistributed by overseeding. Rimed ice particles have relatively large terminal fall speeds ($\sim 1 \text{ m s}^{-1}$), therefore they follow fairly steep trajectories as they fall to the ground. If clouds on the windward side of a mountain are artificially over-seeded, supercooled droplets can be virtually eliminated and growth by riming significantly reduced (Fig. 6.49). In the absence of riming, the ice

Fig. 6.48 Causality or chance coincidence? Explosive growth of cumulus cloud (a) 10 min; (b) 19 min; 29 min; and 48 min after it was seeded near the location of the arrow in (a). [Photos courtesy of J. Simpson.]

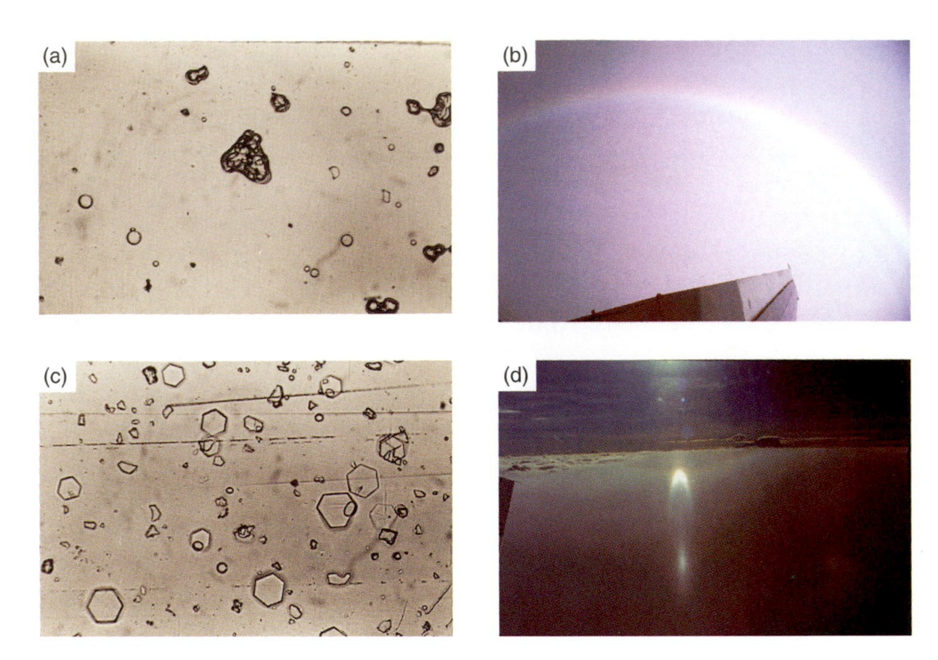

Fig. 6.49 (a) Large rimed irregular particles and small water droplets collected in unseeded clouds over the Cascade Mountains. (b) Cloud bow produced by the refraction of light in small water droplets. Following heavy seeding with artificial ice nuclei, the particles in the cloud were converted into small unrimed plates (c) which markedly changed the appearance of the clouds. In (d) the uniform cloud in the foreground is the seeded cloud and the more undulating cloud in the background is the unseeded cloud. In the seeded cloud, optical effects due to ice particles (portion of the 22° halo, lower tangent arc to 22° halo, and subsun) can be seen. [Photographs courtesy of Cloud and Aerosol Research Group, University of Washington.]

particles grow by deposition from the vapor phase and their fall speeds are reduced by roughly a factor of 2. Winds aloft can then carry these crystals farther before they reach the ground. In this way, it is argued, it might be possible to divert snowfall from the windward slopes of mountain ranges (where precipitation is often heavy) to the drier leeward slopes.

6.6.3 Inadvertent Modification

Some industries release large quantities of heat, water vapor, and cloud-active aerosol (CCN and ice nuclei) into the atmosphere. Consequently, these effluents might modify the formation and structure of clouds and affect precipitation. For example, the effluents from a single paper mill can profoundly affect the surrounding area out to about 30 km (Fig. 6.50). Paper mills, the burning of agricultural wastes, and forest fires emit large numbers of CCN ($\sim 10^{17}$ s^{-1} active at 1% supersaturation), which can change droplet concentrations in clouds downwind. High concentrations of ice nuclei have been observed in the plumes from steel mills.

Fig. 6.50 The cloud in the valley in the background formed due to effluents from a paper mill. In the foreground, the cloud is spilling through a gap in the ridge into an adjacent valley. [Photograph courtesy of C. L. Hosler.]

Large cities can affect the weather in their vicinities. Here the possible interactions are extremely complex since, in addition to being areal sources of aerosol, trace gases, heat, and water vapor, large cities modify the radiative properties of the Earth's surface, the moisture content of the soil, and the surface roughness. The existence of urban "heat islands," several degrees warmer than adjacent less populated regions, is well documented. In the summer months increases in precipitation of 5–25% over background values occur 50–75 km downwind of some cities (e.g., St. Louis, Missouri). Thunderstorms and hailstorms may be more frequent, with the areal extent and magnitude of the perturbations related to the size of the city. Model simulations indicate that enhanced upward air velocities, associated with variations in surface roughness and the heat island effect, are most likely responsible for these anomalies.

6.4 Holes in Clouds

Photographs of holes (i.e., relatively large clear regions) in thin layers of supercooled cloud, most commonly altocumulus, date back to at least 1926. The holes can range in shape from nearly circular (Fig. 6.51a) to linear tracks (Fig. 6.51b). The holes are produced by the removal of supercooled droplets by copious ice crystals (~100–1000 per liter) in a similar way to the formation of holes in supercooled clouds by artificial seeding (see Fig. 6.41). However, the holes of interest here are formed by natural seeding from above a supercooled cloud that is intercepted by a fallstreak containing numerous ice particles (see Fig. 6.45) or by an aircraft penetrating the cloud.

In the case of formation by an aircraft, the ice particles responsible for the evaporation of the supercooled droplets are produced by the rapid expansion, and concomitant cooling, of air in the vortices produced in the wake of an aircraft (so-called *aircraft produced ice particles* or APIPs). If the air is cooled below about −40 °C, the ice particles are produced by homogeneous nucleation (see Section 6.5.1). With somewhat less but still significant cooling, the ice crystals may be nucleated heterogeneously (see Section 6.5.2). The crystals so produced are initially quite small and uniform in size, but they subsequently grow fairly uniformly at the expense of the supercooled droplets in the cloud (see Fig. 6.36). The time interval between an aircraft penetrating a supercooled cloud and the visible appearance of a clear area is ~10–20 min.

APIPs are most likely at low ambient temperatures (at or below −8 °C) when an aircraft is flown at maximum power but with gear and flaps extended; this results in a relatively low airspeed and high drag. Not all aircraft produce APIPs.

Fig. 6.51 (a) A hole in a layer of supercooled altocumulus cloud. Note the fallout of ice crystals from the center of the hole. [Copyright A. Sealls.] (b) A clear track produced by an aircraft flying in a supercooled altocumulus cloud. [Courtesy of Art Rangno.]

The shape of the hole in a cloud depends on the angle of interception of the fallstreak or the aircraft flight path with the cloud. For example, if an aircraft descends steeply through a cloud it will produce a nearly circular hole (Fig. 6.51a), but if the aircraft flies nearly horizontally through a cloud it will produce a linear track (Fig. 6.51b).

6.7 Thunderstorm Electrification

The dynamical structure of thunderstorms is described in Chapter 10. Here we are concerned with the microphysical mechanisms that are thought to be responsible for the electrification of thunderstorms and with the nature of lightning flashes and thunder.

6.7.1 Charge Generation

All clouds are electrified to some degree.[41] However, in vigorous convective clouds sufficient electrical charges are separated to produce electric fields that exceed the dielectric breakdown of cloudy air ($\sim$1 MV m^{-1}), resulting in an initial *intracloud* (i.e., between two points in the same cloud) lightning discharge.

The distribution of charges in thunderstorms has been investigated with special radiosondes (called altielectrographs), by measuring the changes in the electric field at the ground that accompany lightning flashes, and with instrumented aircraft. A summary of the results of such studies, for a relatively simple cloud, is shown in Fig. 6.52. The magnitudes of the lower negative charge and the upper positive charge are $\sim$10–100 coulombs (hereafter symbol "C"), or a few nC m^{-3}. The location of the negative charge (called the *main charging zone*) is rather well defined between the -10 °C and about -20 °C temperature levels. The positive charge is distributed in a more diffuse region above the negative charge. Although there have been a few reports of lightning from warm clouds, the vast majority of thunderstorms occur in cold clouds.

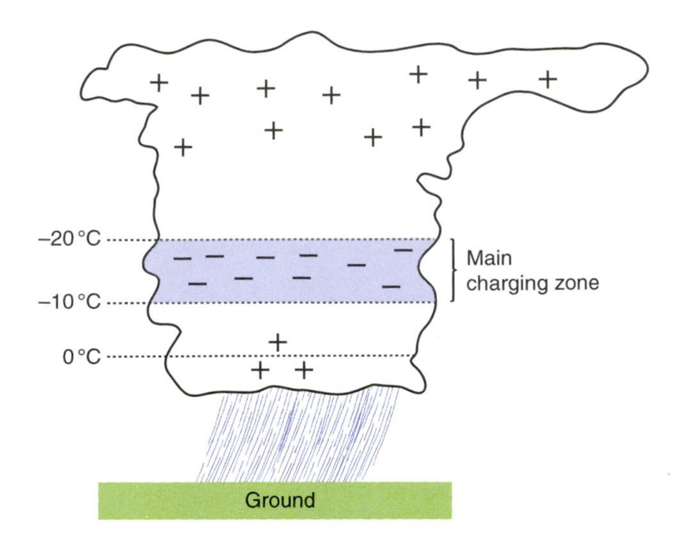

Fig. 6.52 Schematic showing the distribution of electric charges in a typical and relatively simple thunderstorm. The lower and smaller positive charge is not always present.

An important observational result, which provides the basis for most theories of thunderstorm

[41] Benjamin Franklin, in July 1750, was the first to propose an experiment to determine whether thunderstorms are electrified. He suggested that a sentry box, large enough to contain a man and an insulated stand, be placed at a high elevation and that an iron rod 20–30 ft in length be placed vertically on the stand, passing out through the top of the box. He then proposed that if a man stood on the stand and held the rod he would "be electrified and afford sparks" when an electrified cloud passed overhead. Alternatively, he suggested that the man stand on the floor of the box and bring near to the rod one end of a piece of wire, held by an insulating handle, while the other end of the wire was connected to the ground. In this case, an electric spark jumping from the rod to the wire would be proof of cloud electrification. (Franklin did not realize the danger of these experiments: they can kill a person—and have done so—if there is a direct lightning discharge to the rod.) The proposed experiment was set up in Marly-la-Ville in France by d'Alibard[42], and on 10 May 1752 an old soldier, called Coiffier, brought an earthed wire near to the iron rod while a thunderstorm was overhead and saw a stream of sparks. This was the first direct proof that thunderstorms are electrified. Joseph Priestley described it as "the greatest discovery that has been made in the whole compass of philosophy since the time of Sir Isaac Newton." (Since Franklin proposed the use of the lightning conductor in 1749, it is clear that by that date he had already decided in his own mind that thunderstorms were electrified.) Later in the summer of 1752 (the exact date is uncertain), and before hearing of d'Alibard's success, Franklin carried out his famous kite experiment in Philadelphia and observed sparks to jump from a key attached to a kite string to the knuckles of his hand. By September 1752 Franklin had erected an iron rod on the chimney of his home, and on 12 April 1753, by identifying the sign of the charge collected on the lower end of the rod when a storm passed over, he had concluded that "clouds of a thundergust are most commonly in a negative state of electricity, but sometimes in a positive state—the latter, I believe, is rare." No more definitive statement as to the electrical state of thunderstorms was made until the second decade of the 20th century when C. T. R. Wilson[43] showed that the lower regions of thunderstorms are generally negatively charged while the upper regions are positively charged.

[42] **Thomas Francois d'Alibard** (1703–1779) French naturalist. Translated into French Franklin's *Experiments and Observations on Electricity*, Durand, Paris, 1756, and carried out many of Franklin's proposed experiments.

[43] **C. T. R. Wilson** (1869–1959) Scottish physicist. Invented the cloud chamber named after him for studying ionizing radiation (e.g., cosmic rays) and charged particles. Carried out important studies on condensation nuclei and atmospheric electricity. Awarded the Nobel Prize in physics in 1927.

electrification, is that the onset of strong electrification follows the occurrence (detected by radar) of heavy precipitation within the cloud in the form of graupel or hailstones. Most theories assume that as a graupel particle or hailstone (hereafter called the *rimer*) falls through a cloud it is charged negatively due to collisions with small cloud particles (droplets or ice), giving rise to the negative charge in the main charging zone. The corresponding positive charge is imparted to cloud particles as they rebound from the rimer, and these small particles are then carried by updrafts to the upper regions of the cloud. The exact conditions and mechanism by which a rimer might be charged negatively, and smaller cloud particles charged positively, have been a matter of debate for some hundred years. Many potentially promising mechanisms have been proposed but subsequently found to be unable to explain the observed rate of charge generation in thunderstorms or, for other reasons, found to be untenable.

Exercise 6.6 The rate of charge generation in a thunderstorm is $\sim 1\ \mathrm{C\,km^{-3}\,min^{-1}}$. Determine the electric charge that would have to be separated for each collision of an ice crystal with a rimer (e.g., a graupel particle) to explain this rate of charge generation. Assume that the concentration of ice crystals is $10^5\ \mathrm{m^{-3}}$, their fall speed is negligible compared to that of the rimer, the ice crystals are uncharged prior to colliding with the rimer, their collision efficiency with the rimer is unity, and all of the ice crystals rebound from the rimer. Assume also that the rimers are spheres of radius 2 mm, the density of a rimer is 500 kg $\mathrm{m^{-3}}$, and the precipitation rate due to the rimers is 5 cm per hour of water equivalent.

Solution: If $\frac{dN}{dt}$ is the number of collisions of ice crystals with rimers in a unit volume of air in 1 s, and each collision separates q coulombs (C) of electric charge, the rate of charge separation per unit volume of air per unit time in the cloud by this mechanism is

$$\frac{dQ}{dt} = \frac{dN}{dt} q \tag{6.39}$$

If the fall speed of the ice crystals is negligible, and the ice crystals collide with and separate from a rimer with unit efficiency,

$$\begin{aligned}\frac{dN}{dt} &= \text{(volume swept out by one rimer in 1 s)}\\ &\quad \text{(number of rimers per unit volume of air)}\\ &\quad \text{(number of ice crystals per unit volume of air)}\\ &= (\pi r_H^2 v_H)(n_H)(n_I)\end{aligned} \tag{6.40}$$

where, r_H, v_H, and n_H are the radius, fall speed, and number concentration of rimers and n_I the number concentration of ice crystals.

Now consider a rain gauge with cross-sectional area A. Because all of the rimers within a distance v_H of the top of the rain gauge will enter the rain gauge in 1 s, the number of rimers that enter the rain gauge in 1 s is equal to the number of rimers in a cylinder of cross-sectional area A and height v_H; that is, in a cylinder of volume $v_H A$. The number of rimers in this volume is $v_H A n_H$. Therefore, if each rimer has mass m_H, the mass of rimers that enter the rain gauge in 1 s is $v_H A n_H m_H$, where $m_H = (4/3)\pi r_H^3 \rho_H$ and ρ_H is the density of a rimer. When this mass of rimers melt in the rain gauge, the height h of water of density ρ_l that it will produce in 1 s is given by

$$hA\rho_l = v_H A n_H \left(\frac{4}{3}\pi r_H^3 \rho_H\right)$$

or

$$v_H n_H = \frac{3h}{4\pi} \frac{\rho_l}{\rho_H} \frac{1}{r_H^3} \tag{6.41}$$

From (6.39)–(6.41)

$$q = \frac{4\rho_H r_H}{3h\rho_l n_I} \frac{dQ}{dt}$$

Substituting,

$\frac{dQ}{dt} = \frac{1}{60 \times 60}\ \mathrm{C\,km^{-3}\,s^{-1}}$, $h = \frac{5 \times 10^{-2}}{60 \times 60}\ \mathrm{m\,s^{-1}}$, $\rho_l = 10^3\ \mathrm{kg\,m^{-3}}$, $n_I = 10^5\ \mathrm{m^{-3}}$, $\rho_H = 500\ \mathrm{kg\,m^{-3}}$ and $r_H = 2 \times 10^{-3}$ m, we obtain $q = 16 \times 10^{-15}$ C per collision or 16 fC per collision. ■

We will now describe briefly a proposed mechanism for charge transfer between a rimer and a colliding ice crystal that appears promising, although it remains to be seen whether it can withstand the test of time.

Laboratory experiments show that electric charge is separated when ice particles collide and rebound. The magnitude of the charge is typically about 10 fC per collision, which, as we have seen in Exercise 6.6, would be sufficient to explain the rate of charge generation in thunderstorms. The sign of the charge received by the rimer depends on temperature, the liquid water content of the cloud, and the relative rates of growth from the vapor phase of the rimer and the ice crystals. If the rimer grows more slowly by vapor deposition than the ice crystals, the rimer receives negative charge and the ice crystals receive the corresponding positive charge. Because the latent heat released by the freezing of supercooled droplets on a rimer as it falls through a cloud will raise the surface temperature of the rimer above ambient temperatures, the rate of growth of the rimer by vapor deposition will be less than that of ice crystals in the cloud. Consequently, when an ice crystal rebounds from a rimer, the rimer should receive a negative charge and the ice crystal a positive charge, as required to explain the main distribution of charges in a thunderstorm.

The charge transfer appears to be due to the fact that positive ions move through ice much faster than negative ions. As new ice surface is created by vapor deposition, the positive ions migrate rapidly into the interior of the ice, leaving the surface negatively charged. During a collision material from each of the particles is mixed, but negative charge is transferred to the particle with the slower growth rate.

In some thunderstorms, a relatively weak positive charge is observed just below the main charging zone (Fig. 6.52). This may be associated with the charging of solid precipitation during melting or to mixed-phase processes.

6.7.2 Lightning and Thunder

As electrical charges are separated in a cloud, the electric field intensity increases and eventually exceeds that which the air can sustain. The resulting dielectric breakdown assumes the form of a lightning flash that can be either (1) within the cloud itself, between clouds, or from the cloud to the air (which we will call *cloud flashes*) or (2) between the cloud and the ground (a *ground flash*).

Ground flashes that charge the ground negatively originate from the lower main negative charge center in the form of a discharge, called the *stepped leader*, which moves downward toward the Earth in discrete steps. Each step lasts for about 1 μs, during which time the stepped leader advances about 50 m; the time interval between steps is about 50 μs. It is believed that the stepped leader is initiated by a local

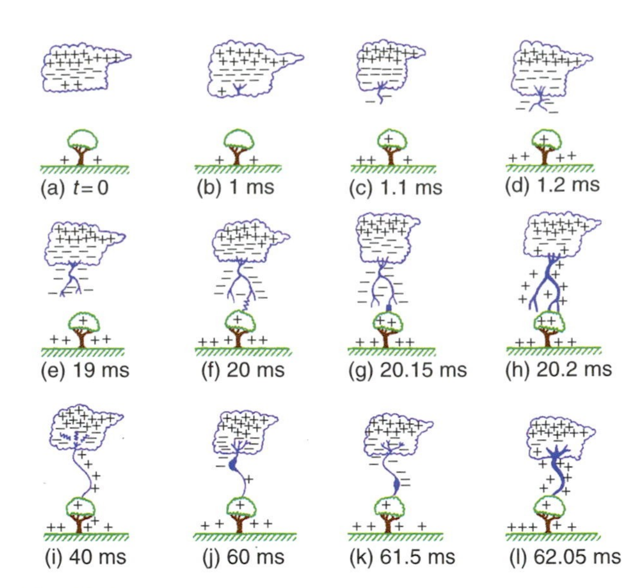

Fig. 6.53 Schematics (not drawn to scale) to illustrate some of the processes leading to a ground flash that charges the ground negatively. (a) cloud charge distribution, (b) preliminary breakdown, (c–e) stepped leader, (f) attachment process, (g and h) first return stroke, (i) K and J processes, (j and k) the dart leader, and (1) the second return stroke. [Adapted from M. Uman, *The Lightning Discharge*, Academic Press, Inc., New York, 1987, p. 12, Copyright 1987, with permission from Elsevier.]

discharge between the small pocket of positive charge at the base of a thundercloud and the lower part of the negatively charged region (Fig. 6.53b). This discharge releases electrons that were previously attached to precipitation particles in the negatively charged region. These free electrons neutralize the small pocket of positive charge that may be present below the main charging zone (Fig. 6.53c) and then move toward the ground (Fig. 6.53c–e). As the negatively charged stepped leader approaches the ground, it induces positive charges on the ground, especially on protruding objects, and when it is 10−100 m from the ground, a discharge moves up from the ground to meet it (Fig. 6.53f). After contact is made between the stepped leader and the upward connecting discharge, large numbers of electrons flow to the ground and a highly luminous and visible *lightning stroke* propagates upward in a continuous fashion from the ground to the cloud along the path followed by the stepped leader (Fig. 6.53g and 6.53h). This flow of electrons (called the *return stroke*) is responsible for the bright channel of light that is observed as a lightning stroke. Because the stroke moves upward so quickly (in about 100 μs), the whole return stroke channel appears to the eye to brighten simultaneously. After the downward flow of electrons, both the return stroke and the ground, to which it is linked, remain positively charged in response to the remainder of the negative charge in the main charging zone.

Following the first stroke, which typically carries the largest current (average 30,000 A), subsequent strokes can occur along the same main channel, provided that additional electrons are supplied to the top of the previous stroke within about 0.1 s of the cessation of current. The additional electrons are supplied to the channel by so-called K or J *streamers*, which connect the top of the previous stroke to progressively more distant regions of the negatively charged area of the cloud (Fig. 6.53i). A negatively charged leader, called the *dart leader*, then moves continuously downward to the Earth along the main path of the first-stroke channel and deposits further electrons on the ground (Figs. 6.53j and 6.53k). The dart leader is followed by another visible return stroke to the cloud (Fig. 6.53l). The first stroke of a flash generally has many downward-directed branches (Fig. 6.54a) because the stepped leader is strongly branched; subsequent strokes usually show no branching, because they follow only the main channel of the first stroke.

Most lightning flashes contain three or four strokes, separated in time by about 50 ms, which can remove 20 C or more of charge from the lower region of a thundercloud. The charge-generating mechanisms within the cloud must then refurbish the charge before another stroke can occur. This they can do in as little as 10 s.

In contrast to the lightning flashes described earlier, most flashes to mountain tops and tall buildings are initiated by stepped leaders that start near the top of the building, move upward, and branch toward the base of a cloud (Fig. 6.54b). Lightning rods[44] protect tall structures from damage by routing the strokes to the ground through the rod and down conductors rather than through the structure itself.

A lightning discharge within a cloud generally neutralizes the main positive and negative charge centers. Instead of consisting of several discrete strokes, such a discharge generally consists of a single, slowly moving spark or leader that travels between the positively and the negatively charged regions in a few tenths of a second. This current produces a low but continuous luminosity in the cloud upon which may be superimposed several brighter pulses, each lasting about 1 ms. Tropical thunderstorms produce about 10 cloud discharges for every ground discharge, but in temperate latitudes the frequencies of the two types of discharge are similar.

The return stroke of a lightning flash raises the temperature of the channel of air through which it passes to above 30,000 K in such a short time that the air has no time to expand. Therefore, the

(a) (b)

Fig. 6.54 (a) A time exposure of a ground lightning flash that was initiated by a stepped leader that propagated from the cloud to the ground. Note the downward-directed branches that were produced by the multibranched stepped leader. [Photograph courtesy of NOAA/NSSL.] (b) A time exposure of a lightning flash from a tower on a mountain to a cloud above the tower. This flash was initiated by a stepped leader that started from the tower and propagated upward to the cloud. In contrast to (a), note the upward-directed branching in (b). [Photograph courtesy of R. E. Orville.]

[44] The use of lightning rods was first suggested by Benjamin Franklin in 1749, who declined to patent the idea or otherwise profit from their use. Lightning rods were first used in France and the United States in 1752. The chance of houses roofed with tiles or slate being struck by lightning is reduced by a factor of about 7 if the building has a lightning rod.

pressure in the channel increases almost instantaneously to 10−100 atm. The high-pressure channel then expands rapidly into the surrounding air and creates a very powerful shock wave (which travels faster than the speed of sound) and, farther out, a sound wave that is heard as *thunder*.[45] Thunder is also produced by stepped and dart leaders, but it is much weaker than that from return strokes. Thunder generally cannot be heard more than 25 km from a lightning discharge. At greater distances the thunder passes over an observer's head because it is generally refracted upward due to the decrease of temperature with height.

Although most ground lightning flashes carry negative charge to the ground, about 10% of the lightning flashes in midlatitude thunderstorms carry a positive charge to the ground. Moreover, these flashes carry the largest peak currents and charge transfers. Such flashes may originate from the horizontal displacement by wind shear of positive charge in the upper regions of a thunderstorm (as depicted in Fig. 6.52) or, in some cases, from the main charge centers in a thunderstorm being inverted from normal.

6.7.3 The Global Electrical Circuit

Below an altitude of a few tens of kilometers there is a downward-directed electric field in the atmosphere during fair weather. Above this layer of relatively strong electric field is a layer called the *electrosphere*, extending upward to the top of the ionosphere in which the electrical conductivity is so high that it is essentially at a constant electric potential. Because the electrosphere is a good electrical conductor, it serves as an almost perfect electrostatic shield.

The magnitude of the fair weather electric field near the surface of the Earth averaged over the ocean ~130 V m^{-1}, and in industrial regions it can be as high as 360 V m^{-1}. The high value in the latter case is due to the fact that industrial pollutants decrease the electrical conductivity of the air because large, slow-moving particles tend to capture ions of higher mobility. Because the vertical current density (which is equal to the product of the electric field and electrical conductivity) must be the same at all levels, the electric field must increase if the conductivity decreases. At heights above about 100 m, the conductivity of the air increases with height and therefore the fair weather electric field decreases with height. The increase in electrical conductivity with height is due to the greater ionization by cosmic rays and diminishing concentrations of large particles. Thus, at 10 km above the Earth's surface the fair weather electric field is only 3% of its value just above the surface. The average potential of the electrosphere with respect to the Earth is ~250 kV, but most of the voltage drop is in the troposphere.

The presence of the downward-directed fair weather electric field implies that the electrosphere carries a net positive charge and the Earth's surface a net negative charge. Lord Kelvin, who in 1860 first suggested the existence of a conducting layer in the upper atmosphere, also suggested that the Earth and the electrosphere act as a gigantic spherical capacitor, the inner conductor of which is the Earth, the other conductor the electrosphere, and the (leaky) dielectric the air. The electric field is nearly constant despite the fact that the current flowing in the air (which averages about 2 to 4×10^{-12} A m^{-2}) would be large enough to discharge the capacitor in a matter of minutes. Thus, there must be an electrical generator in the system. In 1920, C. T. R. Wilson proposed that the principal generators are thunderstorms and electrified shower clouds, and this idea is now almost universally accepted. As we have seen, thunderstorms separate electric charges in such a way that their upper regions become positively charged and their bases negatively charged. The upper positive charges are leaked to the base of the electrosphere through the relatively highly conducting atmosphere at these levels. This produces a diffuse positive charge on the electrosphere, which decreases with height (as does the fair weather electric field) with a scale height of ~5 km. Below a thunderstorm the electrical conductivity of the air is low. However, under the influence of the very large electric fields, a current of positive charges, the *point discharge current*,[47] flows upwards from the Earth (through trees and other pointed obstacles). Precipitation particles are polarized by the fair weather electric field, and by the electric field

[45] This explanation for thunder was first given by Hirn[46] in 1888.

[46] **Gustave Adolfe Hirn** (1815–1890) French physicist. One of the first to study the theory of heat engines. Established a small network of meteorological stations in Alsace that reported observations to him.

[47] Point discharges at mastheads, etc., are know as *St. Elmo's fire*.

beneath thunderstorms, in such a way that they tend to preferentially collect positive ions as they fall to the ground. A positive charge equivalent to about 30% of that from point discharges is returned to the Earth in this way. Finally, ground lightning flashes transport negative charges from the bases of thunderstorms to the ground.

A schematic of the main global electrical circuit is shown in Fig. 6.55. A rough electrical budget for the Earth (in units of C km^{-2} $year^{-1}$) is 90 units of positive charge gained from the fair weather conductivity, 30 units gained from precipitation, 100 units of positive charge lost through point discharges, and 20 units lost due to the transfer of negative charges to the Earth by ground lightning flashes.

Monitoring of lightning flashes from satellites (Fig. 6.56) shows that the global average rate of ground flashes is ~12–16 s^{-1}, with a maximum rate of ~55 s^{-1} over land in summer in the northern hemisphere. The global average rate of total lightning flashes (cloud and ground flashes) is 44 ± 5 s^{-1}, with a maximum of 55 s^{-1} in the northern hemisphere summer and a minimum of 35 s^{-1} in the northern hemisphere winter. About 70% of all lightning occurs between 30 °S and 30 °N, which reflects the high incidence of deep, convective clouds in this region. Over the North American continent ground flashes occur about 30 million times per year!

Over the United States there is a ground network that detects ground flashes. By combining counts of ground flashes from this network with counts of total flashes from satellite observations, the ratio of cloud flashes to ground cloud can be derived. This ratio varies greatly over the United States, from a maximum of ~10 over Kansas and Nebraska, most of Oregon, and parts of northwest California to a ratio of ~1 over the Appalachian Mountains, the Rockies, and the Sierra Nevada Mountains.

Because lightning is associated with strong updrafts in convective clouds, measurements of lightning can serve as a surrogate for updraft velocity and severe weather.

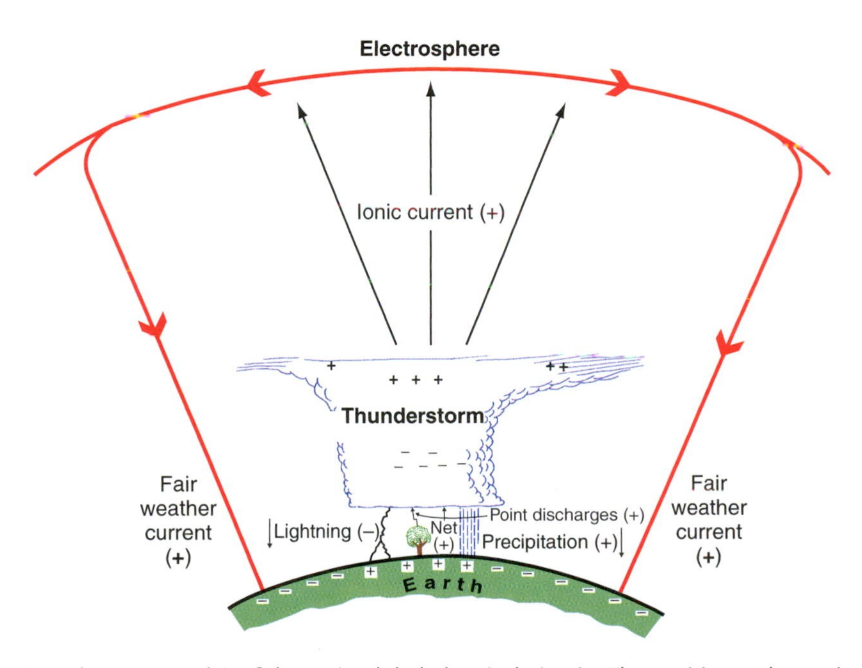

Fig. 6.55 Schematic (not drawn to scale) of the main global electrical circuit. The positive and negative signs in parentheses indicate the signs of the charges transported in the direction of the arrows. The system can be viewed as an electrical circuit (red arrows) in which electrified clouds are the generators (or batteries). In this circuit positive charge flows from the tops of electrified clouds to the electrosphere. Thus, the electrosphere is positively charged, but it is not at a sharply defined height. In fact most of the positive charge on the electrosphere is close to the Earth's surface. The fair-weather current continuously leaks positive charge to the Earth's surface. The circuit is completed by the transfer of net positive charge to the bases of electrified clouds due to the net effect of point discharges, precipitation, and lightning. In keeping with the normal convention, the current is shown in terms of the direction of movement of positive charge, but in fact it is negative charge in the form of electrons that flows in the opposite direction. See text for further details.

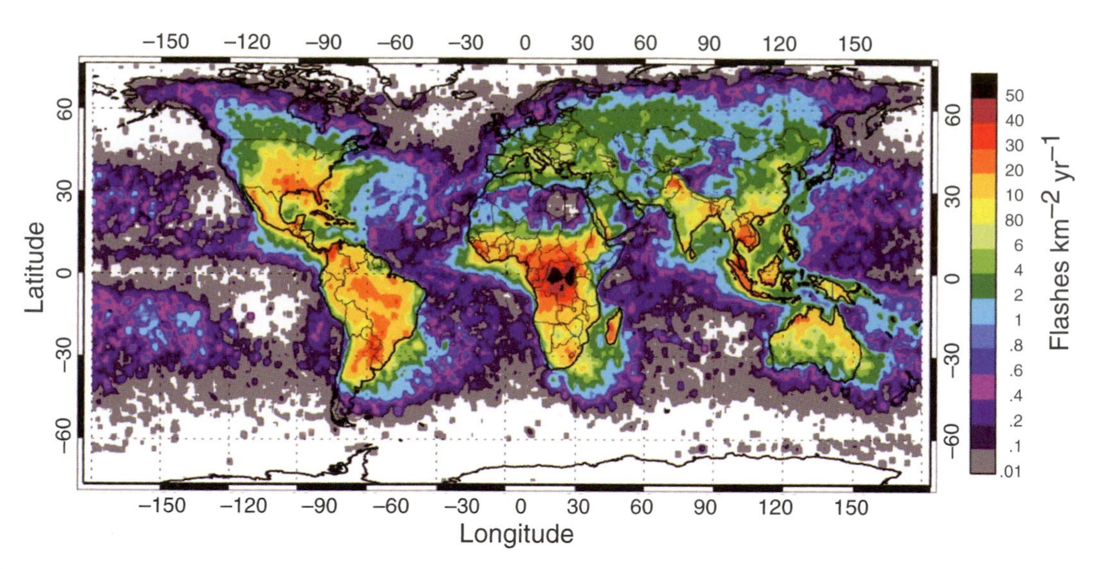

Fig. 6.56 Global frequency and distribution of total lightning flashes observed from a satellite. [From H. J. Christian et al., "Global frequency and distribution of lightning as observed from space by the Optical Transient Detector," *J. Geophys. Res.* **108**(D1), 4005, doi:10.1029/2002JD002347 (2003). Copyright 2003 American Geophysical Union. Reproduced by permission of American Geophysical Union.]

6.5 Upward Electrical Discharges

In 1973 a NASA pilot, in a surveillance aircraft flying at 20 km, recorded the following: "*I approached a vigorous, convective turret close to my altitude that was illuminated from within by frequent lightning. The cloud had not yet formed an anvil. I was surprised to see a bright lightning discharge, white-yellow in color, that came directly out of the center of the cloud at its apex and extended vertically upwards far above my altitude. The discharge was very nearly straight, like a beam of light, showing no tortuosity or branching. Its duration was greater than an ordinary lightning flash, perhaps as much as five seconds.*"

Since then numerous types of lightning-related, transient luminous phenomena in the stratosphere and mesosphere have been documented, which go under the names of *sprites, elves*, and *blue jets* (Fig. 6.57).

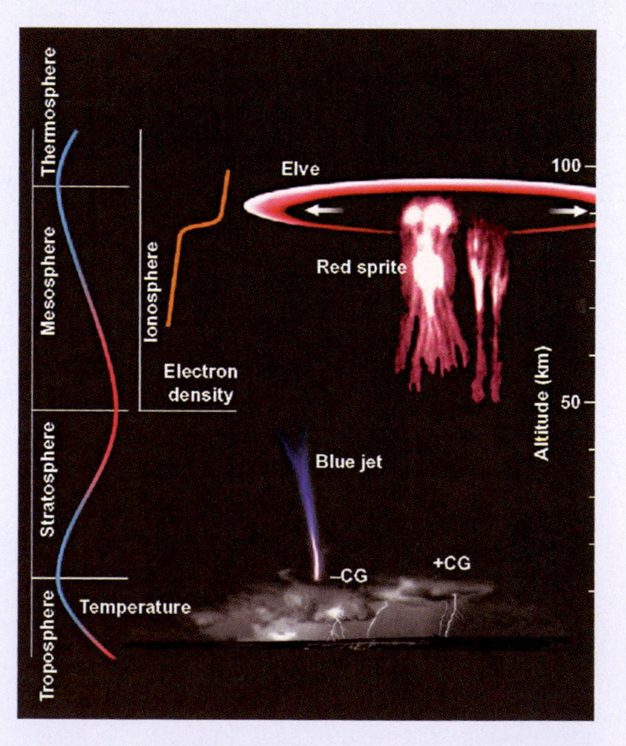

Fig. 6.57 Transient luminous emissions in the stratosphere and mesosphere. [Reprinted with permission from T. Neubert, "On Sprites and their exotic Kin." *Science* **300**, 747 (2003). Copyright 2003 AAAS.]

Sprites are luminous flashes that last from a few to a few hundred milliseconds. Sprites may extend from ~90 km altitude almost down to cloud tops and more than 40 km horizontally. They are primarily red, with blue highlights on their lower regions; they can sometimes be seen by eye. Sprites are believed to be generated by an electric

Continued on next page

6.5 Continued

field pulse when particularly large amounts of positive charge are transferred from a thunderstorm to the ground by a lightning stroke, often from the stratiform regions of large mesoscale convective systems discussed in Section 8.3.4. In contrast to the fully ionized channels of a normal lightning stroke, sprites are only weakly ionized.

Elves are microsecond-long luminous rings located at ~90 km altitude and centered over a lightning stroke. They expand outward horizontally at the speed of light and are caused by atmospheric heating produced by the electromagnetic pulse generated by a lightning stroke. They are not visible by eye.

Blue jets are partially ionized, luminous cones that propagate upward from the tops of thunderstorms at speeds of ~100 km s^{-1} and reach altitudes of ~40 km. On occasions, blue jets trigger sprites, thereby creating a direct, high-conductivity electrical connection from a thunderstorm to the ionosphere. These rare events do not appear to be directly associated with cloud-to-ground lightning flashes. They last only ~100–200 ms and are difficult to see by eye even at night.

Other less well-documented upward propagating discharges have been reported. Figure 6.58 shows an upward-extending column of white light, about 1 km in length, from the top of a thunderstorm.

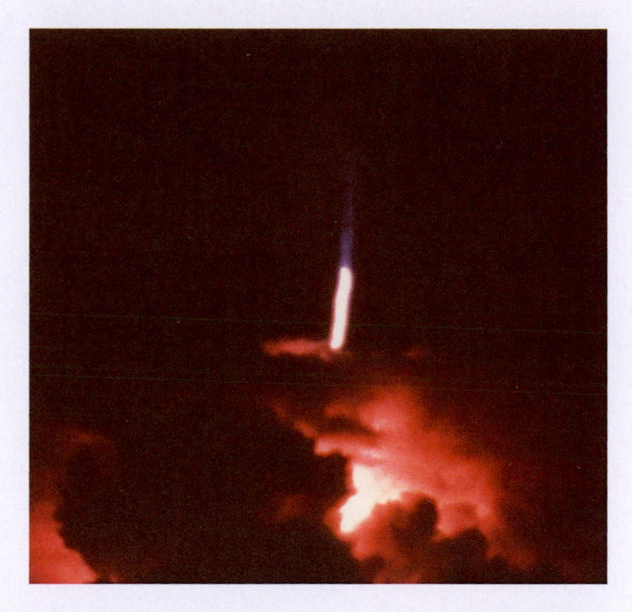

Fig. 6.58 Upward discharge from a thunderstorm near Darwin, Australia. There is a blue "flame" at the top of the white channel that extends upward another kilometer or so. [From *Bull. Am. Meteor. Soc.* **84**, 448 (2003).]

These various phenomena likely play a role in the global electrical circuit and perhaps also in the chemistry of the stratosphere and mesosphere in ways yet to be elucidated.

6.8 Cloud and Precipitation Chemistry[48]

In Chapter 5 we discussed trace gases and aerosols in the atmosphere. This section is concerned with the roles of clouds and precipitation in atmospheric chemistry. We will see that clouds serve as both sinks and sources of gases and particles, and they redistribute chemical species in the atmosphere. Precipitation scavenges particles and gases from the atmosphere and deposits them on the surface of the Earth, the most notable example being *acid precipitation* or *acid rain*.

6.8.1 Overview

Some important processes that play a role in cloud and precipitation chemistry are shown schematically in Fig. 6.59. They include the transport of gases and particles, nucleation scavenging, dissolution of gases into cloud droplets, aqueous-phase chemical reactions, and precipitation scavenging. These processes, and their effects on the chemical composition of cloud water and precipitation, are discussed in turn in this section.

6.8.2 Transport of Particles and Gases

As depicted on the left side of Fig. 6.59, gases and particles are carried upward on the updrafts that feed clouds. Some of these gases and particles are transported to the upper regions of the clouds and are ejected into the ambient air at these levels. In this way, pollutants from near the surface of the Earth (e.g., SO_2, O_3, particles) are distributed aloft. Solar radiation

[48] See footnote 1 in Chapter 5.

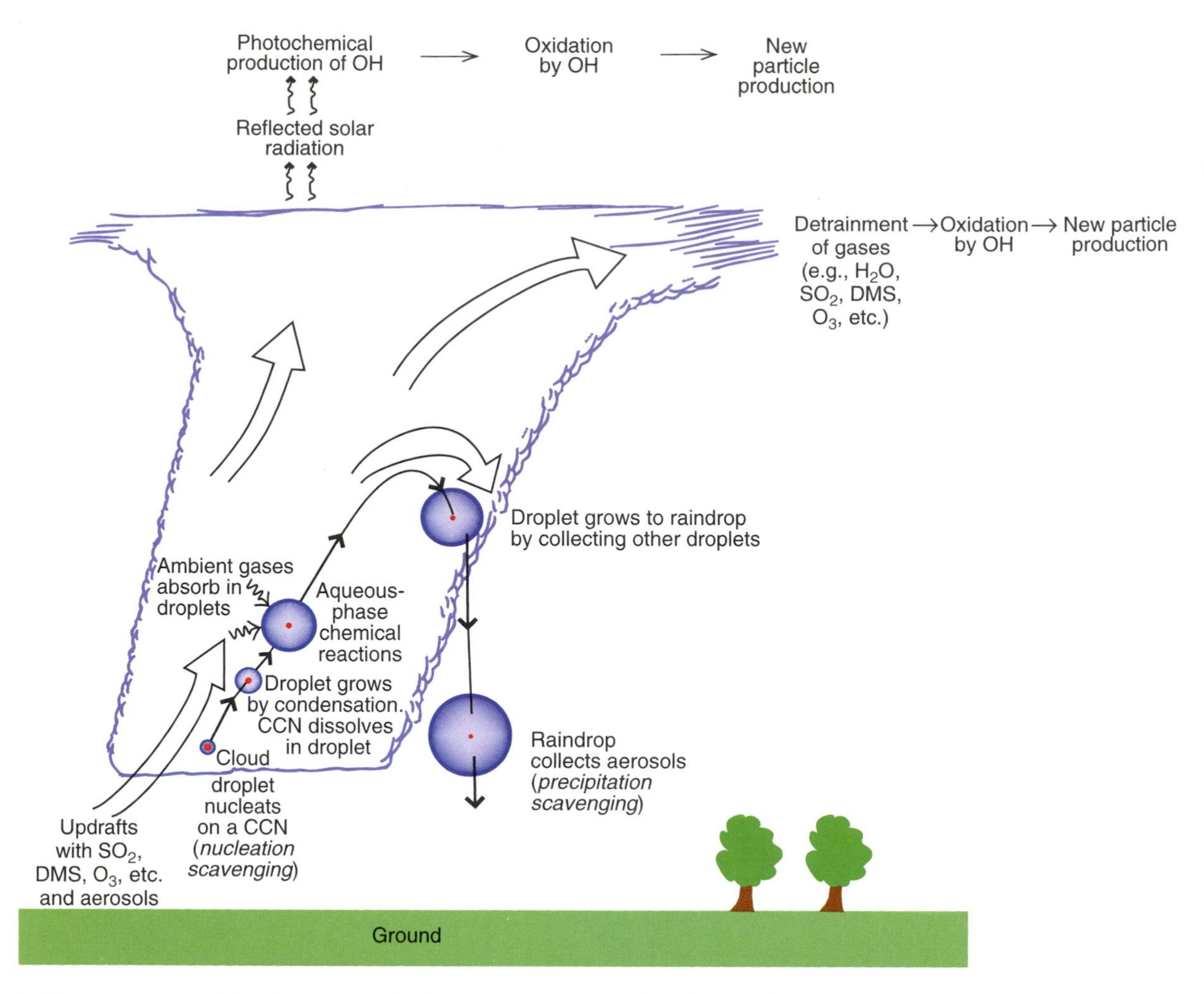

Fig. 6.59 Schematic of cloud and precipitation processes that affect the distribution and nature of chemicals in the atmosphere and the chemical compositions of cloud water and precipitation. The broad arrows indicate airflow. Not drawn to scale.

above the tops of clouds is enhanced by reflection from cloud particles, thereby enhancing photochemical reactions in these regions, particularly those involving OH. In addition, the evaporation of cloud water humidifies the air for a considerable distance beyond the boundaries of a cloud. Consequently, the oxidation of SO_2 and DMS by OH, and the subsequent production of new aerosol in the presence of water vapor (see Section 5.4.1.d), is enhanced near clouds.

6.8.3 Nucleation Scavenging

As we have seen in Section 6.1.2, a subset of the particles that enter the base of a cloud serve as cloud condensation nuclei onto which water vapor condenses to form cloud droplets. Thus, each cloud droplet contains at least one particle from the moment of its birth. The incorporation of particles into cloud droplets in this way is called *nucleation scavenging*. If a CCN is partially or completely soluble in water, it will dissolve in the droplet that forms on it to produce a solution droplet.

6.8.4 Dissolution of Gases in Cloud Droplets

As soon as water condenses to form cloud droplets (or haze or fog droplets), gases in the ambient air begin to dissolve in the droplets. At equilibrium, the number of moles per liter of any particular gas that is dissolved in a droplet (called the *solubility*, C_g, of the gas) is given by Henry's law (see Exercise 5.3). Because the liquid phase is increasingly favored over the gas phase as the temperature is lowered, greater quantities of a gas become dissolved in water droplets at lower temperatures.

Exercise 6.7 Determine the value of the Henry's law constant for a gas that is equally distributed (in terms of mass) between air and cloud water if the liquid water content (LWC) of the cloud is 1 g m^{-3} and the temperature is 5 °C. (Assume that 1 mol of the gas occupies a volume of 22.8 liters at 1 atm and 5 °C.)

Solution: Let the gas be X, and its solubility in the cloud water C_g. Then

$$\begin{array}{c}\text{Amount of } X \text{ (in moles)}\\ \text{in the cloud water}\\ \text{contained in 1 m}^3\text{ of air}\end{array} = \begin{array}{c}C_g \text{ (number of liters}\\ \text{of cloud water in}\\ \text{1 m}^3\text{ of air)}\end{array}$$

$$= C_g \frac{LWC}{\rho_l} \qquad (6.42)$$

where LWC is in kg m^{-3} and ρ_w is the density of water in kg $liter^{-1}$. From (5.3) and (6.42)

Amount of X (in moles) in the cloud water contained in 1 m^3 of air

$$= k_H p_g \frac{LWC}{\rho_l} \qquad (6.43)$$

where p_g is the partial pressure of X in atm and k_H is the Henry's law constant for gas X. Since 1 mol of the gas at 1 atm and 5 °C occupies a volume of 22.8 liters (or 0.0228 m^3), $(0.0228)^{-1}$ moles of the gas occupies a volume of 1 m^3 at 1 atm and 5 °C. Therefore, for a partial pressure p_g of X,

$$1 \text{ m}^3 \text{ of air contains } \frac{p_g}{0.0228} \text{ moles of } X \text{ at } 5\,^\circ\text{C} \qquad (6.44)$$

For the same number of moles of X (and therefore mass of X) in the air and in the water, we have from (6.42) and (6.40)

$$k_H p_g \frac{LWC}{\rho_l} = \frac{p_g}{0.0228}$$

or

$$k_H = \frac{\rho_l}{0.0228(LWC)}$$

Since, $\rho_l = 10^3$ kg m^{-3} = 1 kg per liter, and for a $LWC = 1$ g $m^{-3} = 10^{-3}$ kg m^{-3},

$$k_H = \frac{1}{0.0228(10^{-3})} = 4.38 \times 10^4 \text{ mole liter}^{-1} \text{ atm}^{-1}$$

This value of k_H corresponds to a very soluble gas, such as hydrogen peroxide (H_2O_2). ■

6.8.5 Aqueous-Phase Chemical Reactions

The relatively high concentrations of chemical species within cloud droplets, particularly in polluted air masses, lead to fast aqueous-phase chemical reactions. To illustrate the basic principles involved, we will consider the important case of the conversion of SO_2 to H_2SO_4 in cloud water.

The first step in this process is the dissolution of SO_2 gas in cloud water, which forms the bisulfite ion [HSO_3^- (aq)] and the sulfite ion [SO_3^{2-} (aq)]

$$SO_2(g) + H_2O(\ell) \rightleftarrows SO_2 \cdot H_2O(aq) \qquad (6.45a)$$

$$SO_2 \cdot H_2O(aq) + H_2O(\ell) \rightleftarrows HSO_3^{2-}(aq) + H_3O^+(aq) \qquad (6.45b)$$

$$HSO_3^-(aq) + H_2O(\ell) \rightleftarrows SO_3^{2-}(aq) + H_3O^-(aq) \qquad (6.45c)$$

As a result of these reactions, much more SO_2 can be dissolved in cloud water than is predicted by Henry's law, which does not allow for any aqueous-phase chemical reactions of the dissolved gas.

After $SO_2 \cdot H_2O$(aq), HSO_3^-(aq), and SO_3^{2-}(aq) are formed within a cloud droplet, they are oxidized very quickly to sulfate. The oxidation rate depends on the oxidant and, in general, on the pH of the droplet.[49] The fastest oxidant in the atmosphere, over a wide range of pH values, is hydrogen peroxide (H_2O_2). Ozone (O_3) also serves as a fast oxidant for pHs in excess of about 5.5.

6.8.6 Precipitation Scavenging

Precipitation scavenging refers to the removal of gases and particles by cloud and precipitation elements

[49] The pH of a liquid is defined as

$$pH = -\log[H_3O^+(aq)]$$

where $[H_3O^+(aq)]$ is the concentration (in mole per liter) of H_3O^+ ions in the liquid. Note that a change of unity in the pH corresponds to a change of a factor of 10 in $[H_3O^+(aq)]$. *Acid* solutions have pH <7 and *basic* solutions pH >7.

(i.e., *hydrometeors*). Precipitation scavenging is crucially important for cleansing the atmosphere of pollutants, but it can also lead to acid rain on the ground.

Section 6.8.3 discussed how aerosol particles are incorporated into cloud droplets through nucleation scavenging. Additional ways by which particles may be captured by hydrometeors are *diffusional* and *inertial collection*. Diffusional collection refers to the diffusional migration of particles through the air to hydrometeors. Diffusional collection is most important for submicrometeor particles, as they diffuse through the air more readily than larger particles. Inertial collection refers to the collision of particles with hydrometeors as a consequence of their differential fall speeds. Consequently, inertial collection is similar to the collision and coalescence of droplets discussed in Section 6.4.2. Because very small particles follow closely the streamlines around a falling hydrometeor, they will tend to avoid capture. Consequently, inertial collection is important only for particles greater than a few micrometers in radius.

6.8.7 Sources of Sulfate in Precipitation

The relative contributions of nucleation scavenging, aqueous-phase chemical reactions, and precipitation scavenging to the amount of any chemical that is in hydrometeors that reach the ground depend on the ambient air conditions and the nature of the cloud. For illustration we will consider results of model calculations on the incorporation of sulfate (an important contributor to acid rain) into hydrometeors that originate in warm clouds.

For a warm cloud situated in heavily polluted urban air, the approximate contributions to the sulfate content of rain reaching the ground are nucleation scavenging (37%), aqueous-phase chemical reactions (61%), and below cloud base precipitation scavenging (2%). The corresponding approximate percentages for a warm cloud situated in clean marine air are 75, 14, and 11. Why do the percentages for polluted and clean clouds differ so much? The principal reason is that polluted air contains much greater concentrations of SO_2 than clean air. Therefore, the production of sulfate by the aqueous-phase chemical reactions discussed in Section 6.8.5 is much greater in polluted air than in clean air.

6.8.8 Chemical Composition of Rain

The pH of pure water in contact only with its own vapor is 7. The pH of rainwater in contact with very clean air is 5.6. The lowering of the pH by clean air is due to the absorption of CO_2 into the rainwater and the formation of carbonic acid.

The pH of rainwater in polluted air can be significantly lower than 5.6, which gives rise to acid rain. The high acidity is due to the incorporation of gaseous and particulate pollutants into the rain by the mechanisms discussed in Sections 6.8.3–6.8.7. In addition to sulfate (discussed in Sections 6.8.5 and 6.8.7), many other chemical species contribute to the acidity of rain. For example, in Pasadena, California, there is more nitrate than sulfate in rain, due primarily to emissions of NO_x from cars. In the eastern United States, where much of the acidity of rain is due to the long-range transport of emissions from electric power plants, sulfuric acid and nitric acid contribute ~60 and ~30%, respectively, to the acidity.

6.8.9 Production of Aerosol by Clouds

We have seen in Section 6.8.5 that due to aqueous-phase chemical reactions in cloud droplets, particles released from evaporating clouds may be larger and more soluble than the original CCN on which the cloud droplets formed. Hence, cloud-processed particles can serve as CCN at lower supersaturations than the original CCN involved in cloud formation. We will now describe another way in which clouds can affect atmospheric aerosol.

In the same way as small water droplets can form by the combination of water molecules in air that is highly supersaturated, that is, by homogeneous nucleation (see Section 6.1.1), under appropriate conditions the molecules of two gases can combine to form aerosol particles; a process referred to as *homogeneous-bimolecular nucleation*. The conditions that favor the formation of new particles by homogeneous-bimolecular nucleation are high concentrations of the two gases, low ambient concentrations of preexisting particles (which would otherwise provide a large surface area onto which the gases could condense rather than condensing as new particles), and low temperatures (which favor the condensed phase). These conditions can be satisfied in the outflow regions of clouds for the following reasons.

As shown in Fig. 6.59, and discussed earlier, some of the particles carried upward in a cloud are removed by

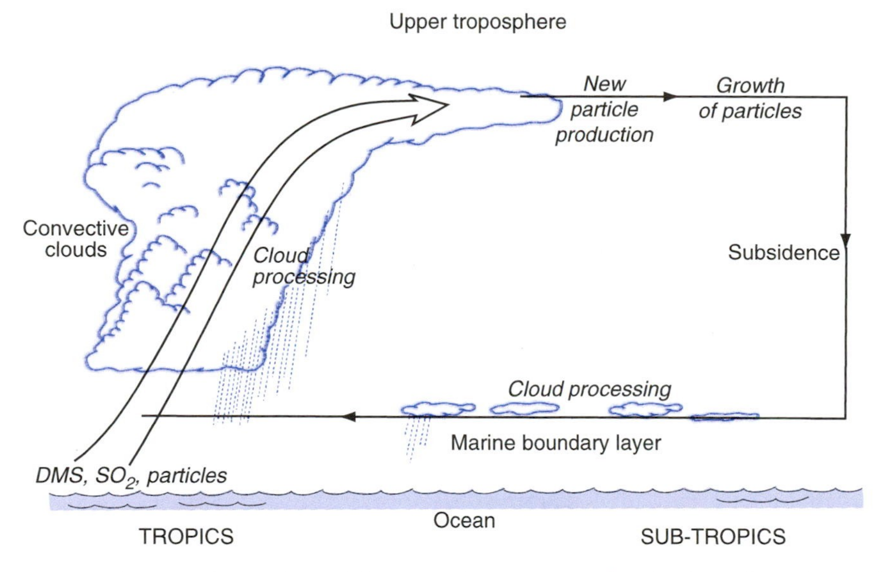

Fig. 6.60 Transport of aerosol between the tropics and the subtropics by means of the Hadley cell circulation. [Adapted from F. Raes et al., "Cloud Condensation nuclei from DMS in the natural marine boundary layer: Remate VS. in-situ production," in G. Restelli and G. Angeletti, eds, *Dimethylsulphide: Oceans, Atmospheres and Climate*, Kluwer Academic Publishers, Dordrecht, The Netherlands, 1993, Fig. 4, p. 317, Copyright 1993 Kluwer Academic Publishers, with kind permission of Springer Science and Business Media.]

cloud and precipitation processes. Therefore, the air detrained from a cloud will contain relatively low concentrations of particles, but the relative humidity of the detrained air will be elevated (due to the evaporation of cloud droplets) as will the concentrations of gases, such as SO_2, DMS, and O_3. Also, in the case of deep convective clouds, the air detrained near cloud tops will be at low temperatures. All of these conditions are conducive to the production of new particles in the detrained air. For example, O_3 can be photolyzed to form the OH radical, which can then oxidize SO_2 to form $H_2SO_4(g)$. The $H_2SO_4(g)$ can then combine with $H_2O(g)$ through homogeneous-bimolecular nucleation to form solution droplets of H_2SO_4.

As shown schematically in Fig. 6.60, the formation of particles in the outflow regions of convective clouds may act on a large scale to supply large numbers of particles to the upper troposphere in the tropics and subtropics and also, perhaps, to the subtropical marine boundary layer. In this scenario, air containing DMS and SO_2 from the tropical boundary layer is transported upward by large convective clouds into the upper troposphere. Particle production occurs in the outflow regions of these clouds. These particles are then transported away from the tropics in the upper branch of the Hadley cell and then subside in the subtropics. During this transport some of the particles may grow sufficiently (by further condensation, coagulation, and cloud processing) to provide particles that are efficient enough as CCN to nucleate droplets even at the low supersaturations typical of stratiform clouds in the subtropical marine boundary layer.

Exercises

6.8 Answer or explain the following in light of the principles discussed in Chapter 6.

(a) Small droplets of pure water evaporate in air, even when the relative humidity is 100%.

(b) A cupboard may be kept dry by placing a tray of salt in it.

(c) The air must be supersaturated for a cloud to form.

(d) Cloud condensation nucleus concentrations do not always vary in the same way as Aitken nucleus concentrations.

(e) CCN tend to be much more numerous in continental air than in marine air.

(f) Growth by condensation in warm clouds is too slow to account for the production of raindrops.

(g) Measurements of cloud microstructures are more difficult from fast-flying than from slow-flying aircraft.

(h) If the liquid water content of a cloud is to be determined from measurements of the droplet spectrum, particular attention should be paid to accurate measurements of the larger drops.

(i) The tops of towering cumulus clouds often change from a cauliflower appearance to a more diffuse appearance as the clouds grow.

(j) Why are actual LWC in clouds usually less than the adiabatic values? Can you suggest circumstances that might produce cloud LWC that are greater than adiabatic values?

(k) Patches of unsaturated air are observed in the interior of convective clouds.

(l) Cloud droplets growing by condensation near the base of a cloud affect each other primarily by their combined influence on the ambient air rather than by direct interactions. [**Hint**: consider the average separation between small cloud droplets.]

(m) After landing on a puddle, raindrops sometime skid across the surface for a short distance before disappearing.

(n) The presence of an electric field tends to raise the coalescence efficiency between colliding drops.

(o) Raindrops are more likely to form in marine clouds than in continental clouds of comparable size.

(p) Collision efficiencies may be greater than unity, but coalescence efficiencies may not.

(q) In some parts of convective clouds the liquid water content may be much higher than the water vapor mixing ratio at cloud base.

(r) In the absence of coalescence, cloud droplet spectra in warm clouds would tend to become monodispersed.

(s) Raindrops reaching the ground do not exceed a certain critical size.

(t) Ice crystals can be produced in a deep-freeze container by shooting the cork out of a toy pop-gun.

(u) Large supersaturations with respect to water are rare in the atmosphere but large supercoolings of droplets are common.

(v) Large volumes of water rarely supercool by more than a few degrees.

(w) The Millipore filter technique may be used to distinguish deposition from freezing nuclei.

(x) Aircraft icing may sometimes be reduced if the aircraft climbs to a higher altitude.

(y) Present techniques for measuring ice nucleus concentrations may not simulate atmospheric conditions very well.

(z) Ice particles are sometimes much more numerous than measurements of the concentrations of ice nuclei would suggest they should be.

(aa) The length of a needle crystal increases relatively rapidly when it is growing by the deposition of water vapor.

(bb) Natural snow crystals are often composed of more than one basic ice crystal habit.

(cc) No two snowflakes are entirely identical.

(dd) Riming tends to be greatest at the edges of ice crystals (see, for example, Figs. 6.40c and 6.40d).

(ee) Riming significantly increases the fall speeds of ice crystals.

(ff) Graupel pellets tend to be opaque rather than transparent.

(gg) Strong updrafts are required to produce large hailstones.

(hh) Aggregated ice crystals have relatively low fall speeds for their masses.

(ii) When snow is just about to change to rain, the snowflakes often become very large.

(jj) The melting level is a prominent feature in radar imagery.

(kk) Cold fogs are easier to disperse by artificial means than warm fogs.

(ll) The charging mechanism responsible for the small pocket of positive charge that exists below the melting level in some thunderstorms must be more powerful than the mechanism responsible for the generation of the main charge centers.

(mm) When the atmospheric electric field between cloud base and the ground is directed downward, precipitation particles reaching the Earth generally carry a negative charge, and when the electric field is in the upward direction the precipitation is generally positively charged.

(nn) The presence of water drops in the air reduces the potential gradient required for dielectric breakdown.

(oo) The ratio of cloud to ground to intercloud lightning flashes is ~10 over Kansas and Nebraska but only ~1 over the Rocky Mountains.

(pp) Lightning is rarely observed in warm clouds.

(qq) Lightning occurs much more frequently over the continents than over the oceans (Fig. 6.56).

(rr) Visibility often improves after a thunderstorm.

(ss) The concentrations of aerosols between cloud droplets (called *interstitial aerosol*) are generally less than they are in ambient air.

(tt) The residence time for particles in the atmosphere is short (~1 min) for particles ~10^{-3} μm in diameter, large (~ months) for particles ~1 μm in diameter, and short (~1 min) for particles in excess of ~100 μm in diameter.

(uu) Feed lots for cattle contribute to the amount of nitrate in rain.

6.9 Figure 6.61 indicates a film of liquid (e.g., a soap film) on a wire frame. The area of the film can be changed by moving a frictionless wire that forms one side of the frame. The surface tension of the liquid is defined as the force per unit length that the liquid exerts on the movable wire (as indicated by the broad black arrow in Fig. 6.61). If the surface energy of the liquid is defined as the work required to create a unit area of new liquid, show that the numerical values of the surface tension and surface energy of the liquid are the same.

6.10 Calculate the relative humidity of the air adjacent to a pure water droplet 0.2 μm in radius if the temperature is 0 °C. (The surface energy of water at 0 °C is 0.076 J m^{-2} and the number density of molecules in water at 0 °C is 3.3×10^{28} m^{-3}.)

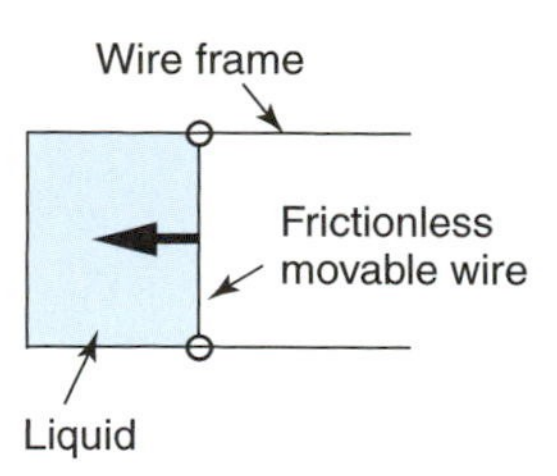

Fig. 6.61

6.11 Use the Köhler curves shown in Fig. 6.3 to estimate:

(a) The radius of the droplet that will form on a sodium chloride particle of mass 10^{-18} kg in air that is 0.1% supersaturated.

(b) The relative humidity of the air adjacent to a droplet of radius 0.04 μm that contains 10^{-19} kg of dissolved ammonium sulfate.

(c) The critical supersaturation required for an ammonium sulfate particle of mass 10^{-19} kg to grow beyond the haze state.

6.12 Show that for a very weak solution droplet ($m \ll \frac{4}{3}\pi r^3 \rho' M_s$), Eq. (6.8) can be written as

$$\frac{e'}{e_s} \simeq 1 + \frac{a}{r} - \frac{b}{r^3}$$

where $a = 2\sigma'/n'kT$ and $b = imM_w/\frac{4}{3} M_s \pi \rho'$. What is your interpretation of the second and third terms on the right-hand side of this expression? Show that in this case the peak in the Köhler curve occurs at

$$r \simeq \left(\frac{3b}{a}\right)^{1/2} \text{ and } \frac{e'}{e_s} \simeq 1 + \left(\frac{4a^3}{27b}\right)^{1/2}$$

6.13 Assuming that cloud condensation nuclei (CCN) are removed from the atmosphere by first serving as the centers on which cloud droplets form, and the droplets subsequently grow to form precipitation particles, estimate the residence time of a CCN in a column extending from the surface of the Earth to an altitude of 5 km. Assume that the annual rainfall is 100 cm and the cloud liquid water content is 0.30 g m^{-3}.

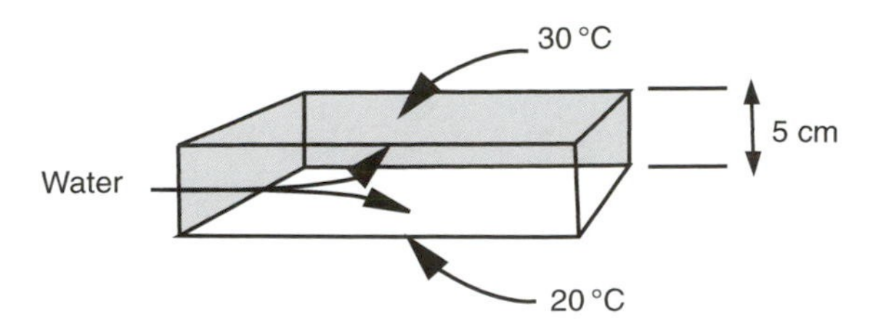

Fig. 6.62

6.14 A thermal diffusion chamber, used for measuring the concentration of CCN active at a given supersaturation, consists of a large shallow box of height 5 cm. The top of the box is maintained at 30 °C, the base of the box at 20 °C, and the temperature in the box varies linearly with height above the base of the box. The inside surfaces of the top and base of the box are covered with water (Fig. 6.62). (a) What is the maximum supersaturation (in %), with respect to a plain surface of water, inside the box? (b) What distance above the base of the box does this maximum supersaturation occur?

To answer this question, use the values given in the following table to plot on a large sheet of graph paper e_s, the saturation vapor pressure over a plain surface of water, as a function of temperature T from the bottom to the top of the box. Plot the actual vapor pressure e in the box as a function of T on the same graph.

T (°C)	e_s (hPa)	T (°C)	e_s (hPa)
20	23.4	26	33.6
21	24.9	27	35.6
22	26.4	28	37.8
23	28.1	29	40.1
24	29.8	30	42.4
25	31.7		

6.15 The air at the 500-hPa level in a cumulonimbus cloud has a temperature of 0 °C and a liquid water content of 3 g m^{-3}. (a) Assuming the cloud drops are falling at their terminal fall speeds, calculate the downward frictional drag that the cloud drops exert on a unit mass of air. (b) Express this downward force in terms of a (negative) virtual temperature correction. (In other words, find the decrease in temperature the air would have to undergo if it contained no liquid water in order to be as dense as the air in question.) [**Hint**: At temperatures around 0 °C the virtual temperature correction due to the presence of water vapor in the air can be neglected.]

6.16 For droplets much larger than the wavelength of visible light, and for a droplet spectrum sufficiently narrow that the effective droplet radius r_e is approximately equal to the mean droplet radius, the optical depth τ_c of a cloud is given approximately by

$$\tau_c = 2\pi h r_e^2 N$$

where h is the cloud depth and N is the number concentration (m^{-3}) of cloud droplets. Derive approximate expressions for (a) the cloud liquid water content (LWC in kg m^{-3}) in terms of r_e, N, and the density of water ρ_l, (b) the LWC in terms of r_e and τ_c, and (c) an expression for the liquid water path (LWP in kg m^{-2}) in terms of r_e, τ_c, and ρ_l.

6.17 Air with a relative humidity of 20% at 500 hPa and −20 °C is entrained into a cloud. It remains at 500 hPa while cloud droplets are evaporated into it. To what temperature can it be cooled by this process? Suppose the same parcel of air is carried down to 1000 hPa in a downdraft. What will be the temperature of the air parcel if it remains saturated? What will be its temperature if its relative humidity is 50% when it reaches the ground? (Use a skew $T - \ln p$ chart to solve this exercise.)

6.18 (a) For the case of no condensation and no entrainment, that is the dry adiabatic case, show that Eq. (6.18) reduces to $d\theta' = 0$. (b) For the case of condensation but no entrainment, show that Eq. (6.18) reduces to the equivalent of Eq. (3.70). (c) Show from Eq. (6.18) that when entrainment is present the temperature of a rising parcel of cloudy air decreases faster than when entrainment is absent.

6.19 Assuming that the radius of a rising parcel of warm air (i.e., a *thermal*), considered as a sphere, is proportional to its height above the ground, show that the *entrainment rate* (defined as dm/mdt, where m is the mass of the thermal at time t) is inversely proportional to the radius of the thermal and directly proportional to the upward speed of the thermal.

6.20 Consider a saturated parcel of air that is lifted adiabatically. The rate of change in the supersaturation ratio $S(S = e/e_s$ where e is the vapor pressure of the air and e_s is the saturated vapor pressure) of the air parcel may be written as

$$\frac{dS}{dt} = Q_1 \frac{dz}{dt} - Q_2 \frac{d(LWC)}{dt}$$

where dz/dt and LWC are the vertical velocity and the liquid water content of the air parcel, respectively.

(a) Assuming that S is close to unity, and neglecting the difference between the actual temperature and the virtual temperature of the air, show that

$$Q_1 \simeq \frac{g}{TR_d}\left(\frac{\varepsilon L_v}{Tc_p} - 1\right)$$

where T is the temperature of the parcel (in degrees kelvin), L_v is the latent heat of condensation, g is the acceleration due to gravity, R_d and R_v are the gas constants for 1 kg of dry air and 1 kg of water vapor, respectively, $\varepsilon = R_d/R_v$, and c_p is the specific heat at constant pressure of the saturated air. [**Hint**: Consider ascent with no condensation. Introduce the mixing ratio of the air, which is related to e by Eq. (3.62).]

(b) Derive a corresponding approximate expression for Q_2 in terms of R_d, T, e, e_s, L_v, p, c_p, and the density of the moist air (ρ). Assume that $e/e_s \simeq 1$, $T \simeq T_v$, and $\varepsilon >> w$ (the mixing ratio of the air).

6.21 Derive an expression for the height h above cloud base of a droplet at time t that is growing by condensation only in a cloud with a steady updraft velocity w and supersaturation S. [**Hint**: Use (6.24) for the terminal fall speed of a droplet together with (6.21).]

6.22 An isolated parcel of air is lifted from cloud base at 800 hPa, where the temperature is 5 °C, up to 700 hPa. Use the skew $T - \ln p$ chart to determine the amount of liquid water (in grams per kilogram of air) that is condensed during this ascent (i.e., the adiabatic liquid water content).

6.23 A drop with an initial radius of 100 μm falls through a cloud containing 100 droplets per cubic centimeter that it collects in a continuous manner with a collection efficiency of 0.800. If all the cloud droplets have a radius of 10 μm, how long will it take for the drop to reach a radius of 1 mm? You may assume that for the drops of the size considered in this problem the terminal fall speed v (in m s^{-1}) of a drop of radius r (in meters) is given by $v = 6 \times 10^3 r$. Assume that the cloud droplets are stationary and that the updraft velocity in the cloud is negligible.

6.24 If a raindrop has a radius of 1 mm at cloud base, which is located 5 km above the ground, what will be its radius at the ground and how long will it take to reach the ground if the relative humidity between cloud base and ground is constant at 60%? [**Hint**: Use (6.21) and the relationship between v and r given in Exercise 6.23. If r is in micrometers, the value of G_l in (6.21) is 100 for cloud droplets, but for the large drop sizes considered in this problem the value of G_l should be taken as 700 to allow for ventilation effects.]

6.25 A large number of drops, each of volume V, are cooled simultaneously at a steady rate $\beta(=dT/dt)$. Let $p(V, t)$ be the probability of ice nucleation taking place in a volume V of water during a time interval t. (a) Derive a relationship between $p(V, t)$ and $\int_0^{T_t} J_{LS}\, dT$, where J_{LS} is the ice nucleation rate (per unit volume per unit time) and T_t is the temperature of the drops at time t. (b) Show that a n-fold increase in the cooling rate produces the same depression in the freezing temperature as an n-fold decrease in drop volume. [**Hint**: For (b) you will need to use some of the expressions derived in Exercise (6.4).]

6.26 A cloud is cylindrical in shape with a cross-sectional area of 10 km^2 and a height of 3 km. All of the cloud is initially supercooled and the liquid water content is 2 g m^{-3}. If all of the water in the cloud is transferred onto ice nuclei present in a uniform concentration of 1 liter^{-1}, determine the total number of ice crystals in the cloud and the mass of each ice crystal produced. If all the ice crystals precipitate and melt before they reach the ground, what will be the total rainfall produced?

6.27 Calculate the radius and the mass of an ice crystal after it has grown by deposition from the vapor phase for half an hour in a water-saturated environment at -5 °C. Assume that the crystal is a thin circular disk with a constant thickness of 10 μm. [**Hint**: Use Eq. (6.37) and Fig. 6.39 to estimate the magnitude of $G_i S_i$. The electrostatic capacity C of a circular disk of radius r is given by $C = 8r\varepsilon_0$, where ε_0 is the permittivity of free space.]

6.28 Calculate the time required for an ice crystal, which starts off as a plane plate with an effective circular radius of 0.5 mm and a mass of 0.010 mg, to grow by riming into a spherical graupel particle 0.5 mm in radius if it falls through a cloud containing 0.50 g m^{-3} of small water droplets that it collects with an efficiency of 0.60. Assume that the density of the final graupel particle is 100 kg m^{-3} and that the terminal fall speed v (in m s^{-1}) of the crystal is given by $v = 2.4M^{0.24}$, where M is the mass of the crystal in milligrams.

6.29 Calculate the time required for the radius of a spherical snowflake to increase from 0.5 mm to 0.5 cm if it grows by aggregation as it falls through a cloud of small ice crystals present in an amount 1 g m^{-3}. Assume the collection efficiency is unity, the density of the snowflake is 100 kg m^{-3}, and that the difference in the fall speeds of the snowflake and the ice crystals is constant and equal to 1 m s^{-1}.

6.30 Determine the minimum quantity of heat that would have to be supplied to each cubic meter of air to evaporate a fog containing 0.3 g m^{-3} of liquid water at 10 °C and 1000 hPa. [**Hint**: You will need to inspect the saturation vapor pressure of air as a function of temperature, as given on a skew $T - \ln p$ chart.]

6.31 If 40 liters of water in the form of drops 0.5 mm in diameter was poured into the top of a cumulus cloud and all of the drops grew to a diameter of 5 mm before they emerged from the base of the cloud, which had an area of 10 km^2, what would be the amount of rainfall induced? Can you suggest any physical mechanisms by which the amount of rain produced in this way might be greater than this exercise suggests?

6.32 Compare the increase in the mass of a drop in the previous exercise with that of a droplet 20 μm in radius that is introduced at cloud base, travels upward, then downward, and finally emerges from cloud base with a diameter of 5 mm.

6.33 If hailstones of radius 5 mm are present in concentrations of 10 m^{-3} in a region of a hailstorm where the hailstones are competing for the available cloud water, determine the size of the hailstones that would be produced if the concentrations of hailstone embryos were increased to 10^4 m^{-3}.

6.34 A supercooled cloud is completely glaciated at a particular level by artificial seeding. Derive an approximate expression for the increase in the temperature ΔT at this level in terms of the original liquid water content w_l (in g kg^{-1}), the specific heat of the air c, the latent heats of fusion L_f and deposition L_d of the water substance, the original saturation mixing ratio with respect to water w_s (in g kg^{-1}), and the final saturation mixing ratio with respect to ice w_i (in g kg^{-1}).

6.35 Ignoring the second term on the right-hand side of the answer given in the previous exercise, calculate the increase in temperature produced by glaciating a cloud containing 2 g kg^{-1} of water.

6.36 On a certain day, towering cumulus clouds are unable to penetrate above the 500-hPa level, where the temperature is -20 °C, because of the presence of a weak stable layer in the vicinity of that level where the environmental lapse rate is 5 °C km^{-1}. If these clouds are seeded with silver iodide so that all of the liquid water in them (1 g m^{-3}) is frozen, by how much will the tops of the cumulus clouds rise? Under what conditions might such seeding be expected to result in a significant increase in precipitation? [**Hint**: From the skew $T - \ln p$ chart, the

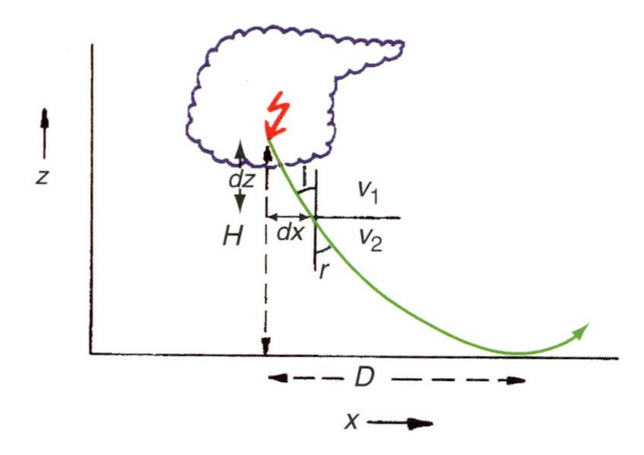

Fig. 6.63

saturated adiabatic lapse rate at 500 hP is about 6 °C km^{-1}.]

6.37 If the velocities of sound in two adjacent thin layers of air are v_1 and v_2, a sound wave will be refracted at the interface between the layers and $\frac{\sin i}{\sin r} = \frac{v_1}{v_2}$ (see Fig. 6.63). Use this relationship, and the fact that $v \propto (T)^{1/2}$, where T is the air temperature (in degrees kelvin), to show that the equation for the path of a sound wave produced by thunder that is heard at the maximum horizontal distance from the origin of the thunder is

$$dx = -(T/\Gamma z)^{1/2}\, dz$$

where x and z are the coordinates indicated in Fig. 6.63, Γ is the temperature lapse rate in the vertical (assumed constant), and T is the temperature at height z.

6.38 Use the expression derived in Exercise (6.37) to show that the maximum distance D at which a sound wave produced by thunder can be heard is given approximately by (see Fig. 6.63 from Exercise 6.37)

$$D = 2(T_0 H/\Gamma)^{1/2}$$

where T_0 is the temperature at the ground. Calculate the value of D given that $\Gamma = 7.50$ °C km^{-1}, $T_0 = 300$ K, and $H = 4$ km.

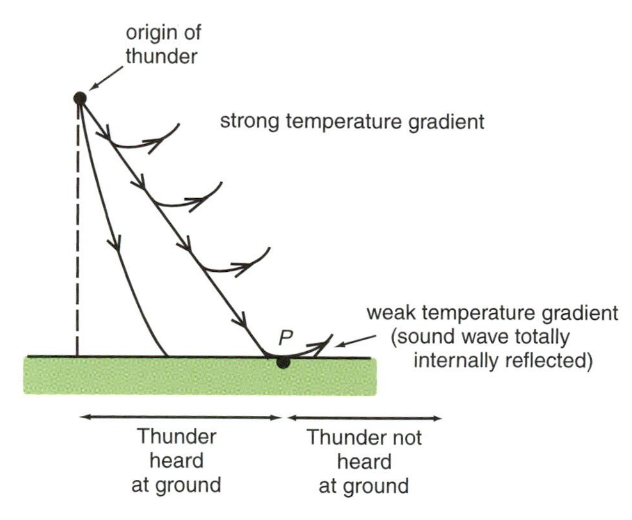

Fig. 6.64

6.39 An observer at the ground hears thunder 10 s after he sees a lightning flash, and the thunder lasts for 8 s. How far is the observer from the closest point of the lightning flash and what is the minimum length of the flash? Under what conditions would the length of the flash you have calculated be equal to the true length? (Speed of sound is 0.34 km s^{-1}.)

6.40 (a) What is the change in the oxidation number of sulfur when HSO_3^-(aq) is oxidized to H_2SO_4(aq)? (b) What are the changes in the oxidation numbers of sulfur in $SO_2{\cdot}H_2O$(aq) and SO_3^{2-}(aq) when they are converted to H_2SO_4?

Atmospheric Dynamics

7

This chapter introduces a framework for describing and interpreting the structure and evolution of large-scale atmospheric motions, with emphasis on the extratropics. We will be considering motions with horizontal scales[1] of hundreds of kilometers or longer, vertical scales on the order of the depth of the troposphere, and timescales on the order of a day or longer. Motions on these scales are directly and strongly influenced by the Earth's rotation, they are in hydrostatic balance, and the vertical component of the three-dimensional velocity vector is more than three orders of magnitude smaller than the horizontal component but is nonetheless of critical importance. The first section introduces some concepts and definitions that are useful in characterizing horizontal flow patterns. It is followed by a more extensive section on the dynamics of large-scale horizontal motions on a rotating planet. The third section introduces the system of *primitive equations*, which is widely used for predicting how large-scale motions evolve with time. The chapter concludes with brief discussions of the atmospheric general circulation and numerical weather prediction.

7.1 Kinematics of the Large-Scale Horizontal Flow

Kinematics deals with properties of flows that can be diagnosed (but not necessarily predicted) without recourse to the equations of motion. It provides a descriptive framework that will prove useful in interpreting the horizontal equation of motion introduced in the next section.

On any "horizontal" surface (i.e., a surface of constant geopotential Φ, pressure p, or potential temperature θ), it is possible to define a set of *streamlines* (arbitrarily spaced lines whose orientation is such that they are everywhere parallel to the horizontal velocity vector $\mathbf{V}$ at a particular level and at a particular instant in time) and *isotachs* (contours of constant scalar wind speed V) that define the position and orientation of features such as jet streams. At any point on the surface one can define a pair of axes of a system of *natural coordinates* (s, n), where s is arc length directed downstream along the local streamline, and n is distance directed normal to the streamline and toward the left, as shown in Fig. 7.1. It follows that at any point in the flow, the scalar wind speed

$$V = \frac{ds}{dt} \text{ and } \frac{dn}{dt} = 0$$

The direction of the flow at any point is denoted by the angle ψ, which is defined relative to an arbitrary reference angle. The unit vector $\mathbf{s}$ is in the local downstream direction and $\mathbf{n}$ is in the transverse direction.

7.1.1 Elementary Kinematic Properties of the Flow

At any point in the flow it is possible to define the kinematic properties listed in Table 7.1, all of which have units of inverse time (s^{-1}). The topmost four rows are elementary properties or building blocks that have precise mathematical definitions in natural coordinates. *Shear* is defined as the rate of change of

[1] The *scale* of the motions denotes the distance over which scalar fields, such as pressure or the meridional wind component, vary over a range comparable to the amplitude of the fluctuations. For example, the horizontal scale of a one-dimensional sinusoidal wave along a latitude circle is on the order of $1/2\pi$ wavelength and the scale of a circular vortex is on the order of its radius.

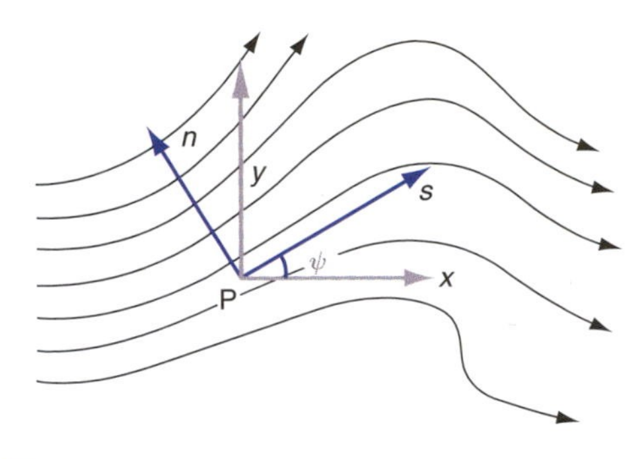

Fig. 7.1 Natural coordinates (s, n) defined at point P in a horizontal wind field. Curved arrows represent streamlines.

the velocity in the direction transverse to the flow. *Curvature* is the rate of change of the direction of the flow in the downstream direction. Shear and curvature are labeled as *cyclonic (anticyclonic)* and are assigned a positive (negative) algebraic sign if they are in the sense as to cause an object in the flow to rotate in the same (opposite) sense as the Earth's rotation Ω, as viewed looking down on the pole. In other words, *cyclonic* means counterclockwise in the northern hemisphere and clockwise in the southern hemisphere. *Diffluence/confluence* is the rate of change of the direction of the flow in the direction transverse to the motion, defined as positive if the streamlines are spreading apart in the downstream direction. *Stretching/contraction* relates to the rate of change of the speed of the flow in the downstream direction, with stretching defined as positive.

7.1.2 Vorticity and Divergence

Vorticity and divergence are scalar quantities that can be defined not only in natural coordinates, but also in Cartesian coordinates (x, y) and for the horizontal wind vector **V**. *Vorticity* is the sum of the shear and the curvature, taking into account their algebraic signs, and *divergence* is the sum of the diffluence and the stretching.

It can be shown that *vorticity* ζ is given by

$$\zeta = 2\omega \tag{7.1}$$

where ω is the rate of spin of an imaginary puck moving with the flow. This relationship is illustrated in the following exercise for the special case of solid body rotation.

Exercise 7.1 Derive an expression for the vorticity distribution within a flow characterized by counterclockwise solid body rotation with angular velocity ω (see Fig. 7.2b).

Solution: Vorticity is the sum of the shear and the curvature. Using the definitions in Table 7.1, the shear contribution is $-\partial V/\partial n = \partial V/\partial r$, where r is radius in polar coordinates, centered on the axis of rotation. In one circuit ψ rotates through angle 2π. Therefore, the curvature contribution $V\partial\psi/\partial s = V(2\pi/2\pi r) = V/r$. For solid body rotation with angular

Table 7.1 Definitions of properties of the horizontal flow

	Vectorial	Natural coords.	Cartesian coords.
Shear		$-\dfrac{\partial V}{\partial n}$	
Curvature		$V\dfrac{\partial \psi}{\partial s}$	
Diffluence		$V\dfrac{\partial \psi}{\partial n}$	
Stretching		$\dfrac{\partial V}{\partial s}$	
Vorticity ζ	$\mathbf{k}\cdot\nabla\times\mathbf{V}$	$V\dfrac{\partial \psi}{\partial s} - \dfrac{\partial V}{\partial n}$	$\dfrac{\partial v}{\partial x} - \dfrac{\partial u}{\partial y}$
Divergence $\mathrm{Div}_H\mathbf{V}$	$\nabla\cdot\mathbf{V}$	$V\dfrac{\partial \psi}{\partial n} + \dfrac{\partial V}{\partial s}$	$\dfrac{\partial u}{\partial x} + \dfrac{\partial v}{\partial y}$
Deformation			$\dfrac{\partial u}{\partial x} - \dfrac{\partial v}{\partial y};\ \dfrac{\partial v}{\partial x} + \dfrac{\partial u}{\partial y}$

velocity ω the contribution from the shear is therefore $d(\omega r)/dr = \omega$ and the contribution from the curvature is $\omega r/r = \omega$. The shear and curvature contributions are in the same sense and therefore additive. Hence, $\zeta = 2\omega$. ■

Divergence $\nabla \cdot \mathbf{V}$ (or $\text{Div}_H\mathbf{V}$) is related to the time rate of change of area. Consider a block of fluid of area A, moving downstream in the flow. In Cartesian (x, y) coordinates, the Lagrangian time rate of change is

$$\frac{dA}{dt} = \frac{d}{dt}\delta x \delta y = \delta y \frac{d}{dt} \delta x + \delta x \frac{d}{dt} \delta y$$

If the dimensions of the block are very small compared to the space scale of the velocity fluctuations, we may write

$$\frac{d}{dt}(\delta x) = \delta u = \frac{\partial u}{\partial x} \delta x$$

where the partial derivative $\partial/\partial x$ in this expression indicates that the derivative is taken at constant y. The time rate of change of δy can be expressed in an analogous form. Substituting for the time derivatives in the expression for dA/dt, we obtain

$$\frac{dA}{dt} = \delta x \delta y \frac{\partial u}{\partial x} + \delta x \delta y \frac{\partial v}{\partial y}$$

Dividing both sides by $\delta x \delta y$ yields

$$\frac{1}{A}\frac{dA}{dt} = \frac{\partial u}{\partial x} + \frac{\partial v}{\partial y} \tag{7.2}$$

where the right-hand side may be recognized as the Cartesian form of the divergence in Table 7.1. Hence, *divergence is the logarithmic rate of expansion of the area enclosed by a marked set of parcels moving with the flow*. Negative divergence is referred to as *convergence*.

Some of the relationships between the various kinematic properties of the horizontal wind field are illustrated by the idealized flows shown in Fig. 7.2.

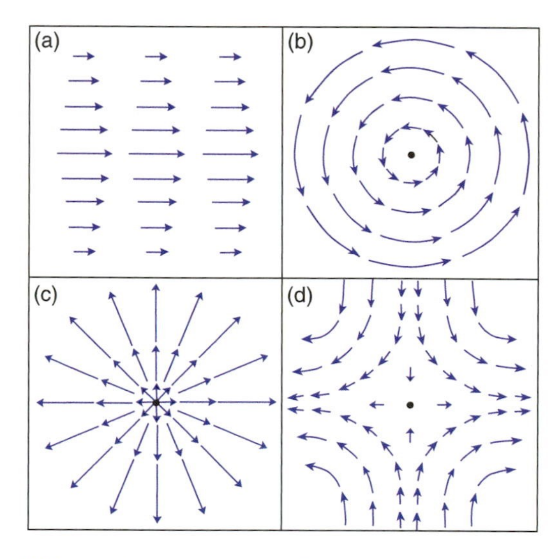

Fig. 7.2 Idealized horizontal flow configurations. See text for explanation.

(a) Sheared flow without curvature, diffluence, stretching, or divergence. From a northern hemisphere perspective, shear and vorticity are cyclonic in the top half of the domain and anticyclonic in the bottom half.
(b) Solid body rotation with cyclonic shear and cyclonic curvature (and hence, cyclonic vorticity) throughout the domain, but without diffluence or stretching, and hence without divergence.
(c) Radial flow with velocity directly proportional to radius. This flow exhibits diffluence and stretching and hence divergence, but no curvature or shear and hence no vorticity.
(d) Hyperbolic flow that exhibits both diffluence and stretching, but is nondivergent because the two terms exactly cancel. Hyperbolic flow also exhibits both shear and curvature, but is irrotational (i.e., vorticity-free) because the two terms exactly cancel.

Vorticity is related to the line integral of the flow along closed loops. From Stokes' theorem, it follows that

$$C \equiv \oint V_s ds = \iint \zeta dA \tag{7.3}$$

where V_s is the component of the velocity along the arc ds (defined as positive when circulating around the loop in the same sense as the Earth's rotation), and C is referred to as the *circulation* around the loop. The term on the right hand side can be rewritten as $\overline{\zeta} A$, where $\overline{\zeta}$ is the spatial average of the vorticity within the loop and A is the area of the loop. The analogous relationship for the component of the velocity V_n outward across the curve is

$$\oint V_n ds = \iint \mathrm{Div}_H \mathbf{V} dA \tag{7.4}$$

which follows from Gauss's theorem.

Exercise 7.2 At the 300-hPa (~10 km) level along 40 °N during winter the zonally averaged zonal wind $[u]$ is eastward at 20 m s^{-1} and the zonally averaged meridional wind component $[v]$ is southward at 30 cm s^{-1}. Estimate the vorticity and divergence averaged over the polar cap region poleward of 40 °N.

Solution: Based on (7.3), the vorticity averaged over the polar cap region is given by

$$\bar{\zeta} = \frac{[u]\oint_{40\,^\circ\mathrm{N}} ds}{\iint_{40\,^\circ\mathrm{N}} dA} = \frac{2\pi R_E [u] \cos 40^\circ}{2\pi R_E^2 \int_{40^\circ}^{90^\circ} \cos\phi d\phi}$$

$$= \frac{[u]}{R_E} \frac{\cos 40^\circ}{(1 - \sin 40^\circ)} = 6.74 \times 10^{-6}\,\mathrm{s}^{-1}$$

In a similar manner, the divergence over the polar cap region is given by

$$\overline{\mathrm{Div}_H \mathrm{V}} = \frac{-[v]_{40\,^\circ\mathrm{N}}\, ds}{\iint_{40\,^\circ\mathrm{N}} dA}$$

$$= \frac{-[v]}{R_E} \frac{\cos 40^\circ}{(1 - \sin 40^\circ)}$$

$$= 1.01 \times 10^{-7}\,\mathrm{s}^{-1}$$

■

7.1.3 Deformation

Deformation, defined in the bottom line of Table 7.1, is the sum of the confluence and stretching terms. If the deformation is positive, grid squares oriented along the (s, n) axes will tend to be deformed into rectangles, elongated in the s direction. Conversely, if the deformation is negative, the squares will be deformed into rectangles elongated in the direction transverse to the flow. In Cartesian coordinates, the *deformation tensor* is made up of two components: the first relating to the stretching and squashing of grid squares aligned with the x and y axes, and the second with grid squares aligned at an angle of 45° with respect to the x and y axes. Figure 7.2(d) shows an example of a horizontal wind pattern consisting of pure deformation in the first component. Here the x axis corresponds to the *axis of dilatation* (or stretching) and the y axis corresponds to the *axis of contraction*. If this wind pattern were rotated by 45° relative to the x and y axes, the deformation would be manifested in the second component: rectangles would be deformed into rhomboids.

Figure 7.3 illustrates how even a relatively simple large-scale motion field can distort a field of passive tracers, initially configured as a rectangular grid, into an elongated configuration in which the individual grid squares are stretched and squashed beyond recognition. Some squares that were initially far apart end up close together, and vice versa.

Deformation can sharpen preexisting horizontal gradients of temperature, moisture, and other scalar

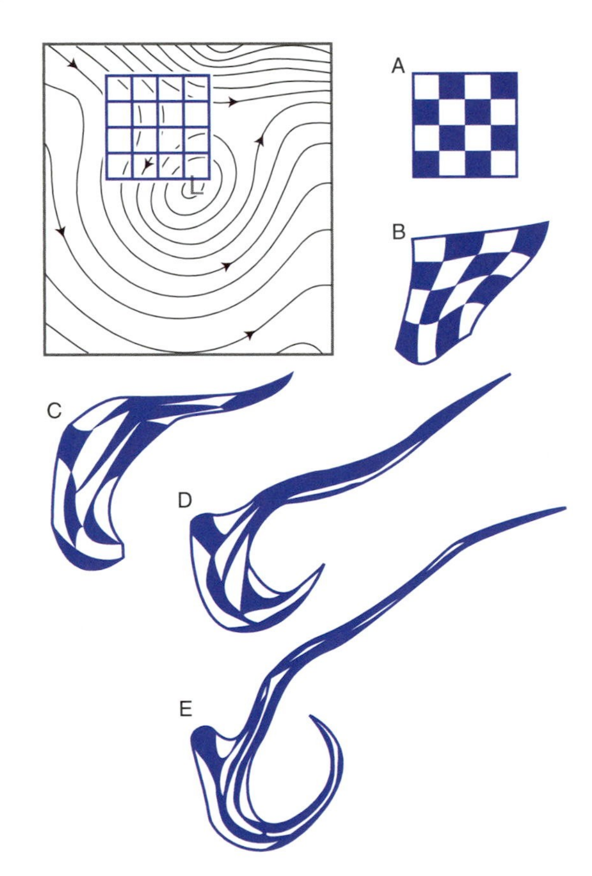

Fig. 7.3 (Top) A grid of air parcels embedded in a steady state horizontal wind field indicated by the arrows. The strength of the wind at any point is inversely proportional to the spacing between the contours at that point. (A–E) How the grid is deformed by the flow as the tagged particles move downstream; those in the upper right corner of the grid moving eastward and those in the lower left corner moving southward and then eastward around the closed circulation. [From *Tellus*, 7, 141–156 (1955).]

variables, creating features referred to as *frontal zones*. To illustrate how this sharpening occurs, consider the distribution of a hypothetical *passive tracer*, whose concentration $\psi(x, y)$ is conserved as it is carried around (or *advected*) in a divergence-free horizontal flow. Because $d\psi/dt = 0$ in (1.3), it follows that the time rate of change at a fixed point in space is given by the horizontal advection; that is

$$\frac{\partial \psi}{\partial t} = -\mathbf{V} \cdot \nabla \psi \qquad (7.5a)$$

or, in Cartesian coordinates,

$$\frac{\partial \psi}{\partial t} = -u\frac{\partial \psi}{\partial x} - v\frac{\partial \psi}{\partial y} \qquad (7.5b)$$

The time rate of change of ψ at a fixed point (x, y) is positive if ψ increases in the upstream direction in which case $\mathbf{V}$ at (x, y) must have a component directed *down* the gradient of ψ; hence the minus signs in (7.5a,b).

Suppose that ψ initially exhibits a uniform horizontal gradient, say from north to south. Such a gradient will tend to be sharpened within regions of negative $\partial v/\partial y$. In the pattern of pure deformation in Fig. 7.2d, $\partial v/\partial y$ is negative throughout the domain. Such a flow will tend to sharpen any preexisting north–south temperature gradient, creating an east–west oriented frontal zone, as shown in Fig. 7.4a. Frontal zones can also be twisted and sharpened by the presence of a wind pattern with shear, as shown in Fig. 7.4b.

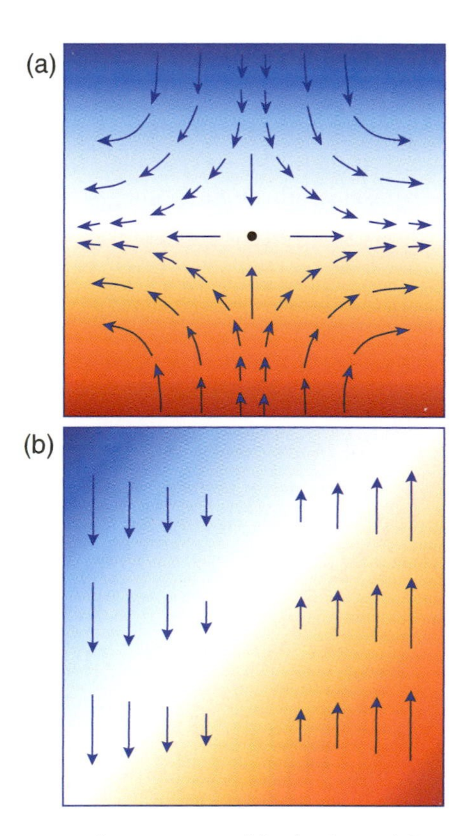

Fig. 7.4 Frontal zones created by horizontal flow patterns advecting a passive tracer with concentrations indicated by the colored shading. In (a) the gradient is being sharpened by deformation, whereas in (b) it is being twisted and sharpened by shear. See text and Exercise 7.11 for further explanation.

7.1.4 Streamlines versus Trajectories

If the horizontal wind field $\mathbf{V}$ is changing with time, the *streamlines* of the instantaneous horizontal wind field considered in this section are not the same as the *horizontal trajectories* of air parcels. Consider, for example, the case of a sinusoidal wave that is propagating eastward with phase speed c, superimposed on a uniform westerly flow of speed U, as depicted in Fig. 7.5. The solid lines represent horizontal streamlines at time t, and the dashed lines represent the horizontal streamlines at time $t + \delta t$, after the wave has moved some distance eastward. The trajectories originate at point A, which lies in the trough of the wave at time t. If the westerly flow matches the rate of eastward propagation of the wave, the parcel will remain in the wave trough as it moves eastward, as indicated by the straight trajectory AC. If the westerly flow through the wave is faster than the rate of propagation of the wave (i.e., if $U > c$), the air parcel will overtake the region of southwesterly flow ahead of the trough and drift northward, as indicated by the

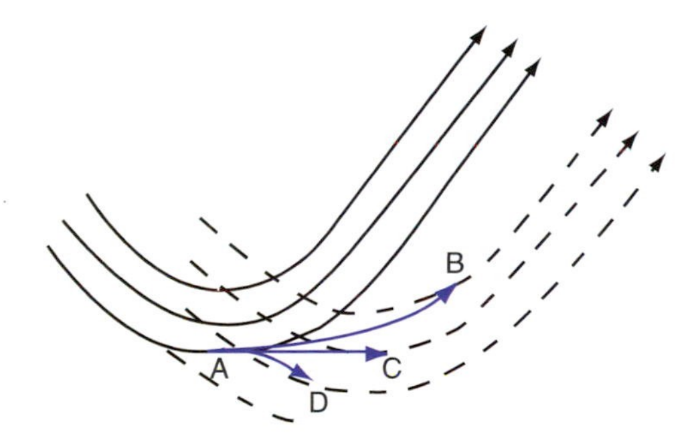

Fig. 7.5 Streamlines and trajectories for parcels in a wave moving eastward with phase velocity c embedded in a westerly flow with uniform speed U. Solid black arrows denote initial streamlines and dashed black arrows denote later streamlines. Blue arrows denote air trajectories starting from point A for three different values of U. AB is the trajectory for $U > c$; AC for $U = c$, and AD for $U < c$.

trajectory AB. Conversely, if $U < c$, the air parcel will fall behind the trough of the wave and curve southeastward, as indicated by AD. By construction, each of the three trajectories is parallel to the initial streamline passing through point A and to the respective later streamlines passing through points B, C, and D. The longest trajectory (AB) corresponds to the highest wind speed.

7.2 Dynamics of Horizontal Flow

Newton's second law states that in each of the three directions in the coordinate system, the acceleration a experienced by a body of mass m in response to a resultant force ΣF is given by

$$a = \frac{1}{m}\sum F \tag{7.6}$$

This relationship describes the motion in an inertial (nonaccelerating) frame of reference. However, it is more generally applicable, provided that *apparent forces* are introduced to compensate for the acceleration of the coordinate system. In a rotating frame of reference two different apparent forces are required: a *centrifugal force* that is experienced by all bodies, irrespective of their motion, and a *Coriolis force* that depends on the relative velocity of the body in the plane perpendicular to the axis of rotation (i.e., in the plane parallel to the equatorial plane).

7.2.1 Apparent Forces

The force per unit mass that is referred to as *gravity* or *effective gravity* and denoted by the symbol g represents the vectorial sum of the true gravitational attraction g* that draws all elements of mass toward Earth's center of mass and the much smaller apparent force called the *centrifugal force* $\Omega^2 R_A$, where Ω is the rotation rate of the coordinate system in radians per second (i.e., s^{-1}) and R_A is the distance from the axis of rotation. The centrifugal force pulls all objects outward from the axis of planetary rotation, as indicated in Fig. 7.6. In mathematical notation,

$$\mathbf{g} = \mathbf{g}^* + \Omega^2 \boldsymbol{R_A}.$$

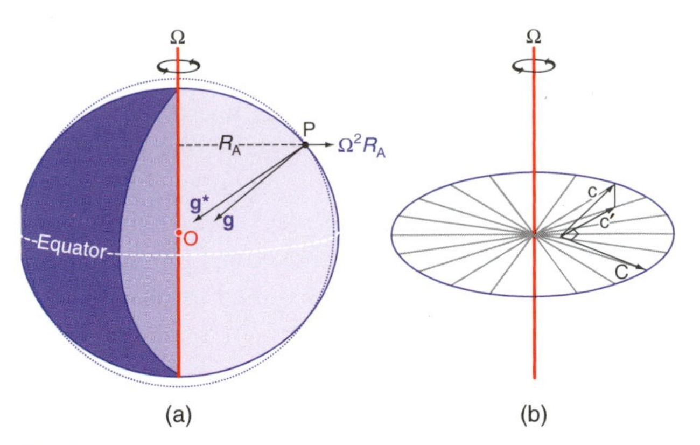

Fig. 7.6 (a) Apparent forces. Effective gravity **g** is the vectorial sum of the true gravitational acceleration **g*** directed toward the center of the Earth O and the *centrifugal force* $\Omega^2 R_A$. The acceleration **g** is normal to a surface of constant geopotential, an oblate spheroid, depicted as the outline of the Earth. The dashed reference line represents a true spherical surface. (b) The Coriolis force C is linearly proportional to c′ the component of the relative velocity c in the plane perpendicular to the axis of rotation. When viewed from a northern hemisphere perspective, C is directed to the right of c′ and lies in the plane perpendicular to the axis of rotation.

Surfaces of constant geopotential Φ, which are normal to **g**, are shaped like oblate spheroids, as indicated by the outline of the Earth in Fig. 7.6. Because the surface of the oceans and the large-scale configuration of the Earth's crust are incapable of resisting any sideways pull of effective gravity, they have aligned themselves with surfaces of constant geopotential. A body rotating with the Earth has no way of separately sensing the gravitational and centrifugal components of effective gravity. Hence, the $\Omega^2 \boldsymbol{R_A}$ term is incorporated into **g** in the equations of motion.

An object moving with velocity **c** in the plane perpendicular to the axis of rotation experiences an additional apparent force called the *Coriolis*[2] *force* $-2\Omega \times \mathbf{c}$. This apparent force is also in the plane perpendicular to the axis of rotation, and it is directed transverse to the motion in accordance with the right hand rule (i.e., if the rotation is counterclockwise when viewed from above, the force is directed to the right of **c** and vice versa).

When the forces and the motions are represented in a spherical coordinate system, the horizontal component of the Coriolis force arising from the

[2] **G. G. de Coriolis** (1792–1843) French engineer, mathmatician, and physicist. Gave the first modern definition of kinetic energy and work. Studied motions in rotating systems.

horizontal motion **V** can be written in vectorial form as

$$\mathbf{C} = -f\mathbf{k} \times \mathbf{V} \tag{7.7}$$

where f, the so-called *Coriolis parameter*, is equal to $2\Omega \sin \phi$ and **k** is the local vertical unit vector, defined as positive upward. The $\sin \phi$ term in f appropriately scales the Coriolis force to account for the fact that the local vertical unit vector **k** is parallel to the axis of rotation Ω only at the poles. Accordingly, the Coriolis force in the horizontal equation of motion increases with latitude from zero on the equator to $2\Omega V$ at the poles, where V is the (scalar) horizontal wind speed. The Coriolis force is directed toward the right of the horizontal velocity vector in the northern hemisphere and to the left of it in the southern hemisphere. On Earth

$$\Omega \equiv 2\pi \text{ rad day}^{-1} = 7.292 \times 10^{-5}\,\text{s}^{-1}$$

where *day* in this context refers to the *sidereal day*,[3] which is 23 h 56 min in length.

7.1 Experiment in a Dish

The role of the Coriolis force in a rotating coordinate system can be demonstrated in laboratory experiments. Here we describe an experiment that makes use of a special apparatus in which the centrifugal force is incorporated into the vertical force called gravity, as it is on Earth. The apparatus consists of a shallow dish, rotating about its axis of symmetry as shown in Fig. 7.7. The rotation rate Ω is tuned to the concavity of the dish such that at any given radius the outward-directed centrifugal force exactly balances the inward-directed component of gravity along the sloping surface of the dish; that is

$$g\frac{dz}{dr} = \Omega^2 r$$

where z is the height of the surface above some arbitrary reference level, r is the radius, and Ω is the rotation rate of the dish. Integrating from the center out to radius r yields the parabolic surface

$$z = \frac{\Omega^2 r^2}{2g} + constant$$

The constant of integration is chosen to make $z = 0$ the level of the center of the dish.

Consider the horizontal trajectory of an idealized frictionless marble rolling around in the dish, as represented in both a fixed (inertial) frame of reference and in a frame of reference rotating with the dish. (To view the motion in the rotating frame of reference, the video camera is mounted on the turntable.)

In the fixed frame of reference the differential equation governing the horizontal motion of the marble is

$$\frac{d^2 r}{dt^2} = -\Omega^2 r$$

It follows from the form of this differential equation that the marble will execute elliptical trajectories, symmetric about the axis of rotation, with

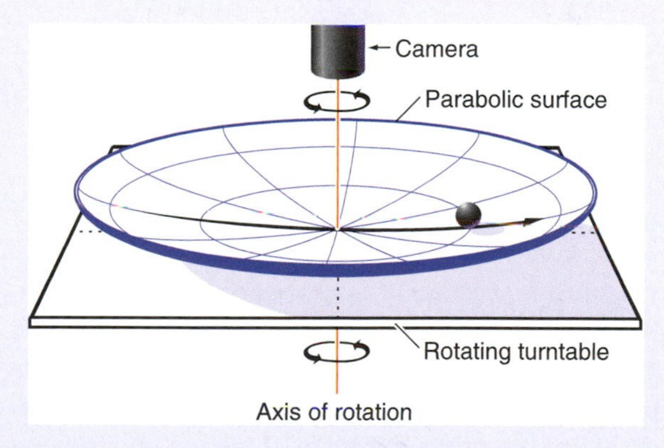

Fig. 7.7 Setup for the rotating dish experiment. Radius r is distance from the axis of rotation. Angular velocity Ω is the rotation rate of the dish. See text for further explanation.

Continued on next page

[3] The time interval between successive transits of a star over a meridian.

7.1 Continued

period $2\pi/\Omega$, which exactly matches the period of rotation of the dish. The shape and orientation of these trajectories will depend on the initial position and velocity of the marble. In this example, the marble is released at radius $r = r_0$ with no initial velocity. After the marble is released it rolls back and forth, like the tip of a pendulum, along the straight line pictured in Fig. 7.7, with a period equal to $2\pi/\Omega$, the same as the period of rotation of the dish. This oscillatory solution is represented by the equation

$$r = r_0 \cos \Omega t$$

where t is time and radius r is defined as positive on the side of the dish from which the marble is released and negative on the other side. The velocity of the marble along its pendulum-like trajectory

$$\frac{dr}{dt} = -\Omega r_0 \sin \Omega t$$

is largest at the times when it passes through the center of the dish at $t = \pi/2,\ 3\pi/2 \ldots$, and the marble is motionless for an instant at $t = \pi,\ 2\pi \ldots$ when it reverses direction at the outer edge of its trajectory.

In the rotating frame of reference the only force in the horizontal equation of motion is the Coriolis force, so the governing equation is

$$\frac{d\mathbf{c}}{dt} = -2\Omega\mathbf{k} \times \mathbf{c}$$

where $\mathbf{c}$ is the velocity of the marble and $\mathbf{k}$ is the vertical (normal to the surface of the dish) unit vector. Because $d\mathbf{c}/dt$ is perpendicular to $\mathbf{c}$, it follows that c, the speed of the marble as it moves along its trajectory in the rotating frame of reference, must be constant. The direction of the forward motion of the marble is changing with time at the uniform rate 2Ω, which is exactly twice the rate of rotation of the dish. Hence, the marble executes a circular orbit called an *inertia circle*, with period $2\pi/2\Omega = \pi/\Omega$ (i.e., half the period of rotation of the dish), with circumference $c\ (\pi/\Omega)$, and radius $c\ (\pi/\Omega)/2\pi = c/2\Omega$. Because $d\mathbf{c}/dt$ is to the right of $\mathbf{c}$, it follows that the marble rolls clockwise, i.e., in the opposite sense as the rotation of the dish.

As in the fixed frame of reference, the trajectory of the marble depends on its initial position and relative motion. Because the marble is released at radius r_0 with no initial motion in the fixed frame of reference, it follows that its speed in the rotating frame of reference is $c = \Omega r_0$, and thus the radius of the inertia circle is $\Omega r_0/2\Omega = r_0/2$. It follows that the marble passes through the center of the dish at the midpoint of its trajectory around the inertia circle.

The motion of the marble in the fixed and rotating frames of reference is shown in Fig. 7.8. The rotating dish is represented by the large circle. The point of release of the marble is labeled **0** and appears at the top of the diagram rather than on the left side of the dish as in Fig. 7.7. The pendulum-like trajectory of the marble in the fixed frame of reference is represented by the straight

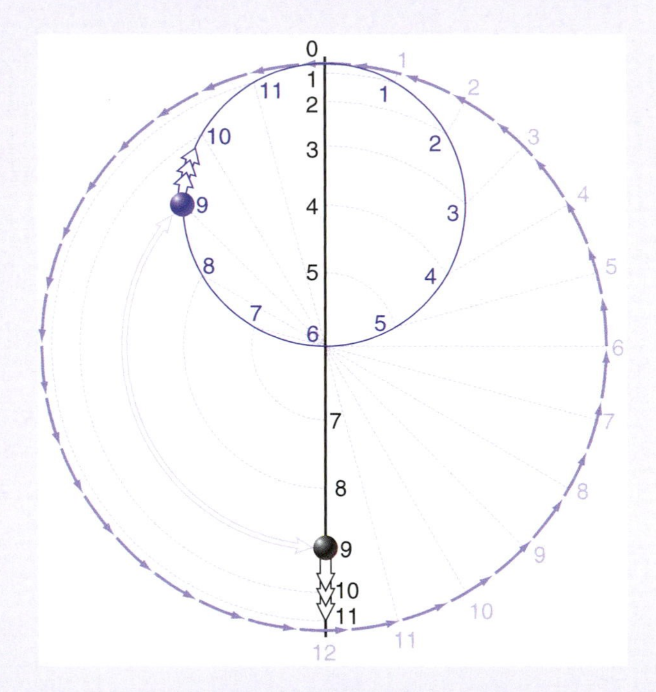

Fig. 7.8 Trajectories of a frictionless marble in fixed (black) and rotating (blue) frames of reference. Numbered points correspond to positions of the marble at various times after it is released at point 0. One complete rotation of the dish corresponds to one swing back and forth along the straight vertical black line in the fixed frame of reference and two complete circuits of the marble around the blue inertia circle in the rotating frame of reference. The light lines are reference lines. See text for further explanation.

7.1 Continued

black vertical line passing through the center of the dish. Successive positions along the trajectory at equally spaced time intervals are represented by numbers: point **1** represents the position of the marble after 1/24 of the period of rotation of the dish, point **12** after one-half period of rotation, etc. Hence, the numerical values assigned to the points are analogous to the 24 h of the day on a rotating planet. Note that the spacing between successive points is largest near the middle of the pendulum-like trajectory, where the marble is rolling fastest.

The position of the marble in the rotating (blue) frame of reference can be located at any specified time without invoking the Coriolis force simply by subtracting the displacement of the dish from the displacement of the marble in the fixed frame of reference. For example, to locate the marble after one-eighth rotation of the dish, we rotate point **3** clockwise (i.e., in the direction opposite to the rotation of the dish) one-eighth of the way around the circle; point **9** needs to be rotated three-eighths of the way around the circle, and so forth. The rotated points map out the trajectory of the marble in the rotating frame of reference: an inertia circle with a period equal to one-half revolution of the dish (i.e., 12 "hours"). In Fig. 7.8 the marble is pictured at 9 o'clock on the 24-hour clock in the two frames of reference.

Alternatively, the inertia circle can be constructed by first rotating the marble backward (clockwise) from its point of release to subtract the rotation of the dish and then moving the marble radially the appropriate distance along its pendulum-like trajectory. Reference lines for performing these operations are shown in Fig. 7.8. Examples of trajectories for two other sets of initial conditions are presented in Exercise 7.14.

7.2.2 Real Forces

The real forces that enter into the equations of motion are *gravity*, the *pressure gradient force*, and *frictional force* exerted by neighboring air parcels or adjacent surfaces.

a. *The pressure gradient force*

The vertical component of the pressure gradient force (per unit mass) $-(1/\rho)\partial p/\partial z$ has already been introduced in the context of the hydrostatic equation. The horizontal component of the pressure gradient force is given by the analogous expression

$$\mathbf{P} \equiv -\frac{1}{\rho}\nabla p \tag{7.8a}$$

or, in component form,

$$P_x = -\frac{1}{\rho}\frac{\partial p}{\partial x}; \quad P_y = -\frac{1}{\rho}\frac{\partial p}{\partial y} \tag{7.8b}$$

The pressure gradient force is directed down the horizontal pressure gradient ∇p from higher toward lower pressure. Making use of the hydrostatic equation (3.17) and the definitions of geopotential (3.20) and geopotential height (3.22), the horizontal pressure gradient force can be expressed in the alternative forms

$$\mathbf{P} \equiv -g\nabla z = -g_0\nabla Z = -\nabla\Phi \tag{7.9}$$

where the gradients of geometric height, geopotential height, and geopotential are defined on sloping pressure surfaces. Hence, the pressure gradient force can be interpreted as the component of effective gravity g in the plane of the pressure surface, analogous to the "downhill" force on a ball rolling on a sloping surface. Pressure surfaces exhibit typical slopes on the order of 100 m per thousand kilometers, or 1 in 10^4. Hence, the horizontal component of the pressure gradient force is roughly four orders of magnitude smaller than the vertical component; i.e., it is on the order of 10^{-3} m s^{-2}.

The pressure field on weather charts is typically represented by a set of contours plotted at regularly spaced intervals as in Fig. 1.19. The lines that are used to depict the distribution of pressure on geopotential height surfaces are referred to as *isobars*, and the lines used to depict the distribution of geopotential height on pressure surfaces are referred to as *geopotential height contours*. Exercise 3.3 demonstrates that, to a close approximation, isobars on a constant geopotential surface (e.g., sea level) can be converted into geopotential height contours on a nearby pressure surface simply by relabeling them using a constant of proportionality based on the hypsometric equation (3.29). For example, near sea level, where atmospheric pressure decreases with height at a rate of ~1 hPa per 8 m, the conventional 4-hPa contour interval for isobars of sea-level pressure is approximately equivalent to a 30-m contour interval for geopotential height contours on a specified pressure surface. It follows that (7.8) and (7.9) yield virtually identical distributions of **P**.

In the oceans the horizontal pressure gradient force is due to both the gradient in sea level on a surface of constant geopotential and horizontal gradients in the density of the overlying water in the column. Near the ocean surface this force is primarily associated with the horizontal gradient in sea level. The topography of sea level is difficult to estimate in an absolute sense from observations because Earth's *geoid* (i.e., geopotential field) exhibits nonellipsoidal irregularities of its own that have more to do with plate tectonics than with oceanography. Temporal variations in sea level relative to Earth's much more slowly evolving geoid are clearly revealed by satellite altimetry.

b. The frictional force

The frictional force (per unit mass) is given by

$$\mathbf{F} = -\frac{1}{\rho}\frac{\partial \tau}{\partial z} \tag{7.10}$$

where τ represents the vertical component of the *shear stress* (i.e., the rate of vertical exchange of horizontal momentum) in units of N m^{-2} due to the presence of smaller, unresolved scales of motion.[4] Vertical exchanges of momentum usually act to smooth out the vertical profile of **V**. The rate of vertical mixing that is occurring at any particular level and time depends on the strength of the vertical wind shear $\partial \mathbf{V}/\partial z$ and on the intensity of the unresolved motions, as discussed in Chapter 9. In the free atmosphere, above the boundary layer, the frictional force is much smaller than the pressure gradient force and the Coriolis force. However, within the boundary layer the frictional force is comparable in magnitude to the other terms in the horizontal equation of motion and needs to be taken into account.

The shear stress τ_s at the Earth's surface is in the opposing direction to the surface wind vector $\mathbf{V}_s$ (i.e., it is a "drag" on the surface wind) and can be approximated by the empirical relationship

$$\tau_s = -\rho C_D \mathbf{V}_s V_s \tag{7.11}$$

where ρ is the density of the air, C_D is a dimensionless *drag coefficient*, the magnitude of which varies with the roughness of the underlying surface and the static stability, $\mathbf{V}_s$ is the surface wind vector, and V_s is the (scalar) surface wind speed. Within the lowest few tens of meters of the atmosphere, the stress decreases with height without much change in direction. Hence, within this so-called *surface layer*, the frictional force $\mathbf{F}_s = -\partial\tau/\partial z$ is directed opposite to $\mathbf{V}_s$ and is referred to as *frictional drag*.

7.2.3 The Horizontal Equation of Motion

The horizontal component of (7.6), written in vectorial form, per unit mass, is

$$\frac{d\mathbf{V}}{dt} = \mathbf{P} + \mathbf{C} + \mathbf{F} \tag{7.12}$$

where $d\mathbf{V}/dt$ is the Lagrangian time derivative of the horizontal velocity component experienced by an air parcel as it moves about in the atmosphere. Substituting for **C** from (7.7) and for **P** from (7.8a) we obtain

$$\frac{d\mathbf{V}}{dt} = -\frac{1}{\rho}\nabla p - f\mathbf{k} \times \mathbf{V} + \mathbf{F} \tag{7.13a}$$

or, in component form on a tangent plane (i.e., neglecting smaller terms that arise due to the curvature of the coordinate system),

$$\frac{du}{dt} = -\frac{1}{\rho}\frac{\partial p}{\partial x} + fv + F_x,$$
$$\frac{dv}{dt} = -\frac{1}{\rho}\frac{\partial p}{\partial y} - fu + F_y \tag{7.13b}$$

The density dependence can be eliminated by substituting for **P** using (7.9) instead of (7.8), which yields

$$\frac{d\mathbf{V}}{dt} = -\nabla\Phi - f\mathbf{k} \times \mathbf{V} + \mathbf{F} \tag{7.14}$$

[4] The contributions of the corresponding horizontal exchanges of momentum to **F** do not need to be considered because the horizontal gradients are much smaller than the vertical gradients.

In (7.13a) the horizontal wind field is defined on surfaces of constant geopotential so that $\nabla\Phi = 0$, whereas in (7.14) it is defined on constant pressure surfaces so that $\nabla p = 0$. However, pressure surfaces are sufficiently flat that the **V** fields on a geopotential surface and a nearby pressure surface are very similar.

7.2.4 The Geostrophic Wind

In large-scale wind systems such as baroclinic waves and extratropical cyclones, typical horizontal velocities are on the order of $10\ \text{m s}^{-1}$ and the timescale over which individual air parcels experience significant changes in velocity is on the order of a day or so ($\sim 10^5$ s). Thus a typical parcel acceleration $d\mathbf{V}/dt$ is $\sim 10\ \text{m s}^{-1}$ per 10^5 s or $10^{-4}\ \text{m s}^{-2}$. In middle latitudes, where $f \sim 10^{-4}\ \text{s}^{-1}$, an air parcel moving at a speed of $10\ \text{m s}^{-1}$ experiences a Coriolis force per unit mass **C** $\sim 10^{-3}\ \text{m s}^{-2}$, about an order of magnitude larger than the typical horizontal accelerations of air parcels.

In the free atmosphere, where the frictional force is usually very small, the only term that is capable of balancing the Coriolis force **C** is the pressure gradient force **P**. Thus, to within about 10%, in middle and high latitudes, the horizontal equation of motion (7.14) is closely approximated by

$$f\mathbf{k} \times \mathbf{V} \simeq -\nabla\Phi$$

Making use of the vector identity

$$\mathbf{k} \times (\mathbf{k} \times \mathbf{V}) = -\mathbf{V}$$

it follows that

$$\mathbf{V} \simeq \frac{1}{f}(\mathbf{k} \times \nabla\Phi)$$

For any given horizontal distribution of pressure on geopotential surfaces (or geopotential height on pressure surfaces) it is possible to define a *geostrophic*[5] *wind field* $\mathbf{V}_g$ for which this relationship is exactly satisfied:

$$\mathbf{V}_g \equiv \frac{1}{f}(\mathbf{k} \times \nabla\Phi) \tag{7.15a}$$

or, in component form,

$$u_g = -\frac{1}{f}\frac{\partial\Phi}{\partial y}, \quad v_g = \frac{1}{f}\frac{\partial\Phi}{\partial x} \tag{7.15b}$$

or, in natural coordinates,

$$V_g = -\frac{1}{f}\frac{\partial\phi}{\partial n} \tag{7.15c}$$

where V_g is the scalar geostrophic wind speed and n is the direction normal to the isobars (or geopotential height contours), pointing toward higher values.

The balance of horizontal forces implicit in the definition of the geostrophic wind (for a location in the northern hemisphere) is illustrated in Fig. 7.9. In order for the Coriolis force and the pressure gradient force to balance, the geostrophic wind must blow parallel to the isobars, leaving low pressure to the left. In either hemisphere, the geostrophic wind field circulates cyclonically around a center of low pressure and vice versa, as in Fig. 1.14, justifying the identification of local pressure minima with cyclones and local pressure maxima with anticyclones. The tighter the spacing of the isobars or geopotential height contours, the stronger the Coriolis force required to balance the pressure gradient force and hence, the higher the speed of the geostrophic wind.

7.2.5 The Effect of Friction

The three-way balance of forces required for flow in which $d\mathbf{V}/dt = 0$ in the northern hemisphere in the presence of friction at the Earth's surface is illustrated in Fig. 7.10. As in Fig. 7.9, **P** is directed normal

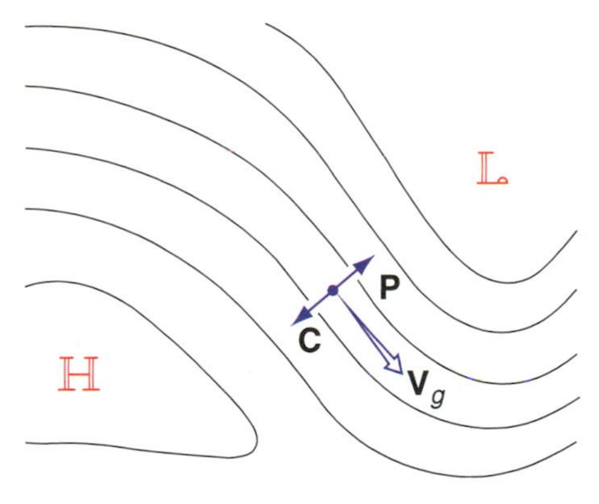

Fig. 7.9 The geostrophic wind V_g and its relationship to the horizontal pressure gradient force P and the Coriolis force C in the northern hemisphere.

[5] From the Greek: *geo* (Earth) and *strophen* (to turn)

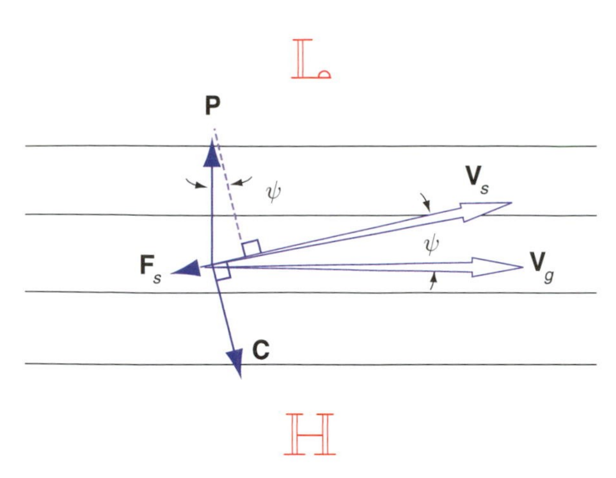

Fig. 7.10 The three-way balance of forces required for steady surface winds in the presence of the frictional drag force F in the northern hemisphere. Solid lines represent isobars or geopotential height contours on a weather chart.

to the isobars, **C** is directed to the right of the horizontal velocity vector $\mathbf{V}_s$, and, consistent with (7.11), $\mathbf{F}_s$ is directed opposite to $\mathbf{V}_s$. The angle ψ between $\mathbf{V}_s$ and $\mathbf{V}_g$ is determined by the requirement that the component of **P** in the forward direction of $\mathbf{V}_s$ must be equal to the magnitude of the drag $\mathbf{F}_s$, and the wind speed V_s is determined by the requirement that **C** be just large enough to balance the component of **P** in the direction normal to $\mathbf{V}_s$; i.e.,

$$fV_s = |\mathbf{P}| \cos \psi$$

It follows that $|\mathbf{C}| < |\mathbf{P}|$ and, hence, the scalar wind speed $V_s = |\mathbf{C}|/f$ must be smaller than $V_g = |\mathbf{P}|/f$. The stronger the frictional drag force $\mathbf{F}_s$, the larger the angle ψ between $\mathbf{V}_g$ and $\mathbf{V}_s$ and the more *subgeostrophic* the surface wind speed $\mathbf{V}_s$. The cross-isobar flow toward lower pressure, referred to as the *Ekman*[6] *drift*, is clearly evident on surface charts, particularly over rough land surfaces. That the winds usually blow nearly parallel to the isobars in the free atmosphere indicates that the significance of the frictional drag force is largely restricted to the boundary layer, where small-scale turbulent motions are present.

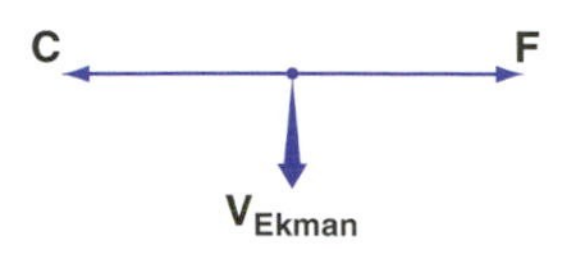

Fig. 7.11 The force balance associated with Ekman drift in the northern hemisphere oceans. The frictional force F is in the direction of the surface wind vector. In the southern hemisphere (not shown) the Ekman drift is to the left of the surface wind vector.

The same shear stress that acts as a drag force on the surface winds exerts a forward pull on the surface waters of the ocean, giving rise to *wind-driven currents*. If the ocean surface coincided with a surface of constant geopotential, the balance of forces just below the surface would consist of a two-way balance between the forward pull of the surface wind and the backward pull of the Coriolis force induced by the Ekman drift, as shown in Fig. 7.11. Although the large scale surface currents depicted in Fig. 2.4 tend to be in geostrophic balance and oriented roughly parallel to the mean surface winds, the Ekman drift, which is directed normal to the surface winds, has a pronounced effect on horizontal transport of near-surface water and sea-ice, and it largely controls the distribution of upwelling, as discussed in Section 7.3.4. Ekman drift is largely confined to the topmost 50 m of the oceans.

7.2.6 The Gradient Wind

The centripetal accelerations observed in association with the curvature of the trajectories of air parcels tend to be much larger than those associated with

[6] **V. Walfrid Ekman** (1874–1954) Swedish oceanographer. Ekman was introduced to the problem of wind-driven ocean circulation when he was a student working under the direction of Professor Vilhelm Bjerknes.[7] Fridtjof Nansen had approached Bjerknes with a remarkable set of observations of winds and ice motions taken during the voyage of the *Fram*, for which he sought an explanation. Nansen's observations and Ekman's mathematical analysis are the foundations of the theory of the wind-driven ocean circulation.

[7] **Vilhelm Bjerknes** (1862–1951) Norwegian physicist and one of the founders of the science of meteorology. Held academic positions at the universities of Stockholm, Bergen, Leipzig, and Kristiania (renamed Oslo). Proposed in 1904 that weather prediction be regarded as an initial value problem that could be solved by integrating the governing equations forward in time, starting from an initial state determined by current weather observations. Best known for his work at Bergen (1917–1926) where he assembled a small group of dedicated and talented young researchers, including his son Jakob. The most widely recognized achievement of this so-called "Bergen School" was a conceptual framework for interpreting the structure and evolution of extratropical cyclones and fronts that has endured until the present day.

the speeding up or slowing down of air parcels as they move downstream. Hence, when $d\mathbf{V}/dt$ is large, its scalar magnitude can be approximated by the centripetal acceleration V^2/R_T, where R_T is the local radius of curvature of the air trajectories.[8] Hence, the horizontal equation of motion reduces to the balance of forces in the direction transverse to the flow, i.e.,

$$\mathbf{n}\frac{V^2}{R_T} = -\nabla\Phi - f\mathbf{k}\times\mathbf{V} \qquad (7.16)$$

The signs of the terms in this three-way balance depend on whether the curvature of the trajectories is cyclonic or anticyclonic, as illustrated in Fig. 7.12. In the cyclonic case, the outward centrifugal force (the mirror image of the centripetal acceleration) reinforces the Coriolis force so that a balance can be achieved with a wind speed smaller than would be required if the Coriolis force were acting alone. In flow through sharp troughs, where the curvature of the trajectories is cyclonic, the observed wind speeds at the jet stream level are often smaller, by a factor of two or more, than the geostrophic wind speed implied by the spacing of the isobars. For the anticyclonically curved trajectory on the right in Fig. 7.12 the centrifugal force opposes the Coriolis force, necessitating a *supergeostrophic* wind speed in order to achieve a balance.

The wind associated with a three-way balance between the pressure gradient and Coriolis and

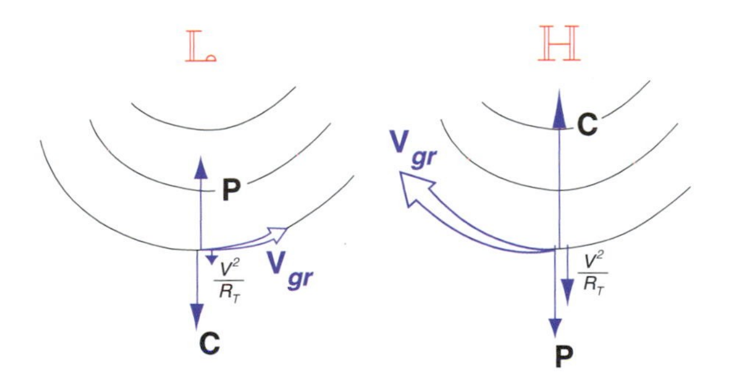

Fig. 7.12 The three-way balance involving the horizontal pressure gradient force P, the Coriolis force C, and the centrifugal force $|V|^2/R_T$ in flow along curved trajectories in the northern hemisphere. (Left) Cyclonic flow. (Right) Anticyclonic flow.

centrifugal forces is called the *gradient wind*. The solution of (7.16), which yields the speed of the gradient wind can be written in the form

$$V_{gr} = \frac{1}{f}\left(|\nabla\Phi| + \frac{V_{gr}^2}{R_T}\right) \qquad (7.17)$$

and solved using the quadratic formula. From Fig. 7.12, it can be inferred that R_T should be specified as positive if the curvature is cyclonic and negative if the curvature is anticyclonic. For the case of anticyclonic curvature, a solution exists when

$$|\nabla\Phi| \le \frac{f^2|R_T|}{4}$$

7.2.7 The Thermal Wind

Just as the geostrophic wind bears a simple relationship to $\nabla\Phi$, the vertical shear of the geostrophic wind bears a simple relationship to ∇T. Writing the geostrophic equation (7.15a) for two different pressure surfaces and subtracting, we obtain an expression for the vertical wind shear in the intervening layer

$$(\mathbf{V}_g)_2 - (\mathbf{V}_g)_1 = \frac{1}{f}\mathbf{k}\times\nabla(\Phi_2 - \Phi_1) \qquad (7.18)$$

In terms of geopotential height

$$(\mathbf{V}_g)_2 - (\mathbf{V}_g)_1 = \frac{g_0}{f}\mathbf{k}\times\nabla(Z_2 - Z_1) \qquad (7.19a)$$

or in component form,

$$(u_g)_2 - (u_g)_1 = -\frac{g_0}{f}\frac{\partial(Z_2 - Z_1)}{\partial y},$$

$$(v_g)_2 - (v_g)_1 = \frac{g_0}{f}\frac{\partial(Z_2 - Z_1)}{\partial x} \qquad (7.19b)$$

This expression, known as the *thermal wind equation*, states that the vertically averaged vertical shear of the geostrophic wind within the layer between any two pressure surfaces is related to the horizontal gradient of thickness of the layer in the same manner in which geostrophic wind is related to geopotential height. For example, in the northern hemisphere the *thermal wind*

[8] In estimating the radius of curvature in the gradient wind equation it is important to keep in mind the distinction between *streamlines* and *trajectories*, as explained in Section 7.1.4.

(namely the vertical shear of the geostrophic wind) "blows" parallel to the thickness contours, leaving low thickness to the left. Incorporating the linear proportionality between temperature and thickness in the hypsometric equation (3.29), the thermal wind equation can also be expressed as a linear relationship between the vertical shear of the geostrophic wind and the horizontal temperature gradient.

$$(\mathbf{V}_g)_2 - (\mathbf{V}_g)_1 = \left(\frac{R}{f}\ln\frac{p_1}{p_2}\right)\mathbf{k} \times \nabla(\overline{T}) \qquad (7.20)$$

where $\overline{T}$ is the vertically averaged temperature within the layer.

To explore the implications of the thermal wind equation, consider first the special case of an atmosphere that is characterized by a total absence of horizontal temperature (thickness) gradients. In such a *barotropic*[9] atmosphere $\nabla T = 0$ on constant pressure surfaces. Because the thickness of the layer between any pair of pressure surfaces is horizontally uniform, it follows that the geopotential height contours on various pressure surfaces can be neatly stacked on top of one another, like a set of matched dinner plates. It follows that the direction and speed of the geostrophic wind must be independent of height.

Now let us consider an atmosphere with horizontal temperature gradients subject to the constraint that the thickness contours be everywhere parallel to the geopotential height contours. For historical reasons, such a flow configuration is referred to as *equivalent barotropic*. It follows from the thermal wind equation that the vertical wind shear in an equivalent barotropic atmosphere must be parallel to the wind itself so the direction of the geostrophic wind does not change with height, just as in the case of a pure barotropic atmosphere. However, the slope of the pressure surfaces and hence the speed of the geostrophic wind may vary from level to level in association with thickness variations in the direction normal to height contours, as illustrated in Fig. 7.13.

In an equivalent barotropic atmosphere, the isobars and isotherms on horizontal maps have the same shape. If the highs are warm and the lows are cold, the amplitude of features in the pressure field and the speed of the geostrophic wind increase with height. If the highs are cold and the lows are warm,

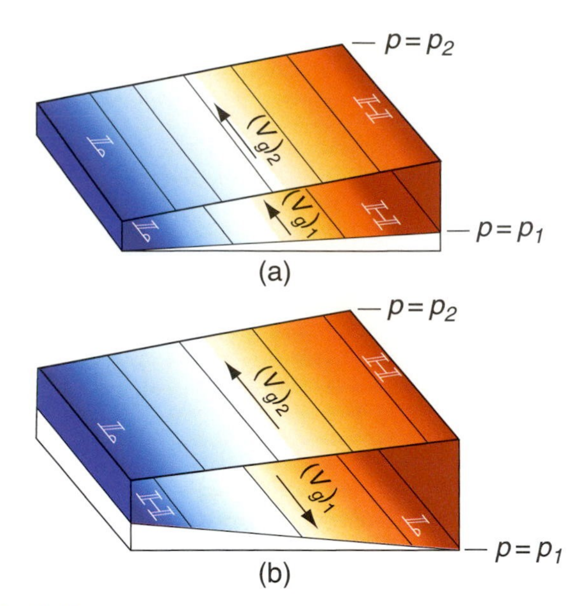

Fig. 7.13 The change of the geostrophic wind with height in an equivalent barotropic flow in the northern hemisphere: (a) V_g increasing with height within the layer and (b) V_g reversing direction. The temperature gradient within the layer is indicated by the shading: blue (cold) coincides with low thickness and tan (warm) with high thickness.

the situation is just the opposite; features in the pressure and geostrophic wind field tend to weaken with height and the wind may even reverse direction if the temperature anomalies extend through a deep enough layer, as in Fig. 7.13b. The zonally averaged zonal wind and temperature cross sections shown in Fig. 1.11 are related in a manner consistent with Fig. 7.13. Wherever temperature decreases with increasing latitude, the zonal wind becomes (relatively) more westerly with increasing height, and vice versa.

Exercise 7.3 During winter in the troposphere ~30° latitude, the zonally averaged temperature gradient is ~0.75 K per degree of latitude (see Fig. 1.11) and the zonally averaged component of the geostrophic wind at the Earth's surface is close to zero. Estimate the mean zonal wind at the jet stream level, ~250 hPa.

Solution: Taking the zonal component of (7.20) and averaging around a latitude circle yields

$$[u_g]_{250} - [u_g]_{1000} = -\frac{R}{2\Omega\sin\phi}\frac{\partial[T]}{\partial y}\ln\frac{1000}{250}$$

[9] The term barotropic is derived from the Greek *baro*, relating to pressure, and *tropic*, changing in a specific manner: that is, in such a way that surfaces of constant pressure are coincident with surfaces of constant temperature or density.

Noting that $[u_g]_{1000} \simeq 0$ and $R \simeq R_d$

$$[u_g]_{250} = -\frac{287 \text{ J deg}^{-1} \text{ kg}^{-1}}{2 \times 7.29 \times 10^{-5} \text{ s}^{-1} \sin 30^\circ} \times \ln 4$$

$$\times \frac{-0.75 \text{ K}}{1.11 \times 10^5 \text{ m}} = 36.8 \text{ m s}^{-1},$$

in close agreement with Fig. 1.11. ■

In a fully *baroclinic atmosphere*, the height and thickness contours intersect one another so that the geostrophic wind exhibits a component normal to the isotherms (or thickness contours). This geostrophic flow across the isotherms is associated with *geostrophic temperature advection. Cold advection* denotes flow across the isotherms from a colder to a warmer region, and vice versa.

Typical situations corresponding to cold and warm advection in the northern hemisphere are illustrated in Fig. 7.14. On the pressure level at the bottom of the layer the geostrophic wind is from the west so the height contours are oriented from west to east, with lower heights toward the north. Higher thickness lies toward the east in the cold advection case (Fig. 7.14a) and toward the west in the warm advection case (Fig. 7.14b).[10]

The upper level geostrophic wind vector $\mathbf{V}_{g2}$ blows parallel to these upper-level contours. Just as the upper level geopotential height is the algebraic sum of the lower level geopotential height Z_1 plus the thickness Z_T, the upper level geostrophic wind vector $\mathbf{V}_{g2}$ is the vectorial sum of the lower level geostrophic wind $\mathbf{V}_{g1}$ plus the thermal wind $\mathbf{V}_T$. Hence, *thermal wind* is to *thickness* as *geostrophic wind* is to *geopotential height.*

From Fig. 7.14 it is apparent that cold advection is characterized by *backing* (cyclonic rotation) of the geostrophic wind vector with height and warm advection is characterized by *veering* (anticyclonic rotation). By experimenting with other configurations of height and thickness contours, it is readily verified that this relationship holds, regardless of the direction of the geostrophic wind at the bottom of the layer or the orientation of the isotherms, and it holds for the southern hemisphere as well.

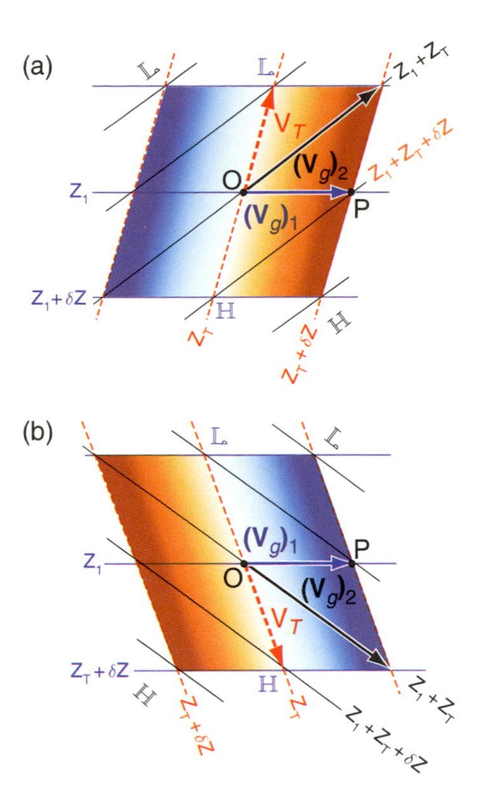

Fig. 7.14 Relationships among isotherms, geopotential height contours, and geostrophic wind in layers with (a) cold and (b) warm advection. Solid blue lines denote the geopotential height contours at the bottom of the layer and solid black lines denote the geopotential height contours at the top of the layer. Red lines represent the isotherms or thickness contours within the layer.

Making use of the thermal wind equation it is possible to completely define the geostrophic wind field on the basis of knowledge of the distribution of $T(x, y, p)$, together with boundary conditions for either $p(x, y)$ or $\mathbf{V}_g(x, y)$ at the Earth's surface or at some other "reference level." Thus, for example, a set of sea-level pressure observations together with an array of closely spaced temperature soundings from satellite-borne sensors constitutes an observing system suitable for determining the three-dimensional distribution of $\mathbf{V}_g$.

[10] Given the distribution of geopotential height Z_1 at the lower level and thickness Z_T, it is possible to infer the geopotential height Z_2 at the upper level by simple addition. For example, the height Z_2 at point O at the center of the diagram is equal to $Z_1 + Z_T$, and at point P it is $Z_1 + (Z_T + \delta Z)$, and so forth. The upper-level height contours (solid black lines) are drawn by connecting intersections with equal values of this sum. If the fields of Z_1, Z_2, and Z_T are all displayed using the same contour interval (say, 60 m), then all intersections between countours will be three-way intersections. In a similar manner, given a knowledge of the geopotential height field at the lower and upper levels, it is possible to infer the thickness field Z_T by subtracting Z_1 from Z_2.

7.2.8 Suppression of Vertical Motions by Planetary Rotation[11]

Through the action of the Coriolis force, Earth's rotation imparts a special character to large-scale motions in the atmosphere and oceans. In the absence of planetary rotation, typical magnitudes of the horizontal and vertical motion components (denoted here as U and W, respectively) would scale, relative to one another, in proportion to their respective length (L) and depth scales (D) of the horizontal and vertical motions, i.e.,

$$\frac{W}{U} \sim \frac{D}{L}$$

Boundary layer turbulence and convection, whose timescales are much shorter then the planetary rotation do, in fact, scale in such a manner. However, the vertical component of large-scale motions is smaller than would be expected on the basis of their aspect ratio. For example, in extratropical latitudes the depth scale is ~5 km (half the depth of the troposphere) and the length scale is ~1000 km: hence, $D/L \sim 1/200$. Typical horizontal velocities in these systems are ~10 m s^{-1}. The corresponding vertical velocities, scaled in accordance with a 1/200 aspect ratio, should be ~5 cm s^{-1}. However, the observed vertical velocities (~1 cm s^{-1}) are almost an order of magnitude smaller than that.

Similar considerations apply to the ratio of typical values of divergence and vorticity observed in large-scale horizontal motions. Because these quantities involve horizontal derivatives of the horizontal wind field, one might expect them both to scale as

$$\frac{U}{L} \sim \frac{10\ \text{m s}^{-1}}{1{,}000\ \text{km}} = \frac{10\ \text{m s}^{-1}}{10^6\ \text{m}} = 10^{-5}\,\text{s}^{-1}$$

just like the elementary properties of the horizontal flow (shear, curvature, diffluence and stretching) and the terms $\partial u/\partial x$, $\partial u/\partial y$, $\partial v/\partial x$, and $\partial v/\partial y$ in Cartesian coordinates. Vorticity does, in fact, scale in such a manner, but typical values of the divergence are $\sim 10^{-6}\,\text{s}^{-1}$, almost an order of magnitude smaller than U/L.

The smallness of the divergence in a rotating atmosphere reflects a high degree of compensation between diffluence and stretching terms, with diffluence occurring in regions in which the flow is slowing down as it moves downstream, and vice versa. This *quasi-nondivergent* character of the horizontal wind field is in sharp contrast to the flow of motor vehicles on a multilane highway, in which a narrowing of the roadway is marked by a slowdown of traffic. A similar compensation is observed between the terms in the Cartesian form of the divergence; that is, to within 10–20%

$$\frac{\partial u}{\partial x} \approx -\frac{\partial v}{\partial y}$$

Because estimates of the divergence require estimating a small difference between large terms, they tend to be highly sensitive to errors in the wind field. The geostrophic wind field also tends to be quasi-nondivergent, but not completely so, because of the variation of the Coriolis parameter f with latitude (see Exercise 7.20).

7.2.9 A Conservation Law for Vorticity

A major breakthrough in the development of modern atmospheric dynamics was the realization, in the late 1930s, that many aspects of the behavior of large-scale extratropical wind systems can be understood on the basis of variants of a single equation based on the conservation of vorticity of the horizontal wind field.[12] To a close approximation, the time rate of change of vorticity can be written in the form

$$\frac{\partial}{\partial t}(f + \zeta) = -\mathbf{V} \cdot \nabla(f + \zeta) - (f + \zeta)(\nabla \cdot \mathbf{V}) \quad (7.21a)$$

or, in Lagrangian form,

$$\frac{d}{dt}(f + \zeta) = -(f + \zeta)(\nabla \cdot \mathbf{V}) \quad (7.21b)$$

In these expressions, ζ, the vorticity of the horizontal wind field, appears in combination with the Coriolis

[11] The remainder of this section introduces the student to more advanced dynamical concepts that are not essential for an understanding of subsequent sections of this chapter. For a more thorough and rigorous treatment of this material, see J. R. Holton, *Introduction to Dynamical Meteorology*, 4th Edition, Academic Press (2004).

[12] We offer a proof of this so-called *vorticity equation* in Exercise 7.32.

parameter f, which can be interpreted as the *planetary vorticity* that exists by virtue of the Earth's rotation: 2Ω (as inferred from Exercise 7.1) times the cosine of the angle between the Earth's axis of rotation and the local vertical (i.e., the colatitude). Hence $(f + \zeta)$ is the *absolute vorticity*: the sum of the *planetary vorticity* and the *relative vorticity* of the horizontal wind field. In extratropical latitudes, where $f \sim 10^{-4}\ \mathrm{s}^{-1}$ and $\zeta \sim 10^{-5}\ \mathrm{s}^{-1}$, the absolute vorticity is dominated by the planetary vorticity. In tropical motions the relative and planetary vorticity terms are of comparable magnitude.

In fast-moving extratropical weather systems that are not rapidly amplifying or decaying, the divergence term in (7.21) is relatively small so that

$$\frac{\partial \zeta}{\partial t} \simeq -\mathbf{V} \cdot \nabla(\zeta + f) \tag{7.22a}$$

or, in Lagrangian form,

$$\frac{d}{dt}(\zeta + f) \simeq 0 \tag{7.22b}$$

In this simplified form of the vorticity equation, absolute vorticity $(\zeta + f)$ behaves as a *conservative tracer*: i.e., a property whose numerical values are conserved by air parcels as they move along with (i.e., are advected by) the horizontal wind field.

This nondivergent form of the vorticity equation was used as a basis for the earliest quantitative weather prediction models that date back to World War II.[13] The forecasts, which were based on the 500-hPa geopotential height field, involved a four-step process:

1. The *geostrophic vorticity*

$$\zeta_g \equiv \frac{\partial v_g}{\partial x} - \frac{\partial u_g}{\partial y} = \frac{1}{f}\left(\frac{\partial^2 \Phi}{\partial x^2} + \frac{\partial^2 \Phi}{\partial y^2}\right) \tag{7.23}$$

   is calculated for an array of grid points, using a simple finite difference algorithm,[14] and added to f to obtain the absolute vorticity.
2. The advection term in (7.22a) is estimated to obtain the geostrophic vorticity tendency $\partial \zeta_g / \partial t$ at each grid point.
3. The corresponding geopotential height tendency $\chi \equiv \partial \Phi / \partial t$ at each grid-point is found by inverting and solving the time derivative of Eq. (7.23).
4. The χ field is multiplied by a small time step Δt to obtain the incremental height change at each grid point and this change is added to the initial 500-hPa height field to obtain the forecast of the 500-hPa height field.

Let us consider the implications of applying this forecast model to an idealized geostrophic wind pattern with a sinusoidal wave with zonal wavelength L superimposed on a uniform westerly flow with velocity U. The geostrophic vorticity ζ_g is positive in the wave troughs where the curvature of the flow is cyclonic, zero at the inflection points where the flow is straight, and negative in the ridges where the curvature is anticyclonic. The vorticity perturbations associated with the waves tend to be advected eastward by the westerly background flow: positive vorticity tendencies prevail in the southerly flow near the inflection points downstream of the troughs of the waves, and negative tendencies in the northerly flow downstream of the ridges (points A and B, respectively, in Fig. 7.15). The positive vorticity tendencies downstream of the troughs induce height falls, causing the troughs to propagate eastward, and similarly for the ridges. Were it not for the advection of planetary vorticity, the wave would propagate eastward at exactly the same speed as the "steering flow" U.

In addition to the vorticity tendency resulting from the advection of relative vorticity ζ, equatorward flow induces a cyclonic vorticity tendency, as air from higher latitudes with higher planetary vorticity is

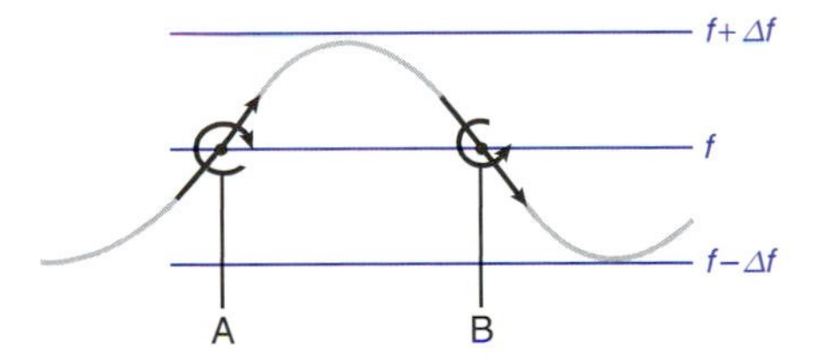

Fig. 7.15 Patterns of vorticity advection induced by the horizontal advection of absolute vorticity in a wavy westerly flow in the northern hemisphere.

[13] The barotropic model was originally implemented with graphical techniques using a light table to overlay various fields. Later it was programmed on the first primitive computers.

[14] See J. R. Holton, *An Introduction to Dynamic Meteorology*, Academic Press, pp. 452–453 (2004).

advected southward. Because absolute vorticity ($f + \zeta$) is conserved, the planetary vorticity of southward moving air parcels is, in effect, being converted to relative vorticity, inducing a positive tendency. In a similar manner, advection of low planetary vorticity by the poleward flow induces an anticyclonic vorticity tendency. In this manner the meridional gradient of planetary vorticity on a spherical Earth causes waves to propagate westward relative to the steering flow.

The relative importance of the tendencies induced by the advection of relative vorticity and planetary vorticity depends on the strength of the steering flow U and on the zonal wavelength L of the waves: other things being equal, the smaller the scale of the waves, the stronger the ζ perturbations, and hence the stronger the advection of relative vorticity. For baroclinic waves with zonal wavelengths ~4000 km, the advection of relative vorticity is much stronger than the advection of planetary vorticity and the influence of the advection of planetary vorticity is barely discernible. However, for *planetary waves* with wavelengths roughly comparable to the radius of the Earth, the much weaker eastward advection of relative vorticity is almost entirely cancelled by the advection of planetary vorticity so that the net vorticity tendencies in (7.24) are quite small. Waves with horizontal scales in this range tend to be quasi-stationary in the presence of the climatological-mean westerly steering flow. The monthly, seasonal, and climatological-mean maps discussed in Chapter 10 tend to be dominated by these so-called *stationary waves*.

The barotropic vorticity equation (7.22a) can be written in the form

$$\frac{\partial \zeta}{\partial t} \simeq -\mathbf{V} \cdot \nabla\zeta - \beta v \tag{7.24}$$

where

$$\beta \equiv \frac{\partial f}{\partial y} = \frac{\partial}{\partial y}(2\Omega \sin\phi) = \frac{2\Omega \cos\phi}{R_E} \tag{7.25}$$

In middle latitudes, $\beta \sim 10^{-11}\,\mathrm{s^{-1}\,m^{-1}}$. The term $-\beta v$ in (7.24) is commonly referred to as the *beta effect*. Wave motions in which the beta effect significantly influences the dynamics are referred to as *Rossby*[15] *waves*.

The divergence term in (7.21) is instrumental in the amplification of baroclinic waves, tropical cyclones, and other large-scale weather systems and in maintaining the vorticity perturbations in the low level wind field in the presence of frictional dissipation. In extratropical latitudes, where f is an order of magnitude larger than typical values of ζ, the divergence term is dominated by the linear term $-f(\nabla \cdot \mathbf{V})$, which represents the deflection of the cross-isobar flow by the Coriolis force. For example, the Coriolis force deflects air converging into a cyclone toward the right in the northern hemisphere (left in the southern hemisphere), producing (or intensifying) the cyclonic circulation about the center of the system. This term is capable of inducing vorticity tendencies on the order of

$$f(\nabla \cdot \mathbf{V}) \sim 10^{-4}\mathrm{s^{-1}} \times 10^{-6}\mathrm{s^{-1}} \sim 10^{-10}\mathrm{s^{-2}}$$

comparable to those induced by the vorticity advection term[16]

$$\mathbf{V} \cdot \nabla(\zeta) \sim 10\,\mathrm{m\,s^{-1}} \times \frac{10^{-5}\,\mathrm{s^{-1}}}{10^{6}\,\mathrm{m}}$$

$$\sim 10^{-10}\mathrm{s^{-2}}$$

The non-linear term $\zeta(\nabla \cdot \mathbf{V})$ in (7.21) also plays an important role in atmospheric dynamics. We can gain some insight into this term by considering the time-dependent solution of (7.21b), subject to the initial condition $\zeta = 0$ at $t = 0$, prescribing that $\nabla \cdot \mathbf{V} = C$, a constant. If the flow is divergent ($C > 0$), vorticity will initially decrease (become anticyclonic) at the rate $-fC$, but the rate of decrease will slow exponentially with time as $\zeta \rightarrow -f$ and $(f + \zeta) \rightarrow 0$. In contrast, if the flow is convergent ($C < 0$) the vorticity tendency

[15] **Carl Gustav Rossby** (1898–1957) Swedish meteorologist. Studied under Vilhelm Bjerknes. Founded and chaired the Department of Meteorology at Massachusetts Institute of Techology and the Institute for Meteorology at the University of Stockholm. While serving as chair of the Department of Meteorology at the University of Chicago during World War II, he was instrumental in the training of meteorologists for the U.S. military, many of whom became leaders in the field.

[16] In the previous subsection the divergence term was assumed to be negligible in comparison to the horizontal advection term, whereas in this subsection it is treated as if it were comparable to it in magnitude. This apparent inconsistency is resolved in Section 7.3.4 when we consider the vertical structure of the divergence profile.

will be positive and will increase without limit as ζ grows.[17] This asymmetry in the rate of growth of relative vorticity perturbations is often invoked to explain why virtually all intense closed circulations are cyclones, rather than anticyclones, and why sharp frontal zones and shear lines nearly always exhibit cyclonic, rather than anticyclonic vorticity.

7.2.10 Potential Vorticity

It is possible to derive a conservation law analogous to (7.22) that takes into account the effect of the divergence of the horizontal motion. Consider a layer consisting of an incompressible fluid of depth $H(x, y)$ moving with velocity $\mathbf{V}(x, y)$. From the conservation of mass, we can write

$$\frac{d}{dt}(HA) = 0 \tag{7.26}$$

where A is the area of an imaginary block of fluid, but it may also be interpreted as the area enclosed by a tagged set of fluid parcels as they move with the horizontal flow. Making use of (7.2), we can write

$$\frac{1}{H}\frac{dH}{dt} = -\frac{1}{A}\frac{dA}{dt} = -\nabla \cdot \mathbf{V}$$

Hence the layer thickens wherever the horizontal flow converges and thins where the flow diverges. Using this expression to substitute for the divergence in (7.21b) yields

$$\frac{d}{dt}(f + \zeta) = \frac{(f + \zeta)}{H}\frac{dH}{dt}$$

In Exercise 7.35 the student is invited to verify that this expression is equivalent to the simple conservation equation

$$\frac{d}{dt}\left(\frac{f + \zeta}{H}\right) = 0 \tag{7.27}$$

The conserved quantity in this expression is called the *barotropic potential vorticity*.

From (7.27) it is apparent that vertical stretching of air columns and the associated convergence of the horizontal flow increase the absolute vorticity $(f + \zeta)$ in inverse proportion to the increase of H. The "spin up" of absolute vorticity that occurs when columns are vertically stretched and horizontally compressed is analogous to that experienced by an ice skater going into a spin by drawing his/her arms and legs inward as close as possible to the axis of rotation. The concept of potential vorticity also allows for conversions between relative and planetary vorticity, as described in the context of (7.22).

Based on an inspection of the form of (7.27) it is evident that a uniform horizontal gradient in the depth of a layer of fluid plays a role in vorticity dynamics analogous to the beta effect. Hence, Rossby waves can be defined more generally as waves propagating horizontally in the presence of a gradient of potential vorticity.

An analogous expression can be derived for the conservation of potential vorticity in large-scale, adiabatic atmospheric motions.[18] If the flow is adiabatic, air parcels cannot pass through isentropic surfaces (i.e., surfaces of constant potential temperature). It follows that the mass in the column bounded by two nearby isentropic surfaces is conserved following the flow; i.e.,

$$\frac{A\,\delta p}{g} = constant \tag{7.28}$$

where δp is the pressure difference between two nearby isentropic surfaces separated by a fixed potential temperature increment $\delta\theta$, as depicted in Fig. 7.16. The pressure difference between the top

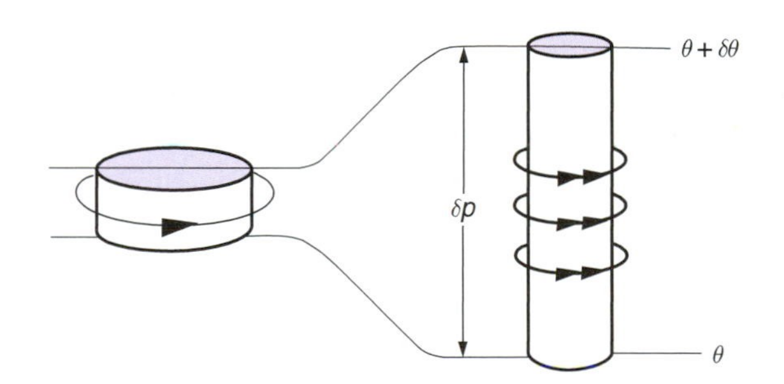

Fig. 7.16 A cylindrical column of air moving adiabatically, conserving potential vorticity. [Reprinted from Introduction to Dynamic Meteorology, 4th Edition, J. R. Holton, p. 96, Copyright 2004, with permission from Elsevier.]

[17] Divergence tends to be inhibited by the presence of both planetary vorticity and relative vorticity. In this sense, prescribing the divergence to be constant while the relative vorticity is allowed to increase without limit is unrealistic.

[18] The conditions under which the assumption of adiabatic flow is justified are discussed in Section 7.3.3.

and the bottom of the layer is inversely proportional to the static stability, i.e.,

$$\delta p = \left(\frac{\partial \theta}{\partial p}\right)^{-1} \delta\theta \tag{7.29}$$

The tighter the vertical spacing of the isentropes (i.e., the stronger the static stability), the smaller pressure increment (and mass per unit area) between the surfaces. Combining (7.28) and (7.29), absorbing $\delta\theta/g$ into the constant, and differentiating, following the large-scale flow, yields

$$\frac{d}{dt}\left[\left(\frac{\partial \theta}{\partial p}\right)^{-1} A\right] = 0 \tag{7.30}$$

Combining (7.26), (7.29), and (7.30), we obtain

$$\frac{d}{dt}\left[(f + \zeta_\theta)\frac{\partial \theta}{\partial p}\right] = 0 \tag{7.31a}$$

The conserved quantity in the square brackets is the isentropic form of *Ertel's*[19] *potential vorticity*, often referred to simply as *isentropic potential vorticity* (PV). The θ subscript in this expression specifies that the relative vorticity must be evaluated on potential temperature surfaces, rather than pressure surfaces. PV is conventionally expressed in units of *potential vorticity units* (PVU), which apply to the quantity in brackets in (7.31) multiplied by g, where 1 PVU = 10^{-6} m^2 s^{-1} K kg^{-1}.

In synoptic charts and vertical cross sections, tropospheric and stratospheric air parcels are clearly distinguishable by virtue of their differing values of PV, with stratospheric air having higher values. The 1.5 PVU contour usually corresponds closely to the tropopause level. Poleward of westerly jet streams PV is enhanced by the presence of cyclonic wind shear, and in the troughs of waves it is enhanced by the cyclonic curvature of the streamlines. Hence the PV surfaces are depressed in these regions. Surfaces of constant ozone, water vapor, and other conservative chemical tracers are observed to dip downward in a similar manner, indicating a depression of the tropopause that conforms to the distortion of the PV surfaces. In a similar manner the PV surfaces and the tropopause itself are elevated on the equatorward flank of westerly jet streams, where the PV is relatively low due to the presence of anticyclonic wind shear, and in ridges and anticyclones, where the curvature of the flow is anticyclonic.

As air parcels move through various regions of the atmosphere it takes several days for their potential vorticity to adjust to the changing ambient conditions. Hence, incursions of stratospheric air into the troposphere are marked, not only by conspicuously low dew points and high ozone concentrations, but also by high values of potential vorticity, as manifested in high static stability $\partial\theta/\partial p$ and/or strong cyclonic shear or curvature.

The distribution of PV is uniquely determined by the wind and temperature fields. A prognostic equation analogous to (7.22a): namely, the Eulerian counterpart of (7.31a)

$$\frac{\partial}{\partial t}\left[(\zeta_\theta + f)\frac{\partial \theta}{\partial p}\right] = -\mathbf{V}_\theta \cdot \nabla\left[(\zeta_\theta + f)\frac{\partial \theta}{\partial p}\right] \tag{7.31b}$$

can be used to update PV field. The updated distribution of PV on potential temperature surfaces, together with the temperature distribution at the Earth's surface, can be inverted to obtain the updated three-dimensional distribution of wind and temperature. Although they are not as accurate as numerical weather prediction based on the primitive equations discussed in the next section, forecasts based on PV have provided valuable insights into the development of weather systems.

7.3 Primitive Equations

This section introduces the complete system of so-called *primitive equations*[20] that governs the evolution of large-scale atmospheric motions. The horizontal equation of motion (7.13) is a part of that system. The other primitive equations relate to the vertical component of the motion and to the time rates of change of the thermodynamic variables p, ρ, and T.

[19] **Hans Ertel** (1904–1971) German meteorologist and geophysicist. His theoretical studies, culminating in his potential vorticity theorem, were key elements in the development of modern dynamical meteorology.

[20] Here the term primitive connotes fundamental or basic; i.e., unrefined by scaling considerations other than the assumption of hydrostatic balance.

7.3.1 Pressure as a Vertical Coordinate

The primitive equations are easiest to explain and interpret when pressure, rather than geopotential height, is used as the vertical coordinate. The transformation from height (x, y, z) to pressure (x, y, p) coordinates is relatively straightforward because pressure and geopotential height are related through the hydrostatic equation (3.17) and surfaces of constant pressure are so flat we can ignore the distinction between the horizontal wind field $\mathbf{V}_p(x, y)$ on a surface of constant pressure and $\mathbf{V}_z(x, y)$ on a nearby surface of constant geopotential height.

The vertical velocity component in (x, y, p) coordinates is $\omega \equiv dp/dt$, the time rate of change of pressure experienced by air parcels as they move along their three-dimensional trajectories through the atmosphere. Because pressure increases in the downward direction, positive ω denotes sinking motion and vice versa. Typical amplitudes of vertical velocity perturbations in the middle troposphere, for example, in baroclinic waves, are $\sim$100 hPa day^{-1}. At this rate it would take about a week for an air parcel to rise or sink through the depth of the troposphere.

The (x, y, p) and (x, y, z) vertical velocities ω and w are related by the chain rule (1.3)

$$\omega \equiv \frac{dp}{dt} = \frac{\partial p}{\partial t} + \mathbf{V} \cdot \nabla p + w\frac{\partial p}{\partial z}$$

Substituting for dp/dz from the hydrostatic equation (3.17) yields

$$\omega = -\rho g w + \frac{\partial p}{\partial t} + \mathbf{V} \cdot \nabla p \tag{7.32}$$

This relation can be simplified to derive an approximate linear relationship between ω and w. Typical local time rates of change of pressure in extratropical weather systems are $\sim$10 hPa day^{-1} or less and the term $\mathbf{V} \cdot \nabla p$ tends to be even smaller due to the quasi-geostrophic character of large-scale atmospheric motions. Hence, to within $\sim$10%,

$$\omega \simeq -\rho g w \tag{7.33}$$

Based on this relationship, 100 hPa day^{-1} is roughly equivalent to 1 km day^{-1} or 1 cm s^{-1} in the lower troposphere and twice that value in the midtroposphere.

Within the lowest 1–2 km of the atmosphere, where w and ω are constrained to be small due to the presence of the lower boundary, the smaller terms in (7.32) cannot be neglected. At the Earth's surface the geometric vertical velocity is

$$w_s = \mathbf{V} \cdot \nabla z_s \tag{7.34}$$

where z_s is the height of the terrain.

7.3.2 Hydrostatic Balance

In (x, y, z) coordinates, Newton's second law for the vertical motion component is

$$\frac{dw}{dt} = -\frac{1}{\rho}\frac{\partial p}{\partial z} - g + C_z + F_z \tag{7.35}$$

where C_z and F_z are the vertical components of the Coriolis and frictional forces, respectively. For large-scale motions, in which virtually all the kinetic energy resides in the horizontal wind component, the vertical acceleration is so small in comparison to the leading terms in (7.35) that it is not practically feasible to calculate it. To within $\sim$1%, the upward directed pressure gradient force balances the downward pull of gravity, not only for mean atmospheric conditions, but also for the perturbations in p and ρ observed in association with large-scale atmospheric motions.[21] Hence, vertical equation of motion (7.35) can be replaced by the hydrostatic equation (3.17) or, in pressure coordinates, by the hypsometric equation (3.29).

7.3.3 The Thermodynamic Energy Equation

The evolution of weather systems is governed not only by dynamical processes, as embodied in Newton's second law, but also by thermodynamic processes as represented in the first law of thermodynamics. In its simplest form the first law is a prognostic equation relating to the time rate of change of temperature of an air parcel as it moves through the atmosphere. These changes in temperature affect the thickness pattern, which, together with appropriate boundary conditions, determines the distribution of

[21] For a rigorous scale analysis of the primitive equations, see J. R. Holton, *An Introduction to Dynamic Meteorology*, 4th Edition, Academic Press, pp. 41–42 (2004).

geopotential Φ on pressure surfaces. Thus if the horizontal temperature gradient changes in response to diabatic heating, the distribution of the horizontal pressure gradient force $(-\nabla\Phi)$ in the horizontal equation of motion (7.9) will change as well.

The first law of thermodynamics in the form (3.46) can be written in the form

$$J\,dt = c_p dT - \alpha\,dp$$

where J represents the diabatic heating rate in J kg^{-1} s^{-1} and dt is an infinitesimal time interval. Dividing through by dt and rearranging the order of the terms yields

$$c_p \frac{dT}{dt} = \alpha \frac{dp}{dt} + J$$

Substituting for α from the equation of state (3.3) and ω for dp/dt and dividing through by c_p, we obtain the *thermodynamic energy equation*

$$\frac{dT}{dt} = \frac{\kappa T}{p}\omega + \frac{J}{c_p} \tag{7.36}$$

where $\kappa = R/c_p = 0.286$.

The first term on the right-hand side of (7.36) represents the rate of change of temperature due to adiabatic expansion or compression. A typical value of this term in °C per day is given by $\kappa T \delta p / p_m$, where $\delta p = \omega \delta t$ is a typical pressure change over the course of a day following an air parcel and p_m is the mean pressure level along the trajectory. In a typical middle-latitude disturbance, air parcels in the middle troposphere (p_m ~500 hPa) undergo vertical displacements on the order of 100 hPa day^{-1}. Assuming $T \sim 250$ K, the resulting adiabatic temperature change is on the order of 15 °C per day.

The second term on the right-hand side of (7.36) represents the effects of diabatic heat sources and sinks: absorption of solar radiation, absorption and emission of longwave radiation, latent heat release, and, in the upper atmosphere, heat absorbed or liberated in chemical and photochemical reactions. In addition, it is customary to include, as a part of the diabatic heating, the heat added to or removed from the parcel through the exchange of mass between the parcel and its environment due to unresolved scales of motion such as convective plumes. Throughout most of the troposphere there tends to be a considerable amount of cancellation between the various radiative terms so that the net radiative heating rates are less than 1 °C per day. Latent heat release tends to be concentrated in small regions in which it may be locally comparable in magnitude to the adiabatic temperature changes discussed earlier. The convective heating within the mixed layer can also be locally quite intense, e.g., where cold air blows over much warmer ocean water. However, throughout most of the troposphere, the sum of the diabatic heating terms in (7.36) is much smaller than the adiabatic temperature change term.

Using the chain rule (1.3) (7.36) may be expanded to obtain the Eulerian (or "local") time rate of change of temperature

$$\frac{\partial T}{\partial t} = -\mathbf{V} \cdot \nabla T + \left(\frac{\kappa T}{p} - \frac{\partial T}{\partial p} \right) \omega + \frac{J}{c_p} \tag{7.37}$$

The first term on the right-hand side of (7.37) may be recognized as the horizontal advection term defined in (7.5), and the second term is the combined effect of adiabatic compression and vertical advection. In Exercise 7.38 the reader is invited to show that when the observed lapse rate is equal to the dry adiabatic lapse rate, the term in parentheses in (7.37) vanishes. In a stably stratified atmosphere, $\partial T/\partial p$ must be less than in the adiabatic lapse rate, and thus the term in parentheses must be positive. It follows that sinking motion (or subsidence) always favors local warming and vice versa: the more stable the lapse rate, the larger the local rate of temperature increase that results from a given rate of subsidence.

If the motion is adiabatic ($J = 0$) and the atmosphere stably stratified, air parcels will conserve potential temperature as they move along their three-dimensional trajectories, carrying the potential temperature surfaces (or isentropes) with them. For steady state (7.37) reduces to

$$\frac{\omega}{V} = \frac{\partial T/\partial s}{(\kappa T/p - \partial T/\partial p)}$$

which requires that the slopes of the three-dimensional trajectories be identical to the slopes of the local isentropes.

Throughout most of the extratropical troposphere the trajectories typically slope about half as steeply as the isentropes. Hence the horizontal advection term tends to dominate and the motion has the effect of flattening out the isentropes. In time, the isentropic surfaces would become completely flat, were it not for the meridional heating gradient, which is continually tending to lift them at high latitudes and depress them at low latitudes. The role of the horizontal advection term in producing rapid local temperature changes observed in association with extratropical cyclones is illustrated in the following exercise.

Exercise 7.4 During the time that a frontal zone passes over a station the temperature falls at a rate of 2 °C per hour. The wind is blowing from the north at 40 km h^{-1} and temperature is decreasing with latitude at a rate of 10 °C per 100 km. Estimate the terms in (7.37), neglecting diabatic heating.

Solution: $\partial T/\partial t = -2\ ^\circ\text{C h}^{-1}$ and

$$-\mathbf{V}\cdot\nabla T = -v\frac{\partial T}{\partial y} = 40\ \text{km h}^{-1} \times \frac{-10\ ^\circ\text{C}}{100\ \text{km}} = -4\ ^\circ\text{C h}^{-1}$$

The temperature at the station is dropping only half as fast as the rate of horizontal temperature advection so large-scale subsidence must be warming the air at a rate of 2 °C h^{-1} as it moves southward. It follows that the meridional slopes of the air trajectories must be half as large as the meridional slopes of the isentropes.

Within the tropical troposphere the relative magnitude of the terms in (7.37) is altogether different from that in the extratropics. Horizontal temperature gradients are much weaker than in the extratropics, so the horizontal advection term is unimportant and temperatures at fixed points vary little from one day to the next. In contrast, diabatic heating rates in regions of tropical convection are larger than those typically observed in the extratropics. Ascent tends to be concentrated in narrow rain belts such as the ITCZ, where warming due to the release of latent heat of condensation is almost exactly compensated by the cooling induced by the vertical velocity term in (7.37). The prevailing lapse rate throughout the tropical troposphere all the way from the top of the boundary layer up to around the 200-hPa level is nearly moist adiabatic so that the lifting of saturated air does not result in large temperature changes, even when the rate of ascent is very large. Within the much larger regions of slow subsidence, warming due to adiabatic compression is balanced by weak radiative cooling. ■

7.3.4 Inference of the Vertical Motion Field

The vertical motion field cannot be predicted on the basis of Newton's second law, but it can be inferred from the horizontal wind field on the basis of the *continuity equation*, which is an expression of the conservation of mass.

Air parcels expand and contract in response to pressure changes. In general, these changes in volume are of two types: those associated with sound waves, in which the momentum of the air plays an essential role, and those that occur in association with hydrostatic pressure changes. When the equation for the continuity of mass is formulated in (x, y, p) coordinates, only the hydrostatic volume changes are taken into account.

Consider an air parcel shaped like a block with dimensions δx, δy, and δp, as indicated in Fig. 7.17. If the atmosphere is in hydrostatic balance, the mass of the block is given by

$$\delta M = \rho\ \delta x\ \delta y\ \delta z = \frac{-\delta x\ \delta y\ \delta p}{g}$$

Because the mass of the block is not changing with time,

$$\frac{d}{dt}(\delta x\ \delta y\ \delta p) = 0 \tag{7.38}$$

or, in expanded form,

$$\delta y\ \delta p \frac{d}{dt}\delta x + \delta x\ \delta p \frac{d}{dt}\delta y + \delta x\ \delta y \frac{d}{dt}\delta p = 0.$$

Expanding the time rates of change of δx, δy, and δp in terms of partial derivatives of the velocity components, as in the derivation of (7.2), it is readily shown that

$$\frac{\partial u}{\partial x} + \frac{\partial v}{\partial y} + \frac{\partial \omega}{\partial p} = 0 \tag{7.39a}$$

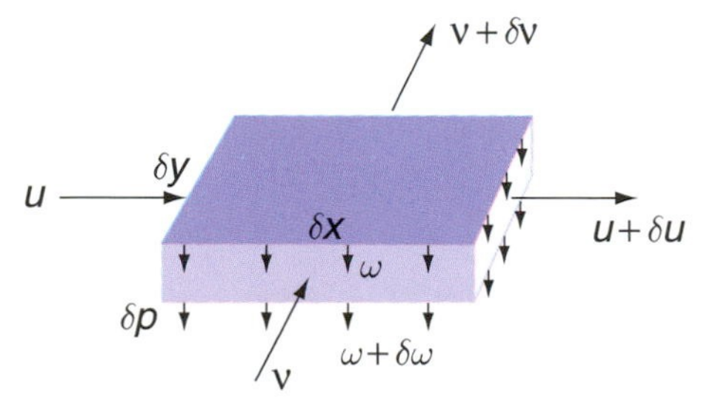

Fig. 7.17 Relationships used in the derivation of the continuity equation.

or, in vectorial form,

$$\frac{\partial \omega}{\partial p} = -\nabla \cdot \mathbf{V} \tag{7.39b}$$

where $\nabla \cdot \mathbf{V}$ is the divergence of the horizontal wind field. Hence, horizontal divergence ($\nabla \cdot \mathbf{V} > 0$) is accompanied by vertical squashing ($\partial\omega/\partial p < 0$), and horizontal convergence is accompanied by vertical stretching, as illustrated schematically in Fig. 7.18.

Within the atmospheric boundary layer air parcels tend to flow across the isobars toward lower pressure in response to the frictional drag force. Hence, the low-level flow tends to converge into regions of low pressure and diverge out of regions of high pressure, as illustrated in Fig. 7.19. The frictional convergence within low-pressure areas induces ascent, while the divergence within high-pressure areas induces subsidence. In the oceans, where the Ekman drift is in the opposite sense, cyclonic, wind-driven gyres are characterized by *upwelling*, and anticyclonic gyres by *downwelling*. In a similar manner, regions of Ekman drift in the offshore direction are characterized by *coastal upwelling*, and the easterly surface winds along the equator induce a shallow Ekman drift away from the equator, accompanied by *equatorial upwelling*.

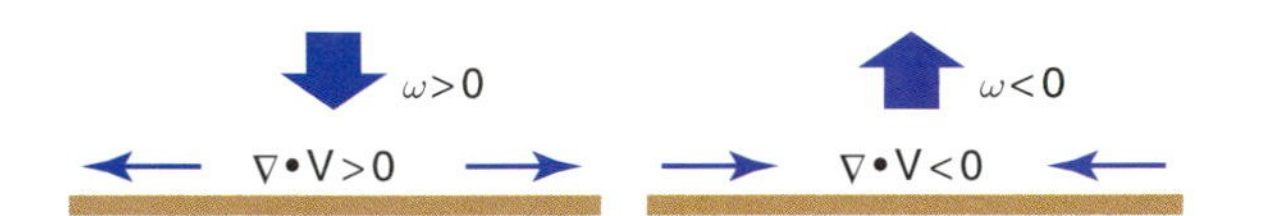

Fig. 7.18 Algebraic signs of ω in the midtroposphere associated with convergence and divergence in the lower troposphere.

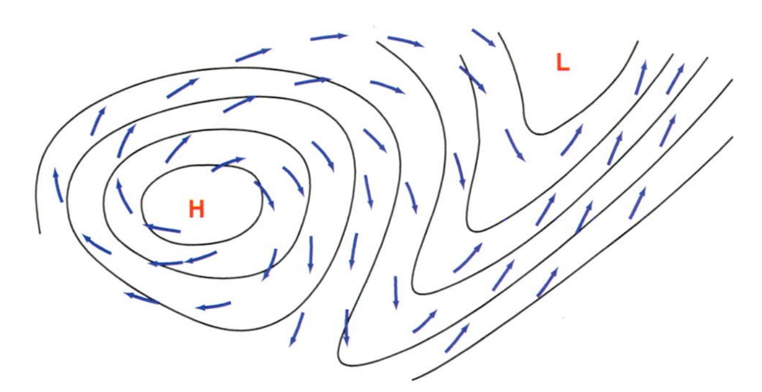

Fig. 7.19 Cross-isobar flow at the Earth's surface induced by frictional drag. Solid lines represent isobars.

The vertical velocity at any given point (x, y) can be inferred diagnostically by integrating the continuity equation (7.39b) from some reference level p^* to level p,

$$\omega(p) = \omega(p^*) - \int_{p^*}^{p} (\nabla \cdot \mathbf{V})\, dp \tag{7.40}$$

In this form, the continuity equation can be used to deduce the vertical velocity field from a knowledge of the horizontal wind field. A convenient reference level is the top of the atmosphere, where $p^* = 0$ and $\omega = 0$. Integrating (7.40) downward from the top down to the Earth's surface, where $p = p_s$ we obtain

$$\omega_s = -\int_0^{p_s} (\nabla \cdot \mathbf{V})\, dp$$

Incorporating this result and (7.34) into (7.32) and rearranging, we obtain Margules'[22] *pressure tendency equation*

$$\frac{\partial p_s}{\partial t} = -\mathbf{V}_s \cdot \nabla p_s - w_s \frac{\partial p}{\partial z} - \int_0^{p_s} (\nabla \cdot \mathbf{V})\, dp \tag{7.41}$$

which serves as the bottom boundary condition for the pressure field in the primitive equations.

At the Earth's surface, the pressure tendency term $\partial p_s/\partial t$ is on the order of 10 hPa per day and the advection terms $-\mathbf{V}_s \cdot \nabla p_s$ and $-w_s(\partial p/\partial z)$ are usually even smaller. It follows that

$$\int_0^{p_s} (\nabla \cdot \mathbf{V})\, dp = p_s \{\nabla \cdot \mathbf{V}\} \sim 10\ \text{hPa day}^{-1}$$

where $\{\nabla \cdot \mathbf{V}\}$ is the mass-weighted, vertically-averaged divergence. Because $p_s \simeq 1000$ hPa and 1 day $\sim 10^5$ s, it follows that

$$\{\nabla \cdot \mathbf{V}\} \sim 10^{-7}\ \text{s}^{-1}$$

[22] **Max Margules** (1856–1920) Meteorologist, physicist, and chemist, born in the Ukraine. Worked in intellectual isolation on atmospheric dynamics from 1882 to 1906, during which time he made many fundamental contributions to the subject. Thereafter, he returned to his first love, chemistry. Died of starvation while trying to survive on a government pension equivalent to $2 a month in the austere post-World War I period. Many of the current ideas concerning the kinetic energy cycle in the atmosphere (see Section 7.4) stem from Margules' work.

That the vertically averaged divergence is typically about an order of magnitude smaller than typical magnitudes of the divergence observed at specific levels in the atmosphere reflects the tendency (first pointed out by Dines[23]) for compensation between lower tropospheric convergence and upper tropospheric divergence and vice versa. Vertical velocity ω, which is the vertical integral of the divergence, tends to be strongest in the mid-troposphere, where the divergence changes sign.[24]

It is instructive to consider the vertical profile of atmospheric divergence in light of the idealized flow patterns for the two-dimensional flow in two different kinds of waves in a stably stratified liquid in which density decreases with height. Fig. 7.20a shows an idealized *external wave* in which divergence is independent of height and vertical velocity increases linearly with height from zero at the bottom to a maximum at the free surface of the liquid. Fig. 7.20b shows an *internal wave* in a liquid that is bounded by a rigid upper lid so that $w = 0$ at both top and bottom, requiring perfect compensation between low level convergence and upper level divergence and *vice versa*. In the Earth's atmosphere, motions that resemble the internal wave contain several orders of magnitude more kinetic energy than those that resemble the external wave. In some respects, the stratosphere plays the role of a "lid" over the troposphere, as evidenced by the fact that the geometric vertical velocity w is typically roughly an order of magnitude smaller in the lower stratosphere than in the midtroposphere.

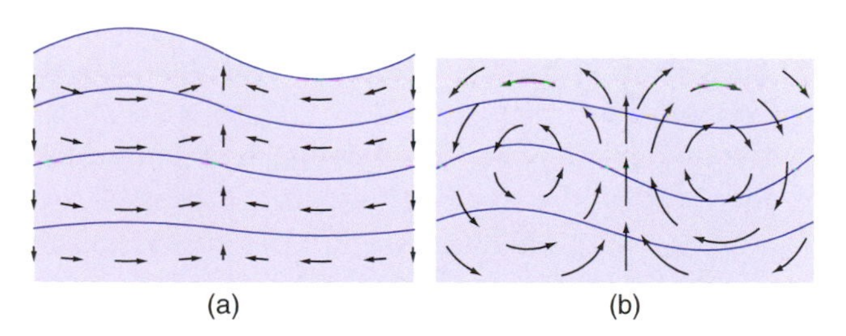

Fig. 7.20 Motion field in a vertical cross section through a two-dimensional wave propagating from left to right in a stably stratified liquid. The contours represent surfaces of constant density. (a) An *external wave* in which the maximum vertical motions occur at the free surface of the liquid, and (b) an *internal wave* in which the vertical velocity vanishes at the top because of the presence of a rigid lid.

7.3.5 Solution of the Primitive Equations

The primitive equations, in the simplified form that we have derived them, consist of the horizontal equation of motion

$$\frac{d\mathbf{V}}{dt} = -\nabla\Phi - f\mathbf{k} \times \mathbf{V} + \mathbf{F} \tag{7.14}$$

the hypsometric equation

$$\frac{\partial\Phi}{\partial p} = \frac{-RT}{p} \tag{3.23}$$

the thermodynamic energy equation

$$\frac{dT}{dt} = \frac{\kappa T}{p}\omega + \frac{J}{c_p} \tag{7.36}$$

and the continuity equation

$$\frac{\partial\omega}{\partial p} = -\nabla \cdot \mathbf{V} \tag{7.39b}$$

together with the bottom boundary condition

$$\frac{\partial p_s}{\partial t} = -\mathbf{V}_s \cdot \nabla p - w_s \frac{\partial p}{\partial z} - \int_0^{p_s} (\nabla \cdot \mathbf{V})\, dp \tag{7.41}$$

Bearing in mind that the horizontal equation of motion is made up of two components, the system of primitive equations, as presented here, consist of five equations in five dependent variables: u, v, ω, Φ, and T. The fields of diabatic heating J and friction $\mathbf{F}$ need to be prescribed or *parameterized* (i.e., expressed as functions of the dependent variables). The horizontal equation(s) of motion, the thermodynamic energy equation, and the equation for pressure on the bottom boundary all contain time derivatives and are

[23] **Willam Henry Dines** (1855–1927) British meteorologist. First to invent a device for measuring both the direction and the speed of the wind (the Dines' pressure-tube anemometer). Early user of kites and ballons to study upper atmosphere.

[24] The midtroposphere minimum in the amplitude of $\nabla \cdot \mathbf{V}$ justifies the use of (7.22) in diagnosing the vorticity balance at that level. Alternatively (7.22) can be viewed as relating to the vertically averaged tropospheric wind field, whose divergence is two orders of magnitude smaller than its vorticity.

therefore said to be *prognostic equations*. The remaining so-called *diagnostic equations* describe relationships between the dependent variables that apply at any instant in time.

The primitive equations can be solved numerically. The dependent variables are represented on an array of regularly spaced horizontal grid points and at a number of vertical levels. In global models, the grid points usually lie along latitude circles and meridians, although the latitudinal spacing is not necessarily uniform. The equations are converted to the Eulerian form (1.3) so that the calculations can be performed on a fixed set of grid points and levels rather than requiring the grid to move with the air trajectories. Initial conditions for the dependent variables are specified at time t_0, taking care to ensure that the diagnostic relations between the variables are satisfied (e.g., the geopotential field must be hydrostatically consistent with the temperature field).

Terms in the equations involving horizontal and vertical derivatives are evaluated using *finite difference techniques* (approximating them as differences between the numerical values of the respective variables at neighboring grid points and levels in the array divided by the distances between the grid points) or as the coefficients of a mutually orthogonal set of analytic functions called *spherical harmonics*.[25] The time derivative terms $\partial \mathbf{V}/\partial t$, $\partial T/\partial t$, and $\partial p_s/\partial t$ are then evaluated and their respective three-dimensional fields are projected forward through a short time step Δt so that, for example,

$$u(t_0 + \Delta t) = u(t_0) + \frac{\partial u}{\partial t}\Delta t$$

The diagnostic equations are then applied to obtain dynamically consistent fields of the other dependent variables at time $t_0 + \Delta t$. The process is then repeated over a succession of time steps to describe the evolution of the fields of the dependent variables.

The time step Δt must be short enough to ensure that the fields of the dependent variables are not corrupted by spurious small scale patterns arising from numerical instabilities. The higher the spatial resolution of the model, the shorter the maximum allowable time step and the larger the number of calculations required for each individual time step. Hence the computer resources required for numerical weather prediction and climate modeling increase sharply with the spatial resolution of the models.

7.3.6 An Application of Primitive Equations

To show how the dynamic processes represented in the primitive equations give rise to atmospheric motions and maintain them in the presence of friction, let us consider the events that transpire during the first few time steps when an atmospheric primitive equation model is "turned on," starting from a state of rest. Initially the atmosphere is stably stratified and the pressure and potential temperature surfaces are horizontally uniform.

Starting at time $t = 0$, the tropics begin to warm and the polar regions begin to cool in response to an imposed distribution of diabatic heating, which is designed to mimic the equator-to-pole heating gradient in the real atmosphere. As the tropical atmosphere warms, thermal expansion causes the pressure surfaces in the upper troposphere to bulge upward, while cooling of the polar atmosphere causes the pressure surfaces to bend downward as depicted in Fig. 7.21a. These tendencies are evident even in the very first time step of the time integration. The sloping of the pressure surfaces gives rise to an equator-to-pole pressure gradient at the upper levels. The pressure gradient force, in turn, drives the poleward flow depicted in Fig. 7.21a, which comes into play in the second time step of the integration.

The poleward mass flux results in a latitudinal redistribution of mass, causing surface pressure to drop in low latitudes and rise in high latitudes. The resulting low level equator-to-pole pressure gradient drives a compensating equatorward low level flow. Hence, the initial response to the heating gradient is the development of an equator-to-pole circulation cell, as shown in Fig. 7.21b. The Coriolis force in the horizontal equation of motion imparts a westward component to the equatorward flow in the lower branch of the cell and an eastward component to the poleward flow in the upper branch, as shown in Fig. 7.21c.

[25] *Spherical harmonics* are products of Legendre polynomials in the latitude domain and sines and cosines in the longtitude domain. In the spherical harmonic representation of the primitive equations, operations such as taking horizontal gradients ($\partial/\partial x$, $\partial/\partial y$), Laplacians (∇^2), and inverse Laplacians (∇^{-2}) reduce to simple algebraic operations performed on the coefficients.

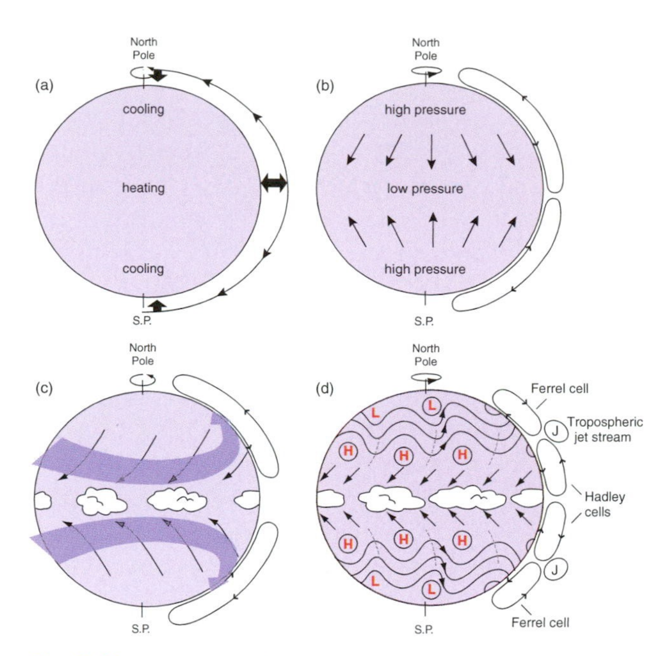

Fig. 7.21 Schematic depiction of the general circulation as it develops from a state of rest in a climate model for equinox conditions in the absence of land–sea contrasts. See text for further explanation.

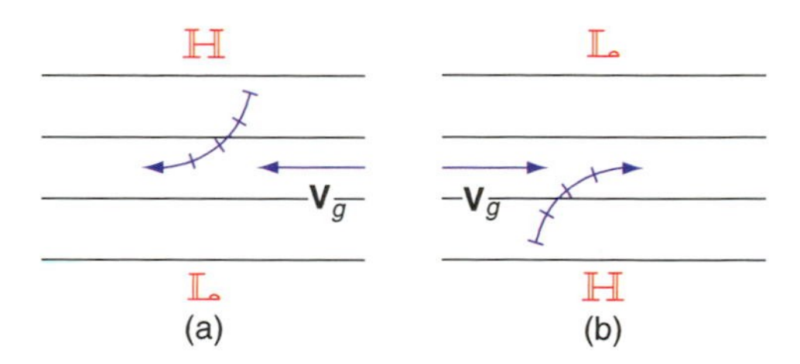

Fig. 7.22 Trajectories of air parcels in the middle latitudes of the northern hemisphere near the Earth's surface (left) and in the upper troposphere (right) during the first few time steps of the primitive equation model integration described in Fig. 7.21. The solid lines represent geopotential height contours and the curved arrows represent air trajectories.

The flow becomes progressively more zonal with each successive time step, until the meridional component of the Coriolis force comes into geostrophic balance with the pressure gradient force, as illustrated in Fig. 7.22. As required by thermal wind balance, the vertical wind shear between the low level easterlies and the upper level westerlies increases in proportion to the strengthening equator-to-pole temperature gradient forced by the meridional gradient of the diabatic heating. The Coriolis force acting on the relatively weak meridional cross-isobar flow is instrumental in building up the vertical shear. Frictional drag limits the strength of the surface easterlies, but the westerlies aloft become stronger with each successive time step. Will these runaway westerlies continue to increase until they become supersonic or will some as yet to be revealed process intervene to bring them under control? The answer is reserved for the following section.

7.4 The Atmospheric General Circulation

Dating back to the pioneering studies of Halley[26] in 1676 and Hadley in 1735, scientists have sought to explain why the surface winds blow from the east at subtropical latitudes and from the west in middle latitudes, and why the trade winds are so steady from one day to the next compared to the westerlies in midlatitudes. Such questions are fundamental to an understanding of the atmospheric *general circulation*; i.e., the statistical properties of large-scale atmospheric motions, as viewed in a global context.

Two important scientific breakthroughs that occurred around the middle of the 20th century paved the way for a fundamental understanding of the general circulation. The first was the simultaneous discovery of *baroclinic instability* (the mechanism that gives rise to baroclinic waves and their attendant extratropical cyclones) by Eady[27] and Charney.[28] The second was the advent of *general circulation models*: numerical

[26] **Edmund Halley** (1656–1742) English astronomer and meteorologist. Best known for the comet named after him. Determined that the force required to keep the planets in their orbits varies as the inverse square of their respective distances from the sun. (It remained for Newton to prove that the inverse square law yields elliptical paths, as observed.) Halley undertook the business and printing ("at his own charge") of Newton's masterpiece *the Principia*. First to derive a formula relating air pressure to altitude.

[27] **E. T. Eady** (1915–1966) English mathematician and meteorologist. Worked virtually alone in developing a theory of baroclinic instability while an officer on the Royal Air Force during World War II. One of us (P. V. H.) is grateful to have had him as a tutor.

[28] **Jule G. Charney** (1917–1981) Made major contributions to the theory of baroclinic waves, planetary-waves, tropical cyclones, and a number of other atmospheric and oceanic phenomena. While working at the Institute for Advanced Studies at Princeton University he helped lay the groundwork for numerical weather prediction. Later served as professor at Massachusetts Institute of Technology where one of us (J. M. W.) is privileged to have had the opportunity to learn from him. Played a prominent role in fostering large international programs designed to advance the science of weather prediction.

models of the global atmospheric circulation based on the primitive equations that can be run over long enough time intervals to determine seasonally varying, climatological mean winds and precipitation, the degree of steadiness of the winds, etc.

The importance of these developments becomes apparent if we extend the numerical integration described in the last section forward in time, based on results that have been replicated in many different general circulation models (GCM's). The ensuing developments could never have been anticipated on the basis of the simple dynamical arguments presented in the previous section. When the meridional temperature gradient reaches a critical value, the simulated circulation undergoes a fundamental change: baroclinic instability spontaneously breaks out in midlatitudes, imparting a wave-like character to the flow. The baroclinic waves create and maintain the circulation in the configuration indicated in Fig. 7.21d. In the developing waves, warm, humid subtropical air masses flow poleward ahead of the eastward moving surface cyclones, while cold, dry polar air masses flow equatorward behind the cyclones. The exchange of warm and cold air masses across the 45° latitude circle results in a net poleward flux of sensible and latent heat, arresting the buildup of the equator-to-pole temperature gradient.

Successive generations of baroclinic waves develop and evolve through their life cycles, emanating from their source region in the midlatitude lower troposphere, first dispersing upward toward the jet stream (~10 km) level and thence equatorward into the tropics. The equatorward dispersion of the waves in the upper troposphere causes the axes of their ridges and troughs to tilt in the manner shown in Fig. 7.21d. As a consequence of this tilt, poleward-moving air to the east of the troughs exhibits a stronger westerly wind component than equatorward moving air to the west of the troughs. The difference in the amount of westerly momentum carried by poleward and equatorward moving air parcels gives rise to a net poleward flux of westerly momentum from the subtropics into middle latitudes. In response to the import of westerly momentum from the subtropics, the surface winds in midlatitudes shift from easterly (Fig. 7.21c) to westerly (Fig. 7.21d), the direction of the midlatitude surface winds in the real atmosphere.

Baroclinic waves drive their own weak northern and southern hemisphere mean meridional circulation cells, referred to as *Ferrel*[29] *cells*, characterized by poleward, frictionally induced Ekman drift at the latitude of the storm tracks (~45°), ascent on the poleward flank, and descent on the equatorward flank, as indicated in Fig. 7.21d. Hence, with the spontaneous development of the baroclinic waves, the Hadley cells withdraw into the tropics, and a region of subsidence develops at subtropical (~30°) latitudes. These regions of subsidence coincide with the subtropical anticyclones (Figs. 1.18 and 1.19), which mark the boundary between the tropical trade winds and the extratropical westerlies. Most of the world's major desert regions lie within this latitude belt.

7.4.1 The Kinetic Energy Cycle

Frictional dissipation observed within the planetary boundary layer and within patches of turbulence within the free atmosphere continually depletes the kinetic energy of large-scale wind systems. Half the energy would be gone within a matter of days were there not some mechanism operating continually to restore it. The ultimate source of this kinetic energy is the release of potential energy through the sinking of colder, denser air and the rising of warmer, less dense air, which has the effect of lowering the atmosphere's center of mass, flattening the potential temperature surfaces, and weakening the existing horizontal temperature gradients. A simple steady-state laboratory analog of the kinetic energy cycle is depicted schematically in Fig. 7.23.

In small-scale convection, the kinetic energy generated by the rising of buoyant plumes is imparted to the vertical component of the motion. The upward buoyancy force disturbs the balance of forces in the vertical equation of motion, inducing vertical accelerations. The vertical overturning that occurs in association with large-scale atmospheric motions cannot be viewed as resulting from buoyancy forces because, in the primitive equations, the vertical equation of

[29] **William Ferrel** (1817–1891) American scientist. A schoolteacher early in his career and later held positions at the Nautical Almanac Office, the Coast Survey, and the Signal Office, which housed the Weather Bureau. Was the first to correctly deduce that gravitational tides retard the earth's rotation, and to explain the profound influence of the Coriolis force on winds and ocean currents that leads to geostrophic flow, as articulated in Buys Ballot's law.

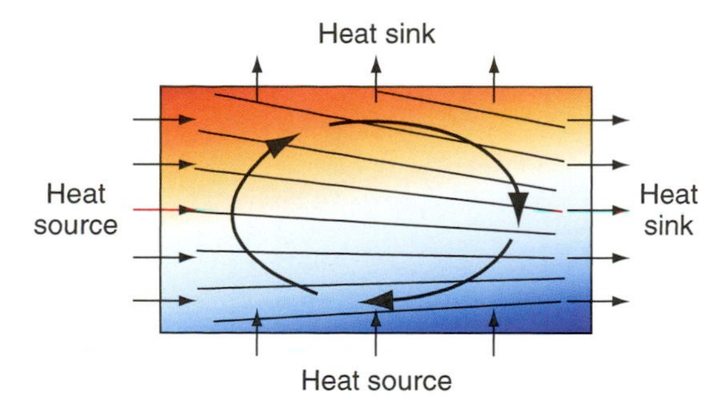

Fig. 7.23 Vertical cross section through a steady-state circulation in the laboratory, driven by the distribution of heat sources and heat sinks as indicated. The colored shading indicates the distribution of temperature and density, with cooler, denser fluid represented by blue. The sloping black lines represent pressure surfaces. Note that the flow is directed down the horizontal pressure gradient at both upper and lower levels.

motion is replaced by the hydrostatic equation, which has no vertical acceleration term. In large-scale motions, kinetic energy is imparted directly to the horizontal wind field. When warm air rises and cold air sinks, the potential energy that is released does work on the horizontal wind field by forcing it to flow across the isobars from higher toward lower pressure. In the equation for the time rate of change of kinetic energy,

$$\mathbf{V} \cdot \frac{d\mathbf{V}}{dt} = \frac{d}{dt}\frac{V^2}{2} = -\mathbf{V} \cdot \nabla\Phi + \mathbf{F} \cdot \mathbf{V} \tag{7.42}$$

the cross-isobar flow $-\mathbf{V} \cdot \nabla\Phi$ is the one and only source term. Flow across the isobars toward lower pressure is prevalent close to the Earth's surface, where the dissipation term $\mathbf{F} \cdot \mathbf{V}$ is most intense. Frictional drag is continually acting to reduce the wind speed, causing it to be subgeostrophic, so that the Coriolis force is never quite strong enough to balance the pressure gradient force. The imbalance drives a cross-isobar flow toward lower pressure (Fig. 7.19), which maintains the kinetic energy and the surface wind speed in the presence of frictional dissipation. This process is represented by the second term on the right-hand side of (7.42). It can be shown that, in the integral over the entire mass of the atmosphere, the generation of kinetic energy by the $-\mathbf{V} \cdot \nabla\Phi$ term in (7.42) is equal to the release of potential energy associated with the rising of warm air and sinking of cold air.

The trade winds in the lower branch of the Hadley cell are directed down the pressure gradient, out of the subtropical high pressure belt and into the belt of low pressure at equatorial latitudes. That the winds in the upper troposphere blow from west to east implies that Φ decreases with latitude at that level, and hence that the poleward flow in the upper branch of the Hadley cell is also directed down the pressure gradient.

Closed circulations like the Hadley cell, which are characterized by the rising of warmer, lighter air and the sinking of colder, denser air and the prevalence of cross-isobar horizontal flow toward lower pressure, release potential energy and convert it to the kinetic energy of the horizontal flow. Circulations with these characteristics are referred to as *thermally direct* because they operate in the same sense as the global kinetic energy cycle. Other examples of thermally direct circulations in the Earth's atmosphere are the large-scale overturning cells in baroclinic waves, monsoons and tropical cyclones. Circulations such as the Ferrel cell, that operate in the opposite sense, with rising of colder air and sinking of warmer air are referred to as *thermally indirect.*

Because thermally direct circulations are continually depleting the atmosphere's reservoir of potential energy, something must be acting to restore it. Heating of the atmosphere by radiative transfer and the release of the latent heat of condensation of water vapor in clouds is continually replenishing the potential energy in two ways: by warming the atmosphere in the tropics and cooling it at higher latitudes and by heating the air at lower levels and cooling the air at higher levels. The former acts to maintain the equator-to-pole temperature contrast on pressure surfaces, and the latter expands the air in the lower troposphere and compresses the air in the upper troposphere, thereby lifting the air at intermediate levels, maintaining the height of the atmosphere's center of mass against the lowering produced by thermally direct circulations. Hence, the maintenance of large-scale atmospheric motions requires both horizontal and vertical heating gradients analogous to those in the laboratory experiment depicted in Fig. 7.23. In contrast, the maintenance of convection requires only vertical heating gradients.

The most important heat source in the troposphere is the release of latent heat of condensation that occurs in association with precipitation. Latent heat release tends to occur preferentially in rising air, which becomes saturated as it cools adiabatically. This preferential heating of the rising air tends to maintain and enhance the horizontal temperature gradients that drive thermally direct circulations, rendering the motions more vigorous than they would be in a dry atmosphere. Condensation heating plays a supporting role as an energy source for extratropical cyclones and it plays a starring role in the energetics of tropical cyclones.

The *kinetic energy cycle* can be summarized in terms of the flowchart presented in Fig. 7.24. Potential energy generated by heating gradients is converted to kinetic energy of both large-scale and smaller scale convective motions. Kinetic energy is drained from the large-scale reservoir by shear instability and flow over rough surfaces, both of which generate small-scale turbulence and wave motions, as discussed in Chapter 9. Kinetic energy in the small-scale reservoir is transferred to smaller and smaller scales until it becomes indistinguishable from random molecular motions and thus becomes incorporated into the atmosphere's reservoir of internal energy. This final step is represented in Fig. 7.24 as "a drop in the bucket" to emphasize that it is a small contribution to the atmosphere's vast internal energy reservoir: only locally in regions of very strong winds such as those near the centers of tropical cyclones is frictional dissipation strong enough to affect the temperature.

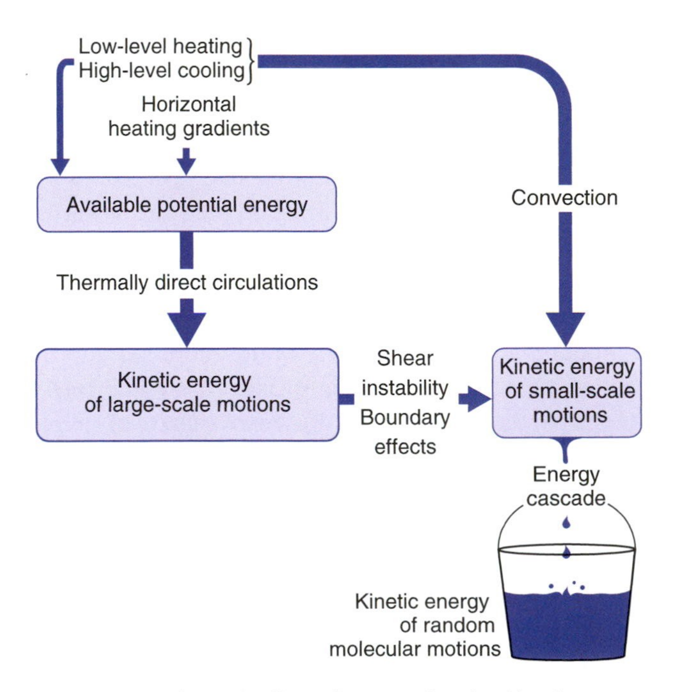

Fig. 7.24 Schematic flow diagram for the kinetic energy cycle in the atmospheric general circulation.

7.4.2 The Atmosphere as a Heat Engine

The primary tropospheric heat sources tend to be concentrated within the planetary boundary layer, as discussed in the previous chapter, and in regions of heavy precipitation. The primary heat sink, radiative cooling in the infrared, is more diffuse. In a gross statistical sense, the center of mass of the heating is located at a slightly lower latitude and at a slightly lower altitude than the center of mass of the cooling, and is therefore at a somewhat higher temperature. The atmospheric general circulation can be viewed as a heat engine that is continually receiving heat at a rate Q_H at that higher temperature T_H and rejecting heat at a rate Q_C at the lower temperature T_C. By analogy with (3.78) for the Carnot cycle, we can write an expression for the thermal efficiency of the atmospheric heat engine

$$\eta = \frac{W}{Q_H} \tag{7.43}$$

where W is the rate at which work is done in generating kinetic energy in thermally direct circulations. The integral of the kinetic energy generation term $-\mathbf{V} \cdot \nabla\Phi$ over the mass of the atmosphere is estimated to be ~1–2 W m^{-2}. Q_H, the net incoming solar radiation averaged over the surface of the Earth, is ~240 W m^{-2}. Hence, based on this definition, the thermal efficiency of the atmospheric heat engine is less than 1%. In contrast to the situation in the Carnot cycle, much of the heat transfer that takes place within the atmosphere is irreversible. For example, cold, polar air masses that penetrate into the tropics are rapidly modified by the heat fluxes from the underlying surfaces and radiative transfer. It follows that relationships derived from the reversible Carnot cycle are not strictly applicable to the atmospheric heat engine.

7.5 Numerical Weather Prediction

Until the 1950s, day-to-day weather forecasting was largely based on subjective interpretation of synoptic charts. From the time of the earliest synoptic networks, it has been possible to make forecasts by extrapolating the past movement of the major features on the charts. As forecasters began to acquire experience from a backlog of past weather situations, they were able to refine their forecasts somewhat by making use of historical analogs of the current weather situation, without reference to the underlying dynamical principles introduced in this chapter. However, it soon became apparent that the effectiveness of this so-called "analog technique" is inherently limited by the unavailability of very close analogs in the historical records. On a global basis the three-dimensional structure of the atmospheric motion field is so complicated that a virtually limitless number of distinctly different flow patterns are possible, and hence the probability of two very similar patterns being observed, say, during the same century, is extremely low. By 1950 the more advanced weather forecasting services were already approaching the limit of the skill that could be obtained strictly by the use of empirical techniques.

The earliest attempts at numerical weather prediction, based on the methodology described in Section 7.2.9, date back to the 1950s and primitive equation models came into widespread use soon afterward. In comparison to the models developed in these pioneering efforts, today's numerical weather prediction models have much higher spatial resolution and contain a much more accurate and detailed representation of the physical processes that enter into the primitive equations. The increase in the physical complexity of the models would not have been possible without the rapid advances in computer technology that have taken place during the past 50 years: to make a single forecast for 1 day in advance by running current operational models on a computer of the type available 50 years ago would require millennia of computer time! The payoff of this monumental effort is the remarkable improvement in forecast skill documented in Fig. 1.1.

The initial conditions for modern numerical weather prediction are based on an array of global observations, an increasing fraction of which are remote measurements from radiometers carried on board satellites. In situ observations include surface reports, radiosonde data, and flight level data from commercial aircraft. In situ measurements of pressure, wind, temperature, and moisture are combined with satellite-derived radiances in dynamically consistent, multivariate *four-dimensional data assimilation systems* like the one described in the Appendix of chapter 8 on the book web site.

The global observing system describes only the resolvable scales of atmospheric variability, and the analysis is subject to measurement errors even at the largest scales. The data assimilation scheme is designed to correct for the systematic errors in the observations and to minimize the impact of random errors on the forecasts. The errors in the initial conditions for numerical weather prediction have been continually shrinking in response to a host of incremental improvements in the observing and data assimilation systems. Nevertheless, there will always remain some degree of uncertainty (or errors) in the initial conditions and, due to the nonlinearity of atmospheric motions, these errors inevitably amplify with time. Beyond some threshold forecast interval the forecast fields are, on average, no more like the observed fields against which they are verified than two randomly chosen observed fields for the same time of year are like one another. For the extratropical atmosphere this so-called *limit of deterministic predictability* is believed to be on the order of 2 weeks.

Figure 7.25 shows a set of forecasts, made on successive days, for the same time, together with the *verifying analysis* (i.e., the corresponding "observed" fields for the time that matches the forecast). This particular example was chosen because the level of skill is typical of wintertime forecasts made with today's state-of-the art numerical weather prediction systems. Forecast skill declines monotonically as the forecast interval lengthens. The 1- and 3-day forecasts replicate the features in the verifying analysis with a high degree of fidelity; the 7-day forecast still captures all the major features in the field, but misses many of the finer details. The 10-day forecast is worthless over much of the hemisphere, but even at this long lead time, the more prominent features in the verifying analysis are already apparent over the Atlantic, European, and western North American sectors.

The increase in the uncertainty of the forecasts with increasing forecast interval is illustrated in Fig. 7.26 in the context of a simplified forecast model based on the *Lorenz attractor* described in Box 1.1. In the first experiment shown in the left panel, the initial conditions that make up the ensemble lie within a region of the attractor for which the forecast uncertainty actually declines for a while as the numerical integration proceeds, as evidenced by the decreasing size of successive forecast ellipses. In the second experiment (middle

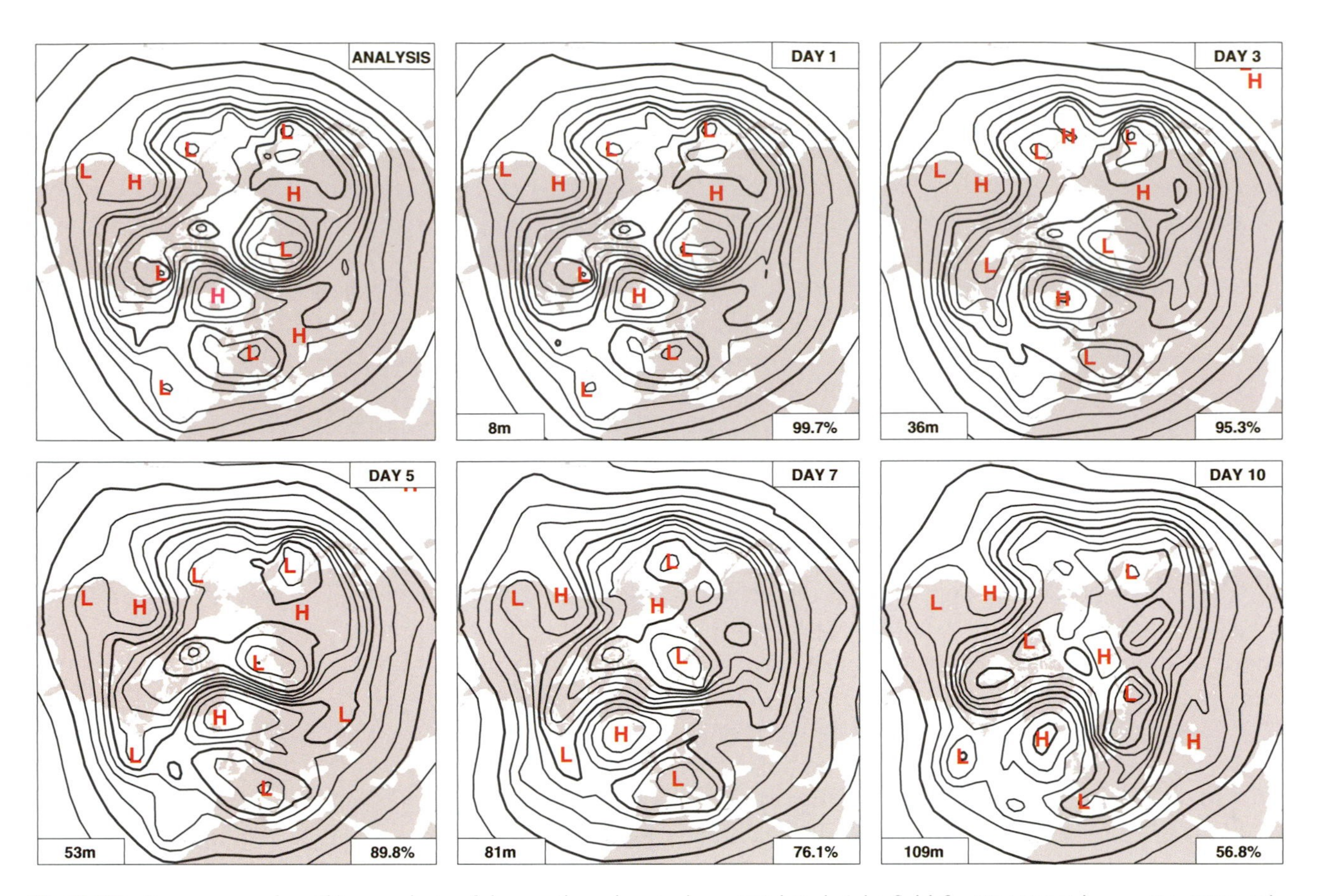

Fig. 7.25 Forecasts and verifying analysis of the northern hemisphere 500-hPa height field for 00 UTC February 24, 2005. The number printed in the lower left corner is the *root mean squared* error and the percentage shown in the lower right corner of each panel is the *anomaly correlation*, a measure of the hemispherically averaged skill of the forecast. [Courtesy of Adrian J. Simmons, European Centre for Medium Range Weather Forecasts (ECMWF).]

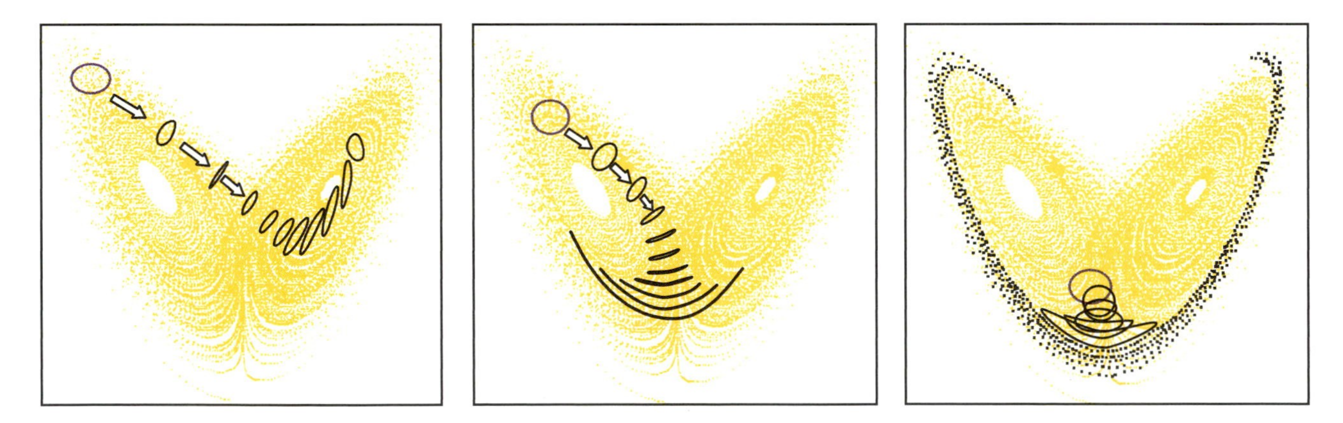

Fig. 7.26 Three sets of *ensemble forecasts* performed using the highly simplified three-variable representation of the atmosphere described in Box 1.1. The time-dependent solution of the governing governing equations traces out the three-dimensional geometric pattern, referred to as the *Lorenz attractor*. Each panel represents a numerical experiment in which the initial conditions are the ensemble of points that lie along the purple ellipse and the forecasts at successive forecast times are represented by the "down-stream" black ellipses. Each point on the ellipse of initial conditions is uniquely identified with a point on each of the successive forecast ellipses. [Courtesy of T. N. Palmer, ECMWF.]

panel), successive forecast ellipses widen markedly as the points representing the individual forecasts that make up the ensemble approach the bottom of the attractor and spread out into the two separate loops. In the third experiment (right panel) the rapid increase in the uncertainty of the forecasts begins earlier in the forecast interval, and the points along the bottom of the attractor spread rapidly around their respective loops as the forecast proceeds. Hence the predictability of this simple model is highly dependent on the initial state.

Ensemble forecasts carried out with the full set of governing equations for large-scale atmospheric motions provide a means of estimating the uncertainties inherent in weather forecasts at the time when they are issued. The results are not as easy to interpret as those for the idealized model based on the Lorenz attractor, but they are nonetheless informative. As in the idealized experiments, the ensemble forecasts also provide an indication of the range of atmospheric states that could develop out of the observed initial conditions. In current operational practice, different forecast models, as well as perturbed initial conditions, are used to generate different members of the ensemble. At times when the entire hemispheric circulation is relatively predictable, members of the ensemble do not diverge noticeably from one another until relatively far into the forecast. Often the errors grow most rapidly over one particular sector of the hemisphere due to the presence of local instability in the hemispheric flow pattern. The rate of divergence of the individual members of the ensemble provides a measure of the credibility of the forecasts in various sectors of the hemisphere and the length of the time interval over which the forecasts can be trusted.

Figure 7.27 shows 7-day ensemble forecasts for a typical winter day. The mean is considerably smoother than its counterpart in Fig. 7.25 because it represents an average over many individual forecasts. Some of the individual forecasts, like the one in the lower left panel, capture the features in the verifying analysis with remarkable fidelity. Unfortunately, there is no way of identifying these highly skillful forecasts at the time that the ensemble forecast is made.

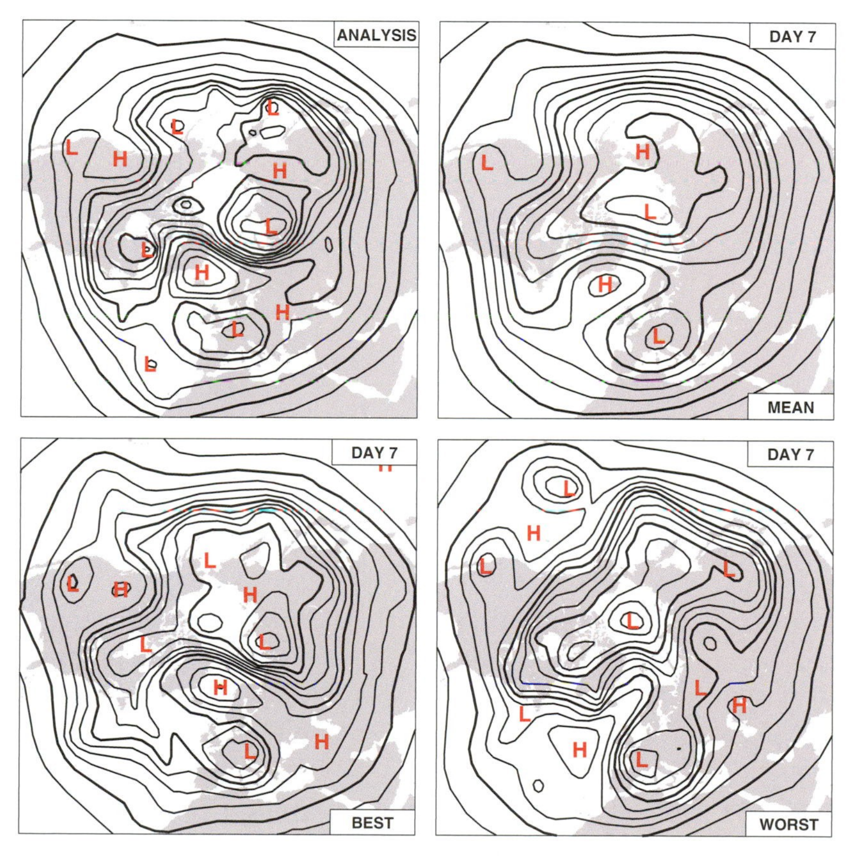

Fig. 7.27 As in Fig. 7.25 but for the 7-day forecasts generated by the ensemble forecasting system in current use at ECMWF. *Mean* is the average of the 50 members of the ensemble *Best* and *Worst* forecasts are selected based on anomaly correlations with the verifying analysis. [Courtesy of Adrian J. Simmons, ECMWF.]

Exercises

7.5 Explain or interpret the following on the basis of the principles discussed in this chapter.

(a) A diffluent flow does not necessarily exhibit divergence.

(b) A flow with horizontal shear does not necessarily exhibit vorticity.

(c) A person of fixed mass weighs slightly less when flying on an eastbound plane than when flying on a westbound plane.

(d) A satellite can be launched in such a way that it remains overhead at a specified longitude directly over the equator in a so-called *geostationary orbit*.

(e) The oblateness of the shape of Jupiter is more apparent than that of the Earth.

(f) The Coriolis force has no discernible effect on the circulation of water going down the drain of a sink.

(g) The vertical component of the Coriolis force is not important in atmospheric dynamics, nor is the Coriolis force induced by vertical motions.

(h) The strong winds encircling hurricanes are highly subgeostrophic.

(i) Cyclones tend to be more intense than anticyclones.

(j) The wind in valleys usually blows up or down the valley from higher toward lower pressure rather than blowing parallel to the isobars.

(k) Surface winds are usually closer to geostrophic balance over oceans than over land.

(l) When high and low cloud layers are observed to be moving in different directions, it can be inferred that horizontal temperature advection is occurring.

(m) Veering of the wind with height within the planetary boundary layer is not necessarily an indication of warm advection.

(n) Areas of precipitation tend to be associated with convergence in the lower troposphere and divergence in the upper troposphere.

(o) The estimation of divergence from wind observations is subject to larger percentage errors than the estimation of vorticity.

(p) The pressure gradient force does not affect the circulation around any closed loop that lies on a pressure surface.

(q) Motions in the middle troposphere tend to be quasi-nondivergent.

(r) The primitive equations assume a simpler form in pressure coordinates than in height coordinates.

(s) The thermodynamic energy equation assumes a particularly simple form in isentropic coordinates.

(t) In middle latitudes, local rates of change of temperature tend to be smaller than the changes attributable to horizontal temperature advection.

(u) Temperatures do not always rise in regions of warm advection.

(v) The release of latent heat in the midtroposphere has the effect of increasing the isentropic potential vorticity of the air in the column below it.

(w) Rising of warm air and sinking of cold air result in the generation of kinetic energy, even in hydrostatic motions.

(x) The Hadley cell does not extend from equator to pole.

(y) A term involving the Coriolis force does not appear in Eq. (7.42).

(z) The easterlies in the lower branch of the Hadley cell are maintained in the presence of friction.

(aa) Baroclinic waves and monsoons tend to be more vigorous in general circulation models that incorporate the effects of moisture than in those that do not.

(bb) Hydroelectric power may be viewed as a by-product of the atmospheric general circulation.

7.6 Describe the vorticity distribution within a flow characterized by counterclockwise circular flow with tangential velocity u inversely proportional to radius r.

Solution: The radial velocity profile is $ur = k$, where k is a constant. Hence the contribution of the shear to the vorticity is $d(k/r)/dr = -k/r^2$, and the contribution of the curvature is

$(k/r)/r = +k/r^2$. As in Exercise 7.1, the contributions are of equal absolute magnitude, but in this case they cancel so $\zeta = 0$ everywhere except at the center of the circle where the flow is undefined and the vorticity is infinite. Such a flow configuration is referred to as an *irrotational vortex*. ■

7.7 At a certain location along the ITCZ, the surface wind at 10 °N is blowing from the east–northeast (ENE) from a compass angle of 60° at a speed of 8 m s^{-1} and the wind at 7 °N is blowing from the south–southeast (SSE) (150°) at a speed of 5 m s^{-1}. (a) Assuming that $\partial/\partial y \gg \partial/\partial x$, estimate the divergence and the vorticity averaged over the belt extending from 7 °N to 10 °N. (b) The meridional component of the wind drops off linearly with pressure from sea level (1010 hPa) to zero at the 900-hPa level. The mixing ratio of water vapor within this layer is 20 g kg^{-1}. Estimate the rainfall rate under the assumption that all the water vapor that converges into the ITCZ in the low level flow condenses and falls as rain.

7.8 Consider a velocity field that can be represented as

$$\mathbf{V}_\Psi = \mathbf{k} \times \nabla\Psi$$

or, in Cartesian coordinates,

$$u_\Psi = -\partial\Psi/\partial y; \quad v_\Psi = \partial\Psi/\partial x$$

where Ψ is called the *streamfunction*. Prove that $\text{Div}_H\,\mathbf{V}$ is everywhere equal to zero and the vorticity field is given by

$$\zeta = \nabla^2\Psi \tag{7.44}$$

Given the field of vorticity, together with appropriate boundary conditions, the inverse of (7.22); namely

$$\Psi = \nabla^{-2}\zeta$$

may be solved to obtain the corresponding streamfunction field. Because the true wind field at extratropical latitudes tends to be quasi-nondivergent, it follows that $\mathbf{V}$ and $\mathbf{V}_\Psi$ tend to be quite similar.

7.9 For streamfunctions Ψ with the following functional forms, sketch the velocity field $\mathbf{V}_\Psi$. (a) $\Psi = my$, (b) $\Psi = my + n\cos 2\pi x/L$, (c) $\Psi = m(x^2 + y^2)$, and (d) $\Psi = m(xy)$ where m and n are constants.

7.10 For each of the flows in the previous exercise, describe the distribution of vorticity.

7.11 Apply Eq. (7.5), which describes the advection of a passive tracer ψ by a horizontal flow pattern to a field in which the initial conditions are $\partial\psi/\partial x = 0$ and $\psi = -my$. (a) Prove that at the initial time $t = 0$,

$$\frac{d}{dt}\left(\frac{\partial\psi}{\partial x}\right) = -m\frac{\partial v}{\partial x} \quad \text{and} \quad \frac{d}{dt}\left(-\frac{\partial\psi}{\partial y}\right) = m\frac{\partial v}{\partial y}$$

Interpret this result, making use of Fig. 7.4. (b) Prove that for a field advected by the pure deformation flow in Fig. 7.4a, the meridional gradient $-\partial\psi/\partial y$ grows exponentially with time, whereas for a field advected by the shear flow in Fig. 7.4b, $\partial\psi/\partial x$ increases linearly with time.

7.12 Prove that for a flow consisting of pure rotation, the circulation C around circles concentric with the axis of rotation is equal to 2π times the angular momentum per unit mass.

7.13 Extend Fig. 7.8 by adding the positions of the marble at points 13–24.

7.14 Consider two additional "experiments in a dish" conducted with the apparatus described in Fig. 7.7. (a) The marble is released from point r_0 with initial counterclockwise motion Ωr_0 in the fixed frame of reference. Show that the orbits of the marble in both fixed and rotating frames of reference are circles of radius r_0, concentric with the center of the dish. (b) The marble is released from point r_0 with initial clockwise motion Ωr_0 in the fixed frame of reference. Show that in the rotating frame of reference the marble remains stationary at the point of release.

7.15 Prove that for a small closed loop of area A that lies on the surface of a rotating spherical planet, the circulation associated with the motion in an inertial frame of reference is $(f + \zeta)A$.

7.16 An air parcel is moving westward at 20 m s^{-1} along the equator. Compute: (a) the apparent acceleration toward the center of the Earth from the point of view of an observer external to the Earth and in a coordinate system rotating with the Earth, and (b) the apparent Coriolis force in the rotating coordinate system.

7.17 A projectile is fired vertically upward with velocity w_0 from a point on Earth. (a) Show that in the absence of friction the projectile will land at a distance

$$\frac{4w_0^3\Omega}{3g^2}\cos\phi$$

to the west of the point from which it was fired. (b) Calculate the displacement for a projectile fired upward on the equator with a velocity of 500 m s^{-1}.

7.18 A locomotive with a mass of 2×10^4 kg is moving along a straight track at 40 m s^{-1} at 43 °N. Calculate the magnitude and direction of the transverse horizontal force on the track.

7.19 Within a local region near 40 °N, the geopotential height contours on a 500-hPa chart are oriented east–west and the spacing between adjacent contours (at 60-m intervals) is 300 km, with geopotential height decreasing toward the north. Calculate the direction and speed of the geostrophic wind.

7.20 (a) Prove that the divergence of the geostrophic wind is given by

$$\nabla \cdot \mathbf{V}_g \equiv \frac{\partial u_g}{\partial x} + \frac{\partial v_g}{\partial y} = \frac{v_g}{f}\frac{\partial f}{\partial y} = -v_g\frac{\cot\phi}{R_E} \quad (7.45)$$

and give a physical interpretation of this result. (b) Calculate the divergence of the geostrophic wind at 45 °N at a point where $v_g = 10$ m s^{-1}.

7.21 Two moving ships passed close to a fixed weather ship within a few minutes of one another. The first ship was steaming eastward at a rate of 5 m s^{-1} and the second northward at 10 m s^{-1}. During the 3-h period that the ships were in the same vicinity, the first recorded a pressure rise of 3 hPa while the second recorded no pressure change at all. During the same 3-h period, the pressure rose 3 hPa at the location of the weather ship (50 °N, 140 °W). On the basis of these data, calculate the geostrophic wind speed and direction at the location of the weather ship.

7.22 At a station located at 43 °N, the surface wind speed is 10 m s^{-1} and is directed across the isobars from high toward low pressure at an angle $\psi = 20°$. Calculate the magnitude of the frictional drag force and the horizontal pressure gradient force (per unit mass).

7.23 Show that if friction is neglected, the horizontal equation of motion can be written in the form

$$\frac{d\mathbf{V}}{dt} = -f\mathbf{k} \times \mathbf{V}_a \quad (7.46)$$

where $\mathbf{V}_a \equiv \mathbf{V} - \mathbf{V}_g$ is the *ageostrophic* component of the wind.

7.24 Show that in the case of anticyclonic air trajectories, gradient wind balance is possible only when

$$\left|\frac{\partial \Phi}{\partial n}\right| \leq f^2R_T/4$$

[**Hint**: Solve for the speed of the gradient wind, making use of the quadratic formula.]

7.25 Prove that the thermal wind equation can also be expressed in the forms

$$\frac{\partial \mathbf{V}_g}{\partial p} = -\frac{R}{fp} \times \nabla T$$

and

$$\frac{\partial \mathbf{V}_g}{\partial z} = \frac{g}{fT} \times \nabla T + \frac{1}{T}\frac{\partial T}{\partial z}\mathbf{V}_g$$

7.26 (a) Prove that the geostrophic temperature advection in a thin layer of the atmosphere (i.e., the rate of change of temperature due to the horizontal advection of temperature) is given by

$$\frac{f}{R\ln(p_B/p_T)} V_{gB}V_{gT}\sin\theta$$

where the subscripts B and T refer to conditions at the bottom and top of the layer, respectively, and θ is the angle between the geostrophic wind at the two levels, defined as positive if the geostrophic wind veers with increasing height.

(b) At a certain station located at 43 °N, the geostrophic wind at the 1000-hPa level is blowing from the southwest (230°) at 15 m s^{-1} while at the 850-hPa level it is blowing from the west–northwest (300°) at 30 m s^{-1}. Calculate the geostrophic temperature advection.

7.27 At a certain station, the 1000-hPa geostrophic wind is blowing from the northeast (050°) at 10 m s^{-1} while the 700-hPa geostrophic wind is blowing from the west (270°) at 30 m s^{-1}. Subsidence is producing adiabatic warming at a rate of 3 °C day^{-1} in the 1000 to 700-hPa layer and diabatic heating is negligible. Estimate the time rate of change of the thickness of the 1000 to 700-hPa layer. The station is located at 43 °N.

7.28 Prove that in the line integral around any closed loop that lies on a pressure surface, the pressure gradient force $-\nabla\Phi$ vanishes, i.e.,

$$\oint \mathbf{P} ds = -\oint \nabla\Phi ds = 0 \qquad (7.47)$$

7.29 Prove that in the absence of friction (i.e., with the only force being the pressure gradient force), the circulation

$$C_a \equiv \oint \mathbf{c} \cdot ds \qquad (7.48)$$

is conserved following a closed loop of parcels as they move where **c** is the velocity in an inertial frame of reference.

7.30 Based on the result of the previous exercise, prove that in an inertial frame of reference

$$[c_s]\frac{dL}{dt} = -L\frac{d[c_s]}{dt}$$

where

$$[c_s] \equiv \frac{\oint \mathbf{c} \cdot ds}{L}$$

$[c_s]$ is the tangential velocity averaged over the length of the loop and $L = \oint ds$ is the length of the loop. Hence, a lengthening of the loop must be accompanied by a proportionate decrease in the mean tangential velocity along the loop and vice versa.

7.31 For the special case of an axisymmetric flow in which the growing or shrinking loop is concentric with the axis of rotation, prove that the conservation of circulation is equivalent to the conservation of angular momentum. [**Hint**: Show that angular momentum $M = C/2\pi$.]

7.32 Starting with the horizontal equation of motion (7.14) and ignoring friction, derive a conservation law for vorticity ζ in horizontal flow on a rotating planet.

Solution: We begin by rewriting the horizontal equation of motion (7.14) in Cartesian form, ignoring the frictional drag term

$$\frac{du}{dt} = -\frac{\partial\Phi}{\partial x} + fv \qquad (7.49a)$$

$$\frac{dv}{dt} = -\frac{\partial\Phi}{\partial y} - fu \qquad (7.49b)$$

Then we expand these expressions in Eulerian form, neglecting the vertical component of the advection, which is about an order of magnitude smaller than the corresponding horizontal advection terms

$$\frac{\partial u}{\partial t} = -u\frac{\partial u}{\partial x} - v\frac{\partial u}{\partial y} - \frac{\partial\Phi}{\partial x} + fv \qquad (7.50a)$$

$$\frac{\partial v}{\partial t} = -u\frac{\partial v}{\partial x} - v\frac{\partial v}{\partial y} - \frac{\partial\Phi}{\partial y} - fu \qquad (7.50b)$$

Differentiating (7.50a) with respect to y and (7.50b) with respect to x, and subtracting the first from the second yields, after rearranging terms

$$\frac{\partial}{\partial x}\frac{\partial v}{\partial t} - \frac{\partial}{\partial y}\frac{\partial u}{\partial t} = -\frac{\partial u}{\partial x}\frac{\partial v}{\partial x} + \frac{\partial u}{\partial x}\frac{\partial u}{\partial x} - u\frac{\partial}{\partial x}\frac{\partial v}{\partial x} + u\frac{\partial}{\partial y}\frac{\partial u}{\partial x}$$
$$-\frac{\partial v}{\partial x}\frac{\partial v}{\partial y} + \frac{\partial v}{\partial y}\frac{\partial u}{\partial y} - v\frac{\partial}{\partial x}\frac{\partial v}{\partial x} + v\frac{\partial}{\partial y}\frac{\partial u}{\partial y}$$
$$-f\left(\frac{\partial u}{\partial x} + \frac{\partial v}{\partial y}\right) - v\frac{\partial f}{\partial y}$$

Reversing the order of differentiation of the terms on the left-hand side yields $\partial\zeta/\partial t$. The four terms involving products of velocity derivatives can be rewritten in the more compact form

$$-\zeta\left(\frac{\partial u}{\partial x} + \frac{\partial v}{\partial y}\right)$$

and the four remaining terms in the two top lines as

$$-u\frac{\partial \zeta}{\partial x} - v\frac{\partial \zeta}{\partial y}$$

With these substitutions, we obtain

$$\frac{\partial \zeta}{\partial t} = -u\frac{\partial \zeta}{\partial x} - v\frac{\partial \zeta}{\partial y} - (f + \zeta)\left(\frac{\partial u}{\partial x} + \frac{\partial v}{\partial y}\right) - v\frac{\partial f}{\partial y}$$

Rearranging slightly (noting that $\partial f/\partial t = 0$) and reintroducing vector notation yields the conservation law

$$\frac{\partial}{\partial t}(f + \zeta) = -\mathbf{V} \cdot \nabla(f + \zeta) - (f + \zeta)(\nabla \cdot \mathbf{V})$$

or, in Lagrangian form,

$$\frac{d}{dt}(f + \zeta) = -(f + \zeta)\,(\nabla \cdot \mathrm{V})$$

These equations appear in the text as Eq. (7.21a,b). ■

7.33 Consider a sinusoidal wave along latitude ϕ with wavelength L and amplitude v in the meridional wind component. The wave is embedded in a uniform westerly flow with speed U. (a) Show that the amplitude of the geopotential height perturbations associated with the wave is $fvL/2\pi g$ where f is the Coriolis parameter and g is the gravitational acceleration. (b) Show that the amplitude of the associated vorticity perturbations is $(2\pi/L)v$. Show that the maximum values of the advection of planetary and relative vorticity are βv and $(2\pi/L)^2 Uv$, respectively, and that they are coincident and of opposing sign. (c) Show that the advection terms exactly cancel for waves with wavelength

$$L_s = 2\pi\sqrt{\frac{U}{\beta}}.$$

L_S is referred to as the wavelength of a stationary Rossby wave.

7.34 Suppose that the wave in the previous exercise propagating is along 45° latitude and has a wavelength of 4000 km. The amplitude of the meridional wind perturbations associated with the wave is 10 m s^{-1} and the background flow U = 20 m s^{-1}. Assume that the velocity field is independent of latitude. Using the results of the previous exercise, estimate (a) the amplitude of the geopotential height and vorticity perturbations in the waves, (b) the amplitude of the advection of planetary and relative vorticity, and (c) the wavelength of a stationary Rossby wave embedded in a westerly flow with a speed of 20 m s^{-1} at 45° latitude.

7.35 Verify the validity of the conservation of barotropic potential vorticity $(f + \zeta)/H$, as expressed in Eq. (7.27). [**Hint**: You might wish to verify and make use of the identity $dy/y = -dx/x$, where $y = 1/x$.]

7.36 Consider barotropic ocean eddies propagating meridionally along a sloping continental shelf, with depth increasing toward the east, as pictured in Fig. 7.28, conserving barotropic potential vorticity in accordance with (7.27). There is no background flow. In which direction will the eddies propagate? [**Hint**: Consider the vorticity tendency at points A and B.]

7.37 During winter in middle latitudes, the meridional temperature gradient is typically on the order of 1° per degree of latitude, while potential temperature increases with height at a rate of roughly 5 °C km^{-1}. What is a typical slope of the potential temperature surfaces in the meridional plane? Compare this result

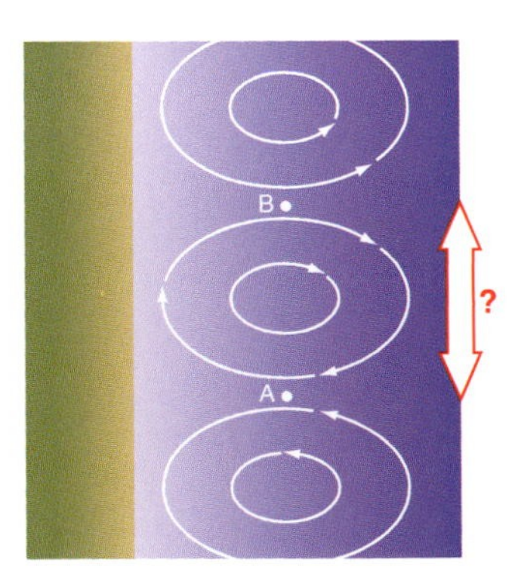

Fig. 7.28 Barotropic ocean eddies propagating along a continental shelf, as envisioned in Exercise 7.36. Lighter blue shading indicates shallower water.

with the slope of the 500-hPa surface in Exercise 7.17.

7.38 Making use of the approximate relationship $\omega \simeq -\rho g w$, show that the vertical motion term

$$\left(\frac{\kappa T}{p} - \frac{\partial T}{\partial p}\right)\omega$$

in Eq. (7.37) is approximately equal to

$$-w(\Gamma_d - \Gamma)$$

where Γ_d is the dry adiabatic lapse rate and Γ is the observed lapse rate.

7.39 Figure 7.29 shows an idealized trapezoidal vertical velocity profile in a certain rain area in the tropics. The horizontal convergence into the rain area in the 1000 to 800-hPa layer is $10^{-5}\,s^{-1}$, and the average water vapor content of this converging air is 16 g kg^{-1}. (a) Calculate the divergence in the 200 to 100-hPa layer. (b) Estimate the rainfall rate using the assumption that all the water vapor is condensed out during the ascent of the air.

Solution: (a) Let $(\nabla \cdot \mathbf{V})_L$ and $(\nabla \cdot \mathbf{V})_H$ refer to the divergences in the lower and upper layers, respectively. Continuity of mass requires that

$$\int_{100}^{200} (\nabla \cdot \mathbf{V})_H \, dp = \int_{800}^{1000} (\nabla \cdot \mathbf{V})_L \, dp$$

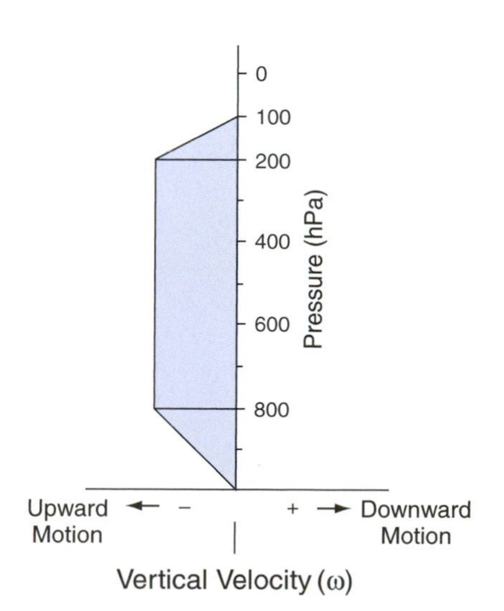

Fig. 7.29 Vertical velocity profile for Exercise 7.39.

Hence,

$$(\nabla \cdot \mathbf{V})_H = (\nabla \cdot \mathbf{V})_L \times \left(\frac{1000 - 800}{200 - 100}\right)$$
$$= 2 \times 10^{-5}\,s^{-1}$$

(b)

$$\begin{aligned}\omega_{800} &= -\int_{1000}^{800} (\nabla \cdot \mathbf{V})_L \, dp + \omega_{1000} \\ &= (\nabla \cdot \mathbf{V})_L \,(1000\text{ hPa} - 800\text{ hPa}) + \omega_{1000} \\ &= (-10^{-5}\,s^{-1})\,(200\text{ hPa}) + 0 \\ &= -2 \times 10^{-3}\text{ hPa s}^{-1}\end{aligned}$$

Making use of the relationship $\omega \approx -\rho g w$, the vertical mass flux ρw can be obtained by dividing ω (in SI units) by g. Therefore, the rate at which liquid water is being condensed out is

$$\frac{2 \times 10^{-1}\text{ Pa s}^{-1}}{(9.8\text{ m s}^{-2})} \times 0.016$$
$$= 3.27 \times 10^{-6}\text{ kg m}^{-2}\,s^{-1}$$

We can express the rainfall rate in more conventional terms by noting that 1 kg in^{-2} of liquid water is equivalent to a depth of 1 mm. Thus the rainfall rate is 3.27×10^{-4} mm s^{-1} or

$$3.27 \times 10^{-1}\text{ mm s}^{-1} \times 8.64 \times 10^4\text{ s day}^{-1}$$
$$= 28.3\text{ mm day}^{-1}$$

which is a typical rate for moderate rain. ■

7.40 In middle-latitude winter storms, rainfall (or melted snowfall) rates on the order of 20 mm day^{-1} are not uncommon. Most of the convergence into these storms takes place within the lowest 1–2 km of the atmosphere (say, below 850 hPa) where the mixing ratios are on the order of 5 g kg^{-1}. Estimate the magnitude of the convergence into such storms.

7.41 The area of a large cumulonimbus anvil in Fig. 7.30 is observed to increase by 20% over a 10-min period. Assuming that this increase in area is representative of the average divergence within the 300 to 100-hPa layer and that the vertical velocity at 100 hPa is zero, calculate the vertical velocity at the 300-hPa level.

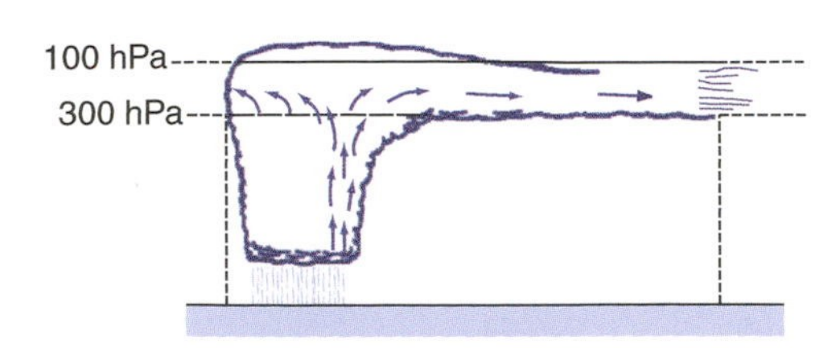

Fig. 7.30 Physical situation in Exercise 7.41.

Solution: From the results of Exercise 7.2,

$$\nabla \cdot \mathbf{V} = \frac{1}{A}\frac{dA}{dt} = \frac{0.20}{600 \text{ s}} = 3.33 \times 10^{-4} \text{ s}^{-1}$$

Making use of (7.39), we have

$$\begin{aligned}\omega_{300} &= \omega_{100} - (\nabla \cdot \mathbf{V})\,(300 - 100) \text{ hPa} \\ &= 0 - 3.33 \times 10^{-4} \text{ s}^{-1} \times 200 \text{ hPa} \\ &= -6.66 \times 10^{-2} \text{ hPa s}^{-1}\end{aligned}$$

To express the vertical velocity in height coordinates we make use of the approximate relation (7.33)

$$\begin{aligned}w_{300} &\simeq -\frac{\omega_{300}}{\rho g} = -\frac{RT}{pg}\omega_{300} = -\frac{H}{p}\omega_{300} \\ &\simeq \frac{7 \text{ km}}{300 \text{ hPa}} \times -6.66 \times 10^{-2} \text{ hPa s}^{-1} \\ &\simeq 1.5 \text{ m s}^{-1}\end{aligned}$$

Note that this is the 300-hPa vertical velocity averaged over the area of the anvil. The vertical velocity within the updraft may be much larger. In contrast, divergences and vertical motions observed in large-scale weather systems are much smaller than those considered in this exercise. ■

7.42 Figure 7.31 shows the pressure and horizontal wind fields in an atmospheric Kelvin wave that propagates zonally along the equator. Pressure and zonal wind oscillate sinusoidally with longitude and time, while $v = 0$ everywhere. Such waves are observed in the stratosphere where frictional drag is negligible in comparison with the other terms in the horizontal equations of motion. (a) Prove that the waves propagate eastward. (b) Prove that the zonal wind component is in geostrophic equilibrium with the pressure field.

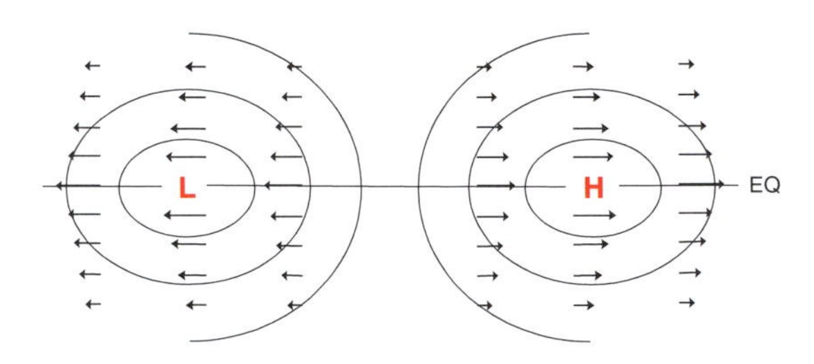

Fig. 7.31 Distribution of wind and geopotential height on a pressure surface in an equatorial Kelvin wave in a coordinate system moving with the wave.

7.43 Averaged over the mass of the atmosphere, the root mean squared velocity of fluid motions is $\sim$17 m s^{-1}. By how much would the center of gravity of the atmosphere have to drop in order to generate the equivalent amount of kinetic energy?

7.44 Suppose that a parcel of air initially at rest on the equator is carried poleward to 30° latitude in the upper branch of the Hadley cell, conserving angular momentum as it moves. In what direction and at what speed will it be moving when it reaches 30°?

7.45 On average over the globe, kinetic energy is being generated by the cross-isobar flow at a rate of $\sim$2 W m^{-2}. At this rate, how long would it take to "spin up" the general circulation, starting from a state of rest?

7.46 The following laboratory experiment provides a laboratory analog for the kinetic energy cycle in the atmospheric general circulation. The student is invited to verify the steps along the way and to provide physical interpretations, as appropriate.

A tank with vertical walls is filled with a homogeneous (constant density) incompressible fluid. The height of the undulating free surface of the fluid is $Z(x, y)$.

(a) Show that the horizontal component of the pressure gradient force (per unit mass) is independent of height and given by

$$\mathbf{P} = -g\nabla Z$$

(b) Show that the continuity equation takes the form

$$\nabla \cdot \mathbf{V} + \frac{\partial w}{\partial z} = 0$$

where $\mathbf{V}$ is the horizontal velocity vector and w is the vertical velocity.

(c) Assuming that $\mathbf{V}$ is independent of height and that the pertubations about the mean height are small show that

$$\frac{dZ}{dt} = -[Z]\,(\nabla \cdot \mathbf{V})$$

where [Z] is the mean height of the free surface of the fluid.

(d) Show that the potential energy of the fluid in the tank is

$$\frac{1}{2}\rho g \iint Z^2\, dA$$

and that the fraction of this energy available for conversion to kinetic energy is

$$\frac{1}{2}\rho g \iint Z'^2\, dA$$

where Z' is the perturbation of the height of the surface from its area averaged height.

(e) Show that the rate of conversion from potential energy to kinetic energy is

$$-\rho g \iint Z' \frac{dZ'}{dt}\, dA$$

(f) If Z' is much smaller than the mean height of the fluid, show that (e) is equivalent to

$$\rho g\,[Z] \iint Z'\,(\nabla \cdot \mathbf{V})\, dA$$

(g) Because the tank is cylindrical and fluid cannot flow through the walls,

$$\iint (\nabla \cdot \mathbf{V}Z')\, dA = 0$$

Making use of this result, verify that (f) can be rewritten as

$$-\rho g\,[Z] \iint (\mathbf{V} \cdot \nabla Z)\, dA$$

Weather Systems

8

With Lynn McMurdie and Robert A. Houze
Department of Atmospheric Sciences
University of Washington

The abundant rainfall that sustains life on Earth is not the bounty of cloud microphysical processes alone. Without vigorous and sustained motions, the atmospheric branch of the hydrological cycle would stagnate. Much of the ascent that drives the hydrologic cycle in the Earth's atmosphere occurs in association with *weather systems* with well-defined structures and life cycles. A small fraction of these systems achieve the status of *storms* capable of disrupting human activities and, in some instances, inflicting damage.

This chapter introduces the reader to the structure and underlying dynamics of weather systems and their associated weather phenomena. The first section is mainly concerned with large-scale extratropical weather systems (i.e., baroclinic waves and the associated *extratropical cyclones*) and their embedded mesoscale fronts. The second section discusses some of the effects of terrain on large-scale weather systems and some of the associated weather phenomena. The third section describes the modes of mesoscale organization of deep cumulus convection. The final section describes a special form of organization in which a mesoscale convective system acquires strong rotation. These so-called *tropical cyclones* tend to be tighter, more axially symmetric, and more intense than their extratropical counterparts.

8.1 Extratropical Cyclones

Extratropical cyclones assume a wide variety of forms, depending on factors such as the background flow in which they are embedded, the availability of moisture, and the characteristics of the underlying surface. This section shows how atmospheric data are analyzed to reveal the structure and evolution of these systems. To illustrate these analysis techniques, we present a case study of a system that brought strong winds and heavy precipitation to parts of the central United States. The particular cyclone system selected for this analysis was unusually intense, but it typifies many of the features of winter storms in middle and high latitudes. Plotting conventions for the synoptic charts that appear in this section are shown in Fig. 8.1. A brief history of synoptic charts and a description of how modern synoptic charts are constructed is presented in the Appendix to Chapter 8 on the book web site.

8.1.1 An Overview

This subsection documents the large-scale structure of the developing cyclone, with emphasis on the 500-hPa height, sea-level pressure, 1000- to 500-hPa thickness (a measure of the mean temperature of the lower troposphere) and vertical velocity fields. The development of the storm is shown to be linked to the intensification of a baroclinic wave.

The hemispheric 500-hPa chart for midnight (00) universal time (UTC: time observed on the Greenwich meridian[1]) November 10, 1998 is shown in Fig. 8.2. At this time, the westerly "polar vortex" is split into two regional cyclonic vortices, one centered over Russia and the other centered over northern Canada. Separating the vortices are pair of *ridges*, where the geopotential height contours bulge poleward. One of the ridges protrudes over Alaska and

[1] At longitudes west of the Greenwich meridian local time (LT) lags universal time (UTC) by 1 h for each ~15° of longitude, less 1 h during daylight savings time. For example, in the United States, 00 UTC corresponds to 19 EST, 20 EDT, and 16 PST of the previous day.

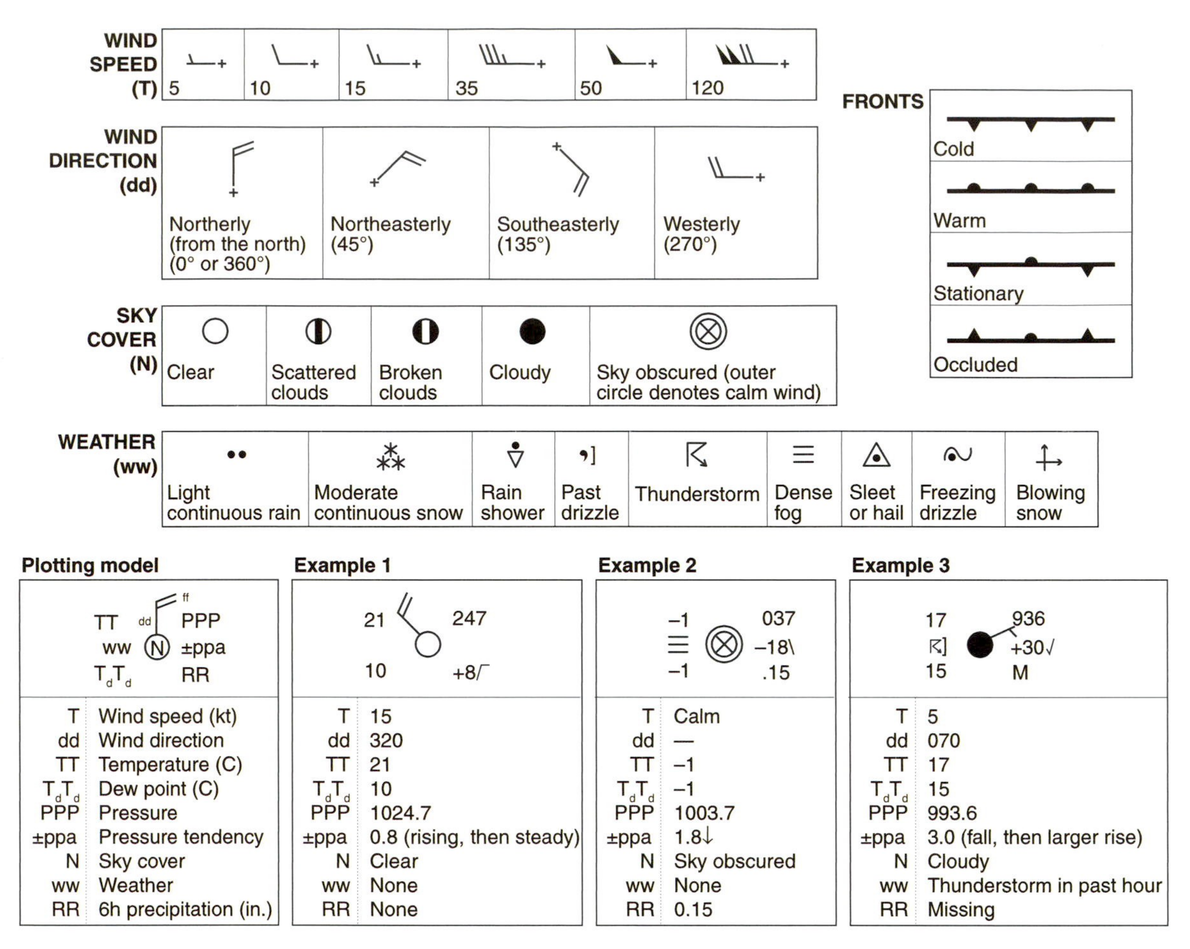

Fig. 8.1 Plotting convections used in synoptic charts.

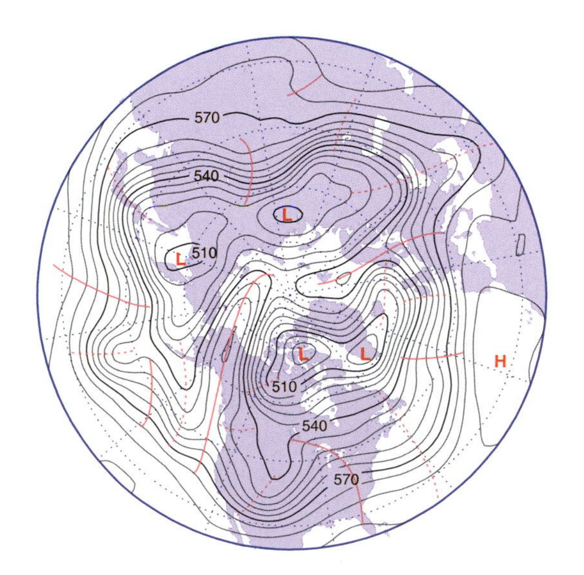

Fig. 8.2 Hemispheric 500-hPa height chart for 00 UTC Nov. 10, 1998. Contours at 60-m intervals. Contours labeled in tens of meters (decameters, dkm). Solid red lines denote the axes of *ridges*, and dashed red lines denote the axes of *troughs* in the 500-hPa wave pattern. [Courtesy of Jennifer Adams, COLA/IGES.]

the other protrudes northward over Scandinavia. Pronounced *troughs* (along which the contours bulge equatorward) are evident over the Black Sea, Japan, the central Pacific, and the United States Great Plains, and several weaker troughs can be identified at other locations. The typical distance between successive troughs (counting the weaker ones) is ~50° of longitude or 4000 km, which corresponds to the theoretically predicted wavelength of baroclinic waves.

Time-lapse animations of weather charts like the one shown in Fig. 8.2 reveal that baroclinic waves move eastward at a rate of $\sim 10 \text{ m s}^{-1}$, which corresponds to the wintertime climatological-mean zonal wind speed around the 700-hPa level. Since the strength of the westerlies generally increases with height within the extratropical troposphere, air parcels above this so-called *steering level* pass through the waves from west to east, while air parcels below that level are overtaken by the waves. Successive ridges (or troughs) typically pass a fixed point on Earth at intervals of roughly 4 days, but they may be only a day or two apart if the steering flow is very strong.

Lapses of a week or longer may occur between wave passages in the sectors of the hemisphere where the westerlies aloft are blocked by strong ridges. The direction of propagation tends to follow the steering flow, which nearly always exhibits a strong eastward component. Baroclinic waves are observed most regularly and tend to be strongest over the oceans, but they can develop over land, as in this case study. Baroclinic wave activity tends to be most vigorous during winter when the meridional temperature gradient across midlatitudes is strongest.

A more detailed view of the 500-hPa height pattern over the North American sector at 00 UTC November 10 is shown in Fig. 8.3, and the charts for

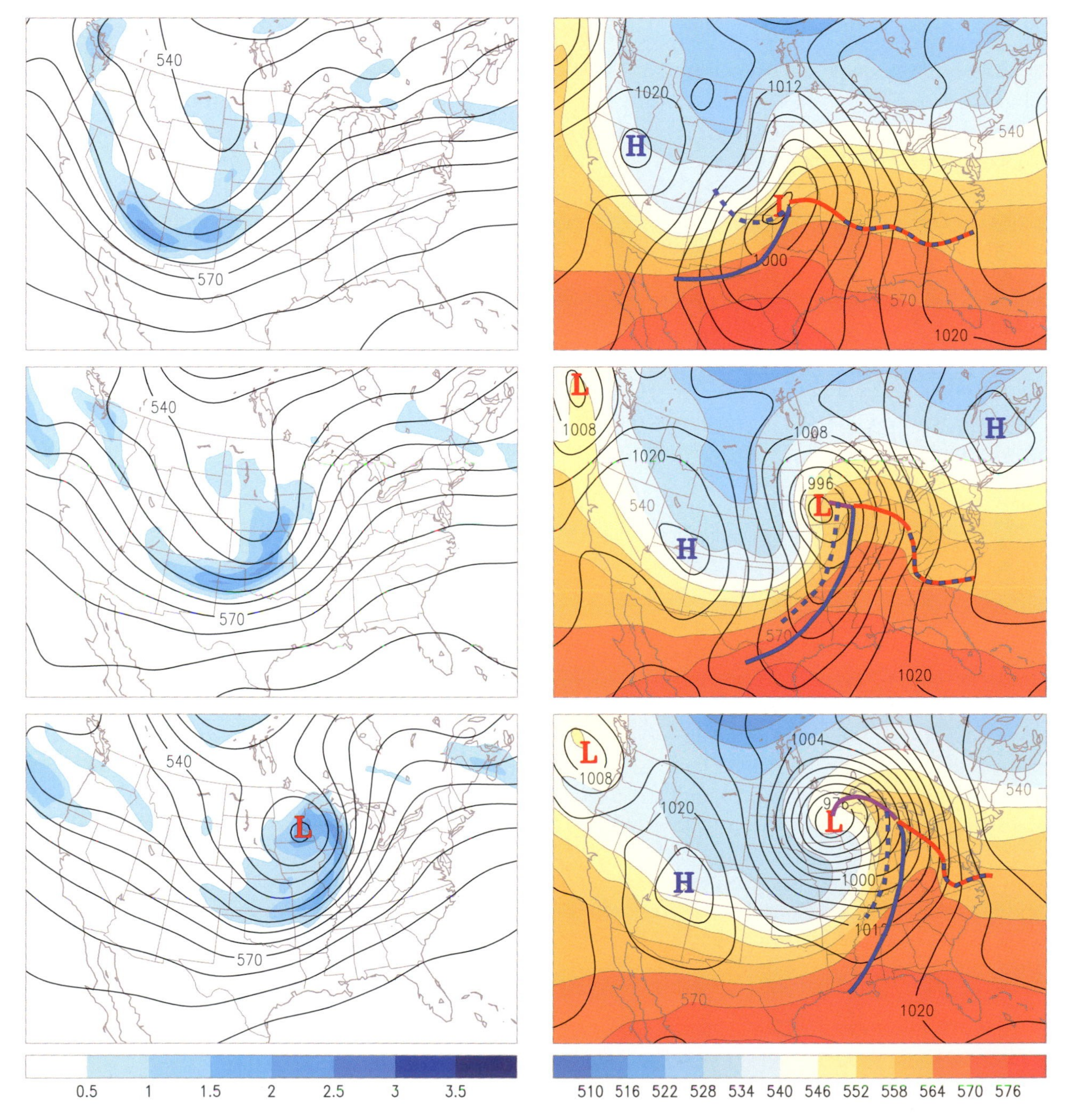

Fig. 8.3 Synoptic charts at 00, 09, and 18 UTC Nov. 10, 1998. (Left) The 500-hPa height (contours at 60-m intervals; labels in dkm) and relative vorticity (blue shading; scale on color bar in units of $10^{-4}\,s^{-1}$). (Right) Sea-level pressure (contours at 4-hPa intervals) and 1000- to 500-hPa thickness (colored shading: contour interval 60 m; labels in dkm). Surface frontal positions, as defined by a skilled human analyst, are overlaid. [Courtesy of Jennifer Adams, COLA/IGES.]

9 and 18 h later are shown below it. Clearly evident in this three-chart sequence is the eastward propagation and intensification of the trough that passes over the United States Great Plains. In the third chart in the sequence, the base of this trough splits off from the westerlies to form a *cutoff low* (i.e., an isolated minimum in the geopotential height field), implying the existence of a closed cyclonic circulation. The dramatic intensification of the winds encircling this feature is reflected in the tightening of the spacing between adjacent 500-hPa height contours.

The intensification of the trough at the 500-hPa level is accompanied by the deepening of the corresponding low pressure center in sea-level pressure field, as shown in the right-hand panels of Fig. 8.3. This surface low marks the center of a closed cyclonic circulation referred to as an *extratropical cyclone*. Also evident in the right-hand panels of Fig. 8.3 is the amplification of the west-to-east gradient in the 1000- to 500-hPa thickness field, indicated by the colored shading. In the first chart of the sequence the developing surface low is located well to the east of the corresponding trough in the 500-hPa height field, but as these features amplify, they come into vertical alignment in subsequent charts of the sequence.

Now let us examine this sequence of events in greater detail. Embedded in the *long-wave trough* over western North America in the first chart in the sequence (Fig. 8.3, upper left panel) are several smaller scale features, which show up clearly in the vorticity field. The vorticity maxima along the coast of British Columbia and over northern Arizona correspond to *short-wave troughs*, in which the horizontal flow exhibits both cyclonic curvature and cyclonic shear. The shear is particularly strong in the Arizona trough. Nine hours later (Fig. 8.3, middle left panel) these vorticity maxima and their associated troughs appear downstream of their previous positions: the former is centered over the state of Washington and the latter has evolved into an elongated comma-shaped band trailing westward from Kansas, across the Texas Panhandle and into New Mexico. In the final chart of the sequence, the head of the comma-shaped feature is centered over southeastern Minnesota.

In the corresponding sequence of surface charts shown in the right-hand panel of Fig. 8.3, the central pressure of the surface low, as analyzed in Fig. 8.3, dropped from 998 hPa at 00 UTC Nov. 10 (top panel) to 978 hPa at 18 UTC (bottom panel), and 968 hPa at 00 UTC Nov. 11 (not shown), a deepening rate of 30 hPa per day, which is three times as rapid as observed in a typical extratropical cyclone. At 00 UTC Nov. 10 (Fig. 8.3, top panel) the center of the extratropical cyclone (as defined by the sea-level pressure field) was located $\sim 1/4$ wavelength downstream of the 500-hPa trough and just about directly underneath the jet stream. In contrast, in the last of the three charts the surface low was situated almost directly beneath the cutoff low in the 500-hPa height field, and on the poleward (cyclonic) side of the jet stream.

The top panel of Fig. 8.4 shows the same information for the same three map times, depicted in a slightly different way. In this case the geopotential height field at the Earth's surface is represented in terms of the geopotential height of the 1000-hPa surface. Contours of 1000-hPa height, 500-hPa height and 1000- to 500-hPa thickness are superimposed on the same set of charts, with the same (60-m) contour interval. The lower panels of Fig. 8.4 show the evolving structure of a typical baroclinic wave, as depicted in a synoptic meteorology textbook written over a generation ago. The high degree of correspondence between the real features observed in this case study and the idealized features in the textbook representation establishes that the case study presented in this section typifies many of the features of baroclinic waves.

The amplification of the wave in the thickness field is due to horizontal temperature advection by the cyclonic circulation around the deepening surface low. The southerly wind component to the east of the low advects warm air northward while the northerly component to the west of the low advects colder air southward. The strengthening of the east–west temperature contrasts in the lower tropospheric temperature field leads to a weakening of the north–south temperature gradient in the background field on which the wave is growing. As the surface low intensifies over the 9-h interval spanned by the first two charts, the winds around it strengthen while the angle between the geopotential height contours and the thickness contours increases, resulting in a dramatic increase in the horizontal temperature advection. However, in the later stage of development during the interval between the second and third charts, the surface low comes into alignment with the 500-hPa trough, and the 1000-hPa height, 500-hPa height, and 1000- to 500-hPa thickness contours come into alignment with each other, resulting in a weakening of the horizontal temperature advection. In the language

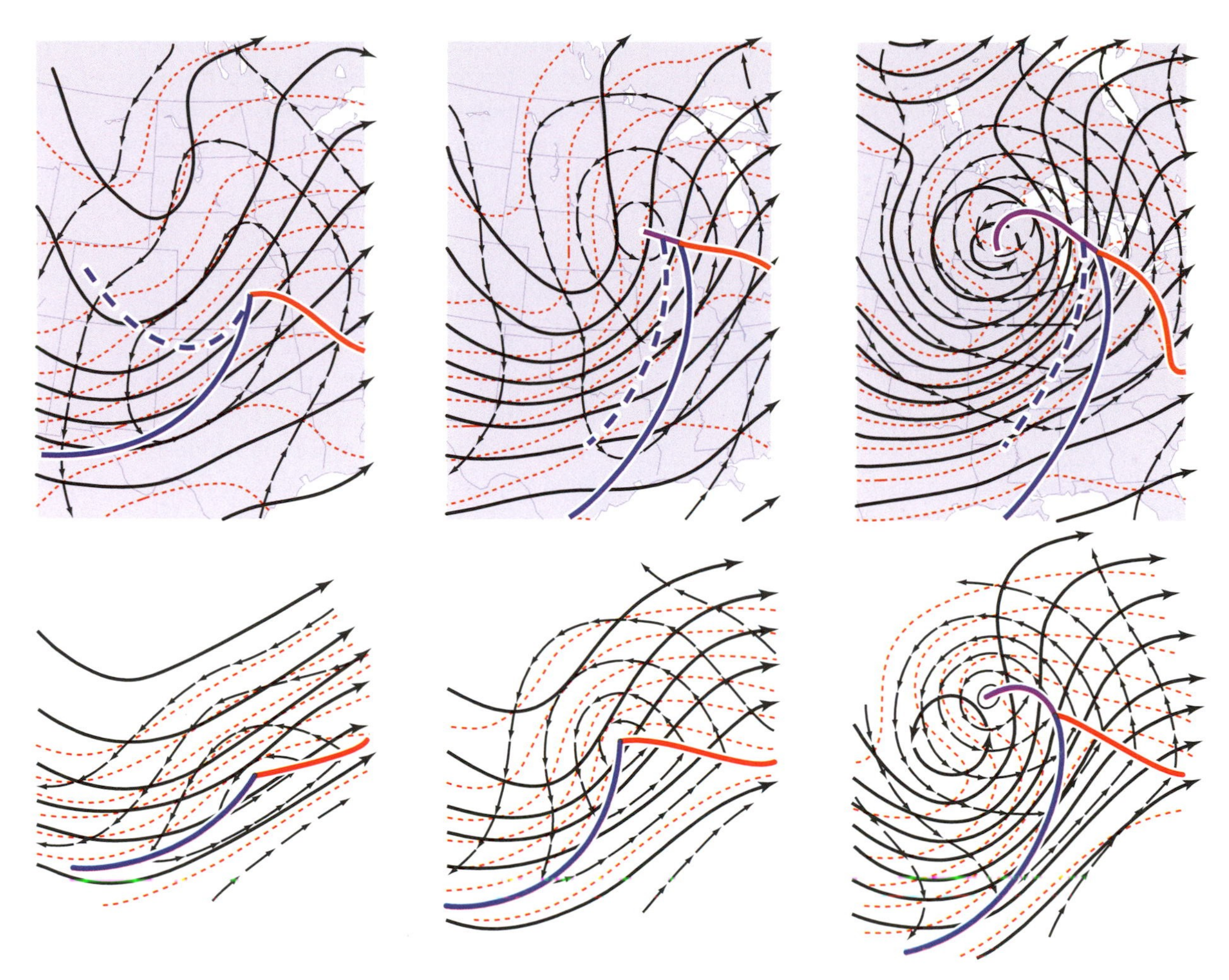

Fig. 8.4 (Top) Fields of 500-hPa height (thick black contours) 1000-hPa height (thin black contours), and 1000- to 500-hPa thickness (dashed red) at 00, 09, and 18 UTC Nov. 10, 1998; contour interval 60 m for all three fields. Arrows indicate the sense of the geostrophic wind. (Bottom) Idealized depictions for a baroclinic wave and its attendant tropical extratropical cyclone in its early (left), developing (center), and mature (right) stages. [Top panel courtesy of Jennifer Adams, COLA/IGES. Bottom panel adapted from *Atmospheric Circulation Systems: Their Structure and Physical Interpretation*, E. Palmén and C.W. Newton, p. 326, Copyright (1969), with permission from Elsevier.]

introduced in Section 7.2.7, the geostrophic wind field evolves from a highly *baroclinic* pattern, with strong turning of the geostrophic wind with height in amplifying baroclinic waves, into a more *equivalent barotropic* pattern, with much less directional shear of the lower tropospheric geostrophic wind field in fully developed baroclinic waves. This transition from a highly baroclinic structure, with strong temperature contrasts in the vicinity of the surface low, to a more barotropic structure with strong winds but weaker temperature gradients, marks the end of the intensification phase in the life cycle of the cyclone.

The vertical velocity field also plays an important role in the development of baroclinic waves. Figure 8.5 shows the vertical velocity field superimposed on the 500-hPa height field. In the left panel, which corresponds to the time when the system is developing most rapidly, the northward moving air in the region of warm advection in advance of the developing surface low is rising, while the southward moving air in the region of cold advection to rear of the cyclone is sinking. It is also apparent from the right-hand panels of Fig. 8.3 that at any given latitude the rising air to the east of the surface low is warmer than the sinking air to the west of it. We recall from Section 7.4.1 that the rising of warm air and sinking of cold air is indicative of a conversion of potential energy into kinetic energy. In the case of baroclinic waves, the potential energy is associated with the east–west temperature gradients and the kinetic energy is primarily associated with the meridional wind component.

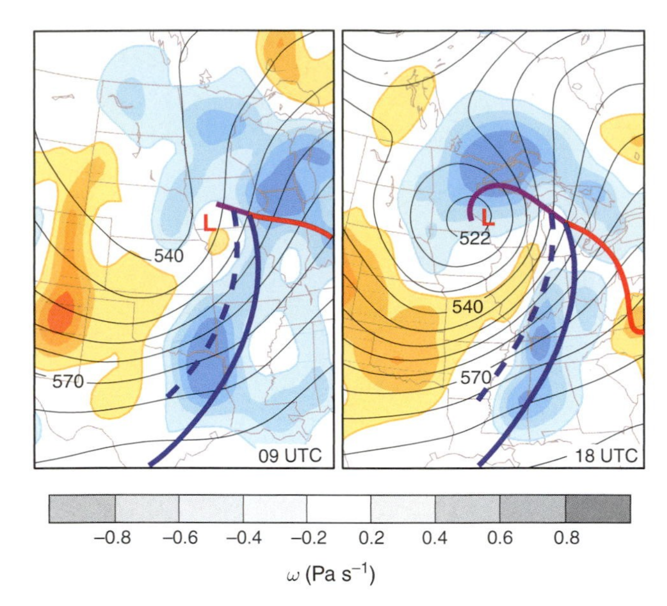

Fig. 8.5 The 500-hPa height (in tens of meters) and vertical velocity (in Pa s^{-1}) fields at the 700-hPa level at 09 and 18 UTC Nov. 10, 1998. Blue shading (negative ω) indicates ascent and tan shading indicates subsidence. [Courtesy of Jennifer Adams, COLA/IGES.]

In the right-hand panel of Fig. 8.5 warmer air to the east of the cyclone is still rising, but the region of ascent wraps around the northern and western flanks of the surface low. In a similar manner, the region of subsidence to the west wraps around the southern and eastern flanks of the cyclone. The juxtaposition of these inward-spiraling rising and subsiding air currents, reminiscent of the "yin-yang pattern" in Asian art, is influential in shaping the cloud and precipitation patterns associated with extratropical cyclones, as shown in the next subsection.

8.1.2 Fronts and Surface Weather

The previous subsection documented the broad outlines of an intense storm that developed over the north central United States. Much of the significant weather observed in association with such systems tends to be concentrated within narrow bands called *frontal zones*, which are marked by sharp horizontal gradients and sometimes by outright discontinuities in wind and temperature. The development of frontal zones (*frontogenesis*, in the vernacular) is initiated by the large-scale horizontal deformation field, as discussed in Section 7.1.3. Mesoscale circulations in the plane perpendicular to the fronts are instrumental in sharpening the temperature contrasts and in organizing the distribution of precipitation into bands oriented parallel to the fronts. This subsection documents the expressions of the November 10, 1998 storm and its attendant frontal zones in (a) wind and pressure, (b) temperature, (c) moisture variables, (d) surface weather, (e) the suite of hourly observations, (f) satellite imagery, and (g) radar imagery.

a. Wind and pressure

Figure 8.6 shows the sea-level pressure and surface winds at 9-h intervals starting at 00 UTC Nov. 10 (note that the field of view is smaller than in the previous charts). At all three map times a pronounced wind-shift line, the expression of the *cold front* in the surface wind field, is evident to the south of the surface low. To the west of the cold front the surface winds exhibit a strong westerly component, whereas to the east of it the southerly wind component is dominant. The isobars bend sharply (and some change direction abruptly or "kink") along the front. Hence, as the front passes, a fixed observer at the Earth's surface would experience a veering (i.e., shifting in an anticyclonic sense) of the wind from southerly to westerly, concurrent with a well-defined minimum in sea-level pressure. Through the three-chart sequence the cold front advances eastward, keeping pace with and showing some tendency to wrap around the surface low as it deepens and tracks northeastward. It appears as though the front is being advected by the intensifying cyclonic circulation.

The wind-shift line extending eastward from the surface low, the expression of the *warm front*, is a more subtle feature, the reality of which becomes clearly evident when the surface charts are analyzed in conjunction with hourly station data, as illustrated later in this subsection. Like the cold front, the warm front shows indications of being advected around the developing surface low. When it passes a station the wind veers from southeasterly to southerly. In the later stages of the development of the cyclone, as represented in the 18 UTC panel in Fig. 8.6, the junction of the cold and warm fronts becomes separated from the center of the surface low and an *occluded front* extends from the center of the surface low to a *triple point* where it meets the junction of the warm and cold fronts. When the occluded front passes a station the surface wind veers from southeasterly to southwesterly.

A fourth wind-shift line, the expression of a *secondary cold front*, rendered in dashed blue, also appears on the charts for 00 and 09 UTC. In the 00 UTC chart the line curves eastward from the eastern

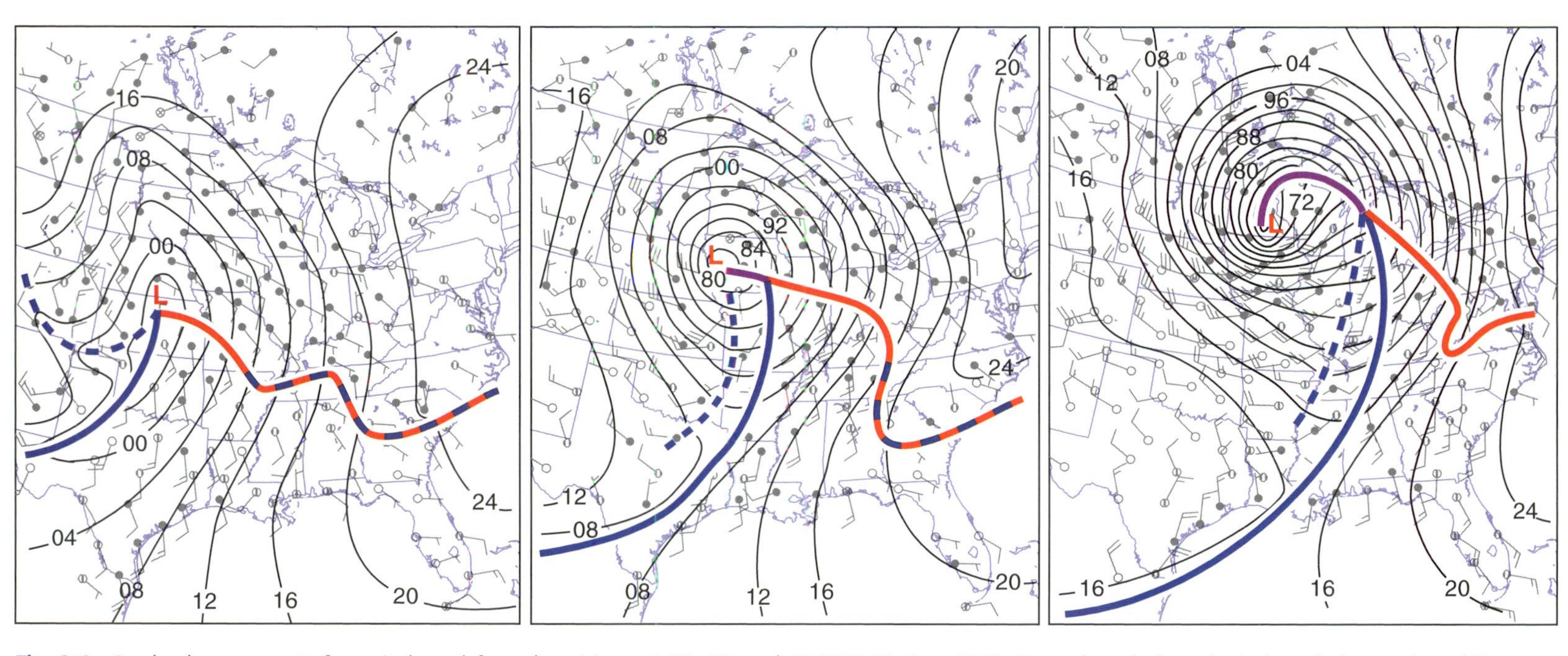

Fig. 8.6 Sea-level pressure, surface winds and frontal positions at 00, 09, and 18 UTC 10 Nov. 1998. Frontal symbols and wind symbols are plotted in accordance with Fig. 8.1. The dashed blue line denotes the secondary cold front. In this figure and in subsequent figures in this section, the frontal positions are defined by a human analyst. The contour interval for sea-level pressure is 4 hPa. [Sea-level pressure and frontal analyses by Lynn McMurdie, figure Courtesy of Jennifer Adams, COLA/IGES.]

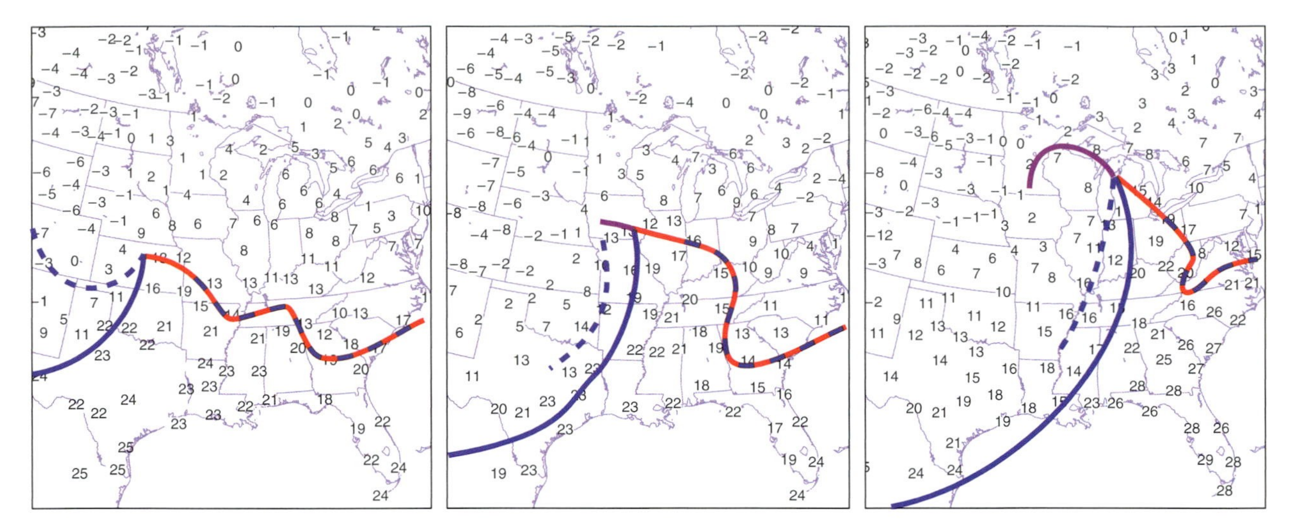

Fig. 8.7 Surface air temperature (in °C) and frontal positions at 00, 09, and 18 UTC 10 Nov. 1998. [Courtesy of Jennifer Adams, COLA/IGES.]

slope of the Colorado Rockies and then northeastward into the center of the surface low. This feature is also embedded in a trough in the sea-level pressure field and causes the surface wind at a fixed station to veer when it passes.

b. Temperature

Figure 8.7 shows the surface air temperatures at the same three map times. The field is represented by raw station data rather than by isotherms, and the positions of the fronts are transcribed from the previous figure. In the southerly flow off the Gulf of Mexico to the east of the cold front, temperatures are relatively uniform, with values in excess of 20 °C extending as far northward as southern Illinois at 09 UTC and values in the teens as far northward as the Great Lakes at 18 UTC. This zone of relatively uniform temperature to the southeast of the surface low is referred to as the *warm sector* of a cyclone. The cold front marks the leading edge of the advancing colder air from the west. In this system, the cold front is not a zero-order discontinuity in the temperature field (i.e., a discontinuity of the temperature itself), but a first-order discontinuity (i.e., a discontinuity in the horizontal temperature gradient). To the east of the cold front the temperatures are relatively homogeneous, while proceeding westward from the front, temperatures drop by 10 °C or more within the first few hundred kilometers. Hence, a cold front can be defined as the warm-air boundary of a *frontal zone* (or *baroclinic zone*) that is advancing in the direction of the warmer air. The passage of a cold front at a station marks the beginning of a period of falling temperatures, heralded by a wind shift.

The November 10, 1998 storm had two cold fronts: a *primary cold front* at the warm air boundary of the frontal zone and a *secondary cold front* within the frontal zone. The two cold fronts show up clearly in the zoomed-in chart for 00 UTC, Nov. 10, 1998, shown in Fig. 8.8. Both fronts are embedded within

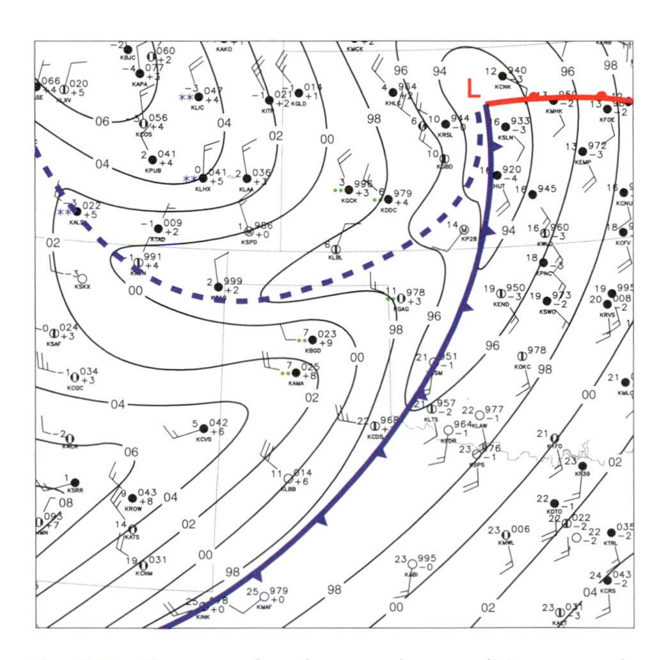

Fig. 8.8 Close-up of surface weather conditions over the southern United States Great Plains at 00 UTC Nov. 10, 1998, showing data plotted using the conventional station model illustrated in Fig. 8.1. [Courtesy of Lynn McMurdie.]

troughs of low pressure and their passage is marked by wind shifts. The passage of the primary cold front marks the onset of the cooling and the passage of the secondary front marks the beginning of an interval of renewed cooling. The secondary cold front marks the leading edge of a band of enhanced baroclinicity (i.e., temperature gradient) within the more broadly defined frontal zone. The passage of such a front marks the onset of renewed cooling.

The more subtle warm front in Fig. 8.7 also marks the warm-air boundary of a baroclinic zone, but in this case the baroclinic zone is advancing northward, displacing the colder air. The passage of a warm front at a fixed station thus is preceded by an interval of rising temperatures. Fronts that exhibit little movement in either direction are labeled as *stationary fronts* and are indicated on synoptic charts as dashed lines with alternating red and blue line segments, as in Figs. 8.6 and 8.7.

From an inspection of Fig. 8.4 it is evident that in the early stages of cyclone development, the cold and warm fronts mark the warm air boundary of the same, continuous baroclinic zone. The cyclone develops along the warm air boundary of the frontal zone, but it subsequently moves away from it, in the direction of the colder air. As this transition occurs, air from within the frontal zone wraps around the cyclone forming the occluded front. It is apparent from Fig. 8.7 that as the occluded front, rendered in purple, approaches a station, surface air temperature rises, and after the front passes the station, the temperature drops. From the standpoint of a stationary observer, experiencing the passage of an occluded front is like experiencing the passage of back-to-back warm and cold fronts except that the temperature changes are usually more subtle because the observer does not experience temperatures as high as those in the warm sector.

Fronts on surface maps are expressions of frontal surfaces that extend upward to a height of several kilometers, sloping backward toward the colder air. Regardless which way the front is moving, air converges toward the front at low levels and the warmer air tends to be lifted up and over the frontal surface along sloping trajectories, as depicted in Fig. 8.9. In the case of a stationary front, warm air may be advancing aloft while the frontal zone air trapped beneath the frontal surface remains stationary. In the case of a cold front, the wind component normal to the front may be in the opposite direction below and above the frontal surface.

Fronts are sometimes pictured as material surfaces, separating air masses characterized by different temperatures and/or humidities, that move about passively in the atmosphere, advected by the winds. This simplistic description ignores the important role of dynamical processes in forming and maintaining fronts. The formation of fronts, a process referred to as *frontogenesis*, involves two-steps. In the first step, the broad, diffuse equator-to-pole temperature gradient tends to be concentrated into frontal zones hundreds of kilometers in width by the large-scale deformation field, as discussed in Section 7.1.3. In the second step, transverse circulations, like those depicted in Fig. 8.9, collapse the low-level temperature gradients within preexisting, still relatively broad frontal zones, down to a scale of tens of kilometers or less.

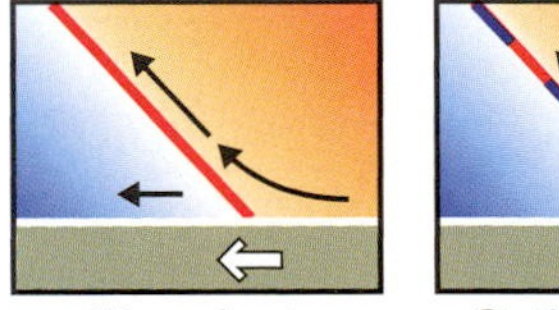

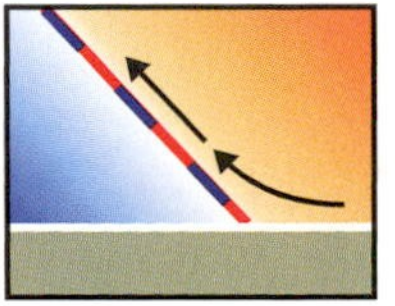

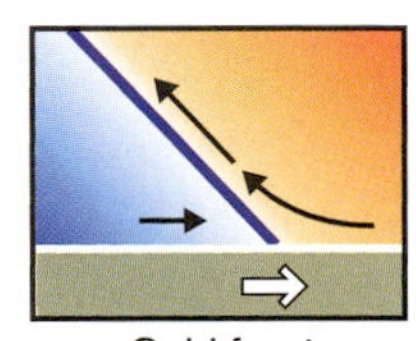

Fig. 8.9 Idealized cross sections through frontal zones showing air motions relative to the ground in the plane transverse to the front. Colored shading indicates the departure of the local temperature from the mean temperature of the air at the same level. (a) Warm front, (b) stationary front with overrunning warm air, and (c) cold front. Heavy arrows at the bottom indicate the sense of the frontal movements.

Lest the role of fronts in mediating surface air temperature be overemphasized, it should be noted that other factors such as time of day, sky cover, altitude of the station, and proximity to large bodies of water can, at times, exert an equally important influence on the temperature pattern. In fact, it is sometimes difficult to locate fronts on the basis of gradients of surface air temperature because

- Over the oceans, surface air temperature is strongly influenced by the temperature of the underlying water, especially in regions where the atmospheric boundary layer is stably stratified.
- In mountainous terrain, large differences in station elevation mask the temperature gradients on horizontal surfaces.
- Unresolved features such as terrain effects, patchy nocturnal inversions, convective storms, and urban heat island effects can raise or lower the temperature at a given station by several degrees relative to that at neighboring stations. Apparent temperature discontinuities associated with these features are sometimes misinterpreted as fronts.

c. Moisture

Frontal zones also tend to be marked by strong gradients in dew point and equivalent potential temperature, especially when the cold air is of continental origin and the warmer air is of marine origin, as is often the case over the eastern United States. In the case study considered in this section, the distributions of temperature and dew point are generally similar. However, during spring and summer, the moisture gradient is often a more reliable indicator of frontal positions than the gradient of surface air temperature because it is less subject to the confounding influence of diurnal variability. For example, during summer over land, the diurnal temperature range at the ground tends to be larger in cool, dry continental air masses than in warm, humid air from off the Gulf of Mexico. Thus, during afternoon it is not uncommon for surface temperatures well behind the cold front to be as high as those on the warm sector of the cyclone, even though there is considerable thermal contrast 1–2 km above the ground. In such situations, the front is more clearly defined in the dew point field than in the temperature field.

Land–sea geometry and terrain features can sometimes give rise to fronts in the moisture field that have no direct relation to extratropical cyclones. For example, during summer, under conditions of southerly low level flow, there often exists a sharp contrast between humid air advected northward from the Gulf of Mexico and much drier air that has subsided along the eastern slopes of the Rockies. The boundary between these marine and continental air masses is referred to as the *dry line*.

d. Hourly observations

Now let us look at the expressions of fronts in hourly surface observations. Hourly pressure, surface wind, temperature, and dew point observations for Gage, Oklahoma, shown in Fig. 8.10, confirm the passage of the primary cold front at 22 UTC (16 LT) Nov. 9, as evidenced by the strong veering of the wind and the onset of an interval of falling temperature and rising sea-level pressure. The passage of the secondary cold front occurred around 03 UTC Nov. 10, when the wind veered and strengthened, the sea-level pressure exhibited a weak minimum, and temperature and dew point began to drop more sharply after having nearly leveled off for several hours.

The time series for Bowling Green, Kentucky, shown in Fig. 8.11, are indicative of a well-defined warm frontal passage around 05 UTC (23 LT), with a

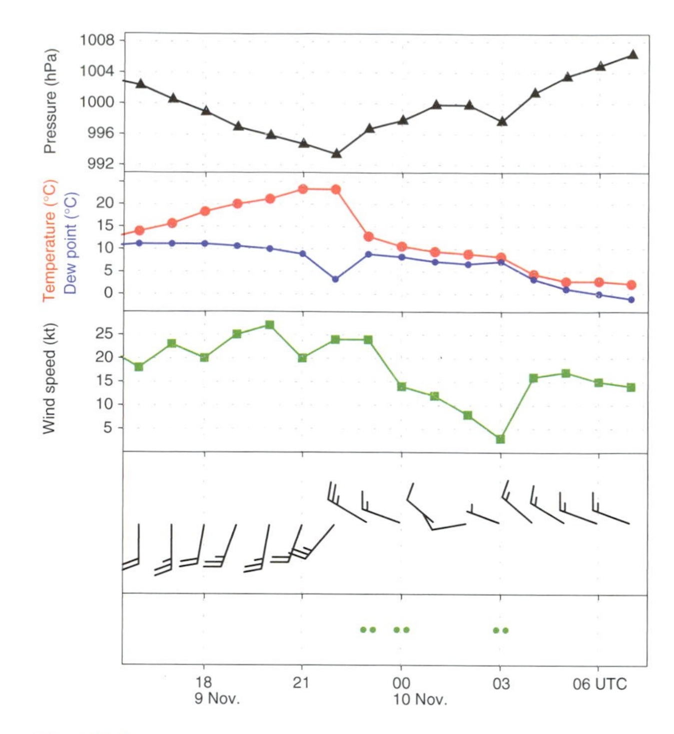

Fig. 8.10 Hourly surface observations at Gage, Oklahoma (KGAG in Fig. 8.36) showing the passage of the primary and secondary cold fronts. The locations of Gage and the other stations for which time series of hourly station observations are shown are indicated in Fig. 8.36 at the end of Section 8.2. [Courtesy of Jennifer Adams, COLA/IGES.]

wind shift from easterly to southerly and a leveling off of the dew point after a prolonged rise. Surface air temperature leveled off an hour later. Pressure continued to drop due to the approach and deepening of the surface low, but the rate of change was smaller than it had been prior to the passage of the front.

As the storm moved northeastward the band of strongest *baroclinicity* (i.e., horizontal temperature gradient) shifted northward into the Great Lakes and the warm front became less distinct. To the east of the Appalachian mountain range the advance of the warm air was delayed by a persistent, topographically induced easterly flow, evident at several of the stations in Fig. 8.6, which advected cooler air southward through the Carolinas and Georgia. By 18 UTC (Fig. 8.6 right panel) the intensifying southerly winds in advance of the approaching cold front scoured out this colder air, resulting in an abrupt northward shift of the warm front at the Earth's surface. Time series of surface variables at Columbia, South Carolina, on the eastern side of the Appalachians (Fig. 8.12) show the warm frontal passage around 16 UTC (11 LT), which was marked by a wind shift and a rapid rise in temperature and dew point. On the west side of the

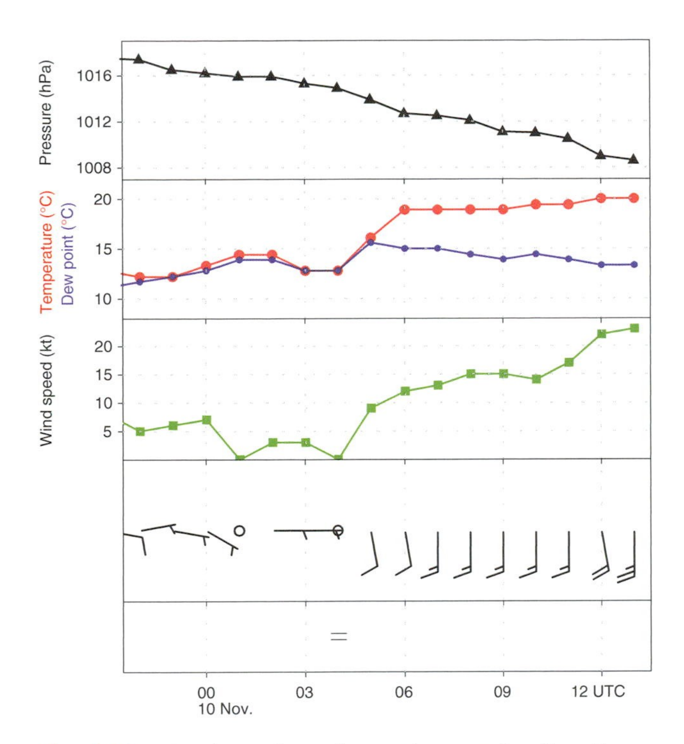

Fig. 8.11 Hourly surface observations at Bowling Green, Kentucky (KBWG in Fig. 8.36) showing the passage of the warm front. [Courtesy of Jennifer Adams, COLA/IGES.]

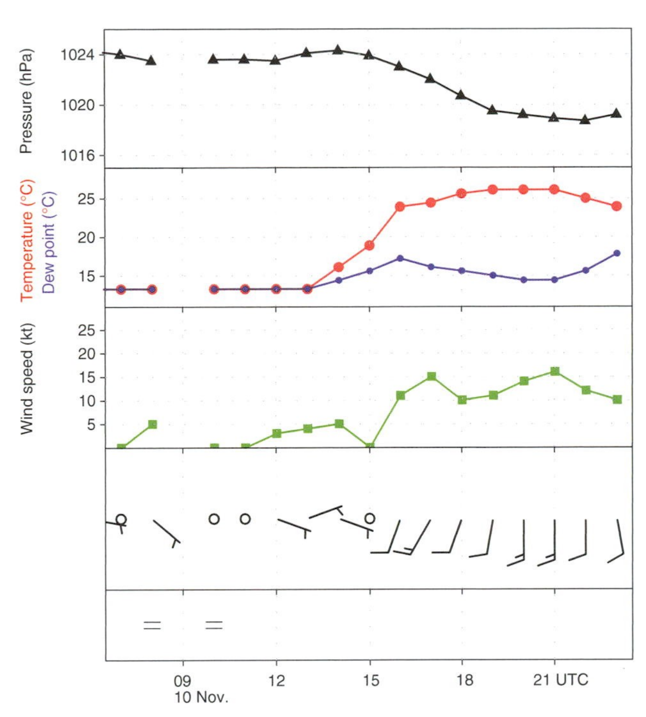

Fig. 8.12 Hourly surface observations at Columbia, South Carolina (KCAE in Fig. 8.36) showing the delayed passage of the warm front. [Courtesy of Jennifer Adams, COLA/IGES.]

Appalachians the northward advance of the warm air occurred 12–18 h earlier.

The time series for Marquette, Michigan (Fig. 8.13) provides an example of the passage of an occluded front. The frontal passage, which occurred around 20 UTC Nov. 20 was attended by a leveling off of the pressure after reaching a remarkably low value of 975 hPa, an abrupt transition from rising to falling temperatures, and a more gradual veering of the wind, from southeasterly to southwesterly. Precipitation ended 3 h before the passage of the front and resumed, in the form of snow showers, 3 h after the frontal passage.

The movement and deepening of the surface low and the advance of the fronts are clearly evident in charts of the 3-h pressure tendency. The example shown in Fig. 8.14 is for the 3-h ending 09 UTC Nov. 10, the time of the middle chart in Figs. 8.6 and 8.7. The falling pressure centered over Iowa reflects both the approach and the deepening of the surface low. The pressure rises behind the cold front reflect the higher density of the colder air that was advancing into territory that was formerly a part of the warm sector of the cyclone. The pressure was falling rapidly ahead of the occluded front, while the pressure was steady behind it, the rising tendency induced by low level cold advection nearly

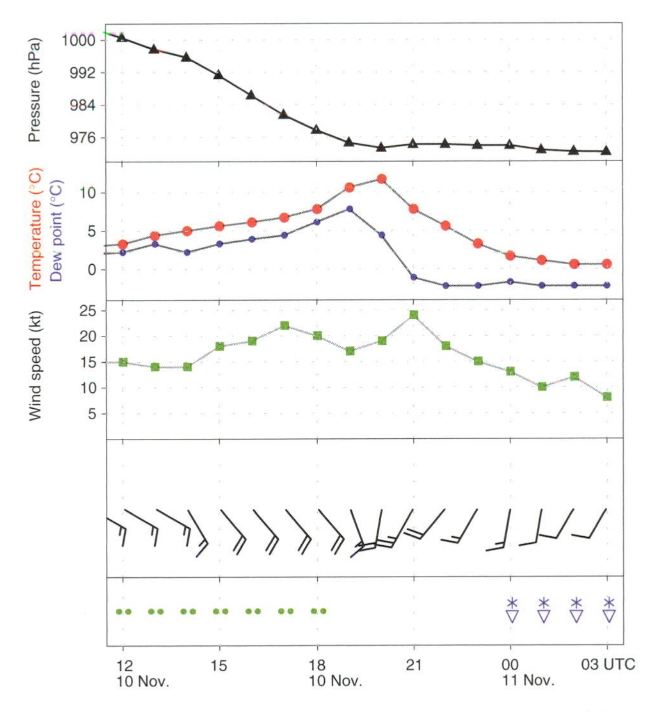

Fig. 8.13 Hourly surface observations at Marquette, Michigan (KMQT in Fig. 8.36) showing the passage of the occluded front. [Courtesy of Jennifer Adams, COLA/IGES.]

balanced by the falling tendency induced by the deepening of the low as it passed to the northwest of the station.

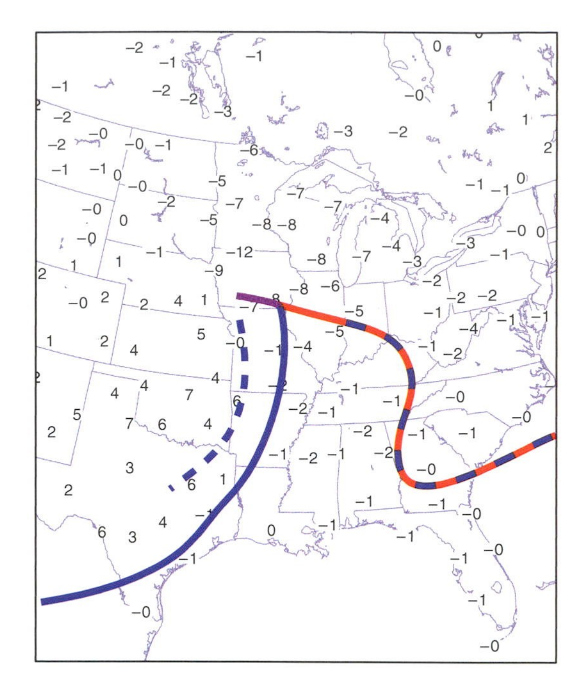

Fig. 8.14 Sea-level pressure tendency (in hPa) for the 3-h interval ending 09 UTC Nov. 10, 1998. Heavy lines denote the frontal positions at this time. [Courtesy of Jennifer Adams, COLA/IGES.]

e. Surface weather

The November 10, 1998 storm produced memorable weather over many parts of the central United States. Figure 8.15 shows the distribution of rain, snow, fog, and thunderstorms at the same times as the charts in Figs. 8.6 and 8.7. At 00 UTC (~18 LT), precipitation was already widespread in the northeast quadrant of the storm, with snow to the north and west and rain to the east and south. With few exceptions, precipitation was light at this time. Many stations to the north of the warm front were reporting fog.

At 09 UTC (03 LT; Fig. 8.15, middle), many of the stations in the Great Lakes region were reporting moderate to heavy rain. Snow reported in southern Minnesota at 00 UTC had changed to rain, reflecting the northwestward advance of the warmer air in the northeast quadrant of the storm, and the approach of the occluded front. The intensity of the snowfall over the Dakotas had increased and rain had changed to snow in eastern Nebraska. With nighttime cooling, fog had become more widespread in the region of the cold air damming over the Carolinas. Although it is not apparent on this map, several of the stations in Illinois and Indiana that reported rain earlier in the evening experienced intermittent fog later in the night, indicative of the passage of the warm frontal zone. Relative to 9 h earlier, more stations along and just behind the cold front were reporting rain at this time.

At 18 UTC (noon LT; Fig. 8.15, right), moderate to heavy snow was falling across much of the northern Great Plains, accompanied by strong winds. Hourly data for Sioux Falls, South Dakota, shown in Fig. 8.16 document blizzard-like conditions prevailing throughout most of the day. Many of the stations farther to the east along the advancing cold front experienced thunderstorms. Although heavy rain continued to be reported at many stations, the broad current of subsiding air circulating around the southern flank of the cyclone (Fig. 8.5) is reflected in the termination of precipitation over Illinois and much of Wisconsin. Marquette, Michigan (Fig. 8.13) experienced a 6-h lapse in precipitation beginning at 18 UTC.

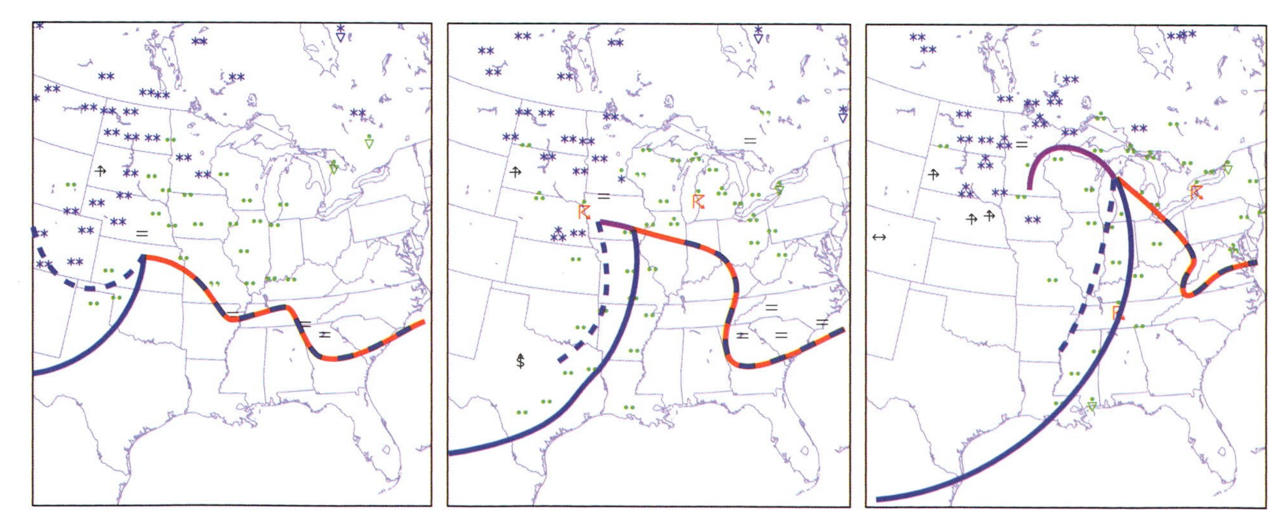

Fig. 8.15 Surface weather observations of rain, snow, fog, and thunderstorms at 00, 09, and 18 UTC 10 Nov. 1998. For plotting conventions see Fig. 8.1. [Courtesy of Jennifer Adams, COLA/IGES.]

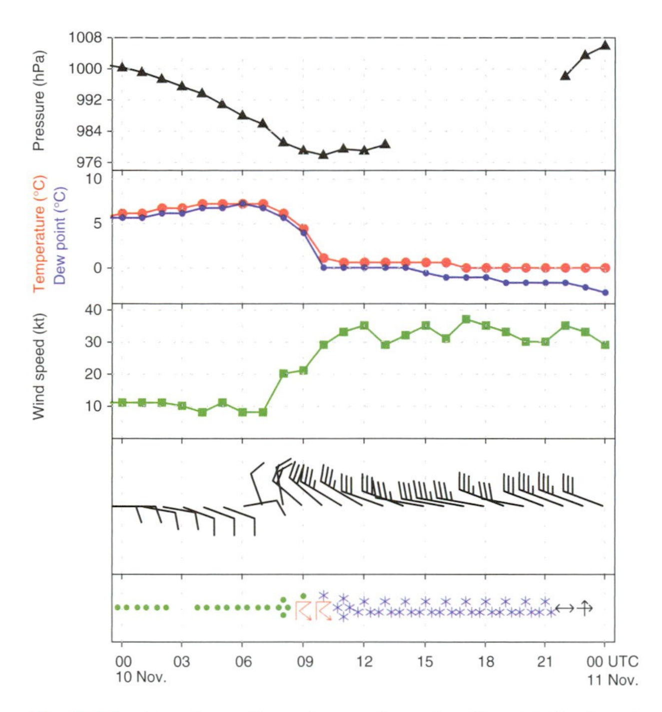

Fig. 8.16 Hourly surface observations for Sioux Falls, South Dakota (KSUX in Fig. 8.36) just to the west of the track of the center of the surface low. Some of the pressure data are missing. [Courtesy of Jennifer Adams, COLA/IGES.]

f. Satellite imagery

Infrared satellite imagery shown in Fig. 8.17 provides a large-scale context for the station observations shown in the previous figures. The first image (0015 UTC) shows a band of clouds with relatively cold tops that accounts for the (mostly light) rain and snow that was falling in the northeastern part of the storm. The warm front at this time corresponds fairly closely to the ragged southern edge of this rain band. Bowling Green, Kentucky, which experienced the passage of the front just a few hours later (Fig. 8.11), was not experiencing rain at this time, but it was located close to the patch of cold cloud tops along the southeastern edge of the band. The narrower and somewhat more coherent band emanating from the cold frontal zone over the Texas Panhandle and extending northward toward the Dakotas was evidently responsible for the light rain at stations in the Texas Panhandle (Figs. 8.8 and 8.15) and Gage, Oklahoma (Fig. 8.10), that was occurring around this time. The well-defined leading edge of this band, which appears as a narrow white line over Texas and as a thin yellow band over Oklahoma, widening into a blue and red "head" near the position of the surface low in Kansas, marks the position of the primary cold front. The patch of colder cloud tops in the northern segment of this band is the embodiment of a broad current of rising air streaming northward above the cold front and wrapping around the developing cyclone to form a comma-shaped "head." It is evident from time-lapse imagery that much of the structure within this air stream can be identified with the spreading of the "anvils" of convective clouds. Over the Texas Panhandle, where the convection along the cold front was shallow at this time, deeper clouds with an associated band of light rain were located, not along the front, but within the frontal zone around 150 km to the northwest of the primary cold front. Hence, these stations experienced

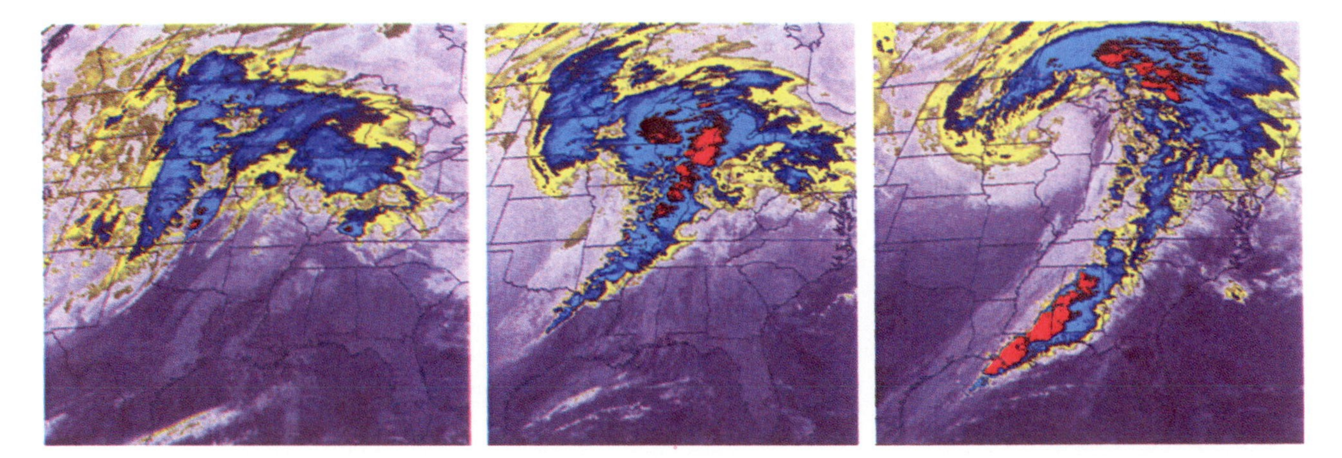

Fig. 8.17 Infrared satellite imagery for 00, 09, and 18 UTC Nov. 10, 1998, based on radiation in the 10.7-μm channel, in which the atmosphere is relatively transparent in the absence of clouds. Radiances, indicative of equivalent black-body temperatures T_E of the Earth's surface or the cloud top, are rendered on a scale ranging from black for the highest values (indicative of cloud-free conditions and a warm surface) with progressively lighter shades of gray indicative of lower temperatures and higher cloud tops. Color is used to enhance the prominence of the coldest (highest) cloud tops in the image.

a period of rain that began a few hours after the frontal passage.

At the time of the second image (09 UTC, Fig. 8.17, middle) the irregularly shaped cloud mass in advance of the warm front has moved northeastward into the southern Great Lakes and has assumed a "comma shape" as it wraps around the northern flank of the intensifying cyclone. The expansion of the area of blue shading over the Dakotas and Nebraska in the "head" of the comma is indicative of a thickening of the cloud deck over that region, consistent with the increase in the rate of snowfall from 00 to 09 UTC (Figs. 8.15 and 8.16). Stations in Illinois and Indiana that were under the cloud deck in the warm frontal zone and experiencing rain at 00 UTC were free of middle and high clouds at 09 UTC, with the clearing coinciding roughly with the passage of the warm front. An important aspect of the development of the cloud pattern in the interval from 00 to the 09 UTC is the pronounced lowering of the cloud top temperatures along the leading edge of the cold frontal cloud band, indicative of the deepening of the convection. As was the case at 00 UTC, this feature coincides with the primary cold front. The remnants of the cloud band that was over the Texas Panhandle at 00 UTC have become aligned with the secondary cold front.

In the final image in Fig. 8.17 at 18 UTC, the "yin-yang" signature in the vertical velocity field (Fig. 8.5, right panel) is clearly evident. The streamer of clouds emanating from the band of convection along the cold front curves cyclonically around the north side of the (now fully developed) cyclone and spirals inward around its western flank, where heavy snow is falling at this time.[2] Meanwhile, the equally pronounced current of darker-shaded subsiding air is wrapping around the southern flank of the cyclone, bringing an end to the precipitation in the areas immediately to the south and east of it. Remnants of the warm frontal cloud band can still be seen advancing northeastward ahead of the system, but they are becoming increasingly detached from the circulation around the cyclone.

Satellite imagery for the water vapor channel, shown in Fig. 8.18, yields additional insights into the structure and evolution of this remarkable storm. At 00 UTC (left) the deep convective clouds in the northern segment of the line of convection along the primary cold front over Kansas are clearly evident. In this respect, this image and the image from the

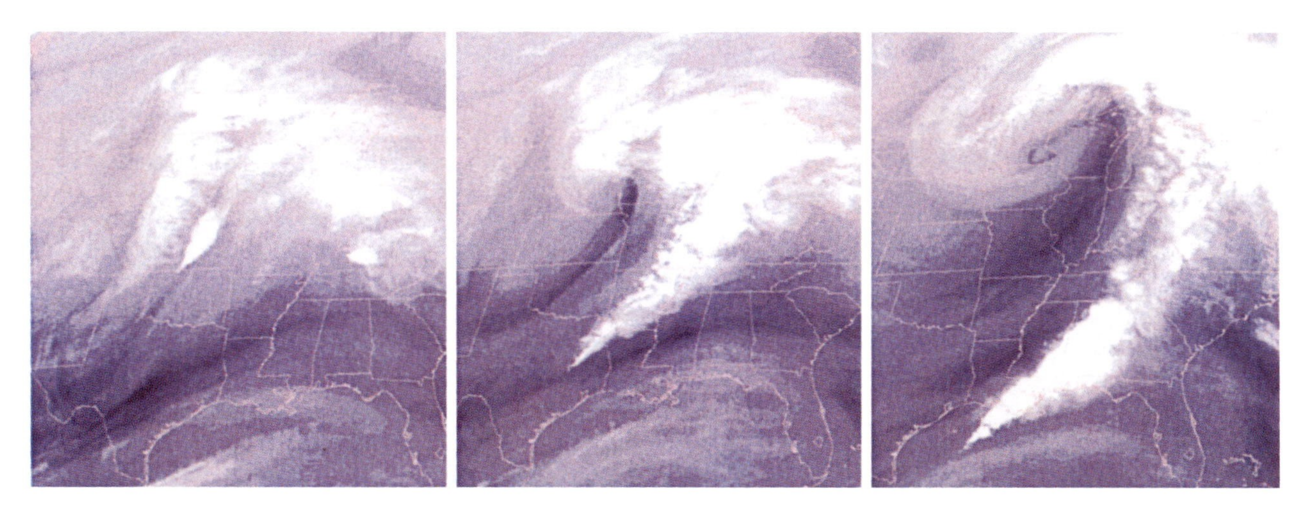

Fig. 8.18 Satellite imagery for 00, 19 and 18 UTC Nov. 10, 1998, based on the 6.7 μm "water vapor channel." The radiances in this band provide a measure of the mid- and upper tropospheric humidity which, in turn, is determined by the air trajectories. Air that has been rising tends to be moist, resulting in a high optical depth, a low equivalent blackbody temperature and a low radiance, and vice versa. Low radiances, indicative of ascent are rendered by the lighter gray shades and high radiances, indicative of subsidence, by the darker shades. The brightest features in the images are clouds with high, cold tops.

[2] The fabled "nor'easters" that bury the eastern seaboard of the United States in half-meter-deep snow from time to time exhibit a structure much like this storm, with the heaviest snowfall in the northwest quadrant of the cyclone. Snowfall tends to be heavier in the coastal storms than in the storm examined in this chapter because much of the ascending air originates over the warm surface waters of the Gulf Stream (Fig. 2.5) where dew points are near 20 °C. The biggest snow producers are storms that slow down or execute tight cyclonic loops during the wrapping-up (or *occlusion*) process, thereby prolonging the interval of heavy snowfall. For an in-depth discussion of nor'easters, see P. J. Kocin and L. W. Uccellini, *Northeast Snowstorms*, Amer. Meteorol. Soc. (2004).

10.7-μm channel, shown in the previous figure, are similar. However, as one follows the front southward through Oklahoma and into Texas, the shallower clouds are masked by the overlying water vapor distribution, which is indicative of a narrow band of subsiding air almost directly above the front.

A prominent feature in the water imagery for the water vapor channel is the so-called *dry slot*, which first becomes apparent in the 09 UTC image and subsequently expands as it wraps around the cyclone. In some storm systems the dry slot is much more prominent in the imagery for the water vapor channel than in that for the 10.7-μm channel. The light gray shading over the Gulf of Mexico is indicative of a deep layer of moist, subtropical air that becomes entrained into the storm as it develops, fueling deep convection along the cold front.

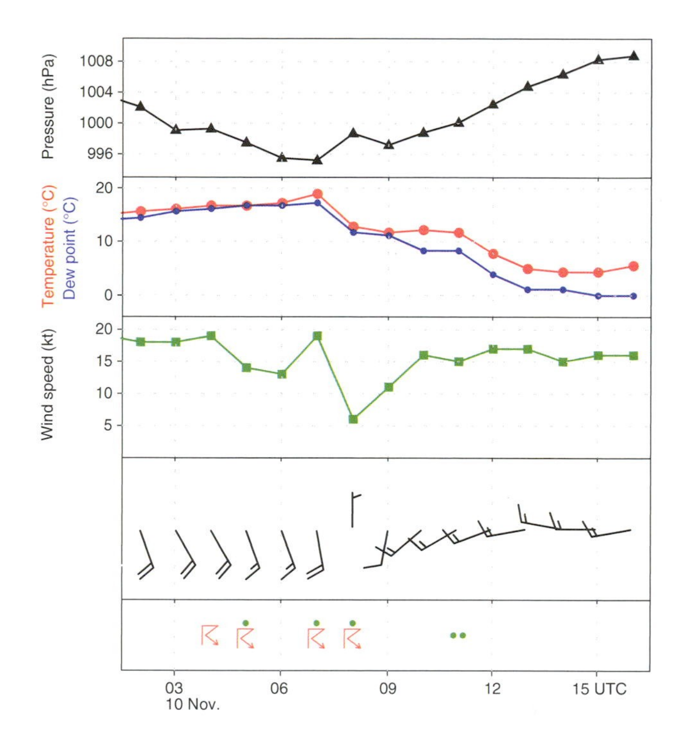

Fig. 8.20 Hourly surface reports for Springfield, Missouri (KSGF in Fig. 8.36) showing the passage of the squall line and primary cold front around 07–08 UTC. [Courtesy of Jennifer Adams, COLA/IGES.]

g. Radar imagery

Composite radar imagery shown in Figs. 8.19 and 8.21 confirms the existence of a narrow, persistent band of deep convection, a feature commonly referred to as a *squall line*, which, in this storm, is coincident with the advancing cold front.[3] Rainfall rates are heaviest along the leading edge of the line and trail off gradually behind it. Figure 8.20 shows hourly surface reports for Springfield, Missouri, located just to the east of the position of the squall line at 0620 UTC, the time of Fig. 8.19. Springfield reported thunder at 04 and 05 UTC and then again at 07 and 08 UTC. The later event marks the passage of the squall line in Fig. 8.19. Some time between the 07 and the 08 observations at Springfield, the temperature dropped by 7 °C and the pressure rose by nearly 4 hPa, signaling a strong cold frontal passage. The most pronounced shift in the wind (from SSW to WSW) did not occur at Springfield until the passage of the secondary cold front around 2 h later, between 09 and 10 UTC, and it was not until that time that the barometer began to rise unequivocally. The drop in temperature and dew point did not resume until between 11 and 12 UTC, when another much weaker rain band passed over the station. The narrow band of dry, subsiding air aloft was also passing over Springfield around 09 UTC (Fig. 8.18, middle).

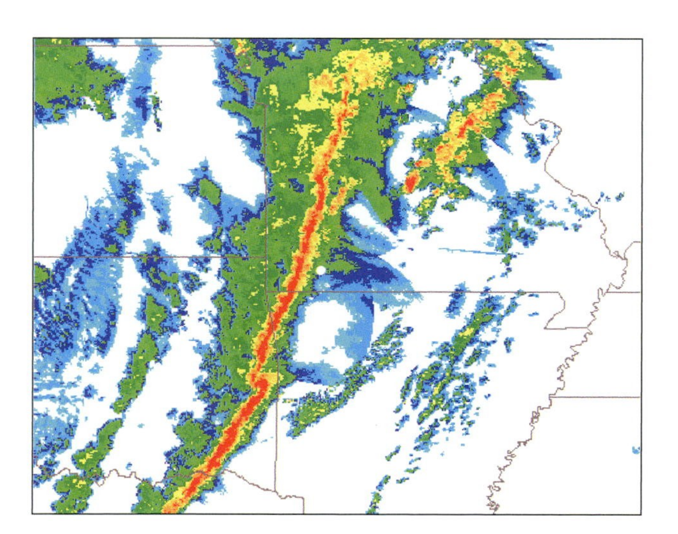

Fig. 8.19 Composite radar image for 0620 UTC Nov. 10, 1998. Estimated rainfall rates increase by about a factor of five from the faintest echoes, rendered in blue, to the strongest echoes, rendered in red. The white circle indicates the location of Springfield, Missouri.

The second radar image shown in Fig. 8.21, based on data taken about 9 h later, still exhibits a well-defined, narrow band of heavy rainfall that is virtually coincident with the position of the primary cold

[3] Squall lines are sometimes observed in the warm sector in advance of, and oriented parallel to, the cold front.

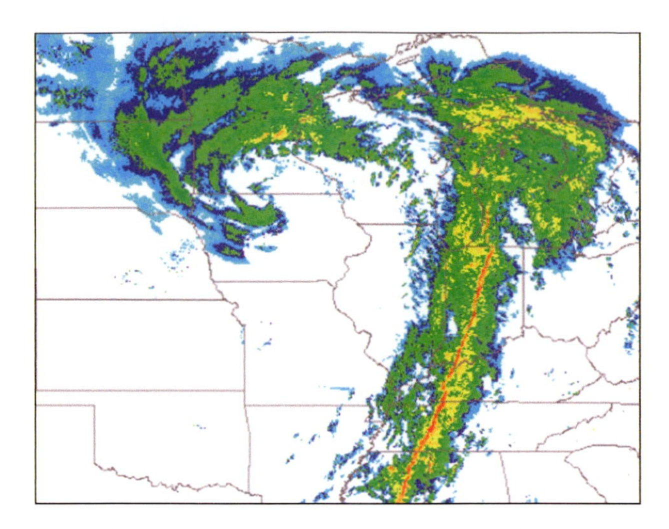

Fig. 8.21 Composite radar image for 1535 UTC Nov. 10, 1998.

front. The major features in the distribution of radar echoes mirror the patterns in the 18 UTC satellite imagery, i.e., the comma-shaped cloud band emerging from the southern tip of the squall line and wrapping around the poleward flank of the cyclone and the slot of dry, relatively cloud-free air intruding from the west and wrapping around the equatorward and eastern flank of the cyclone. This "yin-yang"-like configuration is the signature of intertwined ascending and descending air currents in the vertical velocity field shown in the right-hand side of Fig. 8.5.

8.1.3 Vertical Structure

This subsection examines the vertical structure of this intense baroclinic wave using data formatted in three different ways: upper level charts at selected pressure levels, vertical soundings for selected radiosonde stations, and vertical cross sections.

a. Upper level charts

Figure 8.22 shows a series of upper level charts for 00 UTC Nov. 10, around the time when the associated extratropical cyclone was beginning to deepen rapidly. The corresponding sea level pressure and surface air temperature patterns have already been shown in Figs. 8.6 and 8.7. The 850-hPa height gradients tend to be stronger than the gradients in sea-level pressure (or 1000-hPa height) at the same location.[4] Stronger height gradients are indicative of higher geostrophic wind speeds. Comparing the numbers of wind barbs on the shafts in Figs. 8.6 and 8.22, it is evident that the actual winds are stronger at the 850-hPa level as well. Based on the thermal wind equation (7.20) we know that the strengthening of the westerly component of the wind from the surface to the 850-hPa level is consistent with the prevailing meridional temperature gradient in this layer, with colder air to the north. When the differences in contour intervals in the charts are taken into account, it is readily verified that the geopotential height gradients and wind speeds increase continuously with height up to the 250-hPa level, which corresponds to the level of the jet stream in Fig. 1.11. From 250 to 100 hPa, the highest level shown, the gradients and wind speeds decrease markedly with height.

The 850-hPa isotherms tend to be concentrated within the frontal zone extending from the Great Plains eastward to the Atlantic seaboard and passing through the surface low. To the east of the surface low, southerly winds are advecting the frontal zone northward, whereas to the south of the surface low, westerly winds are advecting it eastward. The frontal zone is particularly tight in the region of cold advection to the south of the surface low, and the temperature is remarkably uniform within a well-defined *warm sector* to the southeast of the surface low. The 850-hPa height contours that pass through the frontal zone exhibit strong cyclonic curvature.[5] Over the Carolinas the warm frontal zone on the 850-hPa chart is positioned quite far to the north of its counterpart on the surface charts. This northward displacement reflects the shallowness of the layer of trapped cool air to the east of the Appalachian mountain range.

Proceeding upward from the 850-hPa to the 250-hPa level, the patterns exhibit notable changes.

[4] From the hypsometric equation it is readily verified that the conventional 4-hPa contour interval for plotting sea-level pressure is roughly comparable to the 30-m contour interval used for plotting the 850-hPa height. Hence, the relative strength of the pressure gradient force (and the geostropic wind) at the two levels can be assessed qualitatively simply by comparing the spacing of the isobars and height contours.

[5] Frontal zones at any level are generally characterized by strong cyclonic vorticity. In stationary frontal zones the vorticity is manifested in the form of shear rather than curvature.

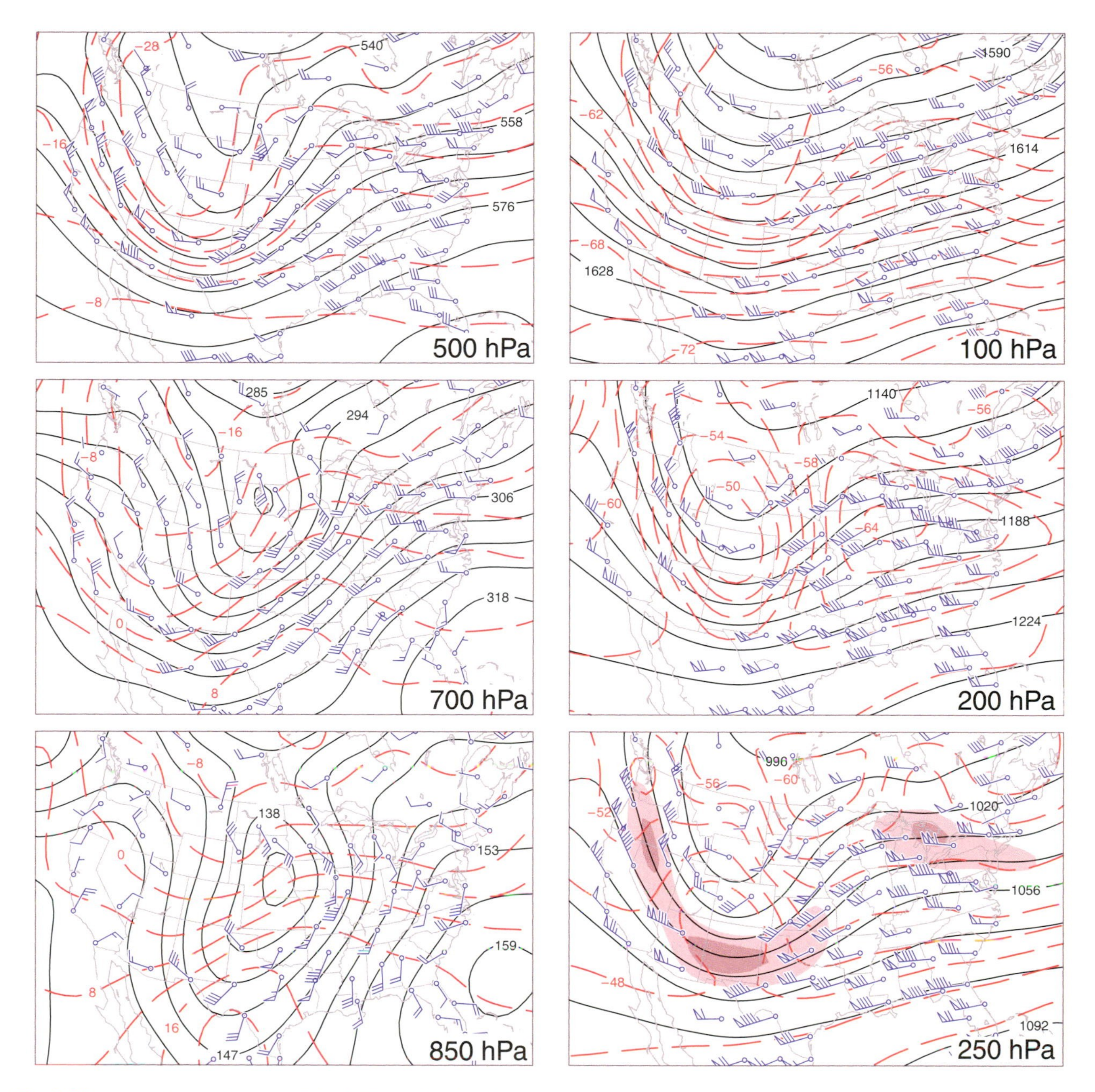

Fig. 8.22 Upper level charts for 00 UTC Nov. 10, 1998, showing geopotential height (black contours), temperature (red contours), and observed winds. Contour interval 30 m for 850- and 700-hPa height, 60 m for 150-hPa height, 120 m for 250- and 200-hPa height, and 60 m for 100-hPa height. The contour interval for temperature is 4 °C in the left panels and 2 °C in the right panels. The shading in the 250-hPa chart are isotachs defining the position of the jet stream. Conventions for plotting wind vectors are shown in Fig. 8.1. [Courtesy of Jennifer Adams, COLA/IGES.]

As noted previously, the geopotential height gradients and the associated geostrophic winds generally increase with height[6] and this tendency is mirrored in the strength of the observed winds. The trough in the geopotential height field tilts westward with height by around 1/4 wavelength from the surface up to the 500-hPa level, but it exhibits relatively little vertical tilt above that level.

[6] In visually comparing the pressure gradients at the various levels, bear in mind that the contour interval doubles from the 700- to the 500-hPa level and doubles again from the 500- to the 250-hPa level.

The temperature contrast between the cold air mass over western Canada and the warm air mass over the subtropics gradually weakens with height. The orientation of the isotherms is much the same at the lowest three levels. Hence the expression of the baroclinic wave in the temperature field does not tilt westward with height. In most extratropical cyclones the baroclinic zones weaken and become progressively more diffuse as one ascends from the Earth's surface to the 500-hPa level. In this particular storm, the warm frontal zone weakens with height but the cold frontal zone remains quite strong up to the 500-hPa level. Upon close inspection it is evident that both warm and cold frontal zones slope backward toward the cold air with increasing height. The horizontal temperature advection within the frontal zones weakens with height as the wind vectors come into alignment with the isotherms. In contrast to the patterns at 850 and 700 hPa, which are highly baroclinic, the structure at the higher levels is more equivalent barotropic.

The temperature patterns in the lower stratosphere are weak and entirely different from those in the troposphere. At these levels (Fig. 8.22, right) the air in troughs in the geopotential height field tends to be warmer than the surrounding air, and the air in ridges tends to be cold. From the hypsometric equation it follows that the amplitudes of the ridges and troughs must decrease with height, consistent with the observations. By the time one reaches the 100-hPa level the only vestige of the baroclinic wave that remains is the weak trough over the western United States.

Now let us examine the structure of the tropopause in this high amplitude baroclinic wave. Vertical temperature profiles for stations in the trough and ridge of the wave are contrasted in Fig. 8.23. The profile for Denver, Colorado, which is located near the center of the 250-hPa trough, is relatively cold throughout the depth of the troposphere. The tropopause is marked by a sharp discontinuity in lapse rate around the 350-hPa (8 km) level, with a transition to more isothermal conditions above. In contrast, the profile for Davenport, Iowa, which is located in the 250-hPa ridge, exhibits a much colder and even sharper tropopause ~180 hPa (12.5 km). The tropopause temperature at this time was 20 °C colder at Davenport than at Denver. Stations such as Amarillo, Texas, which lie close to the axis of the jet stream, exhibit a more gradual decline in the lapse rate as one ascends from the troposphere into the stratosphere. The tropopause is not as well-defined in the Amarillo sounding as it is in the other two soundings.

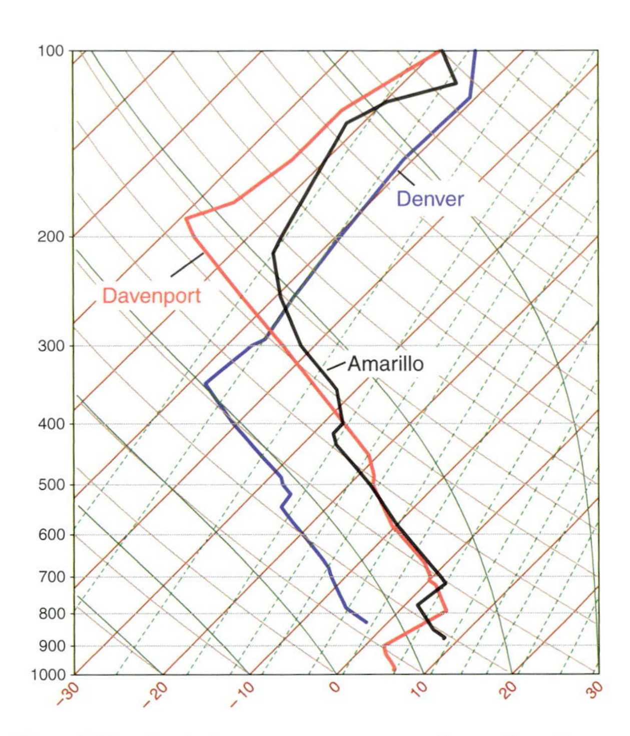

Fig. 8.23 Vertical temperature soundings for Denver, Colorado (blue line), Amarillo, Texas (black line), and Davenport, Iowa (red line) at 00 UTC Nov. 10, 1998, plotted on a skew T - $ln\ p$ diagram. [Courtesy of Jennifer Adams, COLA/IGES.]

Figure 8.24 shows how the tropopause structure relates to the lower tropospheric temperature

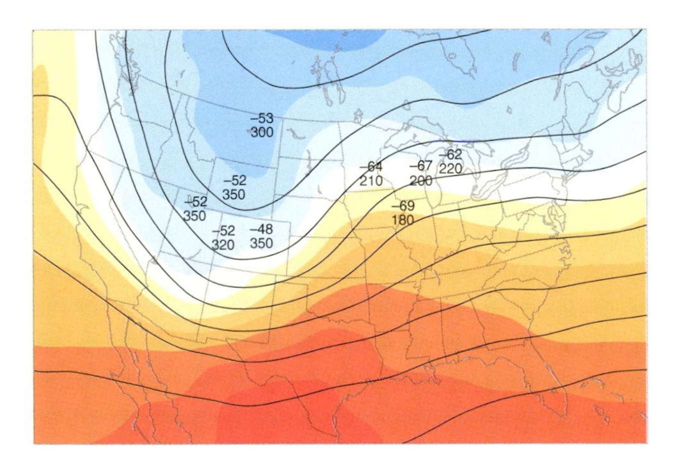

Fig. 8.24 Height contours for the 250-hPa surface superimposed on 1000- to 500-hPa thickness (indicated by colored shading) as in Fig. 8.3 for 00 UTC Nov. 10, 1998. For selected stations, tropopause temperatures (TT in °C) and pressures (PPP in hPa) are plotted (TT/PPP). [Courtesy of Jennifer Adams, COLA/IGES.]

pattern and the flow at the jet stream level (250 hPa). The ridge and trough in the 250-hPa height pattern correspond, respectively, to the axes of the warmest and coldest air in the 1000- to 500-hPa thickness pattern, and the jet streams overlie the baroclinic zones, with colder air lying to the left. The depression of the tropopause in the vicinity of the 250-hPa trough, and directly above the cold air mass in the lower troposphere, is indicative of large-scale subsidence, as required by the continuity of mass [Eq. (7.39), Fig. 7.18]; i.e., as the cold air mass in the lower troposphere spreads out horizontally (as evidenced by the rapid advance of the surface cold front), the air above it must sink. The relatively high tropopause temperatures observed at stations deep within the cold air mass are due to the adiabatic warming of the subsiding air. At the 250-hPa level, relative humidities at these stations (not shown) were in the 25–40% range, consistent with a recent history of subsidence. In contrast, within the relatively warm, ascending air stream over the northern Great Plains, the tropopause is elevated; tropopause temperatures are relatively cold, and relative humidities are ~80%. Figure 8.24 also suggests a possible explanation of why the tropopause in the Amarillo sounding is not as clear as in the soundings for the other two stations shown in Fig. 8.23. Note that Amarillo lies along the axis of the jet stream, where the tropopause is like a vertical wall, with tropospheric air on the anticyclonic side and stratospheric air on the cyclonic side.

b. Frontal soundings

This subsection examines vertical profiles of wind, temperature, and dew point in the lower troposphere at representative stations in different sectors of the developing cyclone. Soundings for two stations within the frontal zone are shown in Fig. 8.25. Amarillo lies within the segment of the frontal zone to the south of the surface low. In the Amarillo sounding the wind *backs* (i.e., turns cyclonically) with increasing height in the layer extending from the surface nearly up to the 700-hPa level. The backing is strongest in the inversion later extending from 780 to 720 hPa. Based on the thermal wind equation [Eq. (7.20)], backing implies cold advection. The layer of strong backing thus corresponds to the cold frontal zone and the cold front intersects the sounding at the top of the layer of strong backing at ~720 hPa. Davenport lies in an analogous position within the frontal zone to the east of the surface low, where the warm air is being advected northward by the southerly component of the wind. In the Davenport sounding the wind *veers* (turns anticyclonically) with increasing height, indicative of warm advection, from the surface up to 800 hPa, which marks the position of the warm front. In both the soundings shown in Fig. 8.25, the frontal zone corresponds to a layer of strong vertical wind

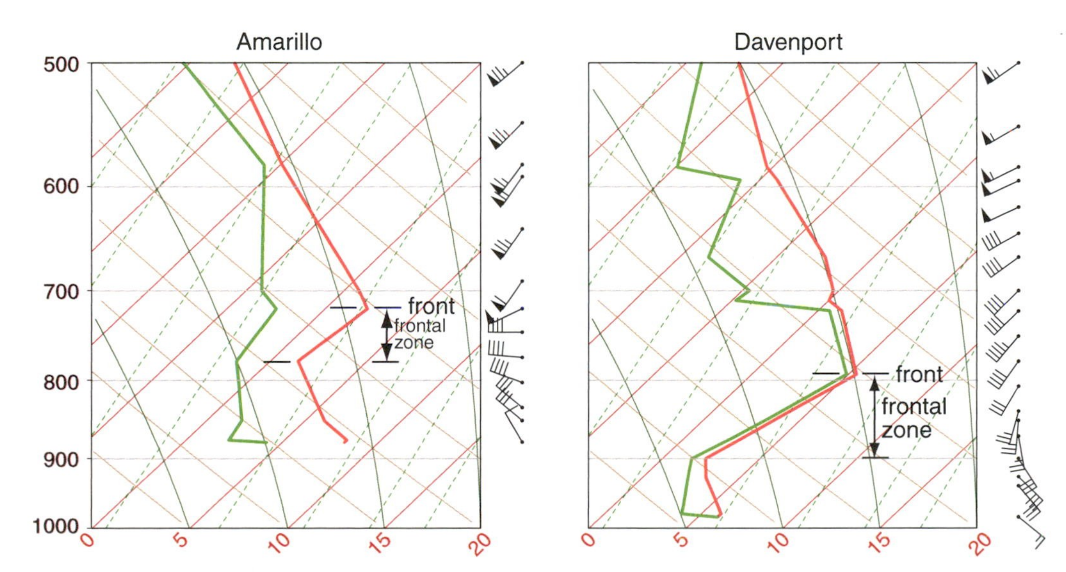

Fig. 8.25 Soundings of wind, temperature (red lines), and dew point (green lines) at 00 UTC Nov. 10, 1998 at Amarillo, Texas (left) in the cold frontal zone and Davenport, Iowa (right) in the warm frontal zone. [Courtesy of Jennifer Adams, COLA/IGES.]

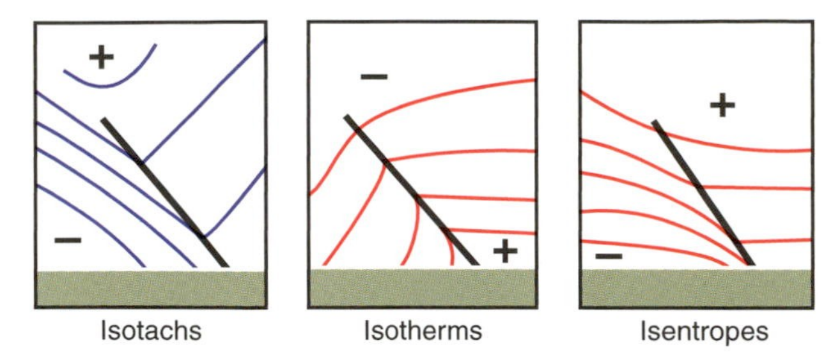

Fig. 8.26 Idealized representations of a sloping frontal zone looking downwind at the jet stream level in the northern hemisphere (or upwind in the southern hemisphere). (Left) The wind component directed normal to the section: values into the section are denoted as positive. (Middle) Temperature. (Right) Potential temperature. Plus (+) and minus (−) signs indicate the polarity of the gradients (e.g., in the left the wind component into the section increases with height).

shear and high static stability, as evidenced by the presence of temperature inversions.

The idealized frontal cross sections shown in Fig. 8.26 are helpful in interpreting the frontal soundings. Consistent with the definition in Section 8.1.2b, at any given level the front marks the warm air boundary of the frontal zone. Consistent with Fig. 8.22, the front slopes backward, toward the colder air, with increasing height. Consistent with Fig. 8.25 the front marks the top of the frontal zone; it is characterized by high static stability and strong vertical wind shear. The frontal zone is depicted as being sharpest at the surface.

Soundings for stations located in the warm sector of the developing cyclone (not shown) exhibit little turning of wind with height other that the frictional veering just above the surface, and relatively little increase of wind speed with height. Stations in the cold sector to the west or northwest of the surface low exhibit relatively little turning of the wind with height, but in some storms they reverse direction, from northeasterly at low levels to southwesterly aloft.

c. Vertical cross sections

Vertical cross sections are the natural complement to horizontal maps in revealing the three-dimensional structure of weather systems. A generation ago, the construction of cross sections was a labor-intensive process that involved blending temperatures and geostrophic winds derived from constant pressure charts with wind and temperature data for intermediate levels extracted from soundings for stations lying along the section. Interpolating fields in the gaping holes between data points could be a formidable challenge, even for the skilled analyst. With today's high-resolution gridded data sets generated by sophisticated data assimilation schemes, all the analyst need do to generate a section is to specify the time and orientation and the fields to be included.

The two most widely used variables in vertical cross sections are temperature (or potential temperature) and geostrophic wind. The sections are usually oriented normal to the jet stream in which case, isotachs of the wind component normal to (or through) the section reveal the location and strength of the jet stream where it passes through the plane of the section, and they often capture the zones of strongest vertical wind shear, where patches of clear air turbulence tend to be concentrated. If the flow through the section is not strongly curved, then the vertical shear of the geostrophic wind component normal to the section and the horizontal temperature gradient along the section are approximately related by the thermal wind equation

$$\frac{\partial V_{gn}}{\partial p} \simeq -\frac{R}{fp}\frac{\partial T_n}{\partial s} \tag{8.1}$$

where V_{gn} is the geostrophic wind component into the section and T_n is temperature in the plane of the section, with the horizontal coordinate s defined as increasing toward the right. Hence, at any point in the section the horizontal temperature gradient is directly proportional to the vertical wind shear. It follows that the horizontal spacing of the isotherms is directly proportional to the vertical spacing of the isotachs plotted in the section. For example, in regions of the section in which the flow is barotropic, the isotherms (or isentropes) are horizontal (i.e., $\partial T/\partial s = 0$) and the isotachs are vertical (i.e., $\partial V_{gn}/\partial p = 0$). These conditions also apply locally in the core of a jet stream. Near the tropopause, the vertical wind shear and the horizontal temperature gradient both undergo a sign reversal at the same level, at which point the isotachs are vertical and the isotherms are horizontal.

The same relationships apply to vertical wind shear and the horizontal gradient of potential temperature. Vertical cross sections for temperature and potential temperature tend to be somewhat different in appearance because temperature in the troposphere usually decreases with height, whereas potential temperature increases with height, and in the stratosphere $\partial\theta/\partial p$ is always strong and negative, whereas $\partial T/\partial p$ is often weak and may be of either sign.

Another variable that is frequently plotted in vertical cross sections is isentropic potential vorticity,

$$PV \equiv -g\,(\zeta_\theta + f)\,\frac{\partial\theta}{\partial p} \tag{8.2}$$

as defined in Section 7.2.10. PV is a conservative tracer that serves as a marker for intrusions of stratospheric air into the troposphere in the vicinity of the jet stream. Air that has resided in the stratosphere for any appreciable length of time acquires high values of static stability $-\partial\theta/\partial p$ by virtue of the vertical gradient of diabatic heating at those levels. Hence, the potential vorticity of stratospheric air tends to be much higher than that of tropospheric air. When a layer of stratospheric air is drawn downward into the troposphere, columns are stretched in the vertical, pulling the potential temperature surfaces apart, thereby causing the static stability to decrease. Conservation of potential vorticity requires that the vorticity of the air within the layer becomes more cyclonic as it is stretched in the vertical.

Now let us consider two examples of vertical cross sections. The first example, shown in Fig. 8.27, is oriented perpendicular to the cold front and jet stream over the southern Great Plains at 00 UTC. The viewer is looking downstream (i.e., northeastward): the colder air is toward the left. In denoting positions along the section, we will be referring to a series of imaginary stations, indicated by letters A, B . . . etc. along the baseline of the section. The front at the Earth's surface is at C and the frontal zone is apparent to the west of station C as a wedge of sloping isotherms (i.e., the red contours) and strong vertical wind shear, as indicated by the close spacing of the isotachs (blue contours) in the vertical. Consistent with the idealized depictions in Fig. 8.26, the front (i.e., the warm air boundary of the frontal zone) slopes backward, toward the cold air, with increasing height. The front becomes less clearly defined at levels above 700 hPa. The jet stream with a maximum wind speed of nearly 50 m s^{-1} passes through the section above station C at the 250-hPa level.

The tropopause is clearly evident in Fig. 8.27 as a discontinuity in the vertical spacing of the isotherms: in the troposphere the isotherms are closely spaced in the vertical, indicative of strong lapse rates, while in the stratosphere, they are widely spaced, indicative of nearly isothermal lapse rates. Consistent with Fig. 8.24, the tropopause is low and relatively warm on the cyclonic (left) side of the jet stream and high

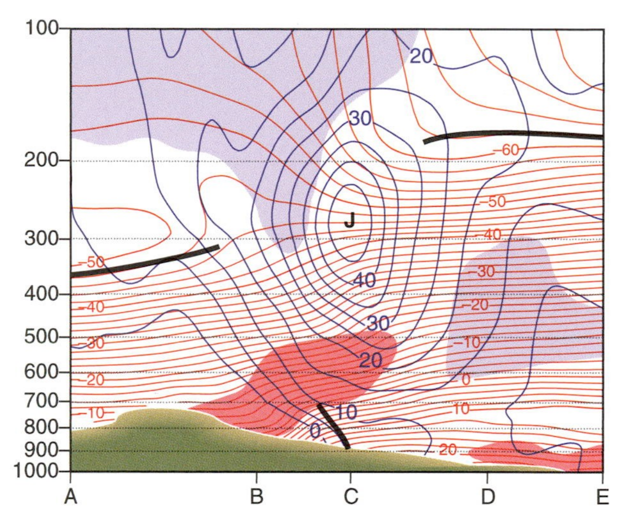

Fig. 8.27 Vertical cross section of wind and temperature for 00 UTC Nov. 10, 1998. This section extends from Riverton, Wyoming to Lake Charles, Louisiana (KRIW to KLCH; see Fig. 8.36). Temperature is indicated by red contours, and isotachs of geostrophic wind speed normal to the section, with positive values defined as southwesterly winds directed into the section, are plotted in blue. Regions with relative humidities in excess of 80% are shaded in red and below 20% in blue. Heavy black lines indicate positions of the surface-based fronts and the tropopause. The orientation of the section relative to the front is indicated in Fig. 8.36 at the end of this section. [Courtesy of Jennifer Adams, COLA/IGES.]

and cold on the anticyclonic (right) side. An aircraft flying along the section at the jet stream (250-hPa) level, passing from the warm side to the cold side of the lower tropospheric frontal zone, would pass from the upper troposphere to the lower stratosphere while crossing the jet stream. Entry into the stratosphere would be marked by a sharp decrease in relative humidity and an increase in the mixing ratio of ozone. One would also observe a marked increase in the PV of the ambient air: a consequence of both the increase in static stability $-\partial\theta/\partial p$ (i.e., compare the lapse rates at the 250-hPa level at stations D and B) in combination with a transition from weak anticyclonic (negative) relative vorticity ζ on the equatorward flank of the jet stream to quite strong cyclonic (positive) relative vorticity on the poleward flank.

Figure 8.28 shows a vertical cross section normal to the frontal zone 12 h later. In this section the red contours are isentropes (rather than isotherms), and high values of PV, indicative of stratospheric air, are indicated by shading. The jet stream is stronger in this section than in the previous one, with peak wind speeds of ~60 m s^{-1}. Immediately beneath the jet

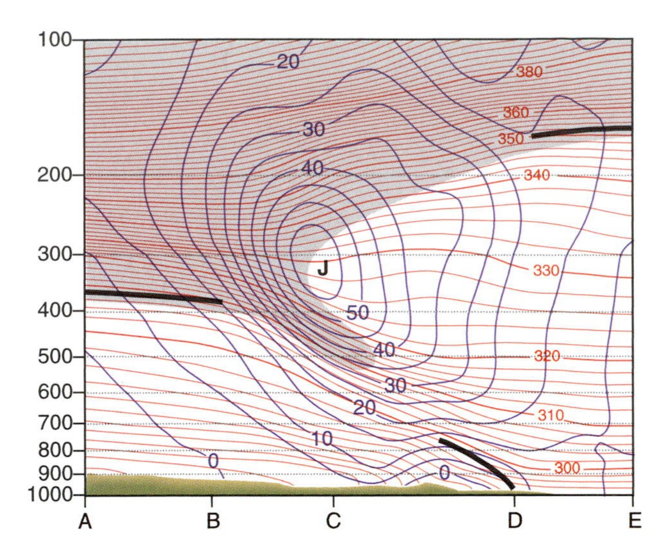

Fig. 8.28 Vertical cross section of wind and potential temperature for 12 UTC Nov. 10, 1998. This section extends from North Platte, Nebraska to Jackson, Mississippi (KLBF to KJAN; see Fig. 8.36). Potential temperature is indicated by red contours, and isotachs of geostrophic wind speed normal to the section are plotted in blue with positive values defined as southwesterly winds directed into the section. The region in which isentropic potential vorticity exceeds 10^{-6} K m^2 s^{-1} kg^{-1} is indicated by shading. Heavy black lines represent the position of the surface-based fronts and tropopause. [Courtesy of Jennifer Adams, COLA/IGES.]

stream is a layer characterized by very strong vertical wind shear. Consistent with the thermal wind equation, the temperature gradient in this layer is very strong, with colder air to the left. The air within this *upper level frontal zone* exhibits strong cyclonic relative vorticity by virtue of its cyclonic shear $\partial V_n/\partial s$ and is also characterized by strong static stability, as evidenced by the tight vertical spacing of the isentropes. It follows that the PV of the air within this upper level frontal zone is much higher than that of typical air parcels at this level and the air within the core of the jet stream. Accordingly, the PV contours are folded backward beneath the jet stream so as to include the upper tropospheric frontal zone within the region of high PV. Since the PV contours define the boundary between tropospheric air and stratospheric air, it follows that the air within the upper part of the frontal zone is of recent stratospheric origin.

Such upper level frontal zones and their associated *tropopause folds* are indicative of extrusions of stratospheric air, with high concentrations of ozone and other stratospheric tracers, into the upper troposphere. Sometimes tropopause folding is a reversible process in which the high PV air within the fold is eventually drawn back into the stratosphere. At other times the process is irreversible: the extruded stratospheric air becomes incorporated into the troposphere, where it eventually loses its distinctively high PV. The extrusion in Fig. 8.28 was evident 12 h earlier in north–south sections through the core of the jet stream over New Mexico, where the jet stream was strongest at that time. With the development of the cyclone, the stratospheric air was drawn downward and northeastward over the cold frontal zone, becoming an integral part of the "dry slot" in the water vapor satellite imagery (Fig. 8.18). The resulting injection of air with high PV into the environment of the cyclone contributed to the remarkable intensification of this system during the later stages of its development.

8.1.4 Air Trajectories

This subsection provides a *Lagrangian* perspective on extratropical cyclones. Lagrangian trajectories are constructed from three-dimensional velocity fields at several successive times separated by an interval of a few hours. The trajectories can either be tracked forward in time from prescribed positions at some initial time t_0 or can be tracked backward in time from prescribed positions at a final time t_f. Since convective motions are not explicitly represented in the synoptic charts, their role in the vertical transport of air parcels must either be ignored, or parameterized in some way.

Figure 8.29 shows a set of trajectories whose end points lie within the cloud shield of a mature extratropical storm. Air parcels ascending along trajectories like these supply most of the moisture that falls as rain and snow in these storms. The trajectories are depicted in coordinates moving northeastward with the center of the surface low, where the coordinate transformation is accomplished by subtracting out the movement of the surface low from the horizontal velocity in each time step of the trajectory calculation. Air parcels such as A that make up the eastern part of the cloud shield can be traced back to low levels in the warm sector of the cyclone; those such as B and C that comprise the northern flank came from the warm frontal zone farther to the north and east. The anticyclonic curvature of trajectories A, B, and C is a consequence of the veering of the wind with height in the region of warm geostrophic temperature advection in advance of the surface low.

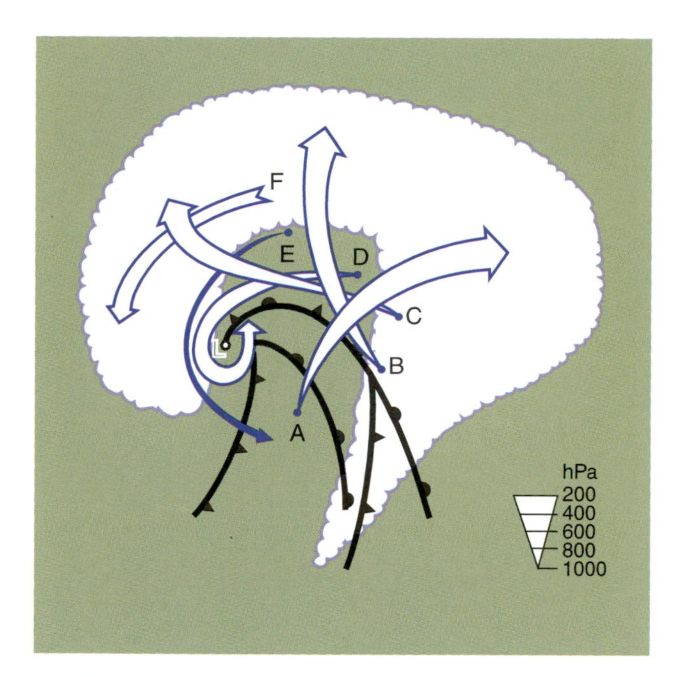

Fig. 8.29 Family of three-dimensional trajectories in an intense extratropical cyclone, as inferred from a high-resolution grid point dataset for an actual storm over the North Atlantic. The trajectories are shown in a coordinate system moving with the cyclone. Two different frontal positions are shown: the lower one is for an earlier time when the configuration is that of an open wave and the upper one is for a later time when the cyclone is in its mature stage and exhibits an occluded front. The configuration of the cloud shield and the position of the surface low correspond to the later time. The width of the arrows gives an indication of the height of the air parcel in accordance with the scale at the lower right. [Adapted from *Mon. Wea. Rev.*, **120** (1995) p. 2295.]

Trajectories D, E, and F are a bit more complicated. D starts out in the warm frontal zone, ascends, and becomes saturated as it circulates around the rear of the surface low within the inner part of the cloud shield. Trajectory D subsequently descends and becomes unsaturated as it recurves northward behind the occluded front. Trajectory E does not ascend appreciably: it passes underneath the head of the comma-shaped feature and into the rear flank of the cold frontal zone. The comma-shaped feature on the western side of the cloud shield is made up of parcels with trajectories such as F, which can be traced back midlevels in the cold air mass to the north of the storm.

Hence, the ascending trajectories describe a fan-like spreading of the rising air. The rising air parcels exhibit a continuum of equivalent potential temperatures, with that for the southernmost trajectory A being highest and those B–F being progressively lower. Wherever the trajectories that start out at the surface cross, the colder one passes beneath the warmer one.

Figure 8.30 shows a bundle of descending trajectories in the cloud-free region to the rear of the cyclone. These air parcels can be traced back to the northwesterly flow in the vicinity of the jet stream level behind the trough of the wave. The trajectories start out vertically stacked and spread out as they descend behind the cold front, with the air parcels warming at a rate close to the dry adiabatic lapse rate. The fan-shaped surface formed by the spreading trajectories slopes upward toward the north. The air parcels that started near the top of the bundle curve cyclonically around the surface low, forming the "dry slot" at the northern part of the fan-shaped surface. The trajectories in the dry slot do not descend all the way to the Earth's surface: they typically level off as they pass over the occluded front and begin to ascend as they approach the cloud shield to the north. However, they are so dry that they remain unsaturated as they cross over the top of the cloud shield.

The air parcels that started near the bottom of the bundle at the jet stream level curve anticyclonically around the surface high, forming the southern part of the fan-shaped surface depicted in Fig. 8.30. The trajectories on this side of the surface descend low

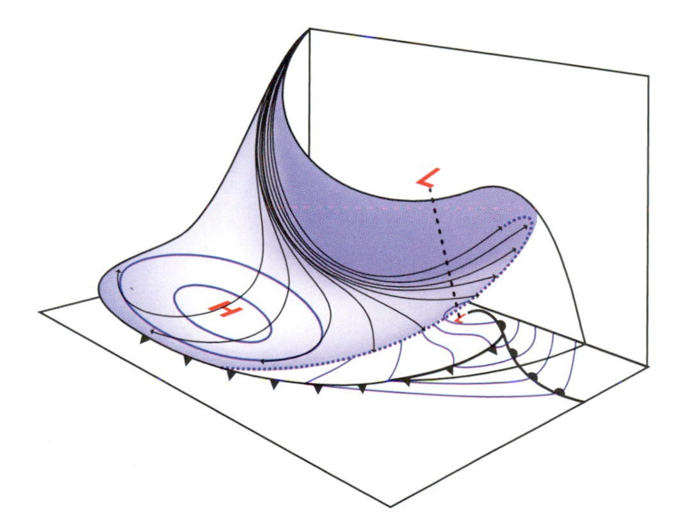

Fig. 8.30 Idealized 24-h trajectories for selected air parcels in the descending branch of an intense extratropical cyclone similar to the one examined in the case study in this section. The trajectories start and end at about the same time. Black arrows are the trajectories and blue contours are isobars of sea-level pressure. [From Project Springfield Report, U.S. Defense Atomic Support Agency, NTIS 607980 (1964).]

enough so that parcels may be entrained into the boundary layer of the modified polar air mass advancing southward behind the cold front. If a pronounced tropopause fold is present at the upstream end of the trajectories at the jet stream level, stratospheric air may be entrained into this anticyclonic air stream. Such relatively rare and brief incursions of stratospheric air into the boundary layer are marked by extremely low relative humidities and high ozone concentrations in surface air.

8.1.5 In Search of the Perfect Storm

For nearly a century meteorologists have argued about what constitutes "the perfect storm": "perfect," not in the sense of most catastrophic, but most typical of the cyclones generated by baroclinic instability in the real atmosphere. The case study featured in this section conforms in most respects to the classical *Norwegian polar front cyclone model* devised by J. Bjerknes and collaborators of the Bergen School during the 1920s for interpreting surface weather observations over the eastern North Atlantic and Europe. Characteristic features of the archetypical Norwegian polar front cyclone, summarized in Fig. 8.31, include the strong cold front, the weaker occluded front, and the comma-shaped cloud shield.

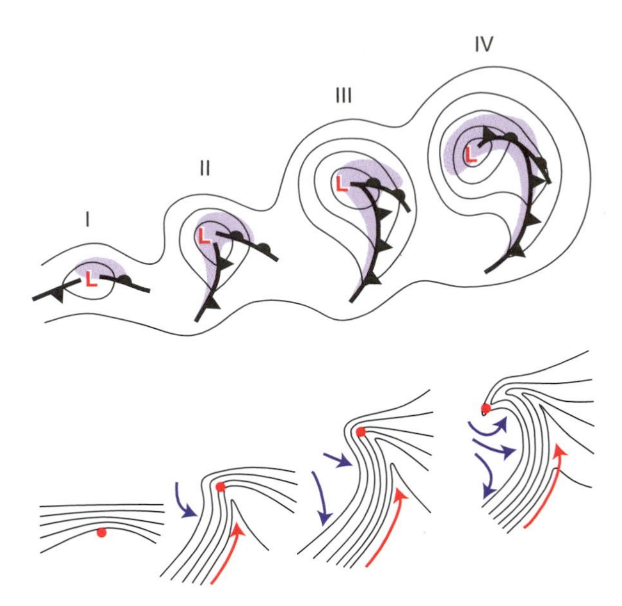

Fig. 8.31 Schematic showing four stages in the development of extratropical cyclones as envisioned in the Norwegian polar front cyclone model. Panels I, II, III and IV represent four successive stages in the life cycle. (Top) Idealized frontal configurations and isobars. Shading denotes regions of precipitation. (Bottom) Isotherms (black) and airflow (colored arrows) relative to the moving cyclone center (red dot). Red arrows indicate the flow in the warm sector, and blue arrows indicate the flow in the cold air mass. Frontal symbols are listed in Table 7.1. [Adapted from *Mon. Wea. Rev.*, **126** (1998) p. 1787.]

Some of the most intense cyclones that develop over the oceans during wintertime exhibit significant departures from this well established paradigm. Their spiral cloud bands, as revealed by satellite imagery, are coiled up more tightly about the center of the surface low than the cloud shield in our case study: see, for example, Fig. 1.12. Unlike the storm in the case study, mature cyclones that exhibit this tightly coiled structure tend to be *warm core*: i.e., the air in the center of the surface low is warmer than the surrounding air on all sides.

Figure 8.32 shows an idealized schematic of the structure and evolution of these tightly coiled storms, as deduced from data from instrumented aircraft flying through them at low levels, as well as numerical simulations with high resolution models based on the primitive equations. The four cyclones represent snapshots of a single cyclone at successive stages of its life cycle as it evolves from a weak frontal wave in (I) to a fully developed cyclone in (IV).

In the early stages of development (I and II) the configuration of the fronts and isotherms resembles the classical, Norwegian polar front cyclone model, with warm and cold fronts beginning to circulate around the center of a deepening low pressure center. The only perceptible difference is the pronounced cyclonic shear of the flow and the relatively greater prominence of the warm front. As the development

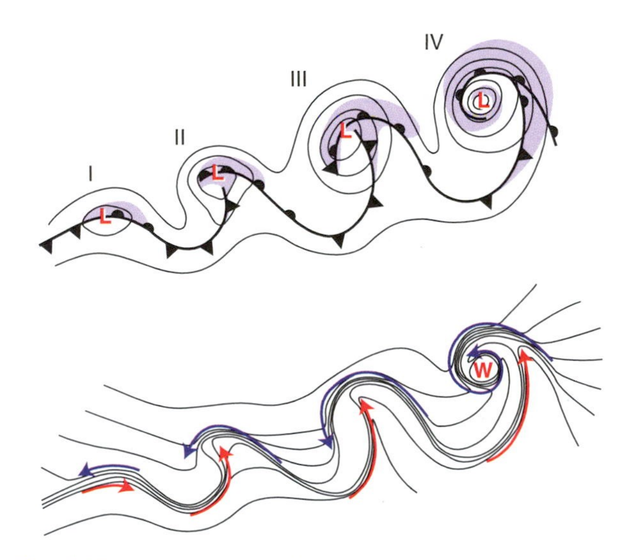

Fig. 8.32 As in Fig. 8.31 but for tightly coiled, warm core storms. [From *Extratropical Cyclones: The Erik Palmén Memorial Volume*, Amer. Meteorol. Soc. (1990) p. 188.]

proceeds, the warm frontal zone continues to sharpen and bridges across the poleward side of the surface low. This zone of sharp thermal contrast maintains its identity as it is advected around the back side of the low in stage III and coiled into a tight, mesoscale spiral in stage IV. This extension of the warm front is sometimes referred to as a *bent back warm occlusion*. This segment of the front is occluded in the sense that the air on the warm side of the frontal zone cannot be traced back to the warm sector of the storm. In this case the label *warm* derives not from the direction of movement, but from the frontal history; depending on the rate of movement of the storm and the direction of the observer relative to the low pressure center, the front may be moving in either direction.

Throughout the development process, the cold frontal zone is less pronounced than the warm frontal zone and the innermost part of it actually weakens as the storm begins to take shape. At stage III, the weakening inner segment of the cold front intersects the stronger warm front at right angles, creating a configuration reminiscent of a T-bone steak. The cold front advances eastward more rapidly than the center of the cyclone and becomes separated from it in stages III and IV.

Cold air spiraling inward along the outer side of the warm front, indicated by the blue arrows in Fig. 8.32, encircles and secludes the relatively warm air in the center of the cyclone, creating the mesoscale warm core. The strongest inflow of warm air, indicated by the red arrow, occurs just ahead of the cold front. Bands of cloudiness and precipitation tend to be located ahead of the cyclonically circulating warm and cold fronts, while drier, relatively cloud-free air spirals inward behind the cold front.

Consistent with the thermal wind equation (as generalized to the gradient wind) the tight cyclonic circulation around the center of the storm weakens rapidly with height above the top of the boundary layer. The wraparound warm front slopes outward, toward the colder air, with increasing height and it diminishes in intensity. Hence, the mesoscale warm air seclusion at the center of the cyclone expands with increasing height, but it also diminishes in intensity.

In the atmospheric dynamics literature, tightly coiled, warm core cyclones are referred to as LC1 storms and cyclones that conform to the Norwegian model as LC2 storms (where LC stands for life cycle). A third category LC3 refers to *open wave cyclones* (i.e., cyclones that never develop occluded fronts) in which the cold front is dominant. One can conceive of an archetypal (or "perfect") storm for each of these three models.

Numerical simulations in which baroclinic waves are allowed to develop on various background flows offer insights as to what conditions favor the development of cyclones that conform to the Norwegian polar front cyclone model versus the tighter, more axially symmetric, warm core cyclones exemplified by Figs. 1.12 and 8.32. The determining factors appear to be the barotropic shear and confluence/diffluence of the background flow.

The three kinds of cyclones (LC1, LC2, and LC3) are different outcomes of the same instability mechanism: baroclinic instability, which can occur even in a dry atmosphere. All three involve the amplification of a wave in the temperature field by horizontal temperature advection and the release of potential energy by the sinking of colder air and the rising of warmer air. In all three, the rising and sinking air flows and their attendant fronts spiral inward toward the center of the cyclone. Even their frontal structures are similar in many respects.

8.1.6 Top-Down Influences

In numerical simulations of baroclinic waves developing on a pure zonal background flow, the disturbances reach their peak amplitude first in the lower troposphere, and a day or so later at the jet stream level. In nature, cyclone development (*cyclogenesis*) is almost always "top-down"; it is initiated and subsequently influenced by dynamical processes in the upper troposphere. To generate a cyclone as intense as the one examined in the case study, conditions in the upper and lower troposphere must both be favorable.

The region of cyclonic vorticity (and potential vorticity) advection downstream of a strong westerly jet is a favored site for cyclogenesis, especially if such a feature passes over a preexisting region of strong low level baroclinicity (e.g., the poleward edge of a warm ocean current, the ice edge, or a weakening frontal zone left behind by the previous storm). Extrusions of stratospheric air, with its high potential vorticity, into frontal zones at the jet stream level can increase the rate of intensification of the cyclonic circulation in the lower troposphere.

Extratropical cyclones sometimes occur in association with long-lived *baroclinic wave packets*, which are more clearly evident at the jet stream level than down at the Earth's surface. The existence and behavior of wave packets are illustrated by the time-longitude plot of the meridional wind component at the jet stream level shown in Fig. 8.33. The pervasive wave-like signature, with a wavelength of around 50° of longitude (~4000 km on the 45 °N latitude circle) is the signature of baroclinic waves. That the individual maxima and minima slope toward the right as one proceeds downward in the diagram is evidence of eastward phase propagation. The average phase speed of the waves in this plot is 7° of longitude per day ($6\,\mathrm{m\,s^{-1}}$). Envelopes comprising several successive waves sectors in which the wave amplitude is relatively large are referred to as *wave packets*. For example, on November 14 a wave packet is passing over the Atlantic sector. Upon close inspection, it is evident that the wave packets propagate eastward with time with a speed of nearly $20\,\mathrm{m\,s^{-1}}$, three times the phase speed of the individual waves embedded within the packets. New waves are continually developing downstream of a wave packet, while mature waves are dying out at the upstream end of it: hence the lifetime of a wave packet transcends the lifetimes of the individual waves of which it is comprised.

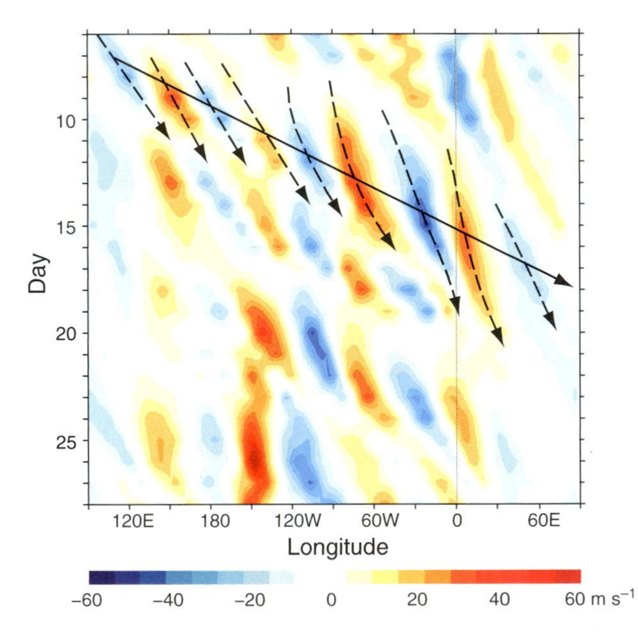

Fig. 8.33 Time-longitude section of the 250-hPa meridional wind component (in $\mathrm{m\,s^{-1}}$) averaged from 35 °N to 60 °N for November 6-28, 2002, a period marked by well-defined baroclinic wave packets and several major northern hemisphere cyclogenesis events. Slopes of the dashed arrows indicate the phase velocities of the waves, and the solid arrow indicates the group velocity of the wave packets. [Courtesy of Ioana Dima.]

The observed tendency for *downstream development* of wave packets is a consequence of the dispersive character of Rossby waves (i.e., the fact that their speed of propagation is a function of their wavelength). The rate of propagation of the packets is closely related to the group velocity of Rossby waves.[7]

8.1.7 Influence of Latent Heat Release

Another factor that contributes to the vigor and diversity of extratropical cyclones is the release of latent heat of condensation in regions of precipitation. Because latent heat release occurs preferentially in warm, rising air masses, it acts to maintain the horizontal temperature gradients within the storm, thereby increasing the supply of potential energy available for conversion to kinetic energy. Numerical simulations of cyclogenesis with and without the inclusion of latent heat release confirm that precipitating storms tend to deepen more rapidly and achieve greater intensities than storms in a dry atmosphere.

Precipitation in extratropical cyclones is often widespread but inhomogeneous in space and time, with much of it concentrated within elongated mesoscale *rain bands* with areas ranging from 10^3 to $10^4\,\mathrm{km^2}$ and with lifetimes of several hours. The axes of the rain bands tend to be aligned with the low level isotherms which, in turn, tend to be aligned with the vertical wind shear and with the fronts, as depicted in Fig. 8.34. The bands that lie along the fronts are fed by ascending air trajectories along the frontal surface, as depicted in Fig. 8.9. Pre- and postfrontal rain bands are the manifestations of instabilities within the broad deformation zone that lead to locally enhanced baroclinicity and upward motion.

Within the rain bands are smaller (10^2 to $10^3\,\mathrm{km^2}$) mesoscale regions in which the precipitation rates are further enhanced by the presence of convective cells, as explained in more detail in the next section. The lifting that occurs in association with a cold front advancing into a warm, humid, convectively unstable air mass can give rise to a line of convective cells forming an intense, narrow

[7] See J. R. Holton, *Dynamic Meteorology*, 4th edition. Academic Press (2004) pp. 185–188.

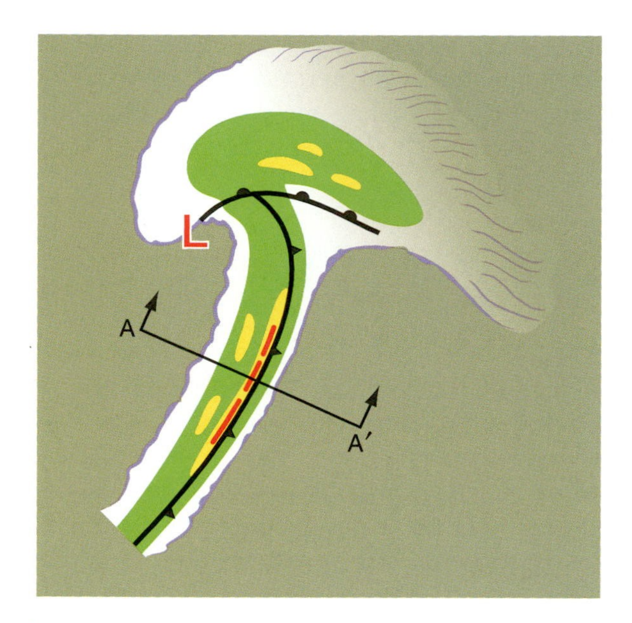

Fig. 8.34 Idealized schematic emphasizing the kinds of mesoscale rain bands frequently observed in association with a mature extratropical cyclone. The green shading within the cloud shield denotes light precipitation, yellow shading denotes moderate precipitation, and red shading denotes heavy precipitation. [Courtesy of Robert A. Houze.]

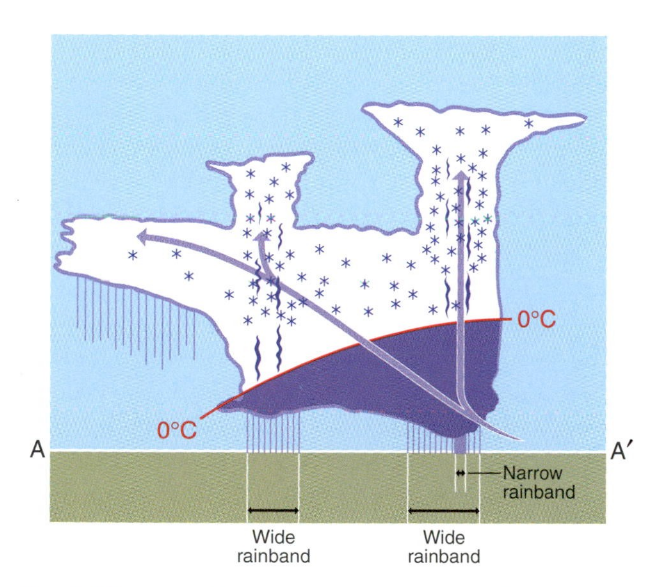

Fig. 8.35 Vertical cross section along AA′ in Fig. 8.34. The position of the cold front at the Earth's surface coincides with the leading edge of the narrow cold frontal rainband, and the frontal surface tilts upward toward the west with a slope comparable to that of the air trajectory. The dark blue shading indicates areas of high liquid water concentration, and the density of the blue asterisks is proportional to the local concentration of ice particles. High liquid water contents are restricted to the layer below the 0 °C isotherm except in regions of strong updrafts in convective cells, as represented by the narrow, dark blue "chimneys." See text for further explanation. [Adapted from *Cloud Dynamics*, R. A. Houze, p. 480, Copyright (1993), with permission from Elsevier.]

cold-frontal rainband, like the ones shown in Figs. 8.19 and 8.21. A vertical cross section through a complex of cold frontal rain bands is shown in Fig. 8.35. The wide frontal and postfrontal rain bands are a consequence of convective cells embedded within the broader rain area. Updrafts in the cells carry cloud liquid water well above the freezing level, where it quickly condenses onto ice particles, enabling them to grow rapidly to a size at which they fall to the ground, enhancing the rainfall rate. The narrow cold frontal rainband coincides with a line of particularly intense convective cells fueled by the lifting action of the front.

Vigorous convective features such as squall lines can sometimes take on a life of their own, modifying the structure of the extratropical cyclones in which they are embedded. Under these conditions, rainfall patterns may depart substantially from those typically associated with baroclinic waves and the frontal configuration may even be modified. Deep convection plays a particularly important role during the warm season, when the equator-to-pole temperature gradient is relatively weak and rainfall rates can be very high.

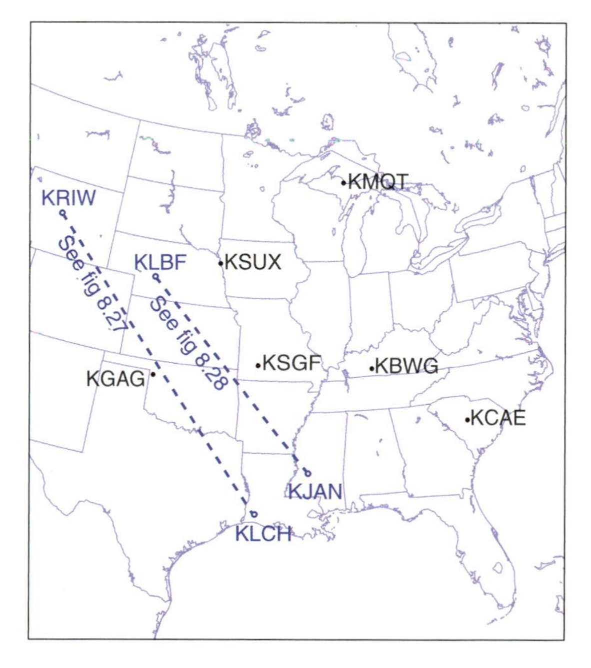

Fig. 8.36 Locations of the stations and vertical cross sections shown in this section. From north to south, KMQT is the station identifier for Marquette, Michigan; KRIW for Riverton, Wyoming; KLBF for North Platte, Nebraska; KSUX for Sioux Falls, South Dakota; KGAG for Gage, Oklahoma; KSGF for Springfield, Missouri; KBWG for Bowling Green, Kentucky; KCAE for Columbia, South Carolina; KJAN for Jackson, Mississippi; and KLCH for Lake Charles, Louisiana.

8.2 Orographic Effects

The structure and propagation of extratropical cyclones may be further influenced by terrain slope and by the blocking of the low level flow by mountain ranges. These influences on the large-scale flow pattern are collectively referred to as terrain effects or *orographic*[8] effects. Just as the large-scale structure of extratropical cyclones can be well simulated and predicted using global models based on the primitive equations, orographic effects can be simulated and, in some cases, predicted using regional mesoscale models that resolve topographic features down to the scale of a few kilometers. Mountains affect the flow on even smaller scales, as discussed in Section 9.5.1.

8.2.1 Lee Cyclogenesis and Lee Troughing[9]

Through the conservation of potential vorticity, as discussed in Section 7.2.10, the vertical stretching and horizontal convergence of a column of air descending along the lee side of a mountain range induces a cyclonic vorticity tendency, which, in turn, gives rise to a local negative pressure tendency. If the downslope flow is strong enough, the pressure fall is reflected in the development of a surface low (*lee cyclogenesis*) or the formation of a trough in the sea-level pressure field on the downwind side of the mountain range. Such lee cyclones or troughs typically exhibit a "warm core" structure, a consequence of the adiabatic warming of the downslope flow, which typically originates from near the level of mountaintops, since the lower level upstream flow is blocked by the mountain range. Hence, lee-side lows and troughs tend to decrease in amplitude with increasing height. When a baroclinic wave passes over a major mountain range, the surface low on the upstream side may decay and be replaced by a new surface low formed by lee cyclogenesis on the downstream side of the range. If the upper level trough moves eastward, across the mountain range, this newly developed surface low will eventually move eastward and deepen under the influence of baroclinic instability, but its track may initially exhibit an equatorward component, for reasons explained in the next subsection.

8.2.2 Rossby Wave Propagation along Sloping Terrain

The flow over and downstream of a mountain range is affected by the presence of a preexisting or newly developed surface low along the lee slope. The warm, westerly, downslope flow in the lee of a north–south-oriented mountain range such as the Rockies tends to be concentrated along the equatorward flank of the surface low, where it is reenforced by the westerly, low level geostrophic wind component circulating around the low. Hence, the vorticity tendency induced by the stretching of air columns in the downslope flow will tend to be most intense, not in the surface low itself, but along its equatorward flank. On the poleward flank of the surface low, the easterly geostrophic wind at the Earth's surface is upslope, in opposition to the flow aloftso the induced vorticity tendency may be weak, or even anticyclonic. In response to the unequal vorticity tendencies to the north and south of it, the propagation of the surface low exhibits an equatorward component. In a similar manner, it can be argued that the low level geostrophic flow circulating around an anticyclone that lies along the eastern slope of a north–south-oriented mountain range must also induce a tendency toward equatorward propagation.[10]

The weather observed in association with a surface high/low couplet, propagating equatorward along the slope of a mountain range in the northern hemisphere, is depicted schematically in Fig. 8.37. The rain or snow tends to occur in the region of upslope flow. Hence, in contrast to the extratropical cyclones over flat terrain, where most of the precipitation occurs in association with falling pressure in advance of the surface low, precipitation in these equatorward propagating systems occurs in association with rising pressure following the passage of the surface low.

---

[8] From the Greek *oros*, mountain.

[9] In the dialect of synoptic meterologists, *ridge* and *trough* are sometimes used as verbs, in the sense of to *form* a ridge or trough.

[10] The tendency for equatorward propagation of surface lows and highs along the eastern slope of a mountain range is a special case of the general tendency for Rossby waves to propagate anticyclonically around large-scale regions of elevated terrain, as required by the conservation of potential vorticity. An application of this same principle to ocean eddies propagating along a sloping continental shelf is considered in Exercise 7.36.

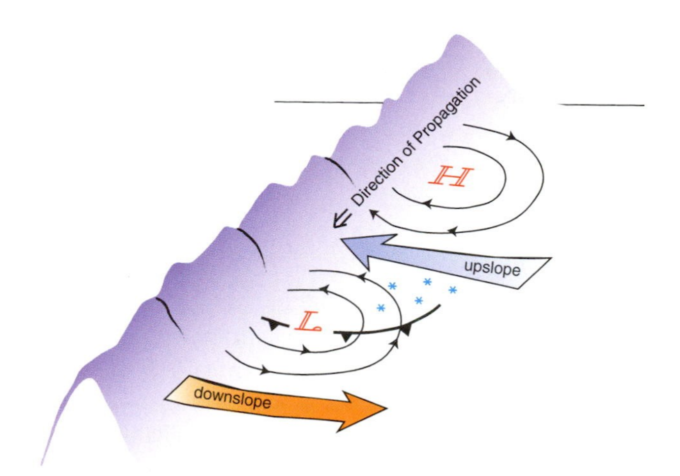

Fig. 8.37 Idealized pressure and geostrophic wind patterns along the eastern slope of a north–south-oriented mountain range in the northern hemisphere. Contours represent geopotential height perturbations on a pressure surface that intersects the terrain, as indicated. Wide arrows indicate regions of upslope and downslope flow along sloping terrain. The effects of cold air damming, considered in the next section, are not represented in this schematic.

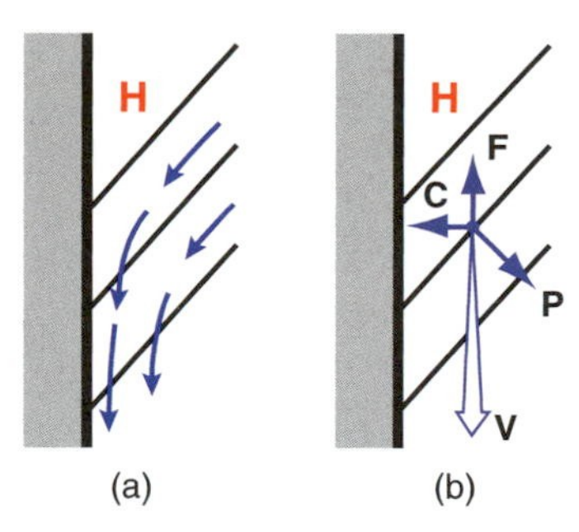

Fig. 8.38 Idealized flow near a north–south-oriented mountain barrier in the northern hemisphere. (a) Deflection of the surface winds as they approach the mountain barrier. The contours represent isobars on a surface of constant geopotential lower than the top of the mountain barrier. (b) Instantaneous balance of forces on an air parcel in idealized flow close to the mountain barrier. **C** is the Coriolis force, directed to the right of the wind, **P** is the pressure gradient force, directed down the pressure gradient, and **F** is the frictional drag force, directed opposite to the wind.

8.2.3 Cold Air Damming

Rossby wave propagation, considered in the previous subsection, involves geostrophic motion along a region of sloping terrain in the absence of discrete boundaries. This subsection considers a phenomenon known as *cold air damming* that involves departures from geostrophic motion due to the presence of a mountain barrier. In describing this phenomenon, the mountain range is represented not as a sloping surface, but as a vertical wall that blocks the flow in the transverse direction.

In the presence of a horizontal pressure gradient along a mountain barrier, the geostrophic wind exhibits a component transverse to the barrier. Because the air cannot flow through the barrier, it follows that the wind must exhibit a substantial ageostrophic component. To illustrate how this ageostrophic component is generated and maintained, let us consider an easterly geostrophic wind impinging on a north–south mountain barrier in the northern hemisphere (e.g., the Front Range of the Colorado Rockies), as depicted schematically in Fig. 8.38a. Trajectories of air parcels curve southward as they approach the barrier, as shown in the left panel. The slowing of the easterly wind component as air parcels approach the barrier implies the existence of an eastward-directed pressure gradient force, as represented by the orientation of the isobars. The higher pressure along the barrier implies that the isentropes are raised there so that at a given pressure level, relatively cold, dense air is banked against the barrier.

The instantaneous balance of forces acting on an air parcel close to the barrier is shown in Fig. 8.38b. In the transverse (east–west) direction, the Coriolis force, directed toward the barrier, is balanced by the "back pressure" exerted by the colder, denser air banked against the barrier. In fact, it is the Coriolis force that maintains the east-to-west slope of the isentropes against the force of gravity. In the direction parallel to the wind, the southward-directed pressure gradient force, unopposed by the Coriolis force, causes the wind speed to increase until it reaches a level at which it is balanced by the frictional drag.

Because the winds aloft nearly always blow from west to east, a region of easterly geostrophic winds at the Earth's surface is typically marked by strong vertical wind shear. On the basis of the thermal wind equation, it can be inferred that such a region must correspond to a frontal zone in which the colder air lies toward the north. Hence, the cold front in Fig. 8.37 is positioned at the southern limit of the easterlies, coincident with the surface low. Propelled by the strong equatorward winds, the cold air along the mountain barrier advances more rapidly than elsewhere along the frontal zone, leading to the formation of a sharp, fast-moving cold front along the southern flank of the zone. Cold fronts accelerated and sharpened in this manner exhibit many of the features of gust fronts described in Section 8.3.3, including strong winds and sharp pressure rises. Wintertime "Texas Northers" that cause abrupt temperature drops in the southern U.S. Great Plains and summertime "Southerly Busters" that cause strong winds in the lee of the coast range in eastern Australia are regional manifestations of this phenomenon, which is a form of cold air damming.

Fast-moving cold fronts propagating along the lee side of mountain ranges in the wake of extratropical cyclones sometimes penetrate deep into the tropics. Although the temperature drops and pressure rises that occur in association with these frontal passages are smaller than those at extratropical latitudes, they are nonetheless large compared to the ordinary low latitude day-to-day variability. These so-called *cold surges* are instrumental in setting the stage for episodes of gap winds discussed in the next subsection and in triggering outbreaks of deep convection. The structure of cold surges is often quite complex, with multiple wind shift s and prefrontal pressure surges preceding the arrival of the slightly cooler, drier air.

8.2.4 Terrain-Induced Windstorms

The presence of terrain gives rise to windstorms of three types: those associated with locally intense *gap winds*, those in which subsidence-induced warming plays an important role, and local windstorms involving mountain lee waves. Although they sometimes occur in combination, each of these phenomena involves somewhat different dynamical mechanisms. Gap winds and regional downslope windstorms are discussed in this subsection; local downslope windstorms associated with mountain lee waves are alluded to briefly in Section 9.5.1.

Windstorms attributable to gap winds occur mainly during wintertime when cold anticyclones build up over the interior of the continents, while sea-level pressure is much lower over the oceans to the south and west. Under these conditions, strong pressure gradients develop across the coastal mountain ranges, and continental air flows southward and/or westward through the gaps in the ranges, producing high winds downstream, over regions such as the Gulf of Alaska and the Mediterranean Sea.

Gap winds are everyday occurrences in favored locations in the trade winds, such as the straits between the Hawaiian islands. They occur frequently enough over Central America to influence the climatology, as illustrated in Fig. 8.39. Following wintertime cold surges, when sea-level pressure on the Atlantic side of the Cordillera is abnormally high, wind speeds downstream of these gaps range as high as $\sim 30 \text{ m s}^{-1}$.

Under conditions of cold air damming, gap winds tend to be strongest not at the narrowest point in the gaps where the flow is most constricted, but on the lee side of the mountain range downwind of the gaps. For lack of a barrier to contain it, cold air emerging from the downwind side of a gap spreads out, subsides, and thins, causing sea-level pressure to drop in the downwind direction. Under the influence of this hydrostatically induced, downwind-directed pressure gradient force, air parcels continue to accelerate for some time after they pass through the gap. Air parcels undergo adiabatic warming as they subside on the downwind side of the gap, but if the temperature differential across the mountain range is large, they may remain colder than the ambient air on the downwind side of the range until they mix with it.

If the elevation of the land surface is substantially higher on the upwind side of a mountain range than on the downwind side, the winds on the downwind side of the range will be warm (or even hot) as a consequence of adiabatic compression. As the air warms, its saturation mixing ratio rises in accordance with the Clausius-Clapeyron equation, while its dew point and mixing ratio remain nearly constant. If the dew point is low to begin with, as is often the case for air masses residing over the continental interior, the relative humidity of the air may be extremely low by the time it completes its descent. The airflow in regional downslope windstorms tends to be complex, with gusty winds in and downstream of the canyons and lighter winds elsewhere, but the extreme warmth and dryness of the downstream airmass are pervasive. A classic example is the so-called *Santa Ana winds* that occur episodically when pressure builds up over the high plateaus in the interior of the western United States during autumn and early winter, driving a hot, dry northeasterly flow through the gaps in the mountain ranges and down the canyons of southern California. A local author described such intervals as "the season of suicide and divorce and prickly dread, wherever the wind blows."[11]

[11] From Joan Didion's collection of essays "Slouching Towards Bethlehem," Noonday Press, 1990.

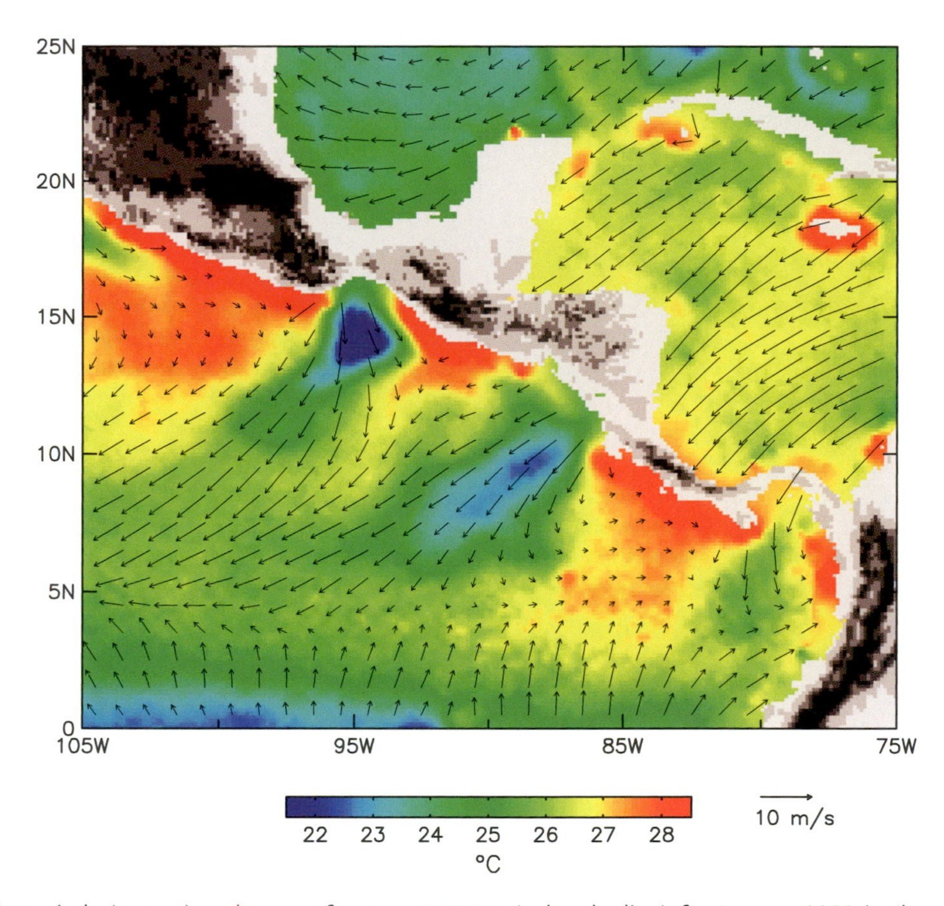

Fig. 8.39 Surface winds (arrows) and sea surface temperatures (color shading) for January 2000 in the vicinity of Central America. Regions of high terrain are indicated by the darker gray and black shading. Note the pronounced signature of strong winds and remarkably low sea surface temperatures in the lee of the (light gray) gaps in the terrain, extending southwestward several hundred kilometers into the Pacific. These regions are also marked by high marine productivity (not shown) due to the mixing of nutrients from below the thermocline. [Winds are from the scatterometer on the NASA QuikSCAT satellite; sea-surface temperature data are from the NASA Tropical Rainfall Measuring Mission satellite. Courtesy of Dudley Chelton.]

During downslope windstorms, existing fires are difficult to control and errant sparks that would ordinarily die out harmlessly can ignite new fires. Figure 8.40 shows smoke from wildfires on October 26–27, 2003, streaming southwest-ward over the Pacific. On these dates, sea-level pressure in excess of 1030 hPa was observed over the Great Basin at an altitude of around 1.5 km, while the pressure in the coastal lowlands of southern California was ~1017 hPa, resulting in a 15-hPa pressure differential and adiabatic warming of ~15 °C. Although few of the regular observing stations (most of which are at airports) experienced wind speeds in excess of 10 m s^{-1}, wind gusts ranged up to 30 m s^{-1} in locations exposed to the winds. Despite the lateness of the season, nighttime temperatures in these exposed locations remained above 25 °C and daytime temperatures throughout the region hovered around 33 °C, with relative humidities on the order of 5%.

8.2.5 Orographic Influences on Precipitation

The general tendency for climatological-mean precipitation to be enhanced on the upwind side of mountain ranges and suppressed on the downwind side, where *upwind* and *downwind* are defined in terms of the prevailing climatological-mean westerly flow, is clearly evident in the distribution of annual mean precipitation over the contiguous United States, shown in Fig. 1.26. This relationship implies that, on average, air tends to flow over mountain ranges without significant deflections. On a day-to-day basis the influence of orography on the distribution of precipitation is more complex and dynamic.

The position and orientation of the rain shadow downwind of a mountain range may change dramatically in response to changes in wind direction and speed that occur in association with the passage of a cyclone. Even within the rain shadow, rain bands can

Fig. 8.40 Dispersion of the smoke plumes from a massive outbreak of California wildfires on October 26, 2003, during a nearly week-long episode of hot, dry, northeasterly Santa Ana winds. Red pixels denote the locations of the fires at this time. Of the three large clusters of fires in the image, the middle one is located east of Los Angeles and the southern one near San Diego. The fires raged out of control until the winds subsided and shifted 2 days later. [NASA Terra MODIS satellite imagery.]

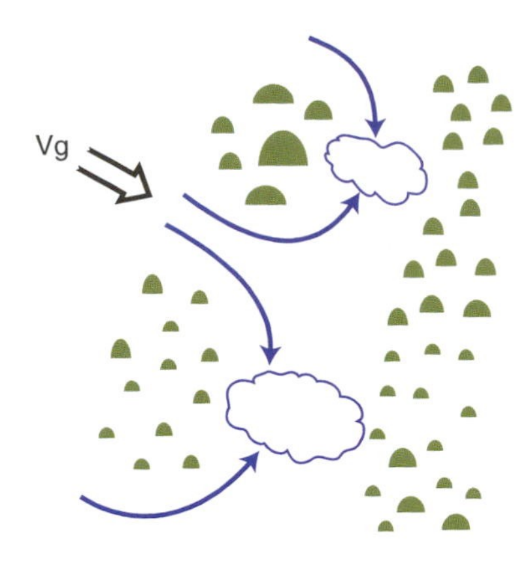

Fig. 8.41 Schematic of low level airflow through gaps on an extended mountain range converging into terrain-induced cyclones on the lee side of the range, giving rise to the development of rain bands. If air simply flowed over the mountains rather than around them, one might expect to find a rain shadow directly downstream of the mountains.

sometimes develop in response to lee cyclogenesis, as air flowing around the two sides of the mountain range converges and lifts the air above it, triggering convection, as illustrated schematically in Fig. 8.41. A necessary condition for the formation of such a *downwind convergence zone* is the presence of conditional instability, which enables the converging air to ascend without being unduly inhibited by the static stability.

Mountain ranges can also give rise to dramatic spatial variations in freezing level. Under conditions of cold air damming, with colder air and higher pressure to the east side of a range, the approach of a cyclone or trough of low pressure from the west increases the pressure gradient across the range, drawing cold air up and over the mountain passes. Freezing levels in this cold, upslope flow tend to be much lower than in the free atmosphere. Hence, under these conditions, frozen precipitation may be falling in the passes while rain is falling at higher elevations. Freezing rain occurs much more frequently in the lowlands along the east side of mountain ranges, in association with cold air damming, than on the west side. The drainage of cold air through a gap in a mountain range can sustain frozen precipitation on the west side (often in the form of sleet and freezing rain) even when the ambient surface air temperature is well above freezing.

8.3 Deep Convection

Among the most striking features in time-lapse satellite imagery are clouds formed by the spreading of updrafts in deep cumulus convection as they approach the tropopause. Deep convection tends to be concentrated within certain preferred regions such as the summer monsoons, persistent bands of low level convergence like the ITCZ, over and along the slopes and crests of mountain ranges, and within the frontal zones and warm sectors of extratropical cyclones. Global surveys indicate that at any given time, cirriform anvil clouds formed in this manner cover only a few percent of the surface area of the Earth, and active convection occupies only a small fraction of the surface area beneath those clouds. Yet despite its small areal coverage, deep convection and its associated stratiform precipitation account for most of the rainfall in the tropics and over the continents of the summer hemisphere.

The horizontal scale of deep convection is much smaller than that of baroclinic waves, the timescale is correspondingly shorter, and the aspect ratio (i.e., the ratio of the characteristic depth scale D to the length scale L) is much larger. As a consequence of the different scaling, the Earth's rotation plays only an indirect role in the dynamics of deep convection, the vertical motions are much stronger than in large-scale motion, and hydrostatic balance does not always prevail.

This section considers deep convection in its various forms:

- individual *convective cells* consisting of a single updraft and downdraft (this category includes some thunderstorms);
- *convective storms* made up of organized groups or sequences of convective cells; and
- *mesoscale convective systems*: bands or zones of clouds and precipitation on a scale of 100 km or larger in at least one direction that are generated by interacting convective cells.

8.3.1 Environmental Controls

In idealized theoretical and modeling studies, deep convection is viewed as developing within a large-scale environment in which temperature, moisture, and horizontal wind are horizontally uniform. The vertical profiles of temperature and moisture in the environment play a critical role in defining the regions of the atmosphere in which deep convection develops spontaneously, and the vertical wind profile determines the direction and rate of movement of convective storms and profoundly influences their structure and evolution.

a. Temperature and moisture stratification

The necessary conditions for the occurrence of deep convection are

- the existence of a conditionally unstable lapse rate (i.e., $\Gamma_w < \Gamma \leq \Gamma_d$)
- substantial boundary-layer moisture, and
- low level convergence (or lifting) sufficient to release the instability.

Convection feeds on the potential energy inherent in the temperature and moisture stratification. The so-called *convective available potential energy* (CAPE), (in J kg^{-1}) of a reference air parcel is given by

$$\text{CAPE} = \int_{\text{LFC}}^{\text{EL}} (F/\rho')\, dz \tag{8.3}$$

where F is the upward buoyancy force per unit volume on the rising air parcel due to the temperature difference between the parcel and its environment, ρ' is the density of the air parcel, LFC is the level of free convection, and EL is the *equilibrium level* above which the parcel is no longer warmer than its environment. By analogy with Exercise 3.11, it is readily shown that the buoyancy force per unit mass (F/ρ') is given by $(\alpha' - \alpha)/\alpha$ times g, where α' is the specific volume of the rising air parcel, α is the specific volume of the environmental air at the same level, and g is the gravitational acceleration. Substituting for gdz from the hydrostatic equation (3.18) and reversing the order of integration yield

$$\text{CAPE} = \int_{\text{EL}}^{\text{LFC}} (\alpha' - \alpha)\, dp$$

Substituting from the equation of state (3.15) yields

$$\text{CAPE} = R_d \int_{\text{EL}}^{\text{LFC}} (T_v' - T_v)\, d\ln p \tag{8.4}$$

If we ignore the small virtual temperature correction, the integral in this expression is simply the area, on the skew-T ln p plot, extending from LFC to EL and bounded by the environmental temperature sounding on the left and a moist adiabat on the right.

The reference air parcel used in computing the CAPE may be an air parcel at the Earth's surface or it may be chosen to be representative of the mean temperature and humidity of the air within the boundary layer. As in the exercises in Chapter 3, the parcel is lifted along a dry adiabat up to its lifting condensation level (LCL) and then along a saturated adiabat as illustrated in Fig. 8.42. The integration begins at the LFC.

Exercise 8.1 Estimate the CAPE for a sounding in which the LFC and EL for the reference air parcel are 700 and 175 hPa, respectively, and within the layer between those levels the reference air parcel is, on average (with respect to ln p), 10 °C warmer than the environmental air at the same level.

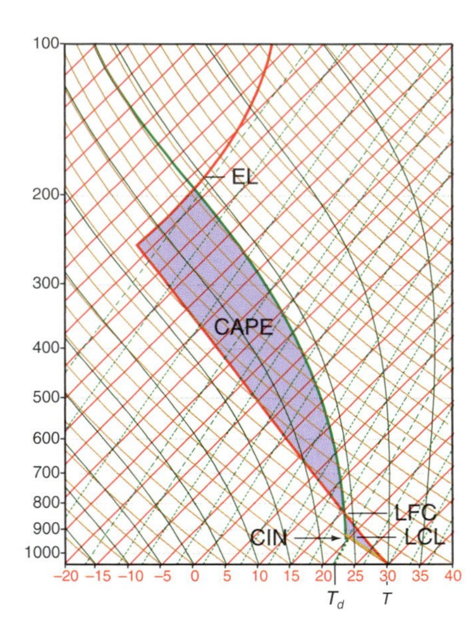

Fig. 8.42 A hypothetical sounding illustrating the concepts of convective available potential energy (CAPE) and convective inhibition (CIN). The CAPE and CIN in this sounding are indicated by shading.

Solution: Substituting values into (8.4), we obtain

$$\begin{aligned}\text{CAPE} &= 287 \times 10\,°\text{C} \times \ln(700/175)\\ &= 3978\ \text{J kg}^{-1}\end{aligned}$$

For reference, values of CAPE ranging from 0 to 1000 J kg^{-1} are considered marginal for deep convection, 1000–2500 adequate to support moderate convection, 2500–4000 adequate to support strong convection, and in excess of 4000 indicative of the potential for extreme convection.[12] ■

If it were a common occurrence for air parcels in ordinary cumulus clouds to reach their level of free convection, the CAPE would never build up to values high enough to support vigorous deep convection. Hence, the degree to which convection is inhibited by the presence of a stable layer or inversion at the top of the planetary boundary layer also plays a role in setting the stage for convective storms. A measure of this so-called *convective inhibition* (CIN or CINH) is the energy, in units of J kg^{-1}, required to lift the reference air parcel to its LFC. So defined, CIN may be viewed as negative CAPE and can be represented as an area on a skew-T ln p plot, as shown in Fig. 8.42. To set the stage for vigorous deep convection it is essential that the CIN be non-zero, but not so large that it precludes the possibility of deep convection altogether. For CIN $>$ 100 J kg^{-1}, deep convection is unlikely to occur in the absence of external forcing, such as might be provided by daytime heating or the approach of a strong front. A return of several thousand units of CAPE on an investment of perhaps several tens of units of "startup costs" required to overcome the CIN rivals the performance of the most successful new enterprises in the world of business!

The geographical distribution of the frequency of lightning flashes shown in Fig. 6.56 provides a measure of the degree to which the vertical stratification of temperature and moisture over various regions of the world is conducive to deep convection. Many of the features in this pattern mirror the distribution of rainfall shown in Fig. 1.25. However, relative to the rainfall, lightning flashes are relatively more frequent over the continents than over the oceans because the heating of the land surface greatly enhances the CAPE during the afternoon hours, giving rise to more vigorous convection.

For the CAPE inherent in the temperature and moisture sounding to be realized, two things need to happen: the environmental air needs to be destabilized (i.e., the CIN needs to be reduced) by lifting and, within this destabilized air mass, air parcels need to be lifted up to their LFC. Lifting destabilizes the environmental sounding by weakening the inversion that caps the mixed layer so that buoyant air parcels from below can break through it. This process is illustrated schematically in Fig. 8.43.

The lifting and associated low level convergence that is responsible for lifting the inversion layer and thereby destabilizing the environmental sounding is usually associated with some large-scale forcing mechanism such as the approach of an extratropical cyclone, which can be anticipated a day or more in advance on the basis of numerical weather prediction. In contrast, the lifting of the air parcel that initiates the deep convection is usually associated with a more

[12] Another widely used indicator of the potential for deep convection is the so-called *lifted index (LI)*, defined as the temperature deficit (relative to the environment) of an air parcel originating at the earth's surface that is lifted dry adiabatically up to its lifting condensation level and then moist adiabatically up to the 500-hPa level. Negative lifted indices are indicative of a potential for deep convection; values below −9 indicate the potential for severe convection. Variants of the lifted index based on air parcels originating at various heights within the planetary boundary layer are also sometimes used.

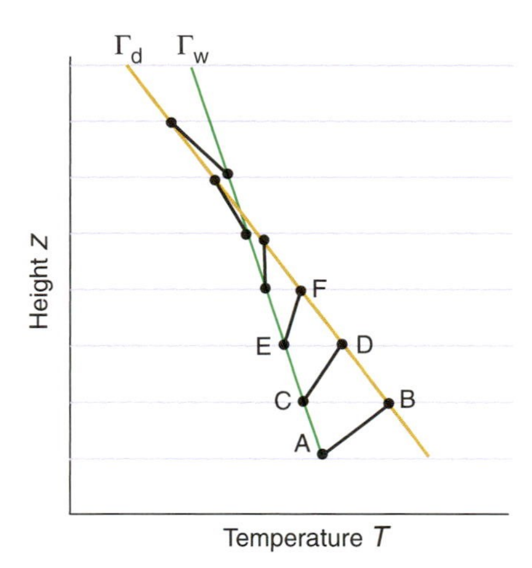

Fig. 8.43 Illustration of the increase in the lapse rate $-dT/dz$ within an inversion layer as the layer is lifted. The black line segment AB represents the temperature profile within the inversion layer before it is lifted; CD represents the temperature profile in the same layer after it is lifted one height increment, EF after it is lifted two height increments, etc. It is assumed that the bottom of the layer is saturated with water vapor and cools at the saturated adiabatic lapse rate as the layer is lifted, while the top of the layer is unsaturated and cools at the dry adiabatic lapse rate. The steepening of the lapse rate due to the differential rate of cooling is partially compensated by the expansion of the air within the layer as it rises. This effect is not represented in the diagram.

localized, short-lived and less predictable forcing mechanism, such as a sea-breeze front, a range of hills, or the leading edge of the outflow from a preexisting convective storm.

b. The vertical wind profile

Convective storms move at a speed approximately equal to the vertically averaged horizontal wind in their environment, where *vertically averaged* denotes a mass (or pressure) weighted average over the depth of the storm. Usually a mid-tropospheric *steering level* can be identified at which the storm motion vector is approximately equal to the wind in the storm's environment. However, it should be understood that the storm is not really *steered* by the wind at any particular level: the steering level is simply the level at which the wind vector most closely matches the layer-mean wind vector. Under some conditions, storms propagate systematically to the left or right of both the vertically averaged wind and the wind at the steering level.

Convective storms often form in an environment in which the *vertical wind shear* vector $\partial \mathbf{V}/\partial z$ is dominated by the increase in scalar wind speed V with height. The strength of the shear affects the vertical tilt of the updrafts and downdrafts within the storm: weak shear favors a structure in which the downdraft ultimately isolates the updraft from its supply of low level moisture, leading to the storm's demise, while strong shear favors a tilted structure with a symbiotic relationship between updraft and downdraft, resulting in more intense, longer lived storms capable of producing hail and strong winds.

Changes in wind direction with height also play an important role in the dynamics of convective storms. Vertical wind profiles that exhibit significant veering and backing are conveniently displayed in terms of a *hodograph*: a plot of the wind components u versus v for a single vertical sounding, with the points representing successive levels in the sounding connected by a curve. At any level in the profile the vertical wind shear vector is tangent to the hodograph curve at that level. In both the idealized hodographs for a hypothetical northern hemisphere station shown in Fig. 8.44, the wind vector $\mathbf{V}$ is rotating clockwise (veering) with height, but the veering is more pronounced in panel (b). The straightness of hodograph in Fig. 8.44a implies that the vertical wind shear vector $\partial \mathbf{V}/\partial z$ does not rotate with height (i.e., that the shear is unidirectional). In contrast, the curvature of the hodograph in Fig. 8.44b implies that both $\mathbf{V}$ and $\partial \mathbf{V}/dz$ rotate with increasing height. The importance of curved hodographs in the dynamics of a class of convective storms called *supercells* is touched on in Section 8.3.2.

In the presence of vertical wind shear, air possesses vorticity that can be visualized as a rolling motion about a horizontal axis. For example, both vertical profiles depicted in Fig. 8.44 exhibit vorticity, in a clockwise sense, about the y axis. In analogy with Table 7.1, the magnitude of the vorticity about the y axis is $(\partial u/\partial z - \partial w/\partial x)$, where $|\partial u/\partial z|$ is several orders of magnitude larger than $|\partial w/\partial x|$. Hence, the vertical shear $\partial u/\partial z$ is, in effect, the vorticity about the y axis.

Exercise 8.2 Compare the vorticity about a horizontal axis due to a vertical wind shear of 3 m s^{-1} per kilometer with the vorticity associated with the Earth's rotation.

Solution: The vorticity is equal to the shear, which is 3 m s^{-1} km^{-1} or 3×10^{-3} s^{-1}. From Exercise 7.1 it can

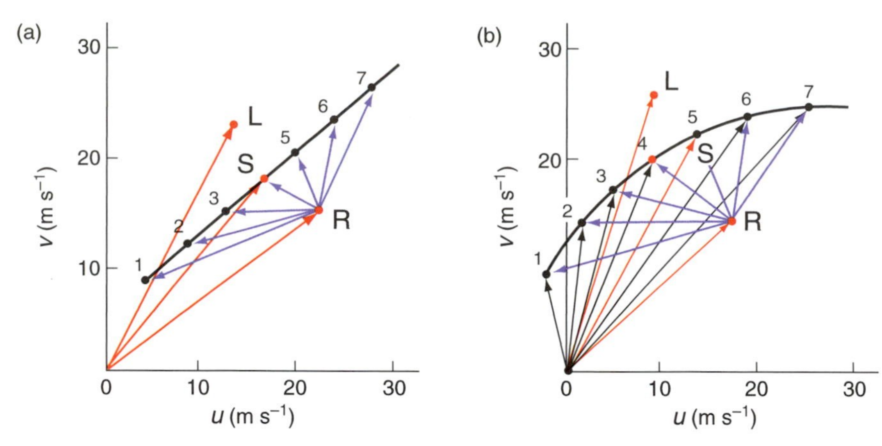

Fig. 8.44 Idealized northern hemisphere vertical wind profiles depicted in the form of *hodographs*. Points on the hodograph indicate the ends of wind vectors radiating out from the origin. The points are numbered in order from the bottom to the top of the profiles, which extend from the ground up to the tropopause. Both profiles exhibit veering of the wind with height. In (a) the vertical wind shear $\partial \mathbf{V}/\partial z$ is unidirectional, while in (b) it rotates clockwise with height, as indicated by the curvature of the hodograph. In both panels a midtropospheric steering level **S** and velocity vectors for hypothetical storms moving toward the left **L** and right **R** of the steering flow are indicated in red. The sense of the curvature of the hodograph determines whether left- or right-moving storms are favored, as explained in the next subsection. Light blue arrows show the relative flow in a coordinate system moving with the right-moving storm.

be inferred that the vorticity associated with the Earth's rotation in the plane perpendicular to the axis of rotation is $2\Omega = 2 \times 7.29 \times 10^{-5}\text{s}^{-1} \simeq 1.5 \times 10^{-4}$. Hence the vorticity inherent in this quite modest vertical wind shear is about 20 times as large as the vorticity associated with the Earth's rotation. ■

When boundary layer air is drawn into the updraft of a convective storm, the vorticity about a horizontal axis may be tilted so that it is transformed into vorticity about a vertical axis, as illustrated in Fig. 8.45. In this schematic the wind, the vertical wind shear, and the storm movement are all envisioned as being in the x direction and the updraft of the storm is centered over the y axis. Note how counterclockwise vorticity about the x axis (as viewed looking in the positive direction along the axis) is tilted into counterclockwise vorticity about the z axis, as viewed from above. This is a powerful mechanism for imparting rotation to convective storms.

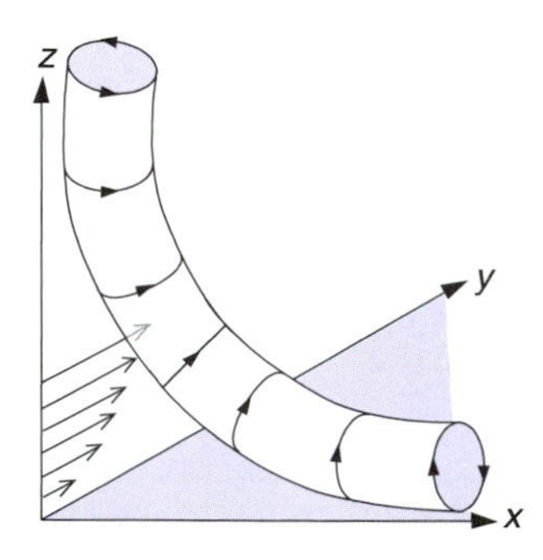

Fig. 8.45 Schematic showing how the updraft of a convective storm can acquire vorticity about a vertical axis by ingesting boundary layer air that possesses vorticity about the x axis by virtue of the vertical shear $\partial u/\partial z$. See text for further explanation.

c. Geographic regions susceptible to convective storms

During springtime, the central United States enjoys the dubious distinction of experiencing more tornadic thunderstorms than any region of comparable size in the world. Storms break out in the warm sectors of extratropical cyclones, where all the essential ingredients are present:

- convectively unstable soundings characterized by warm, humid, boundary-layer air flowing northward from the Gulf of Mexico, capped by dry, conditionally unstable air flowing eastward from the Rockies;
- strong vertical wind shear, as evidenced by the presence of a jet stream at the 250-hPa level;
- strong veering of the wind with height, with southerly low flow underlying westerly flow aloft; and
- strong synoptic-scale lifting associated with the passage of cyclones and their associated fronts.

8.3.2 Structure and Evolution of Convective Storms

Weather-related deaths and injuries, property damage and transportation disruptions occur much more frequently in association with localized *convective storms* than with extratropical cyclones: so much so, in fact, that the terms *severe storm* and *severe weather* are implicitly understood as applying to convective storms, unless otherwise specified.[13]

A great deal has been learned about the structure and dynamics of convective storms during the past few decades. Dual-Doppler analysis[14] has made it possible to map the three-dimensional wind field in the updrafts and downdrafts of storms that are too intense to probe safely with research aircraft. Using numerical models, it has been possible to simulate many of the observed structural features of convective storms and to determine the types of storms that are favored in various large-scale environments. This section presents a brief synopsis of these results.

Thunderstorms are made up of *cells*, which develop, mature, and decay, as illustrated in Fig. 8.46. The left panel shows young, rapidly growing cells in the foreground, fueled by midtropospheric CAPE, with deeper, more fully developed cells in the background. The top of the cell at the upper right is approaching its equilibrium level, beyond which rising air parcels are no longer buoyant, so they spread out to form an anvil. The right panel shows a fully developed anvil, which is still being fed by new convective cells rising out of the boundary layer.

Fig. 8.46 Deep convective cells in various stages in the development of a tornadic storm near Anthony, Kansas, May 25, 1997. The photo in the left panel was taken about an hour and a half before sunset, looking eastward. The photo in the right panel was taken around sunset, looking eastward toward the departing storm. [Courtesy of Brian Morganti.[15]]

[13] Most human fatalities attributable to convective storms are caused by lightning, which is an everyday occurrence during certain seasons in some parts of the world. In an average year ~50 lightning-related deaths occur in the United States, with highest rates per capita in the Rocky Mountain states and Florida, where lightning occurs most frequently. An overwhelming percentage of lightning deaths involve males. Hail is the major source of property damage. In a typical year over the United States hail causes two to three times as much property damage as tornadoes. The highest frequency of occurrence of damaging hail is over the western Great Plains, where it causes annual crop losses amounting to ~5% of total crop value. Slow-moving convective storms are capable of producing flash floods, characterized by abrupt rises in water levels in rivers and streams, that may pose a serious threat to life and property downstream. Winds that form in an environment with strong vertical wind shear are often accompanied by damaging winds, as described in the next subsection.

[14] Analysis of radial velocity measurements from two or more Doppler radars to determine the three-dimensional velocity field in a region of precipitation.

[15] Most of the ground-based photographs shown in this section were provided by amateur or professional meteorologists who maintain or share Web sites: *www.stormeffects.com* (Brian Morganti); *www.twisterchasers.com* (Kathryn Piotrowski); *www.mesoscale.ws* (Eric Nguyen); *skydiary.com* (Chris Kridler); and *www.dblanchard.net* (David Blanchard). These sites provide numerous additional examples of observed convective storms and related phenomena.

Convective storms can be classified as

- relatively small, benign, *single cell storms* that form under conditions of weak vertical wind shear;
- more dangerous *multicell storms* that develop under conditions of strong vertical wind shear; and
- intense, robust, long-lived *supercell storms* with rotating updrafts formed from the splitting of multicell storms.

Both multi- and supercell storms are capable of producing hail and strong winds. Most damaging tornadoes are associated with supercell storms.

a. Ordinary thunderstorms

The term *airmass thunderstorm* or simply *ordinary thunderstorm* is used to describe small isolated cumulonimbus clouds (Fig. 8.47) produced by local convection in an unstable airmass rather than by fronts or instability lines. These systems generally develop just one main precipitation shower (*a single cell storm*), and the pressure field is entirely determined by buoyancy of the warm updraft.

Fig. 8.47 A small isolated cumulonimbus cloud. [Courtesy of Art Rangno.]

Ordinary thunderstorms were the subject of an intensive field program, known as the Thunderstorm Project, which was carried out over Florida and Ohio during the late 1940s. Data on a large number of storms were composited together to construct an idealized model of the life cycle of a typical single-cell thunderstorms. In this model, which is depicted in Fig. 8.48, the life cycle of a single cell within a multicell storm is shown in terms of three stages: *cumulus*, *mature*, and *dissipating*. Multicell storms consist of several such cells, which grow and decay in succession, each having a lifetime of about half an hour.

In the cumulus stage (Fig. 8.48a) the cloud consists entirely of a warm, buoyant plume of rising air. The updraft velocity increases rapidly with height within the cloud and there is considerable entrainment through the lateral cloud boundaries. The top of the cloud moves upward with a velocity of up to $\sim 10\,\mathrm{m\,s^{-1}}$. Because of the large updraft velocities, supercooled raindrops may be present well above the freezing level (a situation potentially hazardous to aircraft because of the possibility of icing). The mature stage (Fig. 8.48b) is characterized by the development of a vigorous downdraft circulation, which coincides with the region of heaviest rain.

The downdraft circulation is initiated by the downward drag force induced by the drops. Dry environmental air entrained into the downdraft (on the right-hand side of Fig. 8.48b) and unsaturated air below cloud base are cooled by evaporation of falling precipitation. In some cases the resulting evaporative cooling is capable of greatly enhancing the negative buoyancy of the downdraft air. In the mature stage, supercooled raindrops still exist well above the freezing level in the updraft, while snowflakes or soft hail pellets may be found below the freezing level in the downdraft. The maximum updraft vertical velocities are in the middle of the cloud, with detrainment above that level. The top of the cloud approaches the tropopause and begins to spread out horizontally as an anvil in the dissipating stage (Fig. 8.48c). As precipitation develops throughout the cloud, the downdraft circulation gradually becomes more extensive until, in the dissipating stage, it occupies virtually the entire cloud. Deprived of a source of supersaturated updraft air, cloud droplets can no longer grow and, as a consequence, precipitation soon ceases. Only about 20% of the water vapor condensed in the updraft actually reaches the ground in the form of precipitation. The remainder evaporates in the downdraft or is left

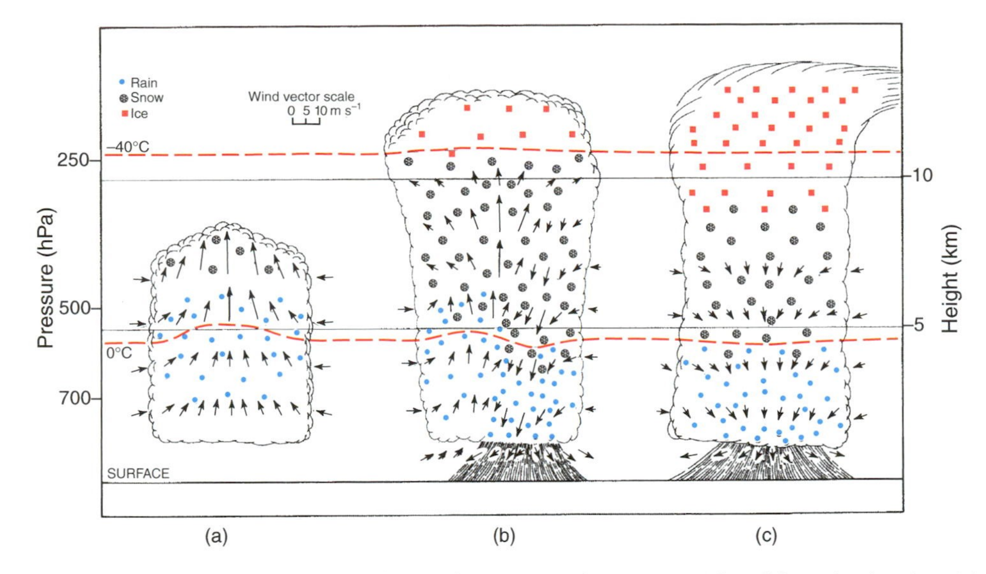

Fig. 8.48 Schematic of a typical ordinary single-cell thunderstorm in three stages of its life cycle showing (a) cumulus stage, (b) mature stage, and (c) dissipating stage. The horizontal scale is compressed by about 30% relative to the vertical scale in the figure. The 0 °C and −40 °C isotherms are indicated in red. [Adapted from *The Thunderstorm*, U.S. Government Printing Office (1949).]

behind as cloud debris (including extensive patches of anvil cirrus), to evaporate into the ambient air.

The single-cell thunderstorm is short lived and rarely produces destructive winds or hail because it contains a built in "self-destruct mechanism" namely, the downdraft circulation induced by the rainshaft (Fig. 8.48c). In the absence of vertical wind shear the thunderstorm has no way of ridding itself of the precipitation it produces without destroying the buoyant updrafts that feed it.

b. Multicell storms

Multicell storms are characterized by a succession of cells, each evolving through its own cycle as in a single cell storm and, in the process, promoting the development of new cells. Under conditions of weak vertical wind shear, the storms tend to be poorly organized and the relationship between the individual cells so weak as to be barely discernible. However, when the shear is strong, the individual cells may be so tightly integrated that they lose their own identity to the larger scale and/or longer lived entity of which they are a part. The mode of organization of multicell storms also depends on the partitioning of the shear between the components aligned with and transverse to the wind itself. A prominent feature of many convective storms is the *gust front*, where warm, moist boundary-layer air is lifted by the leading edge of the evaporatively cooled (and therefore relatively dense) air diverging from the base of the downdraft. New cells tend to form along the advancing gust front, sustaining the multicell storm, while older cells die out as they fall behind the gust front and become surrounded by cooler, denser downdraft air.

A schematic of an idealized symmetric multicell storm is shown in Fig. 8.49 in coordinates moving with the storm. New convective cells are shown forming in the air that is lifted by the approaching gust front. When air lifted by the gust front reaches its level of free convection, it begins to rise spontaneously under the force of its own buoyancy. Water vapor condenses onto cloud droplets and ice particles in the updraft and, when the particles grow sufficiently heavy, their fall speeds exceed the updraft velocity and they fall out. Dry environmental air, with low equivalent potential temperature is shown entering the storm from the rear at middle levels. As precipitation particles fall out of the updraft into this dry air, they partially evaporate and, in so doing, they cool the air toward its wet bulb temperature. As the air cools it becomes negatively buoyant relative to its surroundings and begins to sink. The frictional drag of the falling precipitation particles produces an additional downward force, intensifying the downdraft. The updraft air is shown exiting the storm on the downwind side, forming an anvil that may extend 100 km or more in advance of the storm. Hence, from the perspective of a ground-based observer, the passage of this storm would be marked by thickening high overcast, followed by the approach of a much

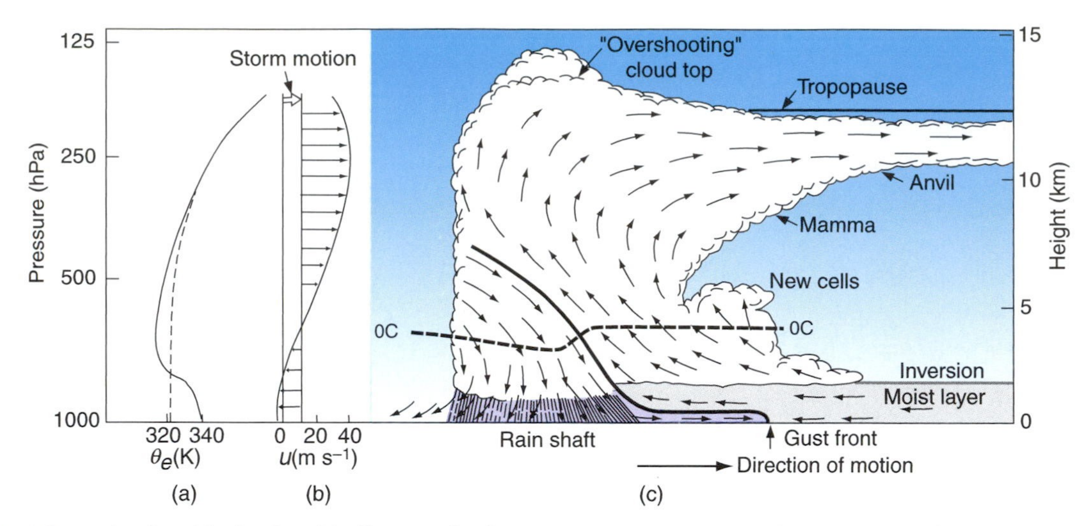

Fig. 8.49 Schematic of an idealized multicell storm developing in an environment with strong vertical shear in the direction of the vertically averaged wind. The vertical profile of equivalent potential temperature θ_e in the environment is shown at the left, together with the wind profile. Arrows in the right panel denote motion relative to the moving storm.

lower, darker cloud base, followed by an abrupt wind shift and temperature drop that marks the arrival of the *gust front*. Heavy precipitation would not begin until a few minutes after the passage of the gust front and might include hail. The shallow pool of cool, moistened downdraft air left behind by the storm may persist for hours, inhibiting the development of further convection.

c. Supercell storms

The distinguishing characteristic of the supercell storm is its rotating updraft that is clearly evident in Fig. 8.50, and even more so in time-lapse photographs and in dual-Doppler radar imagery. Rotation renders the supercell storm more robust by inducing the formation of a *mesolow* (i.e., a pressure minimum) within the updraft, which is superimposed upon the hydrostatically balanced pressure field that exists by virtue of the density gradients. The mesolow forms as the rotating air is pulled outward from the center of the updraft by the centrifugal force. Pressure at the center of the updraft drops until the inward-directed pressure gradient force and the outward-directed centrifugal force come into in a state of *cyclostrophic balance*, as illustrated schematically in Fig. 8.51. Under these conditions,

Fig. 8.50 Supercell thunderstorm over north-central Kansas on May 8, 2001, with a rotating updraft and a shaft of heavy rain and hail. [Courtesy of Chris Kridler.]

$$\frac{v^2}{r} = \frac{1}{\rho}\frac{\partial p}{\partial r} \tag{8.5}$$

where v is the speed of the air circulating around the updraft at radius r.[16]

[16] Geostrophic balance and cyclostrophic balance are special cases of the more general, three-way gradient wind balance discussed in Section 7.2.7. In the case of geostrophic balance, the centrifugal force is neglected, while in the case of cyclostrophic balance the Coriolis force is neglected. Which forces need to be retained and which ones can be neglected depend on whether the vorticity of the system under consideration is much smaller than, comparable to, or much larger than the planetary vorticity; for the specific case of circular vortices, it depends on the rotation rate, as shown in Exercise 8.9. In contrast to flow that is in geostrophic balance, cyclostrophic flow can circulate in either direction around a mesolow and, indeed, both clockwise and counterclockwise circulations have been observed.

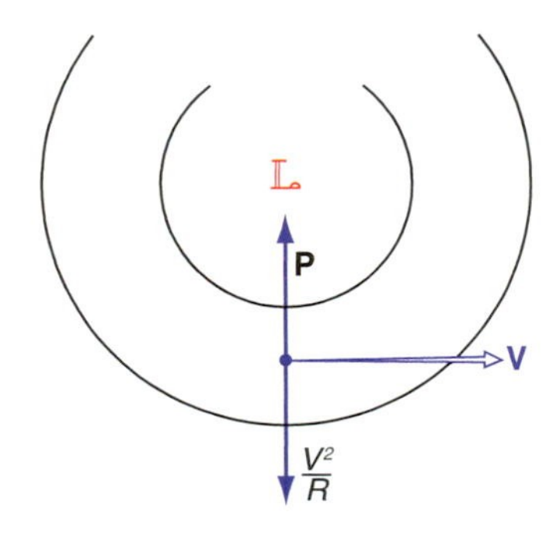

Fig. 8.51 Balance of forces in flow that is in a state of cyclostrophic balance. **P** denotes the pressure gradient force, **V** the horizontal wind vector, V the scalar wind speed, and R the radius of curvature.

Exercise 8.3 Air at cloud base in a supercell updraft is observed to be in solid body rotation out to a radius of 2 km with a period of 15 minutes. Estimate the amplitude of the dynamically-induced radial pressure gradient. The density of the air at cloud base is 1 kg m^{-3}.

Solution: Let p_D be the dynamically induced pressure perturbation (in Pa), r the radius, and ω the rotation rate. Assuming cyclostrophic balance,

$$\frac{1}{\rho}\frac{\partial p_D}{\partial r} = \omega^2 r = \left(\frac{2\pi}{15 \times 60\text{s}}\right)^2 \times 2 \times 10^3\,\text{m} \simeq 0.1\,\text{m s}^{-2}.$$

Multiplying by the density of the air, we obtain $\partial p_D/\partial r \sim 0.1$ Pa m^{-1} = 1 hPa km^{-1}. ■

The intensity of a dynamically induced mesolow is thus proportional to the square of the rotation rate, which tends to be small at the Earth's surface, but increases rapidly with height in the lower part of the updraft. Hence the dynamically induced pressure deficit $-\delta p_D$ at the center of the updraft amplifies with height, giving rise to a hydrostatically unbalanced, upward-directed pressure gradient force $(1/\rho)\partial p_D/\partial z$ below the updraft, which reenforces the upward acceleration of the air beneath the updraft. It is this dynamic (nonhydrostatic) contribution to the three-dimensional pressure distribution that renders super-cell storms more intense and longer lasting than ordinary thunderstorm cells.

Exercise 8.4 Estimate the magnitude of the upward acceleration at the base of the updraft due to the presence of a dynamically induced pressure perturbation p_D that increases from zero at the ground to 1 hPa at an altitude of 1 km above the ground. The density of the air is ~1 kg m^{-3}.

Solution:

$$\frac{dw}{dt} = -\frac{1}{\rho}\frac{\partial p_D}{\partial z} \simeq \frac{1}{1} \times \frac{100\text{ Pa}}{1000\text{ m}} = 0.1\text{ m s}^{-2}$$

Over an interval of 1 min, a force of this magnitude would impart a vertical velocity of 6 m s^{-1} to the air beneath the updraft. The importance of the dynamically induced pressure gradient force in maintaining the updraft is further illustrated in the following exercise. ■

Exercise 8.5 How large a temperature perturbation would be required to impart a hydrostatically balanced vertical acceleration equivalent to that in the previous exercise?

Solution: In Exercise 3.11 it was shown that

$$\frac{\partial^2 z}{\partial t^2} = g\frac{T' - T}{T} = 0.1\text{ ms}^{-2}$$

where T' is the perturbed temperature. Substituting $\rho = 1$ kg m^{-3} and $T = 288$ K and solving for $T' - T$, we obtain a value of ~3 °C.[17] ■

How do supercell storms acquire their rotation? To answer this question, consider a storm developing in an environment in which wind speed is increasing with height, but without any turning of the wind with height. Suppose that an isolated, deep convective updraft develops in this environment and that it initially moves downstream with the vertically averaged steering flow. Boundary layer air is blowing into the updraft from the right and left flanks as well as from the front, as shown in the top panel of Fig. 8.52. As in the schematic shown in Fig. 8.45, the low level air flowing into the right flank of the updraft exhibits counterclockwise vorticity as viewed looking toward the left. As imaginary tubes of this inflow air bend upward into the right side of the updraft, they spin in a counterclockwise sense (looking downward). By the same

[17] This value is an overestimate because the pressure perturbation induced by the rotation of the air in the updraft extends downward, with reduced amplitude, all the way to the ground. The horizontal pressure gradient force due to the presence of the mesolow at the ground strengthens the low level horizontal inflow into the updraft.

line or reasoning, it can be argued that the left side of the updraft, which is the mirror image of the flow in the right side, will acquire a clockwise rotation.

These counterrotating vortices at midlevels of the updrafts induce negative pressure perturbations, which intensify the upward pressure gradient at the base of the updraft. That this reenforcement occurs, not in the middle of the updraft, but along its right and left flanks causes the updraft to widen and eventually split into counterrotating right- and left-moving *supercell* storms, as shown in subsequent panels of Fig. 8.52.

A casual inspection of Fig. 8.52b might suggest that the splitting of the storm is due to the spontaneous formation of a strong downdraft along its axis of symmetry. However, splitting is observed in numerical simulations of this phenomenon, even when the microphysical processes that enhance the downdraft are neglected. It is a consequence of the pressure perturbations induced by counterrotating vortices in the updraft. This storm-splitting process does not, in and of itself, require veering or backing of the environmental wind profile with height.

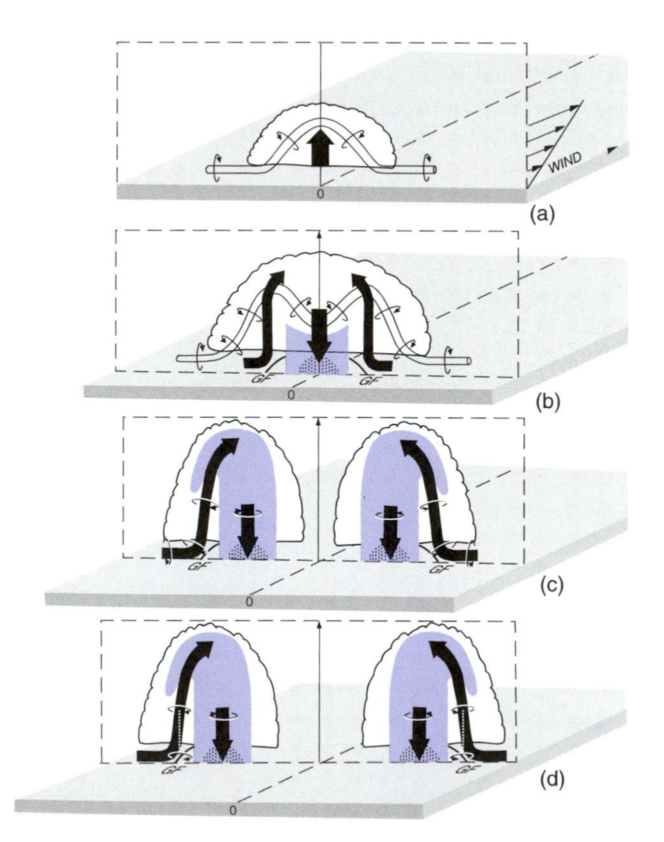

Fig. 8.52 Schematic showing the splitting of a multicell storm into right and left moving supercell storms. Bold black arrows denote updrafts and downdrafts, shading denotes radar echoes, the thin tube in the upper two panels represents a cylinder of marked boundary layer air parcels, and thin circular arrows denote the sense of the rotation of the parcels. [Reprinted from *Adv. Geophys.*, **24**, R. A. Houze and P. V. Hobbs, "Organization and Structure of Precipitating Cloud Systems, p. 263, Copyright (1982), with permission from Elsevier.]

Nearly all intense supercell storms in the northern hemisphere exhibit right-moving and counterclockwise (i.e., cyclonically) rotating updrafts. Until the 1980's it was widely believed that this bias was due to the influence of the Coriolis force, as explained in Section 7.2.9. However, it is now well established, on the basis of numerical experiments, that the contribution of the planetary vorticity to the spin of the updraft of a supercell storm is negligible. The prevalence of right-moving storms is due to the prevalence of clockwise-turning hodographs, like the example in Fig. 8.44b, in large-scale settings conducive to the formation of supercell storms. Hodographs with unidirectional wind shear, like the example in Fig. 8.44a, are equally conducive to left- and right-moving storms. Why veering of the wind with height should favor counterclockwise-rotating storms that move to the right of the steering flow in the northern hemisphere (and vice versa in the southern hemisphere) is beyond the scope of this text. Suffice it to say that veering perturbs the pressure field in and around convective updrafts in a manner that reinforces right-moving storms and suppresses left-moving storms.

Figure 8.53 shows a composite hodograph formed by averaging the hodographs of soundings made in the vicinity of 62 tornadic supercell thunderstorms over the central United States. The average movement of the storms is toward the east–northeast (ENE), well to the right of the vertically averaged steering flow. Wind speed increases rapidly with height, especially in the lower troposphere, and wind direction veers continuously from SSE at the ground to WSW in the upper troposphere, a change in direction of almost 90°. The lower portion of the hodograph exhibits strong curvature in the same sense as the idealized profile in Fig. 8.44b.

The prevalence of cyclonically rotating updrafts in supercell storms is thus a consequence of the facts that

- thermodynamic conditions conducive to deep convection occur more frequently under conditions of warm advection than under conditions of cold advection,
- warm advection occurs in association with veering of the wind with height,
- veering of the wind with height is reflected in anticyclonically curved hodographs that favor cyclonically rotating updrafts.

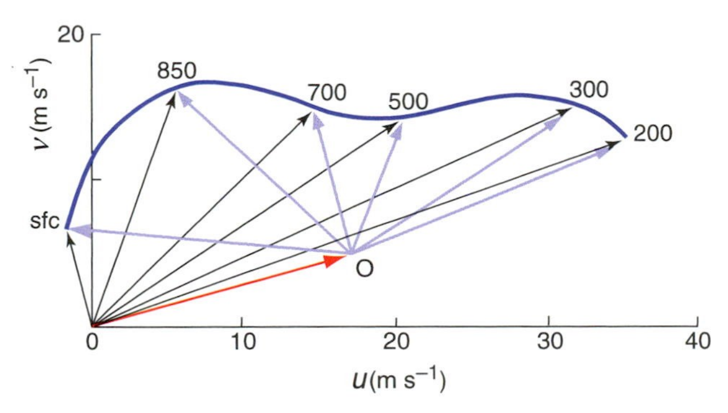

Fig. 8.53 Composite hodograph formed by averaging the hodographs of soundings made in the vicinity of 62 tornadic supercell thunderstorms over the central United States. Labels represent pressure levels in hPa. The red vector from the origin to point **O** indicates the average movement of the storms, and blue arrows represent the winds at various levels as viewed in a coordinate system moving with the storms. This plot was the basis for constructing the idealized hodograph in Fig. 8.44b. [Based on data in *Mon. Wea. Rev.*, **104**, 133–142 as adapted in *Cloud Dynamics*, R. A. Houze, p. 291, Copyright (1993), with permission from Elsevier.]

That warm advection occurs in association with veering of the wind with height is a consequence of the thermal wind equation [Eq. (7.20)], in which the Coriolis parameter $f = 2\Omega \sin\phi$ appears as a linear factor. The veering of the wind with height in the boundary layer due to frictional drag, as explained in Section 7.2.5, also contributes to the prevalence of anticyclonically curved hodographs. Hence, it can be said that the Coriolis force is indirectly responsible for the prevalence of cyclonically rotating supercell storms.

The structure of a typical supercell storm, as revealed by radar imagery, is shown in Fig. 8.54. The strong asymmetry of the patterns at the lower levels reflects the prevalence of right-moving storms. The *bounded weak echo region (BWER)* or *echo free vault* at the 4- and 7-km levels corresponds to the updraft, which is coincident with the axis of rotation. The heaviest rain and hail tend to occur in the downdraft wrapping around the northwest flank of the axis of rotation in a cyclonic sense.

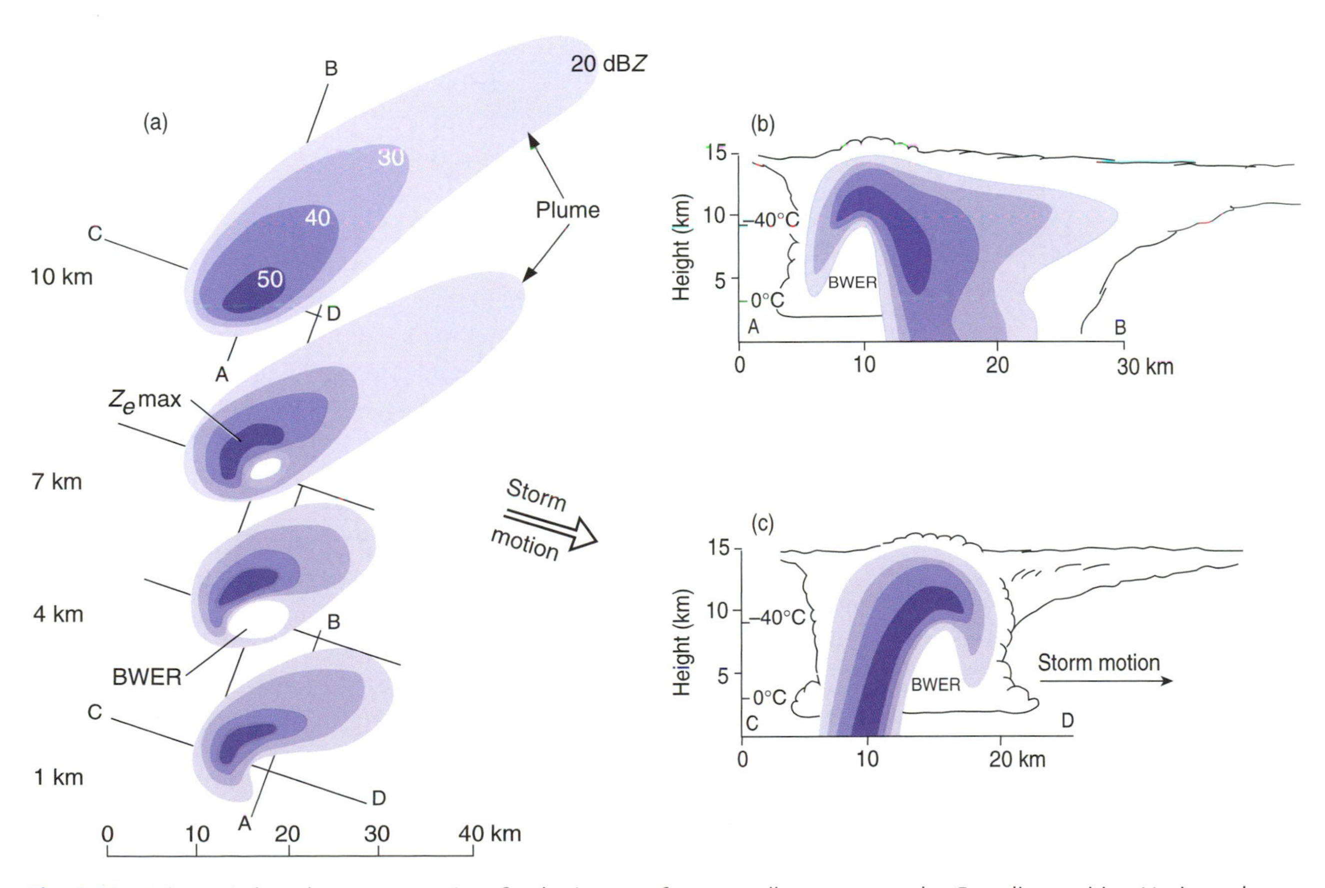

Fig. 8.54 Schematic based on a composite of radar imagery for supercell storms over the Canadian prairies. Horizontal cross sections at left and vertical cross sections at right. The reflectivity scale in dB*Z* is a logarithmic measure of rainfall intensity. BWER denotes the bounded weak echo region that is the signature of the updraft and Z_e max denotes the strongest echoes. [Reprinted from *Cloud Dynamics*, R. A. Houze, p. 293, Copyright (1993), with permission from Elsevier. Adapted from A. J. Chisholm and J. H. Renick, Preprints, International Cloud Physics Conference, London (1972).]

The structure of the storm at the Earth's surface is shown in Fig. 8.55. The mesocyclone marked by the **M** symbol centered on the tornado and the configuration of the gust front, depicted in cold front symbols, resemble a miniature extratropical cyclone. Most of the low level inflow into the rotating updraft takes place along the stationary segment of the gust front to the east of the mesocyclone. Relative to the moving storm, the inflow is westward, parallel to the front. Along section AA′, warm, moist boundary-layer is streaming northward up and over the gust front, while evaporatively-cooled downdraft air remains in place and, in some cases, even advances southward below it. The vertical shear of the air flowing through AA′ into the updraft is thus enhanced by a factor of 3 or more relative to that of the undisturbed environment due to the presence of the gust front, and the rate of vorticity generation by the tilting of the boundary layer air into the updraft is correspondingly increased. Hence, under favorable environmental conditions, the rotation of the right-moving storm formed by the splitting of a multicell storm can be self-amplifying.

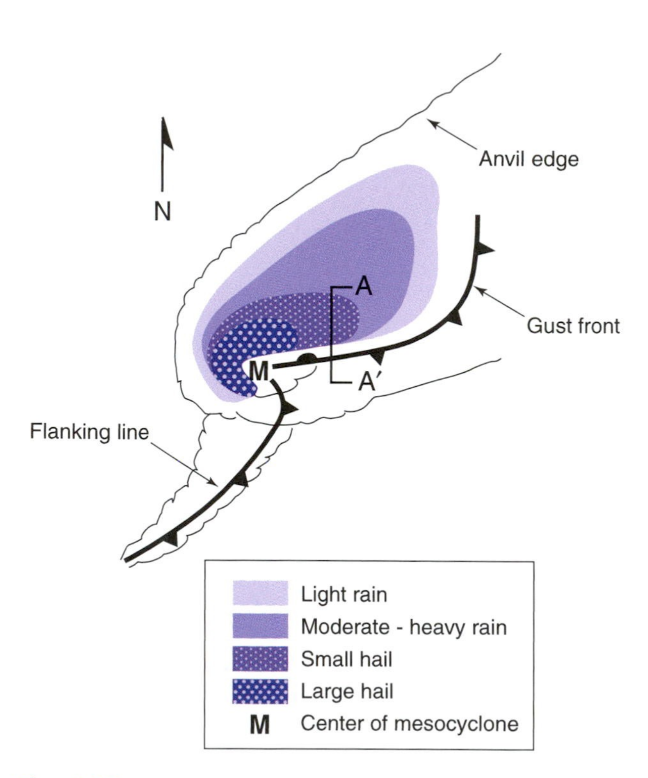

Fig. 8.55 Idealized structure of a right-moving supercell storm. [Based on NOAA National Severe Storms Laboratory publications. Reprinted from *Cloud Dynamics*, R. A. Houze, p. 279, Copyright (1993), with permission from Elsevier.]

- The vertical shear of the environmental air flowing into the storm strengthens the rotation of the updraft and the associated mesocyclone.
- The existence of the mesocyclone, in turn, amplifies the updraft and strengthens the gust front.

Distinctive features of the cloud configuration of a right-moving supercell storm are illustrated schematically in Fig. 8.56, and examples of the associated cloud formations are shown in Figs. 8.57–8.60.

8.3.3 Damaging Winds Associated with Convective Storms

The severe weather that occurs in association with convective storms is often accompanied by, and sometimes dominated by, strong winds associated with tornadoes, gust fronts, and downbursts.

a. Tornadoes

A *tornado*[18] is a rapidly rotating column of air, in contact with the ground, either hanging from or positioned beneath the base of a deep convective cloud. Tornadoes are rendered visible by the lowered lifting condensation level and by the ground debris that they raise, both of which are evident in the examples shown in Figs. 8.61 and 8.62. The distinctive funnel cloud that often appears in association with tornadoes is an indicator of the low atmospheric pressure that prevails in the interior of the vortex. The value of atmospheric pressure that defines the outline of the cloud corresponds roughly to the pressure at the lifting condensation level of the environmental air, although it may be somewhat higher if appreciable quantities of premoistened downdraft air are being entrained into the updraft. Damaging winds associated with tornadoes extend outside the outline of the *funnel cloud* and may be present even in the absence of a well-defined funnel cloud. The airflow within and around a tornado is more complex than

[18] From the Spanish verb *tornar*, to twist.

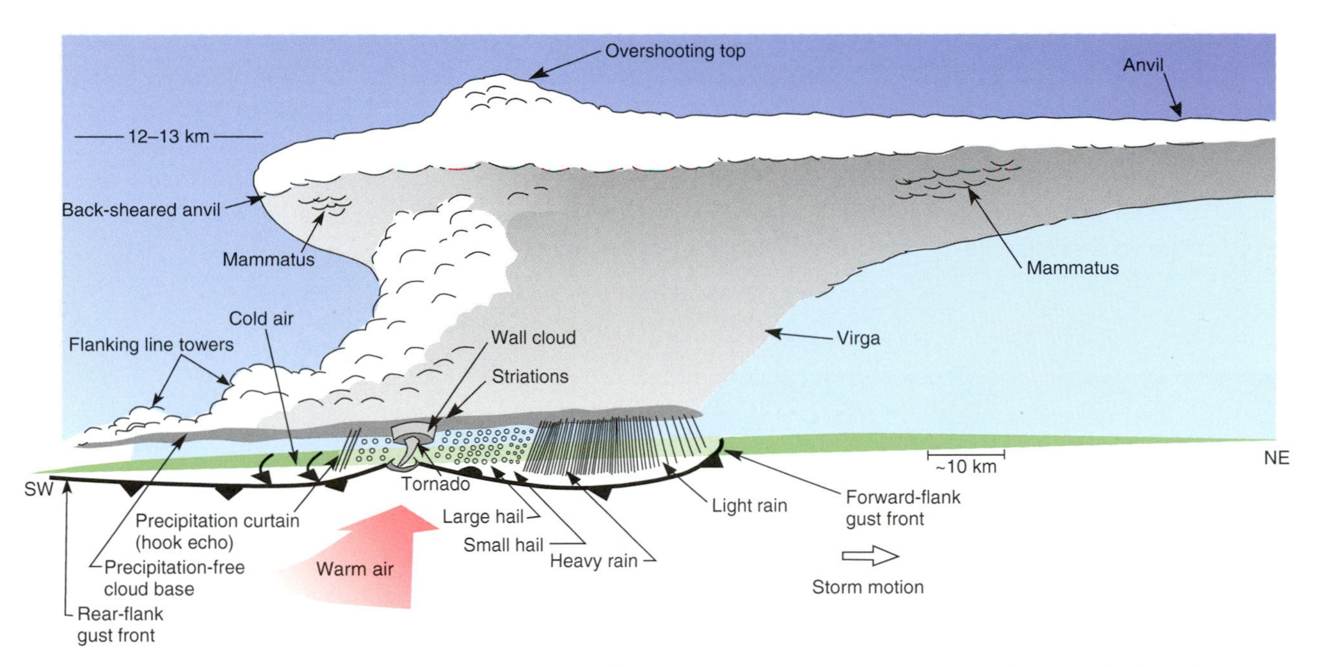

Fig. 8.56 Structure of a typical tornadic supercell storm.[19] Motion of the warm air is relative to the ground. [Based on NOAA National Severe Storms Laboratory publications and an unpublished manuscript by H. B. Bluestein. Reprinted from *Cloud Dynamics*, R. A. Houze, p. 279, Copyright (1993), with permission from Elsevier.]

would be suggested by the compact smooth shape of the funnel cloud. Depending on the ambient lighting, funnel clouds may be dark or light in color. Tornadoes come in a variety of sizes and shapes and a single tornado may have multiple vortices.

The radial profile of tangential wind speed and vertical velocity in tornadoes are not well known, but for assumed profiles, the pressure at the center of a tornado can be estimated on the basis of measurements of the maximum tangential wind speed, assuming that the vortex is in cyclostrophic balance, as demonstrated in the following exercise.

Exercise 8.6 The circulation around many tornadoes resembles a *Rankine vortex*, which is characterized by solid body rotation for radius $r < r_0$ and irrotational flow with tangential velocity v inversely proportional to radius for $r > r_0$. Assuming that the flow is in cyclostrophic balance, estimate the pressure deficit δp (relative to the ambient air pressure) in a tornado with a maximum velocity of 100 m s^{-1}. The density of the air is 1.25 kg m^{-3}. Ignore the variation of air density with pressure.

Solution: The flow is in cyclostrophic balance, so that

$$\frac{1}{\rho}\frac{dp}{dr} = \frac{v^2}{r}$$

Hence, if we ignore the variation of density

$$\delta p = \rho \int_0^\infty \frac{v^2}{r}\, dr$$

Substituting $v = v_0(r/r_0)$ for $r < r_0$ and $v = v_0(r_0/r)$ for $r > r_0$, where v_0 is the tangential velocity at radius r_0, we obtain

$$\begin{aligned}
\delta p &= \frac{\rho v_0^2}{r_0^2}\int_0^{r_0} r dr + \rho v_0^2 r_0^2 \int_{r_0}^{\infty} \frac{dr}{r^3}\, dr \\
&= \frac{\rho v_0^2}{2} - \left(-\frac{\rho v_0^2}{2}\right) \\
&= \rho v_0^2
\end{aligned}$$

[19] Supercell storms can be divided into three categories: classic supercells that resemble this schematic; low precipitation (LP) supercells in which the anatomy of the rotating updraft is often clearly visible; and high precipitation (HP) supercells in which heavy rain (often accompanied by hail) wraps around the trailing side of the mesocyclone, obscuring any embedded tornadoes. In contrast to most classic supercells, HP storms develop strong rotation along the forward (eastern) flank of the gust front.

Substituting $v_0 = 100$ m s^{-1}, $\rho = 1.25$ kg m^{-3}, then $\delta p = 1.25 \times 10^4$ Pa $= 125$ hPa. This estimated value is consistent with the observation that funnel clouds often extend all the way down to the ground. Pressure deficits in excess of 40 hPa have been documented at the Earth's surface, and it is conceivable that even larger ones would be observed if it were possible to obtain measurements in the middle of tornadoes. The lifting of dense objects by tornadoes is suggestive of the existence of upward pressure gradient forces many times the gravitational acceleration g (i.e., pressure decreasing with height at a rate far in excess of the hydrostatic value of ~1 hPa per 8 meters). ■

Fig. 8.57 *Wall cloud* developing along the base of a bell-shaped rotating updraft at the center of the mesocyclone of a supercell storm. The view is looking SW. Observers located to the south of the storm at this time observed a tornado that rapidly became enshrouded in rain. [Courtesy of Brian Morganti.]

Fig. 8.59 Looking north toward the approaching forward flank of a strong gust front marked by a distinctive *arcus cloud* with strong counterclockwise rotation about a horizontal axis pointing into the page. [Courtesy of Kathryn Piotrowski.]

Fig. 8.58 *Shelf cloud* at the base of the updraft of a supercell storm looking southward along the forward flank of the gust front in the direction of the mesocyclone. The gust front is advancing rapidly toward the left (E), propelled by rain-cooled downdraft air. When viewed from a distance looking from left to right, rising cloud motion often can be seen in the leading (outer) part of the shelf cloud, while the underside often appears turbulent, and wind-torn, as in this image. [Courtesy of Brian Morganti.]

Fig. 8.60 Thunderstorms over the midwestern United States, just before sunset July 10, 1994, as revealed in satellite imagery. The low sun angle accentuates the overshooting cloud tops. [NASA-GSFC GOES Project.]

Fig. 8.61 Supercell tornadoes come in many different sizes, shapes and colors. (Top left) Near Sharon, Kansas, May 29, 2004, looking east. The whirling orange cloud is ground debris illuminated by the setting sun. The thinner, light gray funnel cloud extending downward from cloud base is condensed water vapor. [Photo courtesy of Kathryn Piotrowski.] (Top middle) Tornado over Attica, Kansas from the same supercell storm, looking north. Much of the ground debris is from bales of hay. (Top right) Rope-like tornado near Mulvane, Kansas, June 12, 2004, looking east. (Lower left) "Stove pipe" tornado with a smooth white condensation funnel near Big Spring, Nebraska, June 10, 2004, looking NW. Towers of power lines, faintly visible just above the horizon, provide a sense of scale. (Lower right) Wedge tornado (informal term for a large tornado with a condensation funnel that appears at least as wide at the ground as the distance from the ground to cloud base) over Argonia, Kansas, May 29, 2004, looking NNW. [Photos courtesy of Eric Nguyen.]

Fig. 8.62 Tornado over western Nebraska, June 10, 2004, in three stages of development. In the left panel the condensation funnel is just beginning to emerge from the wall cloud, but the circulation at the ground is already raising dust. The condensation funnel is more fully developed in subsequent panels. The wall cloud is clearly visible in the center and right-hand images, together with a *tail cloud* entering from the right and wrapping around the near flank of the wall cloud. [Courtesy of Kathryn Piotrowski.]

To generate a tornado the vorticity in the rotating updraft needs to reach values on the order of 10^{-2} s^{-1}, two orders of magnitude stronger than the Earth's rotation. Then the rotating air column needs to be stretched in the vertical (or more generally, in the direction parallel to its axis of rotation) to further enhance the vorticity.

The effect of the stretching in concentrating the vorticity can be understood as follows. As a closed ring of air converges toward the axis of rotation, its circulation C (or angular momentum) tends to be conserved, as explained in Section 7.2.9. Hence, v increases in inverse proportion to r. Since, from (7.3), the average vorticity within the area enclosed by the loop is equal to the circulation divided by the area enclosed by the loop, vorticity must increase in inverse proportion to the area A of the loop, i.e.,

$$\frac{1}{\zeta}\frac{d\zeta}{dt} = -\frac{1}{A}\frac{dA}{dt}$$

Making use of (7.2) and the continuity equation (7.39), this expression can be rewritten as

$$\frac{1}{\zeta}\frac{d\zeta}{dt} = \frac{d}{dt}\ln\zeta = \frac{\partial\omega}{\partial p} \tag{8.6}$$

Invoking (7.33), and bearing in mind that the vertical stretching that leads to the formation of tornadoes takes place within the lowest 1–2 km above the ground, where density is constant to within 10–20%, we can write

$$\frac{d}{dt}\ln\zeta \simeq \frac{\partial w}{\partial z}$$

Integrating and taking the antilog, we obtain

$$\zeta = \zeta_0 e^{t/T} \quad \text{where } T = (\partial w/\partial z)^{-1} \tag{8.7}$$

The stretching that ultimately leads to tornado formation is due to the upward acceleration of the air at the base of the updraft. In a typical supercell storm the rate of ascent w increases from near zero at the ground to ~3 m s^{-1} at 1 km above the ground so that $\partial w/\partial z \sim 3 \times 10^{-3}$ s^{-1}. Hence, the e-folding time T for the amplification of the vorticity is ~300 s.

Exercise 8.7 Given an ambient vorticity of 10^{-2} s^{-1} and vertical stretching that leads to the concentration of the vorticity with an e-folding time of 5 min, how long would it take to generate an axially symmetric tornado with a maximum wind speed v = 100 m s^{-1} at a radius r_0 of 200 m. Assume that the region of the tornado inside the radius of peak wind speed is characterized by solid body rotation.

Solution: From Exercise 7.1, the vorticity of the fully developed tornado is equal to 2ω, where ω is the rotation rate v/r. Hence

$$\zeta = 2\frac{v}{r} = 2 \times \frac{100 \text{ m s}^{-1}}{200 \text{ m}} = 1 \text{ s}^{-1}$$

To attain this value, the vorticity must amplify by a factor of 100 relative to the ambient value. Taking the natural log of (8.7)

$$\ln\frac{\zeta}{\zeta_0} = \frac{t}{T}$$

and substituting $\zeta/\zeta_0 = 100$ and $T \sim 5$ min, we obtain $t = 23$ min. ■

Rather than being spread uniformly (as would be the case for solid body rotation) most of the vorticity and the ascent within the interior of the tornado vortex tend to become concentrated within a narrow ring, just inside the radius of strongest winds, as shown in Fig. 8.63. Under certain conditions this ring of extremely high vorticity can break down into multiple vortices, whose signatures are clearly evident in aerial photos of debris.

The rotation in a tornadic supercell storm builds up gradually over a period of several hours, but the development of the tornado itself takes place much more quickly. Typical tornado lifetimes are on the order of tens of minutes, during which time they move with the storm. Most tornadoes eventually become surrounded by cooler, less buoyant downdraft air as the *flanking line* or rear flank gust front (Figs. 8.55 and 8.56) wraps around the mesocyclone, reminiscent of the way in which the cold air wraps around an occluding extratropical cyclone. As the mature tornado and its associated mesocyclone weaken and die, a new mesocyclone may form along the gust front, setting the stage for the formation of a second tornado. Over its lifetime a single supercell storm may spawn a family of tornadoes.

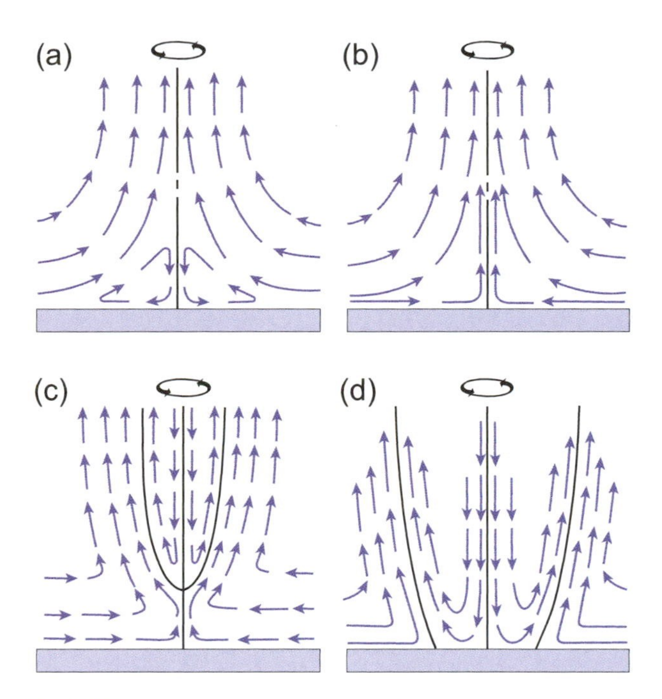

Fig. 8.63 Schematic of the flow in the interior of a tornado during various stages of development. [Based on diagrams in articles in *Thunderstorm Morphology and Dynamics*, University of Oklahoma Press, Norman (1986) 197–236 and in *Proceedings of the Symposium on Tornadoes*, Institute for Disaster Research, Texas Tech University, Lubbock (1976) 107–143, as adapted by *Cloud Dynamics*, R. A. Houze, p. 308, Copyright (1993), with permission from Elsevier.]

Non-supercell tornadoes (Fig. 8.64) form when a patch of boundary layer air with circulation about a vertical axis comes into vertical alignment with a vigorous convective-scale updraft. The source of the vorticity may be a gust front, a convergence line, or wind shear induced by flow around a topographical feature. Over a matter of 15 min or so, the vertical stretching under a 3-m s^{-1} updraft can amplify an initial vorticity perturbation by two orders of magnitude. Waterspouts and dust devils (Fig. 8.65) are smaller, less dangerous vortices that form in a similar manner.

Fig. 8.64 Non-supercell tornado that developed in thunderstorms near Denver, CO, June 15, 1988. [© 1988, courtesy of David Blanchard.]

Fig. 8.65 When sufficient low level rotation is present, dust devils can form beneath strong updrafts in dry convection. This was one of numerous dust devils observed over a recently plowed field late on an August morning near Corvallis, OR. [© 1988, courtesy of David Blanchard.]

b. Downbursts

Downward penetrating downdrafts known as *downbursts* (or, in their most concentrated form, as *microbursts*, Fig. 8.66) pose major hazards to aircraft and can sometimes produce damaging winds at the Earth's surface. In contrast to the surface wind pattern in a tornado, the wind in a microburst does not circulate: it is almost purely divergent. The remarkable strength of the winds associated with microbursts is due not to the concentration of vorticity, but to the negative buoyancy of the downdraft

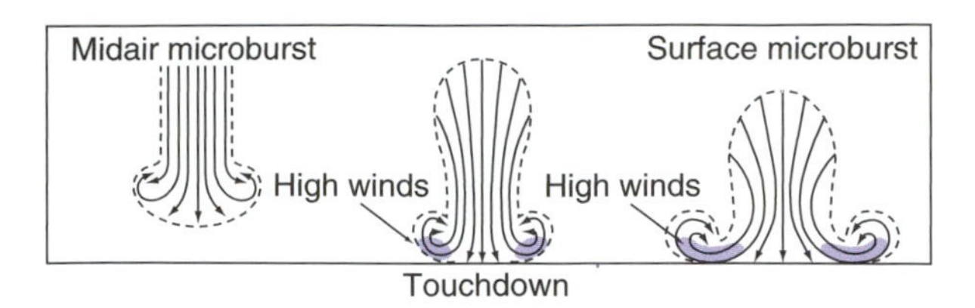

Fig. 8.66 Conceptual model of a downburst. [From T. T. Fujita,[20] *The Downburst–Microburst and Macroburst.* Reports of Projects NIMROD and JAWS, SMRP, University of Chicago, Chicago (1985).]

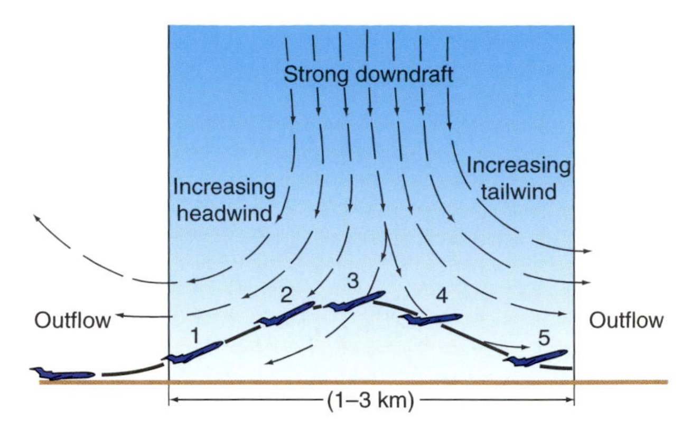

Fig. 8.67 Idealized depiction of an aircraft taking off facing into a microburst. In passing from 2 to 4, the aircraft loses its headwind and begins to experience a tailwind, causing a loss of lift, and consequently a sudden loss of altitude. [From *J. Clim. Appl. Meteorol.*, **25** (1986), p. 1399.]

air, which enables it to penetrate to within a few hundred meters of the Earth's surface before it begins to decelerate. Modeling results indicate that this negative buoyancy is due mainly to evaporative cooling and, in some cases, the melting of hail. In contrast to tornadoes, which form beneath echo-free vaults in the three-dimensional pattern of radar reflectively, microbursts tend to coincide with shafts of heavy precipitation. Vertical accelerations in microbursts are on the order of $0.01g$, which is indicative of a temperature depression of several degrees C relative to the surrounding air, and downward wind speeds of several meters per second extending down to just a few hundred meters above the ground.

The aviation hazard posed by microbursts is illustrated by Fig. 8.67. The danger lies not so much in the loss of altitude due to the downdraft itself, but in the sudden loss of lift that occurs in response to the rapid change in the horizontal velocity when the plane passes through the downdraft, during which time the wind speed relative to the plane may drop by 10 m s^{-1} or more in a matter of 20 s.

c. Gust fronts and derechos

The outflow from downdrafts in convective storms forms a pool of cool, dense air, which displaces the warmer, more buoyant environmental air as it spreads out at the Earth's surface. The advance of the downdraft air tends to be concentrated in the so-called *gust front* on the forward side of the storm relative to the direction of propagation. In coordinates moving with the storm, the gust front is more or less stationary, with warm environmental air approaching it and riding up over it as shown in Fig. 8.68. Gust fronts, sharp cold fronts, and incursions of salt water advancing into estuaries on the incoming tide are examples of a class of fluid flows referred to as *density currents*, which have been extensively investigated in laboratory and numerical modeling simulations. In the laboratory, and sometimes in nature, the interface between the denser and the lighter fluids is rendered visible by suspended particles, as in Fig. 8.69.

A distinctive feature of the gust front (Fig. 8.68) is the bulbous *head*, with its overturning circulation in which surface wind speeds exceed the rate of advance of the front itself. At the leading edge of the head is a pressure maximum at the Earth's surface, with an amplitude on the order of 1 hPa and a horizontal scale of just a few kilometers.[21] An observer at a fixed point at the Earth's surface would experience a pressure surge coincident with the passage of the gust front. In accordance with the horizontal equation of motion, the observer would note a rapid increase in wind speed (toward the right) as the front approaches, peak winds coincident with the passage of the front, and a more gradual decline in wind speed thereafter. The lifting of the warm, moist environmental air above

[20] **Tetsuya Theodore Fujita** (1920–1998). Professor at the University of Chicago, noted for his innovative research on tropical cyclones, tornadoes, and downbursts. Coined the term *microburst* and, based on evidence from aerial photographs of tornado debris, developed a scale for classifying the intensity of tornadoes.

[21] The pressure distribution at the earth's surface in the vicinity of the gust front is made up of a hydrostatic contribution from the excess weight of the cool downdraft air in the head and a dynamical contribution induced by the abrupt deceleration of the downdraft air as it approaches the gust front.

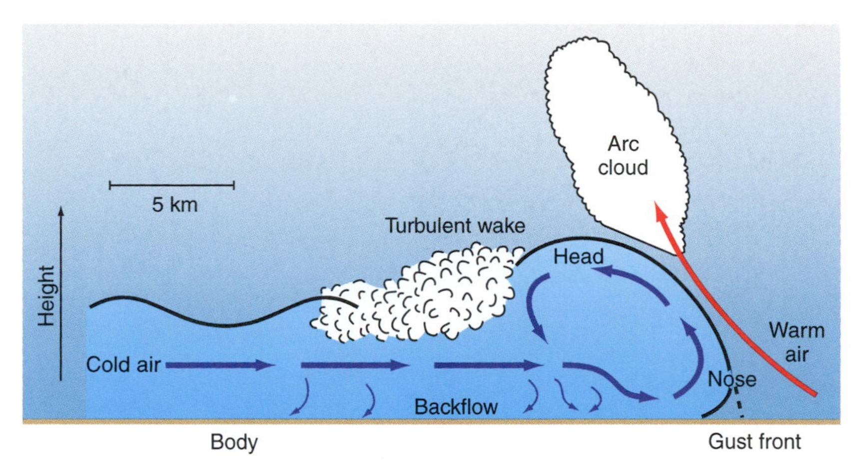

Fig. 8.68 Schematic cross section through a gust front, in coordinates moving with the front. [From *J. Atmos. Sci.*, **44** (1987) p. 1181.]

Fig. 8.69 The nose of a gust front from a thunderstorm cell near Denver, Colorado, as revealed by a cloud of blowing dust. [© 1991, courtesy of David Blanchard.] Convectively generated sand and dust storms are common in the Middle East and sub-Saharan Africa, where they are referred to as *haboobs*.

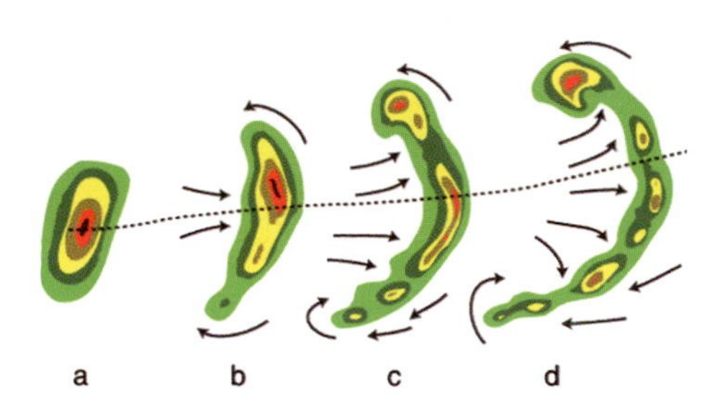

Fig. 8.70 Schematic showing the radar echo of a thunderstorm (a) evolving into a bow echo (b,c) and finally into a comma echo (d) in the northern hemisphere as it moves eastward. Arrows indicate surface winds relative to the moving system. The regions of cyclonic and anticyclonic rotation at the ends of the echo are favored sites for tornado development and the axis of strongest winds is indicated by the dashed line. [Adapted from *J. Atmos. Sci.*, **38** (1981) p. 1528.]

the head is sometimes revealed by a laminar *arcus cloud*, which moves with the gust front (Fig. 8.59). When an arcus cloud becomes detached from the deep convective cloud that spawned the gust front, it is referred to as a *roll cloud*.

Surface winds observed in association with gust fronts are especially strong when the leading edges of the downdrafts from a series of strong thunderstorms coalesce to form an unbroken line of high winds. These intense, long-lived wind-storms, known as *derechos*,[22] tend to develop along forward-curving segments of lines of convection whose signatures in radar imagery are referred to as *bow echoes* (Fig. 8.70).

[22] From the Spanish word for straight ahead.

8.3.4 Mesoscale Convective Systems

Like living organisms, mesoscale convective systems have an identity that transcends the cells (i.e., the individual single-, multi- and supercell convective storms) of which they are made up: in particular, they exhibit larger spatial dimensions and longer lifetimes. Another distinguishing characteristic of mesoscale convective systems is the coexistence of convective and stratiform precipitation. The convective elements within mesoscale convective systems exhibit varying degrees and types of organization, depending on the large-scale environment in which they form. The most distinctive forms of mesoscale organization are *squall*

Fig. 8.71 A squall line over the Gulf of Mexico April 7, 1984 in NOAA GOES satellite imagery (inset) and a high-resolution photograph taken by the crew of the NASA *Challenger* space shuttle.

lines composed of groups of cells arranged in a line behind a long, continuous gust front, with or without trailing stratiform precipitation, and long-lived *mesoscale convective vortices*. Even the weakly organized mesoscale convective systems can be important as rain producers, especially when they occur in association with recurrent climatological features such as the *Mei-yi front* over China or *Baiu front* over Japan or stalled synoptic-scale weather systems.

a. Squall lines

Squall lines are frequently observed over convectively active midlatitude regions such the central United States during spring and summer and they occur over parts of the tropics as well. Tropical squall lines usually propagate westward, whereas midlatitude squall lines propagate eastward. The examples shown in this section pertain to midlatitude squall lines, but many of the same features appear, with reverse symmetry, in their tropical counterparts.

In satellite imagery (Fig. 8.71) squall lines appear as cone-shaped cloud masses, with clouds emanating from the apex. Several overshooting cloud tops, the signature of thunderstorm updrafts, are visible in the high-resolution image. Radar imagery (Fig. 8.72) reveals the internal structure of a typical squall line in greater detail. A line of heavy convective precipitation is observed along the leading edge of the storm where the gradient of radar reflectivity is particularly strong. Serrations in the line with a spacing of 5–10 km and

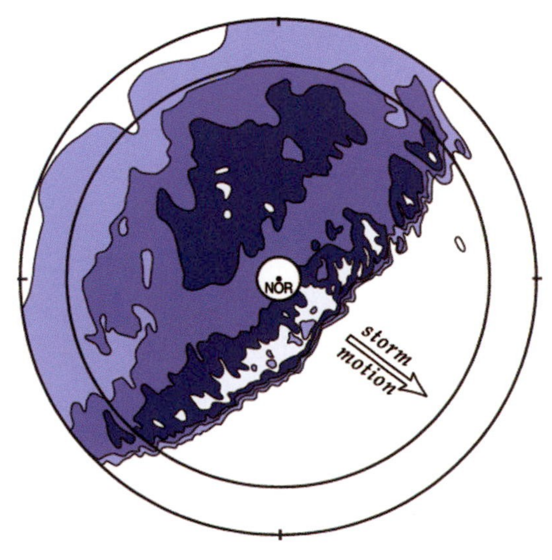

Fig. 8.72 Pattern of radar reflectivity in a squall line over Oklahoma. [From *Mon. Wea. Rev.*, **118** (1990) p. 622.]

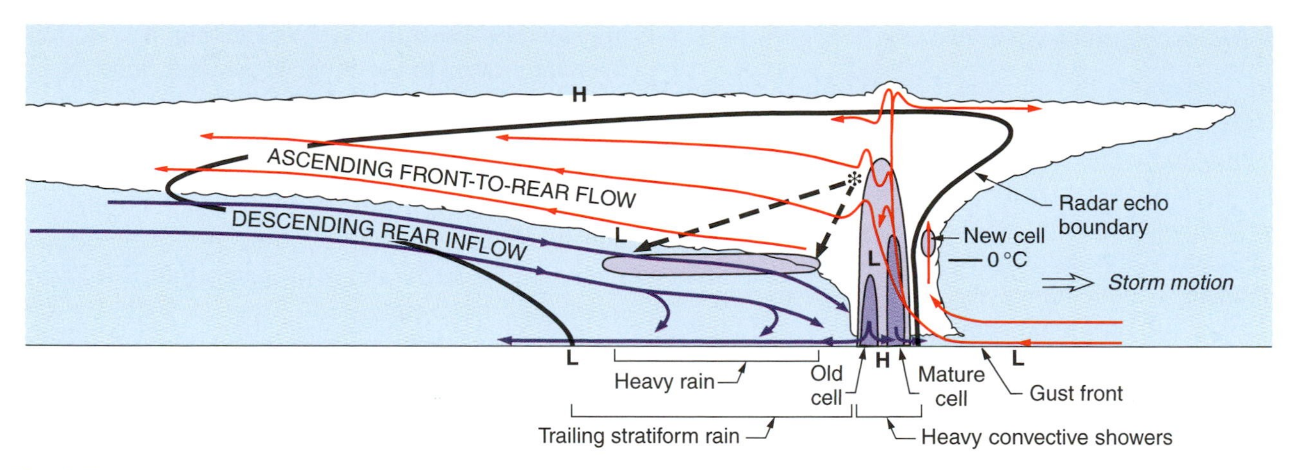

Fig. 8.73 Cross section through an idealized squall line. [Reprinted from *Cloud Dynamics*, R. A. Houze, p. 349, Copyright (1993), with permission from Elsevier.]

oriented at an angle relative to the direction of propagation of the line are reminiscent of the rolls or cloud streets shown in Fig. 9.27 and 9.34, but on a somewhat larger scale. The rainfall rate drops off immediately behind the line, but a secondary maximum is observed in an amorphous region trailing behind the line. The secondary maximum corresponds to a region of stratiform precipitation falling from the older convective cells, which merge to form a more uniform rain area characterized, in the radar reflectivity, by a bright band at the melting level, which is clearly apparent in vertical cross sections (not shown).

A conceptual model of the airflow, microphysical properties, and radar echoes in a midlatitude squall line is shown in Fig. 8.73. The aging convective cells are embedded in the ascending front-to-rear flow, which was once part of a more vigorous convective updraft. Evaporatively cooled air entering from the rear of the line occupies the middle and lower levels to the rear of the convective updrafts. The break between convective and stratiform precipitation corresponds to the region of weak subsidence behind the active convective cells.

The partitioning between convective and stratiform rainfall determines the character of the vertical profiles of vertical velocity and divergence in squall lines. Convective rainfall occurs in association with intense low level convergence into deep updrafts and equally strong divergence in the anvils, where the updrafts encounter their equilibrium level. In accordance with the continuity equation, the divergence tends to be small in the middle troposphere, where the vertical velocity changes relatively little with height, as shown schematically in Fig. 8.74. In contrast, the stratiform rainfall in squall lines is associated with midlevel convergence into the evaporatively cooled downdrafts and low-level divergence. The combined profiles, which are a linear combination of the convective and stratiform profiles, tend to be characterized by elevated peak vertical velocities and more concentrated upper level divergence than in the pure convective profiles.

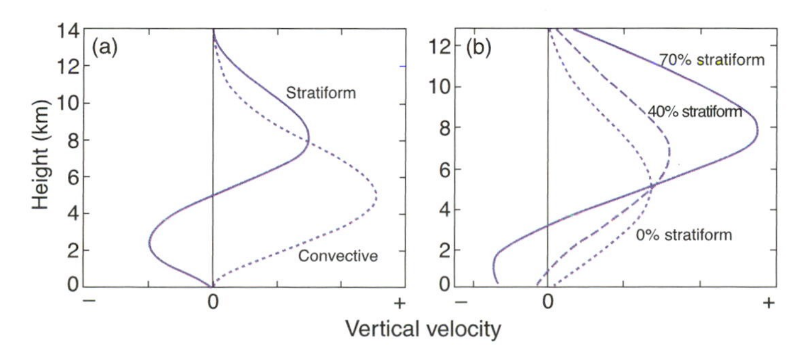

Fig. 8.74 Idealized vertical profiles of vertical velocity observed in association with (a) convective and stratiform precipitation and (b) prescribed linear combinations of convective and stratiform precipitation. [From *Rev. Geophys.*, **42**, 10.1029/2004RG000150 (2004) p. 3. Adapted from *J. Atmos. Sci.*, **61** (2004) p. 1344.]

b. Mesoscale convective vortices

The structure of mesoscale convective vortices (MCVs), depicted in Fig. 8.75, is similar in many respects to that of squall lines, with coexisting regions of convective and stratiform precipitation. The distinguishing characteristic of MCVs is that they exhibit significant rotation, which acts like a flywheel, enhancing their longevity. As in extratropical cyclones, the rotational component of the wind field is in thermal wind balance with the temperature field. However, MCVs differ from the systems associated with baroclinic waves in the sense that the convective system comes first. The rotation, which is virtually always in a cyclonic sense, develops gradually, over the course of a day or so, in response to the concentration of planetary vorticity in the air converging into a preexisting convective system. The rotation is strongest at midtropospheric levels. Anticyclonic outflow is observed in the upper troposphere and the flow at the Earth's surface also tends to be anticyclonic. In accordance with the thermal wind equation, MCVs are warm relative to their surroundings in the upper troposphere and "cold core" in the lower troposphere. The surface winds in MCVs tend to be weak, but the rainfall rates can be quite appreciable. Long-lived, slow-moving MCVs have been implicated in a number of summer flood events.

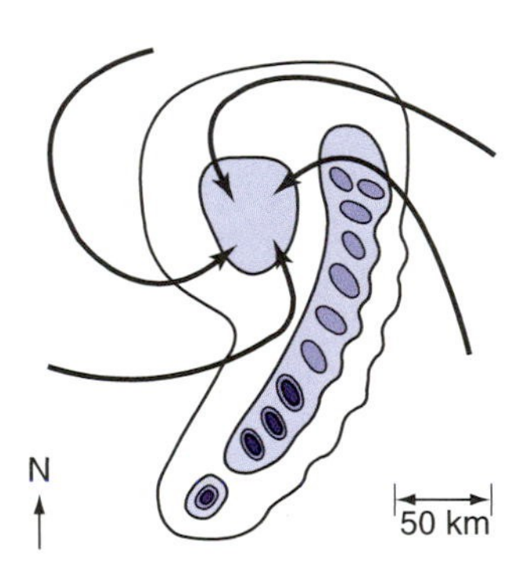

Fig. 8.75 Idealized distribution of radar reflectivity in a mesoscale convective system with rotation. [From *Bull. Amer. Meteorol. Soc.*, **70** (1989) p. 611.]

8.4 Tropical Cyclones

Tropical cyclones are smaller in scale and more axially symmetric than their extratropical counterparts (compare Figs. 1.12 and 1.13) and tend to be longer lived and more intense.[23] Tropical cyclones that achieve peak winds in excess of 32 m s^{-1} are popularly referred to as *hurricanes* or *typhoons*,[24,25] depending on where they form[26]; those with peak wind speeds between 17 and 32 m s^{-1} are referred to as *tropical storms*; and those with weaker, but identifiable circulations are referred to as *tropical depressions*. Here we use the term "tropical cyclone" in a generic sense for all of the above.

At the center of the warm core of a tropical cyclone, surrounded by a rapidly rotating wall of clouds, is the relatively calm *eye* of the storm, which is free of deep convective clouds. Most of the wind damage wrought by tropical cyclones is confined to a narrow swath a few tens of kilometers in width that experiences the passage of the *eyewall* (Figs. 8.76 and 8.77).

8.4.1 Structure, Thermodynamics, and Dynamics

In contrast to extratropical cyclones, which derive their potential energy from the ambient meridional temperature gradient, tropical cyclones derive their potential energy through the fluxes of latent and sensible heat at the air–sea interface. Like the mesoscale convective vortices (MCVs) considered in the previous section, tropical cyclones develop out of mesoscale convective systems and acquire a rotational component in cyclostrophic balance with the pressure field. However, in contrast to MCVs, tropical cyclones exhibit a warm core structure that enables them to draw energy from the fluxes at the air–sea interface.

[23] The lowest sea-level pressure recorded this far in a tropical cyclone was 870 hPa, in the eye of Typhoon Tip, in the western Pacific in 1979. Central pressures as low as 926 hPa have been documented in extratropical cyclones.

[24] In the Beaufort wind scale, winds in excess of 32 m s^{-1} are referred to as being of *hurricane force*. In current operational practice, tropical cyclones with hurricane-force winds are classified as being of *Category 1* ($33 \leq V_m \leq 42$), *Category 2* ($42 \leq V_m \leq 49$), *Category 3* ($49 \leq V_m \leq 58$), *Category 4* ($58 \leq V_m \leq 69$), or *Category 5* ($V_m > 69$ m s^{-1}) on the basis of the peak wind speed V_m.

[25] **Francis Beaufort** (1774–1857). Served as Hydrographer of the Admiralty of the British Royal Navy after being injured by sniper fire while on a patrol mission against pirates. Adaptations of his wind scale have been widely used since 1838.

[26] Tropical cyclones that form over the Atlantic and eastern North Pacific are referred to as *hurricanes*, those that form over the western North Pacific as *typhoons*, and those that form in the Indian Ocean and the South Pacific as tropical cyclones.

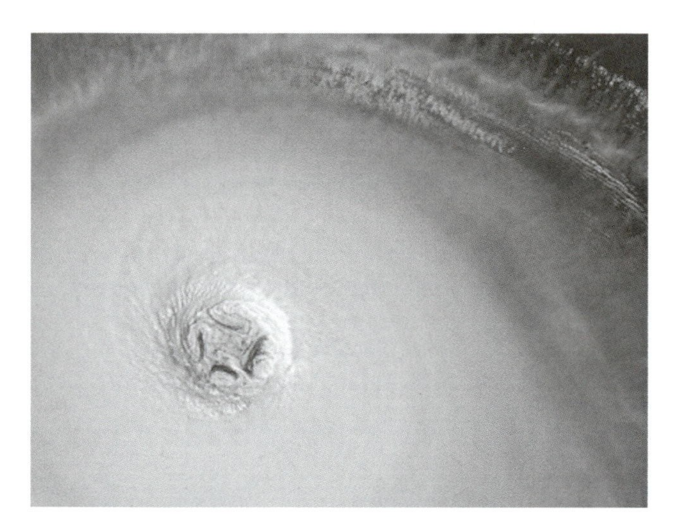

Fig. 8.76 The eye of Hurricane Isabel, passing to the northeast of Puerto Rico at 1315 UTC September 12, 2003. At this time Isabel was a category 5 storm with sustained winds of ~70 m s^{-1}. The eyewall cloud slopes radially outward with increasing height. Lower clouds within the eye itself are arranged in a symmetric pattern. [NOAA GOES-12 Satellite imagery.]

Fig. 8.77 The eye of Category-5 Hurricane Katrina late afternoon August 28, 2005 as viewed from an aircraft flying at an altitude ~3.5 km. The photograph was taken looking toward the east just to the south of the center of the eye. The darker clouds are shaded by the western eyewall cloud. The top of the eyewall cloud slopes radially outward with increasing height like the bleachers in a stadium. [Courtesy of Bradley F. Smull and RAINEX.]

The extraordinarily low pressure in the eye of a tropical cyclone at lower tropospheric levels is due to the warmth (and consequently the low density) of the overlying air. The centripetal acceleration of the flow circulating around the low pressure center is several orders of magnitude larger than the Coriolis force. It follows that the azimuthal wind and temperature fields are in cyclostrophic balance. In such flows, the radial temperature gradient and the vertical shear of the azimuthal wind component are related by an expression analogous to the thermal wind equation, which requires that at any given radius the square of the azimuthal wind speed must decrease with height at a rate proportional to the radially inward temperature gradient. Consistent with the warm core structure of these storms, the intensity of the cyclonic circulation decreases with height.

Figure 8.78 shows a radial cross section through an intense tropical cyclone. The eyewall slopes

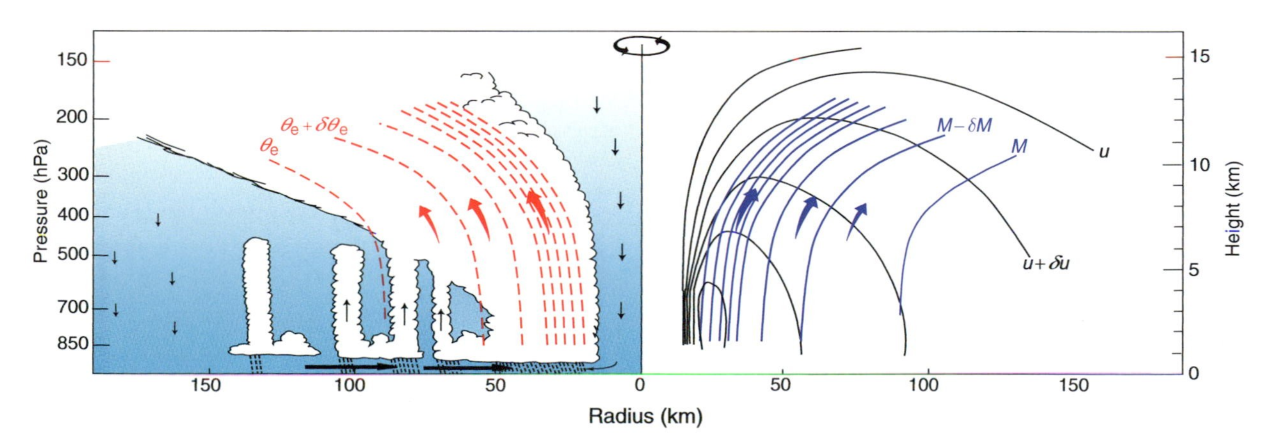

Fig. 8.78 Idealized radial cross section through an intense tropical cyclone showing the distributions of clouds, rain, radial flow and equivalent potential temperature (θ_e) on the left and azimuthal wind speed and angular momentum on the right. The θ_e contours are congruent with angular momentum contours. [Adapted from *Atmospheric Circulation Systems: Their Structure and Physical Interpretation*, E. Palmén and C. W. Newton, p. 481, Copyright (1969), with permission from Elsevier. Modifications based on figures in *Mon. Wea. Rev.*, **104** (1976) 418–442.]

radially outward with height at an angle shallow enough that the slope is visible in satellite imagery, as in the examples shown in Figs. 1.13, 8.76 and 8.77. The eyewall represents a discontinuity in the azimuthal component of the wind: a singularity in the radial shear and the vorticity. Within the eye, wind speeds are light and an inversion is often present at the top of the boundary layer, with warm, moist air below and hot, dry air above, with temperatures as high as 30 °C at the 2–3 km level and relative humidities well below 50%. The extraordinary warmth and dryness of the air within the eye indicate that the air has subsided through a depth of several kilometers. That hot air within the eye is sinking, rather than rising, implies the existence of a thermally indirect circulation in the interior of intense tropical cyclones. However, these storms derive their kinetic energy from the much more extensive thermally direct circulation outside the eyewall.

Air spirals into the storm in the boundary layer and rises in rain bands and in the ring of deep convection surrounding the eyewall, where rainfall rates range up to 5 cm h^{-1}. The low-level radial inflow is strongest just outside the ring of clouds surrounding the eyewall and it drops to near zero at the eyewall itself, while the tangential component of the flow is strongest right at the eyewall. The outflow from the interior of the storm is strongest near the level of the cloud tops and is clearly evident in time-lapse satellite imagery. Away from the eyewall the circulation at cloud-top level is predominantly weakly anticyclonic and it often exhibits pronounced axial asymmetries.

The properties of the boundary-layer air spiraling into an intense tropical cyclone are profoundly influenced by the fluxes of sensible heat, moisture, and momentum at the air–sea interface.[27]

- Much of the warming of the inflowing boundary layer air by the sensible heat flux is compensated by the adiabatic cooling induced by the cross-isobar flow toward lower pressure. Hence, the surface air temperature in the eye is typically only a few degrees higher than that of the surface air in the undisturbed tropical environment, but the potential temperature may be 5–10 °C higher.
- The flux of water vapor at the air–sea interface raises the specific humidity of the inflowing air by up to 5 g kg^{-1}, resulting in an increase in equivalent potential temperature of up to an additional 10–15 K by the time the air reaches the eyewall. The extraordinary warmth of the eyewall in the mid- and upper troposphere is due to the release of this latent heat when the air rises moist-adiabatically within the ring of forced convection just outside the outward-sloping eyewall.
- The downward transfer of azimuthal momentum to the ocean surface acts as a brake on the surface wind speeds. Were it not for this frictional torque, the circulation of the inflow would be conserved, and azimuthal velocity would increase inversely with radius. It is evident from the right-hand side of Fig. 8.78 that this is not the case: the radially inward rate of increase of azimuthal wind speed, although quite substantial, is more gradual than in a constant angular momentum profile. In the presence of frictional drag, the surface wind speed is subcyclostrophic: i.e., the radially outward centrifugal force acting on the air is not quite strong enough to balance the inward directed pressure gradient force. It is this imbalance that drives the low-level radial inflow.
- The frictional dissipation of kinetic energy in the boundary layer inflow is so intense that it constitutes a significant source of diabatic heating.
- The downward transfer of momentum through the air–sea interface induces a cyclonic circulation in the oceanic boundary layer. The centrifugal force acting on this spinning water induces a radial outflow, accompanied by upwelling near the center of the cyclone. If the mixed layer is shallow, water from below the thermocline upwells to the surface, forming a swath of cool surface water along the cyclone track. Some slow-moving storms are weakened by their self-induced pools of cool surface water.

[27] In the presence of hurricane-force winds the air–sea interface is not a continuous surface. According to the Beaufort wind scale, "The air is filled with foam and spray. Sea completely white with driving spray."

Consistent with the decrease in azimuthal wind speed with height near the eyewall, surfaces of constant angular momentum M tilt radially outward with increasing height, as indicated on the right side of Fig. 8.78. In the absence of azimuthal forces, air parcels ascending in convective updrafts within tropical cyclones conserve angular momentum once they leave the boundary layer. Hence, they must spiral outward as they rise, following the M surfaces. Were an air parcel to rise straight up rather than along a surface of constant M, it would arrive at the higher level with a higher azimuthal velocity than the ambient cyclostrophically balanced azimuthal flow and would therefore be subject to an outward-directed, unbalanced, centrifugal force that would push it outward toward the M surface. It follows that in rotating storms fueled by deep convection, surfaces of θ_e and M must be mutually parallel, with θ_e increasing radially inward and M increasing radially outward.[28]

The winds in tropical cyclones are a superposition of the forward motion of the storm and the cyclonic circulation around it. Hence, wind speeds tend to be higher to the right (left) of cyclone tracks in the northern (southern) hemisphere and low-level convergence and ascent tend to be enhanced in the right (left) forward quadrant. Beyond the ring of slantwise deep convection encircling the eyewall, much of the convection in tropical cyclones is concentrated in spiral rain bands that exhibit many of the features of squall lines.

8.4.2 Genesis and Life Cycle

Nearly all tropical cyclones form in preferred regions of the outer tropics (Fig. 8.79) where

- the Coriolis force is appreciable. That virtually all tropical storms spin cyclonically and very few of them form in the equatorial belt proves that the Earth's rotation is instrumental in their formation. The deflection of the low-level inflow by the Coriolis force is responsible for the cyclonic circulation.
- sea surface temperature T_s is sufficiently high. This restriction can be understood as follows. For tropical cyclones to intensify, the boundary layer inflow air must be able to absorb large quantities of latent heat from the underlying ocean surface in the process of becoming saturated with water vapor. The faster the rate of increase of saturation mixing ratio with temperature, the more moisture the air can pick up along its inward trajectory toward the eyewall. Due to the nonlinearity of the Clausius-Clapeyron equation (3.95), dw_s/dT increases with T. Only for temperatures above a threshold value ~26.5 °C is the rate of increase strong enough to support the development of tropical cyclones.
- the vertical wind shear between upper and lower troposphere is weak enough to enable the developing vortex to remain intact.

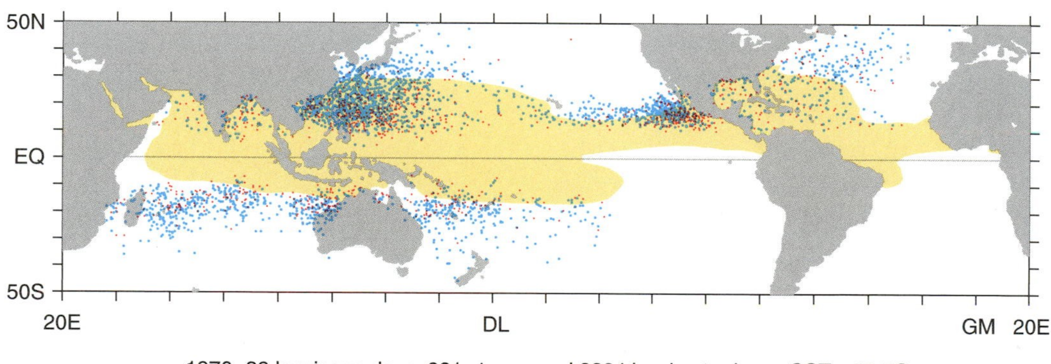

Fig. 8.79 A climatology of tropical cyclones with peak surface wind velocities in excess of 32 m s^{-1}, based on the years 1970–1989. Red dots indicate the genesis regions (i.e., the positions of the cyclones on the first day during which their peak winds exceeded 32 m s^{-1}) and blue dots indicate the positions of the same cyclones on all subsequent days on which their peak winds exceeded 32 m s^{-1}. The shading indicates oceanic regions with sea surface temperatures in excess of 27 °C. [Courtesy of Todd P. Mitchell.]

[28] Similar considerations apply to deep convection that occurs in association with strong frontal zones. The process by which deep convection along sloping trajectories fuels wind systems with strong vorticity is known as *convective symmetric instability*.

The transformation of a nonrotating mesoscale convective system into a tropical cyclone requires time for the converging low-level inflow to concentrate the ambient vorticity. A deep column in the interior of the system also needs to be moistened, eliminating the midtropospheric minimum in θ_e, thereby inhibiting the development of evaporatively cooled downdrafts. As the core finally begins to warm in response to the heating and moistening of the inflow air by the air–sea fluxes, pressure surfaces bulge upward in the upper troposphere, inducing an unbalanced outward pressure gradient force at that level. Upper tropospheric divergence, in turn, induces outflow, which causes sea-level pressure to drop. The radial sea-level pressure gradient drives an inflow of boundary-layer air, which acquires rotation due to the action of the Coriolis force. The fluxes of latent and sensible heat from the underlying ocean surface increase in response to the increasing surface wind speed. Stronger fluxes lead to stronger heating in the interior of the storm, forcing stronger upper level divergence, and so on.

Just where and when tropical cyclones develop depends on many factors. By temporarily enhancing the low-level convergence and vorticity and/or cancelling the planetary-scale vertical wind shear, a preexisting synoptic-scale disturbance (e.g., an *easterly wave* that forms over sub-Saharan Africa and propagates west-ward across the tropical North Atlantic) can create an environment in which a cyclonically rotating MCV can be transformed into a warm core tropical cyclone. Tropical planetary-waves provide week-long "windows of opportunity" for tropical cyclogenesis, which correspond to intervals in which the vertical wind shear over the region of interest is suppressed. By exploiting these relationships with the synoptic- and planetary-scale flow, weather forecasters are often able to anticipate the development of tropical cyclones well in advance of the appearance of their embryonic mesoscale convective systems.

Tropical cyclones require several days in order to reach their peak intensity, at which time the dissipation of kinetic energy in the belt of high winds just outside the eyewall approximately equals the rate of kinetic energy production in the low level, cross-isobar inflow.

Steered by the mass-weighted, vertically averaged flow, tropical cyclones typically tend to track westward for a week or so and then recurve poleward around the western flanks of the subtropical anticyclones. Typical rates of movement are on the order of 5–10 m s^{-1}. As a storm drifts westward and poleward it may vary in intensity from one day to the next as it encounters different environmental conditions and undergoes changes in its own internal structure. Some storms exhibit a pronounced *eyewall-replacement cycle* in which the ring of convection encircling the eyewall contracts with time over the course of a few days and is eventually replaced by an outer eyewall. Storms intensify during the contraction stage and weaken during the replacement stage when the decaying inner and developing outer eyewall are in competition.

Eventually, tropical cyclones either drift into higher latitudes where the sea surface temperatures are too cool to sustain them or they encounter land. Storms that make landfall are radically transformed within a matter of hours. In the absence of latent heat fluxes from the air–sea interface, the extreme low pressure within the eye cannot be sustained. As the depression in the pressure field fills, the azimuthal circulation spins down under the influence of friction. However, the extraordinarily moist "core" of a remnant tropical cyclone may retain its identity for up to a week, posing the risk of flooding should it become stalled over one particular watershed for an extended period of time. Some remnant tropical cyclones that drift poleward toward the extratropical storm track take on characteristics of extratropical cyclones and subsequently reintensify. Others become entrained into existing extratropical cyclones, sometimes leading to rapid deepening.

8.4.3 Storm Surges

The damage wrought by tropical cyclones tends to be concentrated in coastal zones where the winds that develop over the tropical oceans strike with full force, sometimes in combination with even more devastating *storm surges* from the sea. Storm surges represent a superposition of several elements:

- A wind-driven onshore current. If the water adjacent to the coast is shallow, the shoreward flow may extend all the way to the bottom, exerting a strong force on fixed objects that stand in the way of the moving water. As water is pushed against the shore by the wind stress, sea level may rise by as much as several meters.

- Wind waves and swells that carry additional shoreward momentum, intensifying shoreward flow.
- River runoff flowing into bays and inlets from rain-swollen rivers.
- A hydrostatic component, in which the depression of sea-level pressure is compensated by a rise in sea level (~1 m sea-level rise per 100-hPa depression of sea-level pressure).
- Tides, which may serve to exacerbate or reduce the severity of the storm surge, depending on the time of day of the storm and the phase of the moon.

Exercises

8.8 Explain or interpret the following.

(a) Without CIN there could be no CAPE.

(b) Tropical thunderstorms exhibit higher cloud tops than midlatitude thunderstorms, but they rarely produce damaging winds.

(c) Cumulonimbus cloud tops sometimes exhibit a bump immediately above the updraft.

(d) Severe thunderstorms occur much more frequently over the central United States during springtime than during any other season.

(e) Ordinary thunderstorm cells exhibit relatively short lifetimes.

(f) Cyclostrophically balanced flow can circulate in either direction around a center of low pressure, but it is not possible to have a cyclostrophically balanced vortex circulating around a center of high pressure.

(g) Strong updrafts in convective storms are characterized by low radar reflectivity.

(h) Tornadoes form under updrafts rather than under downdrafts.

(i) Left-moving supercell thunderstorms are relatively uncommon in the northern hemisphere, but are the prevalent form of supercell storm in the southern hemisphere.

(j) Most, but not all, tornadoes rotate cyclonically.

(k) The funnel clouds associated with large tornadoes tend to be smooth and perfectly circular, yet the patterns of debris are indicative of chaotic wind patterns that sometimes exhibit multiple vortices.

(l) In contrast to the surface wind pattern in a tornado, the wind in a microburst does not circulate: it is almost purely divergent.

(m) At mid- and upper tropospheric levels θ_e is substantially lower in the middle of the eye of tropical cyclones than in the eyewall cloud.

(n) As the "warm core" of a tropical cyclone develops, sea-level pressure in the eye drops.

(o) The centrifugal force plays a relatively more important role in the balance of forces in tropical cyclones than in extratropical cyclones.

(p) As air parcels spiral inward toward the center of a tornado or tropical cyclone, their azimuthal wind speed increases.

(q) Given two tropical cyclones with the same central pressure but different sizes, the peak wind speed is likely to be higher in the smaller storm.

(r) Coastlines with shallow waters offshore are more vulnerable to coastal surges than those with deep waters offshore.

(s) In the northern hemisphere coastal surges tend to occur to the right of the location at which a tropical cyclone makes landfall.

(t) At radii larger than the eyewall of tropical storms the vorticity tends to be relatively small, but at the eyewall itself, the vorticity is very large and difficult to define. Angular momentum exhibits a discontinuity at the eyewall.

(u) Tropical cyclones rarely if ever form within the equatorial belt or cross the equator.

(v) Tropical cyclones are almost never observed over the South Atlantic and the eastern South Pacific.

8.9 If $\omega = V/R_T$, the local angular velocity of the air moving along a cyclonically curving trajectory, and Ω is the angular velocity of the Earth's rotation, show that the flow is locally geostrophic if $\omega << f$ and cyclostrophic if $\omega >> f$.

8.10 Show that the cyclonic shear of the flow in Fig. 8.31 contributes to the relative prominence of the warm frontal zone relative to the cold frontal zone. [**Hint**: rotate Fig. 7.4b counterclockwise by 90° and apply it to the warm frontal zone.]

8.11 Derive an expression for the process by which transverse component of the vertical wind shear is converted into vorticity about a vertical axis.

Solution: Repeat the derivation of the vorticity equation (7.21) in Exercise 7.32 retaining the vertical advection term $-w(\partial \mathbf{V}/\partial z)$ in the horizontal equation of motion. Show that retention of this term leads to the additional terms

$$\frac{\partial u}{\partial z}\frac{\partial w}{\partial y} - \frac{\partial v}{\partial z}\frac{\partial w}{\partial x} - w\left(\frac{\partial v}{\partial x} - \frac{\partial u}{\partial y}\right)$$

on the right hand side. The first two terms are commonly referred to as the *tilting terms*. In the first of these terms, $-\partial u/\partial z$ represents the clockwise vorticity about the y axis and $\partial w/\partial y$ represents the tilting of the axis of spin of lines of air parcels initially aligned with the y axis into the vertical. The second term has an analogous interpretation for vorticity about the x axis and the third term represents the vertical advection of relative vorticity. ■

8.12 In Exercise 8.6 estimate the pressure deficit at the radius of maximum wind speed.

8.13 Suppose that Dorothy's house[29] had a cross sectional area of 200 m^2, a mean height of 5 m and a mass (including Dorothy and her dog) of 5 metric tons and that when the tornado passed overhead, the decrease in pressure with height was just enough to gently lift the house off the ground. Estimate the rate of decrease of pressure with height at ground level under the tornado. Assume an ambient air density of 1 kg m^{-3}.

8.14 The vertical velocity in a downdraft 500 m above the Earth's surface is 4 m s^{-1} and the radius of the downdraft is 3 km. Estimate the speed of the outflow from the base of the downdraft, averaged over the lowest kilometer. Neglect the vertical variation of density with height and use a value of 1 kg m^{-3}.

8.15 If x is the direction of movement of the gust front and u is the velocity component in that direction, making use of (1.3) and (7.11) show that if the Coriolis force and friction are neglected, $\partial/\partial y = 0$, and the vertical velocity is zero,

$$\frac{d}{dx}\left(\frac{u^2}{2} + \frac{p}{\rho}\right) = 0$$

If U_e is the speed of the u component of the wind in the undisturbed environmental air in advance of the gust front, U_f is the speed of the u component at the gust front (i.e., at the peak of the pressure surge), and δp is the amplitude of the pressure surge, show that

$$\delta p = \rho_0 (U_f - U_e)^2$$

If $\delta p = 1$ hPa, $\rho = 1.25$ kg m^{-3}, and $U_e = 5$ m s^{-1}, estimate the wind speed at the gust front, ignoring the effect of friction.

8.16 As boundary layer air spirals inward into the eye of a tropical cyclone, its temperature remains fixed at 27 °C while the water vapor mixing ratio increases from 15 to 21 g kg^{-1} and the pressure drops from 1012 to 940 hPa. Estimate the resulting increase in its equivalent potential temperature θ_e.

8.17 The lowest sea-level pressure ever observed in a tropical cyclone was 870 hPa in the center of Typhoon Tip in the western Pacific in 1979. Suppose that the 200-hPa pressure surface in this storm was flat. Estimate the ratio of the vertically (with respect to ln p) averaged virtual temperature of the air in the eye of the storm to that in the large-scale environment, and the corresponding temperature difference.

8.18 Just outside the eyewall of an intense tropical cyclone, at a radius of 10 km, the azimuthal

[29] Dorothy is the heroine in L. Frank Baum's childrens' story, *The Wonderful Wizard of Oz*, published in 1900. In the story, Dorothy, with her dog Toto in her arms, is running across the bedroom, desperately trying to reach the trap door of the storm cellar.

"When she was halfway across the room there came a great shriek from the wind, and the house shook so hard that she lost her footing and sat down suddenly upon the floor.

"Then a strange thing happened.

"The house whirled around two or three times and rose slowly through the air. Dorothy felt as if she were going up in a balloon.

"The north and south winds met where the house stood, and made it the exact center of the cyclone. In the middle of a cyclone the air is generally still, but the great pressure of the wind on every side of the house raised it up higher and higher, until it was at the very top of the cyclone; and there it remained and was carried miles and miles away as easily as you could carry a feather."

wind speed is 60 m s^{-1}. Estimate the radial pressure gradient. Assume an air density of 1.1 kg m^{-3}.

8.19 If the azimuthal wind speed in the previous exercise decreases with height from 50 m s^{-1} at the 700-hPa level to zero at the 200-hPa level, estimate the mean radial gradient of virtual temperature within this layer.

8.20 Consider an axially symmetric tropical cyclone that forms at a latitude of 15° in a large-scale environment in which the air is initially at rest. From how far out would the low-level inflow have to come in order to develop an azimuthal wind speed of 40 m s^{-1} at a radius of 20 km in the absence of frictional drag?

8.21 In the eye of an intense tropical cyclone the sea-level pressure is 60 hPa lower than in the large-scale environment. Estimate the elevation of sea level due to the hydrostatic adjustment to the low pressure.

The Atmospheric Boundary Layer

9

by Roland Stull
University of British Columbia, Vancouver, Canada

The Earth's surface is the bottom boundary of the atmosphere. The portion of the atmosphere most affected by that boundary is called the *atmospheric boundary layer* (*ABL*, Fig. 9.1), or *boundary layer* for short. The thickness of the boundary layer is quite variable in space and time. Normally ~1 or 2 km thick (i.e., occupying the bottom 10 to 20% of the troposphere), it can range from tens of meters to 4 km or more.

Turbulence and static stability conspire to sandwich a strong stable layer (called a *capping inversion*) between the boundary layer below and the rest of the troposphere above (called the *free atmosphere*). This stable layer traps turbulence, pollutants, and moisture below it and prevents most of the surface friction from being felt by the free atmosphere.

During fair weather (associated with high-pressure centers), we are accustomed to the diurnal (daily) cycle of changes in temperature, humidity, pollen, and winds that are governed by boundary-layer physics and dynamics. It is cool and calm at night; warm and gusty during daytime. The boundary layer is said to be *unstable* whenever the surface is warmer than the air, such as during a sunny day with light winds over land, or when cold air is advected over a warmer water surface. This boundary layer is in a state of *free convection*, with vigorous thermal updrafts and downdrafts. The boundary layer is said to be *stable* when the surface is colder than the air, such as during a clear night over land, or when warm air is advected over colder water. *Neutral* boundary layers form during windy and overcast conditions, and are in a state of *forced convection*.

Turbulence is ubiquitous within the boundary layer and is responsible for efficiently dispersing the pollutants that accompany modern life. However, the capping inversion traps these pollutants within the boundary layer, causing us to "stew in our own waste." Turbulent communication between the surface and the air is quite rapid, allowing the air to quickly take on characteristics of the underlying surface. In fact, one definition of the boundary layer is that portion of the lower troposphere that feels the effects of the underlying surface within about 30 min or less.

Air masses[1] are boundary layers that form over different surfaces. Temperature differences between neighboring air masses cause baroclinicity that drives extratropical cyclones. Heat and humidity trapped in the boundary layer are important fuels for convective clouds. The capping inversion inhibits thunderstorm formation, allowing the buildup of convective available potential energy (CAPE) in the free atmosphere. Wind shear in the boundary layer, caused by drag near the ground, generates horizontal vorticity that can be tilted by the updrafts in convective clouds to form tornadoes. Dissipation of kinetic energy within the boundary layer serves as a brake on large-scale wind systems.

Turbulence is inspiringly complex, consisting of a superposition of swirls called *eddies* that interact nonlinearly to create quasi-random, chaotic motions. An infinite number of equations is required to fully describe these motions. Hence, a complete solution has not been found. But when averaged over many eddies, we can observe persistent patterns

[1] The term *air mass* refers to an expanse of air with distinctive properties that derive from its residence over a specific source region and are still recognizable for some time after the air has moved into a different geographical setting. For example, air that has resided over a high latitude continent during winter tends to be cold and dry.

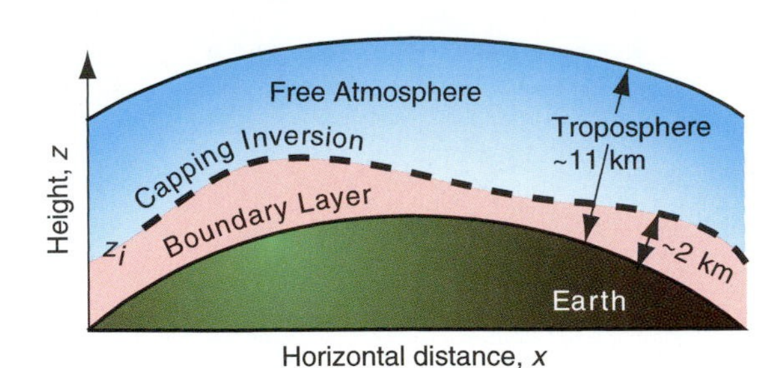

Fig. 9.1 Vertical cross section of the Earth and troposphere showing the atmospheric boundary layer as the lowest portion of the troposphere. [Adapted from *Meteorology for Scientists and Engineers*, A Technical Companion Book to C. Donald Ahrens' Meteorology Today, 2nd Ed., by Stull, p. 65. Copyright 2000. Reprinted with permission of Brooks/Cole, a division of Thomson Learning: www.thomsonrights.com. Fax 800-730-22150.]

and similarities that can be measured and described. In this chapter we explore the fascinating behavior of the boundary layer and the turbulent motions within it.

9.1 Turbulence

Atmospheric flow is a complex superposition of many different horizontal scales of motion (Table 9.1), where the "scale" of a phenomenon describes its typical or average size. The largest are *planetary-scale* circulations that have sizes comparable to the circumference of the Earth. Slightly smaller than planetary scale are *synoptic scale* cyclones, anticyclones, and waves in the jet stream. Medium-size features are called *mesoscale* and include frontal zones, rain bands, the larger thunderstorm and cloud complexes, and various terrain-modulated flows.

Table 9.1 Scales of horizontal motion in the atmosphere

Larger than	Scale	Name
20,000 km		Planetary scale
2,000 km		Synoptic scale
200 km	Meso-α	Mesoscale
20 km	Meso-β	
2 km	Meso-γ	
200 m	Micro-α	Boundary-layer turbulence
20 m	Micro-β	Surface-layer turbulence
2 m	Micro-γ	Inertial subrange turbulence
2 mm	Micro-δ	Fine-scale turbulence
Air molecules	Molecular	Viscous dissipation subrange

Smaller yet are the *microscales*, which contain boundary-layer scales of about 2 km, and the smaller *turbulence* scales contained within it and within clouds. The mesoscale and microscale are further subdivided, as indicated in Table 9.1. This chapter focuses on the microscales, starting with the smaller ones.

9.1.1 Eddies and Thermals

When flows contain irregular swirls of many sizes that are superimposed, the flow is said to be *turbulent*. The swirls are often called *eddies*, but each individual eddy is evanescent and quickly disappears to be replaced by a succession of different eddies. When the flow is smooth, it is said to be *laminar*. Both laminar and turbulent flows can exist at different times and locations in the boundary layer.

Turbulence can be generated *mechanically, thermally*, and *inertially. Mechanical turbulence*, also known as *forced convection*, can form if there is shear in the mean wind. Such shear can be caused by *frictional drag*, which causes slower winds near the ground than aloft; by *wake turbulence*, as the wind swirls behind obstacles such as trees, buildings, and islands (Fig. 9.2); and by *free shear* in regions away from any solid surface (Fig. 9.3).

Thermal or *convective turbulence*, also known as *free convection*, consists of *plumes* or *thermals* of warm air that rises and cold air that sinks due to buoyancy forces. Near the ground, the rising air is often in the form of intersecting curtains or sheets of updrafts, the intersections of which we can identify as *plumes* with diameters about 100 m. Higher in the boundary layer, many such plumes and updraft curtains merge to form larger diameter (~1 km) *thermals*. For air containing sufficient moisture, the tops of these thermals contain cumulus clouds (Fig. 9.4).

Small eddies can also be generated along the edges of larger eddies, a process called the *turbulent cascade*, where some of the *inertial energy* of the larger eddies is lost to the smaller eddies, as eloquently described by Richardson's poem (see Chapter 1). Inertial turbulence is just a special form of shear turbulence, where the shear is generated by larger eddies. The superposition of all scales of eddy motion can be quantified via an *energy spectrum* (Fig. 9.5), which indicates how much of the total turbulence kinetic energy is associated with each eddy scale.

(a)

(b)

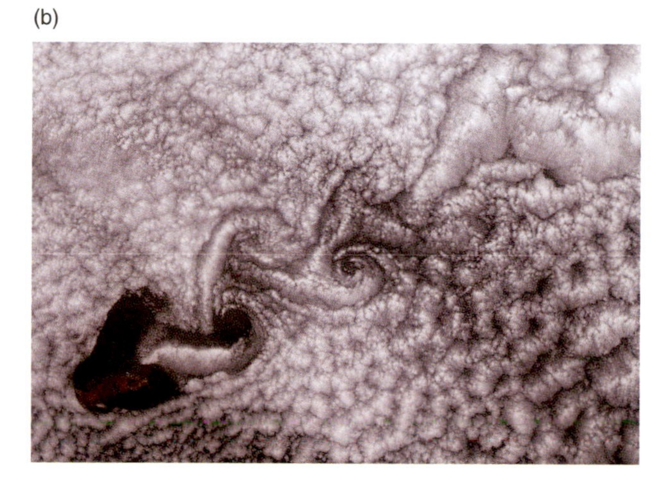

Fig. 9.2 Karman vortex streets in (a) the laboratory, for water flowing past a cylinder [From M. Van Dyke, *An Album of Fluid Motion*, Parabolic Press, Stanford, Calif. (1982) p. 56.], and (b) in the atmosphere, for a cumulus-topped boundary layer flowing past an island [NASA MODIS imagery].

Fig. 9.3 Water tank experiments of a jet of water (white) flowing into a tank of clear, still water (black), showing the breakdown of laminar flow into turbulence. [Photograph by Robert Drubka and Hassan Nagib. From M. Van Dyke, *An Album of Fluid Motion*, Parabolic Press, Stanford, CA. (1982), p. 60.]

Turbulence kinetic energy (TKE) is not conserved. It is continually *dissipated* into internal energy by molecular viscosity. This dissipation usually happens at only the smallest size (1 mm diameter) eddies, but it affects all turbulent scales because of the turbulent cascade of energy from larger to smaller scales. For turbulence to exist, there must be continual generation of turbulence from shear or buoyancy (usually into the larger scale eddies) to offset the transfer of kinetic energy down the spectrum of ever-smaller eddy sizes toward eventual dissipation. But why does nature produce turbulence?

Fig. 9.4 Cumulus clouds fill the tops of (invisible) thermals of warm rising air. [Photograph courtesy of Art Rangno.]

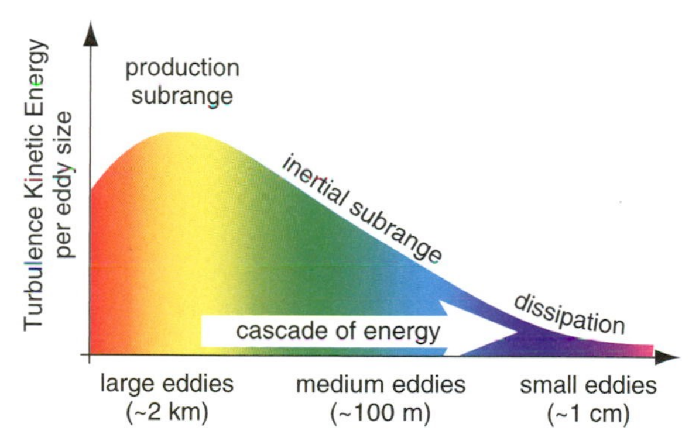

Fig. 9.5 The spectrum of turbulence kinetic energy. By analogy with Fig. 4.2, the total turbulence kinetic energy (TKE) is given by the area under the curve. Production of TKE is at the large scales (analogous to the longer wavelengths in the electromagnetic spectrum, as indicated by the colors). TKE cascades through medium-size eddies to be dissipated by molecular viscosity at the small-eddy scale. [Courtesy of Roland B. Stull.]

Turbulence is a natural response to instabilities in the flow—a response that tends to reduce the instability. This behavior is analogous to *LeChatelier's principle* in chemistry. For example, on a sunny day the warm ground heats the bottom layers of air, making the air statically unstable. The flow reacts to this instability by creating thermal circulations, which move warm air up and cold air down until a new equilibrium is reached. Once this *convective adjustment* has occurred, the flow is statically neutral and turbulence ceases. The reason why turbulence can persist on sunny days is because of continual destabilization by external forcings (i.e., heating of the ground by the sun), which offsets continual stabilization by turbulence.

Similar responses are observed for forced turbulence. Vertical shear in the horizontal wind is a *dynamic instability* that generates turbulence. This turbulence mixes the faster and slower moving air, making the winds more uniform in speed and direction. Once turbulent mixing has reduced the shear, then turbulence ceases. As in the case of convection, persistent mechanical turbulence is possible in the atmosphere only if there is continual destabilization by external forcings, such as by the larger scale weather patterns.

Although the human eye and brain can identify eddies via pattern recognition, the short life span of individual eddies renders them difficult to describe quantitatively. The equations of thermodynamics and dynamics described in Chapters 3 and 7 of this book can be brought to bear on this problem, but the result is an ability to *deterministically* simulate and predict the behavior of each eddy for only exceptionally short durations. The larger diameter thermals can be predicted out to about 15 min to half an hour, but beyond that the predictive skill approaches zero. For smaller eddies of order 100 m, the forecast skill diminishes after only a minute or so. The smallest eddies of order 1 cm to 1 mm can be predicted out to only a few seconds. This inability to deterministically forecast turbulence out to useful periods of days is a result of the highly non-linear nature of turbulent fluid dynamics.

Despite the difficulties of deterministic descriptions of turbulence, scientists have been able to create a *statistical* description of turbulence. The goal of this approach is to describe the net effect of many eddies, rather than the exact behavior of any individual eddy.

9.1.2 Statistical Description of Turbulence

When fast-response velocity and temperature sensors are inserted into turbulent flow, the net effect of the superposition of many eddies of all sizes blowing past the sensor are temperature and velocity signals that appear to fluctuate randomly with time (Fig. 9.6). However, close examination of such a trace reveals that for any half-hour period, there is a well-defined mean temperature and velocity; the range of temperature and velocity fluctuations measured is bounded (i.e., no infinite values); and a statistically robust standard deviation of the signal about the mean can be calculated. That is to say, the turbulence is not completely random; it is *quasi-random*.

Suppose that the velocity components (u, v, w) are sampled at regular time intervals Δt and then digitized and recorded on a computer to form a time series

$$u_i = u(i \cdot \Delta t) \tag{9.1}$$

where i is the index of the data point (corresponding to time $t = i\Delta t$) for $i = 1$ to N in a time series of duration $T = N \cdot \Delta t$. The u-component of the *mean wind* $\overline{u}$,

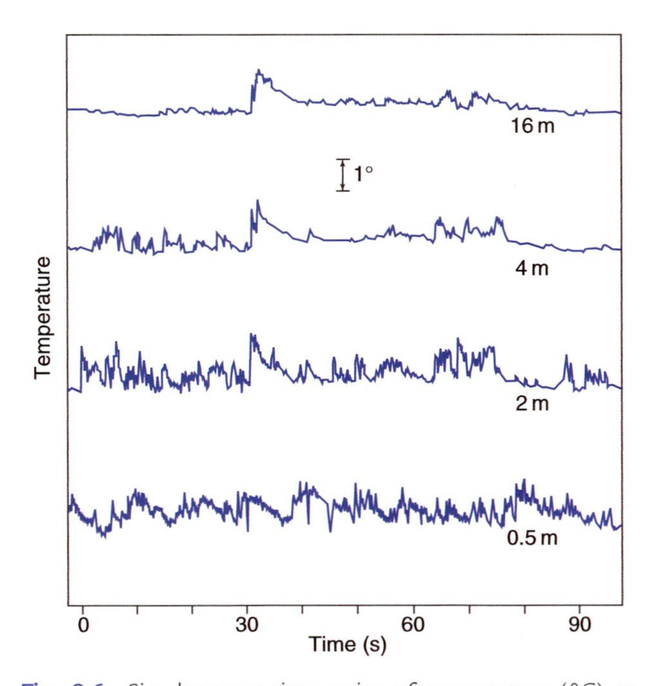

Fig. 9.6 Simultaneous time series of temperature (°C) at four heights above the ground showing the transition from the *surface layer* (bottom 5–10% of the mixed layer) toward the *mixed layer* boundary layer (upper levels of the boundary layer). Observations were taken over flat, plowed ground on a clear day with moderate winds. The top three temperature sensors were aligned in the vertical; the 0.5 m sensor was located 50 m away from the others. [Courtesy of J. E. Tillman.]

based on an average over a time period T (say half an hour), is thus

$$\overline{u} = \frac{1}{N}\sum_{i=1}^{N} u_i \tag{9.2}$$

In the atmosphere, this mean value can change from one half-hour period to the next, resulting in a slow variation of the mean-wind components with time. The velocities referred to in Chapters 7 and 8 are mean velocities (namely, the gusts are averaged out), even though the overbar was not shown. Also, Eq. (9.2) can be generalized: $1/N$ times the sum of N samples of any variable is the average of those samples, and can be indicated by an overbar [e.g., Eq. (9.4)].

Subtracting the mean from the instantaneous component u_i gives just the *fluctuating (gust) portion* of the flow (indicated with a prime)

$$u_i' = u_i - \overline{u} \tag{9.3}$$

which varies rapidly with time. The intensity of turbulence in the u direction is then defined by the *variance*[2]

$$\sigma_u^2 = \frac{1}{N}\sum_{i=1}^{N}[u_i - \overline{u}]^2 = \frac{1}{N}\sum_{i=1}^{N}[u_i']^2 = \overline{[u']^2} \tag{9.4}$$

Again, this is the variance averaged over a half-hour period so this variance value can vary slowly with time over subsequent averaging periods. Similar equations can be defined for the other velocity components.

For situations in which σ_u^2 is relatively constant with time (e.g., the same now as an hour ago), the turbulent nature of the flow is said to be *stationary*. When σ_u^2 is relatively uniform in space (e.g., the same value in one town as in a neighboring town), the flow is said to be *homogeneous*. For situations in which the turbulence intensity at any one point is the same in all directions ($\sigma_u^2 = \sigma_v^2 = \sigma_w^2$), the flow is said to be *isotropic*.

In the atmosphere, fluctuations in velocity are often accompanied by fluctuations in scalar values such as temperature, humidity, or pollutant concentration. For example, in a field of thermals there are regions where warm air is rising (positive potential temperature θ' accompanies positive vertical velocity w'), surrounded by regions where cold air is sinking (negative θ' accompanies negative w'). One measure of the amount that θ and w vary together is the *covariance* (cov)

$$\text{cov}\,(w,\theta) = \frac{1}{N}\sum_{i=1}^{N}[(w_i - \overline{w}) \cdot (\theta_i - \overline{\theta})]$$
$$= \frac{1}{N}\sum_{i=1}^{N}[(w_i') \cdot (\theta_i')] = \overline{w'\theta'} \tag{9.5}$$

If warm air parcels are rising and cold parcels are sinking, as in a thermally direct circulation, then $\overline{w'\theta'} > 0$. Covariances can also be negative or zero for different situations in the atmosphere.

The power of the statistical approach is that the velocity variance is more than just a statistic—it represents the *kinetic energy* associated with the motions on the scale of the turbulence. Similarly, the covariance is a measure of *flux* due to these motions, such as the vertical heat flux in Eq. (9.5). Such interpretations are explained next.

9.1.3 Turbulence Kinetic Energy and Turbulence Intensity

Recall from basic physics that kinetic energy is $KE = \frac{1}{2}mV^2$, where m is mass and V is velocity. In meteorology we often use *specific kinetic energy*, namely KE/m, or the kinetic energy per unit mass. By extension, we can focus on just the portion of specific kinetic energy associated with turbulent fluctuations

$$\frac{TKE}{m} = \frac{1}{2}\left[\overline{u'^2} + \overline{v'^2} + \overline{w'^2}\right]$$

or, using (9.4),

$$\frac{TKE}{m} = \frac{1}{2}[\sigma_u^2 + \sigma_v^2 + \sigma_w^2] \tag{9.6}$$

where TKE is *turbulence kinetic energy*. For laminar flow, which contains no microscale motions, $TKE = 0$, even though $\overline{u}, \overline{v}, \overline{w}$ are not necessarily zero. Larger values of TKE indicate a greater *intensity of the microscale turbulence*. We see now that the three components of velocity variance represent three contributions to the scalar TKE.

[2] Although this is the "biased" variance in statistics terminology, it is negligibly different from the unbiased variance because N is typically very large—~1000 or more. The unbiased variance uses $N - 1$ instead of N in the denominator of Eq. (9.4).

Using what we already know about mechanical and thermal generation of turbulence, and of viscous dissipation, we can write in descriptive form an *Eulerian* (i.e., fixed relative to the ground) forecast equation for turbulence kinetic energy:

$$\frac{\partial (TKE/m)}{\partial t} = Ad + M + B + Tr - \varepsilon \tag{9.7}$$

where

$$Ad = -\overline{u}\frac{\partial (TKE/m)}{\partial x} - \overline{v}\frac{\partial (TKE/m)}{\partial y} - \overline{w}\frac{\partial (TKE/m)}{\partial z}$$

is the *advection* of *TKE* by the mean wind, *M* is *mechanical generation* of turbulence, *B* is *buoyant generation or consumption* of turbulence, *Tr* is *transport* of turbulence energy by turbulence itself, and ε is the viscous *dissipation rate*. The terms *Ad* and *Tr* neither create nor destroy *TKE*, they just *redistribute* it by moving it from one location to another. *M* is usually positive (or zero if there is no shear) and therefore *generates* turbulence, while *B* can be positive or negative. Dissipation is always negative and can be approximated by $\varepsilon = (TKE/m)^{3/2}/L_\varepsilon$, where L_ε is a *dissipation length scale*.

In the absence of the terms *Ad, M, B*, and *Tr*, we see that as long as *TKE* is nonzero, the last term will always cause *TKE* to decrease toward zero. For this reason, turbulence is said to be *dissipative*.

In statically stable environments the buoyancy term can reduce *TKE* by converting it to potential energy by moving cold air up and warm air down. In such situations, the existence of turbulence depends on the relative strengths of mechanical generation (*M*) by wind shear versus buoyant consumption (*B*) by static stability. The ratio of these two terms defines the dimensionless *Richardson number, Ri*, which can be approximated by the vertical gradients of wind and potential temperature

$$Ri = \frac{-B}{M} = \frac{\dfrac{g}{\overline{T}_v}\dfrac{\partial \overline{\theta}_v}{\partial z}}{\left(\dfrac{\partial \overline{u}}{\partial z}\right)^2 + \left(\dfrac{\partial \overline{v}}{\partial z}\right)^2} \tag{9.8}$$

where the term in the numerator is equivalent to the square of the Brunt Väisälä frequency, as defined in (3.75). Laminar flow becomes turbulent when *Ri* drops below the critical value $Ri_c = 0.25$. Turbulent flow often stays turbulent, even for Richardson numbers as large as 1.0, but becomes laminar at larger values of *Ri*. The presence or absence of turbulence for $0.25 < Ri < 1.0$ depends on the history of the flow: a behavior analogous to hysteresis. Flows for which $Ri_c < 0.25$ are said to be *dynamically unstable*.

When the shear in laminar flow across a density interface (e.g., between cold air below and warm air above) increases to the point at which the flow becomes dynamically unstable, the turbulence onset grows as a *Kelvin-Helmholtz (KH) instability* on the interface. First, small waves appear that grow in amplitude and curl over on themselves. If sufficient moisture is present in the atmosphere, a cloud can form in the rising portions of each curl, giving rise to a pattern that looks like breaking waves at the beach when viewed from the side (Fig. 9.7a). When viewed from above or below, these features appear as closely spaced parallel bands of clouds, called *KH billows* or *billow clouds* (Fig. 9.7b), which are perpendicular to the vertical shear vector $\partial \mathbf{V}/\partial z$. The overturning of the billows introduces static instabilities (i.e., locally unstable lapse rates) that further accelerate the transition to turbulence within the shear layer.

The shapes of turbulent eddies are also modulated by the static stability. Under statically unstable conditions with rising thermals, the largest eddies are strongly *anisotropic*, with much greater turbulent energy in the vertical motion component than in the horizontal component. For continuous emissions of smoke from a smoke stack, smoke plumes *loop* up and down and spread more rapidly in the vertical than in the horizontal. Under statically neutral conditions, turbulence is almost isotropic, and smoke plumes spread equally in the vertical and in the horizontal, yielding a conical envelope, a behavior referred to as *coning*. When the flow is statically stable but dynamically unstable, the vertical component of turbulence is partly suppressed by the negative buoyancy of the rising air and the positive buoyancy of the sinking air—a process referred to as *buoyant consumption*—resulting in anisotropy with moderate TKE in the horizontal motion component but very little energy in the vertical component. Smoke plumes in such an environment *fan* out horizontally. In extremely stable conditions, turbulence is completely suppressed, and smoke blows downwind with almost no dispersion, although the plume centerline can oscillate up and down as a laminar wave.

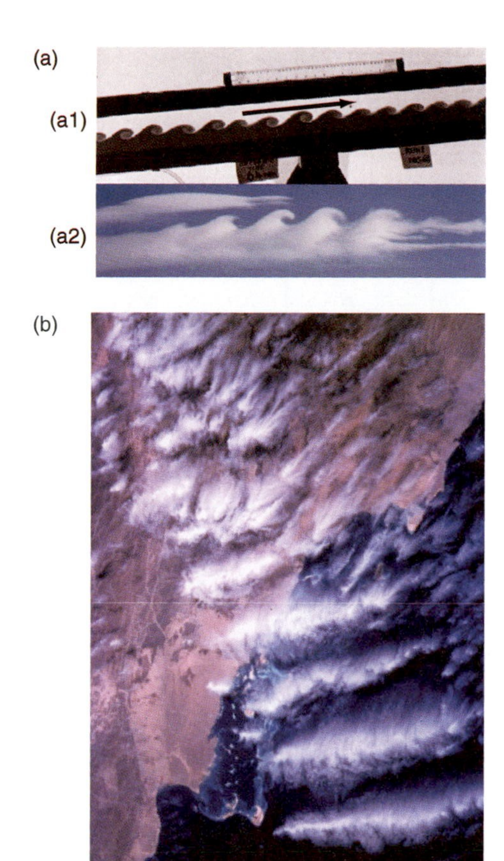

Fig. 9.7 (a) Kelvin-Helmholtz waves (b) Kelvin-Helmholtz billows in clouds. Kelvin-Helmholtz (KH) breakdown of shear flow. (al) A long narrow water tank is filled with a layer of salt brine (dyed a dark color) in the bottom half and clear, fresh water in the top half. When the tank is tilted, the heavy brine flows downslope (to the left in this photo) and the lighter fresh water flows upslope, creating a shear across the density interface. [From *J. Fluid Mech.*, 46 (1971) p. 299, plate 3.] (a2) Similar breaking of KH waves at an atmospheric density interface, by chance made visible by clouds. [Courtesy of Brooks Martner.] (b) KH billow clouds in the atmosphere as viewed from above. [NASA MODIS imagery.]

9.1.4 Turbulent Transport and Fluxes

Covariances can be interpreted as fluxes using the following concepts. Consider a portion of the atmosphere with a constant gradient of potential temperature, as sketched in Fig. 9.8. Consider an idealized eddy circulation consisting of an updraft portion that moves an air parcel from the bottom to the top of the layer, and a compensating downdraft that moves a different air parcel downward. Air mass is conserved (i.e., mass up = mass down). However, the air parcels carry with them small portions of the air from their starting points, and these portions preserve their potential temperatures as they move, resulting in a flux as will now be shown.

In Fig. 9.8a, the thick line represents a statically unstable mean environment $[\overline{\theta}(z)]$. For this case when the rising air parcel reaches its destination, its potential temperature is warmer than the surrounding environment at that altitude. Namely, its deviation from its new environment is $\theta' = (+)$. This air parcel had to move upward to get to its destination so $w' = (+)$. The contribution of this rising parcel to the total covariance is w' times $\theta' = w'\theta' = (+) \cdot (+) = (+)$. Similarly, for the downward-moving $[w' = (-)]$ portion of this eddy, the cold air from aloft finds itself colder $[\theta' = (-)]$ than its new surrounding environment at its final low altitude. Thus, its contribution to the covariance is $w'\theta' = (-) \cdot (-) = (+)$.

The average of these two air parcels represents the covariance, and since each contribution is positive, the average (indicated by the overbar) is also positive: $\overline{w'\theta'} = (+)$. Thus, positive $\overline{w'\theta'}$ covariance is associated with warm air moving up and/or cold air moving down, namely a positive *heat flux* $F_H(=\overline{w'\theta'})$. This form of flux is called a *kinematic heat flux* and has units of (K m s^{-1}). It is related to the traditional heat flux Q_H (W m^{-2}) by

$$Q_H = \rho c_p F_H = \rho c_p \overline{w'\theta'} \tag{9.9}$$

where ρ is the mean air density and c_p is the specific heat of air at constant pressure.

Figure 9.8b shows the contrasting behavior observed in a statically stable environment. In this case, both the upward and downward moving parcels contribute negatively to the covariance. Thus, a downward heat flux is associated with cold air moving up or warm air moving down. Hence, the covariance $F_H = \overline{w'\theta'}$ is negative.

These fluxes can contribute to the warming and cooling of layers of air via the first law of thermodynamics (see Chapter 3), which, in the absence of other heat sources, can be rewritten as

$$\frac{\partial\overline{\theta}}{\partial t} = -\frac{\partial\overline{w'\theta'}}{\partial z} + \cdots \tag{9.10}$$

Note also the analogy with radiative fluxes in the expression for radiative heating rates (4.52).

The net result of this turbulence is that warmer and colder layers are mixed to yield an intermediate potential temperature. In a similar manner, one can

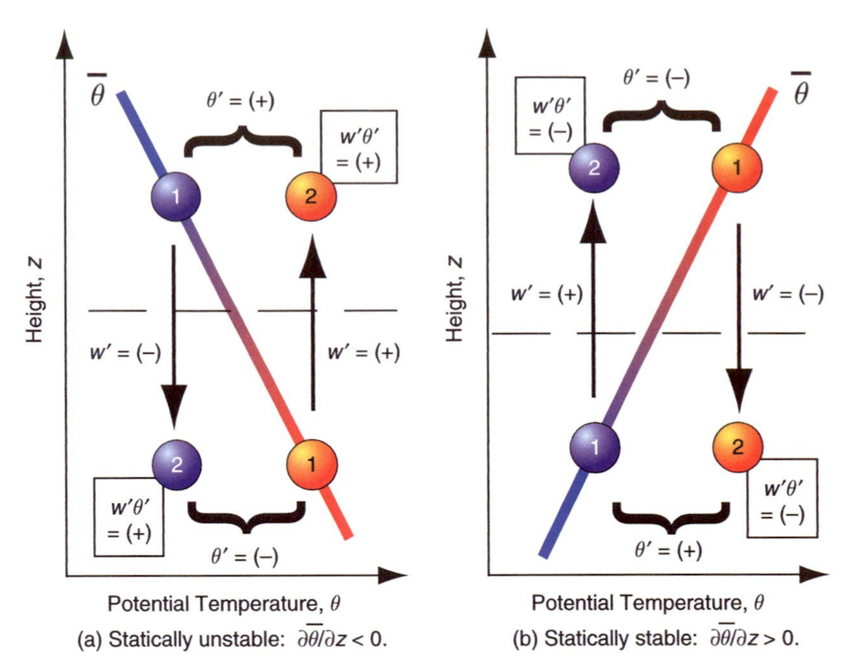

Fig. 9.8 Illustration of how to anticipate the sign of turbulent heat fluxes for small-eddy (local) vertical mixing across a region with a linear gradient in the mean potential temperature θ (thick colored line). Assuming an adiabatic process (no mixing), air parcels (sketched as spheres) preserve their potential temperature (as indicated by their color) of the ambient environment at their starting points (1), even as they arrive at their destinations (2). (a) Statically unstable lapse rate. (b) Statically stable lapse rate. [Adapted from *Meteorology for Scientists and Engineers*, A Technical Companion Book to C. Donald Ahrens' Meteorology Today, 2nd Ed., by Stull, p. 87. Copyright 2000. Reprinted with permission of Brooks/Cole, a division of Thomson Learning: www.thomsonrights.com Fax 800-730-2215.]

conceive of turbulent mixing of moisture, pollutants, and even momentum. In each case, turbulence tends to *homogenize* a fluid.

Turbulence is an extremely efficient mixer. For example, when milk is added to coffee or tea, most people prefer not to wait hours for molecular diffusion to homogenize their drink. Instead, they stir the fluid to generate turbulence, which homogenizes their drink within a few seconds. Atmospheric turbulence is equally efficient at causing mixing—so much so that molecular diffusion and molecular viscosity can be neglected for all motions except the tiniest eddies. In fact, during the daytime over land, convective turbulence is so effective at mixing that the boundary layer is also known as the *mixed layer* because pollutants are distributed so quickly in the vertical.

9.1.5 Turbulence Closure

Equation (9.10) is a forecast equation for potential temperature. The overbar on all the terms in this equation is associated with a process called *Reynolds averaging*—an applied mathematical method that eliminates small *linear* terms such as those associated with nonbreaking *waves*, but retains the *non-linear* terms associated with, or affected by, *turbulence*. There are many other terms that appear on the right-hand side of the full Reynolds-averaged forecast equation, but for now we will focus on just the heat-flux divergence term. To forecast how the mean potential temperature will change with time, we need to know the kinematic heat flux $\overline{w'\theta'}$.

A Reynolds-averaged forecast equation can also be derived for kinematic heat flux $\overline{w'\theta'}$, which is of the form

$$\frac{\partial \overline{w'\theta'}}{\partial t} = -\frac{\partial \overline{w'w'\theta'}}{\partial z} + \cdots \tag{9.11}$$

This new equation yields a forecast of the second-order statistic $\overline{w'\theta'}$, but it introduces a new third-order statistic $\overline{w'w'\theta'}$, which is the turbulent flux of a heat flux. If we write a forecast equation for this third-order statistic, we introduce even higher order unknowns.

This is the *turbulence closure problem*. Mathematically speaking, the equations are not closed. There are always more unknowns than equations. In other words, we need an infinite number of equations to describe turbulence, even if we want only to forecast the mean potential temperature.

To mitigate this difficulty, we can make *closure assumptions*. Namely, we can retain a finite number of equations and then approximate the remaining unknowns as a function of the knowns. The resulting *parameterization* will not give a perfect answer, but it will give an approximate answer that often is good enough.

Turbulence closure assumptions are categorized both by their *statistical order* and by the amount of *nonlocalness* that is included. For the statistical order, if we keep Eq. (9.10) and approximate the unknown $\overline{w'\theta'}$ as a function of the known variables, the result is called *first-order closure*, named after the highest order forecast equation retained. Second-order closure retains both Eqs. (9.10) and (9.11) and parameterizes the third-order statistics $\overline{w'w'\theta'}$.

A common local, first-order closure is called *gradient-transfer theory, K-theory, eddy-diffusivity theory*, or *mixing-length theory*. Analogous to molecular diffusion, it assumes that the flux is linearly proportional to and directed down the local gradient, i.e.,

$$F_H = \overline{w'\theta'} = -K\frac{\partial\overline{\theta}}{dz} \tag{9.12}$$

where an eddy diffusivity, K, is used instead of the molecular diffusivity. The parameter K is prescribed to increase with the intensity of the turbulence, which varies with height above ground, mean wind shear, and surface heating by the sun. Prandtl's[3] mixing length approach was one of the first parameterizations for eddy diffusivity: $K = l^2|\partial V/\partial z|$, where V is mean horizontal wind speed and $l = \left(\overline{z'^2}\right)^{1/2}$ represents an average size or *mixing length* for the eddies. The parameter l is often approximated by $l = kz$ in the *surface layer* (the bottom 5 to 10% of the boundary layer), where $k = 0.4$ is the von Karman[4] constant and z is height above ground level. The wind-shear term in K parameterizes the effects of mechanical generation of turbulence.

The closure in Eq. (9.12) is a *local closure* in the sense that the heat flux at any altitude depends on the local gradient of potential temperature at that same altitude. Namely, it implicitly assumes that only small-size eddies exist. Similar first-order closures can be written for moisture, pollutant, and momentum fluxes.

While local first-order closures often work nicely in laboratory settings, they frequently fail in the unstable atmospheric boundary layer. Under these conditions, thermals cause such intense mixing and homogenization as to eliminate the vertical gradient of mean potential temperature in the middle of the boundary layer, yet there are strong positive heat fluxes caused by the rising thermals. For this situation, *nonlocal* first-order closures have been developed, where the flux across any one altitude depends on transport by all eddy sizes, including the large eddies that move heated air from just above the Earth's surface all the way to the top of the boundary layer.

Finally, a large body of useful results have been complied for statistical *zeroth-order closure*. In this case, neither Eqs. (9.10) nor (9.11) is retained. Instead, the mean flow state $\overline{\theta}$ is parameterized directly. This approach, called *similarity theory*, is illustrated in the next subsection.

9.1.6 Turbulence Scales and Similarity Theory

Some zero and first-order closure schemes rely on simple empirical[5] relationships derived from *dimensional analysis*. Variables that frequently appear in combination with one another are grouped to form new variables that may be nondimensional, such as the Richardson number defined in Eq. (9.8), or may have simple dimensional units such as velocity, length, or time that in some cases relate to the most important scales of motion in the eddies.

A velocity scale that is useful for characterizing the turbulent mixing due to free convection in an

[3] **Ludwig Prandtl** (1874–1953) German aerodynamicist and accomplished pianist. Developed theories for the boundary layer, airfoils, lift vs. drag, and supersonic flow for rocket nozzles. Educated in Munich in mechanics, became professor in Hannover, and later directed the Institute for Technical Physics and the Kaiser Wilhelm Institute for Flow Investigation, University of Göttingen, Germany.

[4] **Theodor von Kármán** (1881–1963) Hungarian aerodynamicist, specializing in supersonic flight. Studied boundary layers and airfoils under Ludwig Prandtl and became professor of aeronautics and mechanics at the University of Aachen, Germany. Worked with Hugo Junkers to help design the first cantilevered wing all metal airplane in 1915. Became director of the Guggenheim Aeronautical Lab at the California Institute of Technology, advancing theoretical aerodynamics and rocket design, and spawning the Jet Propulsion Lab. Was the first recipient of the U.S. National Medal of Science, awarded by John F. Kennedy.

[5] Based on observed relationships between variables.

unstably stratified boundary layer is the *Deardorff velocity scale*

$$w_* = \left[\frac{g \cdot z_i}{T_v} \overline{w'\theta'_s}\right]^{1/3} \tag{9.13}$$

where z_i is the depth of the boundary layer and the subscript s denotes at the surface. Values of w_* have been determined from field measurements and numerical simulations under a wide range of conditions. Typical magnitudes of w_* are $\sim 1 \text{ m s}^{-1}$, which corresponds to the average updraft velocities of thermals.

Another scale u_*, the *friction velocity*, is most applicable to statically neutral conditions in the surface layer, within which the turbulence is mostly mechanically generated. It is given by

$$u_* = \left[\overline{u'w'}^2 + \overline{v'w'}^2\right]^{1/4} = \left|\frac{\tau_s}{\rho}\right|^{1/2} \tag{9.14}$$

where ρ is air density, τ_s is *stress* at the surface (i.e., drag force per unit surface area), and covariances $\overline{u'w'}$ and $\overline{v'w'}$ are the *kinematic momentum fluxes* (vertical fluxes of u and v horizontal momentum, respectively).

The altitude of the capping inversion, z_i, is the relevant length scale for the whole boundary layer for statically unstable and neutral conditions. Within the bottom 5% of the boundary layer (referred to as the *surface layer*), an important length scale is the *aerodynamic roughness length,* z_0, which indicates the roughness of the surface (see Table 9.2). For statically nonneutral conditions in the surface layer, there is an additional length scale, called the *Obukhov length*

$$L \equiv \frac{-u_*^3}{k \cdot (g/T_v) \cdot \left(\overline{w'\theta'}\right)_s}, \tag{9.15}$$

where $k = 0.4$ is the von Karman constant. The absolute value of L is the height below which mechanically generated turbulence dominates.

Typical timescales for the convective boundary layer and the neutral surface layer are

$$t_* = \frac{z_i}{w_*} \qquad t_{*SL} = \frac{z}{u_*} \tag{9.16}$$

where z is height above the surface.

For the convective boundary layer, t_* is of order 15 min, which corresponds to the turnover time for the largest convective eddy circulations, which extend from the Earth's surface all the way up to the capping inversion.

In summary, for convective boundary layers (i.e., unstable mixed layers), the relevant scaling parameters are w_* and z_i. For the neutral surface layer, u_* and z_0 are applicable. Scaling parameters for surface

Table 9.2 The Davenport classification, where z_o is aerodynamic roughness length and C_{DN} is the corresponding drag coefficient for neutral static stability[a]

z_0 (m)	Classification	Landscape	C_{DN}
0.0002	Sea	Calm sea, paved areas, snow-covered flat plain, tide flat, smooth desert.	0.0014
0.005	Smooth	Beaches, pack ice, morass, snow-covered fields.	0.0028
0.03	Open	Grass prairie or farm fields, tundra, airports, heather.	0.0047
0.1	Roughly open	Cultivated area with low crops and occasional obstacles (single bushes).	0.0075
0.25	Rough	High crops, crops of varied height, scattered obstacles such as trees or hedgerows, vineyards.	0.012
0.5	Very rough	Mixed farm fields and forest clumps, orchards, scattered buildings.	0.018
1.0	Closed	Regular coverage with large size obstacles with open spaces roughly equal to obstacle heights, suburban houses, villages, mature forests.	0.030
≥2	Chaotic	Centers of large towns and cities, irregular forests with scattered clearings.	0.062

[a] From Preprints *12th Amer. Meteorol. Soc. Symposium on Applied Climatology*, 2000, pp. 96–99.

layers are u_*, z_0, and L, provided that the stratification is not neutral.

As an example, when dimensional analysis is used to describe the vertical profile of the variance of vertical velocity $\overline{w'^2}$ through the entire depth of the convective boundary layer, the observational data are fit by the function

$$\frac{\overline{w'^2}}{w_*^2} = a\left(\frac{z}{z_i}\right)^b\left(1 - c\frac{z}{z_i}\right)^d$$

in which a, b, c, and d are constants. This expression for dimensionless velocity variance as a function of dimensionless height is applicable to convective boundary layers of any depth and for any surface heat flux. That is to say, vertical profiles of $\overline{w'^2}$ exhibit similar shapes and collapse onto a single curve when plotted on the same pair of dimensionless coordinate axes ($\overline{w'^2}/w_*^2$ versus z/z_i): hence the name *similarity theory*.

When applied to observational data for the statically stable surface layer, dimensional analysis yields the vertical profile of horizontal wind speed

$$\frac{V}{u_*} = 2.5 \ln\left(\frac{z}{z_0} + 8.1\frac{z}{L}\right)$$

This expresson for dimensionless speed as a function of dimensionless height is applicable for any wind speed, height, roughness, and static stability in the surface layer (i.e., the vertical profiles exhibit similar shapes and collapse onto a single similarity curve).

As we have seen, the nature of turbulence is strongly modulated by heat fluxes and drag (momentum fluxes) at the surface. The next section describes the daily evolution of these surface fluxes under fair weather (anticyclonic) conditions.

Exercise 9.1 (a) What is the relationship between the Obukhov length and the Deardorff velocity? (b) For a 1-km-thick boundary layer with a surface kinematic heat flux of 0.2 K m s^{-1} and surface kinematic stress of 0.2 m^2 s^{-2}, over what portion of the boundary layer does mechanical turbulence dominate?

Solution: (a) Combining Eqs. (9.13) and (9.15) yields

$$L/z_i = -u_*^3/(k \cdot w_*^3)$$

(b) $u_*^3 = (0.2 \text{ m}^2 \text{ s}^{-2})^{3/2} = 0.089 \text{ m}^3 \text{ s}^{-3}$. Assuming $T_v = 300$ K as a typical value, then $w_*^3 = 6.5 \text{ m}^3 \text{ s}^{-3}$. Thus, $|L/z_i| = 0.089/(0.4)(6.5) = 0.034$. The mechanically driven surface layer is only 3.4% of the convective boundary layer depth for this example. It is often the case that the surface layer is much thinner than 10% of the boundary-layer depth under free-convective conditions. ■

9.2 The Surface Energy Balance

9.2.1 Radiative Fluxes

The solar radiation incident on the Earth's surface is modulated by the rotation of the Earth, causing a daily (*diurnal*) cycle of incoming solar radiation with reference to local sunrise, noon, and sunset over any point on the surface.

Let $F_s\downarrow$ be the magnitude of the flux of downwelling solar (shortwave) radiation that reaches the surface, integrated over all wavelengths in and near the visible spectrum. The surface reflects some of the sunlight back upward, of magnitude $F_s\uparrow$. Also, the atmosphere emits longwave radiation, some of which $F_L\downarrow$ reaches the Earth's surface. The Earth's surface emits longwave radiation upward, with flux magnitude $F_L\uparrow$. The sum of the inputs to the surface minus the outputs yields the *net radiation flux* F^* absorbed at the surface

$$F^* = F_s\downarrow - F_s\uparrow + F_L\downarrow - F_L\uparrow \tag{9.17}$$

During fair weather with clear skies, the surface radiation fluxes vary with time as sketched in Fig. 9.9. During daytime, the incoming solar radiation is proportional to the sine of the elevation angle of the sun, which varies with time of day, latitude, and season. The solar radiation reflected from the Earth's surface mirrors the incoming direct solar radiation, but with reduced amplitude. At night, there is obviously no shortwave radiation.

The upward and downward longwave fluxes nearly cancel. The $F_L\uparrow$ curve has a slight modulation, which reflects the changing skin temperature [via the Stefan-Boltzmann law (4.12)] of the Earth's surface as it warms and cools in response to variations in solar radiation. Because of their small heat capacity, land surfaces respond almost instantaneously: thus, $F_L\uparrow$ is virtually in phase with the $F_s\downarrow$ flux. However, the $F_L\downarrow$ curve depends on the air temperature, which reaches its maximum in late afternoon before sunset and reaches its minimum just after sunrise.

The algebraic sum of all these fluxes, F^*, is nearly constant and slightly negative during the night and

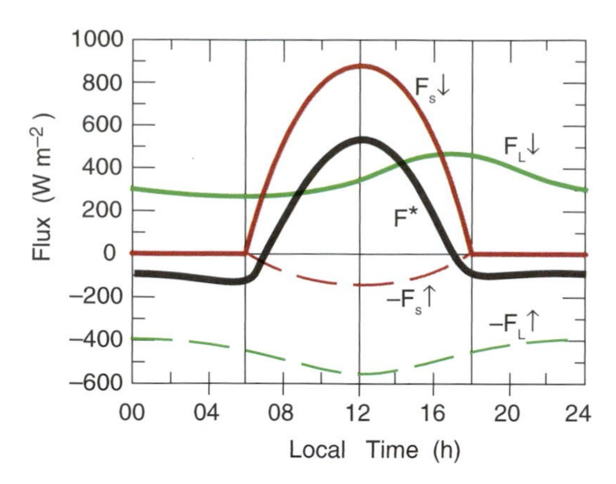

Fig. 9.9 Sketch of contribution of radiative fluxes at the Earth's surface toward the net flux F^* during a daily cycle under clear skies. Positive values represent inputs toward the surface; negative are fluxes away. [Adapted from *Meteorology for Scientists and Engineers,* A Technical Companion Book to C. Donald Ahrens' Meteorology Today, 2nd Ed., by Stull, p. 37. Copyright 2000. Reprinted with permission of Brooks/Cole, a division of Thomson Learning: www.thomsonrights.com. Fax 800-730-2215.]

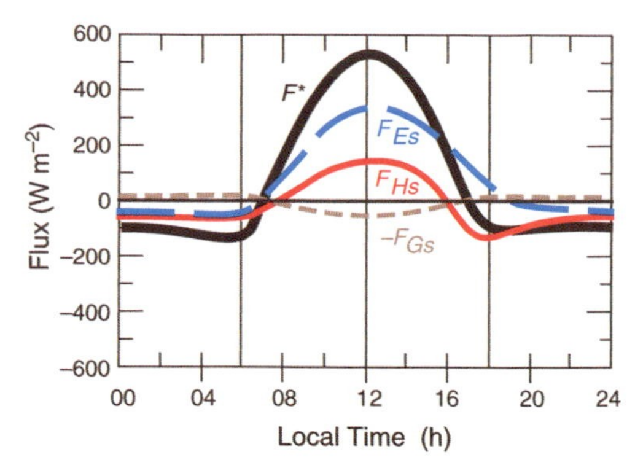

Fig. 9.10 Sketch of the disposition of the net flux F^* into turbulent sensible F_H and latent F_E heat fluxes into the atmosphere at the surface (subscript s), and conduction of heat into the ground F_G during a daily cycle under clear skies. Except for F^* (which is the same as in Fig. 9.9), positive displacements from zero along the ordinate represent upward fluxes, and negative downward. [Adapted from *Meteorology for Scientists and Engineers,* A Technical Companion Book to C. Donald Ahrens' Meteorology Today, 2nd Ed., by Stull, p. 57. Copyright 2000. Reprinted with permission of Brooks/Cole, a division of Thomson Learning: www.thomsonrights.com. Fax 800-730-2215.]

becomes positive with peak near solar noon during daytime, where the sign is defined such that positive implies an input to the surface. This is the "*external*" *forcing* that drives the diurnal variations in the surface heat budget.

9.2.2 Surface Energy Balance over Land

In addition to the radiative fluxes at the Earth's surface, the fluxes of sensible and latent heat also need to be taken into account. The sensible heat flux heats the air in the boundary layer directly. The latent heat flux (i.e., the flux of water vapor times L, the latent heat of vaporization) is not converted to sensible heat and/or potential energy until the water vapor condenses in clouds.

If we imagine the land surface as an infinitesimally thin surface that has zero heat capacity, then the heat flux coming in must balance the heat leaving. Given the net radiation F^* from the previous section, Fig. 9.10 shows how energy gain or loss is partitioned among *sensible heat flux,* F_{Hs}, into the air (positive upward, for flux away from the surface), *latent heat flux* F_{Es} into the air (positive upward), and the *conduction* of heat down into the ground, F_{Gs} (positive downward, away from the surface), where the extra subscript s denotes near the surface. Therefore,

$$F^* = F_{Hs} + F_{Es} + F_{Gs} \tag{9.18}$$

In fair weather conditions with light to calm winds, Fig. 9.11 shows the direction of the fluxes during day and night. Over moist lawns, crops, and forests, most of the sun's energy goes into evaporation during daytime. However, during daytime over a dry desert or unvegetated land (Fig. 9.11c), most of the sun's energy goes into sensible heat flux.

During windy conditions if dry, hot air is advected over a cool, moist surface, such as at a desert oasis (Fig. 9.11d), then the sensible heat flux can be downward from the warm air to the cool surface, even though there is also solar heating of the surface. These two inputs combine to create very large evaporation and associated latent heat flux, known as the *oasis effect*. As a basis for understanding why the surface has this response over desert oases, the next subsection explains bulk aerodynamic methods for estimating surface fluxes.

The magnitude of flux into the ground, F_{Gs}, is ~10% of the net radiation magnitude during daytime, increasing to ~50% at night. The amplitude of the diurnal cycle of surface skin temperature, T_s, is inversely proportional to the conductivity of the soil. The annual cycle of soil temperature as a function of depth, shown in Fig. 9.12, is qualitatively similar to its counterpart for the diurnal cycle (not shown), but it penetrates to a greater depth. Amplitude decreases with increasing depth and the phase becomes progressively later. Bearing in mind that heat is always conducted down

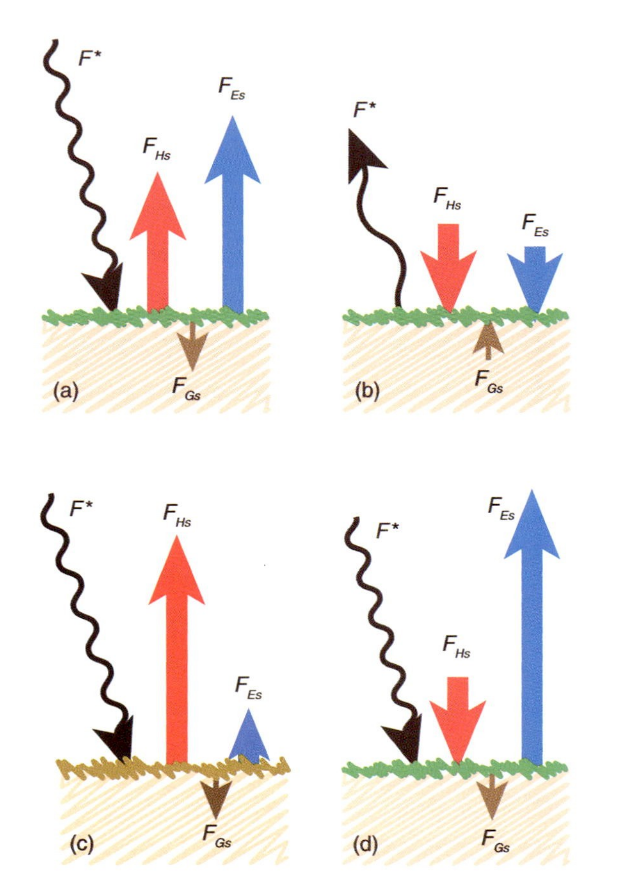

Fig. 9.11 Vertical cross-section sketch of net radiative input to the surface flux, F^*, and resulting heat fluxes into the air and ground for different scenarios (a) Daytime over a moist vegetated surface. (b) Nighttime over a moist vegetated surface. (c) Daytime over a dry desert. (d) Oasis effect during the daytime, with hot dry wind blowing over a moist vegetated surface. (See Fig. 9.10 for explanation of symbols.) [Adapted from *Meteorology for Scientists and Engineers*, A Technical Companion Book to C. Donald Ahrens' Meteorology Today, 2nd Ed., by Stull, p. 57. Copyright 2000. Reprinted with permission of Brooks/Cole, a division of Thomson Learning: www.thomsonrights.com. Fax 800-730-2215.]

the gradient from higher temperature toward lower temperature, it is evident that the downward conduction of heat reduces the warming of the land surface during periods of strong heat input, when the soil temperature decreases with depth, while the upward conduction of heat from below reduces the cooling of the surface during periods of weak insolation.

Ocean surfaces exhibit much larger values of F_{Gs} because turbulence in the ocean can quickly mix heat throughout the top layer of the ocean, called the *ocean mixed layer*, which ranges in depth from a few meters to hundreds of meters. The specific heat of liquid water is also larger than that of soil. These two effects conspire to give the ocean a much larger heat capacity

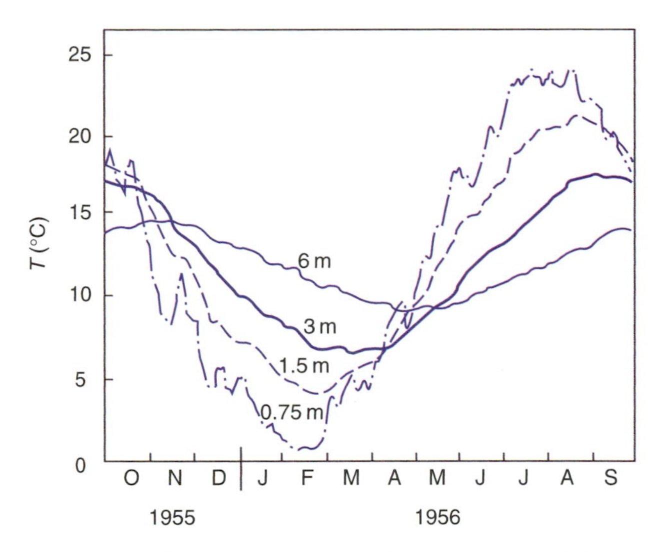

Fig. 9.12 Soil temperatures recorded at an exposed site at levels below the ground. [Adapted from *Trans. Amer. Geophys. Union* 37, 746 (1956).]

than the land surface, enabling it to absorb and store solar energy during the day and release it at night, resulting in nearly constant ocean surface temperatures through the diurnal cycle, and allowing only small temperature changes through the annual cycle.

9.2.3 The Bulk Aerodynamic Formulae

This subsection describes a method that can be used to estimate the fluxes of sensible and latent heat at the Earth's surface. This so-called *bulk aerodynamic method* enables us to estimate the frictional drag on the surface winds as well.

Sensible heat flux between the surface and the overlying air is driven by two processes. Within the bottom few millimeters of the atmosphere, very large vertical gradients of temperature form, causing *molecular conduction* of heat away from the surface into the air. At the bottom of this molecular layer (e.g., at the surface of the ground), there is zero turbulent flux because clods of soil do not usually "dance the eddy dance." But from the top of this *molecular layer* or *microlayer* to the top of the boundary layer, molecular conduction is negligible while *turbulent convection* takes over, moving the warm air upward to distribute sensible heat throughout the boundary layer. Because the microlayer is so thin compared to the boundary-layer depth and because the heat flux across the microlayer is nearly constant and equal to the turbulent eddy flux at the top of the microlayer, it is possible to define an *effective turbulent flux* that is the sum of the molecular and true turbulent components. In practice, the word "effective" is often omitted, and this quantity is referred to simply as the *turbulent flux at the surface*.

The effective sensible heat flux is often parameterized by the temperature difference between the surface and the air. If the surface skin temperature, T_s, is known, then the sensible heat flux (in kinematic units of K m s^{-1}) from the ground to the air can be parameterized as

$$F_{Hs} = C_H |V| (T_s - T_{air}) \qquad (9.19a)$$

where C_H is a dimensionless *bulk transfer coefficient* for heat and $|V|$ and T_{air} are the wind speed and air temperature at standard surface measurement heights (10 and 2 m, respectively). To convert from kinematic to dynamic heat flux (W m^{-2}), F_H must be multiplied by air density times the specific heat at constant pressure ($\rho\, c_p$).

Under statically neutral conditions over flat land surfaces there exists a moderate amount of turbulence that exchanges slow moving air near the ground with faster moving air in the boundary layer, yielding values for C_H in the 0.001 – 0.005 range (designated as C_{H_N} to indicate neutral conditions). The exact value of C_{H_N} depends on surface roughness, similar to the roughness dependence of C_{D_N} shown in Table 9.2.

Under statically unstable conditions, the vigorous turbulence communicates surface drag information more quickly to the boundary layer, causing C_H to be two to three times as large as C_{H_N}. Conversely, as the air becomes more statically stable, the Richardson number increases toward its critical value and the turbulence kinetic energy decreases toward zero, causing C_H to also decrease toward zero.

To estimate the vertical heat flux, one might have expected Eq. (9.19a) to be function of a vertical turbulent-transport velocity w_T times the temperature difference. But for this first-order closure, w_T is parameterized as $C_H|V|$, where it is assumed that stronger winds near the ground generate stronger turbulence, which causes stronger turbulent fluxes.

By combining Eqs. (9.17–9.19a), we see that the surface skin temperature over land on sunny days is really a response to solar heating rather than an independent driving force for the heat flux. For example, on a day with light winds, the net radiation budget causes a certain energy input to the ground, which causes the surface skin temperature to rise according to the first law of thermodynamics. As the skin warms, the sensible heat flux increases in accordance with Eq. (9.19a), as does the evaporation and the conduction of heat into the ground. Since the winds are light, (9.19a) shows that the skin temperature must become quite a bit warmer than the air temperature to drive sufficient sensible heat flux F_{Hs} to help balance the surface heat budget. However, on a windier day, the required heat flux is achieved with a surface skin temperature that is only slightly warmer than the air temperature.

When warmer air is advected over a cooler surface or when the ground is cooled by longwave radiation at night, then $T_s < T_{air}$, and the heat flux becomes downward. This cools the bottom of the boundary layer, and leads to *sub-adiabatic lapse rates* and a reduction or suppression of turbulence. Because turbulence is reduced, the cooling is limited to the bottom of the boundary layer, creating a shallow *stable boundary layer* embedded within the old, deeper boundary layer.

Similar equations, called *bulk aerodynamic relationships*, can be derived for the moisture flux over oceans, lakes, and saturated soil. One can assume that the specific humidity near the surface q_s is equal to its saturation value, as defined by the Clausius-Clapeyron equation, based on the air temperature near the sea surface. Namely, the moisture flux F_{water} [in kinematic units of ($kg_{water\ vapor}/kg_{air}$) (m s^{-1})] from the surface is

$$F_{water} = C_E |V| [q_{sat}(T_s) - q_{air}] \qquad (9.19b)$$

where C_E is a dimensionless bulk transfer coefficient for moisture ($C_E \approx C_H$). This moisture flux is directly related to the latent heat flux (F_{Es}, in kinematic units of K m s^{-1}) at the surface and to the evaporation rate E of water (mm/day) by

$$F_{water} = \gamma F_{Es} = (\rho_{liq}/\rho_{air})\, E \qquad (9.20)$$

where $\gamma = c_p/L_v = 0.4$ [($g_{water\ vapor}/kg_{air}$)/K] is the *psychometric constant* and ρ_{liq} is the density of pure liquid water (not sea water).

The ratio of sensible to latent heat fluxes at the surface is called the *Bowen ratio*[6]: $B = F_{Hs}/F_{Es}$. Due to

[6] **Ira S. "Ike" Bowen** (1898–1973) American physicist and astronomer. Studied under Robert A. Millikan as a graduate student at the University of Chicago and as a research assistant at the California Institute of Technology, where his Ph.D. was on evaporation from lakes and associated heat losses. Identified ultraviolet spectral lines from nebulae. Directed Mt. Wilson and Palomar observatories and the construction of the Hale and Schmidt telescopes. Worked with the Jet Propulsion Laboratory on photography from rockets.

the nonlinearity inherent in the Clausius-Clapeyron equation, the Bowen ratio over the oceans decreases with increasing sea surface temperature. Typical values range from around 1.0 ± 0.5 along the ice edge to less than 0.1 over the tropical oceans where latent heat fluxes dominate due to the warmth of the sea surface. Over land, the evaporation rate, and therefore the Bowen ratio, depends on the availability of water in the soil and on the makeup of the vegetation that transports water from the soil via osmosis. Plants release water vapor into the air via *transpiration* through the open stomata (pores) of leaves. Thus, the Bowen ratio ranges from about 0.1 over tropical oceans, through 0.2 over irrigated crops, 0.5 over grassland, 5.0 over semiarid regions, and 10 over deserts.

For momentum, the bulk aerodynamic approach gives a *drag law*

$$u_*^2 = C_D\,|V|^2 \tag{9.19c}$$

where C_D is the dimensionless *drag coefficient*, ranging in magnitude from 10^{-3} over smooth surfaces to 2×10^{-2} over rough ones (Table 9.2), and u_*^2 is the magnitude of momentum flux lost downward into the ground. C_D is affected not only by *skin friction* (viscous drag), but also by *form drag* (pressure gradients upwind and downwind of obstacles such as trees, buildings, and mountains) and by *mountain-wave drag*. Hence, C_D can be larger than C_H. The drag coefficient C_D varies with stability relative to its neutral value C_{D_N} in the same manner as C_H; namely, $C_D\ C_{D_N}$ for unstable boundary layers, and $C_D\ C_{D_N}$ for stable boundary layers.

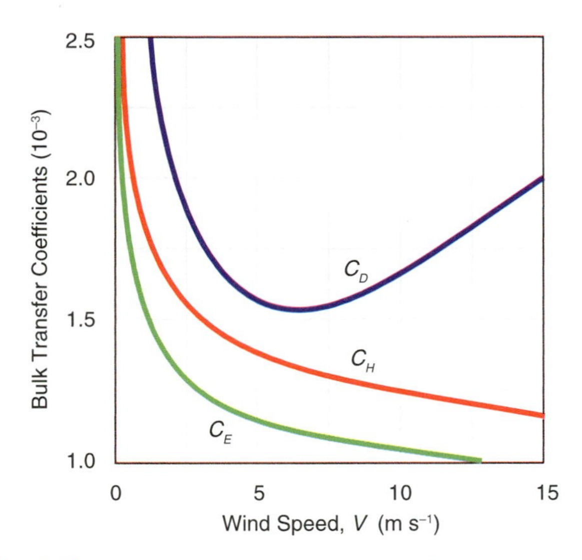

Fig. 9.13 Variation of bulk transfer coefficients for drag (C_D), heat (C_H), and moisture (C_E) with wind speed over the ocean. [Adapted from an unpublished manuscript by M. A. Bourassa and J. Wu (1996).]

Over oceans, an increase in the wind speed leads to an increase in the wave height, which also increases the drag (see Box 9.1). Figure 9.13 shows how the bulk transfer coefficients for momentum, heat, and humidity vary with wind speed measured at height $z = 10$ m over the oceans. For wind speeds larger than 5 m s^{-1}, heat and moisture transfer coefficients gradually decrease with increasing wind speed, whereas the drag coefficient C_D increases. For wind speeds much less than 5 m s^{-1} the bulk formulae are inapplicable, because the vertical turbulent transport between the surface and the air depends more on convective thermals than on wind speed.

9.1 Winds and Sea State

A surface wind gust passing over a water surface produces a discernible patch of tiny *capillary waves* with crests aligned perpendicular to the surface wind vector. Since capillary waves are short lived, their distribution at any given time reflects the current distribution of surface wind. Remote sensing of capillary waves by satellite-borne instruments, called *scatterometers*, provides a basis for monitoring surface winds over the oceans on a global basis.

When forced by surface winds over periods ranging from hours to days, waves with different wavelengths and orientations interact with each other to produce a continuous spectrum of ocean waves extending out to wavelengths of hundreds of meters. The stronger and more sustained the winds, the larger the amplitude of the longer wavelengths. *Wind waves* with the shorter wavelengths tend to propagate in the same direction as the winds. In contrast, the faster propagating long wavelengths tend to radiate outward from regions of strong winds to become *swells* and may thus provide the first sign of an approaching storm. The incidence of wave breaking increases with wind speed. At speeds in excess of 50 m s^{-1}, wave breaking becomes so intense and extensive that the air-sea interface becomes diffuse and difficult to define.

Chapter 7 showed how winds can be forecast by considering the sum of all forces acting on the air. Turbulent drag, as just discussed, is one such force, which always acts opposite to the wind direction (i.e., it slows the wind). More importantly, we see from (9.19c) that the strength of the drag force is proportional to the square of the wind speed, so doubling the wind speed quadruples the drag.

Through these fluxes at the bottom of the atmosphere, the characteristics of the underlying surface are impressed upon the air within the atmospheric boundary layer, but not upon the air in the overlying free atmosphere. Over land the diurnal variations in these fluxes are spread by turbulence throughout the depth of the boundary layer, causing diurnally varying vertical profiles, as described in the next section.

Exercise 9.2 Consider a column of air initially of vertically uniform θ over cold land, capped by a very strong temperature inversion that prevents boundary layer growth. This air column advects with speed U over a warmer ocean surface with potential temperature θ_s. (a) How does temperature vary with distance x from shore? (b) At any fixed distance x from shore, how does the air temperature vary with wind speed?

[**Hint**: Use Taylor's[7] hypothesis: $\frac{\partial \theta}{\partial t} = U\frac{\partial \theta}{\partial x}$.]

Solution: If the only heat into the air column is from the surface, then the change of air temperature with time is found from the heat budget Eq. (9.10) integrated over the boundary layer depth: $\partial\theta/\partial t = F_{Hs}/z_i$, where z_i is constant. Combining this with Taylor's hypothesis gives

$$\frac{\partial \theta}{\partial x} = \frac{1}{U}\frac{F_{Hs}}{z_i}$$

Then, estimating the surface heat flux with bulk aerodynamic methods (Eq. 9.19a) and approximating $(T_s - T)$ by $(\theta_s - \theta)$ gives

$$\frac{\partial \theta}{\partial x} = C_H \frac{\theta_s - \theta}{z_i}$$

(a) Separate variables and integrate: $\theta = \theta_s - (\theta_s - \theta_0)\exp[-C_H x/z_i]$, where θ_0 is the initial potential temperature of the air over land. Thus, the air temperature increases with downwind distance x from the shoreline, rapidly at first, but more gradually further downstream as the air temperature asymptotically approaches the sea-surface temperature.

(b) Surprisingly, air temperature at a fixed distance from shore is independent of wind speed. The reason is that while faster winds cause larger heat fluxes and faster warming of the boundary layer, the faster wind also reduces the time available for warming before the air reaches any distance x from shore. ■

9.2.4 The Global Surface Energy Balance

By applying the bulk aerodynamic formulae to global data sets in which fields of surface air temperature, sea- and land surface temperature, incident solar radiation, and downwelling longwave radiation are derived from assimilation of in situ and space-based observations into state-of-the-art numerical weather prediction models, it is possible to estimate the global distribution of the various terms in the surface energy balance. The net upward transfer of energy through the Earth's surface is

$$F_{net}^{\uparrow} = -F^* + F_{Hs} + F_{Es} \tag{9.21}$$

where F^* is the net downward radiative flux. The sum of the three terms on the right-hand side of (9.21) may be recognized as being equivalent to the term F_{Gs} in (9.18).

The geographical distribution of the annual mean $F_{net}^{\uparrow}$ is shown in Fig. 9.14. As in the net radiation balance at the top of the atmosphere discussed in Section 4.6, the global mean of $F_{net}^{\uparrow}$, averaged over the year, is very close to zero. However, there are local imbalances in excess of 100 W m^{-2}. Because of the small heat capacity of the land surfaces, the net fluxes over the continents are small. The largest imbalances are over regions of the oceans in which the sea

[7] **Geoffrey Ingram Taylor** (1886–1975) British mathematician, physicist, and meteorologist. Studied shock waves, quantum theory, and atmospheric turbulence. Served as meteorologist on an iceberg patrol ship, deployed after ocean liner Titanic collided with an iceberg and sank. Lectured on dynamical meteorology at Cambridge University, devised a statistical method to study turbulent dispersion, examined deformation of crystals, and studied fluid dynamics. Enjoyed boating and flying.

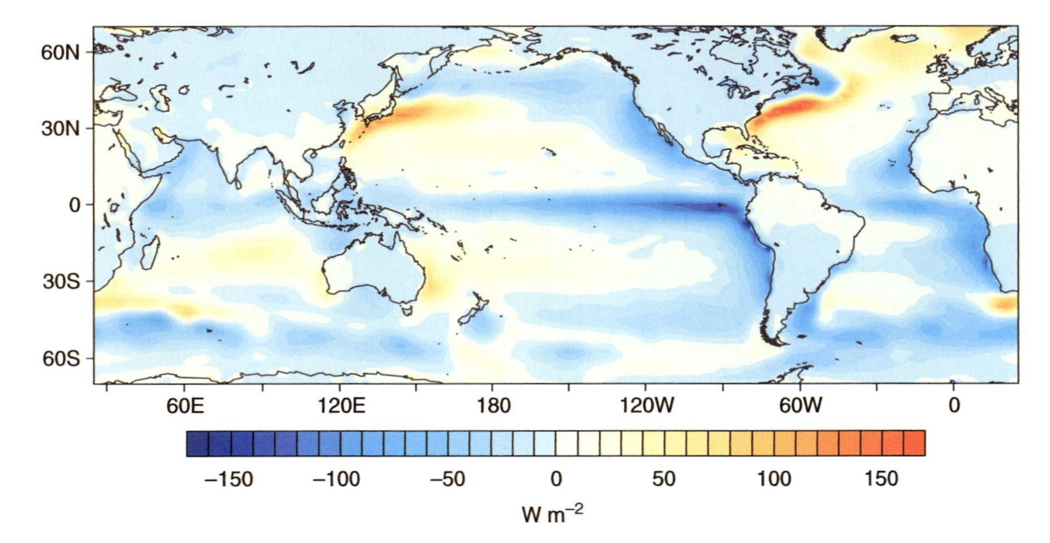

Fig. 9.14 Annual-mean net upward energy flux at the Earth's surface as estimated from Eq. (9.21) based on a reanalysis of 1958–2001 data by the European Centre for Medium Range Weather Forecasting. [Courtesy of Todd P. Mitchell.]

surface is anomalously warm or cold relative to the mean temperature at that latitude (Fig. 2.11). The net flux is upward over the warm waters of the Gulf Stream and the Kuroshio current and it is downward over the regions of coastal and equatorial upwelling where cold water is being brought to the surface.

9.3 Vertical Structure

This section considers the interplay between turbulence and the vertical profiles of wind, temperature and moisture within the boundary layer, drawing heavily on the diurnal cycle over land as an example.

9.3.1 Temperature

Depending on the vertical temperature structure within the boundary layer, turbulent mixing can be suppressed or enhanced at different heights via buoyant consumption or production of T. In fact, it is ultimately the temperature profile that determines the boundary-layer depth.

Recall that the troposphere is statically stable on average, with a potential temperature gradient of 3.3 °C/km (Fig. 9.15). Solar heating of the ground causes thermals to rise from the surface, generating turbulence. Also, drag at the ground causes near-surface winds to be slower than winds aloft, creating wind shear that generates mechanical turbulence. Turbulence generated by processes near the ground mixes surface air of relatively low values of potential temperature, with higher potential temperature air

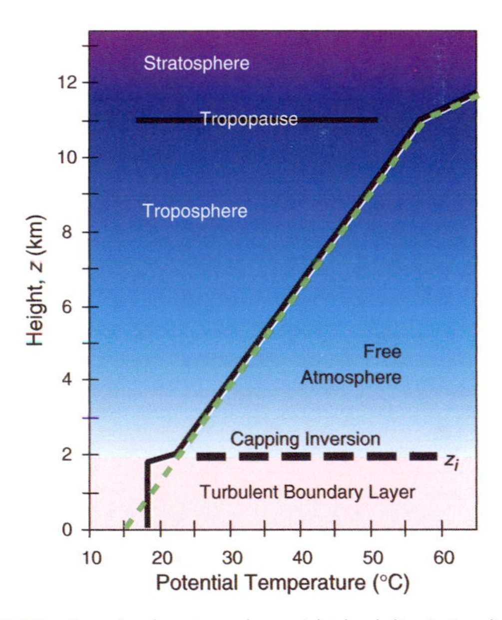

Fig. 9.15 Standard atmosphere (dashed line) in the troposphere and lower stratosphere, and its alteration by turbulent mixing in the boundary layer (solid line). [Adapted from *Meteorology for Scientists and Engineers*, A Technical Companion Book to C. Donald Ahrens' Meteorology Today, 2nd Ed., by Stull, p. 67. Copyrigt 2000. Reprinted with permission of Brooks/Cole, a division of Thomson Learning: www.thomsonrights.com. Fax 800-730-2215.]

from higher altitudes. The resulting mixture has an intermediate potential temperature that is relatively uniform with height (i.e., homogenized within the boundary layer). More importantly, this low altitude mixing has created a temperature jump between the boundary-layer air and the warmer air aloft. This temperature jump corresponds to the capping inversion.

The capping inversion is characterized by high static stability, which suppresses turbulence within it. Turbulence from below has difficulty penetrating the capping inversion and is thus confined within the boundary layer. Hence, the net result is a feedback: boundary-layer turbulence helps create the capping inversion, and the capping inversion tends to trap the turbulence in the boundary layer.

Compared to the mid- and upper troposphere, the fair-weather boundary layer over land exhibits a much larger temperature response to the diurnal cycle because of the rapid turbulent transport forced by alternating heating and cooling of the underlying surface. This effect appears in upper-air soundings as rapid diurnal changes of the vertical profiles within the boundary layer, with slower synoptic scale changes aloft in the free atmosphere. Figure 9.16 contrasts typical vertical profiles of potential temperature and other variables in the boundary layer over land, for day and night. During the daytime, the θ profile is nearly uniform with height over most of the middle of the boundary layer. Because of this extreme homogenization, the daytime boundary layer is also known as the mixed layer, as previously mentioned. Near the bottom is a surface layer (roughly 5% of the depth of the mixed layer), with a superadiabatic temperature gradient, as required to cause a positive heat flux into the air. Near the top is the statically stable capping inversion, which during the daytime is called the *entrainment zone*. Above that is the free atmosphere, illustrated here with the statically stable *standard atmosphere*.

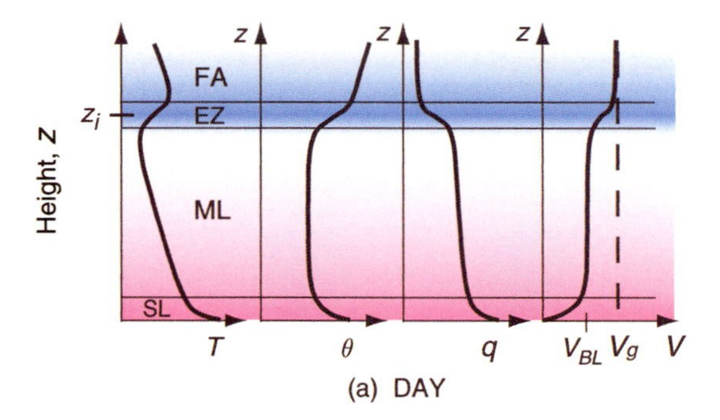

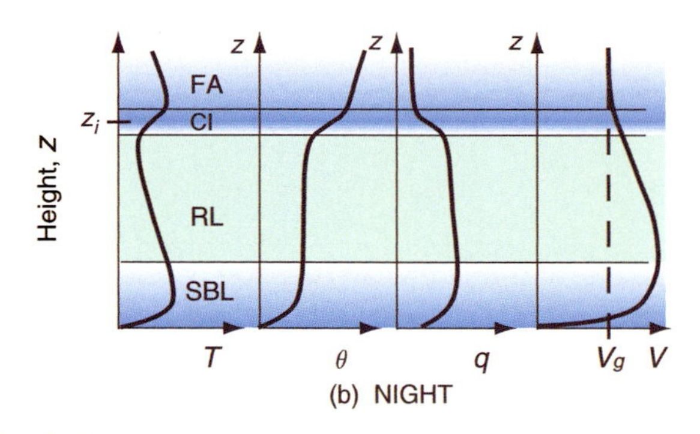

Fig. 9.16 Sketch of typical vertical profiles of temperature (T), potential temperature (θ), specific humidity (q), and wind speed (V) in the bottom of the troposphere. FA, free atmosphere; EZ, entrainment zone; ML, mixed layer; SL, surface layer; CI, capping inversion; RL, residual layer; SBL, stable boundary layer; z_i, height of the capping inversion, which equals top of the boundary layer (BL); V_g, geostrophic wind speed. [Adapted from *Meteorology for Scientists and Engineers*, A Technical Companion Book to C. Donald Ahrens' Meteorology Today, 2nd Ed., by Stull, p. 70. Copyright 2000. Reprinted with permission of Brooks/Cole, a division of Thomson Learning: www.thomsonrights.com. Fax 800-730-2215.]

As mentioned previously, a second stable layer (called the *stable boundary layer* or the *nocturnal boundary layer*) forms at night near the ground in response to the cooling of the air by the radiatively cooled surface. Aloft, the capping inversion formed the previous day is still present. The stable boundary layer near the ground consumes *TKE*, resulting in weak and sporadic turbulence there. Between these two stable layers is the *residual layer*, which contains dying or zero turbulence and also the residual heat, moisture, and pollutants that were mixed there during the previous day.

9.3.2 Humidity

Figure 9.16 also shows the profile of specific humidity, q. During fair weather, the free atmosphere is relatively dry because of the subsiding air in anticyclones (i.e., highs). Evaporation from the surface during daytime adds moisture to the boundary layer. The net result is that there is a rapid decrease of specific humidity with height in the surface layer, as expected by the bulk aerodynamic formula to drive moisture fluxes from the ground into the boundary layer. The moisture added from the ground causes the mixed layer to be more humid than the free-atmosphere air aloft and leads to a humidity jump across the capping inversion.

At night, humidities in most of the middle and top of the boundary layer do not change due to turbulence because turbulence has diminished. However, the radiatively cooled surface can cause dew or frost formation, which reduces humidity in the very bottom of the boundary layer. On other occasions, when dew or frost do not occur, the specific humidity is relatively uniform throughout the bottom and middle of the boundary layer.

9.3.3 Winds

Drag at the ground always causes the wind speed to be reduced, while aloft the winds are stronger (Fig. 9.16). In general, wind speed in the surface layer exhibits a nearly logarithmic profile, as approximated by

$$\overline{V} = \frac{u_*}{k} \ln\left(\frac{z}{z_0}\right) \tag{9.22}$$

where $k = 0.4$ is the von Karman constant and z_0 is the *aerodynamic roughness length* (Table 9.2). The roughness length is defined as the height of zero wind speed as extrapolated down logarithmically from the stronger winds in the surface layer.

Exercise 9.3 By analogy with (9.12) it is possible to define an *eddy viscosity coefficient*

$$K \equiv \frac{-\overline{w'u'}}{\partial \overline{U}/\partial z}$$

that relates the intensity of the vertical mixing of zonal momentum to the local vertical gradient of zonal wind speed, and similarily for the meridional wind component. As in the first-order turbulence closure scheme discussed in Section 9.1.6, K is a local measure of the intensity of the turbulent eddies that are responsible for mixing horizontal momentum downward through the surface layer to the Earth's surface. (a) What is the relationship between eddy viscosity K and height z that yields the logarithmic wind profile (9.22), and what does this suggest about surface-layer similarity theory? (b) What is the relationship between roughness length and drag coefficient? [**Hints**: Let the horizontal mean winds be $(\overline{U}, 0)$ and consider Reynolds stress in only the x-direction.]

Solution: (a) Use Eq. (9.22) for the logarithmic wind profile and assume total wind speed = $\overline{U}$,

$$\overline{U} = (u_*/k)\, ln(z/z_0) = (u_*/k)\, [\ln(z) - \ln(z_0)]$$

Take the partial derivative with respect to height z

$$\partial \overline{U}/\partial z = u_*/(kz)$$

Substituting

$$u_*^2 = -\,\overline{w'u'} = K\frac{\partial \overline{U}}{\partial z}$$

and solving for K we obtain

$$K = k\, z\, u_*$$

It follows that K must increase linearly with height in the surface layer to yield a logarithmic wind profile. K increases with u_*, the square root of the wind stress at the surface, consistent with the notion that windier conditions should be marked by stronger turbulence, which leads to more vigorous eddy mixing. Finally, since the logarithmic wind profile is consistent with a K-theory approach, the fact that it is observed implies that the dominant form of turbulence in the surface layer is small-eddy mixing (i.e., local mixing).

(b) Combine Eq. (9.19c) with the logarithmic wind profile equation

$$C_D = [k/\ln(z/z_0)]^2$$

Thus, rougher surfaces (large z_0) are associated with larger values of the drag coefficient C_D for statically neutral conditions. ■

Equation (9.22) is an example of zeroth-order turbulence closure. It is based on the similarity theory that all wind profiles under statically neutral conditions collapse to one common logarithmic curve (i.e., the curves look similar to each other) when the dimensionless wind speed V/u_* is plotted versus the dimensionless height z/z_0. Often, the expression for wind in the surface layer is written in terms of a dimensionless wind shear

$$\Phi_M = \frac{kz}{u_*}\frac{\partial V}{\partial z} \tag{9.23}$$

where $k = 0.4$ is the von Karmam constant. Thus, for statically neutral conditions, the vertical derivative of Eq. (9.22) can be written as

$$\Phi_M = 1 \tag{9.24}$$

The wind-profile shape varies slightly with static stability, but in general is logarithmic in the surface layer (Fig. 9.17). Under statically stable conditions that are still turbulent (i.e., when $z/L > 0$), the profile is well described by the empirical relationship

$$\Phi_M = 1 + 8.1\frac{z}{L} \tag{9.25}$$

where L is the Obukhov length. For statically unstable conditions, surface-layer similarity theory yields

$$\Phi_M = 1 - 15\left(\frac{z}{L}\right)^{-1/4} \qquad (9.26)$$

Exercise 9.4 Integrate Eq. (9.25) to derive an expression for wind speed as a function of height under statically stable conditions, assuming that $V = 0$ at $z = z_0$ and u_* and L are constant.

Solution: Separate variables:

$$dV = \frac{u_*}{k}\left[\frac{dz}{z} + \frac{8.1}{L}dz\right]$$

Then integrate to obtain

$$V - 0 = \frac{u_*}{k}\left[\ln z - \ln z_0 + \frac{8.1}{L}(z - z_0)\right]$$

Upon rearranging, this can be written as a dimensionless wind speed versus a dimensionless height

$$\frac{V}{u_*} = \frac{1}{k}\left[\ln\frac{z}{z_0} + 8.1\,\frac{(z - z_0)}{L}\right]$$

Thus, wind varies both logarithmically and linearly with height; hence, the vertical wind profile observed under stable conditions is said to be *log-linear*. ■

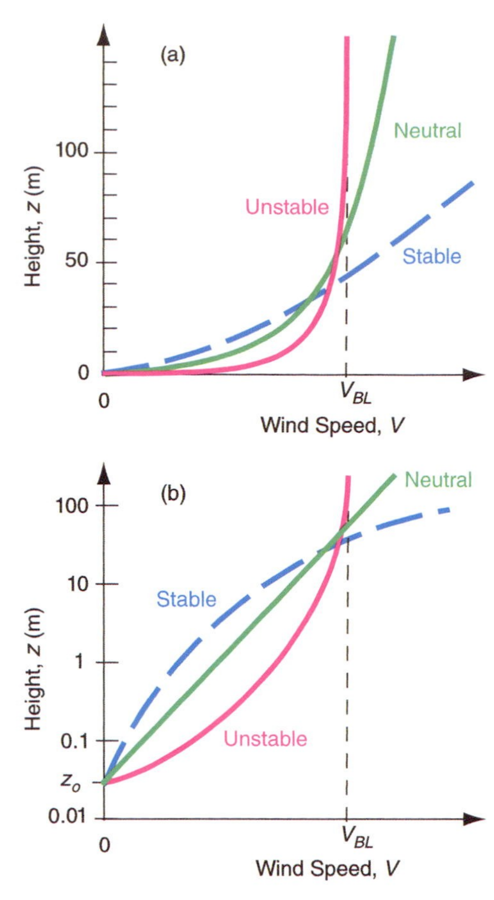

Fig. 9.17 Typical variation of wind speeds with height in the surface layer for different static stabilities, plotted on (a) linear and (b) semi-log graphs. From *Meteorology for Scientists and Engineers*, A Technical Companion Book to C. Donald Ahrens' Meteorology Today, 2nd Ed., by Stull, p. 77. Copyright 2000. Reprinted with permission of Brooks/Cole, a division of Thomson Learning: www.thomsonrights.com. Fax 800-730-2215; and from R. B. Stull, An Introduction to Boundary Layer Meteorology, Kluwer Academic Publishers, Dordrecht, The Netherlands, 1988, Fig. 9.5, p. 377, Copyright 1988 Kluwer Academic Publishers, with kind permission of Springer Science and Bussiness Media.]

Vertical profiles of wind speed for stable, neutral, and unstable profiles are plotted in Fig. 9.17 on linear and logarithmic height scales. Just above the Earth's surface wind speed increases more rapidly with height in the unstable profile than in the stable profile because the more vigorous turbulence that occurs under unstable conditions is more effective at mixing momentum downward. For statically stable surface layers, the turbulence is not strong enough to homogenize the surface layer: in effect, the winds above the surface layer are decoupled from the drag at the ground.

Turbulence communicates drag from the ground throughout the middle part of the mixed layer during daytime, resulting in homogenized, uniform wind speed with height that is slower than geostrophic (i.e., *subgeostrophic*) and which crosses the isobars at a small angle toward lower pressure. Above the capping inversion or entrainment zone, winds quickly recover to their full *geostrophic* values because there is no turbulence there to communicate surface drag information. Thus, wind shear tends to be concentrated in both the surface layer (SL) and the entrainment zone (EZ), as sketched in Fig. (9.16).

At night, the suppression of turbulence by the stable boundary layer causes the air in the residual layer to suddenly be frictionless. This residual-layer air accelerates toward geostrophic, but due to the *Coriolis force* undergoes an *inertial oscillation* in which the wind vector oscillates around the geostrophic wind speed. The winds become faster than geostrophic (*supergeostrophic*) during a portion of the inertial oscillation, causing a low-altitude wind-speed maximum, called the *nocturnal jet*, which is a type of *low-level jet*. Meanwhile, closer to the

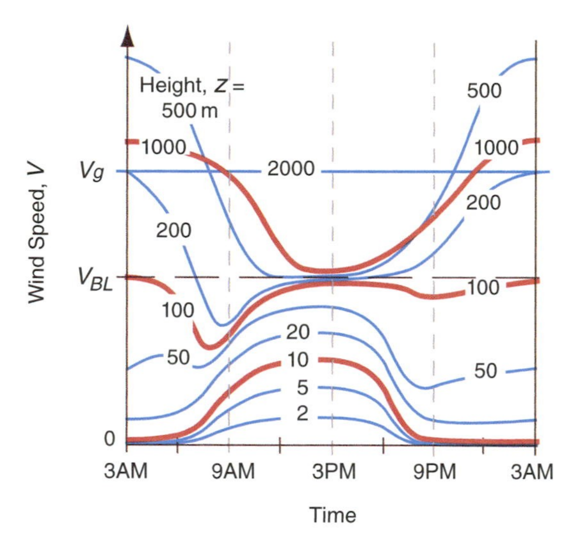

Fig. 9.18 Sketch of variation of wind speed (V) with local time on a sunny day over land, as might be measured at different heights (2 m, 5 m, 10 m . . . . 2000 m) in the boundary layer. V_g is the geostrophic wind. V_{BL} is the mixed layer wind speed, as also sketched in Fig. 9.16. [Adapted from *Meteorology for Scientists and Engineers*, A Technical Companion Book to C. Donald Ahrens' Meteorology Today, 2nd Ed., by Stull, p. 77. Copyright 2000. Reprinted with permision of Brooks/Cole, a division of Thomson Learning: www.thomsonrights.com. Fax 800-730-2215.

ground, the winds become nearly calm because the air is affected by drag at the ground, but is no longer subject to turbulent mixing of stronger winds from aloft because turbulence has diminished. Figure 9.18 shows the diurnal variation of wind speed at various heights within an idealized boundary layer.

9.3.4 Day-to-Day and Regional Variations in Boundary-Layer Structure

We have seen that over land the structure of the boundary layer exhibits a pronounced diurnal cycle in response to the alternating heating and cooling of the underlying surface. Superimposed upon these diurnal variations are day-to-day and longer timescale variations associated with changing weather patterns. Examples include: the flow of cold air over a warmer land surface following the passage of a cold front renders the bottom of the boundary layer more unstable; cloud cover suppresses the diurnal temperature range; and the passage of an anticyclone favors strong nighttime inversions. Day-to-day variations in boundary-layer structure are most pronounced during winter, when the daytime insolation is relatively weak, and during periods of unsettled weather.

Over the oceans, where the diurnal temperature variations are much weaker than over land, the air temperature is closer to being in equilibrium with the underlying surface. With a few notable exceptions,[8] air-sea temperature differences are limited to 1–2 °C. The distribution of the fluxes of latent and sensible heat at the air-sea interface is determined not by the radiation budget, but by the orientation of the low-level wind field relative to the isotherms of sea surface temperature. Over most of the area of the oceans, the flow is across the isotherms from colder toward warmer water so that the air is colder than the underlying surface and boundary layer tends to be weakly unstable, giving rise to a well-defined mixed layer similar to the one in the daytime profiles in Fig. 9.16.

Large expanses of cold advection are evident in the January climatological-mean chart shown in Fig. 9.19. The most prominent of these are off the coasts of Japan and the eastern United States, where

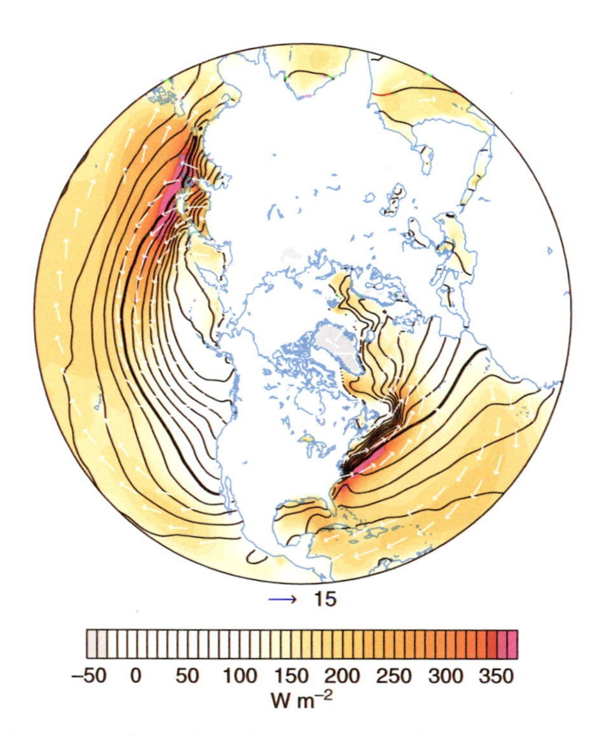

Fig. 9.19 Climatological-mean January latent plus sensible heat fluxes leaving the surface. [Based on data from data by the European Center for Medium Range Weather Forecasting 40-Year Reanalysis. Courtesy of Todd P. Mitchell.]

[8] During wintertime cold air outbreaks, cold continental air flows over the warm waters of the Gulf Stream and Kuroshio currents, giving rise to locally strong fluxes of latent and sensible heat, as discussed in connection with Fig. 9.14.

cold continental air flows over the warm western boundary-currents (i.e., the Kurishio current and the Gulf Stream, respectively). The combined sensible and latent heat fluxes in these zones (indicated by the colored shading in Fig. 9.19) are ~300 W m^{-2} in the climatological mean and they are even larger and during cold air outbreaks. The trade wind belts, where cool air is flowing equator-ward and westward over progressively warmer water, also exhibit a weakly unstable boundary-layer stratification and enhanced sensible and latent heat fluxes.

Stably stratified marine boundary layers, with vertical temperature moisture and wind profiles analogous to the nighttime land profiles in Fig. 9.16, are observed in regions of warm advection, where warm air is flowing over colder water. For example, as air that has resided over the Gulf Stream flows northward over the cold Labrador Current to the southeast of Nova Scotia (see Fig. 2.5) it becomes stably stratified. Under these conditions, the air just above the surface is often chilled to its dew point, resulting in widespread fog. *Advection fog* is also common when air approaching the coast of California passes over the narrow zone of coastal upwelling just offshore, and when warm, humid air masses off the Gulf of Mexico flow northward over snow-covered ground in the midwestern United States during winter.

Over the Arctic and Antarctic, the boundary layer becomes highly stratified in response to the uninterrupted radiative cooling of the surface during the extended polar night. Under conditions of light winds, boundary-layer turbulence virtually disappears and the capping inversion settles to the ground. Under these conditions the surface air temperature may be more than 20 °C lower than the air temperature at the top of the inversion just tens or hundreds of meters above the ground. The surface air temperature may rise sharply whenever the wind speed picks up, mixing warmer air downward from above the inversion or when a cloud layer moves overhead, increasing the downward flux of longwave radiation incident on the Earth's surface.

During cold, calm intervals, Arctic cities and towns such as Fairbanks, Alaska, experience episodes of ice fog when water vapor emitted by automobiles and wood stoves is trapped below the inversion. The frost point of the air is so low that even the emissions from a small urban complex are sufficient to supersaturate the air, resulting in the formation of tiny, sparsely distributed ice crystals referred to by local residents as "diamond dust" because of the way they sparkle in the sunlight.

9.3.5 Nonlocal Influence of Stratification on Turbulence and Stability

Based on the environmental sounding it is possible to make inferences about the vertical distribution of turbulence. In the analysis in Section 3.6, layers were examined individually and judged to be convective or nonconvective depending on the local stratification of temperature and moisture. This approach yields useful information in some cases, but it can sometimes yield misleading results because it fails to take account of the nonlocal character of turbulence.

For example, the local method would incorrectly identify the middle of the mixed layer as statically neutral (implying only moderate, isotropic turbulence and coning of smoke plumes), as opposed to statically unstable (strong turbulence, highly anisotropic, with looping smoke plumes), as demonstrated in Fig. 9.20.

The analysis of soundings can be extended to include non-local influences of stratification in the following manner.

1. First, locate any statically *unstable* regions. This is done by first plotting the θ profile. Then, from every relative maximum in θ, a

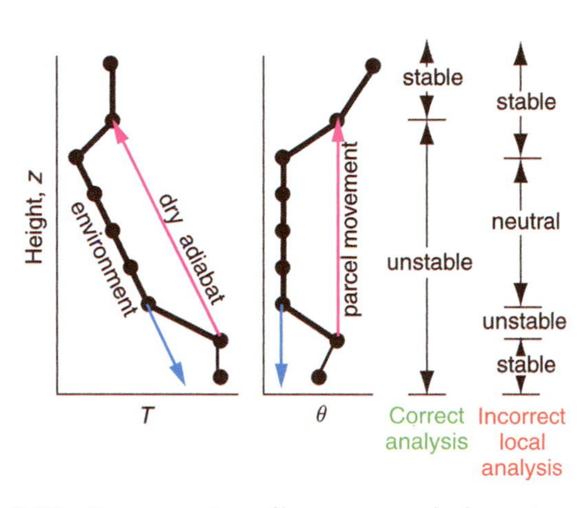

Fig. 9.20 Demonstration of how to properly determine static stability by considering nonlocal air-parcel movement from the relative maxima and minima in the potential temperature θ sounding. T is air temperature. [Adapted from *Meteorology for Scientists and Engineers,* A Technical Companion Book to C. Donald Ahrens' Meteorology Today, 2nd Ed., by Stull, p. 131. Copyright 2000. Reprinted with permission of Brooks/Cole, a division of Thomson Learning: www.thomsonrights.com. Fax 800-730-2215.]

conceptual air parcel is lifted adiabatically until it hits the sounding again (or reaches the highest altitude plotted, whichever is lower). The altitude range from start to end of this lift is statically unstable. Similarly, every relative minimum in θ is located, and a conceptual air parcel is lowered until it hits the sounding again (or hits the ground, whichever is higher). This delineates additional regions of unstable air. The total domain of unstable air is the superposition of all the unstable regions (see Fig. 9.20 for example).

2. Next, for only those subdomains outside of the unstable regions, identify as statically *stable* those regions in which $\partial\theta/\partial z > 0$.
3. Finally, designate any remaining subdomains as statically *neutral* (i.e., regions in which $\partial\theta/\partial z \simeq 0$).

In cloudy regions, the nonlocal method is applicable by changing from dry to moist adiabats when following rising air parcels above their LCL, and changing from moist to dry adiabats when following descending parcels below their LCL.

To determine regions of turbulence, regions of statically unstable air must first be estimated, using nonlocal methods as outlined earlier. Then the Richardson number criterion discussed in Section 9.1.3 is used to identify regions of dynamically unstable air. Finally, turbulence is expected in regions that are statically *or* dynamically unstable. Laminar flow is expected only where the flow is both statically *and* dynamically stable.

Exercise 9.5 A rawinsonde sounding through the lower troposphere gives the following profile information. Which layers of air are turbulent, and why?

z (km)	T (°C)	U (m s^{-1})
13	−58	30
11	−58	60
8	−30	25
5	−19	20
3	−3	18
2.5	1	9
2	2	8
1.6	0	5
0.2	13	5
0	18	0

Assume $V = 0$, $q = 0$, and $T_v = T$.

Solution: To determine the regions in which turbulence is occurring we need to examine both dynamic stability and nonlocal static stability. For dynamic stability, when the derivatives in Eq. (9.8) are approximated with finite differences, the result is called the *bulk Richardson number*

$$R_B = \frac{\frac{g}{\langle T_v \rangle} \Delta\theta_v \cdot \Delta z}{(\Delta U)^2 + (\Delta V)^2}$$

where the angle brackets $\langle T_v \rangle$ represent the average across the whole layer. Also, use $\theta = T + \Gamma z$, where Γ = 9.8 °C km^{-1} is the dry adiabatic lapse rate.

The results are in the following tables:

z (km)	T (°C)	U (m s^{-1})	θ (°C)	T_{avg} (k)	Δz (m)	ΔU (m s^{-1})	$\Delta\theta$ (K)
13	−58	30	69.4				
11	−58	60	49.8	215.15	2000	−30	19.6
8	−30	25	48.8	229.15	3000	35	1.4
5	−19	20	30	248.65	3000	5	18.4
3	−3	18	26.4	262.15	2000	2	3.6
2.5	1	9	25.5	272.15	500	9	0.9
2	2	8	21.6	274.65	500	1	3.9
1.6	0	5	15.68	274.15	400	3	5.92
0.2	13	5	14.96	279.65	1400	0	0.72
0	18	0	18	288.65	200	5	−3.04

For static stability, plot the profile, and lift parcels from every relative maximum and lower from every relative minimum to identify statically unstable regions. The layer $z = 0$ to 1.8 km is statically unstable.

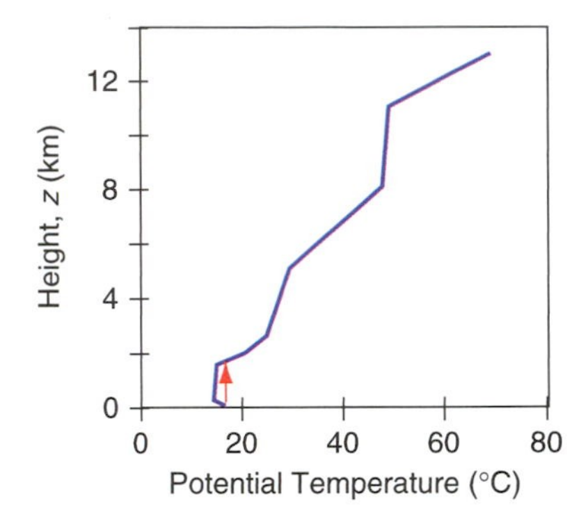

Now we consider the dynamic stability using the bulk Richardson number criterion.

Layer (km)	R_B	Dynamically	Statically	Turbulent
11 to 13	1.98	Stable	Stable	no
8 to 11	0.15	Unstable	Stable	yes
5 to 8	87.02	Stable	Stable	no
3 to 5	67.29	Stable	Stable	no
2.5 to 3	0.20	Unstable	Stable	yes
2 to 2.5	69.58	Stable	Stable	no
1.6 to 2	9.41	Stable	Unstable to 1.8 km	yes to 1.8 km
0.2 to 1.6	$+\infty$	(undefined)	Unstable	yes
0 to 0.2	-0.83	Unstable	Unstable	yes

Summary: The bottom turbulent region 0–1.8 km is the boundary layer. *Clear air turbulence* (CAT) exists near the jet stream, from 8 to 11 km. The other turbulent region is 2.5 to 3 km. ■

9.4 Evolution

This section considers the processes that control the depth of the boundary layer and cause it to evolve in response to changing environmental conditions.

9.4.1 Entrainment

The capping inversion is not a solid boundary. Hence, when rising thermals and turbulent eddies from the mixed layer reach the capping inversion, the inertia of the thermals and eddies causes them to overshoot a small distance through the capping inversion before sinking back into the mixed layer. During this overshoot, the air inside these thermals and eddies has temporarily left the mixed layer, and the pressure gradient created by the incursion of the thermal into the capping inversion drives wisps of free atmosphere air downward through the capping inversion to take the place of the missing air in the mixed layer.

But this exchange is asymmetric. The thermals overshoot into a laminar region of air in the free atmosphere, where nothing prevents these thermals (in which the air is lower in potential temperature than the air in the free atmosphere) from sinking back into the mixed layer. However, the wisps of free atmosphere air that were pushed down into the mixed layer find themselves immediately torn and mixed into the mixed layer by the strong turbulence there. These air parcels become one with the mixed layer and never return to the free atmosphere. This process is called *entrainment*, and the layer in which it takes place is called the *entrainment zone*. Entrainment occurs whenever air from a nonturbulent region is drawn into an adjacent turbulent region. It is a one-way process that adds air mass to the turbulent mixed layer. It can be thought of as a mixed layer that gradually eats its way upward into the overlying air.

Figure 9.21 shows the typical evolution of the boundary layer during fair weather over land in summer. By the end of the night, the boundary layer often consists of a stably stratified shallow boundary layer near the ground (called the *noctural boundary layer*), above which is a nearly neutral layer called the *residual layer*. Above that is the capping inversion as was shown in Figs 9.15 and 9.16.

After sunrise, the warmed ground heats the air touching the ground, creating shallow thermals that

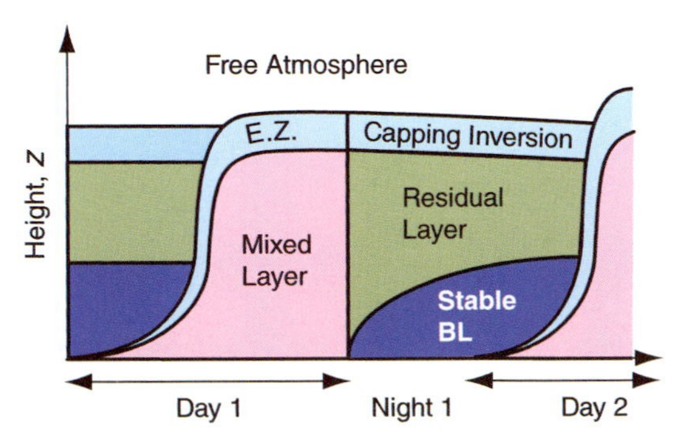

Fig. 9.21 Vertical cross section of boundary-layer structure and its typical evolution during summer over land under fair-weather, cloud-free conditions. E.Z. indicates the entrainment zone. [Adapted from *Meteorology for Scientists and Engineers*, A Technical Companion Book to C. Donald Ahrens' Meteorology Today, 2nd Ed., by Stull, p. 69. Copyright 2000. Reprinted with permission of Brooks/Cole, a division of Thomson Learning: www.thomsonrights.com. Fax 800-730-2215.]

rise and cause intense mixing (creating the *mixed layer*), and which cause entrainment at the top of the mixed layer (in the *entrainment zone*). As the mixed layer (or *convective boundary layer*) grows by entrainment, it can "burn off" the nocturnal inversion and then rapidly rise through the residual layer. Once it hits the capping inversion, the entrainment zone becomes the new capping inversion during the rest of the day.

Around sunset (Fig. 9.21), longwave radiation cools the ground to temperatures less than the overlying air temperature, and two things happen.

1. Thermals cease, allowing turbulence to decay in the former mixed layer. This former mixed layer is now called the *residual layer* because it contains the residual moisture, heat, and pollutants, as was discussed in Fig. 9.16.
2. The cold surface cools the air near the ground, transforming the bottom of the residual layer into a gradually deepening, nocturnal stable boundary layer.

Then the cycle repeats itself, day after day, as long as the weather remains fair. In winter, when the nights are longer than the days, the stable nocturnal boundary layer is much thicker, and the daytime mixed layer is so shallow that the top of the stable boundary layer persists day and night as the capping inversion.

9.4.2 Boundary-Layer Growth

The mixed layer is not a closed system with a fixed mass per unit area. It can incease in depth as air is entrained into it from above, and it can increase or decrease in depth in response to the large-scale vertical velocity; that is

$$\frac{dz_i}{dt} = w_e + w_i \tag{9.27}$$

where w_e is called the *entrainment velocity* (always nonnegative), defined as the volume of air entrained per unit horizontal area per unit time, and w_i is the vertical velocity of the large-scale motion field at the top of the boundary layer (negative for subsidence).

Exercise 9.30 shows that

$$w_i \simeq -z_i\{\nabla \cdot \mathbf{V}\} \tag{9.28}$$

where $\{\nabla \cdot \mathbf{V}\}$ is the mass-weighted divergence within the boundary layer. Over regions of fair weather, w_i is usually downward (negative) and the horizontal flow within the boundary layer is divergent. Divergence may be viewed as thinning the boundary layer over a given region by removing air laterally. Conversely, in regions of large-scale ascent, boundary layer convergence thickens the boundary layer.

A strengthening of the turbulence in the mixed layer causes greater entrainment, whereas a strengthening of the capping inversion $\Delta\theta$ reduces entrainment by reducing the inertial overshoot of the thermals and eddies. The rate of generation of *TKE* by buoyancy is proportional to the *surface sensible heat flux* F_{Hs} (in kinematic units) so the entrainment rate for a free-convective mixed layer is well approximated by

$$w_e = \frac{A\,F_{Hs}}{\Delta\theta} \tag{9.29}$$

where

$$A = -F_{Hzi}/F_{Hs} \tag{9.30}$$

The so-called *Ball parameter A* is on the order of 0.2 for free convection and increases as more mechanical turbulence adds to the entrainment.

The sensible heat flux across the capping inversion F_{Hzi} is usually negative because air with higher potential temperature from the free atmosphere is drawn downward by entrainment. An idealized sensible heat-flux profile under conditions of free convection is shown in Fig. 9.22. The average potential temperature

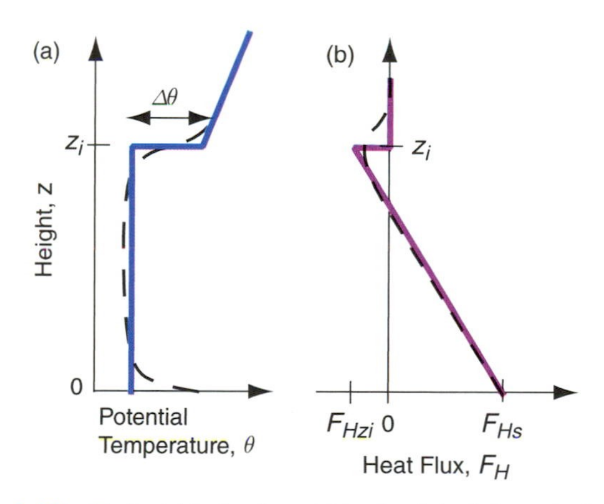

Fig. 9.22 Typical (dashed) and idealized (solid) vertical profiles of potential temperature θ and turbulent sensible heat flux F_H across the mixed layer. The idealized version is sometimes called a *slab model* (because of the uniform θ in the mixed layer) or a *jump model* (because of the discontinuity $\Delta\theta$ in the entrainment zone). [Adapted from *Meteorology for Scientists and Engineers*, A Technical Companion Book to C. Donald Ahrens' Meteorology Today, 2nd Ed., by Stull, p. 69. Copyright 2000. Reprinted with permission of Brooks/Cole, a division of Thomson Learning: www.thomsonrights.com. Fax 800-730-2215.]

in the mixed layer, $\langle\theta\rangle$, is warmed by both the upward sensible heat flux from the surface and from the downward entrainment heat flux at the top

$$\frac{d\langle\theta\rangle}{dt} = \frac{F_{Hs} - F_{Hzi}}{z_i} = \frac{(1 + A)F_{Hs}}{z_i} \qquad (9.31)$$

The entrained air brings with it the heat, moisture, pollutants, and momentum from the free atmosphere. The rates at which these scalar variables are added to the boundary-layer air (per unit area) can be represented as fluxes at the top of the mixed layer. For example, for moisture

$$\overline{w'q'} = -w_e \Delta q \qquad (9.32)$$

where q is specific humidity and Δq is the humidity step across the mixed-layer top ($\Delta q = q_{zi+} - q_{zi-}$). Subscripts $zi+$ and $zi-$ refer to just above and below the entrainment zone, respectively. Equations analogous to (9.32) can be written for any scalar variable, including potential temperature and pollutants, and is also used for the horizontal momentum components u and v. When existing pollutants are mixed down from elevated layers in the residual layer or free atmosphere, the process is called *fumigation*.

Under light wind conditions it is possible to predict the growth of z_i in the diurnal cycle over land based on the early morning temperature sounding and a prediction of the surface sensible kinematic heat flux. Figure 9.23a shows the idealized surface sensible heat-flux evolution, which can be predicted based on a knowledge of the day of the year (which determines sun angles) and the expected cloud coverage. Between the time shortly after sunrise when heat flux becomes positive and the time of day of interest for determining z_i, the amount of accumulated heating is the area under the curve between those two times (the hatched area in Fig. 9.23a). If advection is negligible, it can be assumed that this heat warms the boundary layer.

Assuming an adiabatic temperature profile in the mixed layer, the amount of heat needed to raise z_i from the surface to the level it reaches at $t = t_1$ is the area under the sounding (the hatched area in Fig. 9.23b). The solution is obtained by finding an average potential temperature in the mixed layer, $\langle\theta\rangle$, such that the hatched area under the sounding curve equals the hatched area under the heat-flux curve. Then, knowing $\langle\theta\rangle$, the mixed-layer depth z_i is the height at which $\langle\theta\rangle$ intersects the morning sounding. This procedure is called the *thermodynamic method* or the *encroachment method*.

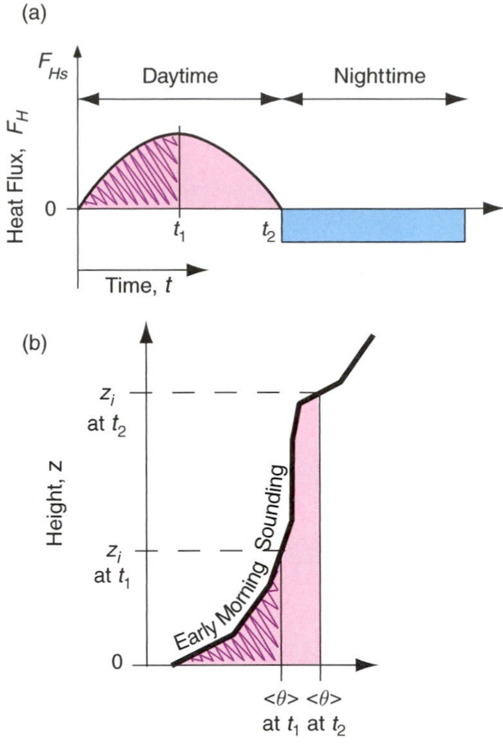

Fig. 9.23 (a) Idealized variation of surface sensible heat flux F_{Hs} (in kinematic units of K m s^{-1}) with time during fair weather over land. The area indicated by the hatched region represents the total amount of heat input into the bottom of the boundary layer from sunrise until time t_1. (b) Typical early morning sounding (heavy line) of potential temperature, θ, showing the change in the mixed-layer average potential temperature, $\langle\theta\rangle$, and depth z_i with time t. The hatched area in (a) equals the hatched area in (b). [Adapted from *Meteorology for Scientists and Engineers*, A Technical Companion Book to C. Donald Ahrens' Meteorology Today, 2nd Ed., by Stull, p. 70, 74. Copyright 2000. Reprinted with permission of Brooks/Cole, a division of Thomson Learning: www.thomsonrights.com. Fax 800-730-2215.]

Exercise 9.6 (a) If the potential temperature profile is linear in the stable boundary layer at sunrise, derive an equation for the growth rate of the mixed layer using the thermodynamic approach. For simplicity, assume the surface heat flux is constant. (b) How does the shape of this curve relate to the initial mixed-layer growth phase?

Solution: (a) If the early morning sounding is $\partial\theta/\partial z = \gamma$, and the surface kinematic heat flux is F_{Hs}, then the area under the heat-flux curve is $A = F_{Hs}\, t$,

where t is time since sunrise. The area under the sounding is a triangle, yielding $A = (\Delta\theta)^2/(2\gamma)$, where $\Delta\theta$ is the amount of warming of the mixed layer since sunrise, assuming, as initial conditions, that $z_i = 0$ at sunrise. Equating the two areas gives $\Delta\theta = (2\gamma F_{Hs}t)^{1/2}$. Because the morning profile is linear, $z_i = \Delta\theta/\gamma$. Thus, $z_i = [(2F_{Hs}/\gamma)\, t]^{1/2}$.
(b) The square root of time dependence from part (a) does not agree with the square of time dependence typically observed in the early morning, as plotted in Fig. 9.24. The difference is due to the fact that during the morning the heat flux is not constant with time, but increases sinusoidally, and also that the initial vertical profile of potential temperature is not linear, but is exponential in shape as in Fig. 9.16b. ■

9.4.3 Cloud-Topped Boundary Layer over Land

The lifting condensation level (LCL) can be below z_i on days when sufficient moisture is present in the boundary layer. In this case, the tops of the thermals that extend above the LCL are filled with cumulus clouds. During fair weather regimes over land, the cloudless boundary layer top typically rises slowly during the morning, while the noctural inversion from the previous night is being "burned off." In contrast, the

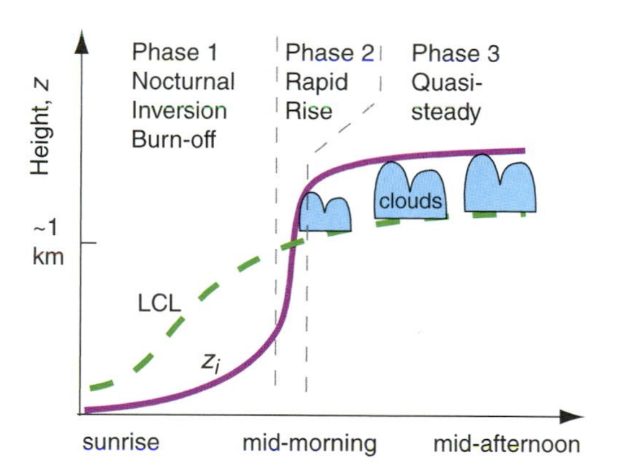

Fig. 9.24 Three-phase growth of mixed layer depth z_i during fair weather in daytime over land showing the height to which thermals can rise. Corresponding variation in the lifting condensation level (LCL) showing the heights needed by thermals to rise to produce cumulus clouds. When $z_i >$ LCL, the LCL marks the cloud base of convective clouds. [Adapted from R. B. Stull, *An Introduction to Boundary Layer Meteorology*, Kluwer Academic Publishers, Dordrecht, The Netherlands, 1988, Fig. 11.10, p. 452, and Fig. 13.12, p. 564, Copyright 1988 Kluwer Academic Publishers, with kind permission of Springer Science and Business Media.]

LCL rises rapidly during the morning due to warming of the air near the surface, which lowers the relative humidity. Thus, the sky typically remains free of boundary-layer clouds through mid-morning because the LCL stays above the top of the rising boundary-layer.

By late morning, the boundary layer typically experiences a rapid-rise phase after the nocturnal inversion has completely disappeared (Fig. 9.24). At around this time, z_i jumps to the height of the capping inversion from the day before (often ~1 or 2 km altitude), and becomes higher than the LCL. This is the time when fair-weather cumulus clouds are most likely to form. The degree of vertical development of the clouds depends on the height difference between the LCL (which marks the cloud base) and z_i (which marks the cloud top). These fair-weather cumulus clouds are distinguishable from the clouds associated with deep convection (see Section 8.2) because they do not penetrate beyond the top of the capping inversion into the free atmosphere. The level of the cloud top is the height to which thermals rise from the surface, and the height to which water vapor and pollutants from the surface are rapidly mixed.

Sometimes, if there is sufficient vertical wind shear $\partial V/\partial z$ in the mostly convective boundary layer, weak parallel counterrotating pairs of circulations form, which are called *horizontal roll vortices* (or "*rolls*" for short). The axes of rotation of the rolls are oriented roughly parallel with the mean wind direction, and the diameter of each roll is of the same order as the boundary-layer depth (Fig. 9.25a). These overturning circulations are almost too weak to measure, but they organize the stronger thermals into parallel lines with a spacing equal to roughly twice the boundary-layer depth. When sufficient moisture is present to support the formation of cumulus clouds, the rolls are visible as parallel rows of cumulus clouds called *cloud streets*, which are apparent in satellite images (Fig. 9.25b) and sometimes from the ground as well.

9.4.4 The Marine Boundary Layer

The cloud-topped boundary layer over the oceans tends to be different from its land counterpart in the following respects,

- Relative humidities of the surface air tend to be higher (usually in excess of 75%).
- Because of the higher relative humidity of the air, cloud cover (much of it in the form of stratus and stratocumulus cloud decks) is much more

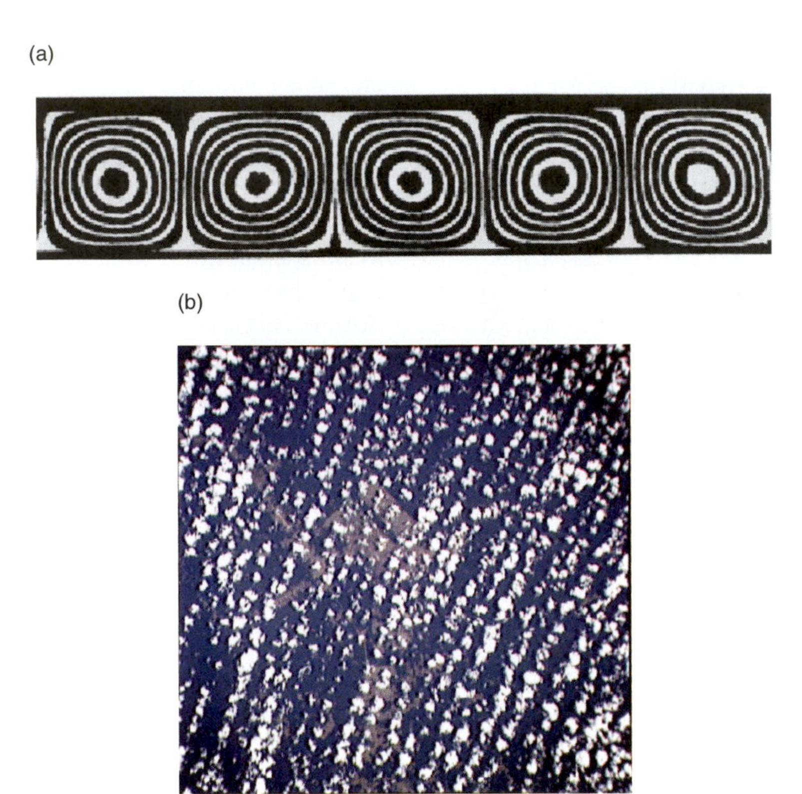

Fig. 9.25 (a) Vertical cross section through the ends of a laboratory simulation of horizontal roll vortices showing the counter-rotating circulations. [From "Influence of Initial and Boundary Conditions on Benard Convection," H.Oertel, Jr., and K. R. Kirchartz, in Recent Developments in Theoretical and Experimental Fluid Mechanics: Compressible and Incompressible Flower, U. Muller, K. G. Roesner, B. Schmidt, eds., Fig. 1, 1979, p. 356, Copyright Springer-Verlag, Berlin, 1979. With kind permission of Springer Science and Business Media.] (b) View looking down on streets of cumulus clouds aligned into rows by the horizontal roll vortices. [NASA MODIS Imagery.]

extensive. Some oceanic regions are cloud covered most of the time.

- With the widespread presence of clouds, radiative transfer plays a more important and more complex role in the boundary-layer heat balance.
- In some regions, drizzle plays a significant role in the boundary-layer heat and water balance.
- The diurnal cycle is not as important and is governed by entirely different physics.

The presence of a cloud layer tends to strengthen the capping inversion through its impact on the radiation balance of the underlying layers (i.e., the ocean plus the atmospheric boundary layer).[9] The strengthening of the capping layer, in turn, reduces the entrainment of dry air into the cloud layer, thereby contributing to the maintenance of the cloud deck. Once formed, marine cloud decks tend to persist because of this positive feedback.

In the cloud-topped boundary layer, convection may be driven by heating from below or by cooling from above. The *heating from below* is determined by the surface buoyancy flux (i.e., the sensible heat flux and the moisture flux to the extent that it affects the virtual temperature) as in the diurnal cycle over land. The *cooling from above* is determined by the flux of longwave radiation at the top of the cloud layer (see Fig. 4.30). Heating from below drives *open cell convection* (Figs. 1.21 and 9.26), with strong, concentrated, buoyant, cloudy plumes interspersed with more expansive regions of weak subsidence that are cloud-free. In contrast, cooling from above drives *closed cell convection* (Figs. 1.7 and 9.26) with strong, negatively buoyant, cloud-free downdrafts, interspersed with more expansive regions of slow ascent

[9] Where the boundary layer is shallow ($z_i \lesssim 1$ km), the reduction in insolation due to the presence of the cloud deck far outweighs the reduction in outgoing longwave radiation from the underlying water because the cloud top is not much cooler than the underlying sea surface. The net effect is radiative cooling of the mixed layer, which helps maintain clouds by keeping the relative humidity high. For boundary-layer depths ~2 km the two effects nearly cancel and the cooling is less efficient.

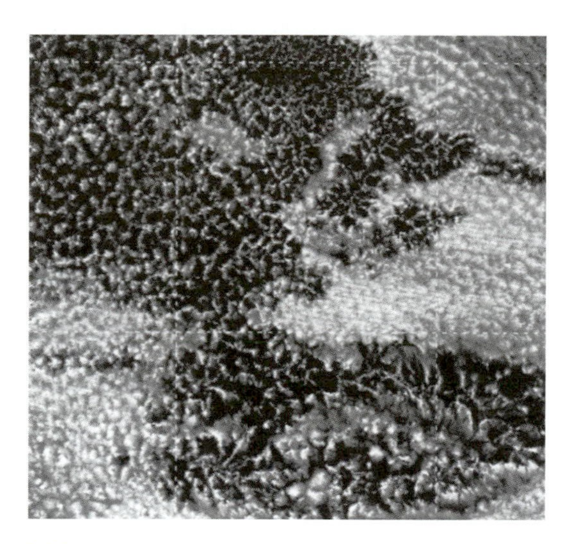

Fig. 9.26 Satellite photograph of mesoscale cellular convection. Open cells are in the upper-left and lower-right quadrants, and closed cells are elsewhere. [Courtesy of NASA MODIS imagery.]

that are cloudy. When satellite images of closed cell convection are rendered as negatives they resemble images of open cell convection, and vice versa. Under conditions of light winds, closed cell convection may assume the form of polygonal cells like those in a honeycomb.

In regions of strong large-scale subsidence where the boundary layer is shallow (i.e., 500 m–1 km in depth), convection-driven heating from below and cooling from above intermingle to form a unified turbulent regime that extends through the depth of the boundary layer. As the boundary layer deepens there is a tendency for the two convective regimes to become decoupled. The lower regime, driven by heating from below, is restricted to the lower part of the boundary layer. As in the daytime boundary layer over land, it may be capped by cumulus clouds.

The capping layer of the upper, radiatively driven regime coincides with the top of a more continuous stratus or stratocumulus cloud deck. The lower and upper regimes are typically separated by an intermediate quiescent layer in which the lapse rate is conditionally unstable. If the cumulus convection in the lower regime becomes sufficiently vigorous that air parcels reach their level of free convection, these buoyant thermals may rise high enough to entrain significant quantities of dry air from above the capping inversion, leading to the dissipation of the stratiform cloud deck and reunification of the turbulent layers.

We can envision the aforementioned sequence of events as occurring if we follow a hypothetical column of boundary-layer air along an equatorward trajectory along the coasts of California, Chile, or Namibia that later curves westward in the trade winds. In response to a weakening of the large-scale subsidence, the boundary layer deepens, and cumulus clouds appear below the base of the stratocumulus deck. The cumulus clouds deepen until they begin to penetrate through the overlying cloud deck. As if by magic, the dreary cloud deck thins and dissipates, leaving behind only picturesque trade wind cumulus.

Drizzle falling from the cloud deck and evaporating into the unsaturated air below it also affects the boundary-layer heat balance. The condensation of water vapor within the cloud deck releases latent heat, and the evaporation of the drizzle drops in the subcloud layer absorbs latent heat. The thermodynamic impact of the downward, gravity-driven flux of liquid water is an upward transport of sensible heat, thereby stabilizing the layer near cloud base.

Low-level cold advection contributes to the maintenance of stratiform cloud decks in two ways: it destabilizes the surface layer, thereby enhancing the flux of water vapor from the sea surface, and it cools the boundary layer relative to the overlying free atmosphere, thereby strengthening the capping inversion and reducing the entrainment of dry air. The subsidence that usually accompanies cold advection also favors a shallow boundary layer with a strong capping inversion. Hence cloud-topped boundary layers prevail in regions of climatological mean cold advection (e.g., to the east of the subtropical anticyclones) and fractional cloud coverage in these regions varies in synchrony with time variations in the strength of the cold advection.

In regions in which the marine boundary layer is topped by cloud decks, fractional cloud coverage tends to be highest around sunrise and lowest during the afternoon. The thinning (and in some cases the breakup) of the overcast during the daytime is due to the absorption of solar radiation just below the cloud tops (see Fig. 4.30). The heating at the top of the cloud deck results in a weakening of the convection within the upper part of the boundary layer, reducing the rate at which moisture is supplied from below. The moisture supply becomes insufficient to replenish the drizzle drops that rain out, and the cloud thins. The warming of the air at the cloud-top level also increases its saturation mixing ratio, thereby allowing some of the liquid water in the cloud droplets to evaporate. After the sun goes down, the air at cloud-top level cools in response to the continuing emission of longwave radiation. In response to the cooling, the convection resumes, renewing the supply of moisture and reestablishing the population of cloud droplets.

A distinctively different kind of marine, cloud-topped boundary layer is sometimes observed during wintertime, when cold, dry continental air crosses the coastline and flows over much warmer water. Under these conditions, sea-air temperature differences may be much larger than those discussed earlier, and the boundary layer may be highly unstable. The strong sensible and latent heat fluxes at the air-sea interface drive intense convection, resulting in a pronounced warming, moistening, and thickening of the boundary layer in a matter of hours or even less over distances of tens of kilometers. This rapid transformation of the boundary layer is a dramatic example of what synoptic meteorologists refer to as *air-mass modification*.

Figure 9.27 shows cold Canadian air streaming southeastward over the Great Lakes, the eastern United States, and the warm Gulf Stream waters of the western Atlantic. As the cold air first moves over unfrozen Lake Ontario, observers standing on the upwind shore of Ontario, Canada, can observe mostly clear skies, but with *steam fog* forming a short distance over the water due to the intense heat and moisture fluxes from the lake surface into the air. As the air crosses the lake, the convective mixed layer deepens and cloud streets develop, aligned with the mean wind direction. The cloud streets widen, deepen, and change into open-cell mesoscale convection as they move downstream. A similar evolution is observed over the Gulf Stream.

Under certain conditions the moisture picked up by the air crossing a large lake can be concentrated within one or more bands of low-level convergence, aligned with the flow, that are capable of producing moderate to heavy snow where they come ashore on the lee side of the lake. Some of the more notable of these *lake-effect snow squall (LES)* events have yielded snowfall amounts in excess of 1 m, which is all the more remarkable considering that the clouds in these systems are typically only 2–3 km deep.

9.4.5 Stormy Weather

In regions of large-scale ascent (e.g., near centers of cyclones and fronts) and in deep convective clouds the mixed-layer top rises to the point where the tropopause, in effect, becomes the capping inversion (Fig. 9.28). It is in these regions that boundary-layer air, with its moisture and pollutants, is most free to mix with the air in the free atmosphere. The character of this mixing is illustrated schematically in Fig. 9.29, in which fronts are depicted as peeling the boundary layer upward along sloping surfaces, and thunderstorms act like vents or chimneys that pump boundary-layer air upward.

After a storm or frontal passage over land, such as the cold front illustrated in Fig. 9.30, the *frontal inversion* (between the top of the cold air mass and the overlying warmer air) is initially the new boundary-layer top. However, after a day or two, when the clouds have cleared and a diurnal cycle of radiative heating and cooling is once again felt by the ground, a new boundary layer forms with depth z_i below the overlying frontal inversion. Such multiple inversion layers (both mixed-layer top and frontal inversion aloft) can often be seen in upper-air soundings.

9.5 Special Effects

In the previous sections the surface underlying the boundary layer was assumed to be flat and either horizontally uniform or (when the effects of horizontal temperature advection were taken into account) slowly varying. This section considers a number of special boundary-layer phenomena that derive from special properties of the underlying surface, in particular,

- terrain,
- discontinuities such as coastines,
- texture, as in the case of forests, and
- human impacts, as in the case of cities.

9.5.1 Terrain Effects

When mountain slopes are heated by the sun during the day in fair-weather conditions, the warm air rises along the slope as an *anabatic wind*[10] (Fig. 9.31). Along a mountain crest, where anabatic winds from the the two sides of the mountain meet, the air rises above the crest and can create *anabatic cumulus clouds* if sufficient moisture is present. Associated with these upslope winds are *return circulations* that include weak downdrafts aloft in the centers of the valley.

[10] *ana*: from the Greek for upward.

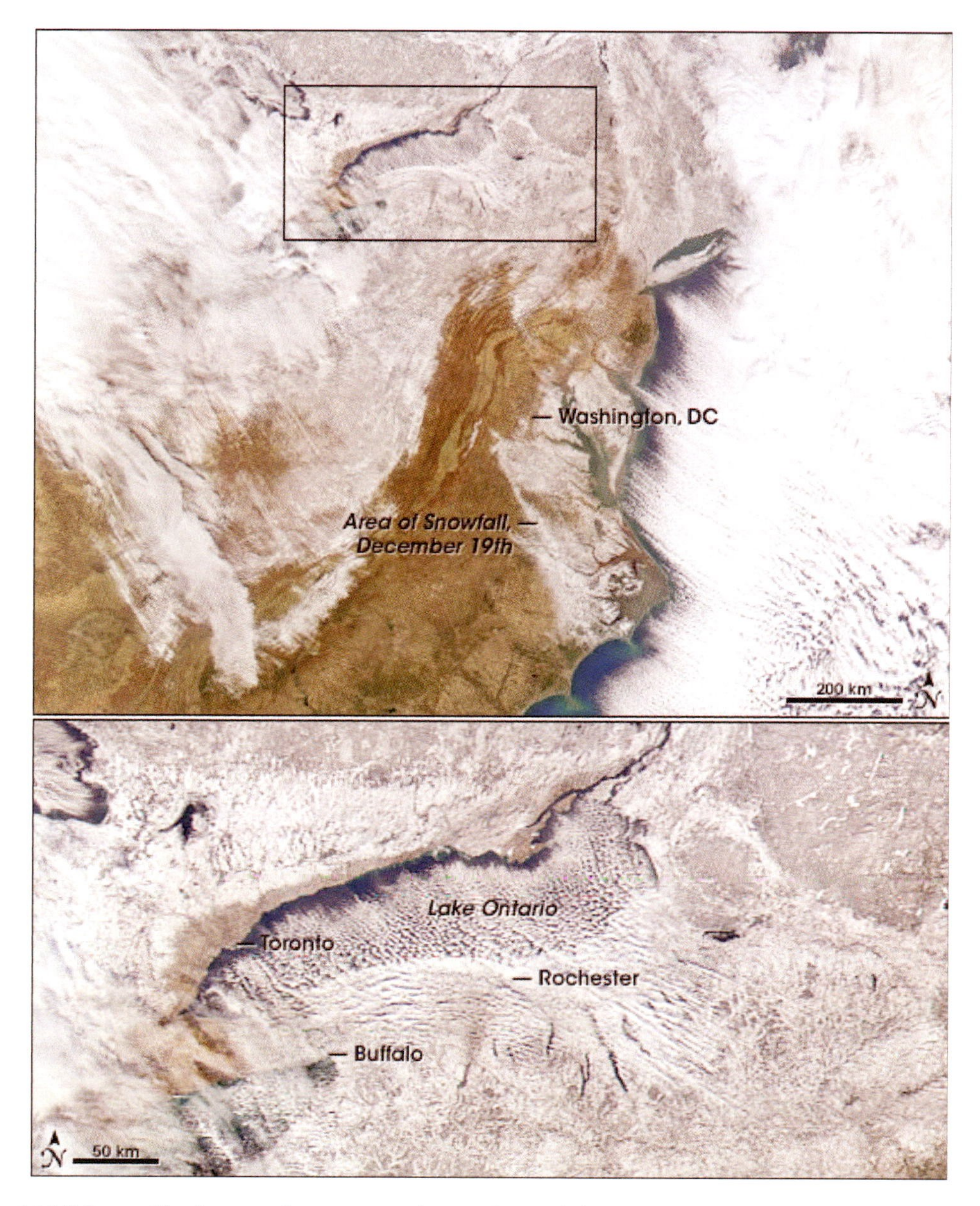

Fig. 9.27 (Top) Visible satellite image of eastern North America and the western North Atlantic showing clouds and snow-covered ground. The region near Lake Ontario outlined by the rectangle is shown in more detail in the bottom figure. Winds are from the northwest, causing cloud streets over the Great Lakes and over the Atlantic Ocean. Water vapor picked up by the air as it crossed Lake Ontario triggered a *lake effect snowstorm* downsteam over upper New York State. [NASA MODIS imagery.]

At night when the mountain slopes are cooled by longwave radiation, the chilled air sinks downslope as a cold *katabatic wind*[11] due to its negative buoyancy. This flow is compensated by a return circulation having weak rising air aloft over the valley center. The chilled air that collects as a *cold pool* in the valley floor is conducive to *fog* formation. Layers of less cooled air do not descend all the way to the valley floor, but spread horizontally at their level of neutral buoyancy, as indicated by the pink arrows in Fig. 9.31. Both the anabatic and katabatic winds are called *cross-valley circulations*.

[11] *kata:* from the Greek for downward.

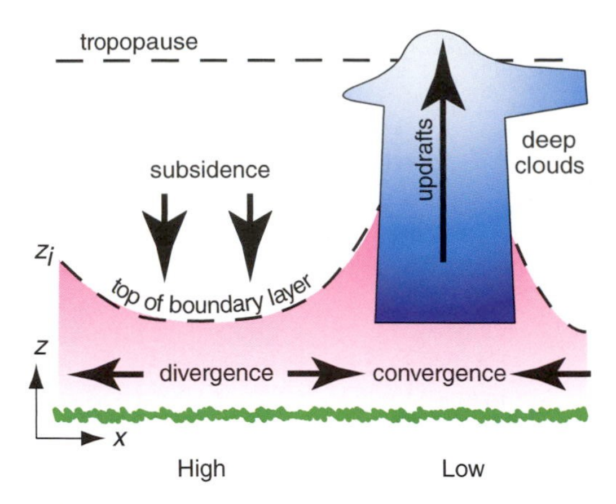

Fig. 9.28 Vertical slice through an idealized atmosphere showing variations in the boundary layer depth z_i in cyclonic (low pressure) and anticyclonic (high pressure) weather conditions. [Adapted from *Meteorology for Scientists and Engineers*, A Technical Companion Book to C. Donald Ahrens' Meteorology Today, 2nd Ed., by Stull, p. 69. Copyright 2000. Reprinted with permission of Brooks/Cole, a division of Thomon Learning: www.thomsonrights.com. Fax 800-730-2215.]

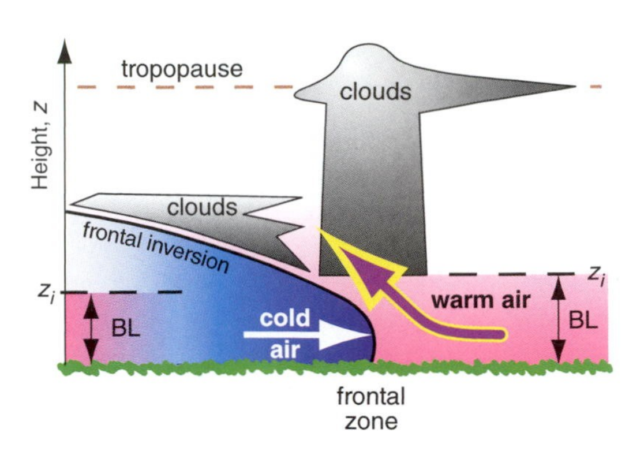

Fig. 9.30 Vertical cross section illustrating the destruction of the capping inversion z_i and venting of boundary layer (BL) air near frontal zones, and subsequent reestablishment of the capping inversion under the frontal inversion after cold frontal passage. [Adapted from *Meteorology for Scientists and Engineers*, A Technical Companion Book to C. Donald Ahrens' Meteorology Today, 2nd Ed., by Stull, p. 69. Copyright 2000. Reprinted with permission of Brooks/Cole, a division of Thomon Learning: www.thomsonrights.com. Fax 800-730-2215.]

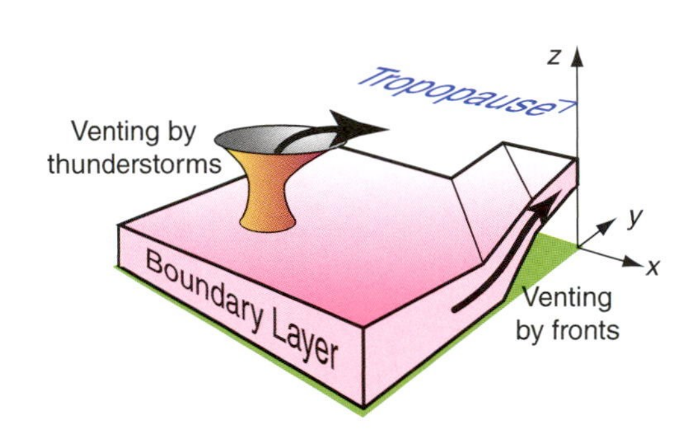

Fig. 9.29 The two dominant mechanisms for boundary-layer air to pass the capping inversion and enter the free atmosphere. [Courtesy of Roland B. Stull.]

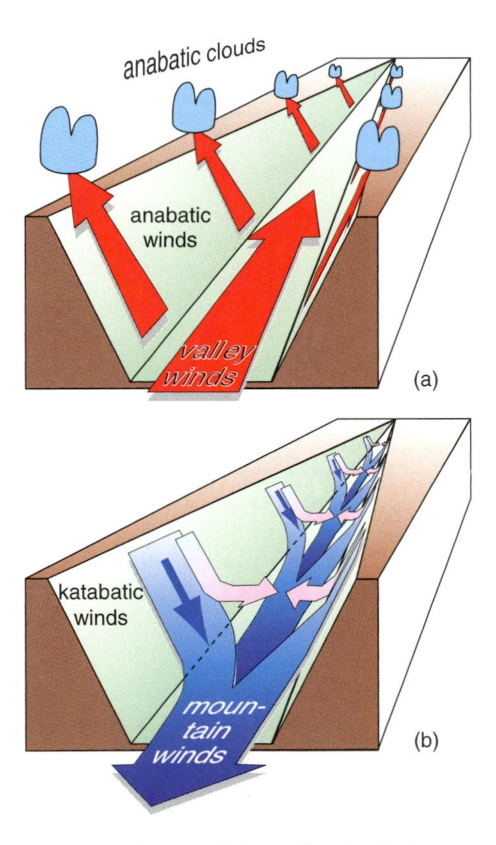

Fig. 9.31 (a) Daytime conditions of anabatic (warm upslope) winds along the valley walls, and associated valley winds along the valley floor. (b) Nighttime conditions of katabatic (cold downslope) winds along the valley walls, and associated mountain winds draining down the valley floor streambed. [Adapted from R. B. Stull, *An Introduction to Boundary Layer Meteorology*, Kluwer, Academic Publishers, Dordrecht, The Netherlands, 1988, Fig. 14.5, p. 592, Copyright 1988 Kluwer Academic Publishers, with kind permission of Springer Science and Business Media.]

Anabatic and katabatic flows are examples of a broader class of diurnally varying thermally driven circulations, which also include land and sea breezes described in the next subsection. Horizontal temperature gradients caused by differential solar heating or longwave cooling cause horizontal pressure gradients, as described by the hypsometric equation and the thermal wind effect. For example, over sloping terrain during the afternoon, a pressure gradient forms normal to the slope, as illustrated in Fig. 9.32. This pressure-gradient force induces an upslope (anabatic) flow during daytime, while the increased density of the air just above the surface at night induces a downslope (katabatic) flow.

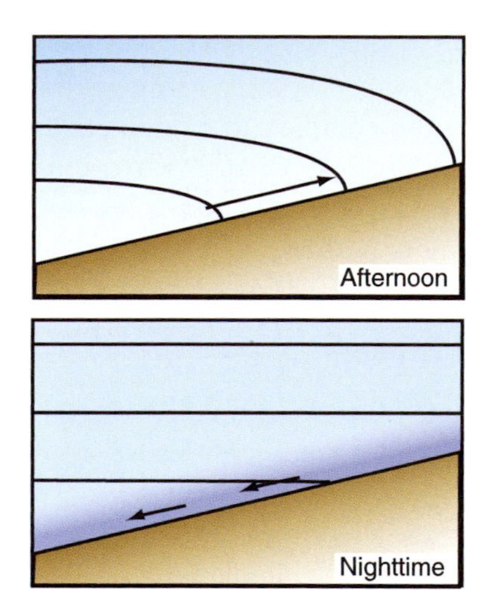

Fig. 9.32 Schematic showing how circulations develop in response to land-sea heating contrasts and heating along sloping terrain. Faded (darkened) blue shading indicates planetary boundary temperatures higher (lower) than free air temperatures at the same level. The contours show how pressure surfaces are distorted by the diurnally varying boundary layer temperatures, and arrows indicate the upslope/downslope accelerations induced by the horizontal temperature gradients.

Along-valley winds also form due to these buoyant forcings (Fig. 9.31). At night, the cold air that pools in a valley follows the streambed downslope, causing cold *mountain winds* to drain out of the mountain valleys onto broader plains. During daytime, there is an up-valley flow called the *valley wind* along the streambed.

There are times when the relatively weak, diurnally dependent, thermally driven circulations discussed in this subsection are over-shadowed by stronger winds that evolve in accordance with the synoptic-scale forcing. Strong synopic-scale pressure gradients in mountainous regions are often attended by a number of distinctive phenomena on smaller scales, including *gap winds* (Section 8.2.4), *mountain waves* and *lenticular clouds*, local accelerations of the wind over mountain crests, *blocking* of low-altitude winds, lee-side "*cavity*" circulations, *rotor clouds, banner clouds, wake turbulence, Karman vortex streets*, downslope wind storms, and *hydraulic jumps* (Fig. 9.33). Even low hills can exert a surprisingly strong influence on cloud patterns (Fig. 9.34) and precipitation amounts.

A nondimensional parameter that is widely used to characterize the flow over mountains and to predict the kinds of flow configurations that might occur is the *Froude*[12] *number*

$$Fr = \frac{V}{NS} \tag{9.33}$$

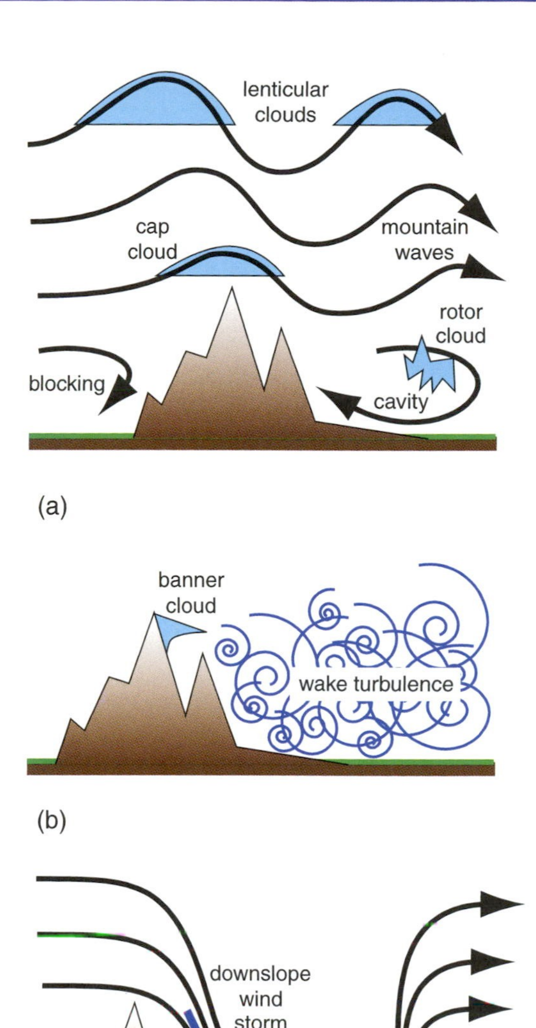

Fig. 9.33 Vertical cross-section sketches of some of the phenomena observed in mountains during windy conditions. [Adapted from R. B. Stull, *An Introduction to Boundary Layer Meteorology*, Kluwer, Academic Publishers, Dordrecht, The Netherlands, 1988, Fig. 14.4, p. 602, and Fig. 14.9, p. 608, Copyright 1988 Kluwer Academic Publishers, with kind permission of Springer Science and Business Media.]

[12] **William Froude** (1810–1879) British hydrodynamicist, engineer, and naval architect who studied water resistance of scale-model ships in towing tanks under contract to the British Navy. Developed an understanding of skin roughness drag versus wave drag and studied ship stability to improve safety and efficiency. Developed the Froude number ratio of inertial to buoyancy forces as a way to scale his model results to full-size ships. Invented a hydraulic dynameter to measure the output of high-powered ship engines.

Fig. 9.34 Cloud streets organized by flow over a nearby ~130-m-high hill. Each photograph was taken on a different day, but the winds on all 3 days were blowing from nearly the same direction. [Photographs courtesy of Art Rangno.]

where V is the wind speed component normal to the mountain, N the Brunt-Väisälä frequency, and S, the vertical (or, for some applications, horizontal) scale of the mountain. The Froude number can be interpreted as the natural wavelength of the buoyancy oscillations, discussed in Exercise 3.13, divided by the wavelength of the mountain. The square of Fr is like a kinetic energy of the incident flow divided by the potential energy needed to lift air over the mountain.

For small values of Fr, the low-level airflow is forced to go around the mountain and/or through gaps. Larger values of Fr are associated with more flow over the top of the mountain. From the results of Exercise 3.13 it is clear that if S is defined as the horizontal wavelength of a sinusoidal mountain range with ridges perpendicular to the flow, the wavelength of the waves induced by flow over the mountain range is proportional to Fr. In the real world, V and N vary with height and the topography of mountain ranges is a two-dimensional function that is often too complex to characterize by a single height or width scale. Hence, to perform realistic simulations of terrain effects requires the use of numerical or laboratory models.

9.5.2 Sea Breezes

Diurnal temperature ranges at the Earth's surface in excess of 15 °C are commonly observed over land in the tropics and in middle latitudes of the summer hemisphere, where daytime insolation is strong. Over land, the low thermal conductivity of the soil tends to concentrate the response within a thin layer just below the surface, as was illustrated in Fig. 9.12. Hence, land surfaces react much more quickly to changes of insolation than the ocean surface. It follows that the continents tend to be warmer than the oceans during the afternoons and they cool off more at night due to the emission of infrared radiation. The resulting land-sea temperature contrasts cause horizontal pressure gradients that drive shallow, diurnally varying, circulations: daytime sea breezes and nighttime land breezes.

A *sea breeze* is a cool shallow wind that blows onshore (from sea to land) during daytime. It occurs during weak background synoptic forcings (i.e., weak or calm geostrophic wind) under generally clear skies and is caused by a 5 °C or greater temperature contrast between the sun-heated warm land and the cooler water. Analogous flows called *lake breezes* form along lake shorelines, while *inland sea breezes* form along boundaries between adjacent land regions with different surface characteristics (e.g., deserts versus vegetated terrain). The sea breeze is a surface manifestation of a thermally driven mesoscale circulation called the *sea-breeze circulation*, which often includes a weak return (land to sea) flow aloft (Fig. 9.35).

A *sea-breeze front* marks the leading edge of the advancing cool marine air and behaves similarly to a weak advancing cold front or a thunderstorm gust front. If sufficient moisture is present in the updraft ahead of the front, a line of cumulus clouds can form along the front, and a line of thunderstorms can be triggered if the atmosphere is convectively unstable. The raised portion of cool air immediately behind the front, called the *sea-breeze head*, is analogous to the head at the leading edge of a gust front (Section 8.3.3c). The sea-breeze head is roughly twice as thick as the subsequent portion of the *feeder* cool onshore flow (which is ~0.5 to 1 km thick). The top of the

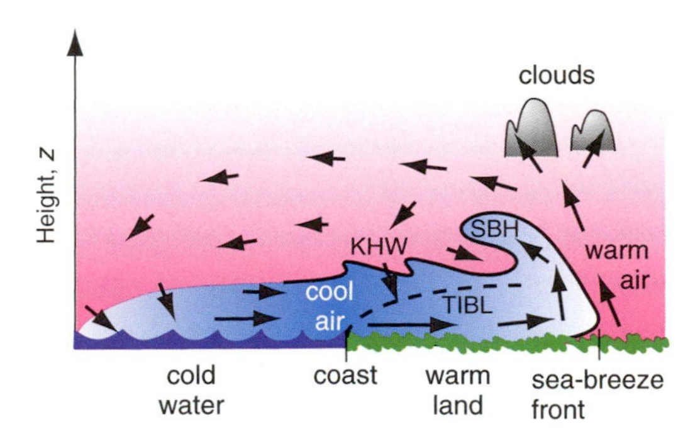

Fig. 9.35 Components of a sea-breeze circulation. SBH, sea-breeze head. KHW, Kelvin-Helmholtz waves. The top of the thermal internal boundary layer (TIBL) is shown by the dashed line. [Adapted from R. B. Stull, *An Introduction to Boundary Layer Meteorology*, Kluwer Academic Publishers, Dordrecht, The Netherlands, 1988, Fig. 14.7, p. 594, Copyright 1988 Kluwer Academic Publishers, with kind permission of Springer Science and Business Media.]

sea-breeze head often curls back over warmer air from aloft in a large horizontal roll eddy.

Vertical wind shear at the density interface between the low-level sea breeze and the return flow aloft can give rise to Kelvin-Helmholtz waves with wavelength of ~0.5 to 1 km along the density interface at the top of the feeder flow behind the head. These waves increase the frictional drag on the sea breeze by entraining low-momentum air from above the interface. A slowly subsiding return flow occurs over water and completes the circulation as the air is again cooled as it blows landward over the cold water. As the cool air flows over the land, a *thermal internal boundary layer* (TIBL) forms just above the ground and grows in depth with the square root of distance from the shore as the marine air is modified by the heat flux from the underlying warm ground.

The initial sea-breeze circulation is not very wide, but spreads out over land and even farther over water as the day progresses. Advancing cold air behind the sea-breeze front behaves somewhat like a *density current* or *gravity current* in which a dense fluid spreads out horizontally beneath a less dense fluid, as has been simulated in water tanks. The sea-breeze front can advance 10 to 200 km inland by the end of the day, although typical advances are ~20 to 60 km unless inhibited by mountains or by opposing synoptic-scale winds. As the front advances, prefrontal waves may cause wind shifts ahead of the front.

When fully developed, surface (10 m height) wind speeds in the marine, inflow portion of the sea breeze at the coast range from ~1 to 10 m s^{-1} with typical values of ~6 m s^{-1}. Wind speed increases with the square root of the land-sea temperature difference. The sea-breeze front usually advances at a speed of ~87% of the surface wind speed. At the end of the day, the sea-breeze circulation dissipates and a weaker, reverse circulation called the *land-breeze* forms in response to the nighttime cooling of the land surface relative to the sea. Sometimes, the now-disconnected sea-breeze front continues to advance farther inland during the night over the growing nocturnal stable boundary layer as a *bore* (a propagating solitary wave with characteristics similar to an hydraulic jump), a phenomenon known in Australia as the *Morning Glory*.

In the plane of a vertical cross section normal to the coastline (as in Fig. 9.35) the surface wind oscillates back and forth between onshore and offshore, reversing directions during the morning and evening hours. The Coriolis force induces an oscillating alongshore wind component that lags the onshore-offshore component by 6 h (or 1/4 cycle). Hence, the diurnal wind vector rotates throughout the course of the day: clockwise in the northern hemisphere and counterclockwise in the southern hemisphere. For example, along a meridional coastline with the ocean to the west in the northern hemisphere, the diurnal component of the surface wind tends to be westerly (onshore) during the afternoon, northerly (alongshore) during the evening, and so on. The diurnal wind component and the mean (24 h average) surface wind are additive. If the mean wind in the above example is blowing, say, from the north, the surface wind speed will tend to be higher around sunset when the mean wind and the diurnal component are in the same direction, than around sunrise, when they are in opposing directions.[13,14]

Many coasts have complex shaped coastlines with bays or mountains, resulting in a myriad of interactions

[13] Irrespective of the effects described here, the wind speed often drops around or just after sunset in response to the weakening of the boundary-layer turbulence, which reduces the downward mixing of momentum.

[14] The diurnal wind vector along sloping terrain rotates in a similar manner, and the daytime strengthening and nighttime weakening of the turbulence in the land boundary layer also contribute to the diurnal cycle in boundary-layer winds. The amplitude of these oscillations tends to be largest around 30° latitude where the diurnal cycle matches the period of an inertia circle.

between local flows that distort the sea breeze and create regions of enhanced convergence and divergence. The sea breeze can also interact with boundary-layer convection, horizontal roll vortices, and urban heat islands, causing complex dispersion of pollutants emitted near the shore. If the onshore synoptic-scale geostrophic wind is too strong, a TIBL develops instead of a sea-breeze circulation.

In regions such as the west coast of the Americas, where major mountain ranges lie within a few hundred kilometers of the coast, sea breezes and terrain effects appear in combination. Figure 9.36 shows an example of how low-level convergence associated with the boundary-layer wind field influences the diurnal cycle in deep convection and rainfall over Central America. Elevated land areas along the crest of the Sierras Madre mountain range experience convection during the afternoon when the sea breeze and upslope flow conspire to produce low-level convergence. In contrast, the offshore waters experience convection most frequently during the early morning hours, in response to the convergence associated with the land breeze, augmented by katabatic winds from the nearby mountains. Morning convection is particularly strong in the Gulf of Panama because of its concave coastline. The prevalence of afternoon convection, driven by the the sea breeze circulation, is also clearly evident over the southeastern United States.

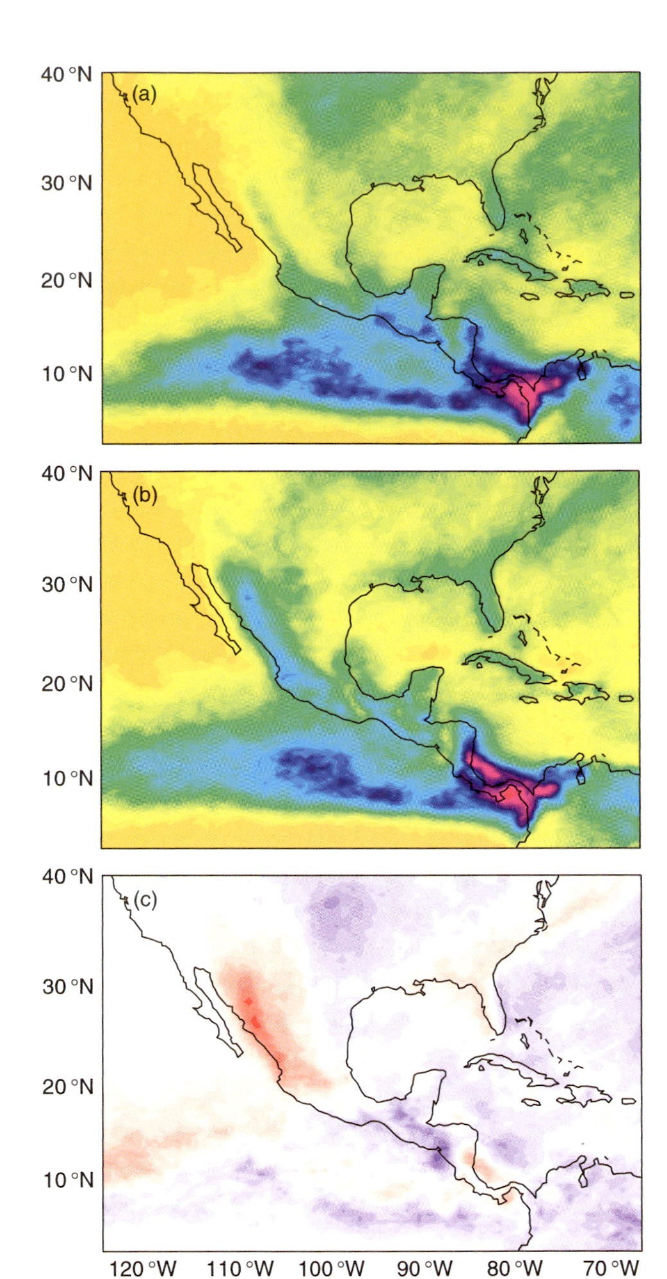

Fig. 9.36 Frequency of deep convection as indicated by clouds with tops colder than −38 °C at two different times of day during July at around (a) 5 AM local time and (b) 5 PM local time. Panel (c) shows the difference (a)-(b). Color scale for (a) and (b) analogous to that in Fig. 1.25; in (c) red shading indicates heavier precipitation at 5 PM. [Courtesy of U.S. CLIVAR Pan American Implementation Panel and Ken Howard, NOAA-NSSL.]

9.5.3 Forest Canopy Effects

The drag associated with *forest canopies* (i.e., the leafy top part of the trees) is so strong that the flow above the canopy often becomes partially separated from the flow below, in the trunk space. In the region above the canopy in the surface layer, the flow exhibits the typical logarithmic wind profile but with the effective aerodynamic surface displaced upward to near the canopy top (Fig. 9.37a). In the trunk space the air often has a relative maximum of wind speed between the ground and the canopy, and often these subcanopy winds flow mostly downslope both day and night if the canopy is dense.

The static stability below the canopy is often opposite to that above the canopy. For example, during clear nights, strong longwave cooling of the canopy will cause a statically stable boundary layer to form above the canopy (Fig. 9.37b). However, some of the cold air from the canopy sinks as upside-down thermals, creating a convective mixed layer in the trunk space at night that gets cooler as the night progresses. During daytime, a superadiabatic surface layer with warm convective thermals forms above the canopy, while in the trunk space the air is warmed from the canopy above and becomes statically stable (Fig. 9.37c). These peculiar configurations of static stability are successfully captured with the nonlocal method of determining static stability.

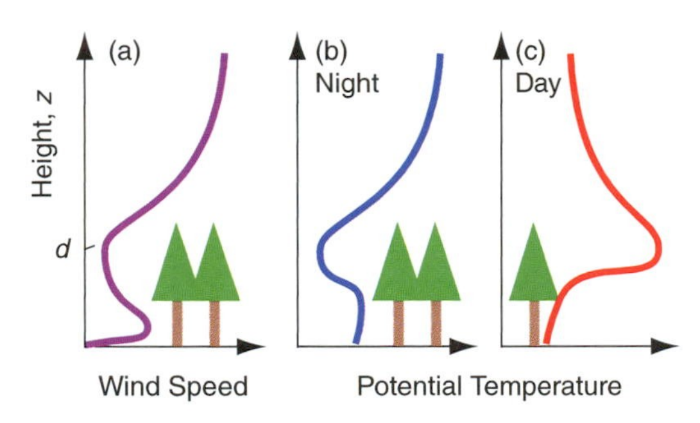

Fig. 9.37 (a) Wind speed in a forest canopy. (b) Potential temperature profile on a clear night with light winds. The level *d* represents the displacement distance of the effective surface for the logarithmic wind profile in the surface layer above the top of the canopy. (c) Same as (b), but on a sunny day. [Courtesy of Roland B. Stull.]

9.5.4 Urban Effects

Large cities differ from the surrounding rural areas by virtue of having larger buildings that exert a stronger drag on the wind; less ground moisture and vegetation, resulting in reduced evaporation; different albedo characteristics that are strongly dependent on the relationship between sun position and alignment of the *urban street canyons*; different heat capacity; and greater emissions of pollutants and anthropogenic heat production. All of these effects usually cause the city center to be warmer than the surroundings—a phenomenon called the *urban heat island* (Fig. 9.38a).

When a light synoptic-scale wind is blowing, the excess warmth and increased pollutants extend downwind as a city-scale *urban plume* (Fig. 9.38b). In some of these urban plumes the length of the growing season (days between last frost in spring and first frost in fall) in the rural areas adjacent to the city has been observed to increase by 15 days compared to the surrounding regions.

Within individual street canyons between strong buildings, winds can be funneled and accelerated. Also, as fast winds from aloft hit tall buildings, the winds are deflected downward to increase wind speeds near the base of these skyscrapers. Behind individual buildings is often a "cavity" of circulation with surface wind direction opposite to the stronger winds aloft, which can cause unanticipated transport of pollutants from the street level up to open windows. In lighter wind conditions, the increased roughness length associated with tall buildings can reduce the wind speed on average over the whole city, allowing pollutant concentrations to build to high levels.

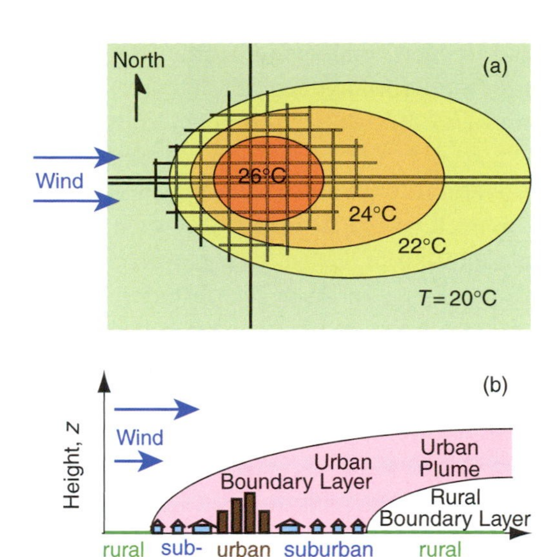

Fig. 9.38 (a) Urban heat island effect, with warmer air temperatures over and downwind of a city. Air temperature excess in the city core can be 2 to 12 °C warmer than the air upstream over rural areas. The grid of lines represents roads. (b) Vertical cross section through a city. The urban plume includes excess heat as well as increased pollution. [Adapted from T. R. Oke, *Boundary Layer Climates*, 2nd Ed., Routledge, New York (1987).]

The largest cities generate and store so much heat that they can create convective mixed layers over them both day and night during fair-weather conditions. This urban heat source is often associated with enhanced thermals and updrafts over the city, with weak return-circulation downdrafts over the adjacent countryside. One detrimental effect is that pollutants are continually recirculated into the city. Also, the enhanced convection over a city can cause measurable increases in convective clouds and thunderstorm rain.

9.6 The Boundary Layer in Context

Boundary-layer meteorology is still a relatively young field of study with many unsolved problems. The basic equations of thermodynamics and dynamics have resisted analytical solution for centuries. The Reynolds-averaging approach resorts to statistical averages of the effects of turbulence in an attempt to circumvent the intrinsic lack of predictability in the deterministic solutions of the eddies themselves. The turbulence closure problem has not been solved, and the chaotic nature of turbulence poses major challenges. Many of the useful empirical relationships, such as the logarithmic wind profile, are founded on similarity theory, which basically ignores physics and dynamics and focuses on the dimensions of the key variables. The nature of turbulence and entrainment of ambient air into clouds is also incompletely understood, as well as the feedbacks between clouds and the fluid motions in the boundary layer.

Despite these formidable obstacles, significant advances are being made in many areas. Boundary-layer models are being coupled with larger-scale models to capture the important effects of multiscale advection. Remote sensing is used to observe the eddies and thermals that cause the turbulence, and very fine resolution computing is being used to simulate boundary-layer turbulence, as explained later.

Horizontal advection of heat, moisture, and momentum by larger-scale mean wind nearly always dominates over turbulent effects in the boundary layer, even during periods of fair weather and light winds. Also, vertical advection by the large-scale motion field is usually the same order of magnitude or larger than the turbulent entrainment at the mixed-layer top. To properly include these larger-scale advective effects, boundary-layer models are often embedded in mesoscale models, which, in turn, are nested in large-scale numerical weather prediction models.

Remote sensing has been used increasingly by researchers since the 1970s to observe the boundary layer. Clear-air radar emits microwaves to observe boundary-layer eddies, as made visible by humidity-related microwave refractivity fluctuations. *Lidar* (light detection and ranging) transmits a beam of radiation at visible or near-visible wavelengths to illuminate aerosols carried aloft by thermals (Fig. 9.39) and uses large telescopes to capture the photons scattered back to the lidar. Sodar (sound detection and ranging) emits loud pulses of sound and uses sensitive microphones to detect faint echoes scattered from regions of strong temperature gradients in and around surface-layer eddies and plumes. *Profilers* monitor the Doppler shift of pulses of off-vertically propagating radio waves, from which the wind profile can be inferred, while *RASS (radio acoustic sounding systems)* use radio waves to measure the speed of sound waves, from which the temperature profile can be inferred.

Realistic numerical simulation of boundary-layer phenomena is inherently more computationally intensive than numerical weather prediction. The range of horizontal scales is larger (in a logarithmic sense), so more grid points are needed. In addition, motions in the boundary layer tend to be fully three dimensional

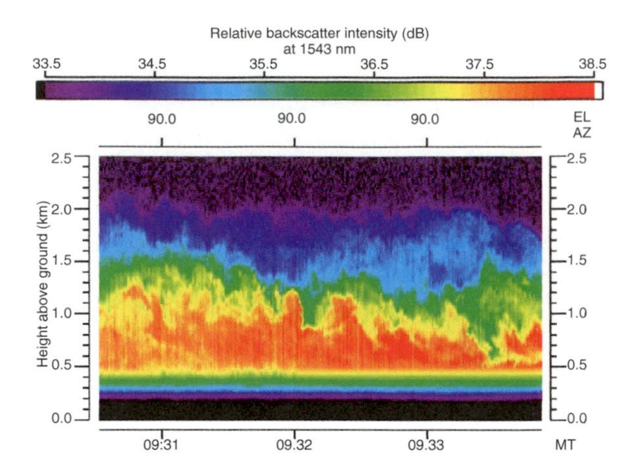

Fig. 9.39 Time-height cross section showing boundary-layer thermals (red and yellow colors) as they drift over a vertically pointing lidar. Pollution particles in the rising thermals, which backscatter light back to the lidar telescope, are rendered in red and yellow colors. Black, purple, and blue indicate air that is relatively clean in the free atmosphere. [Courtesy of Shane Mayor, National Center for Atmospheric Research.]

and nonhydrostatic so additional layers are required in the vertical as well. The timescales of microscale turbulence are orders of magnitude shorter than the timescale for phenomena such as baroclinic waves so much shorter time steps are required.

Despite these formidable computational requirements, some notable advances have been made in recent years. A particularly productive approach has been *large eddy simulation (LES)*, in which numerical weather prediction models are run on high-performance computers with ~10-m grid spacing, which is sufficient to resolve the larger, more energetic eddies. Finer resolution *direct numerical simulation (DNS)* computer models have been run over very small domains with grid spacings as fine as millimeters, which is sufficient to resolve all eddy sizes down to scales on which molecular conduction and diffusion dominate. As computer power continues to increase, DNS and LES are merging to provide direct computation of turbulent flow over arbitrarily complex terrain.

As more of these new observing and numerical-modeling capabilities come on line, it is becoming increasingly possible to transcend the limitations of statistical approaches and similarity theory by pursuing a more phenomenogically based approach to boundary-layer studies. Part of the thrill of boundary-layer research is that there are still many important phenomena and processes waiting to be discovered.

Exercises

9.7 Explain and interpret the following:

(a) Air pollution is often trapped within the boundary layer.

(b) Clear-air turbulence (CAT) experienced by airplane passengers is widespread, but usually not very strong in the boundary layer. It tends to be more sporadic and also more intense near the tropopause.

(c) Air within the free troposphere exhibits only a small diurnal temperature range.

(d) Boundary layers over deserts are often much deeper than over vegetated terrain.

(e) The warmest and highest-rising thermals often have the highest LCLs.

(f) Boundary-layer turbulence in the Earth's atmosphere decays during a solar eclipse.

(g) Birds soar during daytime over land, not usually at night.

(h) Turbulence kinetic energy is dissipated into internal energy.

(i) Convective overturning creates fat thermals with a mean diameter of order of the boundary-layer depth, as opposed to thin thermals of the width of smoke-stack plumes.

(j) The theoretical height of the LCL based on surface temperature and dew point is a very accurate predictor of the height of cloud base for convective clouds.

(k) The LCL is a poor predictor for the base height of stratus clouds.

(l) Deterministic forecasts of any scale of turbulent phenomenon are accurate out to duration roughly equal to the lifetime of individual eddies.

(m) The covariance between u and w is both the vertical flux of horizontal momentum and the horizontal flux of vertical momentum.

(n) The covariance of θ and q does not represent a flux.

(o) During day, boundary layers are often shallow near shorelines and increase with distance inland.

(p) Kelvin-Helmholtz breaking waves and turbulence occur much more often than the occurrence of billow clouds.

(q) The method of estimating turbulent transport demonstrated in Fig. 9.8 does not necessarily work for larger eddies.

(r) Zeroth-order similarity theories give useful important information, even though they contain no dynamics.

(s) Gradient-transfer theory fails in the interior of the mixed layer.

(t) What would be the opposite of the oasis effect, and could it occur in nature?

(u) Significant vertical turbulent fluxes can exist near the surface even in the limit of zero wind speed.

(v) The drag coefficient is related to the roughness length.

(w) During some nights dew will form on the ground, but on others fog forms instead.

(x) Hot-air balloonists like to take off in near-calm conditions at the surface, but can find themselves in fast-moving air a short distance above the surface.

(y) Hot-air balloonists who take off in the near-calm winds of early morning, find it much more difficult to land by mid morning because of faster winds.

(z) On a sunny day, the air in the trunk space below a forest canopy is stably stratified.

(aa) Turbulence in the residual layer decays quickly after sunset.

(bb) Greater subsidence does not inject more air into the top of the mixed layer, but has the opposite effect of making the mixed layer more shallow.

(cc) The rapid-rise phase of mixed-layer growth happens only after the nocturnal inversion has been "burned off."

(dd) The boundary layer is poorly defined near fronts.

(ee) When air flows over a change in roughness, an internal boundary layer develops. The depth of this layer grows with increasing distance from the change.

(ff) Latent and sensible heat fluxes over land surfaces tend to be larger on clear days than on cloudy days.

(gg) Daytime temperatures tend to be higher over deserts than over vegetated land surfaces.

(hh) Other conditions being the same, alpine glaciers and snow-fields lose more mass on a humid summer day than on a dry summer day.

(ii) On a clear, calm day, the surface sensible heat flux into the air does not usually become positive until 30 to 60 min after sunrise.

(jj) In fair weather, the heat and momentum fluxes at the top of the mixed layer due to entrainment are usually downward, but the moisture flux is positive.

(kk) You can estimate the static stability of the boundary layer by looking at the shape of the smoke plume from a smoke stack.

(ll) In Fig. 9.40, why do the surface wind speed and the cloudiness increase as the air flows northward across the sharp front in the sea-surface temperature field that lies along 1 °N?

9.8 Estimate the temperature variance for the velocity trace at the bottom of Fig. 9.6.

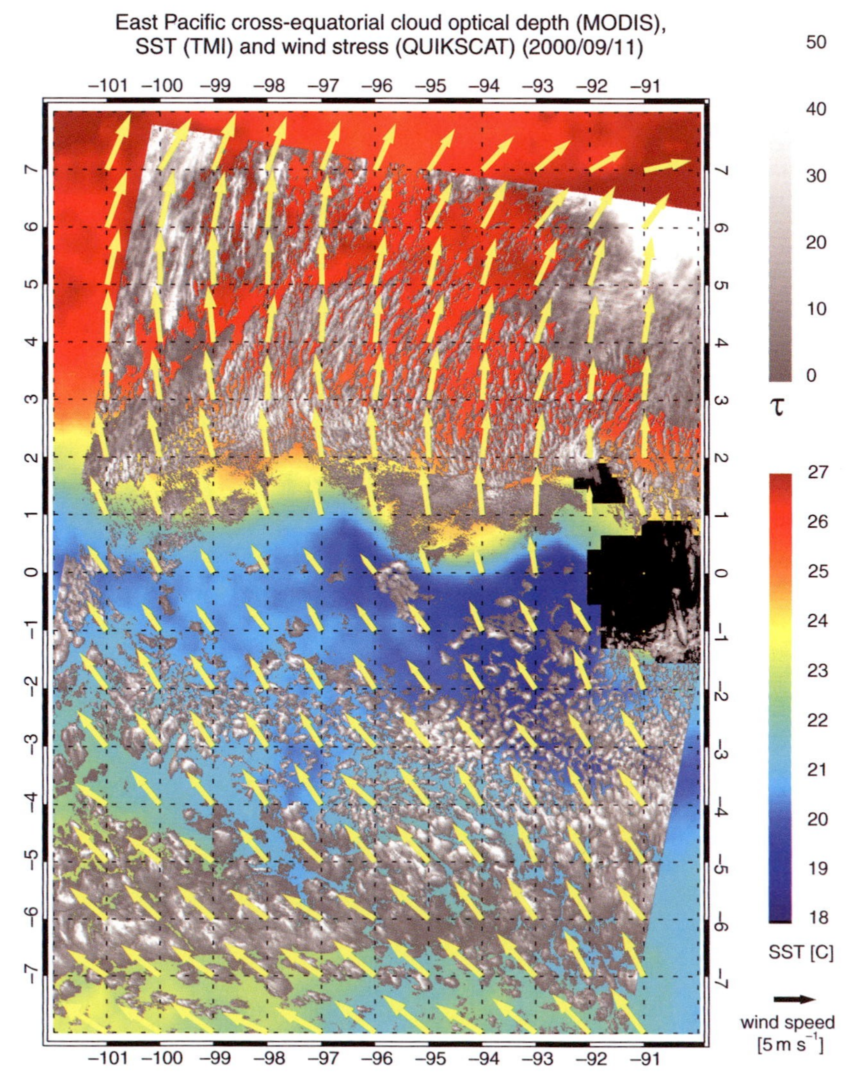

Fig. 9.40 Sea surface temperature (colored shading) surface winds (arrows) and clouds (gray shading) over the equatorial Pacific at a time when the equatorial front in the sea surface temperature field is well defined along 1 °N. The scalloped appearance of the front is due to the presence of tropical instability waves in the ocean. [Based on NASA QUIKSCAT, TMI, and MODIS imagery. Courtesy of Robert Wood.]

9.9 Prove that the definition of covariance reduces to the definition of variance for the covariance between any variable and itself.

9.10 Given the following variances in $m^2 s^{-2}$

Where:	Location A		Location B	
When (UTC):	1000	1100	1000	1100
σ_u^2	0.50	0.50	0.70	0.50
σ_v^2	0.25	0.50	0.25	0.25
σ_w^2	0.70	0.50	0.70	0.25

When, where, and for which variables is the turbulence (a) stationary, (b) isotropic, and (c) homogeneous?
Some, but not all, of the answers Homogeneous for u wind at 1100 UTC. Stationary for v wind at location B. Isotropic at location A at 1100 UTC.

9.11 Given the following synchronous time series for T(°C) and w(m/s), find (a) mean temperature, (b) mean velocity, (c) temperature variance, (d) velocity variance, and (e) kinematic heat flux.

T	21	22	20	25	25	15	18	23	21	24	16	12	19	22
w	1	−2	0	−3	2	−2	−3	3	0	0	1	4	−2	−3

9.12 Under what conditions is the Richardson number not likely to be a reliable indicator for the existence of turbulence? Why?

9.13 If the dissipation length scale is L_ϵ, what is the e-folding time for the decay of turbulence (i.e., time for TKE/m to equal $1/e$ of its initial value), assuming you start with finite TKE but there is no production, consumption, transport, or advection?

9.14 The upward vertical heat flux in soil and rock due to molecular conduction is given by

$$F = -K\frac{\partial T}{\partial z} \qquad (9.34)$$

where K is the *thermal conductivity* of the medium, in units of W m^{-1} K^{-1}, which ranges in value from 0.1 for peat to 2.5 for wet sand. From the first law of thermodynamics, we can write

$$C\frac{\partial T}{\partial t} = -\frac{\partial F}{\partial z} = \frac{\partial}{\partial z}K\frac{\partial T}{\partial z} \qquad (9.35)$$

where C is the heat capacity per unit volume in J m^{-3} K^{-1} (i.e., the product of the specific heat times the density). If K is assumed to be independent of depth

$$\frac{\partial T}{\partial t} = D\frac{\partial^2 T}{\partial z^2} \qquad (9.36)$$

where $D = K/C$ is called the *thermal diffusivity*. Consider the response of the subsurface temperature to a sinusoidal variation in surface temperature with amplitude T_s and period P. (a) Show that the amplitude of the response drops off exponentially with depth below the surface with an e-folding depth of

$$h = \sqrt{\frac{DP}{2\pi}}$$

(b) The *e*-folding depth of the annual cycle, as estimated from Fig. 9.12, is ~2 m. Estimate the *e*-folding depth of the diurnal cycle at the same site.

9.15 For the situation shown in Fig. 9.12 in the text: (a) Making use of Eq. (9.36), derive a relationship for the phase lag Δt between the temperature wave at two depths differing by height Δz, given P the wave period and D the thermal diffusivity of the soil as defined in the previous exercise. (b) Using the annual cycle data given in Fig. 9.12 in the text, estimate the phase lag between the temperature curves at 1.5 and 6 m and then use it to find the value of D.

9.16 Given the heat-flux profile of Fig. 9.22, extend the method of Fig. 9.8 to estimate the sign of the triple correlation $\overline{w'w'\theta'}$ in the middle of the mixed layer, which is one of the unknowns in Eq. (9.11).

9.17 Given: $F_{HS} = 0.2$ K m s^{-1}, $z_i = 1$ km, $u_* = 0.2$ m s^{-1}, $T = 300$ K, $z_0 = 0.01$ m, find and explain the significance of the values of the (a) Deardorff velocity scale; (b) Obukhov length; (c) convective time scale; and (d) wind speed at $z = 30$ m.

9.18 The *flux form of the Richardson number* is

$$R_f = \frac{(g/\overline{T_v})\,\overline{w'\theta'_v}}{\overline{w'u'}\dfrac{\partial \overline{u}}{\partial z} + \overline{w'v'}\dfrac{\partial \overline{v}}{\partial z}}$$

Use gradient transfer theory to show how R_f is related to R_i.

9.19 If the wind speed is 5 m s^{-1} at $z = 10$ m and the air temperature is 20 °C at $z = 2$ m, then (a) what is the value of the sensible heat flux at the surface of unirrigated grassland if the skin temperature is 40 °C? (b) What is the value of the latent heat flux? [**Hint**: $\rho c_p \approx 1231$ (W m^{-2})/(K m s^{-1}) for dry air at sea level.]

9.20 (a) Confirm that the following expression is a solution to Eq. (9.26), and (b) that it satisfies the boundary condition that $V = 0$ at $z = z_0$.

$$\frac{V}{u_*} = \frac{1}{k}\left\{\ln\frac{z}{z_0} - 2\ln\left[\frac{1+x}{2}\right] - \ln\left[\frac{1+x^2}{2}\right] + 2\arctan x - \pi/2\right\}$$

where

$$x = \left[1 - 15\frac{(z - z_0)}{L}\right]^{1/4}$$

9.21 Use the expressions for surface-layer wind speed from Eq. (9.22), from Exercise 9.4 in Section 9.3.3, and from the previous exercise to plot curves of V vs. z for (a) neutral ($L = \infty$), (b) stable ($L = 100$ m), and (c) unstable ($L = -10$ m) stratifications and confirm that their relative shapes are as plotted in Fig. 9.17a and 9.17b. Use $z_0 = 0.1$ m.

9.22 (a) If boundary-layer divergence is constant during fair weather, the mixed layer can stop growing during midafternoon even though there are strong sensible heat fluxes into the boundary layer at the ground. Why? (b) If w_e = constant, derive an equation showing the growth of z_i with time in a region where divergence β is constant.

9.23 If F^* is known, and if $|F_G| = 0.1 \cdot |F^*|$, then show how knowledge of time-averaged temperature at two heights in the surface layer and of mean humidity at the same two heights is sufficient to estimate the sensible and latent heat fluxes in the surface layer.

9.24 As a cold, continental air mass passes over the Gulf Stream on a winter day, the temperature of the air in the atmospheric boundary layer rises by 10 K over a distance of 300 km. Within this interval the average boundary layer depth is 1 km and the wind speed is 15 m s^{-1}. No condensation is taking place within the boundary layer and the radiative fluxes are negligible. Calculate the sensible heat flux from the sea surface.

9.25 (a) If drag at the ground represents a loss of momentum from the mean wind, determine the sign of $\overline{u'w'}_s$ if the mean wind is from the west. Justify your result.

(b) Do the same for a wind from the east, remembering that drag still represents a momentum loss.

9.26 Given the following temperature profile, determine and justify which layers of the atmosphere are (a) statically stable, (b) neutral, and (c) unstable.

z (km)	θ (°C)
2	21
1.8	23
1.6	19
1.4	19
1.2	13
1.0	16
0.2	16
0	10

9.27 If the total accumulated surface heat flux from sunrise through sunset is 5100 km, then use the sunrise sounding in the previous exercise to estimate the depth and potential temperature of the mixed layer just before sunset.

9.28 Given a smoke stack half the height of a valley: (a) describe the path of the centerline of the smoke plume during day and night during fair weather and (b) describe the centerline path of the smoke on a strongly windy day.

9.29 If you know the temperature and humidity jumps across the top of the mixed layer and if you know only the surface heat flux (but not the surface moisture flux), show how you can calculate the entrained heat and moisture fluxes at the top of the mixed layer.

9.30 Show that over flat terrain the large-scale vertical velocity at the top of the boundary layer is approximately equal to

$$w_i \simeq -z_i\{\nabla \cdot \mathrm{V}\} \tag{9.28}$$

where

$$\{\nabla \cdot \mathrm{V}\} \equiv \frac{\int_{p_i}^{p_s} (\nabla \cdot \mathrm{V})\, dp}{(p_s - p_i)}$$

is the mass-weighted divergence in the boundary layer and p_s and p_i are the pressures at the Earth's surface and at the top of the boundary layer.

Climate Dynamics

10

The term *climate* refers to the mean state of the atmosphere and related components of the Earth system and it is also used in reference to atmospheric variability on timescales longer than the 2- to 3-week limit of most deterministic atmospheric predictability. The mean state, including diurnal and seasonal variations, as defined by some prescribed averaging period, is referred to as the *climatological mean*,[1] and departures from this mean state or *normal* are referred to as *climate anomalies*. For example, a 4 °C temperature anomaly denotes a temperature that is 4 °C above the climatological mean for that particular location and time of year. The term *climate variability* refers to long-term variations or changes in the mean state, i.e.,

- *intraseasonal climate variability* denotes month-to-month variations about the seasonally varying climatological mean that occur within the same season (e.g., the distinction between an abnormally warm January and an abnormally cold February),
- *interannual variability* denotes year-to-year variations of annual or seasonal averages (e.g., between the mean temperatures observed in successive winter seasons), and
- *decadal, century-scale*, etc., denotes decade-to-decade, century-to-century, etc. scale variations.

The distinction between *climate variability* and *climate change* is largely semantic: if the variations of interest take place within some specified interval (e.g., the 20th century), they are referred to as the *variability* of the climate within that interval, whereas if the variations involve differences between two successive epochs (e.g., the first and second halves of the 20th century), they are referred to as the *change* in climate from one epoch to the next.

The next section describes the present-day climate and the processes that shape it. The nature and causes of climate variability on various timescales is then discussed and related to the history of past climate. The section on climate variability is followed by a discussion of the climate sensitivity and climate feedbacks. The chapter concludes with a section summarizing some of the scientific issues related to greenhouse warming.

10.1 The Present-Day Climate

The mean state of the climate system is determined by the emission of radiation from the sun, the Earth's rotation rate and orbital characteristics, the composition of the atmosphere, and the interactions between the atmosphere and the other components of the Earth system that determine the fluxes of mass, energy, and momentum at the Earth's surface. Some of the characteristics of that mean state were described in Chapters 1 and 2. This section provides further specifics concerning the diurnally and seasonally varying mean climate and the processes that maintain it.

10.1.1 Annual Mean Conditions

The global mean surface air temperature $T_s = 15$ °C or 288 K and the Earth's equivalent blackbody temperature $T_E = 255$ K are *equilibrium values*, determined by the *balance requirements* that the globally averaged net radiation through the top of the atmosphere, as well as the net energy flux (i.e., the sum of

[1] Conventional climatological means are based on prescribed sets of sequential 30-year periods, the most recent of which ended in December 1990.

the short and longwave radiative fluxes and the fluxes of latent and sensible heat), at the Earth's surface, as prescribed by (9.18) must vanish.

If these balance requirements were not fulfilled, the temperatures would change until an equilibrium was achieved. For example, if the net radiation at the top of the atmosphere were downward, the Earth's equivalent blackbody temperature would have to rise until the outgoing longwave radiation increased enough to eliminate the imbalance. Based on global measurements, such as those shown in Figs. 4.35 and 9.14, it is possible to construct the global energy balance shown in Fig. 10.1.

The 100 units of *insolation* (a synonym for the seasonally varying, climatological-mean solar radiation) incident on the top of the atmosphere represent a flux density of 342 W m^{-2} integrated over the Earth's surface as shown in Fig. 4.8. Of the 100 units incident on the top of the atmosphere, 3 are absorbed by stratospheric ozone and 17 by water vapor and clouds in the troposphere. A total of 30 units are reflected back to space: 20 from clouds and aerosols, 6 from air molecules, and 4 from the Earth's surface. The total reflection (30 out of 100 units of incoming radiation) is the Earth's *albedo*. The remaining 50 units are the net downward shortwave flux through the Earth's surface.

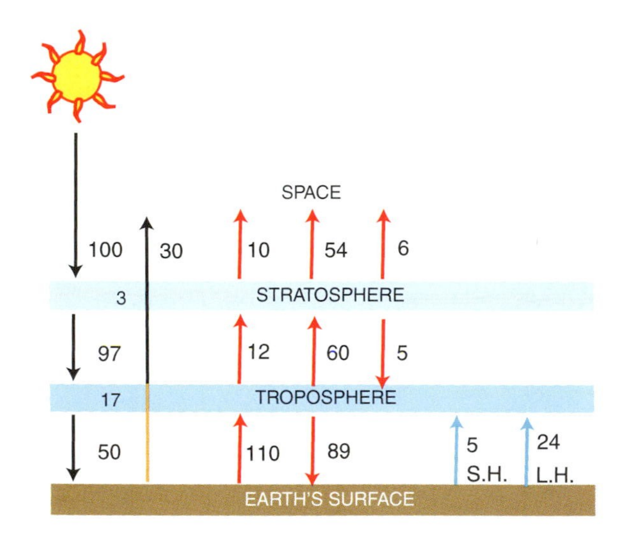

Fig. 10.1 The global energy balance. The 100 units of incoming energy represent the 342 W m^{-2} of incident solar radiation averaged over the area of the Earth. Black arrows represent shortwave radiation, red arrows represent longwave radiation, and blue arrows represent the (nonradiative) fluxes of sensible heat (SH) and latent heat (LH). The absorption and emission by the Earth's surface, the troposphere, and the stratosphere each sum to zero, and the net flux through each of the interfaces sums to zero. [Adapted from Dennis L. Hartmann, *Global Physical Climatology*, p. 28 (Copyright 1994), with permission from Elsevier.]

The Earth disposes of the energy that it absorbs by a combination of longwave radiation and latent and sensible heat fluxes as indicated by the red and blue arrows in Fig. 10.1. The 110 units of longwave radiation emitted from the Earth's surface is equal to effective emissivity of the surface, which ranges from 0.92 to 0.97 locally, times σT_s^4, averaged over the surface of the Earth, where T_s is the surface temperature. The net upward longwave flux through the Earth's surface (i.e., the difference between the upward emission from the surface and the downward emission from clouds and greenhouse gases in the atmosphere) amounts to only 21 units. Were it not for the presence of the downward longwave fluxes (i.e., the greenhouse effect) the Earth's surface would be in equilibrium with the incident solar radiation at a temperature close to T_E and the latent and sensible heat fluxes would be much smaller than observed.

Exercise 10.1 On the basis of the data given in Fig. 10.1, describe the energy balance of the troposphere.

Solution: The troposphere is heated by the absorption of solar radiation, mostly by water vapor and clouds (17 units), the absorption of longwave radiation emitted by the Earth's surface ($110 - 12 = 98$ units), the absorption of longwave radiation emitted by the stratosphere (5 units), the sensible heat transfer through the Earth's surface (5 units), and latent heat release (24 units). Hence, the troposphere must emit a total of $17 + 98 + 5 + 5 + 24 = 149$ units of radiation in the longwave part of the spectrum in order to achieve a balance. It emits 89 units in the downward direction and 60 units in the upward direction. ■

The annual mean latitudinal distribution of net radiation at the top of the atmosphere, shown in Fig. 10.2, was generated by zonally averaging the distributions shown in Fig. 4.35. Because the temperature of the Earth system is changing only very slowly, the globally averaged insolation must be in balance with the outgoing longwave radiation. Hence, there must be a surplus of insolation relative to outgoing longwave radiation at low latitudes and a deficit at high latitudes. It is this imbalance that maintains the observed equator-to-pole temperature contrast in the presence of the strong poleward heat fluxes produced by baroclinic waves, as explained in Section 7.3 and 7.4.

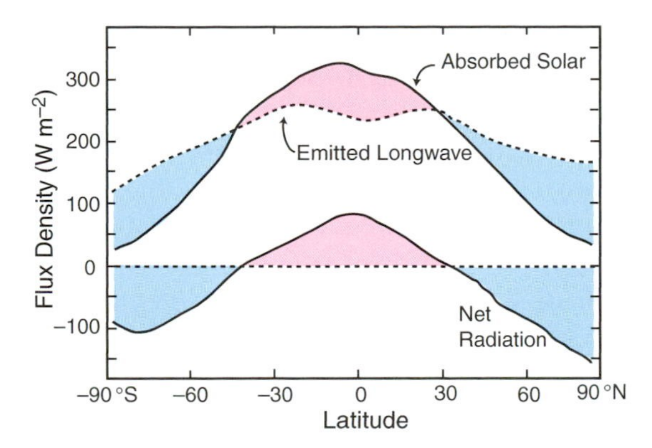

Fig. 10.2 Annual average flux density of absorbed solar radiation, outgoing terrestrial radiation, and net (absorbed solar minus outgoing) radiation as a function of latitude in units of W m^{-2}. Pink (blue) shading indicates a surplus (deficit) of incoming radiation over outgoing radiation. [Adapted from Dennis L. Hartmann, *Global Physical Climatology*, p. 31 (Copyright 1994), with permission from Elsevier.]

Fig. 10.3 Schematic of air parcels circulating in the atmosphere. The Colored shading represents potential temperature or moist static energy, with pink indicating higher values and blue lower values. Air parcels acquire latent and sensible heat during the time that they reside within the boundary layer, raising their moist static energy. They conserve moist static energy as they ascend rapidly in updrafts in clouds, and they cool by radiative transfer as they descend much more slowly in clear air.

The vertical distribution of temperature within the troposphere is determined by the interplay among radiative transfer, convection and large-scale motions. The radiative equilibrium temperature profile is unstable with respect to the dry adiabatic lapse rate. The pronounced temperature minimum near the 10-km level that defines the tropopause (Fig. 1.9) corresponds roughly to the level of unit optical depth for outgoing longwave radiation. Below this level, the repeated absorption and reemission of outgoing longwave radiation render radiative transfer a relatively inefficient mechanism for disposing of the energy absorbed at the Earth's surface. Convection and large-scale motions conspire to maintain the observed lapse rate near a value of 6.5 K km^{-1}.

The observed lapse rate is stable, even with respect to the saturated adiabatic lapse rate, because most of the volume of the troposphere is filled with slowly subsiding air, which loses energy by emitting longwave radiation as it sinks, as documented in Fig. 4.29. As the air loses energy during its descent, its equivalent potential temperature and moist static energy decrease, creating a stable lapse rate, as shown schematically in Fig. 10.3. It is only air parcels that have resided for some time within the boundary layer, absorbing sensible and latent heat from the underlying surface, that are potentially capable of rising through this stably stratified layer. Thermally direct large-scale motions, which are characterized by the rising of warm air and the sinking of cold air, also contribute to the stable stratification. It is possible to mimic these effects in simple *radiative-convective equilibrium* models by artificially limiting the lapse rate, as shown in Fig. 10.4. The tropospheres of Mars and Venus and the photosphere of the sun[2] can be modeled in a similar manner.

The concept of radiative-convective equilibrium is helpful in resolving the apparent paradox that greenhouse gases produce radiative cooling of the atmosphere (Fig. 4.29), yet their presence in the atmosphere renders the Earth's surface warmer than it would be in their absence. An atmosphere entirely transparent to solar radiation and in pure radiative equilibrium would neither gain nor lose energy by radiative transfer in the longwave part of the spectrum. However, it is apparent from Fig. 10.4 that under conditions of radiative-convective equilibrium, temperatures throughout most of the depth of the troposphere are above radiative equilibrium. It is because of their relative warmth (maintained mainly by latent heat release and, to a lesser extent, by the absorption of solar radiation and the upward transport of sensible heat by atmospheric motions) that greenhouse gases in the troposphere emit more longwave radiation than they absorb. Because the tropospheric lapse rate is determined not

[2] The sun's *photosphere* is defined as the layer from which sunlight appears to be emitted; i.e., the level of unit optical depth for visible radiation. The photosphere marks the transition from a lower, optically dense layer in which convection is the dominant mechanism for transferring heat outward from the nuclear furnace in the sun's core to a higher, optically thin layer in which radiative transfer is the dominant energy transfer mechanism. Like the tropopause in planetary atmospheres, the photosphere is marked by a decrease in the lapse rate from the convectively controlled layer below to the radiatively controlled layer above.

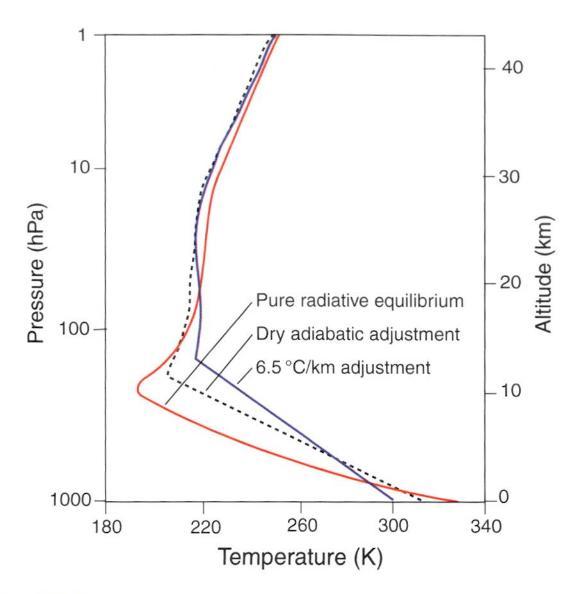

Fig. 10.4 Calculated temperature profiles for the Earth's atmosphere assuming pure *radiative equilibrium* (red curve) and *radiative convective equilibrium,* in which the lapse rate is artificially constrained not to exceed the dry adiabatic value (dashed black curve) and the observed global-mean tropospheric lapse rate (blue curve). [Adapted from *J. Atmos. Sci.*, **21**, p. 370 (1964).]

by radiative transfer, but by convection, it follows that greenhouse gases warm not only the Earth's surface, but the entire troposphere.[3]

In contrast to the troposphere, the stratosphere is close to radiative equilibrium. Heating due to the absorption of ultraviolet solar radiation by ozone is balanced by the emission of longwave radiation by greenhouse gases (mainly CO_2, H_2O, and O_3) so that the net heating rate (Fig. 4.29) is very close to zero. Raising the concentration of atmospheric CO_2 increases the emissivity of stratospheric air, thereby enabling it to dispose of the solar energy absorbed by ozone while emitting longwave radiation at a lower temperature. Hence, while the troposphere is warmed by the presence of CO_2, the stratosphere is cooled.

10.1.2 Dependence on Time of Day

As the Earth rotates on its axis, fixed points on its surface experience large imbalances in incoming and outgoing radiative fluxes as they move in and out of its shadow. As a point rotates through the sunlit, day hemisphere, the atmosphere above it and the underlying surface are heated more strongly by the absorption of solar radiation than they are cooled by the emission of longwave radiation. The energy gained during the daylight hours is lost as the point rotates through the shaded night hemisphere.

Over land, the response to the alternating heating and cooling of the underlying surface produces diurnal variations in temperature, wind, cloudiness, precipitation, and boundary-layer structure, as discussed in Chapter 9. Here we briefly discuss the direct atmospheric response to the hour-to-hour changes in the radiation balance that would occur, even in the absence of the interactions with the underlying surface. This response is often referred to as the *thermal* (i.e., thermally driven) *atmospheric tide*.[4]

Because of the atmosphere's large "thermal inertia," diurnal temperature variations within the free atmosphere are quite small, as illustrated in the following exercise. The thin Martian atmosphere reacts much more strongly to the diurnal cycle in insolation (Table 2.5).

Exercise 10.2 If the Earth's atmosphere emitted radiation to space as a blackbody and if it were completely insulated from the underlying surfaces, at what mass-averaged rate would it cool during the night?

Solution: The cooling rate in degrees K per unit time is equal to the rate of energy loss divided by the heat capacity per unit area of the free atmosphere. During the night the atmosphere continues to emit infrared radiation to space at its equivalent blackbody temperature of 255 K; hence it loses energy at a rate of

$$E = \sigma T^4 = 5.67 \times 10^{-8} \times (255)^4 = 239 \text{ W m}^{-2}$$

The heat capacity of the atmosphere (per m^2) is equal to the specific heat of dry air c_p times the mass per unit area (p/g) or

$$\frac{1004 \text{ J kg}^{-1} \text{ K}^{-1} \times 10^5 \text{ Pa}}{9.8 \text{ m s}^{-2}} \simeq 10^7 \text{ J K}^{-1} \text{m}^{-2}$$

[3] Throughout the tropics the observed lapse-rate Γ is close to the saturated adiabatic value Γ_w. As the Earth's surface warms, the latent heat released in moist adiabatic ascent increases, and so the numerical value of Γ_w decreases. Hence, if Γ remains close to Γ_w as the tropical troposphere warms, temperatures in the upper troposphere will warm more rapidly than surface air temperature, as demonstrated in Exercise 3.45.

[4] The gravitational attraction of the moon and sun also induce atmospheric tides, but these gravitational tides are much weaker than the thermal tides.

The cooling rate is therefore

$$\frac{dT}{dt} = \frac{239\ Wm^{-2}}{10^7 Jk^{-1}m^{-2}} = 2.39 \times 10^{-5}\ Ks^{-1}$$

Night lasts 12 h or 4.32×10^4 s. Hence, the overnight cooling is only ~1 K. ■

The diurnal variation in incident solar radiation is not a pure sine wave: it is proportional to the cosine of the solar zenith angle during the day and zero at night (see Fig. 9.9). Because of the nonsinusoidal time dependence of the forcing, atmospheric tides are made up of an ensemble of frequencies that are integral multiples of one cycle per day. The most important of these are the *diurnal* (one cycle per day) and *semidiurnal* (two cycles per day) tides. The semidiurnal component, a response to the periodic heating and cooling of the stratospheric ozone layer, is clearly evident in tropical sea-level pressure records. The corresponding diurnal and semidiurnal variations in the horizontal wind field increase with height from ~1–2 m s^{-1} at tropospheric levels to more than 10 m s^{-1} in the mesosphere and thermosphere.

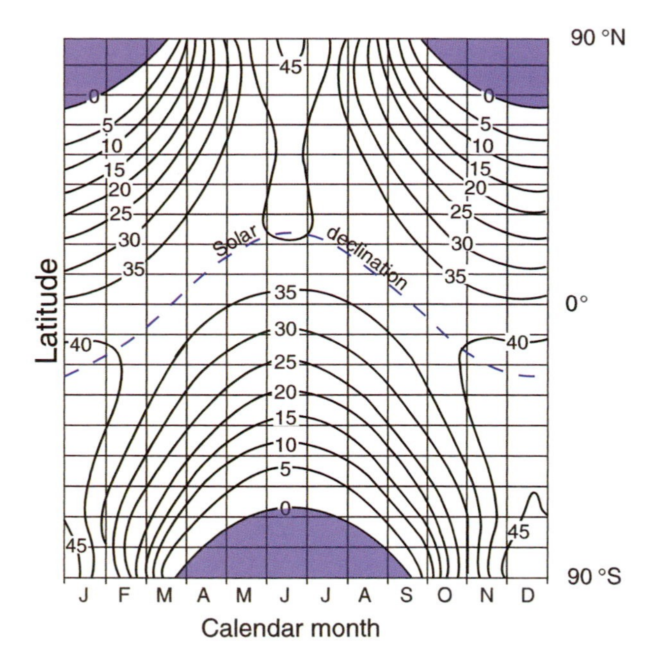

Fig. 10.5 Insolation incident on a unit horizontal surface at the top of the atmosphere, expressed in units of MJ m^{-2} integrated over the 24-h day, as a function of latitude and calendar month. Within the shaded regions the insolation is zero. To convert to units of W m^{-2} multiply the contour labels by 11.57. *Solar declination* refers to the latitude at which the sun is overhead at noon. [Adapted from *Meteorological Tables* (R. J. List, Ed.), 6th Ed., Smithsonian Institute (1951), p. 417.]

10.1.3 Seasonal Dependence

Each year, as Earth revolves around the sun, the extratropical continents experience large temperature swings and many regions of the tropics experience dramatic changes in rainfall. These periodic climate fluctuations are largely a response to the obliquity of the Earth's axis of rotation relative to the plane of the ecliptic.

The seasonally varying latitudinal distribution of insolation incident upon the top of the atmosphere is shown in Fig. 10.5. The equator-to-pole gradient is strongest during winter, when the polar cap is in darkness. In the summer hemisphere the increasing length of daylight with latitude more than offsets the increasing solar zenith angle so that insolation increases slightly with latitude. The large seasonal variations in insolation give rise to seasonally varying imbalances in net radiation at the top of the atmosphere (Fig. 10.6), which range up to 100 W m^{-2} over the subtropical and midlatitude oceans. Net radiation at the surface of the ocean (not shown) exhibits a similar distribution: upward in the winter hemisphere and downward in the summer hemisphere. Most of the excess insolation in the summer hemisphere is stored in the ocean mixed layer and the cryosphere and is released during the following winter, thereby moderating the seasonal contrast in atmospheric temperature.

The storage of heat in the ocean mixed layer is reflected in the warming of a relatively shallow layer of water near the surface during late spring and early summer, which leads to the formation of the *seasonal thermocline*, as shown in Fig. 10.7. During autumn and early winter reduced insolation and enhanced latent and sensible heat fluxes cool the mixed layer, releasing the heat that was stored during the previous summer. The warming of high latitude regions during winter by the release of this stored heat is larger than the poleward heat transport by the western boundary currents and (in the North Atlantic) by the thermohaline circulation. As the seasonal thermocline cools during autumn and winter, it also deepens by entraining water from below. The ocean mixed layer reaches its minimum temperature and maximum depth in spring, setting the stage for the redevelopment of the thermocline much closer to the ocean surface a few months later.

The large seasonal variations in the extent of polar northern hemisphere pack ice (Fig. 2.12) and other elements of the cryosphere also contribute to

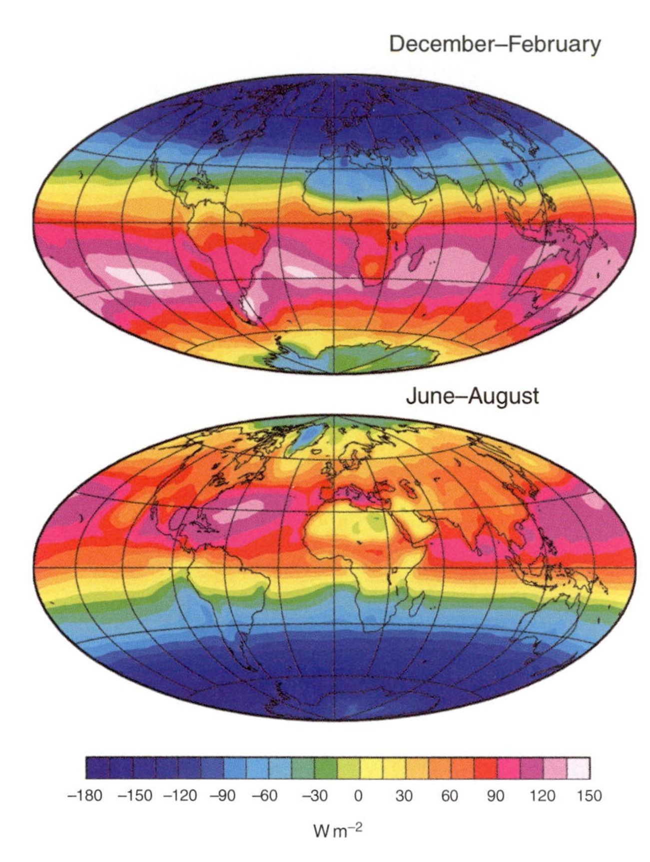

Fig. 10.6 Net radiation at the top of the atmosphere in December-February and June-August in units of W m^{-2}. [Based on data from the NASA Earth Radiation Budget Experiment. Courtesy of Dennis L. Hartmann.]

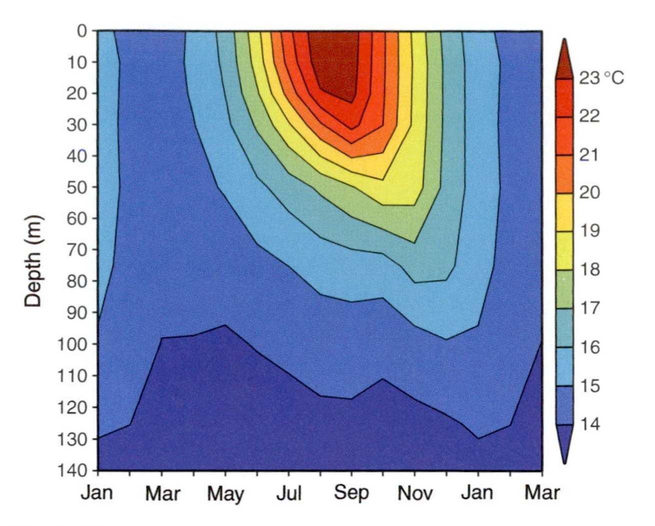

Fig. 10.7 Temperature (in °C) averaged over a region in the central North Pacific (28°−42 °N, 180°−160 °W) as a function of calendar month and depth showing the climatological-mean annual cycle. [Based on data from *World Ocean Atlas*, NOAA National Oceanographic Data Center (1994). Courtesy of Michael Alexander.]

moderating the contrast between winter and summer temperatures. Heat of fusion is taken up by the cryosphere when ice melts during the warm season. A comparable quantity of heat is released during the cold season when sea ice freezes and thickens and snow and ice particles that freeze within the atmosphere precipitate onto ice sheets and glaciers.

The large difference in the annual range of surface air temperature over the continents and oceans shown in Fig. 10.8 reflects widely differing heat capacities of the underlying surfaces. The largest ranges (~50 K) are observed over the interior of Eurasia, North America, and Antarctica, in areas far away from (or shielded by mountain ranges) from the moderating influence of the oceans.

Land–sea temperature contrasts at low latitudes of the summer hemisphere force continental-scale monsoon circulations. In analogy with Fig. 7.21, the warmth of the tropical and subtropical continents relative to the surrounding oceans in the summer hemisphere lifts the pressure surfaces and induces horizontal divergence at upper tropospheric levels, maintaining relatively low sea-level pressure over the continents relative to the surrounding oceans. The pressure contrast between land and sea drives an onshore flow of moist, boundary-layer air, triggering deep convection over land, as depicted schematically in Fig. 10.9.

In the real atmosphere the seasonally varying distribution of precipitation (Fig. 1.25) is also influenced by the land–sea geometry, the distribution of mountain ranges, and the underlying sea-surface temperature distribution. These combined influences account for distinctive regional features such as the summer rainfall maximum over the Bay of Bengal and the extensive subtropical desert regions that experience very little summer rainfall.

The tropospheric jet stream (Fig. 1.11) and the baroclinic wave activity associated with the midlatitude storm tracks tend to be strongest during wintertime, when the equator-to-pole temperature gradient is strongest.[5] At the longitude of Japan, where stationary planetary waves generated by the land–sea contrasts and the flow over the Himalayas reinforce the

[5] A notable exception is the North Pacific, where baroclinic wave activity tends to be suppressed during midwinter relative to late autumn and early spring.

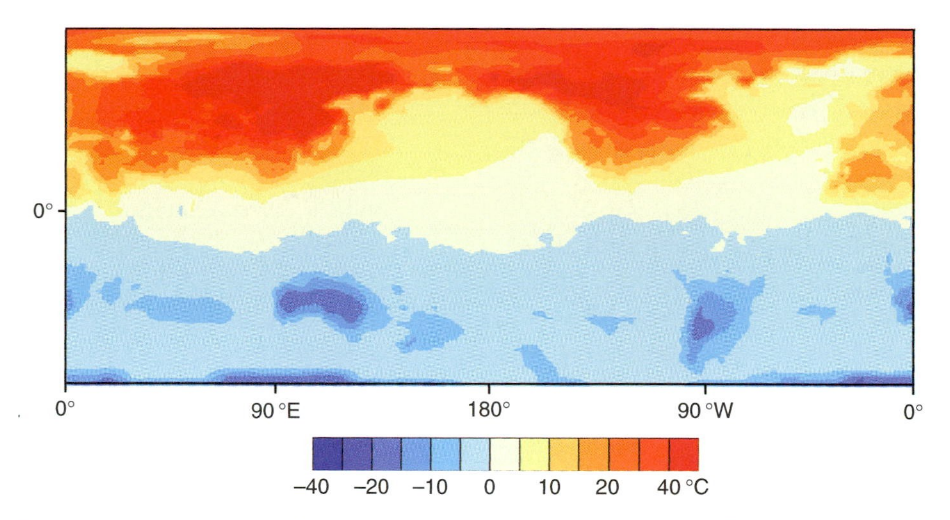

Fig. 10.8 The "Find the Continents" game! Climatological-mean July minus January surface air temperature in °C. [Courtesy of Todd P. Mitchell.]

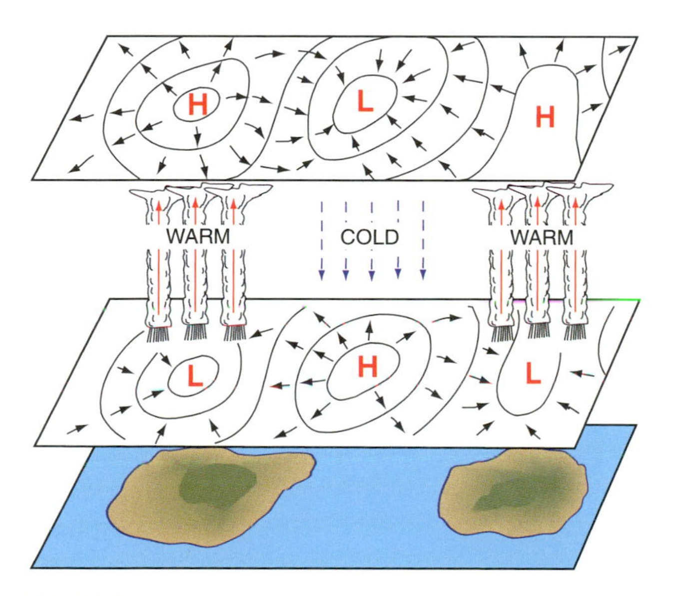

Fig. 10.9 Idealized representation of the monsoon circulations. The islands represent the subtropical continents in the summer hemisphere. Solid lines represent isobars or height contours near sea level (lower plane) and near 14 km or 150 hPa (upper plane). Short solid arrows indicate the sense of the cross-isobar flow. Vertical arrows indicate the sense of the vertical motions in the middle troposphere. Regions that experience of summer monsoon rainfall are also indicated.

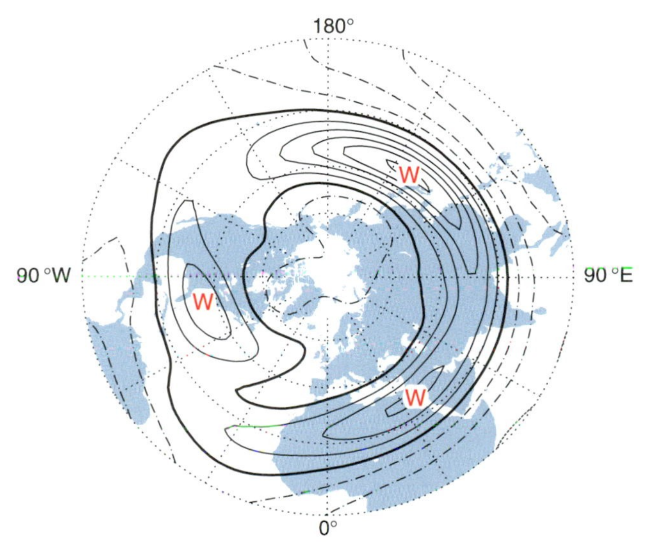

Fig. 10.10 January climatological mean zonal wind speed at the jet stream (250-hPa) level. Contour interval 15 m s^{-1}. The zero contour is bold; positive contours, indicative of westerlies, are solid and negative contours, indicative of easterlies, are dashed. [Based on data from the NCEP-NCAR Reanalysis. Courtesy of Todd P. Mitchell.]

tropospheric jet stream, the climatological mean westerly wind component at the 250-hPa level exceeds 70 m s^{-1} (Fig. 10.10). The Aleutian and Icelandic lows in the sea-level pressure field (Fig. 1.19) also tend to be most pronounced during wintertime, when the high latitude oceans are much warmer than the continents. In contrast, subtropical anticyclones tend to be strongest during summer, when they are reinforced by the monsoon-related land–sea pressure contrasts.

10.2 Climate Variability

Much of what we know about the causes of year-to-year climate variability is based on numerical experiments with atmospheric models like those described in Section 7.3. Imagine a pair of experiments: one run in which the bottom boundary conditions (e.g., sea-surface temperature, sea ice extent, soil moisture) are prescribed to vary from one year to the next in accordance with historical data over, say, the 20th century and the other a run of equal length in which these boundary conditions are prescribed to vary in accordance with seasonally varying, climatological-mean values, year after year. Because weather is inherently unpredictable beyond a time frame of a few weeks, any resemblance between the observed and simulated 100-year sequence daily synoptic charts in the two simulations must be viewed as entirely fortuitous. However, the first run may exhibit climate variability attributable to the year-to-year variations in the boundary conditions, whereas the variability in the second run is generated exclusively by dynamical processes operating within the atmosphere. The ratio of the standard deviation of the year-to-year variability as simulated in the two experiments provides a measure of the relative importance of the *boundary forced* variability relative to the *internally generated* variability.

Numerical experiments of this kind have been conducted using many different models. The results can be summarized as follows.

(1) Most of the year-to-year variability of the tropical atmosphere is boundary forced (i.e., attributable to year-to-year variations in the prescribed boundary conditions, particularly sea-surface temperature over the tropical oceans). In the more realistic models, the simulated year-to-year variations in tropical climate bear a strong resemblance to the observed variations.

(2) At extratropical latitudes boundary forcing and internal atmospheric dynamics both make important contributions to the observed year-to year climate variability. Of the various contributors to the boundary forcing, tropical sea-surface temperature appears to be of primary importance for the northern hemisphere winter climate. Variations in soil moisture and vegetation contribute to the month-to-month persistence of summertime climate anomalies. If the observed values of these fields are prescribed, the simulated year-to-year variations in extratropical climate are correlated with the observed variations, but not as strongly as in (1).

(3) The influences of year-to-year variations in sea-ice extent and extratropical sea-surface temperature are more subtle. By running ensembles of simulations, in which each member is started from different initial conditions but is forced with the same prescribed sequence of boundary conditions, it is possible to identify weak bondary-forced "signals" that stand out above the internally generated "sampling noise."

(4) Most of the intraseasonal variability of the extratropical wintertime circulation appears to be generated internally within the atmosphere.

Variability generated by the interactions between the atmosphere and more slowly varying components of the Earth system is referred to as *coupled climate variability*. Climate variability may also be *externally forced*, e.g., by volcanic eruptions, variations in solar emission, or changes in atmospheric composition induced by human activities.

10.1 Some Basic Climate Statistics[6]

Consider the climatic variable x, which could represent monthly-, seasonal-, annual-, or even decadal-mean temperature at a prescribed latitude, longitude, and height above the Earth's surface. Let X be the climatological-mean value of x. The departure of x from its (seasonally varying) climatological mean value, namely

$$x' = x - X \tag{10.1}$$

is referred to as the *anomaly* in x. For example, a temperature 3 °C below normal is equivalent to a temperature anomaly of −3 °C.

The *variance* of x about the climatological mean[7] is

$$\overline{x'^2} = \overline{(x - X)'^2} \tag{10.2}$$

where the $\overline{(\ )}$ denotes a time mean over the reference period upon which the climatology is based.

Continued on next page

[6] The formalism developed in Box 10.1 is also applicable to boundary-layer quantities. The overbars represent time-averaged quantities and the primed quantities represent fluctuations about the mean that occur in association with boundary-layer turbulence.

[7] $\overline{x'^2}$ is the *temporal variance*. The *spatial variance*, defined in an analogous manner, is a measure of the variability of x about its spatial mean.

10.1 Continued

Variance, a positive definite quantity with units of the square of the variable under examination (e.g., °C^2 for temperature), is a measure of the amplitude of the variability (or dispersion) of x about its climatological-mean value. The *standard deviation* or *root mean squared (r.m.s.) amplitude* of the variations in x about the time mean

$$\sigma(x) \equiv \sqrt{\overline{x'^2}} = \sqrt{\overline{(x - X)^2}} \qquad (10.3)$$

is widely used as a measure of the dispersion. Note that variance and standard deviation do not carry algebraic signs.

The *standardized anomaly*

$$x^* \equiv \frac{x'}{\sigma(x)}$$

is a dimensionless measure of the amplitude of the departure from the mean. For variables such as monthly-mean temperature, sea-level pressure and geopotential height, which tend to be *normally distributed*,[8] ~64% of randomly selected standardized anomalies exhibit absolute values less than 1.0, ~95% of them less than 2.0, and ~99.9% of them less than 3.0. Hence, for such distributions, a standardized anomaly x^* with a value of +1.0 or −1.0 can be considered typical in terms of r.m.s. amplitude.

Figure 10.11 shows an example of a seasonal mean 500-hPa height field together with the corresponding anomaly and standardized anomaly fields. In contrast to the baroclinic waves discussed in Section 8.1.1, which exhibit typical zonal wavelengths of ~40° of longitude, the anomalies in monthly and seasonal mean fields like the ones shown in this example are larger in scale and they are not particularly wave like.

Now let us consider the relationship between two time series $x(t)$ and $y(t)$, which might represent a series of values of the same climatic variable at two different geographical locations or might represent two different variables at the same location. It is assumed that $x(t)$ and $y(t)$ span a common period of record. The dimensionless statistic

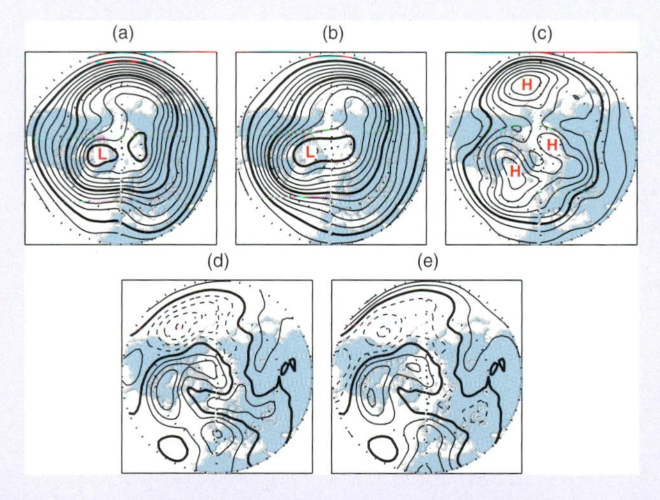

Fig. 10.11 (a) The mean 500-hPa height field averaged over the late winter season January–March 1998. (b) The climatological-mean wintertime (January–March) 500-hPa height field based on the period of record 1958–1999; contour interval 60 m, the 5100-, 5400- and 5700-m contours are bold. (c) The climatological-mean standard deviation of the wintertime-mean (January–March) 500-hPa height field based on the same period of record; contour interval 9 m, the 54-m contour is bold. (d) The anomaly field for January–March 1998, calculated by subtracting (b) from (a); contour interval 30 m, the zero contour is bold, and dashed contours denote negative values. (e) The standardized anomaly field, calculated by dividing (d) by (c); contour interval 0.6 standard deviations, the zero contour is bold, and dashed contours denote negative values. [Based on data from the NCEP-NCAR Reanalyses. Courtesy of Roberta Quadrelli.]

Continued on next page

[8] Possessing a bell-shaped histogram centered on the climatological-mean value, with a width (defined as the distance between the center and the inflection points) of 1 standard deviation.

10.1 Continued

$$r \equiv \overline{x^* y^*} \equiv \frac{\overline{x' y'}}{\sigma(x)\,\sigma(y)} \tag{10.4}$$

called the *correlation coefficient* between x and y, is a measure of the degree to which x and y are *linearly* related (i.e., that one is simply a linear multiple of the other). Values of r range from -1 to $+1$.

To illustrate how the correlation coefficient can be used to describe the structure of a spatial field, we show, in Fig. 10.12, scatter plots of standardized wintertime mean sea-level pressure anomalies over Iceland versus sea-level pressure at three different locations. In plot A, y Iceland sea-level pressure is plotted against sea-level pressure at a nearby location; in this case x^* and y^* tend to be of like sign so that $r = \overline{x^* y^*} > 0$ and the data points in the scatter plot lie preferentially in the first and third quadrants of the plot. In such situations x and y are said to be *positively correlated.* Plot B shows pressure over Iceland versus sea-level pressure over southern England; in this case the correlation is weak (i.e., $r \sim 0$). Plot C shows that sea-level pressure anomalies over Iceland and Portugal tend to be of opposing sign (i.e., negatively correlated: $r = \overline{x^* y^*} < 0$ and the points in the plot lie preferentially in the second and fourth quadrants. Statistically significant linear correlations between climatic variables such widely spaced locations are evidence of what are referred to as *teleconnections*.

In general, the higher the value of $|r|$, the tighter the clustering of the dots along the 45° (or 135°) axis in scatter plots of x^* versus y^*. It can be shown that in the limit, as $|r| \rightarrow 1$, the points line up perfectly, in which case, $y^* = rx^*$ or, alternatively, $x^* = ry^*$. If $r = 0$, there is no linear relation between x and y and the least-squares best fit in the scatter plot is the horizontal line that passes through the centroid of the data points. Exercise 10.9 shows that the fraction of the variance of x and y that is common to the two variables (in a linear sense) is given by r^2.[9]

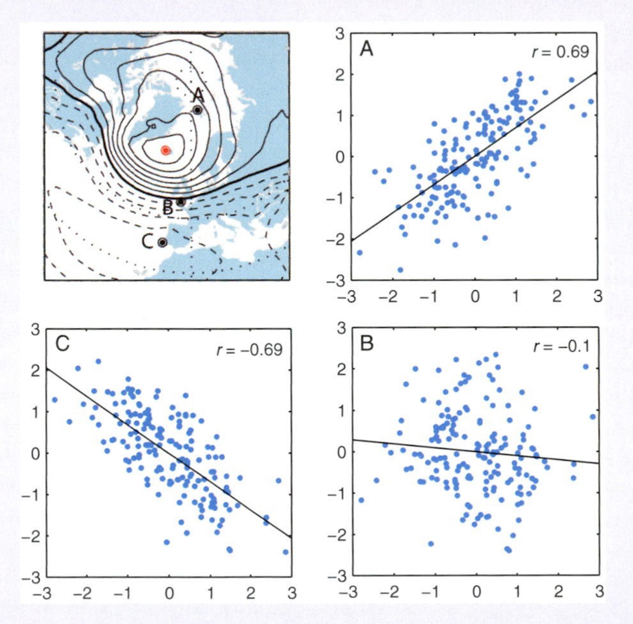

Fig. 10.12 Scatter plots of standardized sea-level pressure anomalies. In all three plots, the x axis refers to anomalies at a grid point over Iceland, indicated by the red dot in the upper left panel. Iceland anomalies are plotted versus anomalies at (**A**) Spitzbergen, (**B**) southern England, and (**C**) Portugal. Numerical values of the correlation coefficients are indicated in the scatter plots, and sloping lines represent the equation $y^* = rx^*$ that corresponds to the least-squares best fit linear regression line. The map in the upper left panel shows the correlation coefficient between sea-level pressure at Iceland and sea-level pressure at every grid point: contour interval 0.15; the zero contour is bold and negative values are indicated by dashed contours. [Based on data from the NCEP-NCAR Reanalyses. Courtesy of Roberta Quadrelli.]

[9] The correlation coefficient r can also be interpreted as the slope of the *least-squares best fit regression line* for y regressed upon x; i.e., the slope of the staight line, passing through the origin, for which the mean squared error $(y_i - rx_i)^2$, summed over all data points in the time series, is minimized (see Exercise 10.9).

10.2.1 Internally Generated Climate Variability

The day-to-day variability of the geopotential height field is characterized by a bewildering array of hemispheric patterns. In contrast, the spatial patterns of month-to-month variability, as manifested in hemispheric or global anomaly fields like those shown in Fig. 10.11 tend to be simpler. Much of the structure in such maps can be interpreted in terms of the superposition of a limited number of *preferred spatial patterns* that may appear with either polarity when they are present. In the northern hemisphere these patterns are much more prominent during winter than during summer.

A prominent pattern in the northern hemisphere wintertime geopotential height field is variously known as the *North Atlantic Oscillation (NAO)*, the *Arctic Oscillation (AO)*, or the *northern hemisphere annular mode (NAM)*. The sea-level pressure signature of this pattern, shown in Fig. 10.13, is marked by anomalies of opposing sign over the Arctic polar cap region and the Atlantic/Mediterranean sector of middle latitudes. The Southern hemisphere annular mode (not shown) is even more symmetric about the pole and is evident not just during the winter, but throughout the year.

When sea-level pressure is below normal over the polar cap region and above normal over the Mediterranean, the annular mode is said to be in its *high index* polarity. At these times the jet streams and storm tracks tend to be displaced poleward of their normal positions, temperatures tend to be unseasonably mild over Eurasia and most the United States (Fig. 10.14), and northern Europe experiences

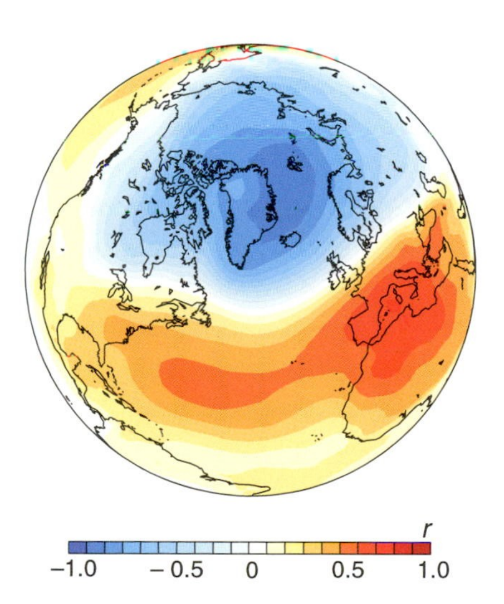

Fig. 10.13 Pattern of sea-level pressure anomalies associated with the *northern hemisphere annular mode* (a.k.a., *North Atlantic Oscillation*). Colored shading indicates the correlation coefficient between the time series of the index of the pattern shown in Fig. 10.17 and the time series of the monthly mean sea-level pressure anomalies at each grid point. The colors thus indicate the sign of the anomalies associated with the high index polarity of the annular mode: the low index polarity is characterized by anomalies of opposite sign. [Based on November–April data from the NCEP-NCAR Reanalysis. Courtesy of Todd P. Mitchell.]

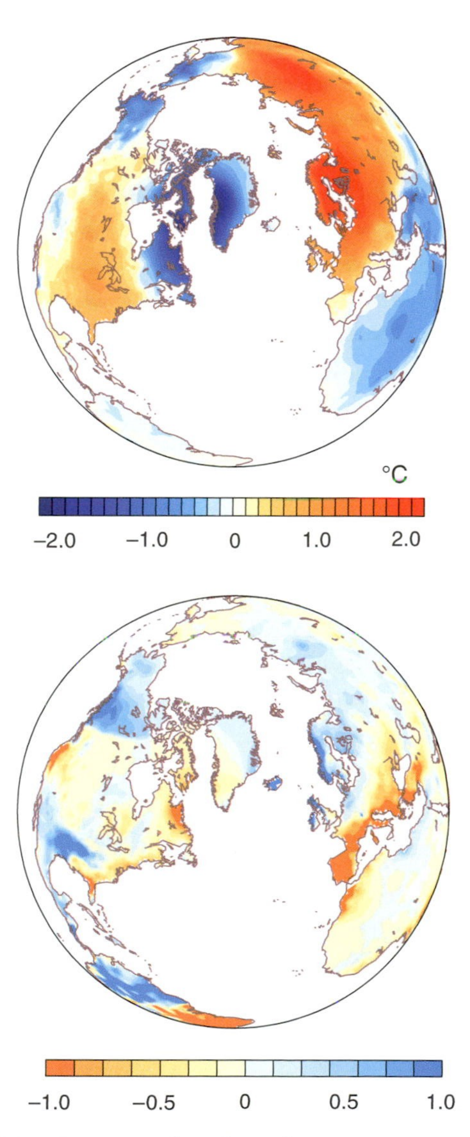

Fig. 10.14 Patterns of surface air temperature anomalies (top) and standardized precipitation anomalies (bottom) observed in months when the northern annular mode is in its *high index polarity* characterized by below normal sea-level pressure over the Arctic. Warm colors indicate above normal temperature and below normal precipitation, and shades of blue indicate anomalies of opposing sign. Typical amplitudes of the temperature and standardized precipitation anomalies are indicated by the color scale. [Courtesy of Todd P. Mitchell.]

heavier than normal rainfall while the Mediterranean basks in sunshine. In contrast, episodes of abnormally high pressure over the Arctic (i.e., the *low index polarity* of the annular mode) tend to be marked by relatively frequent occurrence of cold-air outbreaks over Eurasia and the United States, raising the demand for heating oil on the world market, and stormy weather over the Mediterranean.

The sea-level pressure signature of another prominent wintertime pattern, the so-called *Pacific/North American (PNA) pattern*, is shown in Fig. 10.15. The PNA pattern, with its strongest sea-level pressure anomalies over the North Pacific, has a strong impact on wintertime climate anomalies downstream over North America, as documented in Fig. 10.15. When pressures over the North Pacific are below normal, temperatures over much of western North America tend to be relatively mild and precipitation tends to be abnormally heavy along the coasts of the Gulf of Alaska and the Gulf of Mexico; Hawaii tends to be dry.

The processes that give rise to the annular modes and PNA pattern are not fully understood. The annular modes involve north–south shifts in the *storm tracks* (i.e., the belts of strongest baroclinic wave activity) and the surface westerlies. When the annular modes are in their high index polarity, the storm tracks and associated westerly wind belts are shifted poleward of their climatological-mean positions, and vice versa. These shifts are believed to occur as a result of the interactions between the waves and the mean westerly winds upon which they are superimposed. The PNA pattern, which has no clearly defined southern hemisphere counterpart, involves the alternating extension and retraction of the downstream end of the jet stream that passes just to the south of Japan in Fig. 10.10. When the PNA pattern is in its high index polarity, the jet extends farther east into the central Pacific than in Fig. 10.10, and vice versa. These extensions and retractions of the jet are believed to be the manifestation of a form of instability of the northern hemisphere wintertime climatological-mean stationary-wave pattern.

The annular modes and the PNA pattern play important roles in the variability of the extratropical circulation on timescales of decades and longer. Of particular interest is the tendency toward the positive polarity of both northern and southern annular modes from the 1970s to the 1990s (Fig. 10.17). The positive trends in the indices of the annular modes

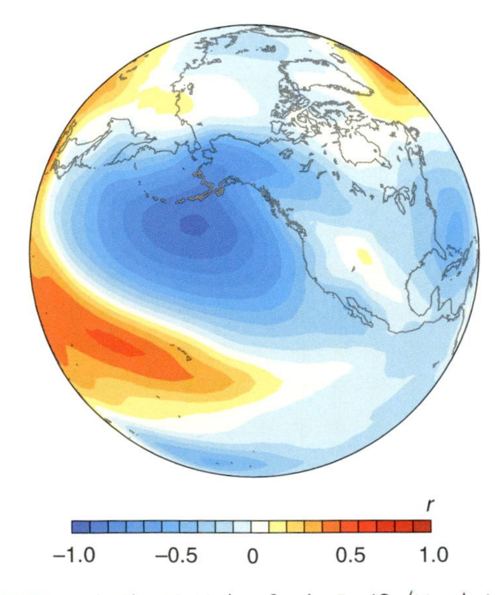

Fig. 10.15 As in Fig. 10.13, but for the Pacific/North American pattern. [Courtesy of Todd P. Mitchell.]

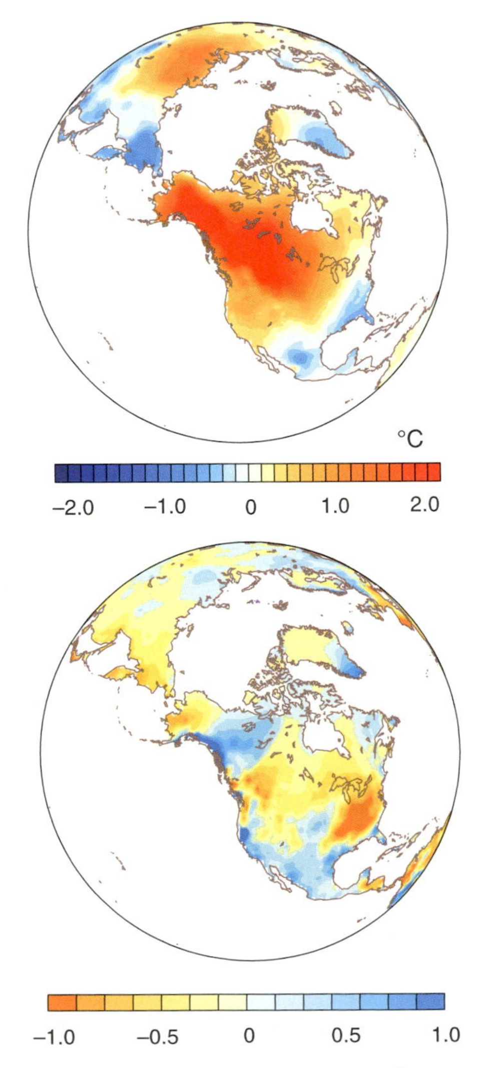

Fig. 10.16 As in Fig. 10.14, but for the Pacific/North American pattern. [Courtesy of Todd P. Mitchell.]

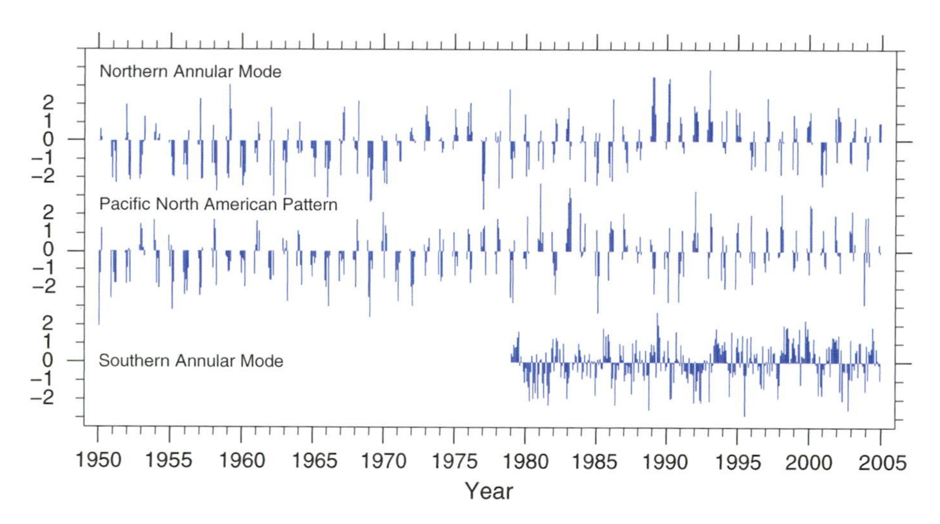

Fig. 10.17 Time series of standardized indices of the annular modes and the PNA pattern. The northern hemisphere indices are shown for the winter months (November–March) only, whereas the index of the southern annular mode is shown for all calendar months. [Courtesy of Todd P. Mitchell.]

are indicative of a poleward shift of the wintertime storm tracks and the westerly wind belt in both hemispheres. During the winters since 1977 the PNA pattern has exhibited a preference for its high index polarity, with below normal sea-level pressure over the Gulf of Alaska and a relative absence of cold air masses over Alaska and western Canada.

10.2.2 Coupled Climate Variability

The northern hemisphere annular mode and the Pacific/North American pattern are believed to be generated by processes within the atmosphere because they emerge as the leading modes of variability in extended runs of atmospheric general circulation models in which the bottom boundaries are held fixed. This section considers *coupled climate variability* involving interactions between the atmosphere and other components of the Earth system. Under certain conditions, these interactions can give rise to modes of variability that are qualitatively different from the kinds of variability that the atmosphere is capable of generating through its own internal dynamical processes.

a. Coupling with the tropical ocean

The most prominent coupled mode involving the atmosphere and ocean is known as *ENSO* (a hybrid acronym in which the first two letters refer to *El Niño*, a recurrent pattern of positive sea-surface temperature anomalies in the equatorial Pacific[10] (Fig. 10.18) and the second two letters refer to the associated global pattern sea-level pressure (Fig. 10.19), which is known as the *Southern Oscillation*[11,12]). Time series of equatorial Pacific sea-surface temperature (an indicator of the status of El Niño) and sea-level pressure at Darwin, Australia (an indicator of the state of the Southern

[10] The name *El Niño*, Spanish for male child, was used by Ecuadorian and Peruvian fishermen to refer to a warm coastal current that often appears Christmas. The name is now more widely used to denote episodic warmings of the entire region of the equatorial Pacific covered by the warm colored shading in Fig. 10.18.

[11] The name *Southern Oscillation* was coined by Walker in the 1920's to denote a global pattern of climate anomalies that included the sea-level pressure anomalies shown in Fig. 10.19.

[12] **Gilbert Thomas Walker** (1869–1958) English mathematician and meteorologist. Attended St. Paul's School, London, where, according to his own account, he spent 2 years on the classical side but was then "sent in disgrace" to the mathematical side when he made some appalling linguistic blunders. While at St. Paul's he was awarded a prize for a gyroscope that he made. In 1897 he published an original paper on the flight of boomerangs, which he was expert at throwing. He also wrote articles on the mathematics of billiards, golf, and bicycle riding. Although he had not published any papers on meteorology, in 1904 he was appointed Director of Observations in India, a post he retained until 1924. Discovered the Southern Oscillation, which turned out to be useful in seasonal forecasting over India. Later became Professor of Meteorology at the Imperial College of Science, London, where he became interested in the forms of clouds and the conditions that produced them. He was an accomplished flute player; some modern flutes are made according to his suggestions.

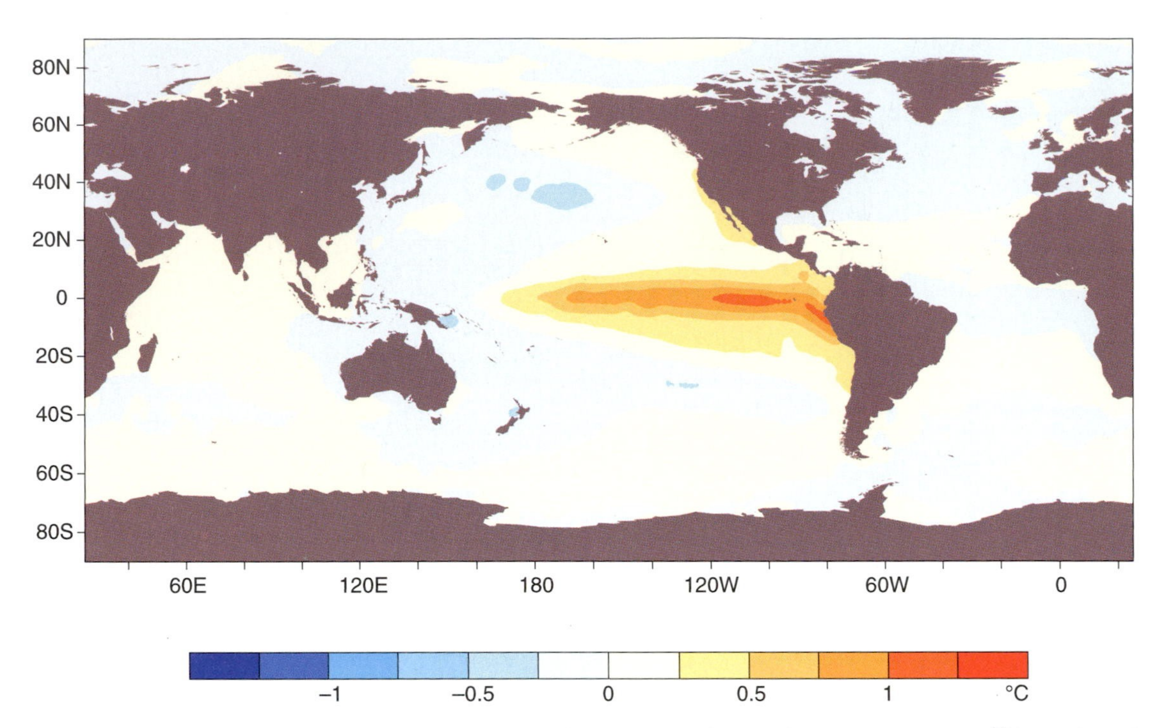

Fig. 10.18 Global pattern of sea surface temperature anomalies observed during El Niño years (in °C).[13] [Based on data from the U. K. Meteorological Office HadISST. Courtesy of Todd P. Mitchell.]

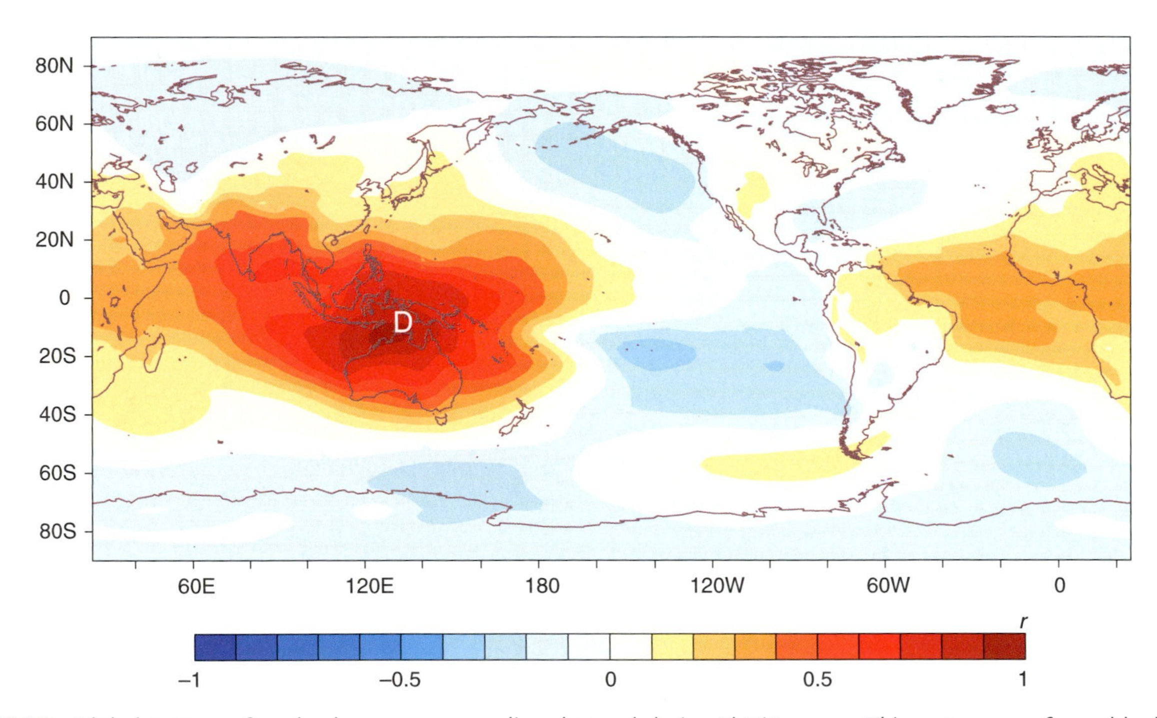

Fig. 10.19 Global pattern of sea-level pressure anomalies observed during El Niño years. This pattern was formed by linearly correlating monthly-mean sea-level pressure at each grid point with sea-level pressure at Darwin, Australia, indicated by the **D** on the map. The Darwin sea-level pressure time series appears on the lower time axis in Fig. 10.20 as an index of the status of the Southern Oscillation. [Darwin time series from the NCAR Climate and Global Dynamics Division and map-form data from the NCEP-NCAR Reanalysis and the Darwin time series from the NCAR Data Library. Courtesy of Todd P. Mitchell.]

[13] Figure 10.18 was constructed by linearly regressing the sea surface temperature anomalies at each grid point onto standardized monthly values of a standardized *El Niño index*, shown in Fig. 10.20, which is defined as the average sea surface temperature anomaly over the region of large equatorial sea-surface temperature anomalies in the eastern equatorial Pacific (6°N−6°S, 180°−90°W) minus the global-mean sea-surface temperature anomaly.

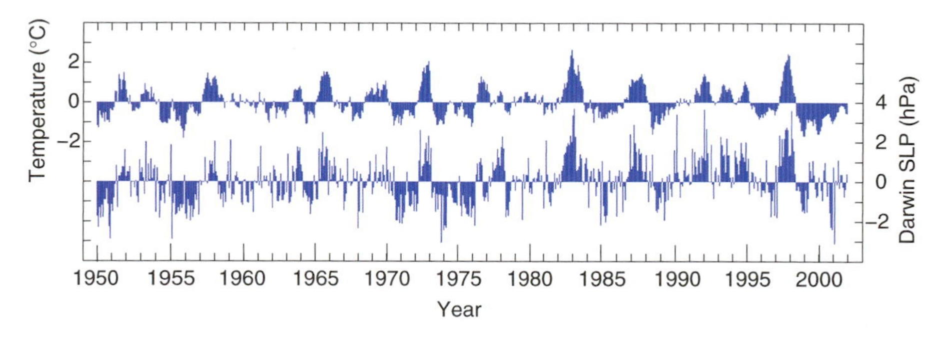

Fig. 10.20 Time series of indices of El Niño (top) and the Southern Oscillation (bottom). Minus the global-mean sea surface temperature anomaly. The Southern Oscillation is represented by Darwin sea-level pressure anomalies. The location of Darwin is indicated by the **D** in Fig. 10.19. Prominent warm episodes of the ENSO cycle occurred during 1957–1958, 1965–1966, 1972–1973, 1982–1983, 1986–1988, and 1997–1998. [Courtesy of Todd P. Mitchell.]

Oscillation), shown together in Fig. 10.20, demonstrate the remarkable strength of the atmosphere–ocean coupling on the interannual timescale that is observed in association with ENSO.[14]

In the long-term climatology (Fig. 1.19), sea-level pressure is higher on the eastern side of the Pacific than on the western side and the easterly surface winds along the equator are directed down the pressure gradient. It is evident from Fig. 10.20 that during El Niño events, sea-level pressure at Darwin tends to be above normal. It follows that El Niño events are maeked by a weakening of the climatological-mean pressure gradient. We will now show, by means of a "case study," that El Niño events are marked by a weakening of the easterly surface winds, as well as the east–west gradients in sea level and thermocline depth.

By most measures, the strongest El Niño event of the 20th century was the one that began in the boreal summer of 1997 and lasted about 9 months. The year that followed was typical of cold events of the ENSO cycle, referred to in the popular press as *La Niña*.[15] Various fields for these contrasting times are shown in the next few figures. Relative to the cold year or to normal conditions, the warm (*El Niño*) year is characterized by:

- a weakening of the easterly trade winds winds along the equator (Fig. 10.21).
- a weakening of the "equatorial cold tongue" in the sea-surface temperature field (Fig. 10.21; compare also with Fig. 2.11).
- reduced productivity of the marine biosphere in the band of upwelling along the equator (Fig. 10.22).
- a rise in sea level in the equatorial eastern Pacific and lowering of sea level in the western Pacific (Fig. 10.23), which is indicative of a weakening of the climatological-mean east-to-west gradient.

These diverse manifestations of El Niño are physically consistent with each other and with the patterns of sea surface temperature anomalies and sea-level pressure anomalies in Figs. 10.18 and 10.19. In accordance with the equation of motion for the zonal wind component on the equator, a reduced east–west pressure gradient during El Niño (i.e., positive sea-level pressure anomalies at Darwin) is consistent with a weakening of the easterly trade winds along the equator. In accordance with the relationships discussed in Section 7.3.4, the weakening of the trade winds is consistent with reduced equatorial upwelling. Reduced equatorial upwelling, in turn, favors higher sea surface temperature and lower marine productivity. The relaxation of the east–west gradient in sea level is a response to the weakening of the surface easterlies along the equator.

The consistency in El Niño-induced variations in wind, sea level, and equatorial upwelling is evident in the schematics shown in Fig. 10.24. The relaxation of the slope of the thermocline occurs in response to

[14] The first convincing proof of the link between El Niño and the Southern Oscillation was in a series of papers by Jacob Bjerknes in the late 1960s. Bjerknes' conceptual model of the life cycle of extratropical cyclones, published nearly 50 years earlier, was highly influential in the development of weather forecasting.

[15] In English, "the girl child."

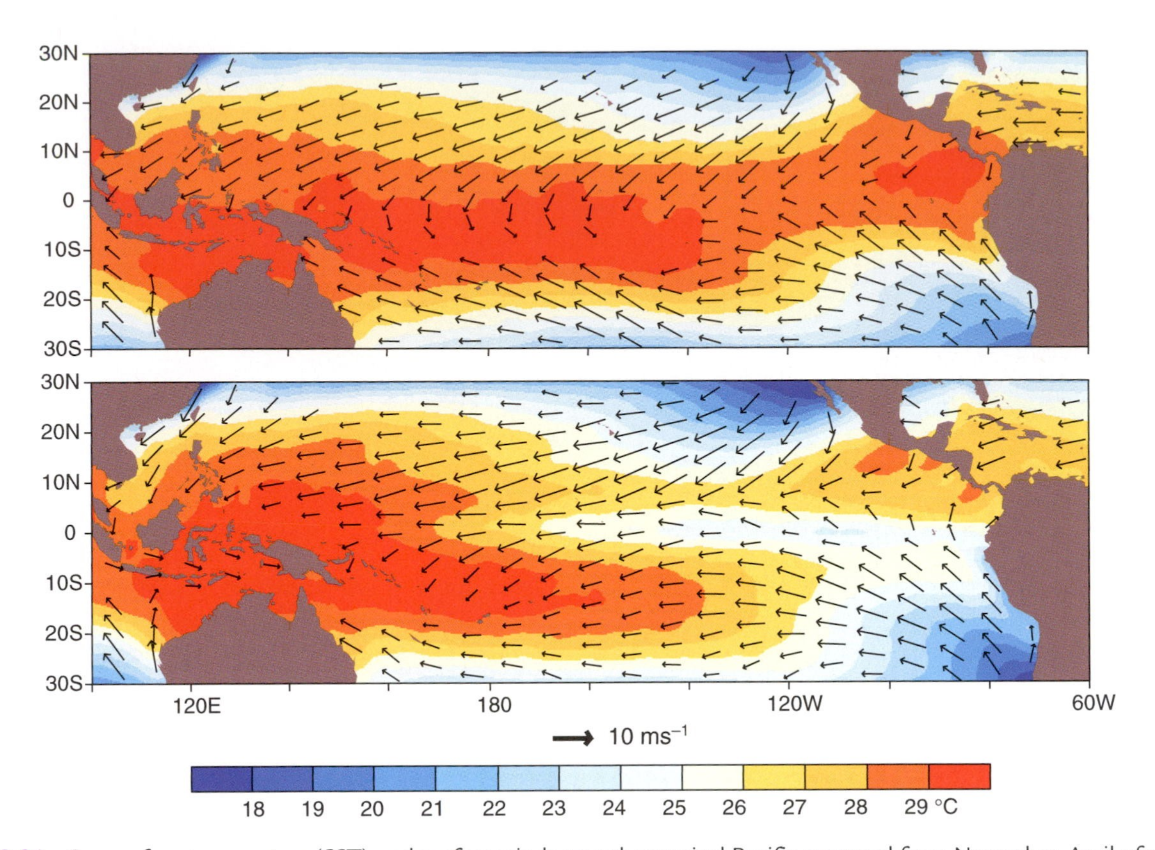

Fig. 10.21 Sea-surface temperature (SST) and surface winds over the tropical Pacific averaged from November–April of a warm year of the ENSO cycle (1997–1998, top) and November-April of a cold year (1998–1999, bottom). During both years the surface winds converge into the region of highest sea-surface temperature, the so-called *warm pool*, encompassing the western Pacific Ocean and Indonesia. During El Niño years the warm pool is shifted eastward. [SST data from the U. K. Meteorological Office HadISST and 10-m winds from the European Remote-sensing Satellite (ERS-2). Courtesy of Todd P. Mitchell.]

the weakening of the trade winds. As a result of the deepening of the thermocline in the eastern Pacific during El Niño events, the upwelled water is not as cold or as rich in nutrients as it is during normal years. Hence, the sea surface temperatures and biological productivity in that region are controlled not so much by the strength of the local wind-induced upwelling as by the stronger wind anomalies in the central Pacific, which regulate the east–west slope of the thermocline.

The difference in the distribution of rainfall over the tropical Pacific between warm and cold years of the ENSO cycle is shown in Fig. 10.25. During El Niño years the ITCZ and the belt of heavy rainfall over the extreme western Pacific intrude into the equatorial dry zone, bringing drenching rains to the Pacific islands that lie close to the equator and to the lowlands of Ecuador, while Indonesia and a number of other areas of the tropics experience drought. The eastward shift in the rain belt during El Niño events is consistent with the eastward shift in the warm pool (Fig. 10.21): it is only at these times that convective updrafts in the atmospheric boundary layer in the central equatorial Pacific are warm and buoyant enough to break through the capping inversion.

The remarkable reproducibility of ENSO-related rainfall anomalies in different El Niño years is demonstrated in Fig. 10.26. That the patterns of tropospheric temperature anomalies shown in these figures are also similar over much of the globe suggests that tropical rainfall anomalies in the equatorial Pacific induce a global response in the tropospheric temperature, geopotential height, and wind fields. This inference is supported by a large body of theory relating to wave propagation in planetary atmospheres and by numerous experiments with global atmospheric models forced with tropical sea-surface temperature anomalies. The response of the global atmosphere to the shifting of the tropical rain belts is reflected in temperature and rainfall anomalies at the Earth's surface at

Fig. 10.22 In January 1998 (top) the 1997–1998 El Niño event was at its height. Because of the weakness of the trade winds at this time, the upwelling of nutrient-rich water was suppressed in the equatorial Pacific. The absence of a green band along the equator in this image is indicative of relatively low chlorophyll concentrations there. By July 1998 (bottom) the trade winds had strengthened and equatorial upwelling had resumed, giving rise to widespread phytoplankton blooms in the equatorial belt. [Imagery courtesy of SeaWiFS Project, NASA/GSFC and ORBIMAGE, Inc.]

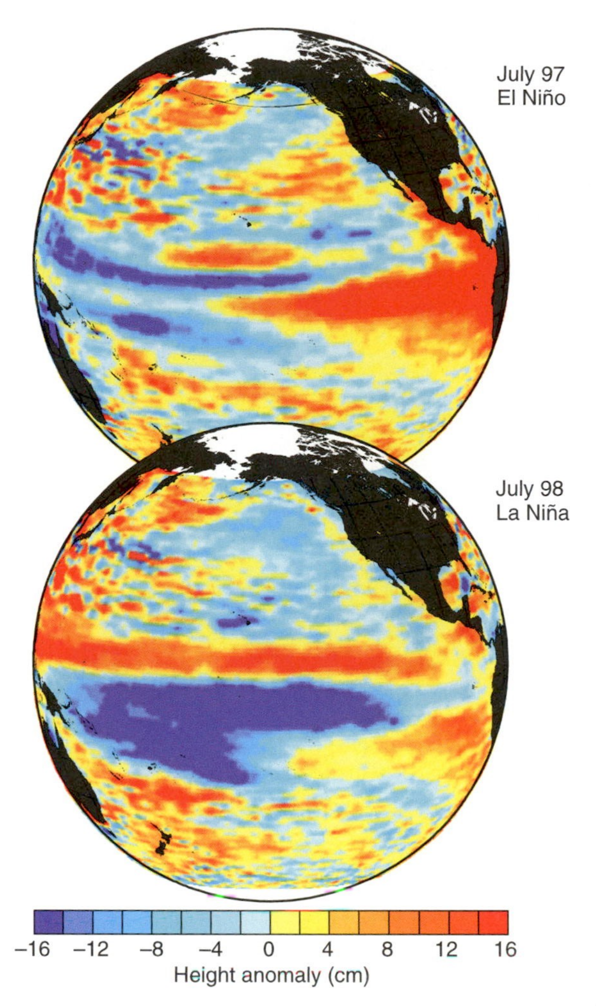

Fig. 10.23 Sea level anomalies measured by the TOPEX/Poseidon altimeter. (Top) July 1997, just after the relaxation of the trade winds following the onset of the 1997–1998 El Niño event. (Bottom) July 1998, a month of unusually strong trade winds. [Courtesy of the NOAA Laboratory for Satellite Altimetry.]

many locations throughout the world. The regional impacts of ENSO are summarized in Fig. 10.27.

Because the ENSO-induced anomalies in the global circulation are due to slowly evolving tropical sea-surface temperature anomalies, they are not subject to the ~2-week predictability limit for weather, which is governed by the atmosphere's capricious internal dynamics. Tropical sea-surface temperature anomalies tend to be highly persistent from the boreal summer through the following winter, and an even better forecast of the wintertime anomalies can be obtained by using statistical prediction schemes or coupled atmosphere–ocean models, both of which exhibit significant skill in ENSO prediction out to a year in advance. Based on a forecast of seasonal mean sea-surface temperature anomalies over the tropical oceans, a variety of statistical and, in some cases, dynamical models are used to predict the corresponding anomalies in rainfall, temperatures, crop yields, oil prices, and other variables.

b. Coupling with the terrestrial biosphere

Atmosphere–ocean coupling affects northern hemisphere climate mainly during its winter season and the ocean plays an important role in it. In contrast, drought and desertification, when they occur in extratropical latitudes, are primarily warm season phenomena. An extended drought popularly known as "the dust bowl" impacted large areas of the United States throughout most of the decade of the 1930s. Over parts of the Great Plains and Midwest many of the 1931–1939 summers were

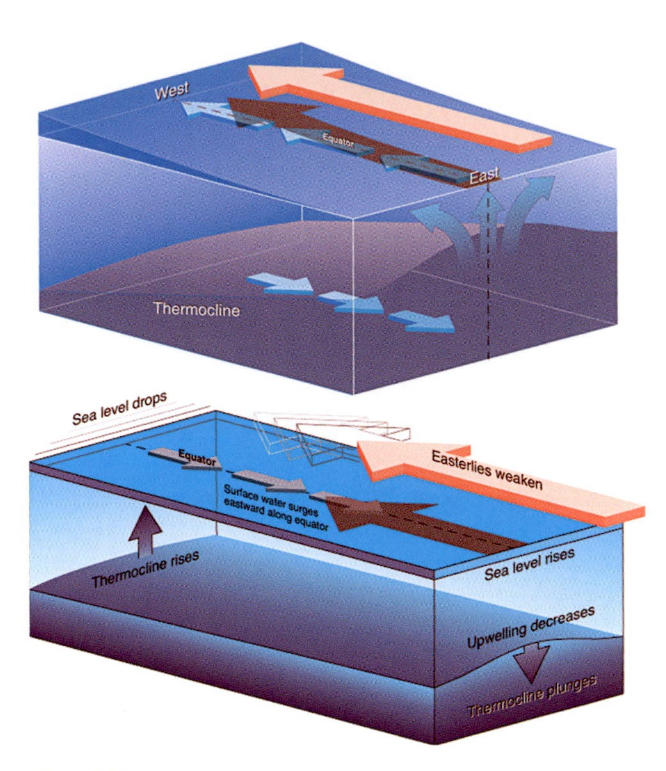

Fig. 10.24 (Top) Schematic showing trade winds (red arrows) maintaining equatorial upwelling and the slope of the equatorial thermocline. (Bottom) Response of equatorial sea-level, currents, and thermocline depth to an abrupt weakening of the trade winds. [From *NOAA Reports to the Nation: El Niño and Climate Prediction*, University Corporation for Atmospheric Research (1994) pp. 12, 14.]

anomalously hot and dry, with daily maxima often in excess of 40 °C, as shown in Fig. 10.28. Large quantities of topsoil were irreversibly lost—blown away in dust storms that darkened skies from the Great Plains as far downstream as the eastern seaboard.

The ENSO cycle modulates the frequency of drought. Over most tropical continents, drought tends to occur more frequently during the warm phase of the ENSO cycle, whereas in the extratropical summer hemisphere it tends to occur somewhat more frequently during the cold phase. The ENSO connection notwithstanding, many of the droughts on timescales of seasons to years appear to be initiated and terminated by random fluctuations in atmospheric circulation, and sustained over long periods of time by positive feedbacks from the biosphere. A few weeks of abnormally hot, dry weather are sufficient to dry the upper layers of the soil, reducing the water available for plants to absorb through their root systems. The plants respond by reducing the rate of evapotranspiration through their leaves during the daylight hours. Reduced evapotranspiration inhibits the ability of the plants to keep themselves and the Earth's surface beneath them cool during the middle of the day, when the incoming solar radiation is strongest, favoring higher afternoon temperatures and it lower humidity within the boundary layer. Because boundary-layer air is the source of roughly half the moisture that condenses in summer rainstorms over the central United States, lower humidity favors reduced precipitation: a positive feedback. Higher daily maximum temperatures, lower humidity, and reduced precipitation all place stress on the plants. If the stress is sufficiently severe and prolonged, the changes in plant physiology become irreversible. Once this threshold is crossed, the earliest hope for the restoration of normal vegetation is the next spring growing season, which may be as much as 9 months away. Throughout the remainder of the summer and early autumn the parched land surface continues to exert a feedback on the atmosphere that acts to perpetuate the abnormally hot, dry weather conditions that caused it in the first place.

The wilting of plants also affects ground hydrology. In the absence of healthy root systems, ground water runs off more rapidly after rain storms, leaving less behind to nurture the plants. Once the water table drops significantly, an extended period of near or above normal precipitation is required to restore it. The remarkable year-to-year persistence of the 1930s drought attests to the remarkable "memory" of the vegetation and the ground hydrology. Hence, once established, an arid climate regime such as the one that prevailed during the dust bowl appears to be capable of perpetuating itself until a well-timed series of rainstorms makes it possible for the vegetation to regain a foothold.

The onset and termination of the 1930s dust bowl are examples of abrupt, but reversible regime shifts between a climate conducive to agriculture and a more desert-like climate. Such shifts have occurred rather infrequently over the United States, but more regularly in semiarid agricultural regions such as the Sahel, northeast Brazil, and the Middle East. If the dry regimes are sufficiently frequent or prolonged, the cumulative loss of topsoil due to wind erosion makes it increasingly difficult for vegetation to thrive, and irreversible *desertification* occurs. The northward expansion of the Sahara desert during the last few centuries of the Roman empire could have involved a series of drought episodes analogous to the dust bowl.

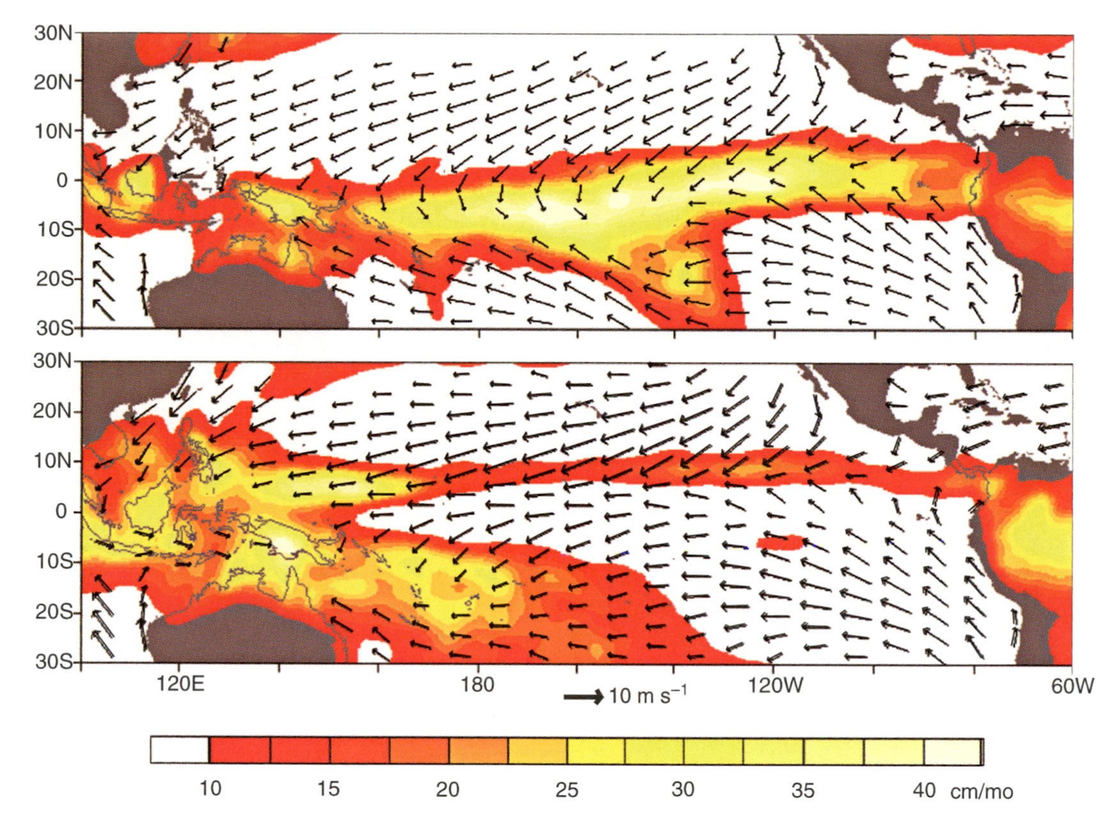

Fig. 10.25 As in Fig. 10.21, but for surface wind and rainfall. The ITCZ and the equatorial dry zone are clearly evident in the lower panel and they are recognizable in a distorted form in the upper panel. In both panels, note the strong correspondence between the regions of heavy rainfall and the regions of convergence in the surface winds. [Ten-meter wind from the European Remote-sensing Satellite (ERs-2) and rainfall from the NCEP Climate Prediction Center Merged Analysis of Precipitation (CMAP). Courtesy of Todd P. Mitchell.]

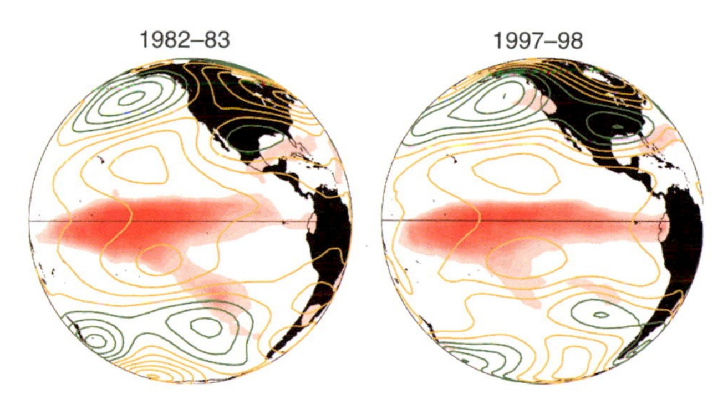

Fig. 10.26 Tropical Pacific rainfall anomalies based on satellite imagery and anomalies in tropospheric temperature as represented by Channel 2 of the microwave sounding unit (MSU) during December–March 1982–1983 and 1997–1998, the strongest El Niño events of the 20th century. Red shading indicates regions of anomalously heavy rainfall, gold (green) contours indicate zero and above (below) normal tropospheric temperatures. Contour interval is 0.5 °C. Note the dipole pattern straddling the region of enhanced rainfall, where the tropospheric temperatures are abnormally high, the prevalence of below normal temperatures over the North Pacific and above normal temperatures over Canada. [Precipitation data is the NCEP Climate Prediction Center Merged Analysis of Precipitation (CMAP). Figure courtesy of Todd P. Mitchell.]

c. Coupling with the cryosphere

The coupling between the atmosphere and the cryosphere exerts a feedback on surface air temperature: that is to say, an increase in surface temperature leads to a decrease in the areal coverage of ice and snow, which tends to lower the albedo and raise the fraction of incoming solar radiation that is absorbed at the Earth's surface, leading to a further increase the surface air temperature. The feedback upon surface air temperature is in the same sense as the temperature perturbation that caused it: hence it is said to be a *positive feedback*. So-called *ice-albedo feedback* contributes to the large variability of surface air temperature over high latitudes, which is evident in the annual cycle (Fig. 10.8) and on timescales ranging from intraseasonal to the scale of the ice ages.

Snow cover and sea ice are the dominant cryospheric participants in climate variability on timescales of centuries or shorter: the more slowly evolving continental ice sheets come into play on millennial and longer timescales. Ice albedo feedback has operated

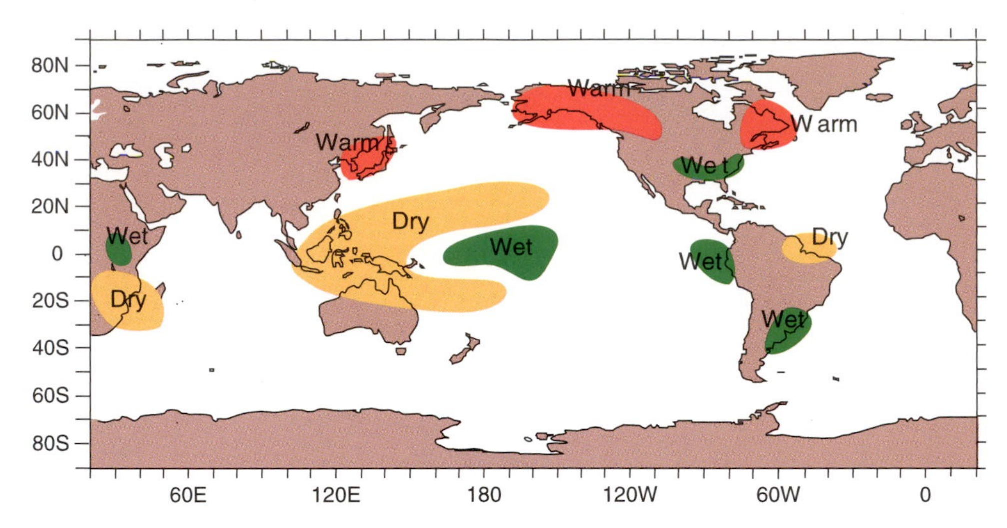

Fig. 10.27 Regional impacts of El Niño on weather and climate during the boreal winter. [Adapted from *Monthly Weather Review*, **115**, p. 1625 (1987) by the NOAA PACS program.]

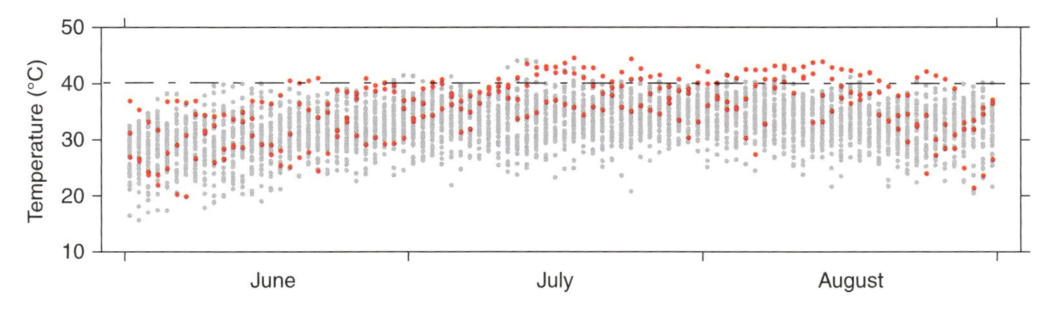

Fig. 10.28 Daily maximum temperature during June–August at Kansas City. Values for "dust bowl" years 1934–1936 are indicated by red dots, and values for all other years by gray dots. [Courtesy of Imke Durre, NOAA National Climatic Data Center.]

only during those relatively short epochs in the Earth's history, such as the last few million years, when the polar regions were glaciated; within those epochs it probably played a more important role during the glacial intervals, when the areal coverage of ice was more extensive, than during interglacials.

Although ice albedo feedback acts locally, its effects are felt globally. Based on estimates of the areal coverage of the continental ice sheets and sea ice at the time of the last glacial maximum ~20,000 years ago, and assuming that the cloud cover was about the same as it is today, the Earth's planetary albedo is estimated to have been higher, by ~0.01 than the current value of 0.305. In Exercise 4.21 the student is asked to show that an increase in planetary albedo from 0.305 to 0.315 would be sufficient to lower the Earth's effective temperature by nearly 1 °C relative to today's value. Ice albedo feedback is believed to have played an essential role in the cycling back and forth between glacial and interglacial epochs in response to the subtle, orbitally induced variations in summer insolation over high latitudes of the northern hemisphere described in Section 2.5.3.

d. Coupling with the Earth's crust

On a time scale of tens of millions of years it is conceivable that the carbonate-silicate cycle discussed in Section 2.3.4, of which volcanic eruptions are a part, could serve to regulate global surface air temperature through the following mechanism. If, for some reason, the temperature of the Earth were to become anomalously warm, weathering of $CaSiO_3$ rocks would be accelerated, making more calcium ions available for carbonate formation [Eq. (2.11)]. Carbonate formation takes up atmospheric carbon dioxide, reducing the greenhouse effect, inducing a temperature response in the opposite sense to the temperature perturbation that initiated the changes. Such a response, called a *negative feedback*, is analogous to a restoring force in a mechanical system, as described in Section 3.6.3.

10.2.3 Externally Forced Climate Variability

Climate variability may also be forced by changes in the sun's emission, by large volcanic eruptions, and by human activities. The amplitude of such externally forced variability depends on the amplitude of the forcing, as well as the sensitivity of the climate system to the forcing, as discussed quantitatively in Section 10.3. If the forcing is sufficiently gradual, like the increase in the sun's emission over the lifetime of the Earth, all the components of the Earth system will remain in equilibrium with it at all times. In contrast, if the forcing changes instantaneously or consists of a short-lived impulse (e.g., as in the case of volcanic eruption), climate will exhibit a transient response with multiple timescales. Discussion of the impact of human activities on climate is reserved for Section 10.4.

a. Solar variability

The intensity of the radiation emitted by the sun varies over a wide range of timescales and the amplitude of these variations is wavelength dependent. Much of this variability is associated with the interrelated phenomena shown in Fig. 10.29 that appear intermittently in *active regions* of the sun.

- *Sunspots* are dark (cool) patches that interrupt the regular pattern of convective cells in the photosphere. Sunspot groups are accompanied by strong magnetic fields with a preferred orientation that reverses between solar hemispheres. Lifetimes of sunspots range from a day up to a few months.
- *Faculae* are bright (hot) spots in the pattern of convective cells that often appear in conjunction with sunspots and are accompanied by strong magnetic fields. Lifetimes of faculae are comparable to those of sunspots.
- *Flares* are intense bursts of ultraviolet and x-ray radiation and high-energy particles emanating from the sun's outer atmosphere within the active regions. They are characterized by strong magnetic fields and violent motions. A typical flare lasts on the order of an hour.

Solar activity is modulated by the ~11-year *solar cycle* illustrated in Fig. 10.30. During the active phase of the cycle, sunspots, faculae and flares are numerous, appearing first at the higher latitudes and then at successively lower latitudes over the course of the next few years. During the quiet phase of the solar cycle, few, if any, of these disturbances are observed. The polarity of the magnetic field in sunspot groups reverses from one cycle to the next. Between successive cycles the polarity of the magnetic field in the sunspot groups reverses; hence a full solar cycle is sometimes regarded as lasting around 22 years.

To understand how disturbances on the sun affect Earth's atmosphere it is necessary to consider in more detail the vertical structure of the sun's outermost layers. Above the photosphere lies the chromosphere, a layer ~2500 km thick in which the temperature increases from a minimum of ~4300 K near the bottom to ~10^5 K at the top. Emanating from the chromosphere in all directions is a stream of extremely hot, ionized particles known as the *solar wind*. In Earth's orbit, particle velocities in the solar

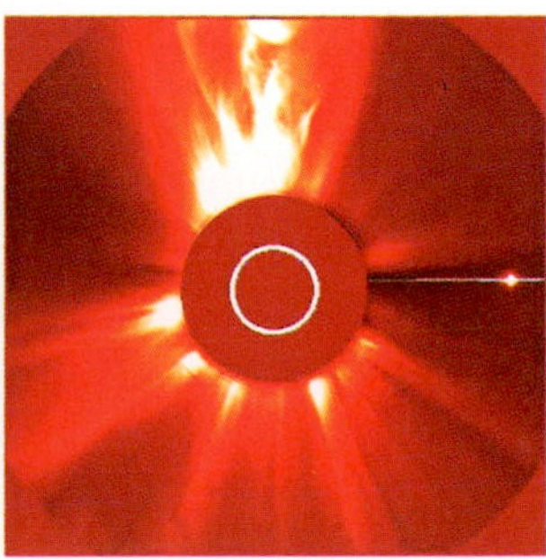

Fig. 10.29 An intense flare in the upper left quadrant of the solar disk that occurred June 6, 2000. The image on the left was taken at the Big Bear Solar Observatory using radiation in the hydrogen alpha line; the second from left is a soft x-ray image obtained from the Yohkoh satellite ("sun beam" in Japanese), the third is another soft x-ray image from the SOHO (Solar and Heliospheric Observatory) satellite, and the panel on the right shows a visible image of the sun taken from the SOHO satellite in which the light coming directly from the sun is blocked (the position of the solar disk is indicated in the image by the white circle). [Assembled by Syun-Ichi Akasofu.]

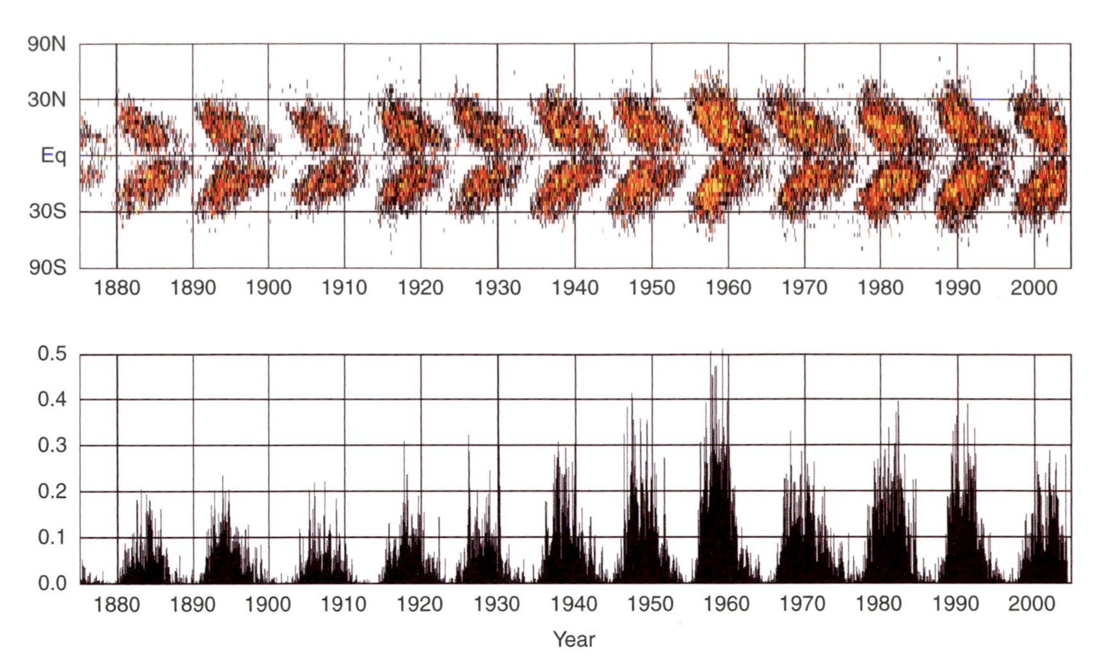

Fig. 10.30 (Top) Fractional area of the photosphere that is covered by sunspots as a function of latitude and time. Yellow shading indicates areal coverage in excess of 1%, red between 0.1 and 1%, and unshaded less than 0.1%. (Bottom) The fractional area of the surface of the entire solar photosphere (in %) covered by sunspots as a function of time. [Courtesy of David Hathaway, NASA Marshall Space Flight Center.]

wind are $\sim$500 km s^{-1} and temperatures sometimes approach 10^6 K.[16] Photospheric radiation scattered by the free electrons in the dense inner portion of the solar wind produces a diffuse luminosity called the *corona* (Fig. 10.29, right), which is visible to the naked eye during solar eclipses. In accordance with Wien's displacement law [Eq. (4.11)], gases in the outer chromosphere and corona emit radiation in the far ultraviolet and x-ray regions of the electromagnetic spectrum. These superheated gases account for virtually all of the sun's emission at wavelengths <0.1 μm.

In contrast to the steadiness of the emission from the photosphere, the radiation from the outer chromosphere and corona increases in response to solar activity and is enhanced greatly at times when flares are in progress. The enhancement is particularly strong in the x-ray region of the spectrum: radiation with wavelengths >0.02 μm is more than an order of magnitude stronger during flares than during quiet periods. However, radiation emitted by the outer chromosphere and corona accounts for less than 1 part in 10^5 of the total solar output. Hence even the strongest flares have no detectable influence on the total flux density of solar radiation reaching the Earth.

In ground-based measurements, such as those described in Box 4.1, temporal variations in the sun's emission are masked by variations in the optical properties of the Earth's atmosphere, which are virtually impossible to quantify. Direct measurements from space-based instruments, which began in 1979, indicate that the sun's total emission varies in synchrony with the solar cycle, with higher values during active intervals than during quiet intervals, as shown in Fig. 10.31. However, the peak-to-peak amplitude of these variations is only $\sim$1 W m^{-2}, less than 0.1% of the mean value of the emission, too small to produce even a 0.1 K change in the Earth's equivalent blackbody temperature.

Upon close inspection of Fig. 10.30 it is evident that the areal coverage of sunspots varies from one solar cycle to the next and that the cycles from 1880 to 1940 were less pronounced than those later in the record. The sun was even less active during the first few decades of the 19th century, and there is compelling historical evidence of an almost complete

[16] High-energy particles emitted by the sun in flares reach the Earth a few days later. The resulting enhancements of the solar wind often produce auroral displays and interfere with radio and electrical power transmission.

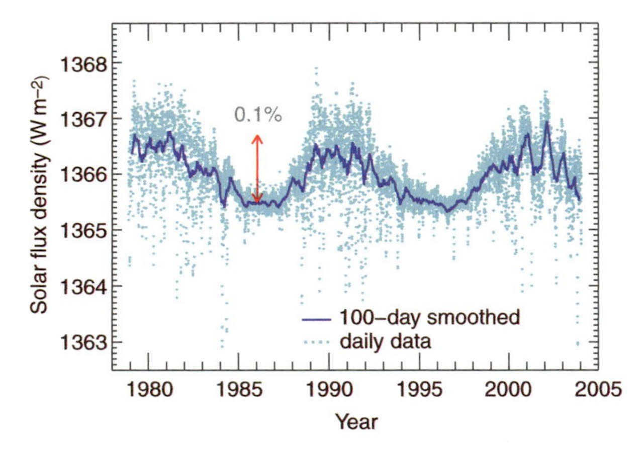

Fig. 10.31 Time variations in the flux density of solar radiation incident on the Earth as inferred from space-based measurements from a variety of instruments. [Courtesy of Judith Lean.]

absence of sunspots from 1645 to 1715, a period referred to as the *Maunder*[17] *minimum* in solar activity, as shown in Fig. 10.32. Solar physicists believe that these extended intervals of reduced solar activity may have been marked by reductions in total emission that were larger, by a factor of 10–30, than those observed in recent decades during the quiet years of the solar cycle. If this interpretation is correct, the consequent decrease in Earth's equivalent blackbody temperature could have been large enough to account for the relatively cold conditions that prevailed during the so-called *Little Ice Age*.

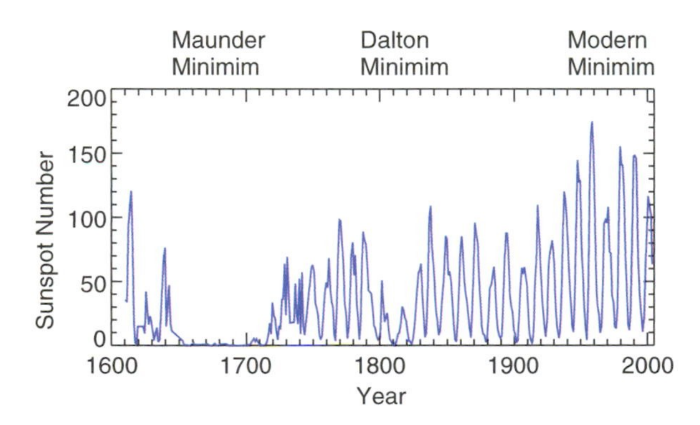

Fig. 10.32 Extended time series of *sunspot number* (defined here as the number of sunspots on the sun's (visible light) disk plus 10 times the number of sunspot groups) showing intervals of low solar activity. [Courtesy of Judith Lean.]

b. Volcanic eruptions

The impact of geothermal energy released in volcanism upon the global heat balance is negligible. The climatic impacts of volcanic eruptions are mainly due to the radiative effects of sulfate aerosols formed from SO_2 emissions. Scavenging by cloud droplets cleanses the troposphere of such volcanic debris within a matter of a few weeks. Hence, it is only major eruptions whose plumes penetrate upward into the lower stratosphere that are capable of significantly perturbing the Earth's climate. A recent example of a volcanic eruption that exerted a descernible influence on global climate is the eruption of Mt. Pinatubo in the Philippines in June 1991 (Fig. 10.33).

Fig. 10.33 The plume from one of a series of eruptive events of Mt. Pinatubo that preceded the massive June 15, 1991 eruption. The event pictured here began at 0841 LT June 13. The photograph was taken 7 min after the start of the eruption. The plume of hot gases is just beginning to spread out at the top as it approaches the tropopause. The cloud from this eruption eventually reached altitudes in excess of 24 km, the maximum range of the weather radar. [Courtesy of Richard Hoblitt, U.S. Geological Survey.]

[17] **Edward Walter Maunder** (1851–1928) English astronomer. After working briefly in a bank he became a photographic and spectroscopic assistant at the Royal Greenwich Observatory, where he carried out his studies of the sun. Also involved in the debate on the canals of Mars.

Figure 10.34 shows the dispersion of the sulfate aerosols that formed in the plume of the Pinatubo eruption. The top panel shows the global distrribution of aerosol concentrations in August 1990, a year before the eruption. Concentrations are enhanced immediately downstream of local source regions such as the Sahara Desert, but they are low elsewhere. This pattern, which is representative of background conditions, is dominated by tropospheric aerosols. The middle panel shows concentrations in August 1991, 2 months after the eruption. Compared to the previous August, concentrations throughout the tropics are strongly enhanced due to the presence of sulfate aerosols in the lower stratosphere. The rapid dispersion of stratospheric aerosols in the east–west direction is due to the strongly sheared zonal flow in the lower stratosphere. The dispersion of the aerosol cloud from the tropics into higher latitudes took place much more slowly. A year later (August 1992, Fig. 10.34, bottom panel), concentrations of aerosols in the tropics have decreased markedly relative to the previous year, but over the globe as a whole, concentrations remain higher than the background levels indicated by the top panel. By this time the stratospheric aerosols are well mixed globally and the gradients in concentration are mainly a reflection of regional sources of tropospheric aerosols.

For a period of several months following the Pinatubo eruption, scattering by the aerosol cloud reduced the flux density of direct solar radiation incident on the Earth's surface by ~30%. Because most of the scattering was in the forward direction, this reduction was compensated by an increase in

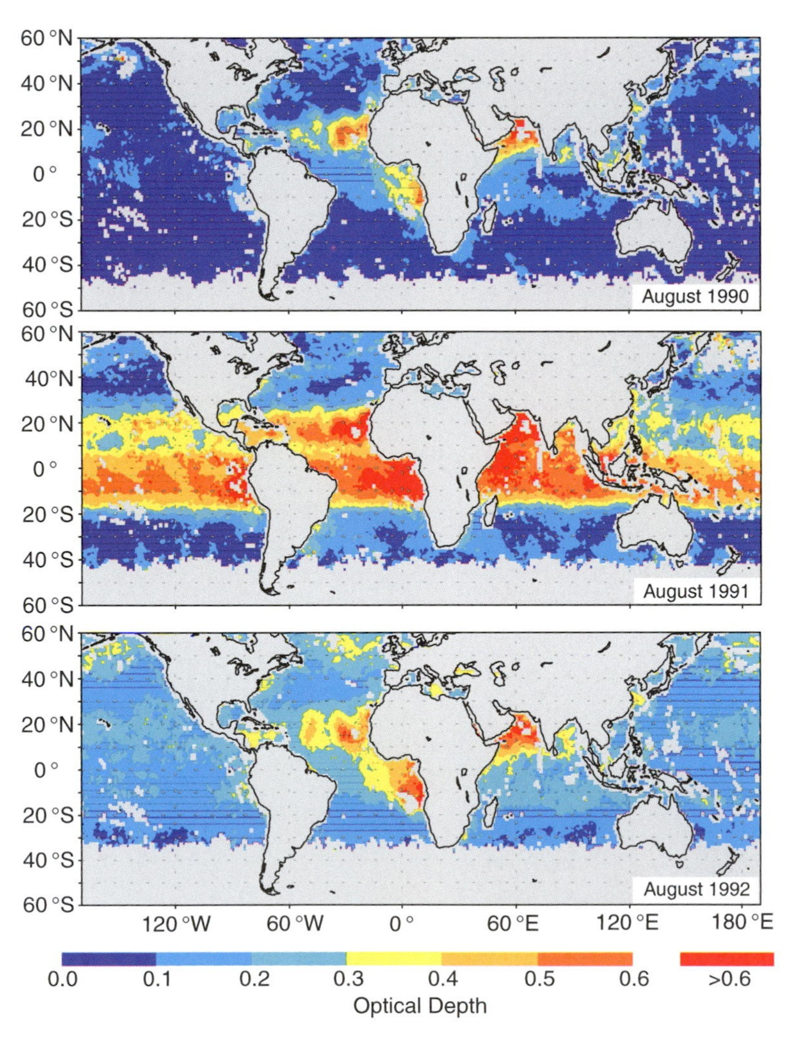

Fig. 10.34 Total concentrations of atmospheric aerosols based on NOAA/AVHRR satellite imagery. The colors represent the monthly averages of optical depth at wavelength of 0.63 μm for cloud-free pixels over the oceans, as inferred from the monochromatic intensity of the reflected radiation at that wavelength. [Data from *J. Geophys. Res.*, **102**(D14), 16923-16934, 1997. Copyright 1997 American Geophysical Union. Reproduced/modified by permission of American Geophysical Union. Courtesy of Alan Robock.]

the flux density of diffuse solar radiation reaching the surface. The absorption and back-scattering by the aerosol layer reduced the total (direct plus diffuse) solar radiation reaching the surface by ~3%. In response to the reduction in total incident solar radiation, global-mean surface air temperature tends to be significantly lower in the wake of major volcanic eruptions than in the long-term average, as shown in Fig. 10.35.

The absorption of incident solar radiation by sulfate aerosols results in a pronounced warming of the lower stratosphere in response to volcanic eruptions, as shown in Fig. 10.36. The widths of the spikes in the temperature time series suggests that the residence time of the aerosols is on the order of 1–2 years, consistent with Fig. 10.34 and the discussion in Section 5.7.3.

The duration of the cooling at the Earth's surface in the wake of volcanic eruptions, as indicated by Fig. 10.35, appears to be somewhat longer than the 1- to 2-year residence time of the stratospheric aerosols. The difference is attributable to the involvement of the ocean mixed layer, with its large heat capacity. Throughout the time interval that the aerosols remain

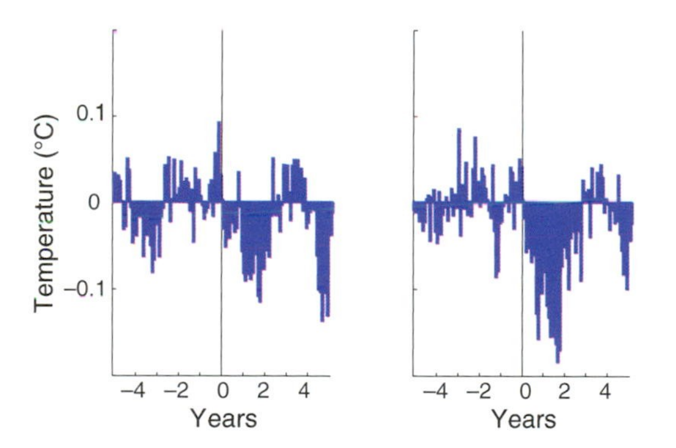

Fig. 10.35 (Left) Composite of global-mean surface air temperature around the time of seven major volcanic eruptions.[18] The time of the eruption corresponds to Year 0 on the x axis and positive values on the x axis denote time after the eruption. The 2- to 3-year-long cool interval beginning at the time of the eruption shows that the eruptions affected global-mean temperature. (Right) The same temperature data, adjusted to remove the temperature variability that is attributable to El Niño, the annular mode and the PNA pattern, so as to more clearly reveal the radiatively induced temperature change following these eruptions. The temperature scales for the two panels are the same. [Courtesy of David W. J. Thompson.]

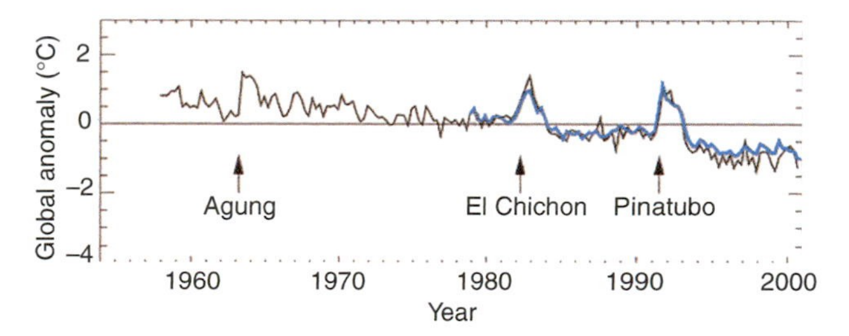

Fig. 10.36 Time series of the globally-averaged temperature in the lower stratosphere. The blue curve is based on the lower stratospheric channel of the microwave sounding unit, which is representative of the layer 15–20 km above the Earth's surface and the black curve is based on radiosonde data for the same layer. [Adapted from Intergovernmental panel on climate change, *Climate Change 2001: The Scientific Basis*, Cambridge University Press, p. 121 (2001).]

in the stratosphere, the Earth's surface and the air immediately above it tends to be anomalously cool. Latent and sensible heat fluxes through the ocean surface tend to be enhanced in response to the enhanced air–sea temperature difference. The enhanced fluxes reduce the amplitude of the atmospheric cooling by extracting heat from the ocean mixed layer and depositing it in the atmosphere. As the ocean mixed layer loses heat it becomes anomalously cool. With the removal of the aerosols the insolation incident upon the troposphere returns to normal and the anomalous fluxes at the air–sea interface reverse direction. An additional year or two is required for the ocean mixed layer to regain the heat that it lost while the aerosols were present in the atmosphere. Not until the ocean comes back into equilibrium does global-mean surface air temperature return to its preeruption level. In a similar manner, the exchange of heat between the atmosphere and the ocean mixed layer damps and delays the response of local sea-surface temperature to the annual cycle and to random month-to-month and year-to-year atmospheric variability.

10.3 Climate Equilibria, Sensitivity, and Feedbacks

To gain a more quantitative understanding of the climate response to a prescribed forcing, it will be useful to place the concepts of climate feedbacks and climate sensitivity into a mathematical framework.

[18] The seven largest volcanic eruptions in the modern record: Krakatoa (August 1883), Tarawera (June 1886), Pelee/Soufriere/Santa Maria (May–October 1902), Katmai (June 1912), Agung (March 1963), El Chichon (April 1982), and Pinatubo (June 1991).

This same framework will provide a basis for understanding the nature of "climate surprises"—instances in which a modest (or gradual) climate forcing could, at least in principle, give rise to a large and abrupt climate change that would be irreversible.

In the interest of brevity we focus on variations in only one climate variable, namely global-mean surface air temperature T_s, which we treat as being governed by three factors: the incident solar radiation, the planetary albedo, and the strength of the greenhouse effect averaged over the Earth's surface. A number of auxiliary variables (e.g., the concentrations of atmospheric water vapor and ozone and the fractions of the surface area of the Earth covered by clouds with tops in various altitude ranges and by ice and snow) enter into the determination of the albedo and the strength of the greenhouse effect. We will treat the forcing and the response as spatially uniform and consider only globally averaged quantities.

The *radiative forcing* F is defined as the net downward flux density or irradiance at the top of the atmosphere[19] that would result if it were applied instantaneously (i.e., without giving the surface air temperature or the vertical temperature profile any time to adjust to it). For example, if the flux density of solar radiation were to increase by the amount dS, the net radiation at the top of the atmosphere would initially be downward and equal to dS. Surface air temperature T_s would gradually rise in response to this imbalance until the outgoing flux of the atmosphere increased by the amount dS, at which time the Earth system can be said to have equilibrated with the forcing. When the system reached its new equilibrium, T_s would have increased by the amount dT_s in response to the solar forcing dS.

In a similar manner, an increase in the greenhouse effect can be expressed as a net downward irradiance dG at the top of the atmosphere, equal to the initial decrease in the upward irradiance of longwave radiation due to the enhanced blanketing effect of the atmosphere. As in the previous example, T_s would rise in response to the imbalance until the outgoing irradiance at the top of the atmosphere became equal to the incoming solar radiation. When the system had fully adjusted to the change, T_s would have increased by the amount dT_s in response to the enhanced greenhouse forcing dG.

The sensitivity of T_s to the radiative forcing F, namely $\lambda \equiv dT_s/dF$, is called the *climate sensitivity*. Consistent with its definition in terms of the total derivative, the climate sensitivity takes into account the changes in the auxiliary variables y_i (the concentration of atmospheric water vapor, the fraction of the Earth's surface covered by snow and ice, low clouds, etc.) that influence T_s. The various processes that play a role in determining the climate sensitivity can be evaluated by expanding the total derivative using the chain rule, which yields

$$\lambda = \frac{dT_s}{dF} = \frac{\partial T_s}{\partial F} + \sum_i \frac{\partial T_s}{\partial y_i}\frac{dy_i}{dF} \tag{10.5}$$

The partial derivative term $\partial T_s/\partial F$ is the climate sensitivity λ_0 that would prevail in the absence of feedbacks involving the auxiliary variables:

$$\lambda_0 \equiv \frac{\partial T_s}{\partial F} \approx \frac{dT_E}{dF}$$

where T_E is the equivalent blackbody temperature, as estimated in Exercise 4.6.

Exercise 10.3 Estimate the sensitivity of the Earth's equivalent blackbody temperature to a change in the solar radiation F_s incident upon the the top of the atmosphere.

[19] The current operational definition of *radiative forcing* uses the tropopause rather than the top of the atmosphere as a reference level with the proviso *"after the stratosphere has come into a new thermal equilibrium state."* This definition is motivated by the following considerations:

1. The tropospheric temperature profile, surface temperature, and the temperature of the ocean mixed layer are strongly coupled because the tropical lapse rate is constrained to be close to moist adiabatic. In contrast, the stratosphere, with its stable stratification, is decoupled from the underlying media. Hence, variations in surface temperature are related more strongly to the radiative forcing at the tropopause than at the top of the atmosphere.
2. The stratosphere equilibrates with an abrupt change in radiative forcing within a matter of months, whereas the troposphere requires decades to fully adjust because of the large thermal inertia of the oceans.

Once the stratosphere has come into thermal equilibrium with the new radiative forcing, the net irradiance through the tropopause and the top of the atmosphere are the same.

Solution: From the Stefan–Boltzmann law (Eq. 4.12)

$$T_E = \left(\frac{F_s}{\sigma}\right)^{1/4}$$

or, taking the natural log of both sides,

$$\ln T_E = \frac{1}{4}\ln F_s - \frac{1}{4}\ln\sigma$$

Taking the differential yields

$$\frac{dT_E}{T_E} = \frac{1}{4}\frac{dF_s}{F_s}$$

Hence,

$$\frac{dT_E}{dF_s} = \frac{1}{4}\frac{T_E}{F_s}$$

Exercise 4.6 shows that for the Earth $F_s = 239.4$ W m^{-2} and $T_E = 255$ K. Hence,

$$\frac{\partial T_s}{\partial F_s} = 0.266\ \mathrm{K(W\ m^{-2})^{-1}}$$

Or, taking the reciprocal, it can be inferred that the Earth's equivalent black body temperature rises 1 K for each 3.76 W m^{-2} of (downward) radiative forcing at the top of the atmosphere. ■

The changes in the auxiliary variables in the last term in (10.5) are a consequence of their temperature dependence. Hence,

$$\frac{dy_i}{dF} = \frac{dy_i}{dT_s}\frac{dT_s}{dF}$$

Substituting into (10.5) we obtain

$$\frac{dT_s}{dF} = \frac{\partial T_s}{\partial F} + \frac{dT_s}{dF}\sum_i f_i \qquad (10.6)$$

where

$$f_i = \frac{\partial T_s}{\partial y_i}\frac{dy_i}{dT_s} \qquad (10.7)$$

are dimensionless *feedback factors* associated with the various feedback processes to be discussed in the following subsections. The feedback factor f_i will be positive if the two derivatives of which it is comprised in (10.7) are of like sign, and it will be negative if they are of opposing sign. [For example, if a rise in T_s results in a decrease in y_i (as is the case for planetary albedo) and if a decrease in y_i favors a further rise in T_s, then the feedback is positive.] The feedback factors are additive, i.e.,

$$f = \sum_i f_i \qquad (10.8)$$

where the algebraic sign of the feedback is taken into account in the summation.

Solving (10.6) for the sensitivity of T_s to the forcing F, we obtain

$$\frac{dT_s}{dF} = \frac{\partial T_s/\partial F}{1-f} \qquad (10.9)$$

Hence, the gain $g \equiv \lambda/\lambda_0$ due to the presence of climate feedbacks is

$$g = \frac{1}{1-f} \qquad (10.10)$$

provided that $f < 1$. A value of $f \geq 1$ corresponds to the case of infinite sensitivity, in which even an infinitesimal forcing may cause the climate system to diverge from its present equilibrium state and seek a new one, as in the mechanical analogs pictured in Fig. 3.17.

Exercise 10.4 Estimate the apparent climate sensitivity $\delta T_s/\delta F$ based on estimates of the differences in T_s and F between current climate and the climate at the time of the last glacial maximum (LGM) around 20,000 years ago, using the relation

$$\frac{\delta T_s}{\delta F} = \frac{T_s\,(\text{current}) - T_s\,(\text{LGM})}{F\,(\text{current}) - F\,(\text{LGM})}$$

Solution: Global-mean surface air temperature is estimated to have been ~5 °C lower at the time of the LGM than it is today. On the basis of ice core data it is known that atmospheric CO_2 concentrations were 180 ppmv, slightly less than half their current values. On the basis of radiative transfer models it is estimated that the climate forcing due to a doubling of the CO_2 concentration is 3.7 W m^{-2}. On the basis of what is known about the coverage of the continental ice sheets and the extent of sea ice at the time of the LGM, it is estimated that the planetary albedo was ~0.01 higher at the time of the LGM than it is today. Assuming that the flux density of solar radiation at the time of the LGM is the same as the current value (342 W m^{-2}), the corresponding increase in the flux of reflected solar radiation at the top of the atmosphere is $342 \times 0.01 = 3.4$ W m^{-2}. Substituting these values into the above expression yields

$$\frac{\delta T_s}{\delta F} = \frac{5\text{ K}}{(3.7 + 3.4)\text{ W m}^{-2}} = 0.70\text{ K per W m}^{-2}$$

Comparing this result with the climate sensitivity in the absence of feedbacks $\lambda_0 = 0.266$ (W m^{-2})$^{-1}$, as computed in Exercise 10.3, and the apparent sensitivity of the climate system is enhanced by a factor of $0.70/0.266 = 2.7$ due to the presence of feedbacks. ■

10.3.1 Transient versus Equilibrium Response

Because of the large heat capacity of the Earth system (in particular, the oceans and the cryosphere), global-mean surface air temperature exhibits a delayed response to climate forcing. An equilibrium response to an abrupt change in the forcing would be achieved only after all components of the system have had adequate time to adjust to the change. The adjustment time is different for different components of the Earth system: the atmosphere adjusts to changes in climate forcing within a matter of a few months, the ocean mixed layer requires a few years, the full depth of the ocean requires centuries, and the continental ice sheets perhaps even longer. The respective adjustment times depend on both the heat capacity of the various components of the Earth system and on the climate sensitivity.

We can gain some insight into the nature of this adjustment process by considering the effect of including a hypothetical ocean mixed layer whose temperature adjusts instantaneously to changes in surface air temperature. With this assumption we can treat the ocean mixed layer as a slab with mean temperature $T = T_0 + T'$, where T_0 is the equilibrium temperature before the forcing F is applied and T' is the time varying (i.e., transient) response to a change Q' in the climate forcing. Based on the energy balance at the top of the atmosphere, we can write

$$c\frac{dT'}{dt} = -\frac{T'}{\lambda} + Q' \qquad (10.11)$$

where c is the heat capacity of the slab in units of J m^{-2} K^{-1} averaged over the surface area of the Earth and λ is the climate sensitivity dT_s/dF. The left-hand side of (10.11) is the rate at which energy is stored in the slab, and the right hand side as the imbalance between the forcing Q' and the increase in outgoing longwave radiation at the top of the atmosphere due to the warming of the slab.

Now let us assume that the perturbation in the forcing Q' is "turned on" at time $t = 0$ and maintained at a constant value after that time. In this case (10.11) can be rewritten as

$$\frac{dT'}{dt} + \frac{T'}{\tau} = \frac{Q'\lambda}{\tau}$$

where $\tau = c\lambda$, and solved to obtain

$$T' = \lambda Q'\left(1 - e^{-t/\tau}\right) \qquad (10.12)$$

Hence, after the forcing is turned on, T' exponentially approaches the equilibrium solution $\lambda Q'$, increasing rapidly at first and then more gradually as the solution is approached. The *e*-folding timescale for the approach to the equilibrium solution is proportional, not only to the heat capacity of the mixed layer, but also to the climate sensitivity. Hence the existence of positive climate feedbacks lengthens the time it takes the climate system to equilibrate with a change in the forcing. It can be shown that for a linear rate of increase in climate forcing, the response of T' is also linear and delayed by the same response time $\tau = c\lambda$.[20]

The time delay in the response to climate forcing due to the heat capacity of the ocean mixed layer is on the order of a decade or less (see Exercise 10.21). The heat capacity of the oceans as a whole is about 50 times larger than that of the ocean mixed layer; the continental ice sheets also possess a high effective heat capacity because of the large magnitude of the latent heat of fusion of water. If the atmosphere exchanged heat freely with these large reservoirs, the response time, as defined in the context of (10.11), would be on the order of centuries or longer. Short-term perturbations of the climate system such as volcanic eruptions would be almost entirely damped out and the impacts of longer term changes in forcing, such as those associated with greenhouse warming, would become apparent only over the course of centuries.

The atmosphere does, in fact, exchange significant quantities of heat with these large reservoirs, but the rates of exchange are much slower than with the ocean mixed layer. The thermohaline circulation ventilates the deep layers of the ocean on timescales of centuries,

[20] See D. L. Hartmann, *Global Physical Climatology*, Academic Press, Section 12.6 (1994).

and the mass of the continental ice sheets responds to imbalances between accumulation and melting on a similar timescale. Exchanges of energy with these reservoirs are reflected in small, but measurable imbalances in the net radiation at the top of the atmosphere. Coincidentally, the warming of the oceans is reflected in a rise of sea level due to the expansion of water as it warms, and the shrinkage of the continental ice sheets causes an additional sea level rise due to the increase of the mass of the oceans. Hence, measurements of sea level change can be used to constrain estimates of the rate of energy exchange with these reservoirs.

Because the exchange of energy between the atmosphere and the larger reservoirs in the Earth system takes place gradually, rather than instantaneously, the response to climate forcing is felt almost immediately. However, the full, *equilibrium response* to a prescribed steady-state forcing is not realized until these large reservoirs have had time to equilibrate with the forcing. For example, if atmospheric CO_2 concentrations were to continue to rise at their present rate until they double (relative to preindustrial concentrations) some time late in the 21st century and remain constant after that, the *transient response* observed at the time of the doubling would not be as large as the *equilibrium response* observed several centuries later.

10.3.2 Climate Feedbacks

It is the sum of the feedback factors f_i that determine the climate sensitivity. Here we consider the contributions of some of the more important individual feedback factors to that sum, including ones that might be negative.

a. Water vapor feedback

As a consequence of the Clausius–Clapeyron equation (see Section 3.7.3), the saturation vapor pressure of water increases exponentially with temperature at a rate of $\sim$7% K^{-1} in the temperature range of interest. If the distribution of relative humidity were to remain constant as the temperature rises, atmospheric water vapor concentrations would increase with temperature at a roughly comparable rate. Higher concentrations of water vapor, the atmosphere's most important greenhouse gas, favor higher surface air temperatures. Based on relatively straightforward radiative transfer calculations under the assumption of constant relative humidity, the feedback factor for this process is estimated to be $\sim$0.5, which, from (10.10), corresponds to a gain of 2 (i.e., a doubling of the response of T_s to a prescribed forcing F) if no other feedback processes were operative.

Due to the nonlinearity of the Clausius–Clapeyron equation (3.92) the strength of the water vapor feedback increases with temperature. If the radiative forcing ever became strong enough to raise the tropical sea-surface temperature from its present value of $\sim$28 °C to above 60 °C, the feedback factor would approach unity, setting the stage for a *runaway greenhouse*. Such a catastrophe, which may have occurred on Venus, would ultimately lead to the evaporation of the entire world's oceans, creating a massive atmosphere consisting mostly of steam, with surface temperatures in excess of 1000 K!

The timescale of the atmospheric branch of the hydrological cycle is so short that the water vapor feedback can be considered to be virtually instantaneous. It sets in rapidly enough to come into play even in the transient response to impulsive forcing such as volcanic eruptions. A by-product of the enhanced water vapor feedback is the intensification of the hydrologic cycle. Climate models indicate that in a warmer world, when it rains it rains harder, and evaporation is also more rapid.

b. Cloud forcing and feedbacks

Clouds reflect a fraction of the solar radiation that would otherwise be absorbed at the Earth's surface and they also contribute to the greenhouse effect. The relative importance of these two competing effects is largely determined by the height and by the optical thickness of the clouds in the shortwave part of the spectrum. For deep cloud layers, such as those typically associated with tropical convection, the two effects nearly cancel one another so that the *net radiative cloud forcing* is small. In contrast, more reflective cloud decks at the top of the planetary boundary layer, whose tops are typically only $\sim$10 °C colder than the underlying water or land surface, contribute much more strongly to the albedo than to the greenhouse effect, and thus produce a large negative net radiative forcing (Fig. 10.37). If the areal coverage of these cloud decks were to increase as the climate warmed, that would constitute a negative feedback on global-mean surface air temperature. However, of the coverage were to decrease as the climate warms, that would constitute a positive feedback.

Stratus and stratocumulus decks tend to occur in regions of subsidence in which the sea surface and

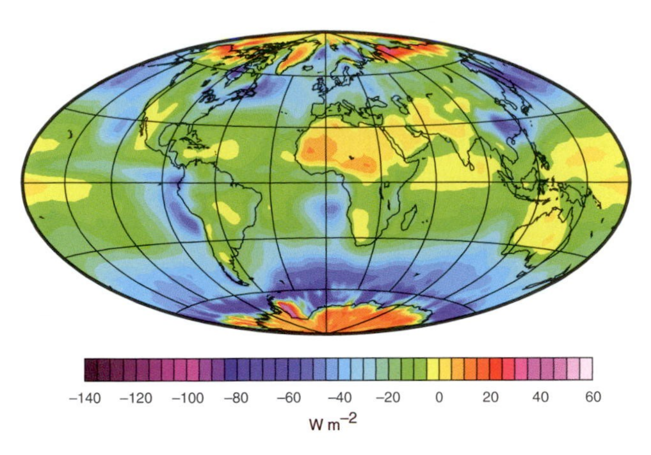

Fig. 10.37 Annual mean net cloud radiative forcing, which is estimated by comparing reflected solar radiation and outgoing solar radiation, as in Fig. 4.35, but computing averages for all pixels and cloud-free pixels separately at each grid point and subtracting the latter from the former. For a more detailed explanation see Exercise 10.22. [Based on data from the NASA Earth Radiation Budget Experiment. Courtesy of Dennis L. Hartmann.]

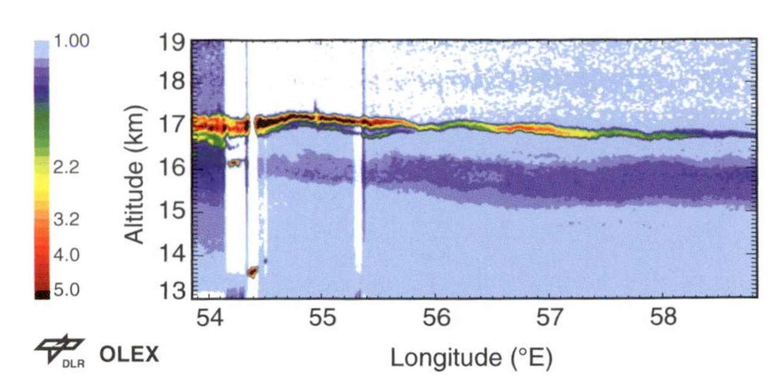

Fig. 10.38 Thin layers of subvisible clouds in the tropical upper troposphere as inferred from Lidar measurements from an aircraft, at a wavelength of 1.064 μm The colored shading indicates the intensity of the light backscattered by cloud droplets light blue shading indicates minimal scattering and white indicates missing data. This observation was taken over the Indian Ocean during a field experiment on board a research aircraft. [From *J. Geophys. Res.*, **107**(D16), 4314–4329, 2002. Copyright 2002 American Geophysical Union. Reproduced/modified by permission of American Geophysical Union. Courtesy of H. Flentje, Deutschen Zentrum für Luftund Raumfahrt (DLR), Germany.]

the atmospheric boundary layer are cool relative to the overlying free atmosphere (i.e., stably stratified). It is not clear how much and in what sense the areal coverage of these clouds would be likely to change in response to global warming.

At the opposite end of the spectrum from boundary layer stratus cloud decks, with respect to cloud forcing, are high thin, *subvisible cirrus* cloud decks with very cold tops, which have been detected in the upper troposphere well above the tops of convective clouds (Fig. 10.38). Subvisible cirrus warm the Earth's surface because they transmit downward solar radiation without significant scattering or absorption, while blocking a larger fraction of the outgoing longwave radiation and reradiating it to space at very low temperatures. The cloud forcing by subvisible cirrus is limited by their small optical thickness which is generally less than 0.1. Their areal coverage remains to be determined.

c. *Ice-albedo feedback*

Ice-albedo feedback has already been discussed in the previous section in the context of coupled climate variability. Its importance depends on the fractional area of the Earth covered by ice. At times like the present, when perennial snow and ice are largely confined to the polar regions, the decrease in fractional areal coverage $\breve{A}$ that would occur in response to an incremental change in global-mean surface temperature [the counterpart of dy_i/dT_s in (10.7)] is relatively modest.

If ice is assumed to cover the area $\breve{A}$ poleward of latitude ϕ in which the Earth's surface is colder than some threshold temperature T^* (presumably near the freezing point), then

$$\frac{d\breve{A}}{dT_s} = m \cos\phi$$

where m is the meridional temperature gradient in the vicinity of the ice edge and the factor $\cos\phi$ reflects the increasing length of latitude circles as the ice edge moves equatorward. Provided that m does not vary strongly with latitude, it follows that the incremental increase in $\breve{A}$ per unit change in T_s decreases with latitude, roughly in proportion to $\cos\phi$. This simple geometric relationship predicts that the strength of the ice-albedo feedback should increase as the ice edge advances toward lower latitudes and vice versa. Simple energy balance model calculations indicate that if the ice line were to advance far enough equatorward, the ice-albedo feedback would become so strong that the denominator in (10.9) would vanish and the climate sensitivity would become infinite. If this situation were to be realized in nature, Earth would abruptly become completely ice covered, as in the *Snowball Earth* scenario described in Box 2.2.

While the sign of the ice-albedo feedback is certainly positive, its magnitude is highly uncertain because it is strongly seasonally dependent and it is coupled to the behavior of clouds, land surface hydrology, and land vegetation over high latitudes. Changes in albedo have the largest impact on the surface energy balance during the summer season, when the insolation is strongest. The presence of extensive stratus cloud decks over the polar oceans during summer reduces the strength of the ice-albedo feedback during this season and renders it more difficult to estimate quantitatively.

The direct effects of changes in snow cover over land are felt mainly in connection with the timing of the spring thaw, but the indirect effects on surface temperature via the ground hydrology linger into summer. An earlier spring thaw increases the possibility that the soil will dry out during summer, allowing for the possibility of higher daytime temperatures. Higher summertime temperatures and longer growing seasons enable shrubs and trees to encroach on areas of tundra, increasing the surface roughness and reducing the albedo. Quantitatively modeling such a diverse array of interactions is a formidable challenge.

d. Carbon dioxide feedback

In the context of greenhouse warming, CO_2 is the dominant radiative forcing, but in the context of glacial–interglacial cycles, changes in radiative forcing due to variations in atmospheric CO_2 concentrations constitute a feedback. Just what caused CO_2 concentrations to increase and decrease nearly in synchrony with temperature on a timescale of thousands of years is still a matter of speculation. Most of the proposed mechanisms involve long-term changes in the oceanic thermohaline circulation that could affect the rate at which the deeper layers of the ocean are ventilated and organic carbon droppings from the "biological pump" discussed in Section 2.3 are recycled back to the surface. If the ventilation rate slowed down significantly, it is conceivable that enough organic carbon could be stored in the deep oceans to account for the ~80 ppmv depression in atmospheric CO_2 concentrations during the ice ages relative to those during interglacial epochs. Whatever their cause, it is clear that the large positive feedback associated with these fluctuations in atmospheric CO_2 concentrations significantly amplify the climate response to orbital forcing.

10.2 Daisyworld

To illustrate the concepts of feedbacks, equilibria, and abrupt climate change, let us consider the climate of *Daisyworld*, an idealized spherical planet with no atmosphere, warmed by a distant sun. The surface temperature of *Daisyworld* is uniform and in radiative equilibrium with the incoming solar radiation. The surface of the planet is black, apart from the part that is covered by perfectly white daisies, which are distributed randomly over the surface of the planet in small enough clumps that the planet looks uniformly gray when viewed from space. Hence, the albedo of *Daisyworld* is numerically equal to the areal coverage of daisies.

Daisies are able to grow only within some finite range of surface temperatures $T_1 \leq T \leq T_2$, and their areal coverage is a maximum at some intermediate temperature, as indicated by the red curve in Figure 10.39. By mediating the planetary albedo, the areal coverage of daisies, in turn, affects the surface temperature of the planet. From the Stefan–Boltzmann law (4.12), under radiative equilibrium conditions

$$\sigma T^4 = (1 - A)F \tag{10.13}$$

where σ is the Stefan–Boltzmann constant, T is the surface temperature of the planet, A is the fractional areal coverage by daisies and F is the flux density of solar radiation averaged over the surface the planet. The effect of the areal coverage of daisies upon the temperature of the planet is indicated by the blue curves in Figure 10.39, each of which corresponds to a different value of F.[21]

Points P and P', where the blue and red curves intersect, represent equilibrium states, for which the fractional area coverage of daisies is consistent with the temperature, and the temperature, in turn, is consistent with the areal coverage of daisies.

Continued on next page

[21] To make the plot as simple as possible, the blue curves are represented as straight lines. In accordance with the Stefan–Boltzmann law, this plotting convention requires that temperature on the x axis be plotted on a non-linear scale with distance from the origin proportional to T^4 and increments of distance δx proportional to $T^3 \delta T$. For purposes of this qualitative discussion, the nonlinearity of the temperature scale is unimportant.

10.2 Continued

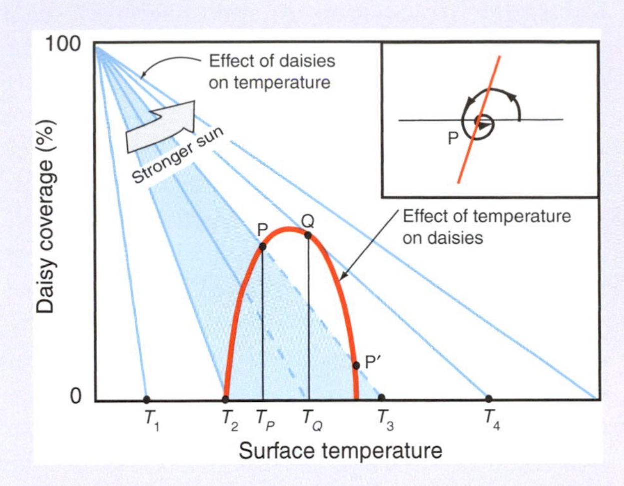

Fig. 10.39 Relationships between the areal coverage of daisies (or, alternatively, the planetary albedo of *Daisyworld*) and temperature, plotted on a non-linear scale. The red curve indicates how the areal coverage of daisies depends on the temperature, and the sloping blue lines indicate how the temperature depends on the areal coverage of daisies. Each blue line corresponds to a different value of the insolation. Points P and P' represent equilibrium states for the same value of the insolation. P represents a stable equilibrium state and P' an unstable equilibrium state. The inset in the upper right corner is an enlargement of the region around P showing qualitatively how the state of the system would respond to a small positive perturbation in temperature, assuming that the insolation is held fixed.

P represents a *stable equilibrium state* in the sense that a small perturbation in either temperature or the areal coverage of daisies would induce a response in the other variable that would tend to reduce the amplitude of the perturbation. For example, if T were raised by a small increment (without changing F), as illustrated in the inset of Fig. 10.39, the areal coverage of daisies and the planetary albedo would increase, resulting in cooling. Cooling, in turn, would reduce the positive amplitude of the perturbation in the areal coverage of daisies, and the state of the system would spiral inward toward P as shown. Alternatively, if the areal coverage of daisies were raised, the temperature of the planet would drop in response to the increased planetary albedo, and the cooling would decrease the amplitude of the perturbation in the areal coverage of daisies. In both cases, perturbations about the equilibrium state are suppressed through the action of negative feedbacks. By the same reasoning it follows that point P' represents an *unstable equilibrium state*, about which small perturbations in temperature or the areal coverage of daisies would spontaneously amplify through the action of positive feedbacks.

Now let us consider how the climate of *Daisyworld* would respond to a slow increase in the sun's emission, such as is believed to have occurred over the lifetime of the solar system. For radiative equilibrium temperatures below T_2 there would be no daisies and temperature would increase with solar emission in accordance with the Stefan–Boltzmann law. However, as soon as F reaches σT_2^4, daisies begin to grow and the sensitivity of *Daisyworld's* climate system to further increases in insolation is radically altered. At temperatures just above T_2 the main response to increasing insolation is an increase in the areal coverage of daisies. The increase in planetary albedo resulting from the proliferation of daisies cancels most of the warming that would have occurred in the absence of the daisies.

This negative feedback process is extraordinarily effective in stabilizing *Daisyworld's* climate. Consider the response to a small increase in F from σT_2^4 to σT_3^4, as indicated by the shaded wedge in Fig. 10.39. In the absence of daisies the temperature of the planet would rise from T_2 to T_3. However, when the daisy-albedo feedback is taken into account, the temperature rises only to T_P. It is - evident from Fig. 10.39 and from Eq. (10.7) that the remarkable stability of the climate is due to the large value of dA/dT at temperatures just above T_2.

As T approaches the optimal temperature for the daisies, the strength of the feedback declines and the climate sensitivity increases. When the insolation reaches the value indicated by the sloping blue line that is tangent to the red curve at Q, a climate catastrophe occurs. Beyond this point any further increase in F, no matter how small, would result in a complete collapse of the daisy population and an abrupt increase in the temperature on the planet from T_Q to T_3. Hence, temperature T_Q with nearly maximal daisy population and the much warmer temperature T_3 with no daisies represent alternative climate states on *Daisyworld* corresponding to the same value of $F = \sigma T_3^4$. Depending on the recent history of the system, the climate could reside in either state, and even an infinitesimally small perturbation in T or A would be sufficient to switch the system from one state to the other. Such states are referred to as *multiple equilibria*.

10.4 Greenhouse Warming[22]

The notion that the burning of fossil fuels could result in a buildup of carbon dioxide in the atmosphere, leading to *greenhouse warming*, dates back over a century.[23,24,25] Yet it was not until the late 1970s that the potential seriousness of this issue began to become widely recognized. In the ensuing years, greenhouse warming has become a lightning rod for scientific and political debate because important policy decisions relating to fossil fuel emissions have to be made in the face of scientific uncertainty. This section reviews what is known, and what is not known, about greenhouse warming, focusing on (1) the buildup of greenhouse gases and the associated forcings, as defined in the previous section, (2) whether human-induced greenhouse warming is already evident, and (3) the predicted extent and impacts of human-induced greenhouse warming.

10.4.1 The Buildup of Greenhouse Gases

Proof that greenhouse gases are actually building up in the atmosphere was not forthcoming until the 1960s, when the CO_2 time series from the first monitoring stations (Fig. 1.3) became long enough to reveal the presence of an upward trend. Averaged over the record extending from 1958 up to the year 2000, the trend is roughly 1.3 ppmv per year; in recent years it has been approaching 2 ppmv (0.5% of the present concentration) per year. The trend is believed to be attributable to fossil fuel burning for the following reasons.

1. Analysis of gas bubbles trapped in ice cores extracted from the Greenland and Antarctic ice sheets indicates that atmospheric CO_2 started to increase around the time of the industrial revolution and it has roughly tracked the rate of growth of fossil fuel consumption since that time (Fig. 5.13).
2. Atmospheric CO_2 concentrations are higher by several ppmv in the northern hemisphere, where most of the strongest carbon sources are located.
3. Atmospheric oxygen is observed to be decreasing at a rate of 3 ppmv per year, consistent with the hypothesis that the CO_2 being added to the atmosphere is a product of combustion.
4. The relative abundance of the radioactive isotope ^{14}C and the stable isotope ^{13}C in atmospheric CO_2 are declining. ^{14}C is virtually absent in fossil fuels and ^{13}C is less abundant in fossil fuels than in atmospheric CO_2 and in carbon dissolved in the oceans.

Figure 10.40 compares the annual rate at which carbon is being released by the consumption of fossil fuels and the rate of increase of the mass of carbon in the atmosphere. On average, the atmosphere is taking up about half the carbon that is being released. Most of the remainder is being taken up by the oceans.

In accordance with Eq. (2.7) and (2.8), the increasing storage of carbon in the oceans in the form of dissolved CO_2 in recent decades is increasing in the concentration in H^+ ions, lowering the pH of sea water. The increasing acidity, in turn, is driving the

[22] The term *greenhouse warming*, as used in this section and in most of the climate literature, connotes human-induced warming due to the buildup of CO_2 and other greenhouse gases in the atmosphere during the period since the industrial revolution.

[23] In 1896 Svante Arrhenius published a seminal paper entitled *On the Influence of Carbonic Acid in the Air upon the Temperature of the Ground* (Proc. Swedish Academy of Sciences, **22**). This was the first attempt to quantify the affects of changes in atmospheric CO_2 concentrations upon the Earth's surface temperature. Although Arrhenius' calculations were based on scanty data, his conclusions turned out to be surprisingly realistic.

Arrhenius was not the first to investigate the greenhouse effect. Fourier introduced the concept, comparing the atmosphere to a glass bowl that is heated not so much by solar radiation as by longwave radiation emitted by the ground.

[24] **Svante August Arrhenius** (1859–1927) Swedish chemist. Best known for his electrolytic theory of dissociation (i.e., that electrolytes, when dissolved in water, become to varying degrees dissociated into electrically opposite positive and negative ions). He was an infant prodigy, teaching himself to read at age three. In addition to his fundamental contributions to chemistry and his recognition of the "greenhouse effect," he suggested that life on Earth might have originated from spores transported from other planets. The story is told that when Arrhenius took the podium to honor guests at the 1926 Nobel Awards dinner, he offered a toast to the guests from each country honored at the ceremony. Because this was during the prohibition era, he toasted the Americans with water. By the end of the toasts, Arrhenius was the only one still sober.

[25] **Jean Baptiste-Joseph, Baron, Fourier** (1768-1830) French mathematician. Helped found mathematical physics and had a strong influence on the theory of functions of a real variable. Accompanied Napoleon on his expedition to Egypt and was also known for his research on antiquities.

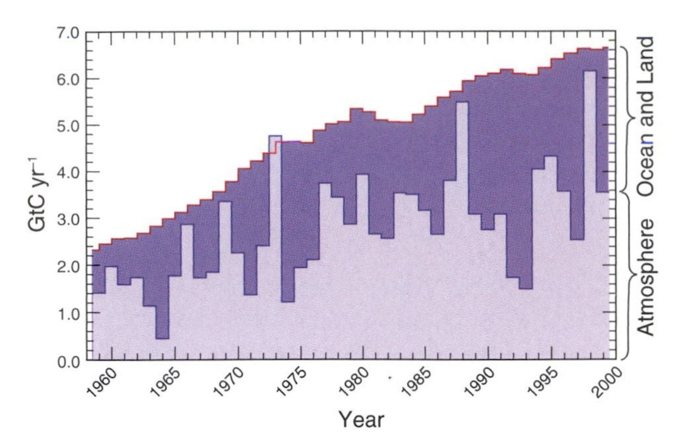

Fig. 10.40 Time series of the annual rate of release of carbon by the consumption of fossil fuels (in PgC or GtC per year) based on industry records (red "staircase"), and the annual rate of increase of the atmospheric carbon inventory (irregular blue line), as inferred from the time derivative of the time series of CO_2 concentration at Mauna Loa (Fig. 1.3). [Reprinted with permission from *Physics Today*, **55**(8), 30–36 (August 2002). Copyright 2002, American Institute of Physics. Courtesy of Nicolas Gruber.]

equilibrium between HCO_3^- and CO_3^{2-} [Eq. (2.9)] toward the left, thereby buffering some of the hydrogen ions.

If it were not replenished, the existing oceanic reservior of CO_3^{2-} ions would be sufficient to buffer roughly three-quarters of the carbon that would be released if the entire fossil fuel reservoir in Table 2.3 were consumed. If the acidity of the oceans increases significantly, the oceanic reservoir of CO_3^{2-} ions could be replenished, at least to some degree, by the dissolution of carbonate sediments from the sea floor.[26] The capacity and responsiveness of the oceans in taking up and buffering CO_2 depend not only on the relevant chemical reaction rates, but also on the rate of ventilation of the deeper layers of the ocean by the thermohaline circulation: the more vigorous the ventilation, the slower the buildup of carbon and acidity in the surface waters and the slower the rate of increase of atmospheric CO_2.

The exchange of carbon between the atmosphere and terrestrial biosphere also affects the rate of buildup of atmospheric CO_2. The prominent "spike" in the uptake by the atmosphere in 1998 (Fig. 10.40) is believed to be due to the CO_2 injected into the atmosphere by the massive forest fire outbreaks that occurred in Indonesia and Amazonia in association with the 1997–1998 El Niño event. Clearing of the tropical rain forests has released additional carbon in recent decades, which has been compensated by an expansion of forests in temperate latitudes.

Based on estimates of the annual rate of consumption of fossil fuels and the rates of change of the atmospheric and oceanic carbon inventories, it is possible to infer, as a residual, the net uptake or release of carbon by the terrestrial biosphere. Current estimates indicate that the terrestrial biosphere, as a whole, has been taking up carbon and that the rate of uptake increased substantially from the 1980s to the 1990s. The identity and likely future behavior of this terrestrial carbon sink are open questions at this point.

Projections of future atmospheric CO_2 concentrations based on three different emissions scenarios are shown in Fig. 10.41. Because it is assumed that the partitioning of the carbon emissions between the atmosphere and the other components of the Earth

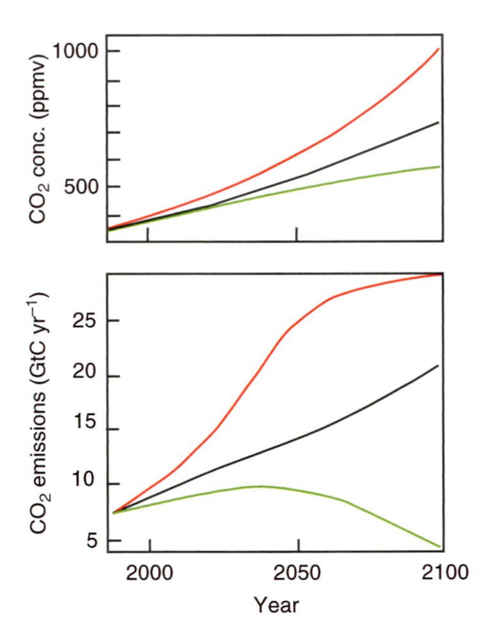

Fig. 10.41 (Top) Projections of future CO_2 concentrations based on the various emissions scenarios plotted in the bottom panel. The red curves are based on the assumption of rapid economic growth, world population peaking in the middle of the 21st century, and continued reliance on fossil fuels. The black curves assume a more gradual increase in fossil fuel consumption, sustained throughout the century. The green curves assume a major shift in the world economy toward sustainability. It is assumed that the oceans and the terrestrial biosphere will continue to take up roughly half the CO_2 injected into the atmosphere by human activities. [Adapted from Intergovernmental Panel on Climate Change, *Climate Change 2001: The Scientific Basis*, Cambridge University Press, p. 11 (2001).]

[26] Concern has been expressed that coral reefs may be vulnerable to an increase in the acidity of the oceans.

system will not change with time, the concentration curves in the top panel are proportional to the time integral of the respective emission curves in the lower panel: hence the monotonic increases in projected CO_2 concentrations, even in the case of the "green" scenario. In all three scenarios a value of 560 ppmv, equivalent to a doubling of preindustrial concentrations, will eventually be attained. Yet it is clear from these projections that current and future decisions on energy policy will play an important role in determining how long it will be before a doubling is reached and how far beyond it CO_2 concentrations will rise before they level off. In the more pessimistic scenarios, concentrations reach 1000 ppmv by the year 2200. Geological evidence indicates that concentrations that high have not occurred since the Eocene epoch 50 million years ago, when tropical animals and plants ranged as far northward as the subarctic.

CO_2 emissions will eventually decline, if for no other reason than fossil fuel reservoirs will become depleted. One might naively predict that the time required for atmospheric concentrations to drop back to preindustrial levels would be roughly comparable to the residence time of CO_2 molecules in the atmosphere, but this clearly is not the case. The ~10-year residence time of atmospheric CO_2 reflects cycling of carbon back and forth between the atmosphere and the marine biosphere and the leafy (as opposed to woody) part of the terrestrial biosphere. These biospheric reservoirs are very small in comparison to the atmospheric reservoir and they are not very expandable. The relaxation of atmospheric CO_2 concentrations back toward preindustrial levels depends on the uptake of carbon by the larger reservoirs in the Earth system, which takes place on a timescale much longer than that of the human-induced buildup of CO_2.

CO_2 is not the only greenhouse gas whose atmospheric concentrations have been increasing during the industrial era. Methane concentrations have exhibited an even larger fractional increase since the industrial revolution and nitrous oxide concentrations have risen as well (Fig. 5.13). CFC concentrations have risen dramatically since the mid-20th century, but are leveling off and are expected to gradually decrease over the next century in response to the Montreal Protocol. HFCs, the replacements for CFCs, are on the rise, along with a number of other long-lived industrial gases. Stratospheric ozone concentrations have declined since the mid-20th century in response to the buildup of CFCs and are just beginning to show signs of recovery. Meanwhile, tropospheric ozone has been increasing in response to worsening urban air pollution.

Even though their concentrations are much lower than that of CO_2, these trace gases contribute substantially to the greenhouse effect because some of their absorption bands occupy regions of the spectrum that would otherwise be *windows* through which terrestrial radiation could pass with relatively little atmospheric absorption. On a per molecule basis, an incremental increase in the concentrations of one of these gases would thus have a much stronger impact on the greenhouse effect than equivalent incremental increase in CO_2, which is so abundant that the atmosphere is relatively opaque in the regions of the spectrum that correspond to its absorption lines.

To compare the relative contributions of various trace gases to the greenhouse effect it is convenient to define a *greenhouse warming potential* (GWP). The GWP is defined as the mass of CO_2 that would need to be instantaneously injected into the atmosphere to produce an incremental increase in the greenhouse effect equivalent to that caused by the injection of 1 kg of a specified gas, integrated over a specified time interval, taking into account the time-dependent decay of the gas in question, as well as that of CO_2. In symbolic form

$$\text{GWP} = \frac{\int_0^T a_x c_x(t)dt}{\int_0^T a_{\text{CO}_2} c_{\text{CO}_2}(t)dt} \tag{10.14}$$

where T is the specified time interval, a_x is the *radiative efficiency* of gas x in units of W m^{-2} kg^{-1} (i.e., the incremental change in radiative forcing that would result from a small change in the concentration of the gas in question, taking into account its present concentration and the concentrations of other greenhouse gases), c_x is the fraction of the mass of gas x injected into the atmosphere at time $t = 0$ that remains in the atmosphere at time t, and a_{CO_2} and c_{CO_2} are the corresponding quantities for CO_2. The numerator of this expression is called the *absolute global warming potential* (AGWP). Greenhouse warming potentials for a few of the important atmospheric greenhouse gases are listed in Table 10.1. CFCs are especially potent in terms of GWP because of their high radiative efficiencies and long lifetimes. As the specified time

Table 10.1 Greenhouse warming potentials for selected greenhouse gases for 20, 100, and 500-year time intervals[a]

Gas	Radiative efficiency	Lifetime (year)	GWP_{20}	GWP_{100}	GWP_{500}
CO_2	0.0155	—	1	1	1
CH_4	0.37	12	62	23	7
N_2O	3.1	114	275	296	156
CFC-12	320	100	10200	10600	5200
HCFC-21	170	2	700	210	65

[a] Radiative efficiency is expressed in W m^{-2} $ppmv^{-1}$. [Adapted from Intergovernmental Panel on Climate Change, *Climate Change 2001: The Scientific Basis*, Cambridge University Press, pp. 388–389 (2001).]

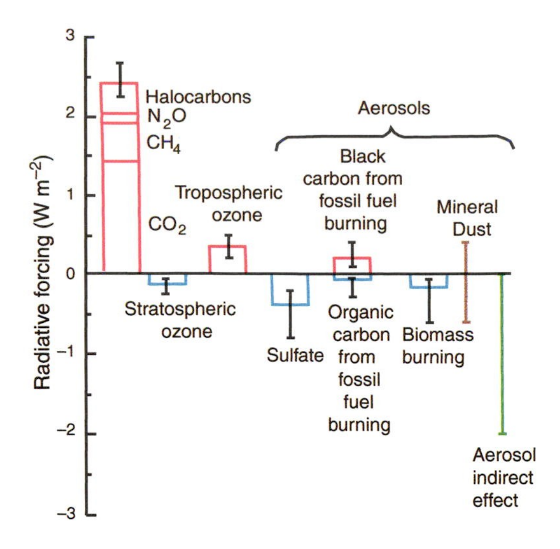

Fig. 10.42 Radiative forcing at the top of the atmosphere due to the human-induced incremental change in atmospheric concentrations of various species of trace gases and aerosols based on present concentrations minus preindustrial concentrations. [Adapted from Intergovernmental Panel on Climate Change, *Climate Change 2001: The Scientific Basis*, Cambridge University Press, p. 8 (2001).]

interval increases beyond the lifetime of the gas, the GWP drops.

In addition to the direct GWP listed in Table 10.1, each gas can be assigned an *indirect GWP* that describes the radiative forcing due to changes in the concentrations of other greenhouse gases or aerosols induced by a small incremental increase in its concentration. For example, the injection of methane into the atmosphere enhances the formation of tropospheric ozone, thereby inducing an indirect warming amounting to ~25% of the direct warming due to the increase in methane itself.

Figure 10.42 shows the contributions of various atmospheric constituents to the radiative forcing at the top of the atmosphere, as defined in the previous section, for the year 2000, relative to conditions prior to the industrial revolution. Hence, these values may be viewed as the contributions of these constituents to the current human-induced greenhouse effect. The left-most bar in Fig. 10.42 shows the contributions of greenhouse gases. The combined warming due to CH_4, N_2O and CFCs and tropospheric ozone is roughly equivalent to that of CO_2 emissions. The contribution from stratospheric ozone is negative because human activity has resulted in decreased concentrations of stratospheric ozone.

Also shown in Fig. 10.42 are estimates of the contributions from aerosols to the human-induced climate forcing. Sulfate aerosols cool because they are highly reflective, whereas black carbon (soot) has the opposing effect. The large and highly uncertain *aerosol indirect effect* represents the effects of increasing concentrations of aerosols upon the properties of clouds. Visual evidence of the whitening of stratus cloud decks by the exhaust plumes from the engines of ships (Fig. 6.9) has fueled speculation that emissions of cloud condensation nuclei from power plants, aviation, biomass burning, and other human activities might be affecting surface air temperature indirectly by changing the optical properties of low clouds and/or by increasing their areal coverage.

There are several ways in which aerosols could conceivably affect the optical properties of clouds. Greater numbers of cloud droplets competing for the same amount of cloud water result in larger numbers of smaller droplets, which favor more reflective clouds. In contrast, carbon (soot) particles injected into the atmosphere by human activities tend to increase the absorptivity of clouds, thereby reducing the fraction of the incident solar radiation that is backscattered to space.

The aerosol indirect effect is believed to be negative and potentially quite important, but its magnitude is highly uncertain. Whether it is or is not an important factor in the human-induced radiative forcing has important implications for estimates of climate sensitivity, as demonstrated in Exercise 10.26.

10.3 Evidence of an Indirect Aerosol Effect

Condensation trails (contrails for short) are cloud-like streamers that can form behind aircraft in cold, humid air (Fig. 10.43). They are produced by water vapor and cloud condensation nuclei (CCN) from aircraft engines. Due to the large concentration of CCN in engine exhausts, recently formed contrails contain large concentrations of small droplets. Therefore, as in the case of continental clouds discussed in Section 6.2, the boundaries of recently formed contrails are usually well defined (e.g., the brighter contrail in the center of the photo). However, as a contrail ages and becomes glaciated, the ice particles survive longer than droplets as they mix with the surrounding subsaturated air. Consequently, the boundaries of older contrails are more diffuse, like the other contrails in the photo.

In the vicinity of large airports, contrails can be very numerous and may spread out to cover large areas of the sky. Because contrails can reduce the transfer of both incoming solar radiation and outgoing infrared radiation, they have the potential to reduce the diurnal temperature range. Following the terrorist attack on the Twin Towers in New York City on 11 September 2001, all commercial aircraft flights in the United States were stopped for 3 days. During this period there was an statistically significant increase of 1.1 °C in the average diurnal temperature range for ground stations across the United States, relative to the climatological mean.

Fig. 10.43 Condensation trails. [Photograph courtesy of Art Rangno.]

Atmospheric concentrations of methane, tropospheric ozone and CFCs respond much more rapidly to changes in source strength than does CO_2. Policy decisions relating to the emissions of these gases could substantially slow the rate of greenhouse warming over the next 50 years. Restrictions mandated by the Montreal Protocol have already halted the rapid buildup of CFCs; changes in land use and oil extraction practices could potentially reduce methane emissions, and tighter air pollution standards could reduce tropospheric ozone.

10.4.2 Is Human-Induced Greenhouse Warming Already Evident?

The 20th century was marked by warming that is reflected in records of surface air temperature over both land and ocean, as shown in Fig. 10.44, in the heat content of the upper oceans, in the retreat of mountain glaciers (Fig. 10.45), the rise in global sea level (Fig. 10.46), and in a variety of other indicators such as melt and freeze dates on lakes and rivers, blooming dates of flowering plants, timing of bird migrations, and the poleward extent of plant, insect, and animal species. To the extent that proxy evidence reflects the true temperature trends during previous centuries, the rate of warming observed during the 20th century appears to be unprecedented in the past millennium.

The warming observed during the 20th century was not uniform either in time or in space. The pronounced warming of the 1920s and 1930s followed by flattening of the temperature trace around the middle of the century, shown in Fig. 10.44, was mainly a northern hemisphere phenomenon. Temperatures over parts of the Arctic rose by several degrees Celsius during the 1920s and dropped by a nearly comparable amount during the 1950s and 1960s. Temperatures over the interior of the Antarctic continent have risen little, if at all, since

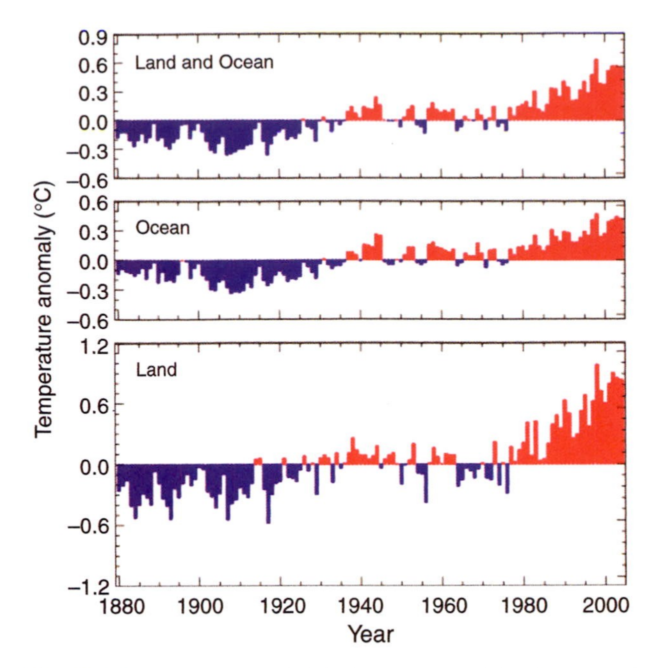

Fig. 10.44 Time series of global-, ocean-, and land-averaged surface temperature anomalies, defined as departures from the 1880–2003 mean. The data are updated through December of 2004. [Courtesy of NOAA's National Climatic Data Center.]

Fig. 10.45 Photographs documenting the retreat of snow and ice on Mt. Kilimanjaro in equatorial Africa during the 20th century. [Courtesy of Lonnie G. Thompson.]

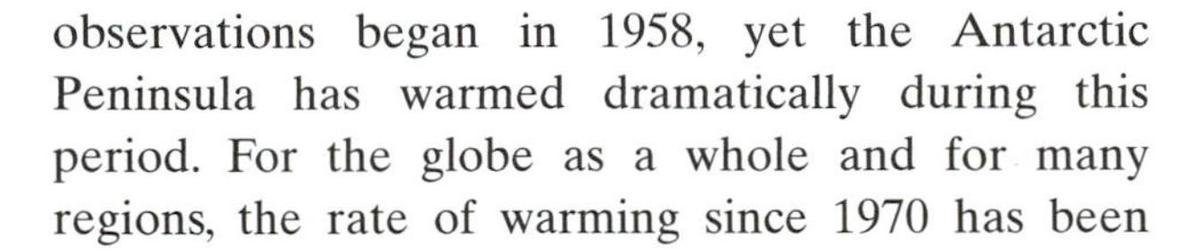

observations began in 1958, yet the Antarctic Peninsula has warmed dramatically during this period. For the globe as a whole and for many regions, the rate of warming since 1970 has been larger than for any period of comparable length earlier in the record.

How much of the warming observed during the 20th century was human induced and how much was

10.4 Global Sea Level Rise

Global-mean sea level varies in response to the expansion and contraction of sea water as the oceans warm and cool (the *thermosteric effect*). Because of the temperature dependence of the coefficient of expansion of sea water, the magnitude of the thermosteric effect depends not only on the net amount of heating of the oceans, but also on the heating distribution in relation to the existing temperature distribution, as illustrated in Exercise 10.28. Global-mean sea level also varies in response to changes in the volume of water that is stored in the oceans (the *eustatic effect*). Changes in the volume of the ice sheets, in the volume of water stored on land in lakes, rivers, reservoirs, soils and aquifers, changes in the shape of the Earth's crust, and (on timescales of millions of years) exchanges of water between the oceans and the mantle all contribute to the eustatic effect.

On timescales of years to decades, local wind-induced variations in sea level, such as the ENSO-related variations shown in Fig. 10.23, are much larger than thermosteric and eustatic changes in global-mean sea level. On the basis of the existing sparse tide-gauge network it is difficult to distinguish between the global greenhouse warming "signal" and the local, wind-induced "noise." On longer timescales, the deformation in the

Continued on next page

10.3 Continued

Earth's crust—in particular, the *isostatic rebound* from the last ice age (i.e., the rising of the crust in areas that were formerly depressed by the weight of the continental ice sheet and the sinking of the crust in the surrounding regions)—further complicates the interpretation of local tide gauge records.

Since 1992, sea level has been monitored on a global basis using satellite altimeters capable of detecting small changes relative to the Earth's *geoid* (surfaces of constant geopotential). These direct measurements, shown in Fig. 10.46, indicate that sea-level is currently rising at a rate of ~3 mm year^{-1} (or 30 cm per century) about twice as rapidly as previous indirect estimates based on measurements of the rate of warming of the world's oceans, trends in salinity over high latitudes (indicative of the the melting of the ice sheets), and changes in the inventory of water on land.

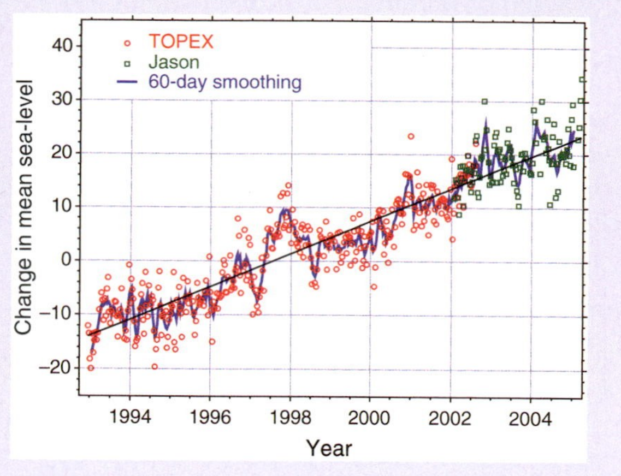

Fig. 10.46 Measurements of 10-day mean and 60-day running mean global-mean sea-level from the NASA/CNES TOPEX Poseidon and Jason satellite-borne altimeters. Changes are with respect to an arbitrary reference level. The data have been corrected to remove the climatological-mean seasonal variations. The sloping black line indicates the linear trend in the data. [Courtesy of Steve Nerem.]

a reflection of the natural variability of the climate system cannot be known with certainty because there is no way of obtaining multiple realizations of the unperturbed climate of the 20th century to serve as a basis for comparison. Natural variability on decadal timescales was probably responsible for some of the regional differences in the trends and the lack of uniformity of the global-mean temperature trend over the course of the century. It may also have contributed to (or detracted from) the net warming. However, the warming of the past 30 years has been so pronounced and the cumulative warming since 1900 so large as to beg for an explanation. Thus far, the only plausible and well-supported interpretation that has been put forward is that they are early indications of a response to the buildup of greenhouse gases.

10.4.3 Projections of Future Human-Induced Greenhouse Warming

There are several ways of estimating how much global-mean surface air temperature would rise in response to a prescribed increase in the concentrations of greenhouse gases. One approach is to rely on the past history of the Earth system to estimate the sensitivity of global mean surface air temperatures to changes in radiative forcing, as illustrated in Exercises 10.4 and 10.26. Empirical estimates of this kind are subject to large uncertainties, as illustrated in Exercise 10.26. Hence, most estimates of climate sensitivity are based on models: either globally averaged, one-dimensional, radiative-convective equilibrium models with assumptions concerning water vapor, cloud and ice-albedo feedbacks; or three-dimensional, coupled atmosphere-ocean-cryosphere-land surface models that explicitly compute water vapor concentrations, ocean temperatures, and currents and include parameterizations of clouds, sea ice, snow cover, land vegetation, and soil wetness.

The advantage of radiative-convective models is that they are simple and they yield results that are easily interpretable. These models predict that a doubling of the atmospheric CO_2 concentration relative to its preindustrial concentration would result in a downward radiative forcing F at the top of the atmosphere of 3.7 W m^{-2}, raising global mean surface air temperature by $\lambda_0 F = 0.96$ K. If water vapor feedback is included under the assumption that relative humidity remains constant, the sensitivity to a prescribed forcing roughly doubles, and if ice-albedo feedback is included as well, the *feedback factor*, as defined in connection with (10.8), increases to around 0.6, yielding an increase in T_s of around 2.5 K for a CO_2 doubling.

In comparison to the simpler radiative convective models, three-dimensional, coupled climate models offer a number of significant advantages for making projections of greenhouse warming,

- They provide a test bed for improving our understanding of boundary layer, cloud, cryospheric, and biospheric processes that have a bearing on climate sensitivity.
- They enable the various components of the atmosphere and the Earth system to interact in ways that cannot be investigated empirically or with the use of radiative-convective models. For example, they allow for the possibility of changes in the tropospheric temperature lapse rate, high or low cloud cover, and surface wind speed in different parts of the world, any of which might have implications for climate sensitivity.
- They yield much more comprehensive predictions of how climate would change in response to a prescribed forcing, with regionally and seasonally specific (as opposed to globally and annually averaged) temperature projections, as well as estimates of patterns of changes in rainfall, snowfall, cloudiness, wind, and the statistics of severe weather events such as tropical cyclones.
- They provide a framework for investigating the impacts of climate change on the Earth system as a whole (e.g., sea level rise, changes in streamflow statistics for various watersheds, desertification of marginally arid regions, and changes in the range of various plant and animal species and diseases).
- Unlike the simpler models, they provide a timetable for the changes that explicitly takes into account the role of the ocean in buffering both the chemical and the thermal response to the burning of fossil fuels.

Numerical simulations of the response to the buildup of concentrations of greenhouse gases have been performed with an array of coupled models. The simulated responses vary somewhat from model to model, but the gains tend to be in the same range as those inferred from the radiative convective models (1.5 to 4.5). The simulated responses share a number of common characteristics that are in accord with simple physical reasoning:

- a polar amplification of the warming, particularly during winter and spring due, in large part, to the positive feedback from the cryosphere. This tendency is also clearly evident in historical temperature records from paleoclimate reconstructions.
- a gradual buildup of the warming due to the large thermal inertia of the oceans.
- more rapid warming over the continents than over the oceans during the early stages of the warming.
- an increase in atmospheric precipitable water, leading to heavier precipitation events (particularly during winter over higher latitudes where the predicted warming is greatest).
- earlier spring snowmelt and enhanced evaporation rates, leading to drying of some continental regions during summer. In such regions, positive feedback from the biosphere amplifies the rise in daytime temperatures.
- a rise in global sea level due to the thermal expansion of sea water as it warms (20–40 cm during the 21st century, compared to ~15 cm during the 20th century). Present models are not capable of predicting the extent of the additional eustatic contribution dominated by the melting of the continental ice sheets and alpine glaciers.
- a reduced rate of formation of North Atlantic deep water in response to the freshening of the surface waters as precipitation increases and ice melts.

There is considerable uncertainty as how the cryosphere will respond to global warming. Will the Arctic pack ice continue to retreat, offering the possibility of new trans-Arctic shipping routes? Will Greenland become ice free, as it apparently was at the time of the last interglacial? Will the melting of the ice sheets be sufficiently rapid to accelerate the rate of sea level rise, as illustrated in Exercise 10.28? How serious is the risk of a "climate surprise" — for example, a sudden disintegration of a major part of one of the ice sheets or a reorganization of the atmospheric and/or ocean general circulation analogous to those that caused the abrupt discontinuities in ice core records? Such complex workings of the Earth system are beyond the capability of the present generation of coupled models to simulate.

10.5 Climate Monitoring and Prediction

Early in the 20th century, understanding of weather phenomena reached the point that many nations of the world deemed it to be in the public interest to establish weather services charged with the task of providing forecasts of changing weather conditions and timely warnings of weather events that might pose risks to life and property. In support of these activities, a global observing system was established under the auspices of the United Nations. Several nations and international consortia established centers that collect atmospheric and other relevant Earth system data and assimilate it into models that diagnose the current state of the atmosphere and predict how it will evolve out to the limit of deterministic weather prediction. This weather information is delivered to the public by a distributed communications system involving both government and private industry.

The global weather observing system is in the process of being augmented to serve the additional task of climate monitoring. Clouds, radiatively active trace gases and aerosols are being added to the suite of atmospheric measurements, and the observing system is being expanded to encompass other components of the Earth system that have a bearing on climate. Many of the needed atmospheric and surface observations can be obtained with complete global coverage by means of remote sensing from satellites. A number of other strategies are being used to monitor subsurface temperatures, currents, and chemical and biological properties of the oceans. Because it deals with relatively weak, slowly evolving climate anomalies, climate monitoring imposes much more stringent requirements on the calibration and long-term stability of measurements than observations in support of numerical weather prediction.

Gridded atmospheric fields used in operational numerical weather prediction are not well suited for climate studies because they contain artificial discontinuities due to changes in instrumentation, changes in the availability of different kinds of measurements, changes in the quality control procedures, and the successive updating of the numerical weather prediction models used to perform the data assimilation. These inhomogeneities can be reduced substantially by performing *reanalysis* of the archived observations, using a single, state-of-the-art numerical weather prediction model and uniform quality control procedures, and correcting the data for known changes in instrument characteristics.

Three-dimensional, coupled climate models were first introduced in the 1970s and have gradually expanded to include additional components of the Earth system. Just as there is more to building a high-performance automobile than merely assembling a set of components designed independently by a number of different manufacturers, there is more to building a high-performance climate model than merely coupling an atmospheric general circulation model (GCM) with an ocean GCM, a sea-ice model, or a land hydrology model.

Atmosphere-only GCMs are designed to produce realistic simulations of the current observed climate when run with climatological-mean sea-surface temperature and sea ice. Ocean-only GCMs are designed to produce realistic simulations of ocean temperatures and currents when run with prescribed surface winds, surface air temperatures, and precipitation, and similar considerations apply to the models of other components of the Earth system. When the models of the various components of the Earth system are coupled, the observational constraints that were used in designing and tuning them are removed. For example, the observed sea-surface temperature is no longer prescribed as part of the lower boundary conditions for the atmosphere: in the coupled model, sea-surface temperature is determined by the mutual adjustment of the atmosphere and ocean GCMs to the insolation and the interactions with the other model components.

The behavior of a coupled model is often quite different from the composite behavior of the component models from which it is constructed. Coupled models may exhibit a pronounced *climate drift*, which reflects the mutual adjustment of the various components as the coupled system erratically wanders away from the imposed initial conditions and establishes its own peculiar climatology. This equilibration process proceeds rapidly at first, but it may take centuries to complete, because of the large thermal inertia of the oceans. Coupled models exhibit greater internal variability than the component models of which they are made up because of their ability to simulated coupled phenomena such as ENSO. Coupled climate models provide a physically consistent framework for data assimilation, not only for the atmosphere, but also for other components of the Earth system.

Although the weather forecaster is still the object of derision from time to time, the general public understands and appreciates the value of weather forecasts and makes routine use of them. The "market" for climate forecasts on timescales ranging from seasons to decades is much more limited: the main "customers" being corporate enterprises in climate-dependent sectors of the world economy, such as water, energy, agriculture, fisheries, forestry, and outdoor recreation. Managers in these fields are already accustomed to using climatological mean data as a basis for decision making, but many are reluctant to use climate forecasts in the absence of a long track record, analogous to the one that exists for weather forecasts. To attract users, climate forecasts need to be tailored to their needs. For example, a public utilities manager might need information on precipitation and snow levels within a particular watershed; an operator of a wind power grid might need statistics of wind speeds at specific sites.

For some applications the relatively coarse resolution atmospheric fields derived from the global models need to be *downscaled*, that is, enhanced by incorporating information on the unresolved scales either statistically or with the help of regional mesoscale models that include a detailed representation of the local terrain. For other applications, variables such as stream flow, crop yields, energy demand, or output from a wind power grid can be related directly to the more predictable indexes of the hemispheric scale circulation, like those used to represent ENSO, using statistical methods based on historical records.

Exercises

10.5 Explain or interpret the following.

(a) Global-mean surface air temperature is below the radiative equilibrium temperature.

(b) In Fig. 10.1, the troposphere emits more longwave radiation in the downward direction than in the upward direction, whereas the stratosphere emits more longwave radiation in the upward direction than in the downward direction

(c) Annual-mean net radiation at the top of the atmosphere is downward at low latitudes and upward at high latitudes.

(d) The annual mean net flux of energy at the air–sea interface is locally downward over the eastern equatorial Pacific.

(e) Mars exhibits a larger diurnal temperature range than Earth; there is almost no diurnal cycle in surface temperature on Venus (Table 2.6).

(f) Dust storms on Mars substantially reduce the diurnal temperature range, but have little influence on the daily average temperature.

(g) Summer insolation (Fig. 10.5) is up to 6% stronger in the southern hemisphere than in the northern hemisphere.

(h) The Earth system is nearly in equilibrium with the annual mean net radiation at the top of the atmosphere, but not with the seasonal mean net radiation.

(i) The net radiation at the Earth's surface averaged over the spring or summer hemisphere is downward.

(j) Climatological-mean insolation is greater at the summer pole than on the equator, yet surface air temperatures in the polar region are lower than over the equator (Fig. 10.5).

(k) The net radiation at the top of the atmosphere over the polar regions is upward, even during summer (Fig. 10.6).

(l) The amplitude of the annual cycle in surface air temperature is larger at high latitudes than at low latitudes (Fig. 10.8).

(m) In middle latitudes the annual cycle in surface air temperature is larger in the northern hemisphere than in the southern hemisphere.

(n) Net radiation at the top of the atmosphere is upward over the Sahara, even during summer, yet the surface is hot.

(o) The phase of the annual cycle in surface air temperature tends to be earlier over the continents than over the oceans.

(p) The ocean mixed layer deepens with time during autumn and winter.

(q) During a single autumn storm the depth of the oceanic mixed layer may deepen dramatically.

(r) During July the equatorial thermocline tilts upward toward the east in both the Pacific and Atlantic Oceans, but not the Indian Ocean. [**Hint**: consider the distribution of surface winds in Fig. 1.18.]

(s) The subtropical oceanic anticyclones tend to be stronger during summer than during winter.

(t) Climate predictability is not subject to the ~2-week limit of deterministic atmospheric predictability.

(u) An atmospheric general circulation model exhibits more interannual variability when it is coupled to a slab mixed layer ocean than when it is run with fixed sea-surface temperature. The enhanced variability is at low frequencies. The deeper the mixed layer, the lower the frequency range of the enhanced variability.

(v) When the easterly surface winds in the central equatorial Pacific weaken, sea-surface temperature rises on the eastern side of the Pacific basin.

(w) A warming of sea-surface temperature over the equatorial eastern Pacific favors a weakening of the easterly surface winds over the equatorial central Pacific.

(x) During summer over continental regions, monthly mean surface air temperature tends to be negatively correlated with the previous month's rainfall. (The correlation derives mainly from daytime temperatures.)

(y) During the 20th century the upward trend in daily minimum temperature was around twice as large as the trend in daily maximum temperature.

(z) A child is sleeping in a cold room and her mother decides she needs an extra blanket. The temperature of the topmost blanket is being monitored with an infrared radiometer mounted on the ceiling. (i) How does the temperature sensed by the infrared thermometer respond to the addition of the extra blanket? Consider both the time-dependent behavior and the new equilibrium temperature after the "system" has adjusted to the addition of the extra blanket. Assume that the initial temperature of the extra blanket is the same as that of the blanket immediately below it and that the child's metabolism is constant. (ii) What sensations of warmth or coolness would the child experience if she were awake during this period? (iii) Is the temperature recorded by the infrared thermometer of any use in diagnosing whether the child is comfortably warm? (iv) In what ways is this situation analogous to the response of global mean surface air temperature to an incremental change in radiative forcing at the top of the atmosphere?

(aa) Describe how the net energy flux at the air–sea interface would evolve in response to an abrupt increase in radiative forcing at the top of the atmosphere.

(bb) Global-mean surface air temperature at high latitudes is more sensitive to external forcing on millennial timescales than on decadal timescales.

(cc) Despite the existence of positive water vapor and ice-albedo feedbacks, the Earth has not experienced a *runaway greenhouse*.

(dd) Cloud-albedo feedback could conceivably be large and of either sign.

(ee) Water vapor is the most important atmospheric greenhouse gas and is a product of fossil fuel burning, yet it is not included among the greenhouse gases in Fig. 10.42.

(ff) The relationship between greenhouse warming and atmospheric CO_2 concentration is expected to be roughly logarithmic, rather than linear (e.g., a quadrupling of CO_2 is expected to produce about twice, rather than four times, as much warming as a CO_2 doubling).

(gg) The atmospheric concentration (by volume) of methane is only roughly 1/200 that of CO_2, yet its present contribution to the greenhouse effect is almost 1/3 as large as that of CO_2.

(hh) The greenhouse warming potential of nitrous oxide is much higher than that of methane, especially over a time interval of a century or longer.

(ii) Human activities are believed to be primarily responsible for the more than 20% increase in the concentration of atmospheric CO_2 that has been observed since monitoring began in the late 1950's and for the ~35% increase since the time of the industrial revolution.

(jj) Most scientists believe that human activities are primarily responsible for the warming observed during the 20th century.

(kk) Even if the rate of fossil fuel consumption were to immediately level off and begin declining, atmospheric CO_2 concentrations would be expected to continue to rise throughout the 21st century.

(ll) Processes within the oceans play an important role in determining the level to which atmospheric CO_2 concentrations will ultimately rise in response to the burning of fossil fuels.

(mm) The sea level rise projected to occur in response to the buildup of greenhouse gases develops more gradually and does not peak until a few hundred years later than the peak CO_2 concentrations.

(nn) The policy dimensions of greenhouse warming are much more complex than those surrounding the ozone hole.

10.6 Repeat Exercise 10.1 for the stratosphere.

10.7 Using data in Fig. 10.7 make a rough estimate of the rate of energy loss through the ocean surface during autumn and early winter, when the ocean mixed later is cooling and deepening. [**Hint**: Consider the top-most 100-m layer, which cools from a vertically averaged temperature of ~18 °C in September to ~15 °C in January, a drop of ~3 °C in ~100 days.]

10.8 If the time series of a climatic variable is perfectly sinusoidal with amplitude A, prove that its root mean squared amplitude or standard deviation is $A/\sqrt{2}$.

10.9 (a) Prove that for standardized variables x and y, the straight line passing through the origin

$$y = rx$$

$$r = \overline{x_i y_i}$$

represents the *least-squares best fit regression line*, where *best fit* is defined in the sense that the quantity

$$Q \equiv \overline{(y_i - rx_i)^2}$$

is minimized. Here, as in the text, the overbar $\overline{(\ )}$ represents the mean over all the data points. In this expression, y_i is the y coordinate of the ith data point in a scatter plot like the ones shown in Fig. 10.12, and rx_i is the y coordinate of the regression line where it passes through $x = x_i$. Hence, $(y_i - rx_i)$ is the error in y_i, as estimated or predicted on the basis of its linear relationship with x.

(b) Prove that r^2 is the variance of y that can be explained in the basis of this least squares best fit and $(1 - r^2)$ is the error or unexplained variance.

Solution: (a) Q can be calculated for any prescribed value of r. At the value of r for which Q is minimized

$$\frac{dQ}{dr} = \frac{d}{dr}\overline{(y_i - rx_i)^2} = 0$$

Expanding, we obtain

$$\frac{d}{dr}(\overline{y_i^2} - 2r\overline{x_i y_i} + r^2\overline{x_i^2}) = 0$$

Differentiating and noting that the data x_i and y_i are independent of r and that $\overline{x_i^2} = 1$ yields

$$r = \overline{x_i y_i}$$

Based on this least squares best fit, we can write

$$y_i = rx_i + \varepsilon_i$$

where ε_i is the error. Squaring and averaging, we obtain

$$\overline{y_i^2} = r^2\overline{x_i^2} + 2\overline{\varepsilon_i x_i} + \overline{\varepsilon_i^2}$$

The error ε_i and x_i must be uncorrelated. If that were not the case, $y = rx$ would not be the line of best fit. Because $\overline{x_i^2} = \overline{y_i^2} = 1$, this expression reduces to

$$1 = r^2 + \overline{\varepsilon_i^2}$$

where r^2 is the explained variance and $\overline{\varepsilon_i^2}$ is the unexplained variance. This relationship can be extended to unstandardized variables, in which case r^2 is the *fraction of the variance* of y that can be explained in the basis of the least-squares best fit and $(1 - r^2)$ is the *fraction of the variance* that remains unexplained. ■

10.10 The standard deviation of monthly mean temperature at a certain station averaged over the winter months December–March is 5.0 °C. The standard deviation of winter seasonal mean temperature is 3.0 °C. What is the standard deviation of monthly mean temperature about the respective seasonal means for individual winters?

10.11 Prove that

$$\overline{x'y'} = \overline{xy} - \overline{x}\,\overline{y}$$

[**Hint**: Make use of the fact that $\overline{\overline{x}y'} = \overline{x'\overline{y}} = 0$.]

10.12 Suppose that the Earth warms by 3 °C during the 21st century and that the entire ocean warms by the same amount. Assume that the cryosphere remains unchanged. In the global energy balance in Fig. 10.1, by how many W m^{-2} would the net incoming solar radiation have to exceed the outgoing Earth radiation at the top of the atmosphere? [**Hint**: The mass of the oceans per unit area of the Earth's surface is given in Table 2.2.]

10.13 On the basis of the global energy balance in Fig. 10.1, estimate what the surface temperature of the Earth would be in the absence of latent and sensible heat fluxes. Assume that the fraction of the longwave radiation emitted from the Earth's surface that is absorbed by the atmosphere and re-emitted back to the surface remains unchanged.

10.14 Using data in Fig. 10.2, estimate the poleward flux of energy by the atmosphere and oceans across 38 °N, where the incoming and outgoing radiation curves intersect. [**Hint**: Assume that the energy transport across the equator is zero. Assume that the energy transport across the equator is zero.]

10.15 Compare the daily insolation upon the top of the atmosphere (a) at the North Pole at the time of the summer solstice and (b) at the equator at the time of the equinox. The *solar declination angle* (the astronomical analog of geographic latitude; i.e., the latitude at which the sun is directly overhead at noontime) at the time of the summer solstice is 23.45° and the Earth–Sun distances are 1.52 and 1.50×10^8 km, respectively.

10.16 Consider the response of the Earth's equivalent blackbody temperature T_E to a volcanic eruption that increases the planetary albedo, resulting in a radiative forcing at the top of the atmosphere of $\delta F = -2$ W m^{-2}. (a) Calculate the equilibrium response. (b) Suppose that the atmosphere is well mixed and thermally isolated from the other components of the Earth system; that the increase in planetary albedo is instantaneous; and that the albedo remains constant at the higher value after the eruption. Show that T_E drops toward its new equilibrium value exponentially, approaching it with an *e*-folding time equal to the atmospheric radiative relaxation time defined and estimated in Exercise 4.29. [**Hint**: Consider the energy balance at the top of the atmosphere. Make use of the fact that that $\delta F/F << 1$ and $\delta T_E/T_E << 1$.]

10.17 Rework Exercise 10.16, but assuming that the radiative flux at the top of the atmosphere abruptly returns to its preeruption value exactly a year after the eruption. Estimate the drop in T_E during the year after the eruption (a) assuming that the atmosphere is well mixed and thermally isolated from the other components of the Earth system and (b) assuming that the atmosphere remains in thermal equilibrium with a 50-m-deep ocean mixed layer that covers the planet. (c) Under which of these scenarios does the Earth system lose more energy during the year that the volcanic debris are present in the atmosphere? (d) Which of these scenarios is more realistic, and why?

10.18 Which of the following has the largest impact on the Earth's equivalent blackbody temperature. (a) The ~0.07% variation sun's emission that is observed to occur in association with the 11-year sunspot cycle. (b) The flux of geothermal energy from the interior of the Earth (0.05 W m^{-2}). (c) The consumption of energy by human activities (10^{13} W).

10.19 Two feedback processes capable of amplifying greenhouse warming by factors of 1.5 and 2.0, respectively, if each were acting in isolation, would be capable of amplifying it by a factor of ________ if they were acting in concert.

10.20 (a) Without using Eq. (10.9) show that for a case of a single auxiliary variable y, the change in global mean surface air temperature T_s resulting from an incremental climate forcing δF is given by

$$\delta T_s = \lambda_0 \delta F\,(1 + f + f^2 + f^3......) \quad (10.16)$$

where f is the feedback factor. (b) Show that (10.9) follows directly from (10.16).

10.21 For a planet with an atmosphere like the Earth's and covered by an ocean 50-m deep, estimate the response time for global-mean surface air temperature to adjust to an instantaneous change in the climate forcing at the top of the atmosphere. Assume that the gain is 2.5.

10.22 Show that the *net cloud radiative forcing* (i.e., the incremental change in downward net radiation at the top of the atmosphere due to the presence of clouds) is given by

$$CF = F(A_{cs} - A) - (OLR - OLR_{cs}) \quad (10.17)$$

where F is the local insolation at the top of the atmosphere, A is the local albedo, OLR is the local outgoing longwave radiation at a point on Earth averaged over an extended period like a season, and A_{cs} and OLR_{cs} are the *clear sky albedo* and *clear sky outgoing longwave radiation* (i.e., the average based on just those instantaneous images that are deemed to be free of clouds) for the same period.

10.23 Based on (10.17) answer the following questions:

Why is the *net cloud radiative forcing* in Fig. 10.37 strongly negative over the regions of persistent stratus cloud decks off the coasts of Peru, Namibia, and Baja California?

Why is the *net cloud radiative forcing* in Fig. 10.37 positive over Antarctica?

10.24 (a) Rework Exercise 10.4 from the perspective of the sensitivity of the climate system to changes in ice cover. This time, instead of including variations in atmospheric CO_2 as part of the forcing, use only the forcing due to the difference in the Earth's albedo. (b) Compare this result with the estimated climate sensitivity in the absence of the CO_2 feedback.

10.25 Using an approach analogous to the one used in Exercise 10.4, estimate the sensitivity of the climate system, comparing global-mean surface air temperature in the year 2000 with conditions at the time of the industrial revolution. Use this value to estimate the rise in surface air temperature that would result from a doubling of the atmospheric carbon dioxide concentration relative to the preindustrial value (280 ppmv), assuming that concentrations of other atmospheric constituents remain fixed at pre-industrial levels and sufficient time has elapsed for the large thermal reservoirs in the Earth system to equilibrate with the forcing.

10.26 (a) Derive a "low end" estimate of the response of global-mean surface air temperature to a doubling of atmospheric CO_2 concentrations relative to preindustrial values, proceeding as in Exercise 10.24, but assuming a storage of 0.3 (rather than 0.7) W m^{-2} and an aerosol forcing of 0.5 (instead of 1.0) W m^{-2} for aerosols. In addition, assume that of the 0.7 K warming of global-mean surface air temperature since 1860, 0.2 K is in response to solar forcing and should therefore not be included in the numerator as part of the temperature change. (b) Derive a "high end" estimate, proceeding as in Exercise 10.24, but assuming an aerosol forcing of −1.4 W m^{-2}.

10.27 Based on the assumptions in Exercise 10.24 concerning the various climate forcings and the estimated climate sensitivity $\lambda = 0.70$ K (W m^{-2})$^{-1}$, calculate the equilibrium response of global mean surface air temperature to the current greenhouse forcing.

10.28 The rise in sea level that would occur in response to a prescribed input of energy depends on how the heating is distributed with respect to the temperature of the water, as illustrated in this exercise. A planet is covered by an ocean 2500 m deep with a 50 m deep mixed layer. The water in the mixed layer is at a temperature of 15 °C and below the mixed layer the water temperature is 5 °C. If the water were mixed to a uniform temperature while conserving energy, (a) what would the final temperature be? (b) By how much would the densities of the upper and lower layers change? (c) By how much would sea level change?

The density of sea water is a complicated function of temperature T, salinity s, and pressure p. The pressure and temperature dependence may be expressed in the polynomial form

$$\rho = \rho_0(p) + c_1(p)\,(T_0(p) - T) + c_2(p)\,(T_0(p) - T)^2 + \cdots\cdots$$

Ignore the rather weak pressure dependence of ρ_0, c_1, and c_2 and assume that the salinity remains constant. With these assumptions, the numerical values of the coefficients are $c_1 = -0.2$ kg m^{-3} K^{-1} and $c_2 = -0.005 \times$ kg m^{-3} K^{-2}. [**Hint**: Consider a two-layer fluid, with temperatures T_1, T_2 and thicknesses d_1

and d_2. If the density difference is not too large, show that the change in the height is

$$dh = -\frac{1}{\rho_0}(d_1 d\rho_1 + d_2 d\rho_2)$$

where

$$d\rho_1 = c_1(T - T_1) + c_2(T - T_1)^2,$$

$$d\rho_2 = c_1(T - T_2) + c_2(T - T_2)^2$$

and

$$T = \frac{d_1 T_1 + d_2 T_2}{d_1 + d_2}]$$

10.29 At present the imbalance in the radiative flux density at the top of the atmosphere due to the increasing storage of energy in the Earth system in response to the buildup of greenhouse gases is estimated to be around 0.8 W m^{-2}. Suppose that this imbalance were to persist over the next century, with 10% the energy storage going into melting the Antarctic and Greenland ice sheets. (a) Estimate the resulting rise in global sea level. (b) At this rate, how long would it take for the Greenland and Antarctic ice sheets to completely melt? [**Hint**: Use the data in Table 2.2. There is no need to consider the areal coverage of the ice sheets.]

Constants and Conversions for Atmospheric Science

Universal Constants

Universal gravitational constant	G	$= 6.67 \times 10^{-11}$ N m^2 kg^{-2}
Universal gas constant in SI units	R^*	$= 8.3143$ J K^{-1} mol^{-1}
Gas constant in chemical units	$(R_c)^*$	$= 0.0821$ L atm K^{-1} mol^{-1}
Speed of light	c	$= 2.998 \times 10^8$ m s^{-1}
Planck's constant	h	$= 6.626 \times 10^{-34}$ J s
Stefan–Boltzmann constant	σ	$= 5.67 \times 10^{-8}$ J s^{-1} m^{-2} K^{-4}
Constant in Wien's displacement law	$\lambda_{max}T$	$= 2.897 \times 10^{-3}$ m K
Boltzmann's constant	k	$= 1.38 \times 10^{-23}$ J K^{-1} molecule^{-1}
Avogadro's number	N_A	$= 6.022 \times 10^{23}$ molecules mol^{-1}
Loschmidt number	L	$= 2.69 \times 10^{25}$ molecules m^{-3}

Air

Typical density of air at sea level	ρ_0	$= 1.25$ kg m^{-3}
Gas constant for dry air	R_d	$= 287$ J K^{-1} kg^{-1}
Effective molecular weight of dry air	M_d	$= 28.97$ kg kmol^{-1}
Specific heat of dry air, constant pressure	c_p	$= 1004$ J K^{-1} kg^{-1}
Specific heat of dry air, constant volume	c_v	$= 717$ J K^{-1} kg^{-1}
Dry adiabatic lapse rate	g/c_p	$= 9.8 \times 10^{-3}$ K m^{-1}
Thermal conductivity at 0 °C (independent of pressure)	K	$= 2.40 \times 10^{-2}$ J m^{-1} s^{-1} K^{-1}

Water Substance

Density of liquid water at 0 °C	ρ_{water}	$= 10^3$ kg m^{-3}
Density of ice at 0 °C	ρ_{ice}	$= 0.917 \times 10^3$ kg m^{-3}
Gas constant for water vapor	R_v	$= 461$ J K^{-1} kg^{-1}
Molecular weight of H_2O	M_w	$= 18.016$ kg kmol^{-1}
Molecular weight ratio of H_2O to dry air	ε	$= M_w/M_d = 0.622$
Specific heat of water vapor at constant pressure	c_{pw}	$= 1952$ J deg^{-1} kg^{-1}
Specific heat of water vapor at constant volume	c_{vw}	$= 1463$ J deg^{-1} kg^{-1}
Specific heat of liquid water at 0 °C	c_w	$= 4218$ J K^{-1} kg^{-1}
Specific heat of ice at 0 °C	c_i	$= 2106$ J K^{-1} kg^{-1}

Latent heat of vaporization at 0 °C	L_v	$= 2.50 \times 10^6$ J kg^{-1}
Latent heat of vaporization at 100 °C		$= 2.25 \times 10^6$ J kg^{-1}
Latent heat of sublimation (H_2O)	L_s	$= 2.85 \times 10^6$ J kg^{-1}
Latent heat of fusion (H_2O)	L_f	$= 3.34 \times 10^5$ J kg^{-1}

Earth and Sun

Acceleration due to gravity at sea level	g_o	$= 9.81$ ms^{-2}
	$m_\oplus$	$= 5.97 \times 10^{24}$ kg
	m_e	$= 5.3 \times 10^{18}$ kg
Radius of the Earth	R_E	$= 6.37 \times 10^6$ m
Area of the surface of the Earth		$= 5.10 \times 10^{14}$ m^2
Mass of an atmospheric column	m_a	$= 1.017 \times 10^4$ kg m^{-2}
Atmosphere to Pascals	1 atm	$= 1.01325 \times 10^5$ Pa
Rotation rate of Earth	Ω	$= 7.292 \times 10^{-5}$ s^{-1}
	$m_\odot$	$= 1.99 \times 10^{30}$ kg
Radius of the sun	$r_\odot$	$= 6.96 \times 10^8$ m
Mean earth–sun distance	d	$= 1.50 \times 10^{11}$ m $= 1.00$ AU
Solar flux	E_s	$= 3.85 \times 10^{26}$ W
Average Intensity of solar radiation	I_s	$= 2.00 \times 10^7$ W m^{-2} sr^{-1}

Units and Conversions

Fahrenheit–Celsius conversion	T_C	$= \frac{5}{9}(T_F - 32)$
Kelvin–Celsius conversion	T_K	$= T_C + 273.15$
Hectopascal conversions	1 hPa	$= 1$ mb $= 10^3$ dynes cm^{-2}
Cubic meters to liters	1 m^3	$= 1000$ L
Days to seconds	1 d	$= 86{,}400$ s
Calories to Joules	1 cal	$= 4.1855$ J
Latitude conversions	1° lat	$= 60$ nautical mi $= 111$ km
		$= 69$ statute mi
Longitude conversions	1° lon	$= 111$ km $\times$ cos(latitude)
Knots to miles per hour	1 knot	$= 1$ nautical mi/h $= 1.15$ statute mi/h
Meters per second to knots	1 m s^{-1}	$= 1.9426$ kt
Sverdrups to m^3 s^{-1}	1 Sv	$= 10^6$ m^3 s^{-1}
Dobson unit	1 DU	$= 2.6 \times 10^{20}$ molecules O_3 cm^{-2}

Abbreviations Used in Captions

AAAS	American Association for the Advancement of Science
AVHRR NDVI	Advanced Very High Resolution Radiometer Normalized Difference Vegetation Index
CLIVAR	International Programme on Climate Variability and Predictability
CMAP	CPC Merged Analysis and Prediction
CNES	Centre National D'Etudes Spatiales (the space agency of France)
COLA	Center for Ocean-Land-Atmosphere Studies
CPC	Climate Prediction Center
DAAC	Distributed Active Archive Center
ECMWF	European Centre for Medium Range Weather Forecasting
ERS-2	European Remote-sensing Satellite
GOES	Geostationary Operational Environmental Satellite (NOAA)
GSFC	Goddard Space Flight Center (NASA)
HadISST	Hadley Centre's Sea-Ice and Sea Surface Temperature data set
IGES	Institute of Global Environment and Society
IPCC	Intergovernmental Panel on Climate Change
JAWS	Joint Airport Windshear Studies project
MODIS	Moderate Resolution Imaging Spectroradiometer (NASA)
NASA	National Aeronautics and Space Administration
NCAR	National Center for Atmospheric Research
NCEP	National Centers for Environmental Prediction
NIMROD	Northern Illinois Research on Downbursts project
NOAA	National Oceanic and Atmospheric Administration
NTIS	National Technical Information Service
PACS	Pan American Climate Studies
PRISM	Pacific Regional Information System
QuikSCAT	Quick Scatterometer
RADARSAT	Canadian remote sensing satellite
RAINEX	Hurricane Rainband and Intensity Change Experiment
SeaWiFS	Sea-viewing Wide Field-of-view Sensor
SMRP	Satellite and Mesometeorology Research project
SST	Sea Surface Temperature
TMI	Tropical Microwave Imager
TOMS	Total Ozone Mapping Spectrometer
TOPEX	Ocean Topography Experiment
UCAR	University Corporation for Atmospheric Research
WHOI	Woods Hole Oceanographic Institute

Index

International Geophysics Series

Volume 1 BENO GUTENBERG. Physics of the Earth's Interior. 1959*

Volume 2 JOSEPH W. CHAMBERLAIN. Physics of the Aurora and Airglow. 1961*

Volume 3 S. K. RUNCORN (ed.). Continental Drift. 1962*

Volume 4 C. E. JUNGE. Air Chemistry and Radioactivity. 1963*

Volume 5 ROBERT G. FLEAGLE AND JOOST A. BUSINGER. An Introduction to Atmospheric Physics. 1963*

Volume 6 L. DEFOUR AND R. DEFAY. Thermodynamics of Clouds. 1963*

Volume 7 H. U. ROLL. Physics of the Marine Atmosphere. 1965*

Volume 8 RICHARD A. CRAIG. The Upper Atmosphere: Meteorology and Physics. 1965*

Volume 9 WILLIS L. WEBB. Structure of the Stratosphere and Mesosphere. 1966*

Volume 10 MICHELE CAPUTO. The Gravity Field of the Earth from Classical and Modern Methods. 1967*

Volume 11 S. MATSUSHITA AND WALLACE H. CAMPBELL (eds.). Physics of Geomagnetic Phenomena (In two volumes). 1967*

Volume 12 K. YA KONDRATYEV. Radiation in the Atmosphere. 1969*

Volume 13 E. PALMÅN AND C. W. NEWTON. Atmospheric Circulation Systems: Their Structure and Physical Interpretation. 1969*

Volume 14 HENRY RISHBETH AND OWEN K. GARRIOTT. Introduction to Ionospheric Physics. 1969*

Volume 15 C. S. RAMAGE. Monsoon Meteorology. 1971*

Volume 16 JAMES R. HOLTON. An Introduction to Dynamic Meteorology. 1972*

Volume 17 K. C. YEH AND C. H. LIU. Theory of Ionospheric Waves. 1972*

Volume 18 M. I. BUDYKO. Climate and Life. 1974*

Volume 19 MELVIN E. STERN. Ocean Circulation Physics. 1975

Volume 20 J. A. JACOBS. The Earth's Core. 1975*

Volume 21 DAVID H. MILLER. Water at the Surface of the Earth: An Introduction to Ecosystem Hydrodynamics. 1977

Volume 22 JOSEPH W. CHAMBERLAIN. Theory of Planetary Atmospheres: An Introduction to Their Physics and Chemistry 1978*

Volume 23 JAMES R. HOLTON. An Introduction to Dynamic Meteorology, 2nd Edition. 1979*

Volume 24 ARNETT S. DENNIS. Weather Modification by Cloud Seeding. 1980*

Volume 25 ROBERT G. FLEAGLE and JOOST A. BUSINGER. An Introduction to Atmospheric Physics, 2nd Edition. 1980*

Volume 26 KUO-NAN LIOU. An Introduction to Atmospheric Radiation. 1980*

Volume 27 DAVID H. MILLER. Energy at the Surface of the Earth: An Introduction to the Energetics of Ecosystems. 1981

Volume 28 HELMUT G. LANDSBERG. The Urban Climate. 1991

Volume 29 M. I. BUDKYO. The Earth's Climate: Past and Future. 1982*

Volume 30 ADRIAN E. GILL. Atmosphere-Ocean Dynamics. 1982

Volume 31 PAOLO LANZANO. Deformations of an Elastic Earth. 1982*

Volume 32 RONALD T. MERRILL AND MICHAEL W. MCELHINNY. The Earth's Magnetic Field: Its History, Origin, and Planetary Perspective. 1983*

Volume 33 JOHN S. LEWIS AND RONALD G. PRINN. Planets and Their Atmospheres: Origin and Evolution. 1983

Volume 34 ROLF MEISSNER. The Continental Crust: A Geophysical Approach. 1986

Volume 35 M. U. SAGITOV, B. BODKI, V. S. NAZARENKO, AND KH. G. TADZHIDINOV. Lunar Gravimetry. 1986

Volume 36 JOSEPH W. CHAMBERLAIN AND DONALD M. HUNTEN. Theory of Planetary Atmospheres, 2nd Edition. 1987

Volume 37 J. A. JACOBS. The Earth's Core, 2nd Edition. 1987*

Volume 38 J. R. APEL. Principles of Ocean Physics. 1987

Volume 39 MARTIN A. UMAN. The Lightning Discharge. 1987*

Volume 40 DAVID G. ANDREWS, JAMES R. HOLTON, AND CONWAY B. LEOVY. Middle Atmosphere Dynamics. 1987

Volume 41 PETER WARNECK. Chemistry of the Natural Atmosphere. 1988*

Volume 42 S. PAL ARYA. Introduction to Micrometeorology. 1988*

Volume 43 MICHAEL C. KELLEY. The Earth's Ionosphere. 1989*

Volume 44 WILLIAM R. COTTON AND RICHARD A. ANTHES. Storm and Cloud Dynamics. 1989

Volume 45 WILLIAM MENKE. Geophysical Data Analysis: Discrete Inverse Theory, Revised Edition. 1989

Volume 46 S. GEORGE PHILANDER. El Niño, La Niña, and the Southern Oscillation. 1990

Volume 47 ROBERT A. BROWN. Fluid Mechanics of the Atmosphere. 1991

Volume 48 JAMES R. HOLTON. An Introduction to Dynamic Meteorology, 3rd Edition. 1992

Volume 49 ALEXANDER A. KAUFMAN. Geophysical Field Theory and Method.

Part A: Gravitational, Electric, and Magnetic Fields. 1992*

Part B: Electromagnetic Fields I. 1994*

Part C: Electromagnetic Fields II. 1994*

Volume 50 SAMUEL S. BUTCHER, GORDON H. ORIANS, Robert J. CHARLSON, AND GORDON V. WOLFE. Global Biogeochemical Cycles. 1992

Volume 51 BRIAN EVANS AND TENG-FONG WONG. Fault Mechanics and Transport Properties of Rocks. 1992

Volume 52 ROBERT E. HUFFMAN. Atmospheric Ultraviolet Remote Sensing. 1992

Volume 53 ROBERT A. HOUZE, JR. Cloud Dynamics. 1993

Volume 54 PETER V. HOBBS. Aerosol-Cloud-Climate Interactions. 1993

Volume 55 S. J. GIBOWICZ AND A. KIJKO. An Introduction to Mining Seismology. 1993

Volume 56 DENNIS L. HARTMANN. Global Physical Climatology. 1994

Volume 57 MICHAEL P. RYAN. Magmatic Systems. 1994

Volume 58 THORNE LAY AND TERRY C. WALLACE. Modern Global Seismology. 1995

Volume 59 DANIEL S. WILKS. Statistical Methods in the Atmospheric Sciences. 1995

Volume 60 FREDERIK NEBEKER. Calculating the Weather. 1995

Volume 61 MURRY L. SALBY. Fundamentals of Atmospheric Physics. 1996

Volume 62 JAMES P. MCCALPIN. Paleoseismology. 1996

Volume 63 RONALD T. MERRILL, MICHAEL W. MCELHINNY, AND PHILLIP L. MCFADDEN. The Magnetic Field of the Earth: Paleomagnetism, the Core, and the Deep Mantle. 1996

Volume 64 NEIL D. OPDYKE AND JAMES E. T. CHANNELL. Magnetic Stratigraphy. 1996

Volume 65 JUDITH A. CURRY AND PETER J. WEBSTER. Thermodynamics of Atmospheres and Oceans. 1998

Volume 66 LAKSHMI H. KANTHA AND CAROL ANNE CLAYSON. Numerical Models of Oceans and Oceanic Processes. 2000

Volume 67 LAKSHMI H. KANTHA AND CAROL ANNE CLAYSON. Small Scale Processes in Geophysical Fluid Flows. 2000

Volume 68 RAYMOND S. BRADLEY. Paleoclimatology, 2nd Edition. 1999

Volume 69 LEE-LUENG FU AND ANNY CAZANAVE. Satellite Altimetry and Earth Sciences: A Handbook of Techniques and Applications. 2000

Volume 70 DAVID A. RANDALL. General Circulation Model Development: Past, Present, and Future. 2000

Volume 71 PETER WARNECK. Chemistry of the Natural Atmosphere, 2nd Edition. 2000

Volume 72 MICHAEL C. JACOBSON, ROBERT J. CHARLSON, HENNING RODHE, AND GORDON H. ORIANS. Earth System Science: From Biogeochemical Cycles to Global Change. 2000

Volume 73 MICHAEL W. MCELHINNY AND PHILLIP L. MCFADDEN. Paleomagnetism: Continents and Oceans. 2000

Volume 74 ANDREW E. DESSLER. The Chemistry and Physics of Stratospheric Ozone. 2000

Volume 75 BRUCE DOUGLAS, MICHAEL KEARNEY, AND STEPHEN LEATHERMAN. Sea Level Rise: History and Consequences. 2000

Volume 76 ROMAN TEISSEYRE AND EUGENIUSZ MAJEWSKI. Earthquake Thermodynamics and Phase Transformations in the Interior. 2001

Volume 77 GEROLD SIEDLER, JOHN CHURCH, AND JOHN GOULD. Ocean Circulation and Climate: Observing and Modelling The Global Ocean. 2001

Volume 78 ROGER A. PIELKE SR. Mesoscale Meteorological Modeling, 2nd Ed. 2001

Volume 79 S. PAL ARYA. Introduction to Micrometeorology. 2001

Volume 80 BARRY SALTZMAN. Dynamical Paleoclimatology: Generalized Theory of Global Climate Change. 2002

Volume 81A WILLIAM H. K. LEE, HIROO KANAMORI, PAUL JENNINGS, AND CARL KISSLINGER. International Handbook of Earthquake and Engineering Seismology, Part A. 2002

Volume 81B WILLIAM H. K. LEE, HIROO KANAMORI, PAUL JENNINGS, AND CARL KISSLINGER. International Handbook of Earthquake and Engineering Seismology, Part B. 2003